T0093544

# INTRODUCTION TO
# ENVIRONMENTAL DATA SCIENCE

## William W. Hsieh

Statistical and machine learning methods have many applications in the environmental sciences, including prediction and data analysis in meteorology, hydrology and oceanography; pattern recognition for satellite images from remote sensing; management of agriculture and forests; assessment of climate change; and much more. With rapid advances in machine learning in the last decade, this book provides an urgently needed, comprehensive guide to machine learning and statistics for students and researchers interested in environmental data science. It includes intuitive explanations covering the relevant background mathematics, with examples drawn from the environmental sciences. A broad range of topics is covered, including correlation, regression, classification, clustering, neural networks, random forests, boosting, kernel methods, evolutionary algorithms and deep learning, as well as the recent merging of machine learning and physics. End-of-chapter exercises allow readers to develop their problem-solving skills, and online datasets allow readers to practise analysis of real data.

WILLIAM W. HSIEH is a professor emeritus in the Department of Earth, Ocean and Atmospheric Sciences at the University of British Columbia. Known as a pioneer in introducing machine learning to environmental science, he has written more than 100 peer-reviewed journal papers on climate variability, machine learning, atmospheric science, oceanography, hydrology and agricultural science. He is the author of the book *Machine Learning Methods in the Environmental Sciences* (Cambridge University Press, 2009), the first single-authored textbook on machine learning for environmental scientists. Currently retired in Victoria, British Columbia, he enjoys growing organic vegetables.

'As a new wave of machine learning becomes part of our toolbox for environmental science, this book is both a guide to the latest developments and a comprehensive textbook on statistics and data science. Almost everything is covered, from hypothesis testing to convolutional neural networks. The book is enjoyable to read, well explained and economically written, so it will probably become the first place I'll go to read up on any of these topics.'

– **Alan Geer**, *European Centre for Medium-Range Weather Forecasts (ECMWF)*

'There is a need for a forward-looking text on environmental data science and William Hsieh's text succeeds in filling the gap. This comprehensive text covers basic to advanced material ranging from timeless statistical techniques to some of the latest machine learning approaches. His refreshingly engaging style is written to be understood and is complemented by a plethora of expressive visuals. Hsieh's treatment of nonlinearity is cutting-edge and the final chapter examines ways to combine machine learning with physics. This text is destined to become a modern classic.'

– **Sue Ellen Haupt**, *National Center for Atmospheric Research*

'William Hsieh has been one of the "founding fathers" of an exciting new field of using machine learning (ML) in the environmental sciences. His new book provides readers with a solid introduction to the statistical foundation of ML and various ML techniques, as well as with the fundamentals of data science. The unique combination of solid mathematical and statistical backgrounds with modern applications of ML tools in the environmental sciences ... is an important distinguishing feature of this book. The broad range of topics covered in this book makes it an invaluable reference and guide for researchers and graduate students working in this and related fields.'

– **Vladimir Krasnopolsky**, *Center for Weather and Climate Prediction, NOAA*

'Dr. Hsieh is one of the pioneers of the development of machine learning for the environmental sciences including the development of methods such as nonlinear principal component analysis to provide insights into the ENSO dynamic. Dr. Hsieh has a deep understanding of the foundations of statistics, machine learning, and environmental processes that he is sharing in this timely and comprehensive work with many recent references. It will no doubt become an indispensable reference for our field. I plan to use the book for my graduate environmental forecasting class and recommend the book for a self-guided progression or as a comprehensive reference.'

– **Philippe Tissot**, *Texas A & M University, Corpus Christi*

# INTRODUCTION TO
# ENVIRONMENTAL DATA SCIENCE

## William W. Hsieh

University of British Columbia

CAMBRIDGE
UNIVERSITY PRESS

Shaftesbury Road, Cambridge CB2 8EA, United Kingdom

One Liberty Plaza, 20th Floor, New York, NY 10006, USA

477 Williamstown Road, Port Melbourne, VIC 3207, Australia

314–321, 3rd Floor, Plot 3, Splendor Forum, Jasola District Centre,
New Delhi – 110025, India

103 Penang Road, #05–06/07, Visioncrest Commercial, Singapore 238467

Cambridge University Press is part of Cambridge University Press & Assessment,
a department of the University of Cambridge.

We share the University's mission to contribute to society through the pursuit of
education, learning and research at the highest international levels of excellence.

www.cambridge.org
Information on this title: www.cambridge.org/9781107065550

First published 2023

Printed in the United Kingdom by TJ Books Limited, Padstow Cornwall

*A catalogue record for this publication is available from the British Library*

*Library of Congress Cataloging-in-Publication data*
Names: Hsieh, William Wei, 1955– author.
Title: Introduction to environmental data science / William W. Hsieh.
Description: New York : Cambridge University Press, 2023. |
Includes bibliographical references and index.
Identifiers: LCCN 2022054278 | ISBN 9781107065550 (hardback)
Subjects: LCSH: Environmental sciences–Data processing. | Environmental
protection–Data processing. | Environmental management–Data processing. |
Machine learning.
Classification: LCC GE45.D37 H74 2023 | DDC 363.700285–dc23/eng20221219
LC record available at https://lccn.loc.gov/2022054278

ISBN 978-1-107-06555-0   Hardback

# Contents

# Preface

Modern data science has two main branches – statistics and machine learning – analogous to physics containing classical mechanics and quantum mechanics. Statistics, the much older branch, grew out from mathematics, while the advent of the computer and computer science in the post–World War II era led to an interest in intelligent machines, henceforth artificial intelligence (AI), and machine learning (ML), the fastest growing branch of AI. As quantum mechanics arrived in the 1920s with a fuzzy, random view of nature, which made many physicists, including Einstein, uncomfortable, the ML models too have been disapprovingly called 'black boxes' from their use of a large number of parameters that are opaque in practical problems. Quantum mechanics was eventually accepted, and a modern physicist learns both classical mechanics and quantum mechanics, using the former on everyday problems and the latter on atomic-scale problems. Similarly, a modern data scientist learns both statistics and machine learning, choosing the appropriate statistical or ML tool based on the particular data problem.

Environmental data science is the intersection between environmental science and data science. Environmental science is composed of many parts – atmospheric science, oceanography, hydrology, cryospheric science, ecology, agricultural science, remote sensing, climate science, and so on. Environmental datasets have their unique characteristics, for example most non-environmental datasets used in ML contain discrete or categorical data (alphabets and numbers in texts, colour pixels in an image, etc.), whereas most environmental datasets contain continuous variables (temperature, air pressure, precipitation amount, pollutant concentration, sea level height, streamflow, crop yield, etc.). Hence, environmental scientists need to assess astutely whether data methods developed from non-environmental fields would work well for particular environmental datasets.

This book is an introduction to environmental data science, attempting to balance the yin (ML) and the yang (statistics) when teaching data science to environmental science students. Written as a textbook for advanced undergraduates and beginning graduate students, it should also be useful for researchers and practitioners in environmental science. The reader is assumed to know multivariate calculus, linear algebra and basic probability.

Sections are marked by the flags $\boxed{\text{A}}$ for core material, $\boxed{\text{B}}$ for generally useful material and $\boxed{\text{C}}$ for more specialized material, and emojis indicating the level of technical difficulty for students – ☺ (easy), ☺ (moderately easy), ☺ (moderate), ☹ (moderately difficult) and ☹ (difficult). For instance, an instructor giving a one-term course would select topics mainly from sections $\boxed{\text{A}}$ and if the students have limited mathematical background, skip topics marked by ☹ and ☹.

The **book website** www.cambridge.org/hsieh-ieds contains downloadable datasets needed for some of the exercises provided in this book, and the solutions to most of the exercises. Readers of the printed book (with only greyscale figures) can also download a file containing coloured figures.

〜〜〜〜〜〜〜〜〜〜〜〜〜〜〜〜〜〜〜〜〜〜〜〜

How this book came about: With an undergraduate degree in mathematics and physics and a PhD (1981) in physical oceanography, I had, prior to 1992, zero knowledge of machine learning and very little of statistics! Through serendipity, I met Dr Benyang Tang, who introduced me to neural network models from ML. After learning this exotic topic (often the hard way) and training some graduate students in this direction, I wrote my first book *Machine Learning Methods in the Environmental Sciences*, published by Cambridge University Press in 2009. This was actually a gruelling ordeal lasting over eight years, so I thought that would be my last book.

However, three things happened on my way to retirement: (i) For a long time, machine learning was a fringe topic in the environmental sciences, but over the last five or six years, it has broken into the mainstream and has been growing exponentially. With so many fascinating new advances, I felt like a young boy unable to leave a toy store. (ii) ML and statistics have been taught separately from different books, which seems unnatural as I gradually view the two as the yin and yang side of a larger data science. Of course, this view is idiosyncratic as every ML/statistics researcher would have his/her own unique view. (iii) At conferences, enthusiastic graduate students told me that they had got into this research area from having read my first book – such comments were heartwarming to an author and made all the hard work worthwhile. So I dropped my serene retirement plans for one more book!

Writing this book has been a humbling learning experience for me. For such a vast, diverse subject, it is impossible to cover all important areas, and contributions from many brilliant scientists have regrettably been omitted.

〜〜〜〜〜〜〜〜〜〜〜〜〜〜〜〜〜〜〜〜〜〜〜〜

I have been fortunate in having supervised numerous talented graduate students, postdoctoral fellows and research associates, many of whom taught me more than I taught them. In particular, Dr Benyang Tang, Dr Aiming Wu and Dr Alex Cannon have, respectively, contributed the most to my research group during the early, mid and late phases of my research career, especially in helping my graduate students with their research projects.

The support from the editorial team led by Dr Matt Lloyd at Cambridge University Press was essential for bringing this book to fruition. The book has also benefitted from the comments provided by many colleagues who carefully read various draft chapters.

Although retired, I remain connected, as professor emeritus, to the Department of Earth, Ocean and Atmospheric Sciences at the University of British Columbia. Having moved from Vancouver to Victoria in 2016, I am grateful to the School of Earth and Ocean Sciences, University of Victoria, for giving me Visiting Scientist status.

Without the loving support from my family (my wife, Jean, and my daughters, Teresa and Serena) and the strong educational roots planted decades ago by my parents and my teachers, especially my PhD supervisor, Professor Lawrence Mysak, I could not have written this book.

# Notation Used

In general, scalars are typeset in italics (e.g. $x$ or $J$), vectors are denoted by lower case bold letters (e.g. $\mathbf{x}$ or $\mathbf{a}$) and matrices by upper case bold letters (e.g. $\mathbf{X}$ or $\mathbf{A}$). The elements of a vector $\mathbf{a}$ are denoted by $a_i$, while the elements of a matrix $\mathbf{A}$ are written as $A_{ij}$ or $(\mathbf{A})_{ij}$. A column vector is denoted by $\mathbf{x}$, while its transpose $\mathbf{x}^{\mathrm{T}}$ is a row vector, for example:

$$
\mathbf{x}^{\mathrm{T}} = [x_1, x_2, \ldots, x_m] \quad \text{and} \quad \mathbf{x} = [x_1, x_2, \ldots, x_m]^{\mathrm{T}} = \begin{bmatrix} x_1 \\ \vdots \\ x_m \end{bmatrix}, \tag{1}
$$

and the inner or dot product of two vectors $\mathbf{a} \cdot \mathbf{x} = \mathbf{a}^{\mathrm{T}}\mathbf{x} = \mathbf{x}^{\mathrm{T}}\mathbf{a}$.

In many environmental problems, $\mathbf{x}$ can denote $m$ different variables or measurements of a variable (e.g. temperature) at $m$ different stations. The measurements are often taken repeatedly at different times up to $n$ times, yielding $\mathbf{x}^{(1)}, \mathbf{x}^{(2)}, \ldots, \mathbf{x}^{(n)}$. The total dataset containing $m$ variables measured $n$ times can be arranged in either of the matrix forms

$$
\begin{bmatrix} x_{11} & \cdots & x_{1n} \\ \vdots & \ddots & \vdots \\ x_{m1} & \cdots & x_{mn} \end{bmatrix} \quad \text{or} \quad \begin{bmatrix} x_{11} & \cdots & x_{1m} \\ \vdots & \ddots & \vdots \\ x_{n1} & \cdots & x_{nm} \end{bmatrix}, \tag{2}
$$

with each matrix being simply the transpose of the other. In my first book (Hsieh, 2009), the first matrix form was used, but the second form has become increasingly widely used, probably due to the way data are typically arranged in spreadsheets. Hence, in this book, the data matrix $\mathbf{X}$ is written as

$$
\mathbf{X} = \begin{bmatrix} x_{11} & \cdots & x_{1m} \\ \vdots & \ddots & \vdots \\ x_{n1} & \cdots & x_{nm} \end{bmatrix} = \begin{bmatrix} \mathbf{x}^{(1)\mathrm{T}} \\ \vdots \\ \mathbf{x}^{(n)\mathrm{T}} \end{bmatrix}. \tag{3}
$$

The probability for discrete variables is denoted by upper case $P$, whereas the probability density for continuous variables is denoted by lower case $p$. The expectation is denoted by $\mathrm{E}[\ldots]$ or $\langle \ldots \rangle$. The natural logarithm is denoted by ln or log.

# Abbreviations

AAO  Antarctic Oscillation

AIC  Akaike information criterion

ANOVA  analysis of variance

ANN  artificial neural network

AO  Arctic Oscillation

AR  auto-regressive

ARIMA  auto-regressive integrated moving average

ARMA  auto-regressive moving average

BIC  Bayesian information criterion

BLUE  best linear unbiased estimator

BMA  Bayesian model averaging

BS  Brier score

CART  classification and regression tree

CCA  canonical correlation analysis

CCDF  complementary cumulative distribution function

CDF  cumulative distribution function

CDN  conditional density network

CI  confidence interval

CNN  convolutional neural network

ConvLSTM  convolutional long short-term memory model

CRPS  continuous ranked probability score

CSI  critical success index

CTFT  continuous-time Fourier transform

DE  differential evolution

DFT  discrete Fourier transform

DL  deep learning

DNN  deep neural network

DNS  direct numerical simulation (in computational fluid dynamics)

DTFT  discrete-time Fourier transform

EA  evolutionary algorithm

ECMWF  European Centre for Medium-Range Weather Forecasts

EDA  exploratory data analysis

EEOF  extended empirical orthogonal function

ELM  extreme learning machine

ENSO  El Niño-Southern Oscillation

EOF  empirical orthogonal function

ES  environmental science

ET  extra trees (extremely randomized trees)

ETS  equitable threat score

FFNN  feed-forward neural network

FFT  fast Fourier transform

GA  genetic algorithm

GAN  generative adversarial network

GBM  gradient boosting machine

GCM  general circulation model or global climate model

GEV  generalized extreme value distribution

GP  Gaussian process model

GSS  Gilbert skill score

HSS  Heidke skill score

IC  information criterion

i.i.d.  independent and identically distributed

IPCC  Intergovernmental Panel on Climate Change

IQR  interquartile range

IR  infrared

KDA  kernel density estimation

KNN  $K$-nearest neighbours

LDA  linear discriminant analysis

LSTM  long short-term memory model

MA  moving average

MAD  median absolute deviation

MAE  mean absolute error

MCA  maximum covariance analysis

MDN  mixture density network

ME  mean error

MJO   Madden–Julian Oscillation
ML   machine learning
MLP   multi-layer perceptron neural
         network
MLR   multiple linear regression
MOS   model output statistics
MSE   mean squared error
MSSA   multichannel singular
         spectrum analysis
NAO   North Atlantic Oscillation
NASA   National Aeronautics and
         Space Administration
         (USA)
NCAR   National Center for
         Atmospheric Research
         (USA)
NCEP   National Centers for
         Environmental Prediction
         (USA)
NLCCA   nonlinear canonical
         correlation analysis
NLCPCA   nonlinear complex PCA
NLPC   nonlinear principal
         component
NLPCA   nonlinear principal
         component analysis
NLSSA   nonlinear singular spectrum
         analysis
NN   neural network
NOAA   National Oceanic and
         Atmospheric
         Administration (USA)
NWP   numerical weather prediction
OSELM   online sequential extreme
         learning machine
PC   principal component
PCA   principal component analysis
PDF   probability density function
         or probability distribution
         function

PI   prediction interval
PNA   Pacific-North American
         pattern
POD   probability of detection
POFD   probability of false detection
PSS   Peirce skill score
QBO   Quasi-Biennial Oscillation
QRNN   quantile regression neural
         network
RBF   radial basis function
RCM   regional climate model
ReLU   rectified linear unit
RF   random forest
RMSE   root mean squared error
RNN   recurrent neural network
ROC   relative operating
         characteristic
RPCA   rotated principal component
         analysis
RPS   ranked probability score
SGD   stochastic gradient descent
SLP   sea level pressure
SOI   Southern Oscillation Index
SOM   self-organizing map
SS   skill score
SSA   singular spectrum analysis
SSE   sum of squared errors
SSR   sum of squares due to
         regression
SST   sea surface temperature; sum
         of squares (total)
SVD   singular value decomposition
SVM   support vector machine
SVR   support vector regression
SWE   snow water equivalent
TS   threat score
UAS   unmanned aerial systems
XGBoost   extreme gradient
         boosting

# 1

# Introduction

*Two roads diverged in a wood, and I –*
*I took the one less traveled by,*
*And that has made all the difference.*

Robert Frost, *The Road Not Taken*

## 1.1  Statistics and Machine Learning  A☺

Back in the early 1970s when the author was starting his undergraduate studies, freshmen interested in studying data analysis would pursue statistics in a mathematics department or a statistics department. In contrast, today's freshmen would most likely study machine learning in a computer science department, though they still have the option of majoring in statistics. Once there was one, now there are two options. Or is machine learning (ML) merely statistics with a fancy new wrapping? In this section, we will try to answer this question by first following the evolution of the two fields.

ML and statistics have very different origins, with statistics being the much older science. Statistics came from the German word *Statistik*, which appeared in 1749, meaning 'collection of data about the State', that is, government data on demographics and economics, useful for running the government. The collected data were analysed and the new science of *probability* was found to provide a solid mathematical foundation for data analysis. Probability itself began in 1654 when two famous French mathematicians, Blaise Pascal and Pierre de Fermat, solved a gambling problem brought to their attention by Antoine Gombaud. Christian Huygens wrote the first book on probability in 1657, followed by contributions from Jakob Bernoulli in 1713 and Abraham de Moivre in 1718. In 1812, Pierre de Laplace published *Théorie analytique des probabilités*, greatly expanding probability from games of chance to many scientific and practical

1

problems. By the time of World War II, statistics was already a well-established field.

The birth of the electronic digital programmable computer by the end of World War II led to the growth of computer science and engineering, and the first successful numerical weather prediction in 1950 (Charney et al., 1950). While computers could compute with enormous speed and solve many problems, it soon became clear that they were very poor at performing simple tasks humans could do easily, such as recognizing a face, understanding speech, and so on.

The term *artificial intelligence* (AI) was invented by John McCarthy when he organized the Dartmouth Summer Research Project on Artificial Intelligence in 1956, an 8-week summer school held at Dartmouth College in Hanover, New Hampshire with about 20 invited attendees. This seminal workshop was considered by many to spark the field of AI research, with AI research mainly pursued by computer scientists/engineers and psychologists.

*Machine learning* (ML) is a major branch of AI[1] that allows computers to learn from data without being explicitly programmed. As for the origin of the term "machine learning", Turing (1950) raised the question 'can computers think?' and introduced the concept of "learning machines". In the 1955 Western Joint Computer Conference in Los Angeles, there was a session on "Learning Machines" (Nilsson, 2009), while the term 'machine learning' appeared later in Samuel (1959).

Meanwhile, the general public has become fascinated with the new genre of science fiction, depicting machines with human intelligence. Under this intoxicating atmosphere, some AI researchers became unrealistically optimistic about how soon it would take to produce intelligent machines; thus, a backlash against overpromises became inevitable. The UK Science Research Council asked the Cambridge Lucasian professor Sir James Lighthill to evaluate the academic research in AI with an outsider perspective, as Lighthill was a fluid dynamicist. The 1973 report was largely negative, stating that 'Most workers in AI research and in related fields confess to a pronounced feeling of disappointment in what has been achieved in the past twenty-five years. Workers entered the field around 1950, and even around 1960, with high hopes that are very far from having been realised in 1972. In no part of the field have the discoveries made so far produced the major impact that was then promised' (Lighthill, 1973). AI was devastated, as the UK closed all academic AI research except at three universities. Around the same time, the US Defense Advanced Research Projects Agency (then known as 'ARPA', now 'DARPA'), which had been the main source of AI funding in the US, also lost faith in AI, leading to drastic funding cuts. There were two major 'AI winters', periods of poor funding in AI, lasting around 1974–1980 and 1987–1993 (Crevier, 1993; Nilsson, 2009).

As AI suffered a tarnished reputation during the long AI winters, many researchers in AI in the mid-2000s referred to their work using other names, such as machine learning, informatics, computational intelligence, soft computing, data

---

[1] AI has other branches besides ML; for example *expert systems* were once very popular but have almost completely disappeared.

driven modelling, data mining and so on, partly to focus on a more specific aspect and partly to avoid the stigma of overpromises and science fiction over-tones associated with the name 'artificial intelligence'. Google Trends reveals how terminology usage changes over time (see Fig. A1 in Appendix A), with 'machine learning' having been searched more often than 'artificial intelligence' on Google since 2015.

The general goal of having computers learning from data without being explicitly programmed was achieved in 1986 with an artificial neural network model called the multi-layer perceptron (Rumelhart et al., 1986a).[2] The rise of the Internet in the mid-1990s meant the connection of numerous computers and datasets, thereby introducing a huge amount of data for ML to extract useful information from. The commercial potential was quickly recognized, leading to the spectacular growth of many high technology companies, which in turn poured massive amounts of funding into ML and AI research.

By the late 1990s, statisticians introduced the new term '*data science*' to broaden statistics by including contributions by computer scientists (C. Hayashi, 1998; Cleveland, 2001). Data science is an interdisciplinary field that tries to extract knowledge from data using techniques from statistics, mathematics, computer science and information science. As such, one could consider statistics and ML as components within data science.

Did the separate paths of evolution taken by statistics and ML bring them to more or less the same domain within data science? Certainly there is par-tial overlap between the two, as it is not uncommon to have similar methods developed independently by statisticians and by ML scientists. Nevertheless, ML and statistics have their own distinct characteristics or cultures (Breiman, 2001b; D. R. Cox, Efron, et al., 2001). In fact, the two cultures are sufficiently different that it would be very difficult for ML to germinate from within statis-tics. For instance, one would expect counterculture art or music movements to germinate from societies with liberal laws rather with rigorous laws. Statis-tics, rooted in mathematics, requires a high standard in mathematical rigour for publications. While rigorous proofs can usually be derived for linear models, they may not even exist for the non-linear models used in ML. Thus, ML can only germinate within a culture that supports a more liberal, heuristic approach to research. Not surprisingly, ML germinated mainly from computer science, psychology, engineering and commerce.

When fitting a curve to a dataset, a statistician would ensure the number of adjustable model parameters is small compared to the sample size (that is, the number of observations) to avoid *overfitting*, that is, the model fitting to the noise in the data as the model becomes too flexible with abundant adjustable parameters (see Section 1.3). This prudent practice in statistics is not strictly followed in ML, as the number of parameters can be greater, sometimes much greater, than the sample size, as ML has developed ways to avoid overfitting while using a large number of parameters. The relatively large number of pa-

---

[2] While multi-layer perceptron neural network models became very popular after the ap-pearance of Rumelhart et al. (1986a), there had been important contributions made by earlier researchers (Schmidhuber, 2015, section 5.5).

rameters renders ML models more difficult to interpret than statistical models; thus, ML models are often regarded somewhat dismissively as 'black boxes'.

The rationale of using large numbers of model parameters in ML is based on AI's desire to develop models following the architecture of the human brain. The following argument is attributed to Geoffrey Hinton, who often used it in his lectures: In the brain, there are more than $10^{14}$ synapses, that is, connections between nerve cells; thus, there are more than $10^{14}$ adjustable parameters in the brain. A human lifetime is of the order of $10^9$ seconds, and learning say 10 data points per second implies a total sample size of $10^{10}$ in a lifetime. Thus, the number of parameters greatly exceeds the sample size for the human brain. In other words, if AI is to model the human brain function it has to explore the domain where the number of model parameters exceeds the sample size. For instance, in the ILSVRC-2012 image classification competition, the winning entry from Hinton's team used 60 million parameters trained with about 1.2 million images (Krizhevsky et al., 2012).

Dualism in nature was noted by the ancient Greek philosopher Heraclitus and in the Chinese philosophy of *yin and yang*, where opposite properties in nature may actually be complementary and interconnected and may give rise to each other as a wave trough gives rise to a wave crest. Yin is the shady or dark side and yang the sunny or bright side. Examples of traditional yin–yang pairs are night–day, moon–sun, feminine–masculine, soft–hard, and so on, and we can now add ML–statistics to the list as the yin and yang sides of data science (Fig. 1.1).

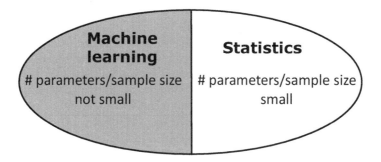

Figure 1.1 ML and statistics tend to occupy different parts of the data science space, as characterized by the number of model parameters to the sample size. In reality, there is overlap and more gradual transition between the two than the sharp boundary shown (see the Venn diagram in Fig. 15.1).

In journeys of discovery, the obscure yin side is often explored after the yang side. For instance, the European maritime exploration to India first proceeded eastward, and only westward from the time of Columbus, as sailing far into the obscure western ocean was not considered sensible nor profitable. In cosmology, the search had originally focused on visible, ordinary matter, but later it was found that ordinary matter accounts for only 4.9% of the Universe, while the rest is the dark universe, containing dark matter (26.8%) and dark energy (68.3%) (Hodson, 2016). Similarly, in data science, the domain where the number of parameters is small relative to sample size was explored first by statisticians, and the seemingly meaningless domain of a large number of parameters was only explored much later by ML scientists, driven by their interest in building models that simulate the human brain. The old constraint requiring the number of parameters to be no larger than the sample size turned out to be breakable, much like the 'sound barrier' preventing supersonic flight. In Fig. 1.1, the yin and yang domains are drawn to be equal in size – in reality, the domain where the number of parameters is restricted to be small is much smaller than the domain without this restriction. The reason ML enjoyed much faster growth than statistics in recent decades is that the solutions of many problems in image and speech recognition, self-driving cars and so on lie in the domain of a large number of parameters.

Another major difference between the statistics and ML cultures lies in their treatment of predictor variables (Breiman, 2001b). Predictor selection, that is, choosing only the relevant predictor variables from a pool of predictors, is commonly practiced in statistics but not often in ML. ML generally does not consider throwing away information a good practice. Furthermore, first selecting predictors based on having high correlation with the response variable then building statistical/ML models leads to overestimation of the prediction skill (DelSole and Shukla, 2009).

In summary, the main tradeoff between statistics and ML is *interpretability* versus *accuracy*. With relatively few parameters and few predictors, statistical models are much more interpretable than ML models. For instance, the parameters in a linear regression model give useful information on how each predictor variable influences the response variable, whereas ML methods such as artificial neural networks and random forests are run as an ensemble of models initialized with different random numbers, leading to a huge number of parameters that are uninterpretable in practical problems. However, as datasets become increasingly larger and more complex, interpretability becomes harder and harder to achieve even with statistical models, while the advantage in prediction accuracy attained by ML models makes them increasingly attractive.

In physics, a similar transition occurred between classical mechanics and quantum mechanics in the 1920s. The clear deterministic view of classical mechanics was replaced by a fuzzy, random picture for atomic particles, thanks to revolutionary concepts like the Heisenberg uncertainty principle, wave-particle duality, and so on. Uncomfortable with the apparent randomness of nature in quantum mechanics, Einstein protested with the famous quote 'God does not play dice with the world' (Hermanns, 1983). A modern physicist learns both classical mechanics and quantum mechanics, using the former on everyday problems and the latter on atomic-scale problems. Similarly, a modern data scientist learns both statistics and machine learning, choosing the appropriate statistical or ML method based on the particular data problem.

## 1.2   Environmental Data Science   [A]☺

Environmental data science is the intersection between environmental science and data science. *Environmental science* (ES) is composed of many branches – atmospheric science, hydrology, oceanography, cryospheric science, ecology, agricultural science, remote sensing, climate science, environmental engineering, and so on, with the data from each branch having their own characteristics. Often, the plural term 'environmental sciences' is used to denote these branches. Statistical methods have long been popular in the environmental sciences, with numerous textbooks covering their applications in climate science (von Storch and Zwiers, 1999), atmospheric science (Wilks, 2011), oceanography (Thomson and Emery, 2014) and hydrology (Naghettini, 2007).

Environmental data tend to have different characteristics from non-environmental data. Most non-environmental datasets in ML applications contain *discrete data* (e.g. intensity of colour pixels in an image) and/or *categorical data* (e.g. alphabets and numbers in texts),[3] whereas most environmental datasets contain *continuous data* (e.g. temperature, air pressure, wind speed, precipitation amount, pollutant concentration, sea level, salinity, streamflow, crop yield, etc.). The discrete/categorical data from ML problems are in general bounded, that is, having a finite domain – for example, a colour pixel normally has intensity values ranging from 0 to 255, while texts are typically composed of 26 alphabets and 10 digits (plus upper cases and some special symbols). In contrast, continuous data are in general not bounded; for example, there are no guaranteed upper limits for variables such as wind speed, precipitation amount and pollutant concentration.

The most common data problem consists of predicting the value of an output variable (also known as a *response* variable or *dependent* variable) given the values of some input variables (a.k.a. *predictors* or *features*).[4] If the output variable is discrete or categorical, this is a *classification* problem, whereas if the output is continuous, it is a *regression* problem. Again, classification is much more common in non-environmental datasets, and many ML methods were developed first for classification and later modified for regression, such as support vector machines (Cortes and V. Vapnik, 1995; V. Vapnik et al., 1997).

After a model has been built or trained with a *training* dataset, its performance is usually evaluated with a separate *test* dataset. If the test input data lie outside the domain of the input data used to train the model, the model will be forced to do *extrapolation*, yielding inaccurate or even nonsensical predictions. Figure 1.2 illustrates why the outlier problem can be much worse with unbounded continuous input data than with finite-domain discrete/categorical data. Thus, making accurate predictions using environmental data could be a much harder problem than typical non-environmental data problems.

---

[3] The difference between categorical data (e.g. water, land, snow, ice) and discrete data (e.g. 1, 2, 3) is that categorical data normally have no natural ordering, though some categorical data (namely *ordinal data*) do have natural ordering (e.g. sunny, cloudy, rainy). See Section 2.1.

[4] "Predictors" are used in the statistics literature while "features" are used in ML.

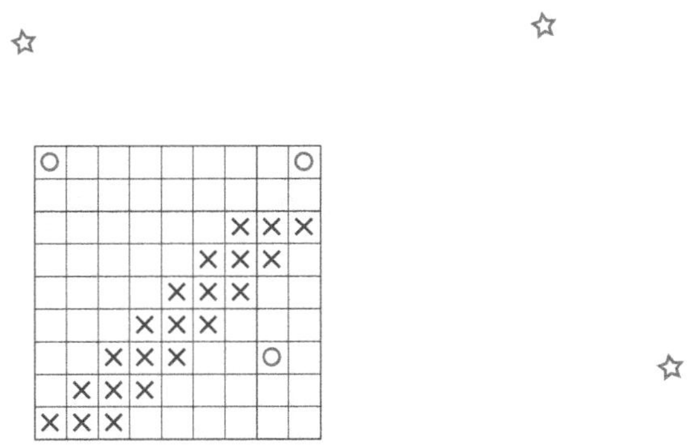

Figure 1.2 Schematic diagram illustrating the problem of outliers in the input data in 2-D. The grid illustrates a finite-domain discrete input data space with crosses indicating training data and circles marking outliers in the test data. For unbounded continuous input variables, the test data can lie well outside the grid and much farther from the training data, as illustrated by the stars.

Let us look at an example of input outliers. A common air quality measure of fine inhalable particles with diameters $\leq 2.5$ $\mu$m is the PM$_{2.5}$ concentration. For predicting the hourly PM$_{2.5}$ concentration in Beijing, an important predictor is the cumulated precipitation (X. Liang et al., 2016), as the pollutant concentration drops after precipitation. When data from 2013 to 2015 were used for training non-linear regression models and data from 2010 to 2012 were used for testing, Hsieh (2020) noticed that the cumulated precipitation of an intense precipitation event reached 223.0 mm in the test data in July 2012, whereas the maximum value in the three years of training data was only 51.1 mm, that is, this input in the test data was over four times the maximum value in the training data, which led to wild extrapolation (Section 16.9).

From the old saying 'climate is what you expect; weather is what you get' (a similar version originated from Mark Twain), it follows that environmental problems also tend to group into 'weather' and 'climate' problems, with the former concerned with short-term variations and the latter concerned with the expected values from long-term records or with longer-term variations. For instance, by averaging daily weather data over three months, one obtains seasonal data and can build models to predict seasonal variations. Farmers, utility companies, and so on have great interest in seasonal forecasts, for example on whether next season will be warm or cool, dry or wet.

The averaging of weather data to form climate data changes the nature of the data through the central limit theorem from statistics. To illustrate this effect, consider the synthetic dataset

$$y = x + x^2 + \epsilon, \tag{1.1}$$

where $x$ is a random variable obeying the Gaussian probability distribution with zero mean and unit standard deviation (see Section 3.4) and $\epsilon$ is Gaussian noise with a standard deviation of 0.5. Averaging these 'daily' data over 30 days reveals a dramatic weakening of the non-linear relation in the original daily data (Fig. 1.3). Thus, in this example, a non-linear regression model will greatly outperform a linear regression model in the daily data but not in the 30-day averaged data. In the real world, tomorrow's weather is not independent from today's weather (i.e. if it is rainy today, then tomorrow will also have higher odds of being rainy). Thus, the monthly data will be effectively averaging over far fewer than 30 independent observations as done in this synthetic dataset, so the weakening of the non-linear relation will not be as dramatic as in Fig. 1.3(c). Nevertheless, using non-linear regression models from ML on climate data will generally be less successful than using them on weather data due to the effects of the central limit theorem (Yuval and Hsieh, 2002).

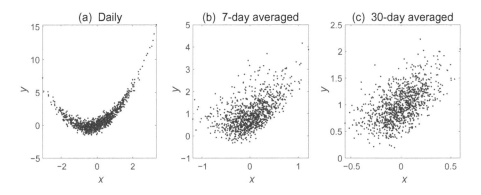

Figure 1.3 Effects of time-averaging on the non-linear relation (1.1). (a) Synthetic 'daily' data from a quadratic relation between $x$ and $y$. The data time-averaged over (b) 7 observations and (c) 30 observations. [Follows Hsieh and Cannon (2008).]

Obtaining climate data from daily weather data by taking the time mean or average is no longer the only statistic used. In the last couple of decades, there has been a growing interest in the climate of extreme weather events (simply called 'climate extremes'), as global climate change may affect the extremes even more than the means. There is now a long list of such climate extreme variables derived from daily data, for example, the annual number of frost days, the maximum number of consecutive days when precipitation is $< 1$ mm, and so on (X. B. Zhang et al., 2011) and ML methods have been used to study climate extremes (Gaitan, Hsieh et al., 2014).

Like seeds broadly dispersed by the wind, ML models landing in numerous environmental fields germinated at different rates depending on the local conditions. If a field already had successful physics-based models, ML models tended to suffer from neglect and slow growth. Meteorology, where dynamical (a.k.a. numerical) models have been routinely used for weather forecasting, has been slower to embrace ML models than hydrology, where by the year 2000 there were already 43 hydrological papers using neural network models (Maier and Dandy, 2000). ML models were readily accepted in hydrology because physical-based hydrological models were not skillful in forecasting streamflow.[5] Remote sensing is another field where ML was quickly adopted (Benediktsson et al., 1990; Atkinson and Tatnall, 1997).

Compared to linear statistical models, non-linear ML models require larger sample sizes to excel. Oceanography, a field where collecting *in situ* observations is far more difficult than in meteorology or hydrology, and climate science, where the long timescales involved preclude large effective sample size, are fields where the adoption of ML have been relatively slow among the environmental sciences. Zwiers and Von Storch (2004) noted: 'much of the work that has had a large impact on climate research has used relatively simple techniques that allow transparent interpretation of the underlying physics'. Nevertheless, in the last few years, ML has grown rapidly even in fields such as oceanography and climate science. Perhaps even more unexpectedly, divergent approaches such as ML and physics have been merging in recent years within environmental science (Chapter 17). The history and practice of AI/ML in the environmental sciences have been reviewed by S. E. Haupt, Gagne et al. (2022) and Hsieh (2022).

## 1.3   A Simple Example of Curve Fitting  $\boxed{\text{A}}$ ☺

In this section, we will illustrate some basic concepts in data science by a simple example of curve fitting, using one independent variable $x$ and one dependent variable $y$. Assume the true signal is a quadratic relation

$$y_{\text{signal}} = x - 0.25\,x^2. \tag{1.2}$$

The $y$ data are composed of the signal plus random noise,

$$y = y_{\text{signal}} + \epsilon, \tag{1.3}$$

where the noise $\epsilon$ obeys a Gaussian probability distribution with zero mean and standard deviation being half that of $y_{\text{signal}}$. The advantage of using synthetic data in this example is that we know what the true signal is.

A polynomial of order $m$,

$$\hat{y} = w_0 + w_1 x + w_2 x^2 + \cdots + w_m x^m, \tag{1.4}$$

has $m + 1$ adjustable model parameters or weights $w_j$ $(j = 0, ..., m)$, with $\hat{y}$ denoting the output value from the polynomial function as opposed to the value $y$ from the data. Polynomials of order 1, 2, 4 and 9 are fitted to the training

---

[5] The difficulty lies in the subsurface flow passing through material, which is not easily observable. The subsurface flow is also complex and non-linear, thereby requiring many parameters to cover for the inexact physics, resulting also in poor model interpretability (Karpatne et al., 2017).

dataset of 11 data points in Fig. 1.4 by minimizing the *mean squared error* (MSE) between the model output $\hat{y}$ and the data $y$,

$$\text{MSE} = \frac{1}{N} \sum_{i=1}^{N} (\hat{y}_i - y_i)^2, \tag{1.5}$$

where there are $i = 1, ..., N$ data points.

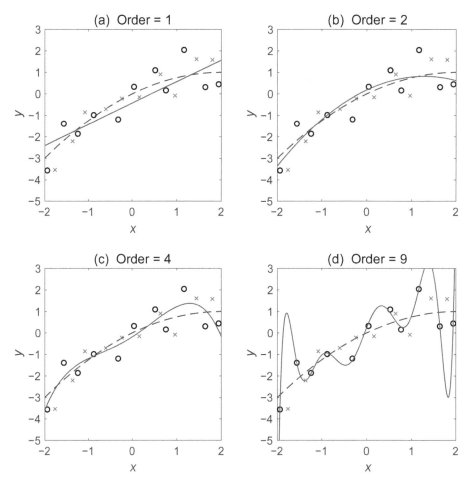

Figure 1.4 Polynomial fit to data using a polynomial of order (a) 1, (b) 2, (c) 4 and (d) 9. The circles indicate the 11 data points used for fitting (i.e. training), the solid curve the polynomial solution $\hat{y}$ and the dashed curve the true signal ($y_{\text{signal}} = x - 0.25\, x^2$). The crosses show 10 new data points used to validate the polynomial fit.

For order 1 (Fig. 1.4(a)), the polynomial reduces to a straight line and the problem is simple linear regression. As the order of the polynomial increases, the curve fit to the 11 training data points improves until the fit is almost perfect in Fig. 1.4(d), where the total number of adjustable parameters is 10, very close to the number of training data points. However, if one compares the polynomial fit (the solid curve) to the true signal (the dashed curve), while there is improvement going from order 1 to order 2, the agreement gets worse at order 4 and is dreadful at order 9.

This example illustrates the concepts of underfitting and overfitting data. At order 1, the model is *underfitting* the data since the true signal is a quadratic but the model is linear. However, as the order increases above 2, the model begins to *overfit* – that is, with more adjustable parameters than is needed to fit the true signal, the model is fitting to the noise in the data. An extra 10 new data points were generated from (1.2) and (1.3). The order 9 polynomial curve predicts these new data (marked by crosses) poorly (Fig. 1.4(d)), albeit the excellent fit to the training data (marked by circles).

With a real world problem, we will not have the luxury of knowing in advance what the true signal is and will be unable to know if our model is overfitting or underfitting. We need some independent *validation* data (sometimes also called test data) not used in the model training to tell us if the model is overfitting.[6] Figure 1.5 plots the mean squared error (MSE) for the training data and the validation data, as the order of the polynomial varies. The order 0 polynomial fit simply fits a constant to the training data, while the order 10 fit to the 11 data points is a perfect fit with zero MSE. While the MSE for the training data keeps dropping as the order increases, the MSE for the validation data drops to a minimum at order 2 then increases as the order increases. This tells us that the model was underfitting for order $< 2$ and overfitting for order $> 2$. Thus, we select the order 2 polynomial as the optimal model for this dataset. This process of using independent validation data to select the optimal model is called *model validation*.

What happens if we have more training data? Figures 1.6(a) and (b) compare the 9th order polynomial fit to training data with 15 and 100 points, respectively. The overfitting in (a) is much reduced in (b); thus, having more data helps in reducing overfitting.

What happens if the data are very noisy? In Fig. 1.6(c), the standard deviation of the added noise in $y$ is four times that of Fig. 1.6(b), indicating noisier data make overfitting worse. However, if we increase the amount of noisy data to 1,000 points in Fig. 1.6(d), the overfitting is much reduced when compared with Fig. 1.6(c), where there are only 100 data points. Thus, one of the main reasons for success in modern data science is that even weak signals imbedded in very noisy data can be successful retrieved if massive amounts of data are available.

---

[6] Strictly speaking, training data, validation data and test data are all separate. Validation data are used to select the best model (e.g. the order 2 polynomial in our example). The performance of the selected model is then evaluated or verified with independent test data. Performance scores from test data are more trustworthy than those from the training and validation data, as the test data have not been used in model training and selection.

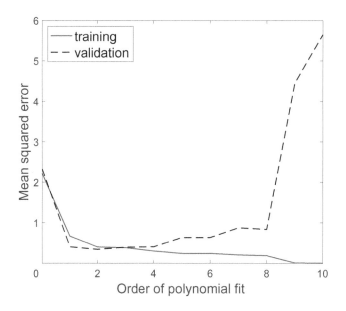

Figure 1.5 Mean squared error of the training and validation data as the order of the polynomial fit varies from 0 to 10.

What happens to the polynomial solution when we extend it outside the training data domain? Within the training domain of $x \in [-2, 2]$, the polynomial solutions for orders 2, 4 and 9 are quite similar to each other (Fig. 1.7), but when they extrapolate outside the training domain, the higher order solutions behave very badly. Furthermore, the wild extrapolation behaviour is irreproducible in that if we generate the synthetic data with a different initialization of the random number generator, we will get a different wild extrapolation picture in Fig. 1.7(d).

Polynomials are notorious for their extrapolation behaviour, since power functions of the form $x^m$ $(m > 1)$ increase in magnitude much faster than $x$ as $x \to \pm\infty$. Thus, modern data science methods such as artificial neural networks use basis functions with less aggressive growth properties than polynomials, which tame but still cannot rule out wild extrapolation (see Section 16.9).

## 1.4   Main Types of Data Problems   [A]☺

This section gives an overview on the basic types of data and data problems. Data are described by (a) discrete or categorical variables, (b) continuous variables and (c) probability distributions. Examples of discrete/categorical variables include binary variables (e.g. [0, 1], [on, off], [true, false]), [1, 2, 3], all integers and, in environmental science, [storm, no storm], [cold, warm], [cold, normal, warm], [dry, normal, wet], [land, water, snow, ice], and so on. Continuous variables in environmental science include temperature, wind speed,

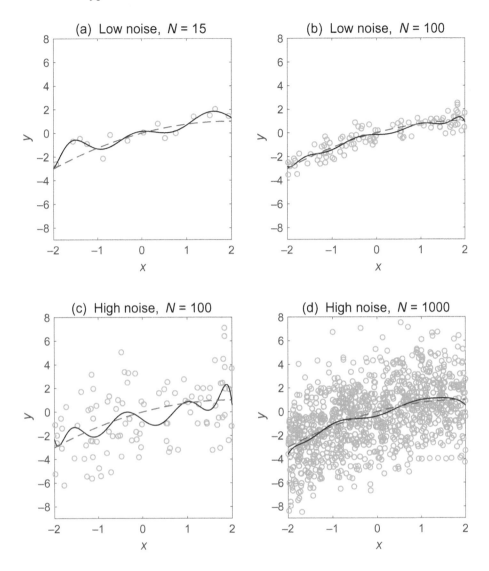

Figure 1.6 The ninth order polynomial fit to data with two noise levels and different numbers of training data points. The circles indicate the data points used for training, the solid curve the polynomial solution $\hat{y}$ and the dashed curve the true signal $y_{\text{signal}}$.

pollutant concentration, and so on. Examples of data described by probability distributions include a Gaussian distribution of given mean and standard deviation describing the temperature, a Weibull distribution describing the wind speed, and so on.

The early applications of ML were almost entirely done with discrete/categorical data; even today, the vast majority of ML applications in commercial

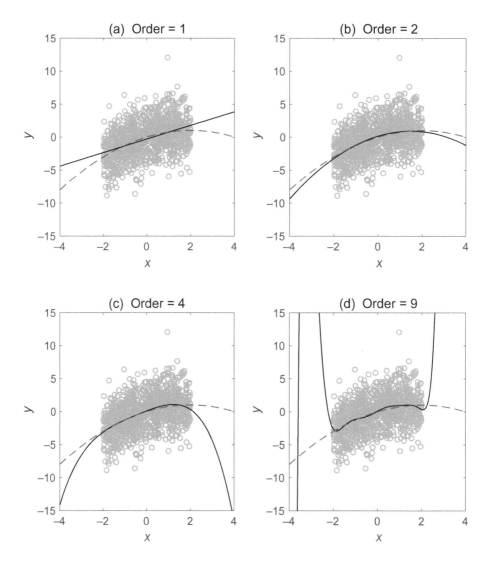

Figure 1.7 Extrapolating the polynomial solution to beyond the training data domain, where 1,000 data points (circles) were used for training and the order of the polynomial was (a) 1, (b) 2, (c) 4 and (d) 9. In (d), extending to the left side, the curve first shot up beyond the top of the plot, then plunged back down.

and engineering fields involve discrete/categorical data. In contrast, environmental scientists prefer continuous data, for example predicting tomorrow's temperature to be 28.4 °C instead of just being 'warm', which is rather vague. Statisticians prefer working with probability distributions, for example, the predicted temperature for tomorrow can be described by a Gaussian distribution with mean of 28.4 °C and a standard deviation of 1.6 °C. Of the three types of data description, the probability distribution approach contains the most information; however, it usually involves assuming the form of the distribution, for example Gaussian, which may or may not be a good assumption. With ML focusing on discrete/categorical data and statistics on probability distributions, there was very little linkage between ML and statistics in the early days. Fortunately, the linkage is much improved – for instance, most ML methods can now be cast in a probabilistic framework (K. P. Murphy, 2012).

What can we learn from the data? There are two main types of learning – *supervised learning* and *unsupervised learning*. An analogy for the former is a student trying to learn the correct French pronunciation being demonstrated by the teacher in a French class. An analogy for the latter is a solitary child playing with a jigsaw puzzle. In unsupervised learning, the student is provided with learning rules, but must rely on self-organization to arrive at a solution, without the benefit of being able to learn from a teacher's demonstration. Besides supervised and unsupervised learning, there is a third and less common type of learning – *reinforced learning*, which is briefly described in Section 1.4.3.

## 1.4.1 Supervised Learning $\boxed{\text{A}}$ ☺

Given some training data $\{\mathbf{x}_i, \mathbf{y}_i\}$, $(i = 1, ..., N)$, supervised learning tries to find a mapping from the input variables $\mathbf{x}$ to the output variables $\hat{\mathbf{y}}$ (with the 'hat' on $\hat{\mathbf{y}}$ distinguishing the model output from the observed data or target data $\mathbf{y}_i$ or $\mathbf{y}$). For a new input $\mathbf{x}'$, one can then use the mapping to predict $\hat{\mathbf{y}}'$. The inputs are also called predictors, independent variables, features, attributes or covariates, and the outputs are also called response variables, dependent variables or predictands. $N$ is called the sample size or the number of observations or data points. Observations are also called examples, cases or patterns in ML. In many applications, the model vector output $\hat{\mathbf{y}}$ reduces to a scalar $\hat{y}$.

Supervised learning is divided into *regression* and *classification* – regression if the output variables are real variables and classification if output variables are discrete/categorical. An example of regression: $\mathbf{x}$ contains three variables, air temperature, humidity and pressure, and $\hat{y}$ is the wind speed for the next day. For regression, the inputs are usually also real variables, though it is possible to include discrete/categorical variables in the inputs.

For classification, the discrete/categorical output $\hat{y} \in \{1, ..., C\}$, with $C$ being the number of classes. If $C = 2$, this is binary classification, whereas for $C > 2$, multi-class classification. The inputs can be real or discrete/categorical variables. For instance, we can again have $\mathbf{x}$ containing air temperature, humidity and pressure, and $\hat{y}$ being 'storm' or 'no storm' for the next day – that is, the trained model will be used to issue storm warnings. Examples of multi-class classification include seasonal temperature forecasts of 'cool', 'normal' or 'warm' conditions, and satellite classifying observed ground pixels as being 'land', 'water', 'snow' or 'ice'.

Environmental scientists may find it surprising that the non-environmental applications of ML are predominantly classification instead of regression. Examples of common non-environmental applications include: (i) spam filters classifying emails into 'spam' and 'not spam', (ii) banks classifying credit card transactions into 'suspicious' and 'not suspicious', (iii) handwriting recognition software using inputs of digitalized pixels of handwriting to classify into alphabets and numerals and (iv) object recognition software using inputs of photo images. In (iv), the number of classes can be very large, since the object in the photo can be a cat, rocket, car, house, and so on.

Some environmental problems involve both classification and regression, for example, in precipitation forecast, it is common to first use a classification model to choose between 'no precipitation' and 'precipitation', and if the output is 'precipitation', then a regression model is used to predict the amount of precipitation.

## 1.4.2   Unsupervised Learning   A☺

In contrast to supervised learning with input data $\mathbf{x}$ and output data $\hat{\mathbf{y}}$, unsupervised learning has only input data $\mathbf{x}$ to work with – the goal here is to find structure within the $\mathbf{x}$ data. For instance, some of the $\mathbf{x}$ data points are similar to each other and are located close together within the $\mathbf{x}$ space, so clustering is used to find the groups or clusters in the $\mathbf{x}$ data. For instance, the large-scale variability in the atmosphere displays several commonly occurring patterns (i.e. teleconnection patterns), and clustering (a.k.a. cluster analysis) (Section 10.1) has been used to extract these patterns from the atmospheric data (M. Bao and Wallace, 2015).

Another application involves reducing the $\mathbf{x}$ space. Suppose $\mathbf{x}$ contains 100 variables, spanning 100 dimensions. The data do not uniformly fill the 100-dimensional space, but may spread mainly along, say, two or three directions. Dimension reduction methods try to find the two or three directions displaying the strongest spread in the data. The essence of the 100-D dataset is now nicely condensed into a dataset with only two or three dimensions. For instance, principal component analysis (Chapter 9) (Jolliffe, 2002; Jolliffe and Cadima, 2016) is commonly used to condense high-dimensional environmental datasets to much lower dimensional datasets.

Which is more important – supervised or unsupervised learning? In Gorder (2006, p. 5), Professor Geoffrey Hinton was quoted on his view that human learning is mostly unsupervised:

> When we're learning to see, nobody's telling us what the right answers are – we just look. Every so often, your mother says "that's a dog", but that's very little information. You'd be lucky if you got a few bits of information – even one bit per second – that way. The brain's visual system requires $10^{14}$ [neural] connections. And you only live for $10^9$ seconds. So it's no use learning one bit per second. You need more like $10^5$ bits per second. And there's only one place you can get that much information – from the input itself.

In their review of deep learning (i.e. neural network models with many processing layers), LeCun, Bengio, et al. (2015, p. 442) concluded:

> Unsupervised learning had a catalytic effect in reviving interest in deep learning, but has since been overshadowed by the successes of purely supervised learning ... we expect unsupervised learning to become far more important in the longer term. Human and animal learning is largely unsupervised: we discover the structure of the world by observing it, not by being told the name of every object.

### 1.4.3 Reinforced Learning $\boxed{\text{A}}$☺

Although much less widely used than supervised learning and unsupervised learning, reinforcement learning is still a notable branch of ML. Mainly used in game theory, control theory and operations research, reinforcement learning is concerned with improving the behaviour of intelligent agents (e.g. robots) to maximize the cumulative reward in an interactive environment. When training a robot soccer team, using supervised learning by teaching the robots to imitate the top human soccer players would be suboptimal, since the robots have different motor skills from humans (e.g. running speed, flexibility, balance, etc.). Instead, reinforced learning would use cumulative reward to gradually improve the performance of the robot soccer team. As the soccer team plays against other teams, rewards for good moves (e.g. controlling the ball, scoring a goal, etc.) and punishments for bad moves are used to gradually change the behaviour of the robots, so the team evolves into a stronger team (Riedmiller et al., 2009).

When building computer game players, using supervised learning by having the computer imitate the top human players would limit the skill of the computer player. To surpass the top human players, reinforced learning is needed. Reinforced learning using neural networks first became famous with the backgammon computer algorithm developed by Tesauro (1994), which improved its skills by playing against itself, eventually attaining the level of top human players.

Reinforced learning is not pursued in this book, as it does not appear to have important applications in the environmental sciences.

## 1.5 Curse of Dimensionality $\boxed{\text{A}}$☺

As modern datasets contain increasingly more variables, their high dimensions introduce a problem known as the 'curse of dimensionality' (Bellman, 1961), where data methods developed for low-dimensional datasets become unusable at high dimensions. Figure 1.8 illustrates the effect of increasing the dimension, where in 1-D, a segment of width 0.5 covers $1/2$ of the unit interval $[0, 1]$, in 2-D, a square of width 0.5 covers $1/4$ of the unit square, while in 3-D, a cube of width 0.5 covers $1/8$ of the unit cube. In a $d$-dimensional space, a hypercube of width 0.5 covers $2^{-d}$ of the unit hypercube. If one samples 100 data points over a hypercube, then in 1-D, each segment of width 0.5 will have an average of 50 data points, in 2-D each square of width 0.5 will have about 25 data points, and so on. In 10-D, there are $2^{10} = 1024$ hypercubes of width 0.5, so each hypercube contains less than 0.1 data point. In other words, sampling becomes very sparse for high-dimensional problems unless one has a huge amount of data. Techniques such as $K$-nearest neighbours (Section 12.4), which makes a prediction for a test input data by looking at the behaviour of its $K$ nearest neighbours in the training data, break down at high dimension since the neighbours are far away. Thus, the problem of test input data being outliers relative to the training input data (as mentioned in Section 1.2) is more serious with high dimensions. The polynomial fit is another example of a method that generalizes poorly to high dimensions (Section 6.3.1).

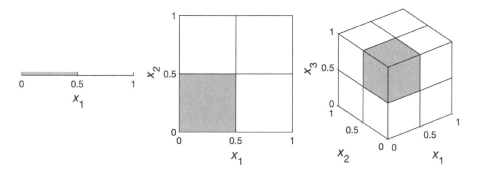

Figure 1.8 The 'curse of dimensionality' effect, as one proceeds from one to three dimensions.

In practice, methods have been successfully developed to work in high-dimensional space. Real high-dimensional data usually concentrate into a much smaller number of effective dimensions, and real data typically have some smoothness properties allowing local interpolation. Methods useful for high-dimensional problems tend to exploit these two properties.

# 2

# Basics

## 2.1 Data Types  A☺

Environmental scientists are familiar with numerical data, especially with continuous numerical data – for example temperature, pressure, pollution concentration, specific humidity, streamflow and sea level height. Numerical data can also be discrete, being recorded in integers instead of real numbers. Figure 2.1 shows there are other types of data, namely categorical data. *Categorical* variables represent data which can be placed into groups or categories. Categorical data can, in turn, be either nominal or ordinal. *Nominal* data do not have order, for example true/false, colour (red, green, blue), country of birth (USA, China, Russia, Liechtenstein, etc.), animals (cats, dogs, elephants, etc.). *Ordinal* data have categories with some natural order, such as weather type (sunny, cloudy, rainy), education level (elementary school graduate, high school graduate, some college and college graduate) and the Likert scale used in customer surveys (strongly disagree, disagree, neutral, agree, strongly agree). Unlike in the environmental sciences, where data are predominantly continuous, data in commercial or computer science application areas tend to be predominantly categorical and/or discrete. This has an important bearing as data methods developed for commercial or computer science applications tend to be predominantly designed for categorical and/or discrete data. Some were later adapted to work with continuous data.

## 2.2 Probability  A☺

Environmental data contain fluctuations – for example, the atmosphere has fluctuations ranging from large-scale weather systems to small-scale turbulence. Thus, an understanding of random variables and probability theory is essential for analysing environmental data.

We start with a simple example for illustrating the basic concepts of probability. Suppose one has 100 days of weather observations, where there two

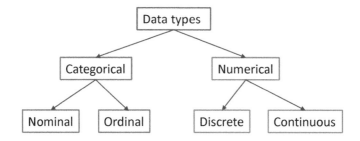

Figure 2.1 Main types of data.

variables in the daily weather, namely temperature and precipitation. For simplicity, the temperature variable has three classes – cold (c), normal (n) and hot
(h) – while the precipitation has two classes – dry (d) and wet (w). Table 2.1
shows the distribution of the weather data – for example, out of 100 days, there
are 15 days of cold dry weather, 5 days of cold wet weather, and so on. The bottom row gives the sum over the numbers in the column above, for example the
total number of cold days is $15 + 5 = 20$. The rightmost column gives the sum
over the row, for example the total number of dry days is $15 + 35 + 10 = 60$.

Table 2.1 Distribution of 100 days of weather observations, with the corresponding probability distribution $P(x, y)$ listed in the table to the right. $P(x)$ and
$P(y)$ are the marginal distributions.

|       | cold | norm | hot | sum |
|-------|------|------|-----|-----|
| dry   | 15   | 35   | 10  | 60  |
| wet   | 5    | 15   | 20  | 40  |
| sum   | 20   | 50   | 30  | 100 |

| $P(x, y)$ | $x = c$ | $x = n$ | $x = h$ | $P(y)$ |
|-----------|---------|---------|---------|--------|
| $y = d$   | 0.15    | 0.35    | 0.10    | 0.60   |
| $y = w$   | 0.05    | 0.15    | 0.20    | 0.40   |
| $P(x)$    | 0.20    | 0.50    | 0.30    | 1      |

Next, we want to obtain the probability distribution $P(x, y)$, where $x$ and
$y$ are the temperature and precipitation variables, respectively. $P(c, d)$, the
probability of cold dry weather, is simply the number of observations with cold
dry weather divided by $N$, the total number of observations, that is, $P(c, d) =$
$15/100 = 0.15$. The probability table is shown on the right side of Table 2.1.
Strictly speaking, probability is defined only in the limit as $N \to \infty$, so our finite
$N$ only allows us to get an estimate of the true probability. $P(x, y)$ is called a
*joint probability* or *joint distribution* as it depends on both $x$ and $y$. The bottom
row $P(x)$ and the rightmost column $P(y)$ are called *marginal distributions*, as
they appear on the margins of probability tables. They are obtained by summing
over $P(x, y)$,

$$P(x) = \sum_y P(x, y), \qquad P(y) = \sum_x P(x, y), \qquad (2.1)$$

as one can check by summing over the rows and columns of $P(x,y)$ in Table 2.1. From $P(y)$, the probability of dry days, $P(d)$, is 0.60, while the probability of wet days is 0.40. Note that

$$\sum_y P(y) = 0.60 + 0.40 = 1, \quad \text{and} \quad \sum_x P(x) = 0.20 + 0.50 + 0.30 = 1, \quad (2.2)$$

as the sum of the probabilities over all the events must equal one.

$P(x|y)$, the *conditional probability* of $x$ given $y$, is the probability of observing $x$ when the value $y$ is already known. For instance, if $x = c$ and $y = d$, $P(c|d)$ is the probability of getting cold temperature under dry conditions. Since the joint probability of getting $x$ and $y$, that is, $P(x,y)$, equals the probability of getting $y$, that is, $P(y)$, multiplied by the conditional probability of getting $x$ given $y$, that is, $P(x|y)$, we can write

$$P(x,y) = P(x|y)P(y). \quad (2.3)$$

Thus,

$$P(x|y) = \frac{P(x,y)}{P(y)}, \quad \text{if } P(y) > 0, \quad \text{otherwise } 0. \quad (2.4)$$

Using the values from Table 2.1, $P(c|d)$, the conditional probability of having cold conditions given it is dry, is $P(c,d)/P(d) = 0.15/0.60 = 0.25$.

Similarly,

$$P(x,y) = P(y|x)P(x), \quad (2.5)$$

where $P(y|x)$ is the conditional probability of $y$ given $x$. Thus,

$$P(y|x) = \frac{P(x,y)}{P(x)}, \quad \text{if } P(x) > 0, \quad \text{otherwise } 0. \quad (2.6)$$

Using Table 2.1, $P(d|c)$, the conditional probability of having dry conditions given it is cold, is $P(c,d)/P(c) = 0.15/0.20 = 0.75$.

Combining (2.3) and (2.5) gives

$$P(y|x)P(x) = P(x|y)P(y). \quad (2.7)$$

Thus,

$$P(y|x) = \frac{P(x|y)P(y)}{P(x)} = \frac{P(x|y)P(y)}{\sum_y P(x,y)} = \frac{P(x|y)P(y)}{\sum_y P(x|y)P(y)}, \quad (2.8)$$

upon invoking (2.1) and (2.3). This equation is called *Bayes theorem* or Bayes rule, having originated from the work of Thomas Bayes (1702–1761), an English mathematician and Presbyterian minister. For more details on Bayes theorem, see Section 2.14.

If the probability of getting $x$ is not affected at all by the given value of $y$, we have

$$P(x|y) = P(x). \quad (2.9)$$

Then, (2.3) simplifies to
$$P(x,y) = P(x)P(y),  \tag{2.10}$$
and $x$ and $y$ are said to be *independent* events. If one computes $P(x)P(y)$ from Table 2.1, one finds the product not equal to $P(x,y)$, so $x$ and $y$ are not independent events in that dataset.

Keeping the same $P(x)$ and $P(y)$ from Table 2.1, one can check that Table 2.2 indeed satisfies (2.10), so $x$ and $y$ are independent. Thus, in this example, the probability of getting dry or wet weather is unaffected by whether the temperature is cold, normal or hot, and similarly, the probability of getting cold, normal or hot weather is unaffected by whether it is dry or wet.

Table 2.2 Probability distribution $P(x,y)$, with $x$ and $y$ being independent.

| $P(x,y)$ | $x = c$ | $x = n$ | $x = h$ | $P(y)$ |
|---|---|---|---|---|
| $y = d$ | 0.12 | 0.30 | 0.18 | 0.60 |
| $y = w$ | 0.08 | 0.20 | 0.12 | 0.40 |
| $P(x)$ | 0.20 | 0.50 | 0.30 | 1 |

## 2.3   Probability Density  $\boxed{\text{A}}$ ☺

Thus far, only probabilities of discrete events have been considered. Next, we extend the concept of probability to continuous variables. Suppose the probability of a real variable $x$ lying within the interval $(x, x + \delta x)$ is denoted by $p(x)\delta x$ for $\delta x \to 0$ (Fig. 2.2(a)), then $p(x)$ is called the *probability density* or

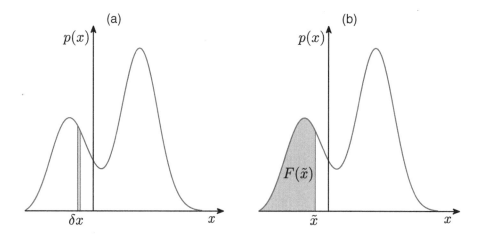

Figure 2.2 (a) The probability of $x$ lying within the interval $(x, x + \delta x)$ is given by the area of the narrow vertical band of height $p(x)$ and width $\delta x$. The two peaks in $p(x)$ indicate the two regions of higher probability. (b) The cumulative distribution $F(\tilde{x})$ is given by the shaded area under the curve.

probability density function (PDF) over $x$. The probability of $x$ lying within the interval $(a, b)$ is obtained by integrating the PDF:[1]

$$P(x \in (a, b)) = \int_a^b p(x)dx. \tag{2.11}$$

As probabilities cannot be negative, it follows that

$$p(x) \geq 0. \tag{2.12}$$

The requirement that the sum of probabilities over all discrete events equals one is replaced in the continuous case by

$$\int_{-\infty}^{\infty} p(x)dx = 1. \tag{2.13}$$

Note that $p(x)$ is not prohibited from exceeding 1.

The *cumulative distribution function* (CDF) $F(\tilde{x})$ is defined to be

$$F(\tilde{x}) = P(x \leq \tilde{x}) = \int_{-\infty}^{\tilde{x}} p(x)dx. \tag{2.14}$$

In Fig. 2.2(b), $F(\tilde{x})$ is seen as the area under the curve $p(x)$, stretching over the interval $-\infty < x \leq \tilde{x}$. It follows from taking the derivative of $F$ that

$$p(x) = \frac{dF(x)}{dx}. \tag{2.15}$$

From (2.14) and (2.11), we have

$$F(b) - F(a) = \int_a^b p(x)dx = P(x \in (a, b)). \tag{2.16}$$

The *complementary cumulative distribution function* (CCDF) or simply the *tail distribution* is

$$\tilde{F}(\tilde{x}) \equiv 1 - F(\tilde{x}) = P(\tilde{x} < x) = \int_{\tilde{x}}^{\infty} p(x)dx. \tag{2.17}$$

If there are two continuous variables $x$ and $y$, the joint probability density distribution is $p(x, y)$. The *marginal probability density* distributions are defined similar to the marginal probability distributions for discrete variables in (2.1), but with integration replacing summation, that is

$$p(x) = \int p(x, y)\, dy, \qquad p(y) = \int p(x, y)\, dx, \tag{2.18}$$

where the integrations are over the domains of $y$ and $x$, respectively.

Similar to (2.6) for discrete variables, the *conditional probability density* distribution can be defined by

$$p(x|y) = \frac{p(x, y)}{p(y)} \tag{2.19}$$

for $p(y) > 0$, and analogously for $p(y|x)$.

---

[1] In this book, we try to follow the convention of using the capital letter $P$ to denote a probability and the small letter $p$ to denote a probability density.

## 2.4 Expectation and Mean  $\boxed{\text{A}}$ ☺

Let $x$ be a random variable that takes on discrete values. For example, $x$ can be the outcome of a die cast, where the possible values are $x_i = i$, with $i = 1, \ldots, 6$. The *expectation* or expected value of $x$ from a population is given by

$$E[x] = \sum_i x_i P_i, \tag{2.20}$$

where $P_i$ is the probability of $x_i$ occurring. If the die is fair, $P_i = 1/6$ for all $i$, so $E[x]$ is 3.5. We also write

$$E[x] = \mu_x, \tag{2.21}$$

with $\mu_x$ denoting the *mean* of $x$ for the population, that is, the *population mean*.

The expectation of a sum of random variables satisfies

$$E[ax + by + c] = a\,E[x] + b\,E[y] + c, \tag{2.22}$$

where $x$ and $y$ are random variables, and $a$, $b$ and $c$ are constants.

For a random variable $x$ that takes on continuous values over a domain $\Omega$, the expectation is given by an integral,

$$E[x] = \int_\Omega x\,p(x)\,\mathrm{d}x, \tag{2.23}$$

where $p(x)$ is the PDF. For any function $f(x)$, the expectation is

$$E[f(x)] = \int_\Omega f(x)p(x)\,\mathrm{d}x \quad \text{(continuous case)},$$

$$E[f(x)] = \sum_i f(x_i)P_i \quad \text{(discrete case)}. \tag{2.24}$$

In real world problems, one normally cannot compute the mean by using the formula for the population mean, that is, (2.20) or (2.23), because one does not know $P_i$ or $p(x)$. One can sample only $N$ measurements of $x$ $(x_1, \ldots, x_N)$ from the population. The *sample mean* $\bar{x}$ or $\langle x \rangle$ is calculated by

$$\bar{x} \equiv \langle x \rangle = \frac{1}{N} \sum_{i=1}^{N} x_i, \tag{2.25}$$

that is, simply taking the average of the $N$ measurements, which is in general different from the population mean $\mu_x$. However, one can show that the expectation of the sample mean equals the population mean. Thus, as the sample size increases, the sample mean approaches the population mean. In general, the sample mean should be regarded as a statistical estimator of the true population mean.

## 2.5   Variance and Standard Deviation   $\boxed{\text{A}}$ ☺

The fluctuations around the mean value are commonly characterized by the variance of the population (i.e. the *population variance*),

$$\text{var}(x) \equiv \text{E}[(x - \mu_x)^2] = \text{E}[x^2 - 2x\mu_x + \mu_x^2]$$
$$= \text{E}[x^2] - 2\mu_x\text{E}[x] + \mu_x^2 = \text{E}[x^2] - \mu_x^2, \tag{2.26}$$

where (2.22) and (2.21) have been invoked. The *population standard deviation* $\sigma$ is the positive square root of the population variance, that is,

$$\sigma^2 = \text{var}(x). \tag{2.27}$$

The *sample standard deviation* $s$ is the positive square root of the *sample variance*, given by

$$s^2 = \frac{1}{N-1}\sum_{i=1}^{N}(x_i - \overline{x})^2. \tag{2.28}$$

The denominator of $N - 1$ (instead of $N$) is a bias correction introduced by Friedrich Bessel (1784–1846) to ensure the expectation of the sample variance equals the population variance. As the sample size increases, the sample variance approaches the population variance. For large $N$, there is negligible difference in the result when using $N - 1$ or $N$ in the denominator of (2.28), so either form can be used. The sample variance is a statistical estimator of the population variance.

Note that the standard deviation has the same unit as the variable $x$. For instance, if $x$ is wind speed measured in $\text{m}\,\text{s}^{-1}$, then its mean and standard deviation will also have units of $\text{m}\,\text{s}^{-1}$, while the variance will have units of $\text{m}^2\text{s}^{-2}$. The mean is a measure of the location of the data, while the standard deviation is a measure of the scale or spread of the data.

Often one would like to compare two very different variables, for example sea surface temperature and fish catch, which have different units and very likely different magnitudes. To avoid 'comparing apples with oranges', one usually standardizes the variables. The *standardized variable*

$$z = (x - \overline{x})/s \tag{2.29}$$

is obtained from the original variable by subtracting the sample mean and dividing by the sample standard deviation.[2] The standardized variable is also called the standard score, $z$-score, $z$-value, normal score, normalized variable or standardized anomaly (where *anomaly* means deviation from the mean value). The advantage of standardizing variables is that now the sea surface temperature and fish catch standardized variables will both have no units and have sample means of zero and sample standard deviations of one.

---

[2] In situations where the population mean $\mu_x$ or standard deviation $\sigma$ are known, they are used instead of the sample mean $\overline{x}$ and standard deviation $s$ in (2.29).

## 2.6   Covariance   $\boxed{A}$☺

For two random variables $x$ and $y$, with mean $\mu_x$ and $\mu_y$, respectively, their *covariance* is defined by

$$\text{cov}(x, y) = \text{E}[(x - \mu_x)(y - \mu_y)]. \tag{2.30}$$

For brevity, we will use 'covariance' instead of 'population covariance' when there is no ambiguity. Note that cov is symmetric with respect to its two arguments, that is, $\text{cov}(x, y) = \text{cov}(y, x)$. The variance in (2.26) is simply a special case of the covariance, with

$$\text{var}(x) = \text{cov}(x, x). \tag{2.31}$$

Covariance is a measure of the joint variability of $x$ and $y$. If high values of $x$ occur together with high values of $y$, then $(x - \mu_x)(y - \mu_y)$ will be positive; similarly, if low values of $x$ occur together with low values of $y$, then $(x - \mu_x)(y - \mu_y)$ will involve multiplying two negative numbers and so will also be positive – leading to a positive covariance. If high values of $x$ occur together with low values of $y$, and vice versa – the covariance will be negative. Thus, a positive covariance indicates a tendency of similar behaviour between $x$ and $y$, whereas a negative covariance indicates opposite behaviour. For instance, if the covariance between temperature and snow amount is negative, then high temperature tends to occur with low snow and low temperature with high snow.

One can show that if $x$ and $y$ are independent, then their covariance is zero. However, the converse is not true in general – for example, if $x$ is uniformly distributed in $[-1, 1]$ and $y = x^2$, one can show that $\text{cov}(x, y)$ is zero, even as $y$ depends on $x$ non-linearly. Thus, covariance only measures the linear joint variability between two variables.

The sample covariance computed from the data by

$$\text{cov}(x, y) = \frac{1}{N-1} \sum_{i=1}^{N} (x_i - \bar{x})(y_i - \bar{y}) \tag{2.32}$$

approaches the population covariance as $N \to \infty$.

The magnitude of the covariance is not too useful since it is not normalized and, therefore, depends on the magnitudes of the variables. For instance, if $x$ and $y$ are measured in units of centimetres instead of metres, the covariance computed using measurements in cm will be $10^4$ times that using metres. The normalized version of the covariance, the correlation coefficient (Section 2.11), is widely used as its magnitude reveals the strength of the linear relation.

If instead of just two variables $x$ and $y$, we have $d$ variables, that is, $\mathbf{x} = x_1, ..., x_d$, then the covariance coefficient generalizes to the *covariance matrix*:

$$\mathrm{cov}(\mathbf{x}) = \mathrm{E}\left[(\mathbf{x} - \mathrm{E}[\mathbf{x}])(\mathbf{x} - \mathrm{E}[\mathbf{x}])^{\mathrm{T}}\right]$$

$$= \begin{bmatrix} \mathrm{var}(x_1) & \mathrm{cov}(x_1, x_2) & \cdots & \mathrm{cov}(x_1, x_d) \\ \mathrm{cov}(x_2, x_1) & \mathrm{var}(x_2) & \cdots & \mathrm{cov}(x_2, x_d) \\ \vdots & \vdots & \ddots & \vdots \\ \mathrm{cov}(x_d, x_1) & \mathrm{cov}(x_d, x_2) & \cdots & \mathrm{var}(x_d) \end{bmatrix}, \tag{2.33}$$

where the superscript T denotes the transpose of a vector or matrix.

Another way to generalize (2.32) is by letting

$$\mathrm{cov}(\mathbf{x}, \mathbf{y}) = \mathrm{E}\left[(\mathbf{x} - \mathrm{E}[\mathbf{x}])(\mathbf{y} - \mathrm{E}[\mathbf{y}])^{\mathrm{T}}\right]$$

$$= \begin{bmatrix} \mathrm{cov}(x_1, y_1) & \mathrm{cov}(x_1, y_2) & \cdots & \mathrm{cov}(x_1, y_d) \\ \mathrm{cov}(x_2, y_1) & \mathrm{cov}(x_2, y_2) & \cdots & \mathrm{cov}(x_2, y_d) \\ \vdots & \vdots & \ddots & \vdots \\ \mathrm{cov}(x_d, y_1) & \mathrm{cov}(x_d, y_2) & \cdots & \mathrm{cov}(x_d, y_d) \end{bmatrix}. \tag{2.34}$$

Clearly, $\mathrm{cov}(\mathbf{x})$ in (2.33) is equivalent to the special case $\mathrm{cov}(\mathbf{x}, \mathbf{x})$ in (2.34). There is no standard nomenclature for $\mathrm{cov}(\mathbf{x})$ and $\mathrm{cov}(\mathbf{x}, \mathbf{y})$, as both are referred to as covariance matrices. Some authors call $\mathrm{cov}(\mathbf{x})$ the variance matrix or the variance-covariance matrix and $\mathrm{cov}(\mathbf{x}, \mathbf{y})$ the covariance matrix, while others call $\mathrm{cov}(\mathbf{x})$ the covariance matrix and $\mathrm{cov}(\mathbf{x}, \mathbf{y})$ the cross-covariance matrix. In this book, we will use 'covariance matrix' to denote either $\mathrm{cov}(\mathbf{x})$ or $\mathrm{cov}(\mathbf{x}, \mathbf{y})$.

## 2.7 Online Algorithms for Mean, Variance and Covariance $\boxed{\text{C}}$☺

In recent decades, there has been increasing interest in *online learning* problems where data become available in a sequential order and the models are to be updated with the continually arriving new data. The traditional *batch learning* approach, which trains the model with the complete training dataset, is very inefficient in the online learning situation – to update the model with one additional data point, the model has to be retrained with the complete dataset containing $N$ points. In contrast, an online learning algorithm would update the model with only the single new data point – and all previous data points can be erased from the computer memory. Obviously, when one has to update a model frequently with newly arrived data, an online learning algorithm would have a huge advantage over a batch learning algorithm in terms of computer time and memory.

First, consider the sample mean. The batch algorithm is given by (2.25), where one has to use all $N$ data points for the computation. To develop an online algorithm, we first define the sample mean computed with $N$ data points to be

$$\bar{x}_N \equiv \frac{1}{N} \sum_{i=1}^{N} x_i, \tag{2.35}$$

which can be rewritten as

$$N\,\overline{x}_N = \sum_{i=1}^{N} x_i = \sum_{i=1}^{N-1} x_i + x_N \qquad (2.36)$$

$$= (N-1)\,\overline{x}_{N-1} + x_N = N\,\overline{x}_{N-1} + x_N - \overline{x}_{N-1}. \qquad (2.37)$$

Thus, the online algorithm for the sample mean is given by

$$\overline{x}_N = \overline{x}_{N-1} + \frac{x_N - \overline{x}_{N-1}}{N}. \qquad (2.38)$$

This means that if one has $\overline{x}_{N-1}$, the sample mean for the first $N-1$ data points, and a new data point $x_N$, then the updated sample mean for the $N$ data points can be obtained simply from $\overline{x}_{N-1}$ and $x_N$. The earlier data points $x_1, \ldots, x_{N-1}$ are not needed in this update of the sample mean. The ability to delete old data can be very helpful as datasets can grow to enormous size as time passes.

Let us count the number of operations in the two approaches. In the batch algorithm (2.35), there are $N-1$ additions followed by one division. In the online algorithm (2.38), there are one subtraction, one division and one addition. When $N$ becomes large, the batch algorithm becomes much slower than the online algorithm.

For online updating of the sample variance, the Welford algorithm (Welford, 1962; Knuth, 1998, vol. 2, p. 232) involves updating the mean with (2.38) and updating the sum of squared differences

$$S_N \equiv \sum_{i=1}^{N} (x_i - \overline{x}_N)^2, \qquad (2.39)$$

by

$$S_N = S_{N-1} + (x_N - \overline{x}_{N-1})(x_N - \overline{x}_N), \qquad N \geq 2, \qquad (2.40)$$

with the sample variance

$$s_N^2 = \frac{S_N}{N-1}. \qquad (2.41)$$

Similarly, for an online algorithm to compute the sample covariance, define

$$C_N \equiv \sum_{i=1}^{N} (x_i - \overline{x}_N)(y_i - \overline{y}_N). \qquad (2.42)$$

With

$$\overline{y}_N = \overline{y}_{N-1} + \frac{y_N - \overline{y}_{N-1}}{N}, \qquad (2.43)$$

one can show that

$$C_N = C_{N-1} + (x_N - \overline{x}_{N-1})(y_N - \overline{y}_N) \qquad (2.44)$$

$$= C_{N-1} + \frac{N-1}{N}(x_N - \overline{x}_{N-1})(y_N - \overline{y}_{N-1}), \qquad (2.45)$$

with the sample covariance being $C_N/(N-1)$.

## 2.8 Median and Median Absolute Deviation
 [A] ☺

In the last few decades, there has been increasing usage of *robust statistics* to alleviate weaknesses in traditional statistical estimators (Wilcox, 2004). Traditional statistical methods commonly make assumptions (e.g. the random variables obey a Gaussian distribution) that may not be valid for some datasets, leading to poor statistical estimates. Statistical methods that perform poorly when the underlying assumptions are not satisfied are called *non-robust*. Robust methods are designed to work well with a broad range of datasets.

Another weakness is referred to as being *non-resistant* to outliers in the data – an *outlier* being an extreme data value arising from a measurement or recording error, or from an abnormal event. For instance, someone entering data by hand may misread '.100' as '100', and ends up entering a number a thousand times larger than the actual value. Non-resistant methods yield poor estimates when given even a small number of outliers. Resistant methods are designed to work well even when the datasets contain outliers.

It is desirable to have methods that are *robust* and *resistant*. Some authors, such as Wilks (2011), make a distinction between robustness and resistance. However, since most methods that are robust are also resistant, and vice versa, many authors do not make a distinction between robustness and resistance and simply refer to all such methods as robust methods.

While the mean and standard deviation are the most common estimators of location and scale (or spread) of the data, they are not resistant to outliers. Suppose student A made seven repeated measurements in a laboratory experiment, recording the values (arranged in ascending order): 1.0, 1.2, 1.2, 1.3, 1.5, 1.7 and 1.8. His lab partner, student B, also recorded the same measurements but mistakenly typed in '18.' instead of '1.8' for the final data point. The mean computed by A was 1.386, but was 3.700 by B. The computed standard deviation was 0.291 by A and 6.310 by B. Clearly, the mean and standard deviation are non-resistant to outliers.

A robust/resistant alternative to the mean is the *median*, defined as the middle value of a population or a sample of measurements sorted in ascending order. In the above example of seven measurements, the middle is the fourth measurement, namely 1.3, as there are three measurements above and three below. What happens if there is an even number of data points? Suppose we drop the seventh data point and are left with six measurements. Then the third and fourth are the two middle points, and we take the average of these two values, that is, $(1.2 + 1.3)/2 = 1.25$, as the median. Thus, the median is defined to be the middle value if $N$, the number of data points, is odd, and to be the average of the two middle values if $N$ is even.

Let us return to the example with the students each recording seven measurements. Student A's mean was 1.386 and his median was 1.3, while student B's mean was 3.700 and his median was 1.3. Thus, with a completely erroneous seventh data point, student B managed to obtain the same median value as student A.

The *breakdown point* of a statistical estimator is the proportion of incorrect data points, for example data points with arbitrarily high or low values, the estimator can handle before giving a completely incorrect result. For the mean, the breakdown point is 0, since the mean cannot handle even one single incorrect data point. In contrast, the median has a breakdown point of 50%. For instance, if the above example has values recorded as 1.0, 1.2, 1.2, 1.3, 999, 999 and 999, the median will still be 1.3.

A robust and resistant substitute for the standard deviation is the *median absolute deviation* (MAD), defined by

$$\text{MAD} = \text{median}(|x - \text{median}(x)|), \tag{2.46}$$

with a breakdown point of 50%. The deviations, $x - \text{median}(x)$, are around the median instead of the mean, and computing the absolute value avoids squaring the deviations in the standard deviation formula (2.28), which amplifies the large deviations.

For student A, with median $= 1.3$, his deviations $x - \text{median}(x) = -0.3$, $-0.1$, $-0.1$, $0.0$, $0.2$, $0.4$ and $0.5$, and the absolute deviations sorted in ascending order are $0.0$, $0.1$, $0.1$, $0.2$, $0.3$, $0.4$ and $0.5$, with MAD $= 0.2$. For student B, the absolute deviations arranged in ascending order are $0.0$, $0.1$, $0.1$, $0.2$, $0.3$, $0.4$ and $16.7$, again with MAD $= 0.2$.

Unlike the mean and standard deviation, there are no simple online learning algorithms for the median and MAD, that is, algorithms where one can erase the old data as new data arrive to update the estimator.

In this book, MAD stands for median absolute deviation around the median. Other estimators with the same acronym MAD can be defined, for example, mean absolute deviation around the mean, mean absolute deviation around the median or even median absolute deviation around the mean (though this final one is not commonly used).

## 2.9   Quantiles  A ☺

Often one is interested in finding a value $x_\alpha$ where $P(x \le x_\alpha) = \alpha$ for a given value of $\alpha$, with $0 \le \alpha \le 1$. For instance, one may want to know the value $x_{0.95}$ where 95% of the distribution lies below the value $x_{0.95}$ – that is, finding the 95th percentile. As the cumulative probability distribution function $F(x_\alpha) \equiv P(x \le x_\alpha)$ from (2.14) is a monotonically increasing function, it has an inverse function $F^{-1}(\alpha)$. $F^{-1}(\alpha)$ is the value of $x_\alpha$ where $F(x_\alpha) = \alpha$. We call $q_\alpha \equiv x_\alpha$ the $\alpha$ *quantile* of $F$.

Figure 2.3 illustrates how to obtain a quantile value from a cumulative distribution function $F(x)$. Along the ordinate axis, we locate the value 0.95. To find $F^{-1}(0.95)$, we simply look at the intersection between the cumulative distribution curve and the horizontal line with ordinate $= 0.95$, leading to the value of the quantile $q_{0.95}$ along the abscissa. $F^{-1}(0.5) = q_{0.5}$ is simply the median, with 50% of the $x$ values distributed above and 50% below the median.

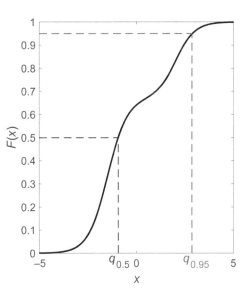

Figure 2.3 A cumulative distribution function $F(x)$. By inverse mapping from the ordinate to the abscissa, one can obtain the quantiles $q_\alpha$. The 95th percentile $q_{0.95}$ and the median $q_{0.5}$ are shown.

The median splits up the cumulative distribution into two equal halves. Other common ways to split up the cumulative distribution into quantiles include *terciles*, with $q_{0.333}$ and $q_{0.667}$ splitting the distribution into three equal parts, and *quartiles*, with $q_{0.25}$, $q_{0.5}$ and $q_{0.75}$ splitting the distribution into four equal parts. The *interquartile range* (IQR), defined by the separation between the third quartile and the first quartile,

$$\text{IQR} = q_{0.75} - q_{0.25}, \qquad (2.47)$$

is often used to characterize the spread or scale of the data, as it measures the spread of the middle 50% of the data. Since it ignores the top and bottom 25% of the data, it is resistant to outliers.

For a five-part split, *quintiles* use $q_{0.2}$, $q_{0.4}$, $q_{0.6}$ and $q_{0.8}$. For a 10-part split, *deciles* use $q_{0.1}$, $q_{0.2}$, ..., $q_{0.9}$. For a 100-part split, *percentiles* use $q_{0.01}$, $q_{0.02}$, ..., $q_{0.99}$.

Next, we examine how quantiles can be computed from a dataset $\{x_1, \ldots, x_N\}$. We first sort the data points into ascending order, that is, $x_{(1)}, \ldots, x_{(N)}$, with $x_{(1)}$ the smallest and $x_{(N)}$ the largest value in the original dataset.

Computing quantiles with observed data is not entirely straightforward. The reason is that the quantile $q_\alpha$ usually falls between $x_{(i)}$ and $x_{(i+1)}$ for some integer $i$. For example, with six data points, the median $q_{0.5}$ falls between $x_{(3)}$ and $x_{(4)}$, so we let $q_{0.5}$ be the average of the two values. In general, $q_\alpha$ need not fall midway between $x_{(i)}$ and $x_{(i+1)}$, so various schemes compute $q_\alpha$ differently. Hyndman and Y. N. Fan (1996) listed nine different schemes for computing quantiles. Fortunately, when $N \geq 100$, the differences between the various schemes become negligible.

## 2.10   Skewness and Kurtosis   $\boxed{\text{B}}$☺

As the mean is computed from the first moment of the data and the variance from the second moment, one can proceed onto *skewness*, a third moment statistic. The population skewness coefficient is traditionally defined by

$$\gamma_{\mathrm{p}} = \mathrm{E}\left[\left(\frac{x - \mu_x}{\sigma}\right)^3\right] = \frac{\mathrm{E}[(x - \mu_x)^3]}{\sigma^3}, \tag{2.48}$$

where $\mu_x$ and $\sigma$ are the population mean and standard deviation, respectively.

The sample skewness is computed from

$$\gamma = \frac{\frac{1}{N}\sum_{i=1}^{N}(x_i - \bar{x})^3}{s^3}, \tag{2.49}$$

where $\bar{x}$ and $s$ are the sample mean and standard deviation, respectively.

The skewness is easily seen to be zero for a symmetric probability distribution, for example the Gaussian distribution (Section 3.4). If the right tail of the Gaussian is made longer or fatter, the skewness becomes positive. If the left tail is longer or fatter, the skewness becomes negative. Note that while symmetry implies zero skewness, the converse is not true, as one can make the left tail fatter and the right tail longer to compensate each other, leaving the skewness at zero. Distributions for variables that are non-negative, for example wind speed, precipitation amount, pollution concentration, and so on tend to have positive skewness.

The cubic power makes the traditional skewness coefficient very non-resistant to outliers, thus rather unreliable to use in practice. A resistant skewness coefficient based on quartiles was introduced in Bowley (1901), generally regarded as the first English-language textbook on statistics, with

$$\gamma_{\mathrm{B}} = \frac{q_{0.75} + q_{0.25} - 2q_{0.5}}{q_{0.75} - q_{0.25}}, \tag{2.50}$$

where the denominator is simply the IQR. Bowley's skewness is also called Yule's coefficient or the Yule–Kendall index (Yule, 1912).

From the fourth moment of the data, the population *kurtosis* is defined by

$$\beta = \frac{\mathrm{E}[(x - \mu_x)^4]}{\sigma^4}. \tag{2.51}$$

For a Gaussian distribution, $\beta = 3$. Distributions with more outliers than the Gaussian has $\beta > 3$, while those with fewer outliers has $\beta < 3$. Many authors use 'kurtosis' to mean 'excess kurtosis' (i.e. the kurtosis of a distribution relative to that of a Gaussian), that is, $\beta' \equiv \beta - 3$, so the Gaussian has $\beta' = 0$. With the fourth power involved, the traditional kurtosis is obviously not resistant to outliers.

For more resistant higher moments, L-moments are often used (Hosking, 1990; von Storch and Zwiers, 1999).

## 2.11 Correlation Ⓐ☺

### 2.11.1 Pearson Correlation Ⓐ☺

The (Pearson) correlation coefficient, widely used to represent the strength of the linear relationship between two variables $x$ and $y$, is defined by

$$\hat{\rho}_{xy} = \frac{\text{cov}(x, y)}{\sigma_x \sigma_y}, \tag{2.52}$$

where $\sigma_x$ and $\sigma_y$ are the population standard deviations for $x$ and $y$, respectively.

For a sample containing $N$ pairs of $(x, y)$ measurements or observations, the *sample correlation* is computed by

$$\rho \equiv \rho_{xy} = \frac{\sum_{i=1}^{N}(x_i - \bar{x})(y_i - \bar{y})}{\left[\sum_{i=1}^{N}(x_i - \bar{x})^2\right]^{\frac{1}{2}}\left[\sum_{i=1}^{N}(y_i - \bar{y})^2\right]^{\frac{1}{2}}}, \tag{2.53}$$

which lies between $-1$ and $+1$. At the value $+1$, $x$ and $y$ will show a perfect straight-line relation with a positive slope, whereas at $-1$, the perfect straight line will have a negative slope. With increasing noise in the data, the sample correlation moves towards 0.

This formula for $\rho$ involves two passes with the data, as it requires a first pass to compute the means $\bar{x}$ and $\bar{y}$. Substituting in the formulas for the means (2.25), one can rewrite (2.53) as

$$\rho = \frac{N\sum_{i=1}^{N}x_i y_i - \sum_{i=1}^{N}x_i \sum_{i=1}^{N}y_i}{\left[N\sum_{i=1}^{N}x_i^2 - \left(\sum_{i=1}^{N}x_i\right)^2\right]^{\frac{1}{2}}\left[N\sum_{i=1}^{N}y_i^2 - \left(\sum_{i=1}^{N}y_i\right)^2\right]^{\frac{1}{2}}}, \tag{2.54}$$

where $\rho$ can be computed by a single pass. For some datasets, this formula can lead to the subtraction of similar numbers, resulting in the loss of significant figures. For instance, consider a number with seven significant figures: 0.1234567. If one is to subtract from it the similar number 0.1234511, one gets 0.0000056, with only two significant figures.

With two variables $x$ and $y$, a *scatterplot* that plots the data points as dots in the $x$-$y$ plane is often useful for visualizing the distribution of the data points. In Fig. 2.4, scatterplots of synthetic $(x,y)$ data are shown, along with the corresponding correlation coefficient. The $x$ variable is from a Gaussian distribution with zero mean and unit standard deviation, while $y = x + $ noise in Figs. 2.4(a), (c) and (e), and $y = -x + $ noise in Figs. 2.4(b), (d) and (f). The added random noise is Gaussian, with increasing noise lowering the magnitude of the correlation from (a) to (f).

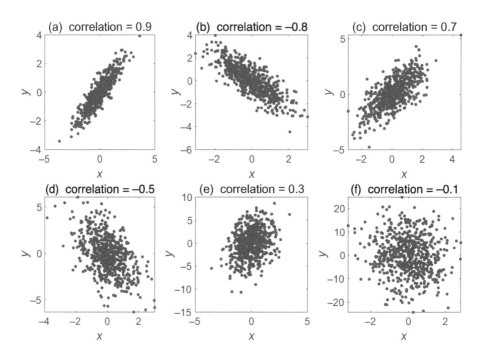

Figure 2.4 Scatterplots showing distribution of $(x, y)$ data and the corresponding Pearson correlation coefficient as the noise level rises from (a) to (f).

It will be instructive to look at scatterplots and correlations with real data. The daily surface air temperature, relative humidity, wind speed and sea level pressure at Vancouver, British Columbia, Canada, from averaged hourly observations by Environment and Climate Change Canada, were downloaded from www.weatherstats.ca. In Fig. 2.5(a), the correlation is $-0.33$ and, indeed, focusing on where the data density is high, we see lower relative humidity concurring with higher temperature, which is not surprising since Vancouver has rainy winters and dry summers. However, when temperature becomes very low, the relative humidity drops as temperature drops, opposite to our expectation from the negative correlation coefficient. The reason for this behaviour is that in winter there are occasional Arctic air outbreaks, bringing very cold, dry air from the Arctic. The strongest correlation of $-0.38$ was found between pressure and wind speed in Fig. 2.5(d), as low pressure systems give rise to stormy weather with high wind speeds.

The Pearson correlation assumes a linear relation between $x$ and $y$; however, the sample correlation is not *robust* to deviations from linearity in the relation, as illustrated in Fig. 2.6(a) where $\rho \approx 0$ though there is a strong (non-linear) relationship between the two variables. Thus, the correlation can be misleading when the underlying relation is non-linear. Furthermore, the sample correlation is not *resistant* to outliers, where in Fig. 2.6(b) $\rho = -0.50$. If the single outlier is

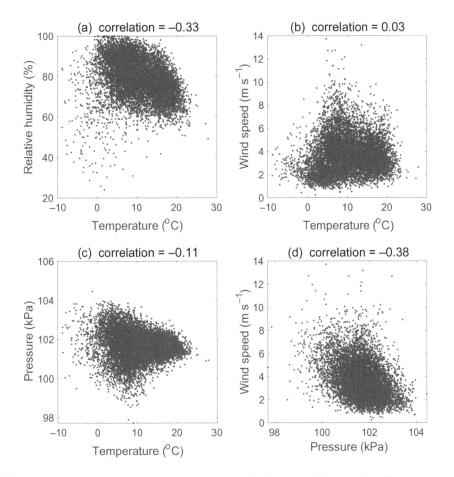

Figure 2.5 Scatterplots and Pearson correlation coefficients of daily weather variables from Vancouver, BC, Canada, with 25 years of data (1993–2017). [Data source: weatherstats.ca based on Environment and Climate Change Canada data.]

removed, $\rho$ changes from $-0.50$ to $+0.70$, that is, the strong positive correlation was completely masked by one outlier. Later in this chapter, we will study more robust/resistant estimates of the correlation.

If there are $d$ variables, for example $d$ stations reporting the air pressure, then correlations between the variables lead to a correlation matrix

$$
\mathbf{C} =
\begin{bmatrix}
\rho_{11} & \rho_{12} & \cdots & \rho_{1d} \\
\rho_{21} & \rho_{22} & \cdots & \rho_{2d} \\
\vdots & \vdots & \ddots & \vdots \\
\rho_{d1} & \rho_{d2} & \cdots & \rho_{dd}
\end{bmatrix},
\tag{2.55}
$$

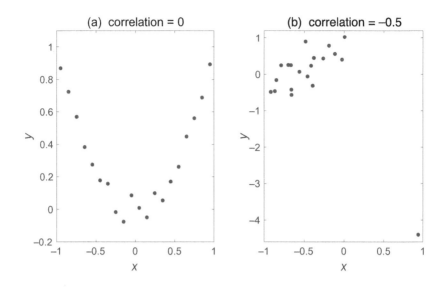

Figure 2.6 (a) An example illustrating that correlation is not robust to deviations from linearity. Here, the strong non-linear relation between $x$ and $y$ is completely missed by the near-zero correlation coefficient. (b) An example showing that correlation is not resistant to outliers. By removing the single outlier on the lower right corner, the correlation coefficient changes from negative to positive.

where $\rho_{ij}$ is the correlation between the $i$th and the $j$th variables. The diagonal elements of the matrix satisfy $\rho_{ii} = 1$, and the matrix is symmetric, that is, $\rho_{ij} = \rho_{ji}$. The $j$th column of $\mathbf{C}$ gives the correlations between the variable $j$ and all other variables. The correlation matrix is simply the normalized version of the covariance matrix $\mathrm{cov}(\mathbf{x})$ in (2.33).

## 2.11.2   Serial Correlation   [A]☺

Often the observations are measurements at regular time intervals, that is, time series, and there is *serial correlation* in the time series – that is, neighbouring data points in the time series are correlated. Serial correlation is well illustrated by *persistence* in weather patterns, for example, if it rains one day, it increases the probability of rain the following day. Serial correlation in a single time series is called *autocorrelation*. Serial correlation can involve more than one time series, for example rainfall today can increase river runoff tomorrow.

  In making statistical estimates, it is common to also estimate the confidence interval (Section 4.4). For instance, for a statistical estimate $z$, we would like to estimate the interval $[z_{\mathrm{lower}}, z_{\mathrm{upper}}]$ containing $z$, where there is 95% chance that the true parameter $z_{\mathrm{true}}$ lies within this confidence interval. In hypothesis testing (Section 4.1), one would like to know if the observed $z$ is enough to reject the null hypothesis at a certain level. In both cases, the answers depend on the sample size $N$, that is, larger sample size makes the confidence intervals narrower, or $z$ significant at a more desirable level.

Unfortunately, traditional confidence interval estimates and significance tests assume the $N$ data points are all independent observations. With serial correlation, this assumption is false, as the number of independent observations is smaller and sometimes much smaller than $N$. For example, suppose the weather is typically three days of rain, alternating with five days of sun, that is, one has a typical rainy event of three days alternating with a sunny event of five days, so over eight days, there are two events. If one has $N = 80$, there are only about 20 events, so the *effective sample size* $N_{eff}$ is only about 20. One needs to use $N_{eff}$ instead of $N$ in the significance tests and confidence interval estimates when there is serial correlation in the data (see Section 4.2.4).

To determine the degree of autocorrelation in a time series, we use the auto-correlation coefficient, where a copy of the time series is shifted in time by a lag of $l$ time intervals and then correlated with the original time series. The lag-$l$ autocorrelation coefficient is given by

$$\rho(l) = \frac{\sum_{i=1}^{N-l}[(x_i - \overline{x})(x_{i+l} - \overline{x})]}{\sum_{i=1}^{N}(x_i - \overline{x})^2}, \tag{2.56}$$

where $\overline{x}$ is the sample mean. There are other estimators of the autocorrelation function, though this version is most commonly used (von Storch and Zwiers, 1999, p. 252). The function $\rho(l)$ has the value 1 at lag 0. From symmetry, one defines $\rho(-l) = \rho(l)$.

The effective sample size can be derived as

$$N_{eff} = N\left[1 + 2\sum_{l=1}^{N-1}\left(1 - \frac{l}{N}\right)\rho(l)\right]^{-1}, \tag{2.57}$$

(von Storch and Zwiers, 1999, p. 372; Thiébaux and Zwiers, 1984). Thiébaux and Zwiers (1984) compared several methods for estimating $N_{eff}$. Their direct estimation approach involves substituting values of $\rho(l)$ into (2.57), Unfortunately, direct estimation involves estimating $\rho(l)$ at large lags (when the true autocorrelation function is effectively zero) and summing over many such terms. Even using the option of truncating the summation at large lags, direct estimation was not among the better methods (Thiébaux and Zwiers, 1984).

A better approach is to assume an auto-regressive (AR) process (Section 11.8). For the simplest AR process of order 1 (abbreviated as AR(1)), when $N$ is large, the effective sample size is approximated by

$$N'_{eff} \approx N\frac{1 - \rho(1)}{1 + \rho(1)}, \tag{2.58}$$

with $\rho(1)$ being the lag-1 autocorrelation coefficient (Zwiers and von Storch, 1995). For $0 \leq \rho(1) < 1$, (2.58) gives $0 < N'_{\text{eff}} \leq N$. If $\rho(1) = 0$, $N'_{\text{eff}} = N$, as expected for independent data. Sallenger et al. (2012) found that $\rho(1)$ from (2.56) did not give stable estimates for noisy time series; instead, they fitted an AR(1) model and substituted the AR(1) coefficient for $\rho(1)$ in (2.58) to obtain $N'_{\text{eff}}$.

It is possible to have $N'_{\text{eff}} > N$ if $\rho(1) < 0$. To keep the effective sample size within a reasonable range, Zwiers and von Storch (1995) recommended using

$$N_{\text{eff}} = \begin{cases} 2 & \text{if } N'_{\text{eff}} \leq 2, \\ N'_{\text{eff}} & \text{if } 2 < N'_{\text{eff}} \leq N, \\ N & \text{if } N < N'_{\text{eff}}, \end{cases} \qquad (2.59)$$

with $N'_{\text{eff}}$ computed from (2.58). How $N_{\text{eff}}$ is used in hypothesis testing is further pursued in Section 4.2.4.

For illustration, the autocorrelation function was computed for the daily temperature at Vancouver, BC in Fig. 2.7. That the autocorrelation function in Fig. 2.7(a) has a strong trough at around 180 days and a strong peak at around 360 days merely indicates that the time series has a strong seasonal cycle. For Gaussian *white noise*,[3] 95% of the distribution falls within the interval $[-1.96/\sqrt{N}, 1.96/\sqrt{N}]$, which is marked by the two horizontal lines in Fig. 2.7, (see Section 3.4) (Box, Jenkins, et al., 2015, Section 2.1.6; von Storch and Zwiers, 1999, pp. 252–253), that is, outside of this interval, there is only 5% chance the true autocorrelation is zero.

For the short record of $N = 90$ days during the winter of 2016–2017 (Fig. 2.7(b)), (2.59) gave $N_{\text{eff}} \approx 7.9$, an order of magnitude smaller than $N$.

## 2.11.3   Spearman Rank Correlation   $\boxed{\text{A}}$ ☺

For the correlation to be more robust and resistant to outliers, the Spearman rank correlation (Spearman, 1904) is often used instead of the Pearson correlation. If the data $\{x_1, \ldots, x_N\}$ are rearranged in order according to their size (starting with the smallest), and if $x$ is the $n$th member, then $\text{rank}(x) \equiv r_x = n$. The Spearman correlation $\rho_{\text{spearman}}$ is simply the Pearson correlation $\rho$ of $r_x$ and $r_y$, that is,

$$\rho_{\text{spearman}}(x, y) = \rho(r_x, r_y). \qquad (2.60)$$

Spearman correlation assesses how well the relationship between two variables can be described by a monotonic function. If the relation is perfectly monotonic, that is, if $x_i < x_j$, then $y_i < y_j$ for all $i \neq j$, then the Spearman correlation takes on the maximum value of 1. The minimum value of $-1$ is attained if $x_i < x_j$, then $y_i > y_j$ for all $i \neq j$. Thus, the Pearson correlation measures if the relation between two variables is linear, while Spearman measures if the relation is monotonic (regardless whether it is linear or non-linear).

---

[3] *White noise* is a random signal having equal intensity at all frequencies. The values at any two times are identically distributed and statistically independent; thus, the autocorrelation $\rho(l) = 0$ for all $l \neq 0$.

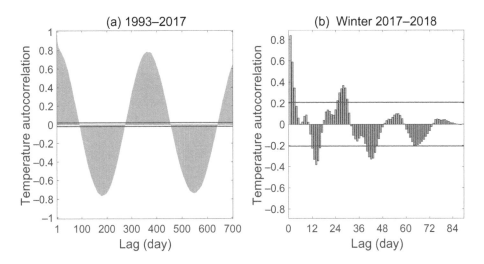

Figure 2.7 Autocorrelation function for the daily temperature at Vancouver, BC during (a) 1993–2017 and (b) winter of 2016–2017 (Dec.-Feb.), with the horizontal lines indicating the 95% confidence interval. [Data source: weather-stats.ca based on Environment and Climate Change Canada data.]

For example, if five measurements of $x$ yielded the values $1, 3, 0, 3$ and $6$, then the corresponding $r_x$ values are $2, 3.5, 1, 3.5$ and $5$ (where the tied values were all assigned an averaged rank). If measurements of $y$ yielded $2, 3, -1, 7$ and $-99$ (an outlier), then the corresponding $r_y$ values are $3, 4, 2, 5$ and $1$. The Spearman rank correlation is $-0.05$, while the Pearson correlation is $-0.79$, which shows the strong influence exerted by an outlier.

In Fig. 2.6(a), the Pearson correlation is $0.00$ while the Spearman correlation is $-0.02$, but in Fig. 2.6(b), it is $-0.50$ for Pearson versus $+0.44$ for Spearman. If the single outlier at the bottom right corner of Fig. 2.6(b) is removed, it is $0.70$ for Pearson and $0.68$ for Spearman. Clearly, the Spearman correlation is much more resistant to the outlier than Pearson.

There are alternative robust and resistant correlations, such as the Kendall rank correlation and the biweight midcorrelation.

## 2.11.4   Kendall Rank Correlation [A]☺

An alternative approach to rank correlation is via Kendall rank correlation or Kendall's tau (after the Greek letter $\tau$) (Kendall, 1938; Kendall, 1945). Given $x_i$ and $y_i$ ($i = 1, \ldots, N$), we say that a pair $(i, j)$, $i < j$, is:
- *concordant* when $x_i - x_j$ and $y_i - y_j$ are both non-zero and have the same sign;
- *discordant* when $x_i - x_j$ and $y_i - y_j$ are both non-zero and have opposite signs.

Let $C$ and $D$ be the number of concordant pairs and discordant pairs, respectively. The total number of pairs $(i, j)$ with $i < j$ is $N_0 = N(N-1)/2$.

Kendall's $\tau_A$ was defined originally in Kendall (1938) as

$$\tau_A = \frac{C - D}{N_0}, \tag{2.61}$$

which has a rather simple interpretation, namely the number of concordant pairs minus the number of discordant pairs, divided by the total number of pairs. If all pairs are concordant, then $C = N_0$, $D = 0$ and $\tau_A = 1$, and if all pairs are discordant, then $\tau_A = -1$.

The denominator was adjusted for ties (i.e. $x_i = x_j$ or $y_i = y_j$) in Kendall (1945), and statistical packages usually implement this modified $\tau$ (often called $\tau_B$), so $\tau_B$ ranges from $-1$ to $+1$ even with tied data. $\tau_B$ is defined by

$$\tau_B = \frac{C - D}{\sqrt{(N_0 - T_x)(N_0 - T_y)}}, \tag{2.62}$$

with

$$T_x = \sum_{j=1}^{g^{(x)}} t_j^{(x)}(t_j^{(x)} - 1)/2, \tag{2.63}$$

$$T_y = \sum_{j=1}^{g^{(y)}} t_j^{(y)}(t_j^{(y)} - 1)/2, \tag{2.64}$$

where $g^{(x)}$ is the number of tied groups in the variable $x$ and $t_j^{(x)}$ is the size of tied group $j$ (e.g. if the value $x = 5.1$ appears twice, $t_j^{(x)} = 2$), and $g^{(y)}$ and $t_j^{(y)}$ are similarly defined for the variable $y$. When there are no ties, $\tau_A = \tau_B$.

In Fig. 2.6(b), the correlation is $-0.50$ for Pearson, $0.44$ for Spearman and $0.35$ for Kendall. If the single outlier at the bottom right corner of Fig. 2.6(b) is removed, it is $0.70$ for Pearson, $0.68$ for Spearman and $0.50$ for Kendall.

The usage of Kendall's $\tau$ has been increasing in recent decades, though whether it is better or worse than the Spearman correlation is problem dependent (W. C. Xu et al., 2013).

## 2.11.5   Biweight Midcorrelation   B☺

We have seen one approach in making correlation more robust and resistant, namely using ranks as in the Spearman and Kendall rank correlations. A different approach involves replacing the non-robust/resistant measures in the Pearson correlation, that is, the mean and deviation from the mean, by the corresponding robust/resistant ones, that is, the median and the deviation from the median. This second approach is used in the biweight midcorrelation (Mosteller and Tukey, 1977; Wilcox, 2004, pp. 203–209).

To calculate the biweight midcorrelation function $\mathrm{bicor}(x, y)$, first rescale $x$ and $y$ by

$$p_i = \frac{x_i - M_x}{9\,\mathrm{MAD}_x}, \quad q_i = \frac{y_i - M_y}{9\,\mathrm{MAD}_y}, \quad i = 1, \ldots, N, \tag{2.65}$$

where $M_x$ and $M_y$ are the median values of $x$ and $y$, respectively, and $\text{MAD}_x$ and $\text{MAD}_y$ (the median absolute deviations) are the median values of $|x_i - M_x|$ and $|y_i - M_y|$, respectively.

Next, define the weights (called 'biweights' by Beaton and Tukey (1974))

$$w_i^{(x)} = \begin{cases} (1 - p_i^2)^2, & \text{if } |p_i| < 1 \\ 0, & \text{if } |p_i| \geq 1, \end{cases} \tag{2.66}$$

$$w_i^{(y)} = \begin{cases} (1 - q_i^2)^2, & \text{if } |q_i| < 1 \\ 0, & \text{if } |q_i| \geq 1. \end{cases} \tag{2.67}$$

The biweight midcorrelation is defined by

$$\text{bicor}(x, y) = \frac{\sum_{i=1}^{N} w_i^{(x)} (x_i - M_x) \, w_i^{(y)} (y_i - M_y)}{\left\{ \sum_{i=1}^{N} [w_i^{(x)} (x_i - M_x)]^2 \right\}^{\frac{1}{2}} \left\{ \sum_{i=1}^{N} [w_i^{(y)} (y_i - M_y)]^2 \right\}^{\frac{1}{2}}}. \tag{2.68}$$

Formally, bicor resembles the Pearson correlation (2.53), except for the presence of the weights $w_i^{(x)}$ and $w_i^{(y)}$ and the use of medians $M_x$ and $M_y$ instead of the means. The weights in (2.66) and (2.66) are set to zero for outliers (large $|p_i|$ or $|q_i|$); thus, bicor is resistant to outliers. The biweight midcorrelation, like the Pearson correlation, ranges from $-1$ (negative association) to $+1$ (positive association).

## 2.12 Exploratory Data Analysis $\boxed{\text{A}}$ ☺

In statistics, exploratory data analysis (EDA) was pioneered by John W. Tukey, who wrote the classic book entitled *Exploratory Data Analysis* (Tukey, 1977). Tukey felt that statistics placed too much emphasis on statistical hypothesis testing, so he advocated EDA, which tries to explore and visualize the data, thereby letting the data suggest what hypotheses to test. Besides using more robust/resistant statistics such as the median and the quartiles to summarize a dataset than using the traditional mean and standard deviation, EDA also uses graphical methods extensively to aid in visualizing the structure of the datasets. Graphical methods include scatterplots (Fig. 2.5), histograms, quantile–quantile plots and boxplots.

### 2.12.1 Histograms $\boxed{\text{A}}$ ☺

*Histograms* (from 'historical diagrams'), introduced by Pearson (1895), present the probability distribution of a given variable by plotting the frequencies of observations occurring over the domain of the variable. To construct a histogram, the domain is partitioned into intervals (called 'bins' or 'buckets'), and the frequency, that is, how many observed values fall into each bin, is counted. The frequencies can be simply the raw counts, or normalized, that is, dividing the counts by the total number of observations. The bins are usually of equal width, but can be of unequal width. With normalized frequencies, the area over each bin gives the probability of occurrence within that interval. The bin width cannot be chosen to be too wide, which smooths out important details in the histogram, nor too narrow, which gives a noisy-looking histogram. (Scott, 2015, p. 78) recommends using a bin width $\leq 2.6\,\mathrm{IQR}/(N^{1/3})$, where IQR is the interquartile range and $N$ the sample size. Most histogram packages will have a reasonable default bin width, so the user does not have to specify the bin width.

Figure 2.8 gives an example of using the histogram method on the weather data for Vancouver, BC. The histogram gives the actual distribution of the data, while a Gaussian distribution curve has also been fitted to the data.[4] Comparing the histogram with the Gaussian curve tells us how close the observed distribution is to a Gaussian distribution. Temperature in Fig. 2.8(a) actually shows a bimodal distribution (i.e. having two humps) in contrast to the unimodal Gaussian, while wind speed in Fig. 2.8(c) is also clearly non-Gaussian. The fit for relative humidity (Fig. 2.8(b)) is poor for the right tail as humidity cannot exceed 100%, and the pressure distribution (Fig. 2.8(d)) is more narrowly distributed than a Gaussian. For precipitation (Fig. 2.8(e)), most days have no precipitation – these dry days are omitted in Fig. 2.8(f). Although the Gaussian distribution is called the 'normal distribution', in reality, environmental variables often do not closely resemble Gaussians.

## 2.12.2   Quantile–Quantile (Q–Q) Plots   B☺

A *quantile–quantile plot* (Q–Q plot) is a probability plot, which provides a graphical tool for comparing two probability distributions by plotting their quantiles against each other. For a chosen set of quantiles, a point $(x, y)$ on the plot corresponds to one of the quantiles of the $y$ distribution plotted against the same quantile of the $x$ distribution.

There are two ways to use a Q–Q plot: (i) to compare observed data with a specified theoretical distribution (e.g. a Gaussian distribution) and (ii) to assess whether two sets of observed data obey the same distribution. If the agreement between the two distributions is perfect, then the plot is a straight line.

There are many ways to choose the quantiles for the plot. If the distribution for the observations $y_i$, $(i = 1, \ldots, N)$, is to be compared with a specified theoretical distribution, one way is to simply use $N$ quantiles. The $y$ data are sorted into ascending order, $y^1, \ldots, y^N$, then the $i$th ordered value $y^{(i)}$ is plotted against the $(i - \frac{1}{2})/N$ quantile of the theoretical distribution along the $x$-axis.

The Q–Q plot for Vancouver's daily temperature versus the standard Gaussian distribution (i.e. Gaussian with zero mean and unit standard deviation) shows the temperature to have weaker tails than the Gaussian, especially for high temperatures (Fig. 2.9(a)), which can also be seen in (Fig. 2.8(a)). However the bimodal structure seen in the histogram is much less noticeable in the Q–Q

---

[4] The Gaussian distribution is specified by two parameters, namely its population mean $\mu$ and variance $\sigma^2$ (see Section 3.4). From the dataset, compute the sample mean and variance and use these as the parameters of the Gaussian distribution.

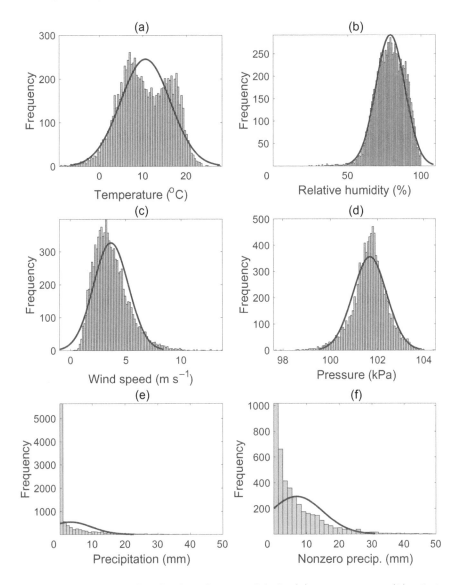

Figure 2.8 Histogram for the distribution of daily (a) temperature, (b) relative humidity, (c) wind speed, (d) sea level pressure, (e) precipitation and (f) non-zero precipitation in Vancouver, BC from 1993 to 2017. A Gaussian distribution curve has also been fitted to the data. Relative humidity is bounded between 0% and 100%, and wind speed is non-negative. Since 53.4% of the days in (e) have no precipitation, the dry days are omitted in (f). [Data source: weatherstats.ca based on Environment and Climate Change Canada data.]

plot. Thus, comparing the histogram with the Q–Q plot, the Q–Q plot tends to be better in revealing the departure of the observations from the theoretical distribution near the tails, but the histogram tends to be better in showing the departure away from the tails.

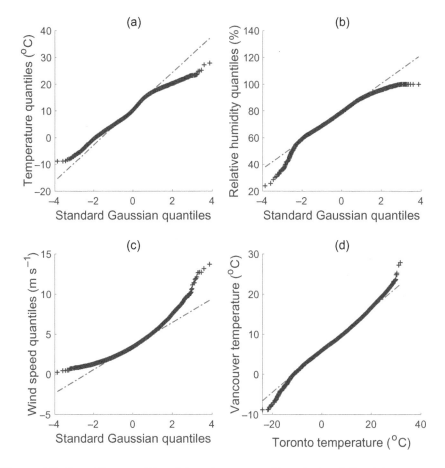

Figure 2.9 Quantile–quantile plots where quantiles of the daily (a) temperature, (b) relative humidity and (c) wind speed in Vancouver, BC from 1993 to 2017 are plotted against the quantiles of the standard Gaussian distribution as indicated by the '+' symbols. If the observed distribution is a perfect Gaussian, the plot will fall on the straight (dot-dashed) line. In (d), the quantiles of the temperature in Toronto, Ontario are plotted against those from Vancouver. [Data source: weatherstats.ca based on Environment and Climate Change Canada data.]

Figure 2.9(b) shows the relative humidity to have a shorter tail than the Gaussian at high values but a longer tail at low values. On the other hand, wind speed (Fig. 2.9(c)) has a longer tail than the Gaussian at high values but a shorter tail at low values.

The Q–Q plot can also be used to assesses whether two sets of observations have the same distribution. One plots the quantile values for the first dataset along the $x$-axis and the corresponding quantile values for the second dataset along the $y$-axis. The two datasets can have different numbers of data points, as a Q–Q plot only plots selected quantiles. If the resulting plot is linear, the two datasets obey the same distribution. Figure 2.9(d) shows the Q–Q plot of Toronto's temperature versus Vancouver's temperature. Toronto's temperature, though having a larger range than Vancouver's, has relatively shorter tails.

### 2.12.3  Boxplots $\boxed{\text{A}}$ ☺

Tukey (1977) advocated using five numbers to summarize a dataset, that is, the median, the lower and upper quartiles (i.e. $q_{0.25}$ and $q_{0.75}$), and the minimum and maximum values. *Boxplots* (or box-and-whisker plots) arose as a visualization tool for the five-number summary (Tukey, 1977).

The top and bottom of each 'box' are the upper and lower quartiles of the sample data, respectively, with the distance between the top and bottom indicating the interquartile range (IQR). The horizontal line or 'waist line' within each box is the sample median. Skewness is present if the median is not centred in the box.

The whiskers are the (dashed) lines extending above and below each box. The most common convention is to have the whisker above the box extending from $q_{0.75}$ to a furthest observation not more than 1.5 IQR above $q_{0.75}$. Any observation beyond is considered an outlier and is plotted as a '+' or 'o' symbol. Similarly, the lower whisker extends from $q_{0.25}$ to a furthest observation not more than 1.5 IQR below $q_{0.25}$, with any observation beyond plotted as an outlier. For a Gaussian distribution, 99.7% of the distribution lies within the interval $[q_{0.25} - 1.5\,\text{IQR},\ q_{0.75} + 1.5\,\text{IQR}]$.

A common variant of the boxplot displays notches on the two sides of the waistline, that is, '>−<' (Fig. 2.10(a)), with the height of the notches indicating the uncertainty of the sample median (McGill et al., 1978). The notches extend

$$\pm\, 1.57\,\text{IQR}/\sqrt{N} \qquad\qquad (2.69)$$

from the sample median. If the notches from two boxes do not overlap, then the two sample medians are considered different at the 5% significance level.

Figure 2.10 shows boxplots of weather variables at three Canadian cities, Vancouver and Victoria in British Columbia on the west coast and Toronto, Ontario to the east. In (a), there are only 90 data points from Dec. 2016 to Feb. 2017, so the notches are much wider than those in (b), where there are 25 years of daily data. Toronto is seen to have a much larger temperature range in (b) and lower relative humidity in (c) than the two west coast cities, while Victoria is seen to have lower wind speed than the other two cities in (d).

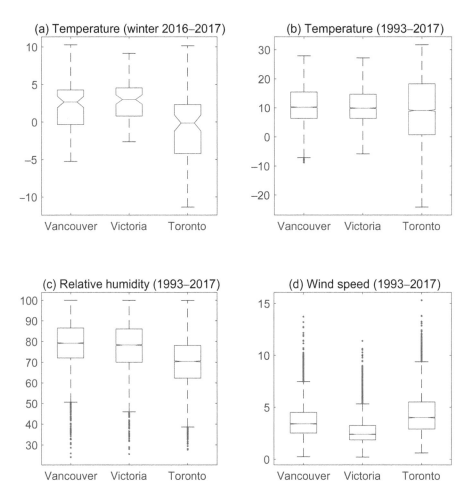

Figure 2.10 Boxplots for the daily weather at three Canadian cities: (a) temperature during the winter of 2016–2017, and (b) temperature, (c) relative humidity and (d) wind speed from 1993 to 2017. [Data source: weatherstats.ca based on Environment and Climate Change Canada data.]

The astute reader may question the notches computed from (2.69) since there is serial correlation in the weather data. Indeed, if we assume the effective sample size $N_{\text{eff}} \approx 8$ for Vancouver temperature (winter 2016–2017) as in Section 2.11.2, then replacing $N$ by $N_{\text{eff}}$ in (2.69) would make the width of the notches 3.4 times as wide. Unfortunately, common boxplot packages do not provide an option for replacing $N$ by $N_{\text{eff}}$ when computing the notches, leading to notch widths that are too narrow for serially correlated data. In such situations, it is best to turn off the option for displaying notches in the boxplot.

# 2.13   Mahalanobis Distance   $\boxed{\text{A}}$☺

For a one-dimensional dataset with mean $\mu$ and standard deviation $\sigma$, and a particular data point $x$, the Mahalanobis distance measures the number of standard deviations between the data point $x$ and the mean $\mu$, that is,

$$d_{\mathrm{M}} = \sqrt{(x - \mu)^2}/\sigma. \qquad (2.70)$$

For higher-dimensional data, the Mahalanobis distance is defined by

$$d_{\mathrm{M}} = \sqrt{(\mathbf{x} - \boldsymbol{\mu})^{\mathrm{T}} \mathbf{C}^{-1} (\mathbf{x} - \boldsymbol{\mu})}, \qquad (2.71)$$

where $\boldsymbol{\mu}$ is the expectation of $\mathbf{x}$, $\mathbf{C}$ is the covariance matrix $\mathrm{cov}(\mathbf{x})$ from (2.33) and $\mathbf{C}^{-1}$ is the inverse of $\mathbf{C}$. Clearly, (2.70) is the special one-dimensional case of the general formula (2.71), as $\mathbf{C}$ reduces to the variance $\sigma^2$ in 1-D. If $\mathbf{C}^{-1}$ is the identity matrix, (2.71) reduces to the familiar Euclidean distance.

Figure 2.11 illustrates why the Mahalanobis distance (and not the Euclidean distance) is the appropriate distance for determining if a data point $\mathbf{x}$ is far from the centre of a dataset. Consider the two points marked by the asterisk and the star. The Euclidean distance (Fig. 2.11(a)) between the asterisk and the centre of the dataset is 7.44, versus 4.48 between the star and the centre. However, in terms of the Mahalanobis distance (Fig. 2.11(d)), the distance from the centre is 2.48 for the asterisk and 4.46 for the star. Thus the Mahalanobis distance correctly indicates the star as being much further from the centre than the asterisk, and should be considered an outlier.

Robust methods have been developed to estimate $\mathbf{C}$ when the data is non-Gaussian. These include the fast-minimum covariance determinant (Fast-MCD) method (Rousseeuw and van Driessen, 1999), the orthogonalized Gnanadesikan–Kettenring (OGK) method (Maronna and Zamar, 2002) and the Olive–Hawkins method (Olive, 2004). Of the three methods, the author's choice is the OGK method, based on some limited tests comparing the accuracy and speed of the three methods.

## 2.13.1   Mahalanobis Distance and Principal Component Analysis   $\boxed{\text{B}}$☹

One can compute the Mahalanobis distance from (2.71), using the sample co-variance matrix for $\mathbf{C}$ and the sample mean for $\boldsymbol{\mu}$. However, for the inquisitive reader, it is illuminating to see how the Mahalanobis distance can be derived from principal component analysis (PCA) (Section 9.1). From (9.29), one can write the centred data $\mathbf{x} - \boldsymbol{\mu}$ (Fig. 2.11(b)) as

$$\mathbf{x} - \boldsymbol{\mu} = \sum_i a_i \mathbf{e}_i, \qquad (2.72)$$

where the summation is over all the PCA modes, that is, over all the principal components (PC) $a_i$ multiplied by their corresponding eigenvectors $\mathbf{e}_i$. The eigenvectors satisfy the eigenequation

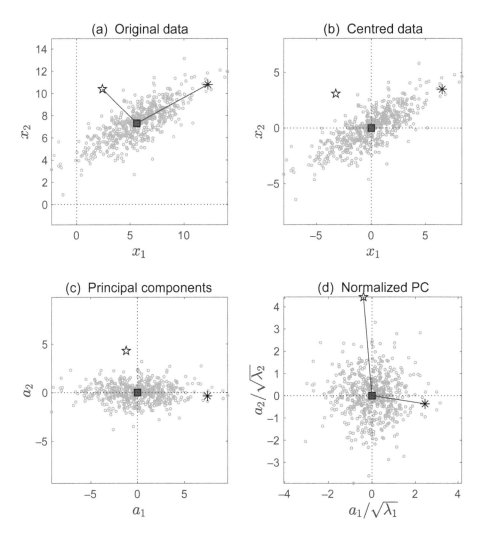

Figure 2.11 Mahalanobis distances versus Euclidean distances, as illustrated by two data points marked by the asterisk and the star. The square marks the centre (i.e. the mean) of the Gaussian dataset containing 500 points. (a) In the original data, the line marking the Euclidean distance from the centre is longer for the asterisk than for the star. (b) Subtracting the mean gives the centred data. (c) Principal components ($a_1$ and $a_2$) are obtained by rotating the centred data, so the direction of the maximum variance is along the horizontal axis. (d) Principal components are normalized to have unit variance in each direction. The line connecting the centre and the asterisk/star gives the Mahalanobis distance. Thus in terms of Euclidean distance, the asterisk is further from the centre than the star, but in terms of Mahalanobis distance, the star is further from the centre than the asterisk.

$$\mathbf{C}\mathbf{e}_i = \lambda_i \mathbf{e}_i, \tag{2.73}$$

where $\lambda_i$ are the eigenvalues. The vector $\mathbf{e}_1$ points in the direction of maximum variance of the dataset, while $\mathbf{e}_2$ points in the direction of maximum variance within the space orthogonal to $\mathbf{e}_1$. In general, $\mathbf{e}_i$ points in the direction of maximum variance within the space orthogonal to $\mathbf{e}_1, \ldots, \mathbf{e}_{i-1}$. The PC $a_1$ is the coordinate in the $\mathbf{e}_1$ direction, while $a_2$ is the coordinate in the $\mathbf{e}_2$ direction (Fig. 2.11(c)). A common convention is to make all eigenvectors of unit length, then $\lambda_i$ is the variance of $a_i$ [see (9.37) and (9.39)].

Left multiplying (2.73) by $\mathbf{C}^{-1}$ gives

$$\mathbf{e}_i = \lambda_i \mathbf{C}^{-1} \mathbf{e}_i, \tag{2.74}$$

$$\lambda_i^{-1} \mathbf{e}_i = \mathbf{C}^{-1} \mathbf{e}_i. \tag{2.75}$$

$$\sum_i a_i \lambda_i^{-1} \mathbf{e}_i = \sum_i a_i \mathbf{C}^{-1} \mathbf{e}_i = \mathbf{C}^{-1} \sum_i a_i \mathbf{e}_i = \mathbf{C}^{-1}(\mathbf{x} - \boldsymbol{\mu}), \tag{2.76}$$

upon invoking (2.72). Thus,

$$d_{\mathrm{M}}^2 = (\mathbf{x} - \boldsymbol{\mu})^{\mathrm{T}} \mathbf{C}^{-1} (\mathbf{x} - \boldsymbol{\mu}) = \sum_j \sum_i a_j \mathbf{e}_j^{\mathrm{T}} a_i \lambda_i^{-1} \mathbf{e}_i. \tag{2.77}$$

With orthonormal eigenvectors, $\mathbf{e}_j^{\mathrm{T}} \mathbf{e}_i = \delta_{ij}$ (the Kronecker delta function, where $\delta_{ij} = 1$ if $i = j$, and 0 otherwise),

$$d_{\mathrm{M}}^2 = \sum_i a_i^2 / \lambda_i. \tag{2.78}$$

The Mahalanobis distance squared is simply the sum over the square of the normalized principal components, as $a_i / \sqrt{\lambda_i}$ can be regarded as the normalized PC. Thus, the distance from the origin in the normalized PC space is the Mahalanobis distance (Fig. 2.11(d)).

## 2.14  Bayes Theorem  A☺

Bayes theorem, named after the Reverend Thomas Bayes (1702–1761), plays a central role in modern statistics (Jaynes, 2003; N. D. Le and Zidek, 2006). Historically, it had a major role in the debate around the foundations of statistics, as the traditional '*frequentist*' school and the *Bayesian* school disagreed on how probabilities should be assigned in applications. Frequentists assign probabilities to random events according to their frequencies of occurrence or to subsets of populations as proportions of the whole. In contrast, Bayesians describe probabilities in terms of beliefs and degrees of uncertainty, similarly to how the general public uses probability. For instance, a sports fan prior to the start of a sports tournament believes that team A, with a stellar historical record, has a probability of 70% for winning a game. However, after watching several mediocre games, the fan may modify his estimate of the winning probability to 50%. Similarly, to estimate the probability of a hypothesis, a Bayesian first specifies/guesses some prior probability, then updates it to a posterior probability by incorporating new data (evidence).

We will use a *classification* problem to illustrate the Bayes approach. Suppose a meteorologist wants to classify the approaching weather state as either storm ($C_1$) or non-storm ($C_2$). Assume there is some *a priori probability* (or simply *prior probability*) $P(C_1)$ that there is a storm, and some prior probability $P(C_2)$ that there is no storm. For instance, from the past weather records, if 15% of the days were found to be stormy during this season, then the meteorologist may assign $P(C_1) = 0.15$, and $P(C_2) = 0.85$. Now suppose the meteorologist has a barometer measuring a pressure $x$ at 6 a.m. The meteorologist would like to obtain an *a posteriori probability* (or simply *posterior probability*) $P(C_1|x)$, that is, the conditional probability of having a storm on that day given the 6 a.m. pressure $x$. In essence, he would like to improve on his simple prior probability with the new information $x$.

The joint probability density $p(C_i, x)$ is the probability density that an event belongs to class $C_i$ and has value $x$. The joint probability density can be written as

$$p(C_i, x) = P(C_i|x)p(x),  \tag{2.79}$$

with $p(x)$ the probability density of $x$. Alternatively, $p(C_i, x)$ can be written as

$$p(C_i, x) = p(x|C_i)P(C_i),  \tag{2.80}$$

with $p(x|C_i)$, the conditional probability density of $x$, given that the event belongs to class $C_i$. Equating the right hand sides of these two equations, we obtain

$$P(C_i|x) = \frac{p(x|C_i)P(C_i)}{p(x)},  \tag{2.81}$$

which is *Bayes theorem*. The previous form of Bayes theorem encountered in (2.8) was for two discrete variables $x$ and $y$, whereas here we have a discrete variable $C$ and a continuous variable $x$. Since $p(x)$ is the probability density of $x$ without regard to which class, it can be decomposed into

$$p(x) = \sum_i p(x|C_i)P(C_i).  \tag{2.82}$$

Substituting this for $p(x)$ in (2.81) yields

$$P(C_i|x) = \frac{p(x|C_i)P(C_i)}{\sum_i p(x|C_i)P(C_i)},  \tag{2.83}$$

where the denominator on the right hand side is seen as a normalization factor for the posterior probabilities to sum to unity. Bayes theorem says that the posterior probability $P(C_i|x)$ is simply $p(x|C_i)$ (the *likelihood* of $x$ given the event is of class $C_i$) multiplied by the prior probability $P(C_i)$ and divided by a normalization factor. The advantage of Bayes theorem is that the posterior probability is now expressed in terms of quantities that can be estimated. For instance, to estimate $p(x|C_i)$, the meteorologist can divide the 6 a.m. pressure record into two classes and estimate $p(x|C_1)$ from the pressure distribution for stormy days, and $p(x|C_2)$ from the pressure distribution for non-stormy days.

For the general situation, the scalar $x$ is replaced by a vector $\mathbf{x}$, and the classes are $C_1, \ldots, C_k$; then, Bayes theorem becomes

$$P(C_i|\mathbf{x}) = \frac{p(\mathbf{x}|C_i)P(C_i)}{\sum_i p(\mathbf{x}|C_i)P(C_i)},$$

(2.84)

for $i = 1, \ldots, k$.

If instead of the discrete variable $C_i$, we have a continuous variable $w$, then Bayes theorem (2.81) for two continuous variables $w$ and $x$ takes the form

$$p(w|x) = \frac{p(x|w)p(w)}{p(x)}.$$

(2.85)

The scalars $x$ and $w$ can be generalized to the vectors $\mathbf{x}$ and $\mathbf{w}$, so Bayes theorem becomes

$$p(\mathbf{w}|\mathbf{x}) = \frac{p(\mathbf{x}|\mathbf{w})p(\mathbf{w})}{p(\mathbf{x})}.$$

(2.86)

Often a model controlled by some parameters $\mathbf{w}$ is used to model the variables $\mathbf{x}$. The model parameters $\mathbf{w}$ are to be estimated using a dataset $D$ containing the observations $\mathbf{x}_1, \ldots, \mathbf{x}_N$. Given a prior distribution $p(\mathbf{w})$ and $p(D|\mathbf{w})$ (i.e. the likelihood of observing $D$ given the parameters $\mathbf{w}$), we can obtain a posterior distribution $p(\mathbf{w}|D)$ from Bayes theorem,

$$p(\mathbf{w}|D) = \frac{p(D|\mathbf{w})p(\mathbf{w})}{p(D)},$$

(2.87)

where $p(D)$ is simply a normalization factor,

$$p(D) = \int p(D|\mathbf{w})p(\mathbf{w})d\mathbf{w}.$$

(2.88)

The likelihood $p(D|\mathbf{w})$ is treated differently by frequentists, who view the parameters $\mathbf{w}$ as being fixed. The frequentists commonly estimate the value of $\mathbf{w}$ by maximizing the likelihood function (Section 3.5). In contrast, Bayesians view the observed data $D$ as fixed, but $\mathbf{w}$ is given by a distribution $p(\mathbf{w}|D)$.

Figure 2.12 illustrates the relation between $p(\mathbf{w}|D)$, $p(D|\mathbf{w})$ and $p(\mathbf{w})$ where, for simplicity, $\mathbf{w}$ is reduced to a scalar $w$. Case (a): Little prior information is available for $w$, that is, a very broad and flat $p(w)$. Case (b): More precise prior information is available from the narrower $p(w)$ distribution.

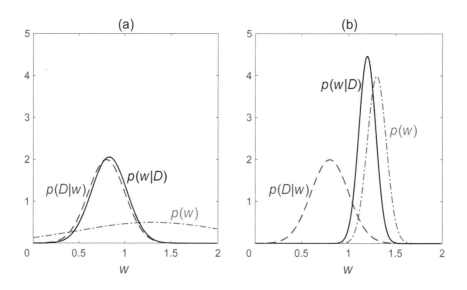

Figure 2.12 Relation between $p(w|D)$, $p(D|w)$ and $p(w)$. (a) A broad and flat distribution of $p(w)$ provides little prior information for estimating $w$, leading to the posterior distribution $p(w|D)$ being very similar to the likelihood $p(D|w)$. (b) A narrow $p(w)$ distribution leads to a larger difference between $p(w|D)$ and $p(D|w)$. If more data are available, $p(D|w)$ will be narrower and more strongly peaked than that shown in (b), and the $p(w|D)$ distribution will be pulled more towards $p(D|w)$. [Follows Cowan (2007)].

## 2.15   Classification  $\boxed{A}$ ☺

Once the posterior probabilities $P(C_i|\mathbf{x})$ have been estimated from (2.84), we can proceed to classification: Given an input or predictor vector $\mathbf{x}$, called a *feature vector* in the ML literature, we choose the class $C_j$ having the highest posterior probability, that is,

$$P(C_j|\mathbf{x}) > P(C_i|\mathbf{x}), \quad \text{for all } i \neq j. \tag{2.89}$$

From (2.84), this is equivalent to

$$p(\mathbf{x}|C_j)P(C_j) > p(\mathbf{x}|C_i)P(C_i), \quad \text{for all } i \neq j. \tag{2.90}$$

In the *feature space* (i.e. the space of the predictor variables $\mathbf{x}$), the pattern classifier has divided the space into *decision regions* $R_1, \ldots, R_k$, so that if a feature vector lands within $R_i$, the classifier will assign the class $C_i$. The decision region $R_i$ may be composed of several disjoint regions, all of which are assigned the class $C_i$. The boundaries between decision regions are called *decision boundaries* or *decision surfaces*.

To justify the decison rule (2.90), consider the probability $P_{\text{correct}}$ of a new pattern being classified correctly:

$$P_{\text{correct}} = \sum_{j=1}^{k} P(\mathbf{x} \in R_j, C_j), \tag{2.91}$$

where $P(\mathbf{x} \in R_j, C_j)$ gives the probability that the pattern that belongs to class $C_j$ has its feature vector falling within the decision region $R_j$, thus classified correctly as belonging to class $C_j$. Note that $P_{\text{correct}}$ can be expressed as

$$P_{\text{correct}} = \sum_{j=1}^{k} P(\mathbf{x} \in R_j | C_j) P(C_j),$$

$$= \sum_{j=1}^{k} \int_{R_j} p(\mathbf{x}|C_j) P(C_j) \mathrm{d}\mathbf{x}. \tag{2.92}$$

To maximize $P_{\text{correct}}$, one needs to maximize the integrand by choosing the decision regions so that $\mathbf{x}$ is assigned to the class $C_j$ satisfying (2.90).

In general, classification need not be based on probability distribution functions, since in many situations, $p(\mathbf{x}|C_i)$ and $P(C_i)$ are not known. The classification procedure is then formulated in terms of *discriminant functions*, which tell us which class we should assign to the given $\mathbf{x}$. For example, in Fig. 2.13(a), $\mathbf{x} = (x_1, x_2)^{\mathrm{T}}$, and the two classes are separated by the line $x_2 = x_1$.

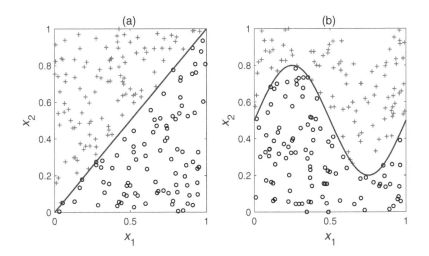

Figure 2.13 (a) A linear decision boundary separating two classes of data denoted by crosses and circles, respectively. (b) A non-linear decision boundary.

The discriminant function can be simply $y(\mathbf{x}) = -x_1 + x_2$, with $C_2$ assigned when $y(\mathbf{x}) > 0$, and $C_1$ otherwise. Thus, the decision boundary is given by $y(\mathbf{x}) = 0$.

When there are more than two classes, the discriminant functions are $y_1(\mathbf{x}), \ldots, y_k(\mathbf{x})$, where a feature vector $\mathbf{x}$ is assigned to class $C_j$ if

$$y_j(\mathbf{x}) > y_i(\mathbf{x}), \quad \text{for all } i \neq j. \tag{2.93}$$

Equation (2.89) can be viewed as a special case of (2.93). An important property of a discriminant function $y_i(\mathbf{x})$ is that it can be replaced by $f(y_i(\mathbf{x}))$ for any monotonic function $f$, since the classification is unchanged as the relative magnitudes of the discriminant functions are preserved by $f$. There are many classical *linear discriminant analysis* methods (Duda et al., 2001), where the discriminant function is a linear combination of the predictor variables $x_l$, that is,

$$y_i(\mathbf{x}) = \sum_l w_{il} x_l + w_{i0} \equiv \mathbf{w}_i^{\mathrm{T}} \mathbf{x} + w_{i0}, \tag{2.94}$$

with parameters $\mathbf{w}_i$ and $w_{i0}$. Based on (2.93), the decision boundary between class $C_j$ and $C_i$ is obtained from setting $y_j(\mathbf{x}) = y_i(\mathbf{x})$, yielding a hyperplane[5] decision boundary described by

$$(\mathbf{w}_j - \mathbf{w}_i)^{\mathrm{T}} \mathbf{x} + (w_{j0} - w_{i0}) = 0. \tag{2.95}$$

Suppose $\mathbf{x}$ and $\mathbf{x}'$ both lie within the decision region $R_j$. Consider any point $\tilde{\mathbf{x}}$ lying on a straight line connecting $\mathbf{x}$ and $\mathbf{x}'$, that is,

$$\tilde{\mathbf{x}} = a\mathbf{x} + (1 - a)\mathbf{x}', \tag{2.96}$$

with $0 \leq a \leq 1$. Since $\mathbf{x}$ and $\mathbf{x}'$ both lie within $R_j$, they satisfy $y_j(\mathbf{x}) > y_i(\mathbf{x})$ and $y_j(\mathbf{x}') > y_i(\mathbf{x}')$ for all $i \neq j$. Since the discriminant function is linear, we also have

$$y_j(\tilde{\mathbf{x}}) = ay_j(\mathbf{x}) + (1 - a)y_j(\mathbf{x}'), \tag{2.97}$$

therefore $y_j(\tilde{\mathbf{x}}) > y_i(\tilde{\mathbf{x}})$ for all $i \neq j$. Thus, any point on the straight line joining $\mathbf{x}$ and $\mathbf{x}'$ must also lie within $R_j$, meaning that the decision region $R_j$ is simply connected and convex. As we shall see later, with non-linear ML methods such as neural networks and support vector machines, the decision boundaries can be curved surfaces (Fig. 2.13(b)) instead of hyperplanes, and the decision regions need not be simply connected nor convex. Discriminant analysis and other classification methods are explained in detail in Chapter 12.

## 2.16   Clustering  $\boxed{\text{A}}$ ☺

In machine learning, there are two general approaches, *supervised learning* and *unsupervised learning*. An analogy for the former is students in a Spanish class

---

[5] A hyperplane is a subspace where the dimension is one less than that of its ambient space.

where the teacher demonstrates the correct Spanish pronunciation. An analogy for the latter is students working on a team project without supervision. In unsupervised learning, the students are provided with learning rules, but must rely on self-organization to arrive at a solution, without the benefit of being able to learn from a teacher's demonstration.

In classification, the training dataset consists of predictors or features $\mathbf{x}_i$ ($\mathbf{x}$ can be made up of continuous and/or discrete and/or categorical variables) and discrete/categorical response variables $C_i$, $i = 1, \ldots, N$, $N$ being the number of observations. Here, $C_i$ serves the role of the teacher or *target* for the classification model output $\tilde{C}_i$, that is, $\tilde{C}_i$ is fitted to the given target data, similar to students trying to imitate the Spanish accent of their teacher; thus, the learning is supervised.

For instance, suppose $\mathbf{x}$ contains three variables – air temperature, humidity and pressure, and $C$ can be 'no precipitation', 'rain' or 'snow' a day later. Such a classification model uses three meteorological inputs to predict whether it will be 'no precipitation', 'rain' or 'snow' a day later.

*Clustering* or cluster analysis is the unsupervised version of classification, that is, we are given the $\mathbf{x}$ data but not the $C$ data. The goal of clustering is to group the $\mathbf{x}$ data into a number of subsets or 'clusters', such that the data within a cluster are more closely related to each other than data from other clusters. After performing clustering on the air temperature, humidity and pressure data, we may indeed find three main clusters. The first cluster of data points may occur where humidity is low and pressure is high, corresponding to days of no precipitation. A second cluster may occur where humidity is high, pressure is low and temperature is high, corresponding to rainy days, while a third cluster may be somewhat similar to the second cluster but occurring at low temperature, corresponding to snowy days. Thus, even without the target $C$ data, we can learn much from the $\mathbf{x}$ data alone.

A simple and popular method for performing clustering is *K-means clustering*. First choose $K$, the number of clusters. Next, start with initial guesses for the mean positions of the $K$ clusters in the $\mathbf{x}$ space (i.e. the position of a cluster centre, a.k.a. *centroid*, is simply the mean position of all the data points belonging to that cluster). Iterate the following two steps until convergence:

(i) For each data point, find the closest centroid [based on the squared Euclidean distance in (10.2)] and assign the data point to be a member of this cluster.

(ii) For each cluster, reassign the centroid to be the mean position of all data points belonging to that cluster.

This is known as Lloyd's algorithm, and is sometimes referred to as 'naive $K$-means' as there are faster algorithms. The initial choice for the $K$ centroids often involves randomly picking $K$ data points from the $\mathbf{x}$ data. The initialization can be improved, for example by using the $K$-means++ method of Arthur and Vassilvitskii (2007), where the centroids are chosen randomly from the data points, but with the probability of choosing a data point being proportional to its squared distance from the closest centroid (among the centres already chosen). See Section 10.2.1 for more details on $K$-means clustering and Chapter 10 for more choices of clustering methods.

Figure 2.14 illustrates $K$-means clustering with $K = 3$ clusters for the daily air pressure and temperature data at Vancouver, BC, Canada during 2013–2017. As air pressure and temperature have different units, clustering was performed on the standardized data. The poor initial choice of the centroids did not hinder the convergence of the clustering algorithm. Upon convergence in Fig. 2.14(d), the three centroids make physical sense: In the Pacific Northwest region of North America, summer tends to be sunny while winter has numerous weather systems passing through, so it makes sense to have one cluster representing summer and two clusters representing winter. In winter, during high pressure days, the clear skies increase outgoing long-wave radiation, leading to colder temperatures (as characterized by the the lower-right centroid) than during low pressure days where the clouds reduce the outgoing long-wave radiation (lower-left centroid). As $K$, the number of clusters, is specified by the user, choosing a different $K$ will in general lead to very different clusters. It turns out $K = 3$ is optimal according to two internal evaluation criteria (see Fig. 10.1).

## 2.17   Information Theory   B☺

Information theory studies the quantification, storage and communication of information. The field started in 1948 with the publication of an article in two parts by Claude E. Shannon (Shannon, 1948a,b), on how best to encode information for transmission. His theory was based on probability theory, and the central concept of information entropy, a measure of the uncertainty in a message, was surprisingly similar to the thermodynamic entropy developed by the physicists Ludwig Boltzmann and J. Willard Gibbs in the 1870s. Information theory has since grown into a large field (Cover and Thomas, 2006) and is connected to machine learning (MacKay, 2003). Information theory has also been applied increasingly to the environmental sciences, especially regarding predictability of dynamical systems (Leung and North, 1990; Kleeman, 2002; DelSole, 2004; DelSole, 2005; Y. M. Tang et al., 2005; DelSole and Tippett, 2007) and the selection of good predictors from a large pool of available predictor variables (Sharma, 2000; May, Dandy, et al., 2008; May, Maier, et al., 2008).

How information is connected to probability can be illustrated by the following example. The statement 'the sun has risen from the east' is much less informative than the statement 'a magnitude 9.0 earthquake has occurred off Japan'. The reason is that the first event has probability 1 (so the statement contains no useful information) while the second event has low probability (so the statement contains potentially life-saving information). Thus, information content is low when the probability of the event occurring is high.

First, start with a discrete random variable $X$, from which we can draw specific values $x$. We want to develop a measure of information content, $h(x)$,

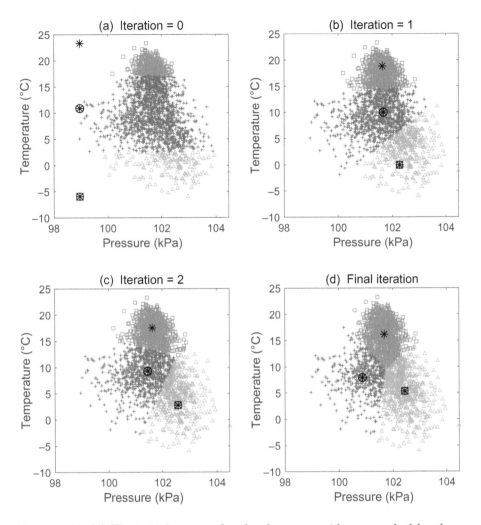

Figure 2.14 (a) The initial guesses for the three centroids are marked by three asterisks. The data points are assigned to clusters based on their nearest centroid. In (b), the centroids have been recalculated based on the mean position of the cluster members in (a), and cluster members in (b) have been reassigned based on their closeness to the centroids in (b). The location of the centroids and their associated cluster members are shown after (c) two iterations and (d) after final convergence of the $K$-means clustering algorithm.

where $h(x)$ is a monotonically decreasing function of the probability $P(x)$. If there are two independent events $x$ and $y$, the information from observing both events should be the sum of the two separate events, that is,

$$h(x, y) = h(x) + h(y). \tag{2.98}$$

Since the probability of observing two independent events obeys $P(x, y) = P(x)\,P(y)$, we can choose $h(x)$ to be

$$h(x) = -\log P(x), \tag{2.99}$$

so that

$$h(x, y) = -\log P(x, y) = -\log[P(x)P(y)] \tag{2.100}$$
$$= -\log P(x) - \log P(y) = h(x) + h(y). \tag{2.101}$$

The minus sign in (2.99) ensures $h(x) \geq 0$ and $h$ is a monotonically decreasing function of $P$. One can choose any base for the logarithm function, but the two most common choices are base 2 and base e, that is, $\log_2$ or $\log_e$ (ln). With base 2, $h(x)$ has units of *bits* (from 'binary digits'), whereas with base e (i.e. using natural logarithm), $h(x)$ has units of *nats*.

## 2.17.1 Entropy B☺

Since $X$ is a random variable, we are more interested in the average amount of information transmitted, that is, the expectation of $h(x)$ than $h(x)$ itself. The expectation of of (2.99) involves summing $h(x)$ weighted by the probability $P(x)$ over all possible states of $x$, that is,

$$H(X) = \mathrm{E}[h(x)] = -\sum_x P(x) \log P(x) = -\sum_i P_i \log P_i, \tag{2.102}$$

where $H(X)$ is called the *entropy* and $P$ at the discrete values of $x$ is also written as $P_i$. For any $x$ with $P(x) = 0$, we will set $P \log P = 0$ since $\lim_{P \to 0} P \log P = 0$.

Readers familiar with the statistical mechanics of Boltzmann and Gibbs will recall that the thermodynamic entropy $S$ is given by

$$S = -k_\mathrm{B} \sum_i P_i \log P_i, \tag{2.103}$$

where $k_\mathrm{B}$ is the Boltzmann constant and the summation is over all $i$ states. By switching to natural units, $k_\mathrm{B}$ becomes unity, and (2.103) is identical in form to (2.102).

Next, consider the simple example of $X$ being a binary variable (0 or 1). If we write $P(1) = \alpha$ and $P(0) = 1 - \alpha$, then (2.102) gives

$$H(X) = -[P(0) \log_2 P(0) + P(1) \log_2 P(1)] \tag{2.104}$$
$$= -[(1 - \alpha) \log_2(1 - \alpha) + \alpha \log_2 \alpha]. \tag{2.105}$$

The entropy is maximized (Fig. 2.15) when $\alpha = 0.5$, that is, $P(1) = P(0) = 0.5$, and minimized ($H = 0$) when $P(1) = 0$ or 1. For the coin flip analogy, $P(1) = 0$ or 1 means the coin is loaded to always come out 'head' (H) or 'tail' (T). Since the outcome is certain, more flips of the coin do not provide any new information, that is, the information content is 0 at $H = 0$. In contrast, when $\alpha = P(1) = P(0) = 0.5$, there is greatest uncertainty since there is equal probability of getting H or T; thus, information on the outcome of coin tosses is most valuable.

Figure 2.15 Entropy $H$ as a function of $\alpha$. When $\alpha = 0.5$, the maximum ($H = 1$) is attained.

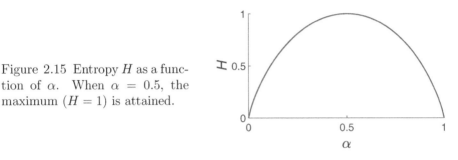

It is straightforward to define entropy for continuous random variables. With $\mathbf{X}$ denoting a random real vector, the entropy is given by

$$H(\mathbf{X}) = -\int p(\mathbf{x}) \log p(\mathbf{x})\, \mathrm{d}\mathbf{x}, \qquad (2.106)$$

where the natural logarithm is commonly used and the summation in (2.102) is replaced by an integration over the entire domain of $\mathbf{x}$. However, when computing with sampled data, continuous variables are commonly discretized or quantized, that is, dividing the domain of each variable into bins and counting how many data points fall within each bin to obtain a histogram and, thereby, a sample discrete probability distribution. We will continue our discussion using the discrete variable formulation.

## 2.17.2 Joint Entropy and Conditional Entropy B☺

Suppose there are two random variables $X$ and $Y$ from which we can draw specific values $x$ and $y$ with joint probability $P(x, y)$. The *joint entropy* between the two random variables is

$$H(X, Y) = \mathrm{E}[-\log P(x, y)] = -\sum_x \sum_y P(x, y) \log P(x, y). \qquad (2.107)$$

If the value of $x$ is already known, then the additional information needed to specify $y$ is $-\log P(y|x)$. The average additional information needed to specify $y$ is the *conditional entropy*

$$H(Y|X) = \mathrm{E}[-\log P(y|x)] = -\sum_x \sum_y P(x, y) \log P(y|x). \qquad (2.108)$$

Using (2.6), it can easily be shown that

$$H(X,Y) = H(X) + H(Y|X),  \tag{2.109}$$

that is, the information needed to describe both $X$ and $Y$ is the information needed to describe $X$ plus the additional information needed to describe $Y$ after $X$ is known.

### 2.17.3   Relative Entropy  B☺

Suppose the unknown true probability distribution is $P(x)$ and we have obtained $Q(x)$, an approximation of the true distribution. A measure of the dissimilarity between the two distributions is the *relative entropy*, also known as the *Kullback–Leibler* (KL) *divergence*, where

$$D_{\mathrm{KL}}(P||Q) = \mathrm{E}\left[\log \frac{P(x)}{Q(x)}\right] = \sum_x P(x)\log \frac{P(x)}{Q(x)}.  \tag{2.110}$$

This quantity is referred to as a divergence instead of a distance as it is asymmetric, that is, $D_{\mathrm{KL}}(P||Q) \neq D_{\mathrm{KL}}(Q||P)$ in general.

$$D_{\mathrm{KL}}(P||Q) = -\sum_x P(x)\log Q(x) + \sum_x P(x)\log P(x) = H_{\mathrm{c}}(P,Q) - H(P),$$

$$\tag{2.111}$$

where $H_{\mathrm{c}}(P,Q)$ is called the *cross-entropy*, with

$$H_{\mathrm{c}}(P,Q) = -\sum_x P(x)\log Q(x).  \tag{2.112}$$

It can be proven that $D_{\mathrm{KL}}(P||Q) \geq 0$, and $D_{\mathrm{KL}}(P||Q) = 0$ if, and only if, $P = Q$ (Cover and Thomas, 2006). In coding theory (Cover and Thomas, 2006), the cross-entropy $H_{\mathrm{c}}(P,Q)$ is the expected number of bits needed to encode data from a source with distribution $P$ while we use model $Q$ to define our codebook. The entropy $H(P)$ is the expected number of bits if we use the true model; thus, the relative entropy $D_{\mathrm{KL}}(P||Q)$ can be interpreted as the expected number of extra bits that must be communicated if a code that is optimal for the incorrect distribution $Q$ is used instead of using a code based on the true distribution $P$.

Relative entropy has been used in studies of predictability of dynamical systems (Kleeman, 2002; DelSole, 2004), including the predictability of the El Niño-Southern Oscillation (ENSO) (see Section 9.1.5), the dominant mode of interannual climate variability in the equatorial Pacific, with global implications (Y. M. Tang et al., 2005). For instance, $Q$ can be the distribution from climatology (i.e. the expected behaviour from historical observed data) while $P$ can be the distribution from the predictions by a numerical model. The relative entropy $D_{\mathrm{KL}}(P||Q)$ is a measure of the additional information provided by the prediction model over the information from climatology (Kleeman, 2002).

## 2.17.4   Mutual Information   B☺

Given two random variables $X$ and $Y$, if we want to detect any linear relation between the two, we can compute the correlation (Section 2.11). But what if the relation is non-linear? Correlation can completely miss a strong non-linear relation. To detect linear and non-linear relations, one can turn to *mutual information* (MI), which determines, via the KL divergence, how dissimilar the joint distribution $P(X, Y)$ is to $P(X) P(Y)$, that is, MI is given by

$$I(X,Y) = D_{\mathrm{KL}}(P(X,Y) \| P(X)P(Y)) = \sum_x \sum_y P(x,y) \log \frac{P(x,y)}{P(x)P(y)}.$$
(2.113)

As $D_{\mathrm{KL}}$ is non-negative, $I(X,Y) \geq 0$. When $X$ and $Y$ are independent variables, $P(X,Y) = P(X) P(Y)$. The logarithm term in (2.113) becomes $\log 1 = 0$, giving $I(X,Y) = 0$. Thus, the minimum MI occurs when $X$ and $Y$ are independent. As $X$ and $Y$ become more dependent, MI increases.

From (2.5), we can substitute $P(x,y) = P(y|x) P(x)$ into (2.113), giving

$$I(X,Y) = - \sum_x \sum_y P(x,y) \log P(y) + \sum_x \sum_y P(x,y) \log P(y|x) \qquad (2.114)$$

$$= - \sum_y P(y) \log P(y) + \sum_x \sum_y P(x,y) \log P(y|x), \qquad (2.115)$$

that is,

$$I(X,Y) = H(Y) - H(Y|X), \qquad (2.116)$$

with $H(Y|X)$ being the conditional entropy. Thus, MI can be viewed as the reduction in the uncertainty in $Y$ when the value of $X$ is known. Similarly,

$$I(X,Y) = H(X) - H(X|Y). \qquad (2.117)$$

A problem with applying the information theory approach to continuous variables is that it is difficult to estimate continuous probability density distributions. When computing with sampled data, continuous variables are often discretized or quantized, that is, dividing the domain of each variable into bins and counting how many data points fall within each bin to obtain a histogram as a discrete approximation of the probability density distribution. The final result unfortunately depends on the choice of the bin width – having a wide bin width gives a crude approximation of a continuous distribution, while having a narrow bin width means there are few data points within each bin, thus a noisy histogram. Reshef et al. (2011) has proposed a way to deal with the bin width problem. An alternative approach is to use kernel density estimation (see Section 3.13) to estimate the distribution, but instead of the bin width there is now an adjustable width parameter of the kernel function. A more recent approach using $K$-nearest neighbour distances to estimate MI (Kraskov et al., 2004) has seen increasing usage.

When there are many available predictors or features, one may want to select the most relevant predictors before building a prediction model. Traditionally, a common approach is to compute the correlation between a predictor variable and the response variable and select the predictors with high correlation. However, predictors non-linearly related to the response variable may be missed using this selection procedure. MI has been proposed as a better measure for predictor selection, as it is not restricted to detecting linear relations (H. C. Peng et al., 2005). In environmental sciences, MI has been proposed for predictor selection in hydrological studies (Sharma, 2000; May, Maier, et al., 2008; May, Dandy, et al., 2008).

# Exercises

*Some exercises involve working with data files, which are downloadable from our book website (web link given in the Preface).*

### 2.1

In a tropical Atlantic region, the number of occurrences when the daily sea surface temperature condition is cool, normal or warm and when the wind condition is calm or stormy have been recorded in the table below. What is the probability of a day being (a) warm and stormy and (b) cool and stormy? What is the probability of the day being (c) stormy if it is a warm day and (d) stormy if it is a cool day?

|        | cool | normal | warm |
|--------|------|--------|------|
| calm   | 1805 | 3661   | 2012 |
| stormy | 32   | 125    | 228  |

### 2.2

A variable $y$ is measured by two instruments placed 50 km apart in the east–west direction. Values are recorded daily for 100 days. The autocorrelation function of $y$ shows the first zero crossing (i.e. the smallest lag at which the autocorrelation is zero) occurring at six days (for both stations). Furthermore, $y$ at one station is correlated with the $y$ at the second station, with the second time series shifted in time by various lags. The maximum correlation occurred with $y$ from the eastern station lagging $y$ from the western station by two days. Assuming a sinusoidal wave is propagating between the two stations, estimate the period, wavelength, and the speed and direction of propagation.

### 2.3

Prove that the expectation of the sample mean in (2.25) equals the population mean.

### 2.4

Given two variables $x$ and $y$ with zero population means: (a) Show that the population covariance $\text{cov}(x, y) = \text{E}[xy]$ is zero if $x$ and $y$ are independent. (b) However, the converse is not true in general. Given $x$ uniformly distributed in $[-1, 1]$ and $y = x^2$, show that $\text{cov}(x, y)$ is zero even though $x$ and $y$ are not independent.

2.5

Using the data file provided on the book website, compare the Pearson correlation with the Spearman and Kendall rank correlations for the time series $x$ and $y$ (each with 40 observations). Repeat the comparison for the time series $x_2$ and $y_2$ (from the same data file as above), where $x_2$ and $y_2$ are the same as $x$ and $y$, except that the fifth data point in $y$ is replaced by an outlier in $y_2$. Repeat the comparison for the time series $x_3$ and $y_3$, where $x_3$ and $y_3$ are the same as $x$ and $y$, except that the fifth data point in $x$ and $y$ is replaced by an outlier in $x_3$ and $y_3$. Make scatterplots of the data points in the $x$–$y$ space, the $x_2$–$y_2$ space and the $x_3$–$y_3$ space. Also plot the linear regression line in the scatterplots.

2.6

Analyse the monthly sea surface temperature anomalies (i.e. deviations from the mean) for the Niño1+2 region in the eastern equatorial Pacific ($0°$–$10°$S, $80°$W–$90°$W) and the Niño 3.4 region in the central equatorial Pacific ($5°$S–$5°$N, $170°$W–$120°$W) (shown in Fig. 9.3 and data downloadable from our book website):
(a) For each variable, compute the histogram and compare to the Gaussian distribution fit to the data. (b) For each variable, compute the quantile–quantile plot relative to the standard Gaussian distribution. (c) Compute the quantile–quantile plot between Niño1+2 and Niño3.4 anomalies. (d) Compute the box-plot for the two variables. [Data source: Climatic Research Unit, University of East Anglia]

2.7

In addition to the Niño1+2 and Niño3.4 sea surface temperature anomalies, analyse the Southern Oscillation Index (SOI) (Tahiti pressure minus Darwin pressure, standardized to zero mean and unit standard deviation): (a) For each of the three time series, plot the Pearson autocorrelation function. (b) Compute the Pearson correlation and the Spearman and Kendall rank correlations between Niño1+2 and Niño3.4 anomalies. (c) Compute the Pearson correlation and the Spearman and Kendall rank correlations between Niño1+2 and SOI. (d) Compute the Pearson correlation and the Spearman and Kendall rank correlations between Niño3.4 and SOI. [Data source: Climatic Research Unit, University of East Anglia]

2.8

Suppose a test for the presence of a toxic chemical in a lake gives the following results: if a lake has the toxin, the test returns a positive result 99% of the time; if a lake does not have the toxin, the test still returns a positive result 2% of the time. Suppose only 5% of the lakes contain the toxin. What is the probability that a positive test result for a lake turns out to be a false positive?

2.9

Use $K$-means clustering to analyse the dataset containing air temperature and humidity at Vancouver, BC, Canada from 2013–2017. Try $K = 2$ and 3, and try to explain what the clusters represent. [Data source: weatherstats.ca based on Environment and Climate Change Canada data]

### 2.10

A biological oceanographer collected 100 water samples at various locations, from which the water temperature $(T)$, the nitrate concentration $(N)$, the silicate concentration $(S)$ and the concentration of a marine microorganism $(M)$ were measured. The measurements were discretized into 'below normal', 'normal' and 'above normal', that is, 1, 2 and 3, respectively. The observed number of occurrences are given in the table below. Which of the three environmental variables has the strongest relation with $M$? Try to determine this using mutual information and using Pearson and Spearman correlation.

| $M\backslash T$ | 1 | 2 | 3 | $M\backslash N$ | 1 | 2 | 3 | $M\backslash S$ | 1 | 2 | 3 |
|---|---|---|---|---|---|---|---|---|---|---|---|
| 1 | 2 | 38 | 3 | 1 | 13 | 9 | 3 | 1 | 3 | 7 | 5 |
| 2 | 5 | 8 | 5 | 2 | 8 | 33 | 7 | 2 | 13 | 26 | 12 |
| 3 | 18 | 4 | 17 | 3 | 4 | 8 | 15 | 3 | 9 | 17 | 8 |

# 3

# Probability Distributions

With a central role in statistics, probability theory also plays a very important role in modern machine learning. In this chapter, we will study some basic types of probability distributions. While these basic distributions are idealistic, they are illuminatingly instructive and can be the building blocks for more complicated models.

The basic probability distributions discussed in this chapter are commonly used to model the probability distribution $p(\mathbf{x})$ of a random variable $\mathbf{x}$, for which we have observed data $x_1, \ldots, x_n$. This is known as *density estimation*. Most of the basic probability distribution functions have only a small number of adjustable parameters. Once the parameters are estimated using the observed data, $p(\mathbf{x})$ is modelled by specifying the small number of parameters, assuming the basic distribution fits the data reasonably well. These distributions with adjustable parameters to control the functional form are called *parametric distributions*.

In this chapter, we will start with parametric distributions for *discrete* variables – the binomial distribution, the Poisson distribution and the multinomial distribution. Then we will move onto parametric distributions for *continuous* variables – the Gaussian (i.e. normal) distribution for a continuous variable over $(-\infty, \infty)$, the gamma distribution over a domain bounded on one side, for example $[0, \infty)$, the beta distribution over a bounded domain, for example $[0, 1]$ and the von Mises distribution over a periodic domain $[0, 2\pi)$. We will also study the multivariate Gaussian distribution, the mixture of Gaussian distributions and extreme value distributions.

A limitation of the parametric approach is that the functional form for the distribution is specified, which may not work well for some datasets. The alternative *non-parametric distribution* approach still has parameters, but the parameters are not used to control the specified functional form; instead, the parameters are used to control the model complexity.

## 3.1  Binomial Distribution  $\boxed{\text{A}}$☺

Let $x$ be a binary variable, that is, it can only take on the value 0 or 1. Examples of environmental binary variables include the prediction of the occurrence of an event, for example a tsunami or a tornado, with $0 = $ 'no occurrence' and $1 = $ 'occurrence'.

Consider the outcome of a coin toss, where $0 = $ 'tail' (T) and $1 = $ 'head' (H). If the probability for getting an H is $p$, then the probability of getting a T is $1 - p$, that is

$$P(x\,|\,p) = \begin{cases} p, & \text{if } x = 1 \\ 1 - p, & \text{if } x = 0. \end{cases} \tag{3.1}$$

We could write the probability $P(x\,|\,p)$ simply as $P(x)$, but the notation $P(x\,|\,p)$ is helpful in reminding us that the parameter $p$ is also needed to specify the probability function. This probability distribution is called the *Bernoulli distribution* (after Jacob Bernoulli) and can be rewritten as

$$\text{Ber}(x\,|\,p) = p^x (1 - p)^{1-x}. \tag{3.2}$$

By substituting in 1 and then 0 for $x$ in (3.2), one recovers (3.1). The mean and variance of this distribution can be easily derived as

$$\text{E}[x] = p, \tag{3.3}$$
$$\text{var}[x] = p(1 - p). \tag{3.4}$$

If one tosses the coin 100 times and gets 65 H, then the sample mean of $x$ is 0.65, and from (3.3), one can estimate $p$ to be 0.65.

Next the coin toss is repeated a total of $N$ times, with the assumption that outcomes are independent events, and the probability $p$ remains constant during the tosses. We want to know the probability of getting $k$ heads out of the $N$ tosses, where $k = 0, 1, \ldots, N$.

For simplicity, start with the case $N = 2$, that is, we toss the coin twice. The possible outcomes are two heads (HH), one head and one tail (HT or TH) and two tails (TT). The probability of getting two heads is $p^2$ and that of two tails is $(1 - p)^2$. The probability of getting one head and one tail is the sum of the probability of getting HT and that of getting TH, that is, $p(1-p) + (1-p)p = 2p(1-p)$.

These probabilities are simply obtained from the *binomial distribution* (introduced by Jacob Bernoulli in 1713),

$$\text{Bin}(k\,|\,N, p) = \binom{N}{k} p^k (1 - p)^{N-k}, \quad k = 0, 1, \ldots, N, \tag{3.5}$$

with two parameters $N$ and $p$, and for our case, $N = 2$ and $k = 0, 1$ and 2. The binomial coefficient is defined by

$$\binom{N}{k} = \frac{N!}{k!\,(N - k)!}, \tag{3.6}$$

where the factorial notation $k! = k(k-1)\ldots 2 \cdot 1$. The binomial coefficient counts the number of ways one can end up with $k$ heads from $N$ tosses. For $N = 2$ and $k = 1$, the binomial coefficient $= 2!/[1!\,(2-1)!] = 2!/(1!\,1!) = (2 \cdot 1)/(1 \cdot 1) = 2$, as there are two ways to get a single head, namely HT and TH.

For example, let us look at the occurrence of category 5 Atlantic hurricanes (category 5 being the strongest), where over 100 years (1919–2018) there have been 25 years with at least one occurrence (Wikipedia contributors, 2018). Some years, there can be more than one occurrence (e.g. the unusually active year of 2005 had four occurrences), with a total of 33 occurrences in those 25 years. If we let $x = 1$ in a year with at least one occurrence and 0 otherwise, then the sample mean of $x = 0.25$. From (3.3) we obtained the estimate $p = 0.25$.

Over a period of 10 years, what is the probability of having $0, 1, 2, \ldots, 10$ years with one or more category 5 Atlantic hurricanes? The binomial distribution (3.5) with $p = 0.25$, $N = 10$ gives the probabilities: 0.056, 0.188, 0.282, 0.250, 0.146, 0.058, 0.016, 0.003, 0.000, 0.000 and 0.000 for $k = 0, 1, 2, \ldots, 10$, respectively. Thus, the probability of having no category 5 hurricane over a 10-year period is 0.056. Figure 3.1 plots the binomial distribution of years with category 5 hurricane(s) per decade and per century.

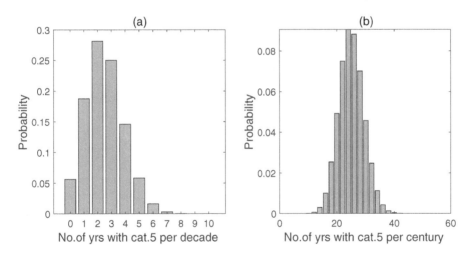

Figure 3.1 Probability distribution of the number of years with category 5 hurricane(s) (a) per decade and (b) per century. The binomial distribution with $p = 0.25$ was used with (a) $N = 10$ and (b) $N = 100$. As $N$ becomes large, the skewness of the distribution disappears, and the binomial distribution can be approximated by the Gaussian (normal) distribution.

As environmental data often have serial correlation (Section 2.11.2) (e.g. tomorrow's air temperature is correlated with today's temperature), the assumption of independent events made by the binomial (and many other) distributions is often not fulfilled. However, if the events typically occur over a time interval larger than the typical time scale where serial correlation is observed, then the assumption of independent events is largely satisfied. For instance, suppose serial correlation is found for precipitation over three days. Then daily precipitation cannot be considered independent events, but weekly averaged and monthly averaged precipitation can.

## 3.2   Poisson Distribution   $\boxed{\text{B}}$☺

Suppose the probability of a single event occurring in a short time interval $\delta t$ is $\nu \delta t$, with $\nu$ being a positive constant. Then the probability of $x$ events ($x = 0, 1, 2, \ldots$) occurring in time $t$ is given by the *Poisson distribution* (first published by Siméon Denis Poisson in 1837),

$$\text{Poi}(x|\lambda) = e^{-\lambda} \frac{\lambda^x}{x!}, \tag{3.7}$$

where the parameter $\lambda = \nu t$ is called the average occurrence rate, as it indicates the average number of events in the given time interval $t$. The Poisson distribution has mean and variance given by

$$E[x] = \lambda, \tag{3.8}$$
$$\text{var}[x] = \lambda. \tag{3.9}$$

From (3.8), $\lambda$ can be estimated from the sample mean of $x$.

Since there were 33 category 5 Atlantic hurricanes observed during 100 years, the sample mean of $x = 33/100 = 0.33$; thus, we use $\lambda = 0.33$. The Poisson distribution then gives the probability of having 0, 1, 2, 3, 4 and 5 category 5 hurricanes in one year as 0.719, 0.237, 0.039, 0.0043, 0.0004 and 0.0000, respectively. Thus, the probability of not having any category 5 hurricane in one year is 0.719; the probability of having four occurrences in one year is 0.0004. Having four occurrences in 2005 was exceedingly rare and not likely to happen again within one's lifetime unless climate change dramatically alters $\lambda$.

## 3.3   Multinomial Distribution   $\boxed{\text{B}}$☺

If instead of tossing a coin with binary outcomes, we are rolling a die with $m$ sides, there are now $m$ possible outcomes $1, 2, \ldots, m$, with probability $p_1, p_2, \ldots, p_m$, respectively. If the die is rolled $N$ times, and we use the vector $\mathbf{x} = [x_1, x_2, \ldots, x_m]$ to denote getting side 1 of the die $x_1$ times, side 2 $x_2$ times and so on, then the probability of getting $\mathbf{x}$ is given by the *multinomial distribution*

$$\text{Mu}(\mathbf{x}|N, p_1, \ldots, p_m) = \binom{N}{x_1 \ldots x_m} p_1^{x_1} p_2^{x_2} \ldots p_m^{x_m}, \tag{3.10}$$

where

$$\sum_{i=1}^{m} x_i = N, \qquad \sum_{i=1}^{m} p_i = 1, \tag{3.11}$$

and the multinomial coefficient

$$\begin{pmatrix} N \\ x_1 \dots x_m \end{pmatrix} = \frac{N!}{x_1! x_2! \dots x_k!} \tag{3.12}$$

is the number of ways to divide a set of $N$ members into $m$ subsets with $x_1, \dots, x_m$ members. The binomial distribution (3.5) is clearly a special case of the multinomial distribution.

For example, consider a river basin where the summer conditions can be classified into drought, flood or normal, with the non-normal summers bad for agriculture. Suppose over the historical data record of 60 summers, there were 10 drought summers and 6 flood summers. We can estimate that $p_1 = 10/60 = 1/6$ for drought, $p_2 = 6/60 = 1/10$ for flood and $p_3 = 1 - p_1 - p_2 = 11/15$ for normal conditions.

Suppose the farmers can survive one non-normal summer, but two consecutive non-normal summers are deadly. What is the probability of having two consecutive non-normal summers for this region? Two consecutive non-normal summers can mean having two droughts, two floods or one drought and one flood.

The probability of having two consecutive drought summers is given by (3.10) with $\mathbf{x} = [2, 0, 0]$ and $N = 2$, that is

$$P_1 = \mathrm{Mu}([2,0,0]\,|\,2, 1/6, 1/10, 11/15) = 0.028. \tag{3.13}$$

The probability of having two consecutive flood summers is

$$P_2 = \mathrm{Mu}([0,2,0]\,|\,2, 1/6, 1/10, 11/15) = 0.010, \tag{3.14}$$

while the probability of having one drought and one flood summer over two consecutive summers is

$$P_3 = \mathrm{Mu}([1,1,0]\,|\,2, 1/6, 1/10, 11/15) = 0.033. \tag{3.15}$$

The probability of having two consecutive non-normal summers is given by $P = P_1 + P_2 + P_3 = 0.071$. Alternatively, $P$ can be calculated from the binomial distribution (3.5), where $P = \mathrm{Bin}(0\,|\,2, 11/15) = 0.071$, that is, zero occurrences of normal events in two tosses.

## 3.4 Gaussian Distribution $\boxed{\text{A}}$ ☺

We now turn to *continuous* probability distributions $p(x)$, where $x$ is a continuous variable. The most commonly used continuous distribution is the Gaussian distribution, named after Carl Friedrich Gauss (though Pierre-Simon Laplace also made major contributions at roughly the same time as Gauss). Since this bell-shaped distribution is so common, it is also called the normal distribution, which is a little excessive since there are many natural phenomena that do not follow the Gaussian distribution.

The probability density function (PDF) $p(x)$ of a Gaussian distribution with population mean $\mu$ and variance $\sigma^2$ is given by

$$\mathcal{N}(x\,|\,\mu,\sigma^2) = \frac{1}{\sqrt{2\pi}\,\sigma}\,\exp\left[-\frac{(x-\mu)^2}{2\sigma^2}\right], \qquad (3.16)$$

as illustrated in Fig. 3.2. Although we use the term 'Gaussian' instead of 'normal distribution', we keep the traditional symbol $\mathcal{N}$ used for the normal distribution.[1] Thus, a Gaussian distribution is specified by two parameters, namely its mean and variance. When given some observed data for $x$, one can compute the sample mean and sample variance for the data. These can then be used as estimates for $\mu$ and $\sigma^2$ for the Gaussian distribution that best fits the given data. This approach to estimating parameters is called the *method of moments* as it involves matching the population and sample moments.

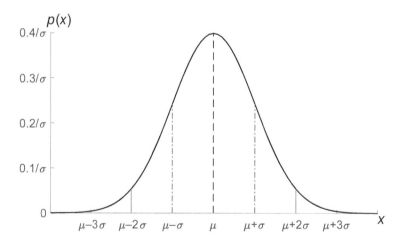

Figure 3.2 Probability density $p(x)$ of a Gaussian distribution with mean $\mu$ and standard deviation $\sigma$.

The probability of having a data point within $\pm\sigma$ of $\mu$ is simply the area underneath the curve $p(x)$ within the interval $\mu - \sigma \le x \le \mu + \sigma$, and is 68.27% for the Gaussian distribution. The probability within $\pm 2\sigma$ of $\mu$ is 95.45% and within $\pm 3\sigma$ is 99.73% for the Gaussian. The Gaussian distribution is symmetric about its mean, so its skewness is zero, and its mean, median and mode are all equal.[2]

---

[1] $\mathcal{N}(\mu,\sigma^2)$ is also commonly used as an abbreviation of $\mathcal{N}(x\,|\,\mu,\sigma^2)$.
[2] The *mode* is the value of $x$ when $p(x)$ is maximum.

The Gaussian distribution with $\mu = 0$ and $\sigma = 1$ is referred to as the *standard Gaussian distribution*,

$$\mathcal{N}(z\,|\,0,1) = \frac{1}{\sqrt{2\pi}} \exp\left(-\frac{z^2}{2}\right). \tag{3.17}$$

One can always convert any Gaussian distribution $\mathcal{N}(x\,|\,\mu,\sigma^2)$ to the standard Gaussian $\mathcal{N}(z\,|\,0,1)$ by introducing a standardized variable $z$ (a.k.a. standard score or $z$-score), where

$$z = \frac{x - \mu}{\sigma} \tag{3.18}$$

(von Storch and Zwiers, 1999, p.35). This ability to convert any Gaussian distribution into the standard Gaussian is useful since tabulated tables in books are provided only for the standard Gaussian. However, looking up statistical tables in books is archaic, as common scientific programming languages all provide functions for standard statistical distributions. For this reason, no statistical tables are provided in this book.

The cumulative distribution function (CDF) of the standard Gaussian distribution is defined by

$$\Phi(z) = \int_{-\infty}^{z} \mathcal{N}(\tilde{z}\,|\,0,1)\,d\tilde{z} = \frac{1}{\sqrt{2\pi}} \int_{-\infty}^{z} \exp\left(-\frac{\tilde{z}^2}{2}\right)d\tilde{z} \tag{3.19}$$

$$= \frac{1}{2}\left[1 + \mathrm{erf}\left(\frac{z}{\sqrt{2}}\right)\right]. \tag{3.20}$$

This is the probability of $\tilde{z}$ taking a value in the interval $(-\infty, z]$. The integral cannot be expressed in terms of elementary functions but can be expressed in terms of the erf or error function (unrelated to the term 'error function' used in machine learning) (von Storch and Zwiers, 1999, p.35).

For the general Gaussian distribution $\mathcal{N}(x\,|\,\mu,\sigma^2)$, the CDF is given by

$$F(x) = \frac{1}{\sqrt{2\pi}\,\sigma} \int_{-\infty}^{x} \exp\left[-\frac{(\tilde{x} - \mu)^2}{2\sigma^2}\right]d\tilde{x} \tag{3.21}$$

$$= \frac{1}{2}\left[1 + \mathrm{erf}\left(\frac{x - \mu}{\sigma\sqrt{2}}\right)\right]. \tag{3.22}$$

For estimating confidence intervals, quantiles of the standard Gaussian distribution are invoked. In Section 2.9 on quantiles, we mentioned that one is often interested in finding a value $x_\alpha$ where the probability $P(x \le x_\alpha) = \alpha$ for a given value of $\alpha$, with $0 \le \alpha \le 1$. For instance, one may want to know the value $x_{0.975}$ where 97.5% of the distribution lie below the value $x_{0.975}$ – that is, finding the 0.975 quantile.

For the standard Gaussian, the CDF $\Phi(z_\alpha)$ from (3.20) has an inverse function $\Phi^{-1}(\alpha)$, that is, $\alpha = \Phi(z_\alpha)$ and $z_\alpha = \Phi^{-1}(\alpha)$. From (3.20), by applying $\mathrm{erf}^{-1}$ to $\mathrm{erf}(z_\alpha/\sqrt{2})$, one gets

$$z_\alpha = \Phi^{-1}(\alpha) = \sqrt{2}\,\mathrm{erf}^{-1}(2\alpha - 1), \tag{3.23}$$

where $\mathrm{erf}^{-1}$ is the inverse error function.

For instance, the quantile $z_{0.975} = 1.96$. Based on the area under the two tails of a standard Gaussian distribution, a standard Gaussian variable will lie outside $\pm 1.96$ in only $2 \times (1 - 0.975) \times 100\% = 5\%$ of cases. Commonly encountered quantiles of the standard Gaussian are as follows: $z_{0.90} = 1.2816$, $z_{0.95} = 1.6449$, $z_{0.975} = 1.9600$, $z_{0.99} = 2.3263$ and $z_{0.995} = 2.5758$.

For the general Gaussian,

$$x_\alpha = F^{-1}(\alpha) = \mu + \sigma \Phi^{-1}(\alpha) = \mu + \sigma\sqrt{2}\,\mathrm{erf}^{-1}(2\alpha - 1), \qquad (3.24)$$

where the Gaussian variable will lie outside $\mu \pm 1.96\,\sigma$ in only 5% of cases.

The Gaussian distribution is found frequently in nature thanks to the central limit theorem (Fischer, 2020). The original version of the theorem, proved by Laplace in 1810, states that the sum or average of a sequence of $N$ independent and identically distributed (i.i.d.)[3] random variables approaches a Gaussian distribution as $N \to \infty$. The original variables need not follow the Gaussian distribution – for example, the variables may be from a distribution where $p(x)$ has multiple maxima. The theorem was generalized later by Aleksandr Lyapunov and others, so the variables no longer have to be identically distributed (i.e. they can follow different distributions). Many variables in nature are the result of summing over many other variables, and hence the common occurrence of the Gaussian distribution.

Let $x_1, \ldots, x_N$ be independent samples from a Gaussian distribution $\mathcal{N}(x\,|\,\mu, \sigma^2)$. The sample mean $\bar{x}$ is computed from

$$\bar{x} = \frac{1}{N} \sum_{i=1}^{N} x_i. \qquad (3.25)$$

It follows that $\bar{x}$ is also a Gaussian random variable with population mean $\mu$ and population variance $\sigma^2/N$ (Zehna, 1970). The term *standard error* is used to denote the standard deviation of the sampling distribution of a statistic (here the statistic is the mean). Thus, the standard error of the mean is

$$\mathrm{SE} = \sigma/\sqrt{N}, \qquad (3.26)$$

where SE can be made much smaller than $\sigma$ by increasing $N$, the number of terms used to compute the sample mean in (3.25). The Gaussian variable $\bar{x}$ will lie outside the interval $\mu \pm 1.96\,\mathrm{SE} = \mu \pm 1.96\,\sigma/\sqrt{N}$ for only 5% of all cases. In practice, the sample standard deviation $s$ from (2.28) is used instead of the population standard deviation $\sigma$ when estimating the SE of the mean in (3.26).

---

[3] In probability theory and statistics, a collection of random variables is *independent and identically distributed* (i.i.d.) if the random variables all obey the same probability distribution and are all mutually independent. In statistics, observations in a sample are often assumed to be i.i.d.

## 3.5 Maximum Likelihood Estimation [A] ☺

When fitting a distribution to a dataset, an alternative and more widely used approach to estimate the parameters of the distribution is the maximum likelihood method. In everyday usage, likelihood is a synonym of probability; however, statisticians make a clear distinction between the two.

Suppose $x$ is a discrete variable with the probability distribution $P(x|\mathbf{a})$, where $\mathbf{a} = [a_1, \ldots, a_M]$ is the set of parameters governing the shape of the distribution function. If we have a dataset $X$ containing the observations $x_1, \ldots, x_N$, the probability of observing $x_j$ ($j = 1, \ldots, x_N$) is $P(x_j|\mathbf{a})$. If we assume the observations are independent of each other, then the probability of observing the dataset $X$ is simply the product of the probabilities of observing each individual data point $x_j$ in $X$, that is

$$P(X|\mathbf{a}) = \prod_{j=1}^{N} P(x_j|\mathbf{a}). \tag{3.27}$$

Suppose we have a function $f(x, y)$. If we decide that this function has one variable and one given parameter controlling the shape of the function, we have two ways to interpret $f$. The first way is to view $x$ as the variable and $y$ as the given parameter; the second way is to view $y$ as the variable and $x$ as the parameter. Thus, $f$ is like a coin with two sides – as we will see in the next paragraph, viewed from one side, $f$ is a probability, and viewed from the other side, $f$ is a likelihood, and the two views are quite distinct.

In $P(X|\mathbf{a})$, $X$ is the set of variables, and $\mathbf{a}$ is the set of given parameters. However, the actual situation is quite the opposite; we have no idea what $\mathbf{a}$ should be, but we are given a set of observations $X$. In other words, $\mathbf{a}$ should be viewed as the set of variables and $X$ the set of given 'parameters'. Thus, the *likelihood* function is defined by

$$L(\mathbf{a}|X) = P(X|\mathbf{a}), \tag{3.28}$$

that is, the mathematical forms of $L$ and $P$ are identical, but they are interpreted very differently, with $L$ treating $\mathbf{a}$ as variables and $X$ as given.

Now that we are given $X$, what is a reasonable way to estimate $\mathbf{a}$? Since we have observed $X$, it seems reasonable that we should find the $\mathbf{a}$ that would maximize the likelihood of observing $X$ – and this is the *maximum likelihood* approach. Of course, if the number of observations is small, this approach would not be very reliable. For mathematical simplicity, instead of maximizing $L$, one maximizes its logarithm,

$$\mathcal{L} \equiv \ln L. \tag{3.29}$$

As $\ln L$ is a monotonically increasing function of $L$, if $\mathbf{a}$ maximizes $\ln L$, it also maximizes $L$. To find the optimal values for $a_1, \ldots, a_M$, we can set

$$\frac{\partial \mathcal{L}}{\partial a_i} = 0, \quad i = 1, \ldots, M \tag{3.30}$$

and solve for the values of $a_1, \ldots, a_M$.

If $x$ is a continuous variable instead of a discrete variable, the probability distribution $P(x\,|\,\mathbf{a})$ is replaced by the probability density distribution $p(x\,|\,\mathbf{a})$, and the likelihood

$$L(\mathbf{a}\,|\,X) = p(X\,|\,\mathbf{a}). \tag{3.31}$$

Next, we illustrate using maximum likelihood to estimate the parameters of a Gaussian distribution $\mathcal{N}(x\,|\,\mu, \sigma^2)$. Equation (3.27) becomes

$$p(X\,|\,\mathbf{a}) = \prod_{j=1}^{N} \mathcal{N}(x_j\,|\,\mu, \sigma^2), \tag{3.32}$$

$$= \prod_{j=1}^{N} \frac{1}{\sqrt{2\pi}\,\sigma} \exp\left[-\frac{(x_j - \mu)^2}{2\sigma^2}\right], \tag{3.33}$$

upon substituting in (3.16). Substituting (3.33) into (3.31) and taking the logarithm, we have

$$\mathcal{L} \equiv \ln L(\mu, \sigma^2\,|\,X) = -\frac{1}{2}N\ln(2\pi) - N\ln\sigma - \frac{1}{2\sigma^2}\sum_{j=1}^{N}(x_j - \mu)^2. \tag{3.34}$$

$$\frac{\partial\mathcal{L}}{\partial\mu} = \frac{1}{\sigma^2}\sum_{j=1}^{N}(x_j - \mu) = 0 \tag{3.35}$$

gives the estimate for the parameter $\mu$,

$$\mu = \frac{1}{N}\sum_{j=1}^{N} x_j, \tag{3.36}$$

which is just the sample mean.

$$\frac{\partial\mathcal{L}}{\partial\sigma} = -\frac{N}{\sigma} + \frac{1}{\sigma^3}\sum_{j=1}^{N}(x_j - \mu)^2 = 0 \tag{3.37}$$

gives the estimate for $\sigma^2$,

$$\sigma^2 = \frac{1}{N}\sum_{j=1}^{N}(x_j - \mu)^2. \tag{3.38}$$

The maximum likelihood estimation for $\sigma^2$ is the sample variance, which is a biased estimator for the population variance. One can correct the bias by changing the denominator from $N$ to $N - 1$, as in (2.28).

## 3.6 Multivariate Gaussian Distribution B☺

Instead of the scalar $x$, we now work with a $D$-dimensional vector $\mathbf{x}$, and the multivariate Gaussian distribution is given by

$$\mathcal{N}(\mathbf{x}|\,\boldsymbol{\mu},\mathbf{C}) = \frac{1}{(2\pi)^{D/2}\,|\mathbf{C}|^{1/2}}\,\exp\left[-\frac{1}{2}(\mathbf{x}-\boldsymbol{\mu})^{\mathrm{T}}\mathbf{C}^{-1}(\mathbf{x}-\boldsymbol{\mu})\right], \qquad (3.39)$$

where the $D$-dimensional vector $\boldsymbol{\mu}$ is the mean of $\mathbf{x}$, $\mathbf{C}$ is a $D \times D$ covariance matrix, as defined in (2.33) and $|\mathbf{C}|$ is the determinant of $\mathbf{C}$. In the exponent, we find the Mahalanobis distance $d_{\mathrm{M}}$ from (2.71) appearing in

$$(\mathbf{x}-\boldsymbol{\mu})^{\mathrm{T}}\mathbf{C}^{-1}(\mathbf{x}-\boldsymbol{\mu}) = d_{\mathrm{M}}^2. \qquad (3.40)$$

As the covariance matrix $\mathbf{C}$ is a real, symmetric matrix, the eigenequation

$$\mathbf{C}\,\mathbf{e}_i = \lambda_i\,\mathbf{e}_i \qquad (3.41)$$

has real eigenvalues $\lambda_i$, and the eigenvectors $\mathbf{e}_i$ can be chosen to be orthonormal, that is,

$$\mathbf{e}_j^{\mathrm{T}}\mathbf{e}_i = \delta_{ij}, \qquad (3.42)$$

with $\delta_{ij}$ the Kronecker delta function (i.e. $\delta_{ij} = 1$ if $i = j$, and 0 otherwise). The matrices $\mathbf{C}$ and $\mathbf{C}^{-1}$ can both be expanded in terms of the eigenvectors,

$$\mathbf{C} = \sum_{i=1}^{D}\lambda_i\mathbf{e}_i\mathbf{e}_i^{\mathrm{T}}, \qquad (3.43)$$

$$\mathbf{C}^{-1} = \sum_{i=1}^{D}\lambda_i^{-1}\mathbf{e}_i\mathbf{e}_i^{\mathrm{T}}. \qquad (3.44)$$

In Section 2.13.1, we have studied the connection between Mahalanobis distance and principal component analysis (PCA). Left multiplying (2.72) by $\mathbf{e}_j^{\mathrm{T}}$ and using the orthonormal property (3.42), one has the principal component

$$a_j = \mathbf{e}_j^{\mathrm{T}}(\mathbf{x}-\boldsymbol{\mu}). \qquad (3.45)$$

With (3.44) and (3.45), we can rewrite (3.40) as

$$d_{\mathrm{M}}^2 = \sum_{j=1}^{D}\lambda_j^{-1}a_j^2. \qquad (3.46)$$

One can interpret $a_j$ as the distance along the $\mathbf{e}_j$ direction upon projecting the vector $\mathbf{x} - \boldsymbol{\mu}$ onto the unit vector $\mathbf{e}_j$. With $\mathbf{a} = [a_1,\ldots,a_D]^{\mathrm{T}}$, we have

$$\mathbf{a} = \mathbf{E}(\mathbf{x}-\boldsymbol{\mu}), \qquad (3.47)$$

where the matrix $\mathbf{E}$ has rows given by $\mathbf{e}_j^{\mathrm{T}}$. From (3.42), $\mathbf{E}$ is an orthogonal matrix satisfying

$$\mathbf{E}\mathbf{E}^{\mathrm{T}} = \mathbf{I} = \mathbf{E}^{\mathrm{T}}\mathbf{E}, \qquad (3.48)$$

with $\mathbf{I}$ the identity matrix.

For easier visualization, let us consider the *bivariate* Gaussian distribution, that is, the multivariate Gaussian distribution in the two-dimensional space $(x_1, x_2)$. The elliptical curve in Fig. 3.3 has semi-major axis of length $\lambda_1^{1/2}$ and semi-minor axis of $\lambda_2^{1/2}$. Equation (3.46) gives $d_M^2 = 1$ on this ellipse, yielding a probability density $\exp(-1/2)$ times that at the centre $\boldsymbol{\mu}$, according to (3.39) and (3.40). The eigenvectors $\mathbf{e}_1$ and $\mathbf{e}_2$ can be regarded as the axes of a new coordinate system, with the semi-major axis of the ellipse pointing in the direction of $\mathbf{e}_1$.

Figure 3.3    The elliptical contour (with semi-major axis of
length $\lambda_1^{1/2}$ and semi-minor axis
of $\lambda_2^{1/2}$) shows where the probability density is $\exp(-1/2)$
times that at the centre $\boldsymbol{\mu}$ for
the Gaussian distribution in a
two-dimensional space $(x_1, x_2)$.
The semi-major and semi-minor
axis are pointed in the directions given by the eigenvectors
$\mathbf{e}_1$ and $\mathbf{e}_2$, respectively, while $a_1$
and $a_2$ are the distances (mea-
sured from $\boldsymbol{\mu}$) along these two
directions.

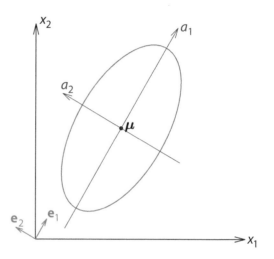

From maximum likelihood estimation, one can obtain the mean and covariance of the multivariate Gaussian distribution,

$$E[\mathbf{x}] = \boldsymbol{\mu}, \tag{3.49}$$

$$\text{cov}(\mathbf{x}) = E\left[(\mathbf{x} - E[\mathbf{x}])(\mathbf{x} - E[\mathbf{x}])^T\right] = \mathbf{C} \tag{3.50}$$

(Bishop, 2006, section 2.3).

The multivariate Gaussian distribution has some disadvantages. First, the distribution can have a large number of parameters, as the symmetric matrix $\mathbf{C}$ has $D(D+1)/2$ parameters in addition to the $D$ parameters in $\boldsymbol{\mu}$. One can reduce the number of parameters by imposing some restriction on $\mathbf{C}$ – for example, by restricting $\mathbf{C}$ to be diagonal, the number of parameters in $\mathbf{C}$ drops to $D$. However, this also limits the model's ability to represent a wide range of observed datasets. Second, as the multivariate Gaussian is unimodal (i.e. having a single maximum), it does not fit observed datasets with multiple humps well.

# 3.7 Conditional and Marginal Gaussian Distributions ☐C☺

Suppose in a multivariate Gaussian distribution, the set of variables $\mathbf{x}$ is partitioned into two subsets $\mathbf{x}_a$ and $\mathbf{x}_b$, where the values of $\mathbf{x}_b$ are specified. The conditional probability density distribution $p(\mathbf{x}_a|\mathbf{x}_b)$, that is, the distribution of $\mathbf{x}_a$ conditional on $\mathbf{x}_b$, can be shown to be a (multivariate) Gaussian distribution (Bishop, 2006, section 2.3.1); that is, a *conditional Gaussian distribution* remains Gaussian.

The order of the variables in $\mathbf{x}$ can be rearranged so the $\mathbf{x}_a$ variables come before the $\mathbf{x}_b$ variables, thus,

$$\mathbf{x} = \left( \begin{array}{c} \mathbf{x}_a \\ \mathbf{x}_b \end{array} \right), \tag{3.51}$$

while the mean $\boldsymbol{\mu}$ and the covariance matrix $\mathbf{C}$ can be written as

$$\boldsymbol{\mu} = \left( \begin{array}{c} \boldsymbol{\mu}_a \\ \boldsymbol{\mu}_b \end{array} \right), \tag{3.52}$$

$$\mathbf{C} = \left[ \begin{array}{cc} \mathbf{C}_{aa} & \mathbf{C}_{ab} \\ \mathbf{C}_{ba} & \mathbf{C}_{bb} \end{array} \right]. \tag{3.53}$$

The symmetry from $\mathbf{C}^{\mathrm{T}} = \mathbf{C}$ renders both $\mathbf{C}_{aa}$ and $\mathbf{C}_{bb}$ symmetric, and $\mathbf{C}_{ba} = \mathbf{C}_{ab}^{\mathrm{T}}$.

For the conditional distribution $p(\mathbf{x}_a|\mathbf{x}_b)$, the mean and the covariance can be shown to be

$$\boldsymbol{\mu}_{a|b} = \boldsymbol{\mu}_a + \mathbf{C}_{ab}\mathbf{C}_{bb}^{-1}(\mathbf{x}_b - \boldsymbol{\mu}_b) \tag{3.54}$$

$$\mathbf{C}_{a|b} = \mathbf{C}_{aa} - \mathbf{C}_{ab}\mathbf{C}_{bb}^{-1}\mathbf{C}_{ba} \tag{3.55}$$

(Bishop, 2006, section 2.3.1).

Next, consider the marginal probability density distribution $p(\mathbf{x}_a)$ defined by

$$p(\mathbf{x}_a) = \int p(\mathbf{x}_a, \mathbf{x}_b) \, d\mathbf{x}_b, \tag{3.56}$$

with $p(\mathbf{x}_a, \mathbf{x}_b)$ being multivariate Gaussian. It can be shown that $p(\mathbf{x}_a)$, the *marginal Gaussian distribution*, is also Gaussian (Bishop, 2006, section 2.3.2), with mean and covariance given by

$$E[\mathbf{x}_a] = \boldsymbol{\mu}_a \tag{3.57}$$

$$\mathrm{cov}[\mathbf{x}_a] = \mathbf{C}_{aa}. \tag{3.58}$$

## 3.8   Gamma Distribution   $\boxed{\text{B}}$ ☺

There are many environmental variables that are not well modelled by the Gaussian distribution. For instance, there are many variables with strongly skewed distributions. The strong skewness is often the result of the variable being incapable of having a negative value. Examples of non-negative variables include precipitation amount, wind speed, streamflow, pollutant concentration, biological population abundance, wildfire burnt area, wave height, concentration of suspended particles in sea water, and so on. Fitting a Gaussian to such variables will give non-zero probability for negative values, which is unphysical.

A common distribution used to fit non-negative variables is the gamma distribution with two non-negative parameters $a$ and $b$,

$$\mathrm{Ga}(x\,|\,a,b) = \frac{(x/b)^{a-1}\exp(-x/b)}{b\,\Gamma(a)}, \quad x > 0,\ a > 0\ \text{and}\ b > 0, \tag{3.59}$$

with $\Gamma$ being the gamma function, defined by

$$\Gamma(z) = \int_0^\infty y^{z-1}\,\mathrm{e}^{-y}\,\mathrm{d}y. \tag{3.60}$$

Since $x$ appears as $x/b$ in the gamma distribution, $b$ is a scale parameter, that is, enlarging $b$ will widen the distribution as well as decrease the amplitude by the factor $b$ found in the denominator of Ga, as the area underneath the distribution curve needs to be unity. The parameter $a$ is a shape parameter, as it modifies the shape of the distribution (Fig. 3.4(a)). For $a = 1$, Ga simply reduces to the exponential distribution. For $a < 1$, Ga $\to \infty$ as $x \to 0$, whereas for $a > 1$, Ga $= 0$ at $x = 0$. As $a \to \infty$, the gamma distribution approaches a Gaussian. The mode (i.e. the maximum) of Ga is located at $(a - 1)\,b$ for $a \geq 1$.

Ga has expected mean $ab$ and variance $ab^2$. One might try to use the method of moments, that is, equate these values to the sample mean and variance, then solve for the parameters $a$ and $b$. Unfortunately, this approach for fitting a gamma distribution to data is not always accurate (Thom, 1958). Instead, the parameters are usually estimated using maximum likelihood, for which there are several options (Thom, 1958; Hahn and Shapiro, 1994; Minka, 2002).

The gamma distribution is fitted to the histograms of three environmental variables (Figs. 3.4 (b), (c) and (d)) – the precipitation amount (from days with non-zero precipitation) in Vancouver, BC, the streamflow of Stave River, BC and the concentration of the atmospheric pollutant $PM_{2.5}$ (fine inhalable particles with diameters $\leq 2.5\ \mu$m) in Beijing. The gamma distribution has $a < 1$ for the precipitation but $a > 1$ for the streamflow and the pollutant.

One way to generalize the gamma distribution is to introduce a location or shift parameter $c$ so the distribution can be shifted along the $x$-axis. This *generalized gamma distribution* is also known as the Pearson type III distribution, as it belonged to a family of continuous probability distributions introduced in a series of papers published by Karl Pearson in 1895, 1901 and 1916. It is defined

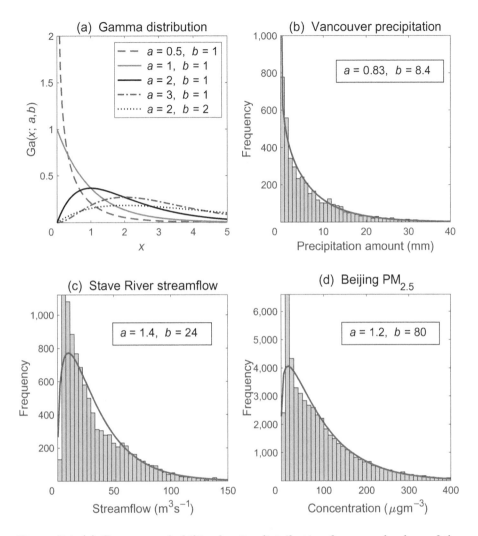

Figure 3.4 (a) Gamma probability density distribution for several values of the shape parameter $a$ and the scale parameter $b$, and histograms of (b) non-zero daily precipitation amount in Vancouver, BC (1993–2017), (c) daily streamflow of Stave River, BC (1985–2011) and (d) hourly concentration of the atmospheric pollutant $PM_{2.5}$ in Beijing (2010–2015). The gamma distribution is fitted to the histograms, with the values of the parameters $a$ and $b$ given in the legends. [Data source: (b) weatherstats.ca based on Environment and Climate Change Canada data, (c) Water Survey of Canada and (d) Machine Learning Repository, University of California Irvine, with data from X. Liang et al. (2016).]

by

$$f(x \,|\, a, b, c) = \frac{[(x - c)/b]^{a-1} \exp[-(x - c)/b]}{|b \, \Gamma(a)|} , \qquad (3.61)$$

with $x > c$ for $b > 0$, or $x < c$ for $b < 0$. Clearly this reduces to the gamma distribution (3.59) when $c = 0$, $x > 0$, $a > 0$ and $b > 0$. When $b < 0$ and $x < c$, the distribution is reflected across the $y$-axis, so it has a long tail to the left. For hydrological studies, the version with long tail to the right ($b > 0$ and $x > c$) is used (Singh, 1998, pp. 231–251).

The gamma distribution (including the generalized version) has been widely applied to environmental variables. In hydrology, it has been used to model streamflow (Botter et al., 2013), transit time estimation in catchments (Hrachowitz et al., 2010) and river catchments (Bastola et al., 2021). In atmospheric sciences, it has been used for modelling precipitation (Groisman et al., 1999; New et al., 2002; Haylock et al., 2006), concentrations of atmospheric pollutants (J. A. Taylor et al., 1986) and wind speed (Carta, Ramirez, et al., 2009), though the Weibull distribution (see Section 3.11) is usually preferred for modelling wind speed. In oceanography, it has been used to model significant wave height (J. A. Ferreira and Soares, 1999) and the size distribution of suspended particles in sea water (important for radiative transfer and light scattering in sea water) (Risovic, 1993). It has been used in ecology/biology to model population abundance (e.g. in fisheries) (Dennis and Patil, 1984) and to model the area burned by wildfire (Littell et al., 2009).

## 3.9 Beta Distribution B☺

For continuous variables, we started with the Gaussian distribution, which is unbounded along the $x$-dimension, then we introduced the gamma distribution, which is bounded on one side, and now we proceed to the beta distribution, which is bounded on both sides, that is, $x \in [0, 1]$. There are environmental variables that are defined only within a bounded interval, such as relative humidity, cloud cover (i.e. fraction of sky covered by clouds), and so on.

The *beta distribution* is defined by

$$\mathrm{Beta}(x \,|\, a, b) = \frac{\Gamma(a + b)}{\Gamma(a)\Gamma(b)} \, x^{a-1}(1 - x)^{b-1}, \quad 0 \le x \le 1, \, a > 0 \text{ and } b > 0, \ (3.62)$$

with $\Gamma$ denoting the gamma function.

The population mean $\mu$ and variance $\sigma^2$ are given by

$$\mu = \frac{a}{a + b}, \qquad (3.63)$$

$$\sigma^2 = \frac{ab}{(a + b)^2(a + b + 1)}. \qquad (3.64)$$

The parameters $a$ and $b$ can be readily estimated using the method of moments, that is, replace the population mean and variance by the sample mean $\bar{x}$ and sample variance $s^2$ in (3.63) and (3.64), then solve for $a$ and $b$ to get

$$a = \bar{x} \left( \frac{\bar{x}(1 - \bar{x})}{s^2} - 1 \right), \tag{3.65}$$

$$b = (1 - \bar{x}) \left( \frac{\bar{x}(1 - \bar{x})}{s^2} - 1 \right), \tag{3.66}$$

if $s^2 < \bar{x}(1 - \bar{x})$ (from the constraint $a > 0$, $b > 0$).

The beta distribution is shown for several choices of the parameters in Fig. 3.5. From (3.62), it is seen that if $a$ or $b$ is $< 1$, the distribution is unbounded at one (or both) of the end points ($x = 0$ and $x = 1$). If $a$ and $b$ are both $> 1$, the distribution is zero at the two end points and has a single hump somewhere in between. The beta distribution is fitted to relative humidity and cloud cover in Fig. 3.6.

Figure 3.5 Beta probability density distribution for several values of the parameters $a$ and $b$. When $a = b = 1$, it turns into the uniform distribution. When the parameters are interchanged, as seen between the cases $a = 4$, $b = 2$ and $a = 2$, $b = 4$, the curves are mirror images of each other.

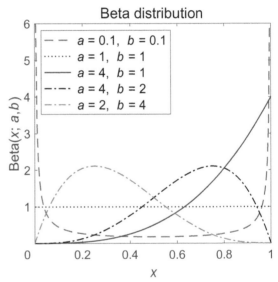

If a variable $y$ is bounded in the interval $[a, b]$, then transforming it by $x = (y - a)/(b - a)$ gives a variable $x$ bounded in the interval $[0, 1]$, so the beta distribution can be fitted to $x$.

The beta distribution has been used to model global cloud cover (Falls, 1974) and atmospheric transmittance for simulating solar irradiation (Graham and Hollands, 1990). It has been used to model relative humidity (Yao, 1974) and in forecasts of threshold relative humidity for plant disease prediction (Wilks and K. W. Shen, 1991). The beta distribution has been used for parameterizing variability of soil properties at the regional level for crop yield estimation (Haskett et al., 1995). As the orientation of leaf angle from being horizontal to vertical changes the interaction between the incoming electromagnetic radiation and the plants, leaf angle distribution has been modelled by the beta distribution (Goel and Strebel, 1984; W. M. Wang et al., 2007). For the biodegradability of

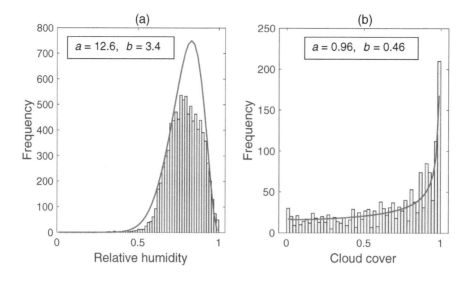

Figure 3.6 Histograms of (a) daily relative humidity and (b) daily cloud cover in Vancouver, BC (1993–2017). The beta distribution is fitted to the histograms, with the values of the parameters $a$ and $b$ given in the legends. [Data source: weatherstats.ca based on Environment and Climate Change Canada data.]

dissolved organic carbon (DOC), the beta distribution has been used to model the probability of biodegradability (Vahatalo et al., 2010).

## 3.10   Von Mises Distribution  C☺

In environmental sciences, variables defined over a periodic or circular domain include wind directions or ocean current directions, as well as variables defined over calendar times (e.g. time of occurrence within a day or a year). For such variables, *circular statistics* or *directional statistics* has been developed (N. I. Fisher, 1995). Instead of the $x$ domain, we now have the periodic or circular domain $0 \leq \theta < 2\pi$. The *von Mises distribution* is essentially the Gaussian distribution over a periodic domain (Bishop, 2006, section 2.3.8) and is defined by

$$p(\theta \,|\, a, b) = \frac{1}{2\pi I_0(b)} \exp[b \cos(\theta - a)], \qquad (3.67)$$

where $I_0(b)$ is the zero-order Bessel function of the first kind (*NIST Digital Library of Mathematical Functions*) defined by

$$I_0(b) = \frac{1}{2\pi} \int_0^{2\pi} \exp(b \cos \theta) \, d\theta. \qquad (3.68)$$

The parameter $a$, analogous to $\mu$ in the Gaussian distribution, gives the mean, while $b$ is analogous to the inverse of the variance $(1/\sigma^2)$ in the Gaussian

(Fig. 3.7). Maximum likelihood can be used to estimate $a$ and $b$ (Bishop, 2006). Like the Gaussian distribution, the von Mises distribution is limited by having a single hump; thus, a mixture (i.e. a sum) of von Mises distributions is often used to deal with multiple maxima in the data.

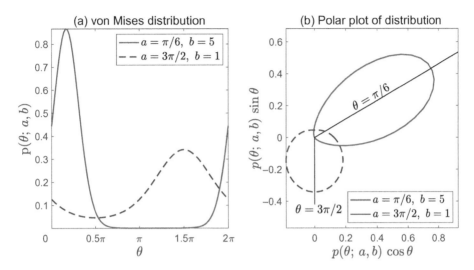

Figure 3.7 (a) von Mises probability density distribution for two sets of the parameters $a$ and $b$. (b) The same distribution shown in a polar plot $(r\cos\theta, r\sin\theta)$, where $r$, the radial distance from the origin, is given by $p(\theta|\, a, b)$.

The von Mises distribution (or a mixture of von Mises distributions) has been used to model wind direction (Carta, Bueno, et al., 2008; L. Bao et al., 2010) and the direction of wildfire (Barros et al., 2012). The distribution has also been used to model the dates of occurrence of floods in a calendar year (L. Chen et al., 2010) and the first date of flowering or fruiting in plants (Morellato et al., 2010).

## 3.11   Extreme Value Distributions ☐C☹

For environmental variables, their extreme values are the ones which cause disasters – for example floods, droughts, heat waves, storms, and so on – leading to human suffering, loss of life and economic damage. The worrying issue of long-term changes in the distribution of extreme events from global climate change has been investigated, for example for storms (Lambert and J. C. Fyfe, 2006) and heavy precipitation events (X. Zhang et al., 2001).

A common measure of extreme value is the highest (or lowest) daily value in one year, such as highest daily precipitation, temperature, wind speed or streamflow in a year. The statistics of such extreme values are particularly important for civil engineers designing infrastructure strong enough to withstand the biggest storm or flood expected once in 50 years or 100 years, or insurance companies trying to assess the risks of storms, floods, wildfires, earthquakes, and so on in order to set their insurance premiums appropriately. In recent decades, there has been increasing interest in applying extreme value theory to the environmental sciences, largely stimulated by the problem of whether the statistics of extreme events has been changing under anthropological climate change (Zwiers and Kharin, 1998; Kharin, Zwiers, X. Zhang, and G. C. Hegerl, 2007; Garrett and Müller, 2008; S. K. Min, X. B. Zhang, et al., 2011; Kharin, Zwiers, X. Zhang, and Wehner, 2013).

The generalized extreme value (GEV) distribution (Coles, 2001; de Haan and A. Ferreira, 2006) has been used to fit extreme data such as the maximum (or minimum) daily value over one year.[4] For a variable $x$, the standardized variable $s = (x - \mu)/\sigma$, with the location parameter $\mu \in \mathbb{R}$ and the scale parameter $\sigma > 0$. The GEV cumulative distribution function (CDF) is given by

$$G(s|\xi) = \exp[-(1 + \xi s)^{-1/\xi}]; \quad \xi \neq 0, \tag{3.69}$$

with $\xi$ being the shape parameter (Coles, 2001). As $(1 + \xi s) \geq 0$ is required, this means $s > -1/\xi$ for $\xi > 0$ and $s < -1/\xi$ for $\xi < 0$. To define GEV for $\xi = 0$, we need to take the limit as $\xi \to 0$ in (3.69). Letting $n = 1/(\xi s)$ and invoking

$$\lim_{n \to \infty} \left(1 + \frac{1}{n}\right)^n = e, \quad \text{and} \quad \lim_{\xi \to 0}(1 + \xi s)^{-1/\xi} = \lim_{n \to \infty} \left(1 + \frac{1}{n}\right)^{-ns} = e^{-s}, \tag{3.70}$$

we obtain

$$G(s|\xi = 0) = \exp[-\exp(-s)]. \tag{3.71}$$

The GEV distribution is divided into three sub-families by $\xi = 0$, $\xi > 0$ and $\xi < 0$, corresponding to the Gumbel (type I), Fréchet (type II) and Weibull (type III) families, respectively.

By differentiating $G(s|\xi)$, we obtain the GEV probability density function (PDF),

$$g(s|\xi) = \begin{cases} (1 + \xi s)^{(-1/\xi)-1} \exp[-(1 + \xi s)^{-1/\xi}], & \xi \neq 0, \\ \exp(-s)\exp[-\exp(-s)], & \xi = 0. \end{cases} \tag{3.72}$$

Again, when $\xi \neq 0$, $(1 + \xi s) \geq 0$ is required, implying $s > -1/\xi$ for $\xi > 0$ and $s < -1/\xi$ for $\xi < 0$. Thus, the PDF is bounded below for $\xi > 0$ and above for $\xi < 0$, and unbounded for $\xi = 0$, as seen in Fig. 3.8.

---

[4] In general, given a time series $z$, we compute the maximum or minimum value over a block of $m$ data points and repeat over consecutive blocks to get a new time series $x$ for the GEV analysis. Strictly speaking, GEV distribution theory is valid only asymptotically, that is, when the extreme values are obtained from blocks with large $m$. Annual extremes are from $m = 365$ (or 366 for leap years). However, serial correlation and the presence of an annual cycle (e.g. the maximum annual temperature can only come from the summer days) may greatly reduce the effective size of $m$.

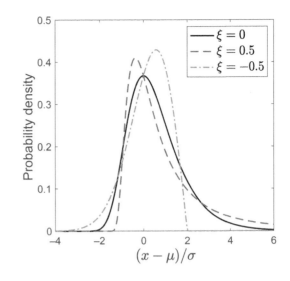

Figure 3.8 GEV probability density curves for the three sub-families: type I ($\xi = 0$), type II ($\xi > 0$) and type III ($\xi < 0$).

The quantile function $q$ is the inverse of the CDF $G$ (see Section 2.9), where

$$q(\alpha \,|\, \mu, \sigma, \xi) = \begin{cases} \mu + \sigma[(-\ln \alpha)^{-\xi} - 1]/\xi, & \xi > 0 \text{ and } \alpha \in [0, 1), \\ & \text{or } \xi < 0 \text{ and } \alpha \in (0, 1], \quad (3.73) \\ \mu - \sigma \ln(-\ln \alpha), & \xi = 0 \text{ and } \alpha \in (0, 1). \end{cases}$$

The quantile function $q(\alpha)$ gives the *return level* associated with the *return period* $1/(1 - \alpha)$, that is, the $q(\alpha)$ level is expected to be exceeded on average once every $1/(1 - \alpha)$ years. For instance, if we choose $\alpha = 0.99$, then $q(0.99)$ gives the 100 year return level.

Figure 3.9 shows the GEV fit to the annual maximum daily flow rate of the River Thames at Kingston, UK, 1882–2017. The record contains 135 hydrological years, with the the hydrological year starting at 09:00 GMT on 1st October. Of the 135 annual values, the minimum and maximum values are 95 and 806 $\mathrm{m}^3\mathrm{s}^{-1}$, respectively. The GEV fit from maximum likelihood yielded the parameter values $\mu = 275$ $\mathrm{m}^3\mathrm{s}^{-1}$, $\sigma = 96$ $\mathrm{m}^3\mathrm{s}^{-1}$ and $\xi = -0.05$. The 10 year, 100 year and 1,000 year return levels are given by the quantile levels $q(0.9)$, $q(0.99)$ and $q(0.999)$, yielding, respectively, 480, 670 and 836 $\mathrm{m}^3\mathrm{s}^{-1}$.

Unfortunately, the GEV model fit is quite sensitive to the few largest extreme values in the data. For instance, if we remove the highest value 806 $\mathrm{m}^3\mathrm{s}^{-1}$ from the data record and repeat the GEV fit, we would obtain the 10 year, 100 year and 1,000 year return levels as being 468, 626 and 750 $\mathrm{m}^3\mathrm{s}^{-1}$, respectively. Furthermore, most annual maximum time series available are far shorter than the Thames at Kingston record of 135 years, rendering the GEV fit even more sensitive to the very few largest extreme values.

As *rogues waves*, that is, extremely large waves on the ocean surface, pose a serious danger in the marine environment, their occurrence rates have been studied using dynamical and statistical approaches. While the GEV model can be used to estimate the distribution of extreme wave heights, Gemmrich and

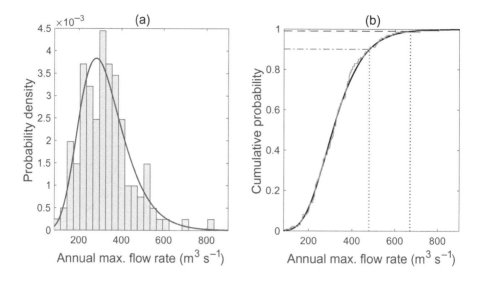

Figure 3.9 GEV fit to the annual maximum daily flow rate of the River Thames at Kingston, UK, with (a) the PDF from the GEV model shown as a curve over the histogram of the observed data, and (b) the CDF from the GEV model shown by the solid black curve, with the empirical CDF shown by the fainter curve. The 0.9 and 0.99 quantiles of the GEV model are indicated by the horizontal dot-dashed and dashed line, respectively, with the corresponding vertical dotted lines indicating, respectively, the 10 year and 100 year return levels along the abscissa. [Data source: National River Flow Archive, UK]

Garrett (2011) noted that such an application requires not only the block size $m$ in the extreme value theory to be large, but $\ln m$ to be large, which is unlikely to be satisfied in practical applications.

## 3.12 Gaussian Mixture Model $\boxed{\text{B}}$ ☺

Although commonly used, Gaussian distributions do not fit many datasets well. For instance, a dataset may be distributed with two peaks instead of a single peak as in a Gaussian distribution. A common approach is to use a *mixture model* to allow more flexibility in fitting the data.

In a mixture model, the probability distribution $p(\mathbf{x})$ is composed of a sum of component distribution functions,

$$p(\mathbf{x}) = \sum_{k=1}^{K} \pi_k f(\mathbf{x}|\,\mathbf{a}_k), \qquad (3.74)$$

where $f(\mathbf{x}|\,\mathbf{a}_k)$ is a specific family of probability distributions with parameters $\mathbf{a}_k$, $\pi_k$ are the *mixing coefficients* and $K$ the number of components. For instance, a common choice for $f$ is the Gaussian family, with

$$p(\mathbf{x}) = \sum_{k=1}^{K} \pi_k \, \mathcal{N}(\mathbf{x}|\,\boldsymbol{\mu}_k, \mathbf{C}_k), \tag{3.75}$$

where $\mathcal{N}$ is the multivariate Gaussian distribution (3.39) with mean $\boldsymbol{\mu}_k$ and covariance $\mathbf{C}_k$ for the $k^{\text{th}}$ component. We will proceed with the *Gaussian mixture model* in this section, although the general mixture model is not restricted to only using the Gaussian family.

For $p(\mathbf{x})$ to be a valid probability density, the mixing coefficients must satisfy two conditions,

$$0 \le \pi_k \le 1, \tag{3.76}$$

$$\sum_{k=1}^{K} \pi_k = 1. \tag{3.77}$$

Suppose a dataset is distributed showing two peaks, so we choose a Gaussian mixture model with two Gaussian components to fit this dataset. Each data point is assumed to have come from one of the two Gaussian distributions. We define a two-dimensional binary *latent variable*[5] $\mathbf{z} = [z_1, z_2]$, where if a data point comes from the first Gaussian distribution, we set $\mathbf{z} = [1, 0]$ and if the point comes from the second Gaussian, $\mathbf{z} = [0, 1]$. The probability of choosing $\mathbf{z} = [1, 0]$ is $\pi_1$, while the probability of choosing $\mathbf{z} = [0, 1]$ is $\pi_2$, with $\pi_1$ and $\pi_2$ being the two mixing coefficients.

In general, instead of two components, we have $K$ components in the mixture model, and the latent variable $\mathbf{z}$ becomes a $K$-dimensional binary variable. The variable $\mathbf{z}$ uses the so-called 1-of-$K$ representation, that is, all its elements are zero except for the $k$th element $z_k$, with $z_k = 1$, when the data point comes from the $k$th Gaussian distribution. The probability

$$P(z_k = 1) = \pi_k, \tag{3.78}$$

can be re-expressed as

$$P(\mathbf{z}) = \prod_{k=1}^{K} \pi_k^{z_k}, \tag{3.79}$$

as $\mathbf{z}$ uses the 1-of-$K$ representation.

The conditional distribution of $\mathbf{x}$ given a particular value of $\mathbf{z}$, where the only non-zero component is $z_k$, is simply the $k$th Gaussian distribution in the mixture model, that is,

$$p(\mathbf{x}|\,z_k = 1) = \mathcal{N}(\mathbf{x}|\,\boldsymbol{\mu}_k, \mathbf{C}_k), \tag{3.80}$$

which can be re-expressed as

$$p(\mathbf{x}|\,\mathbf{z}) = \prod_{k=1}^{K} \mathcal{N}(\mathbf{x}|\,\boldsymbol{\mu}_k, \mathbf{C}_k)^{z_k}. \tag{3.81}$$

---

[5] Latent variables are variables that are not directly observed but can be inferred from other observed variables via a mathematical model.

From the joint distribution $p(\mathbf{x}, \mathbf{z})$, the marginal distribution

$$p(\mathbf{x}) = \sum_{\mathbf{z}} p(\mathbf{x}, \mathbf{z}) = \sum_{\mathbf{z}} p(\mathbf{x}|\mathbf{z}) P(\mathbf{z}), \qquad (3.82)$$

upon invoking (2.3). Substituting in (3.79) and (3.81), we get

$$p(\mathbf{x}) = \sum_{k=1}^{K} \pi_k \mathcal{N}(\mathbf{x}|\boldsymbol{\mu}_k, \mathbf{C}_k), \qquad (3.83)$$

as the summation over $\mathbf{z}$ is equivalent to a summation over $k = 1, \ldots, K$. Thus, we have again arrived at the same Gaussian mixture model (3.75), but via a formulation using a latent variable $\mathbf{z}$. This latent variable formulation will lead to simplification when we try to derive the expectation–maximization (EM) algorithm later.

From Bayes theorem (2.84), the conditional probability of $\mathbf{z}$ given $\mathbf{x}$ is given by

$$P(z_k = 1|\mathbf{x}) = \frac{p(\mathbf{x}|z_k = 1) P(z_k = 1)}{\sum_{j=1}^{K} p(\mathbf{x}|z_j = 1) P(z_j = 1)}. \qquad (3.84)$$

We can regard $\pi_k$ from (3.78) as a prior probability and $P(z_k = 1|\mathbf{x})$ as a posterior probability. Introducing the shorthand $q(z_k)$ to denote the posterior probability and substituting in (3.78) and (3.80), we have

$$q(z_k) \equiv P(z_k = 1|\mathbf{x}) = \frac{\pi_k \mathcal{N}(\mathbf{x}|\boldsymbol{\mu}_k, \mathbf{C}_k)}{\sum_{j=1}^{K} \pi_j \mathcal{N}(\mathbf{x}|\boldsymbol{\mu}_j, \mathbf{C}_j)}. \qquad (3.85)$$

Next, we turn to maximum likelihood for estimating the parameters of the Gaussian mixture model from a set of observations $\{\boldsymbol{x}_1, \ldots, \mathbf{x}_N\}$, with each observation $\mathbf{x}_n \in \mathbb{R}^D$. The dataset can be written as an $N \times D$ matrix $\mathbf{X}$, where the $n$th row is given by $\boldsymbol{x}_n^{\mathrm{T}}$. If we assume the observations are independent of each other, then the probability density for the dataset $\mathbf{X}$ is simply the product of the probability density of the individual observations, that is,

$$p(\mathbf{X}|\boldsymbol{\pi}, \boldsymbol{\mu}, \mathbf{C}) = \prod_{n=1}^{N} \left( \sum_{k=1}^{K} \pi_k \mathcal{N}(\mathbf{x}_n|\boldsymbol{\mu}_k, \mathbf{C}_k) \right). \qquad (3.86)$$

The log likelihood $\mathcal{L}$ is given by

$$\mathcal{L} = \ln L(\boldsymbol{\pi}, \boldsymbol{\mu}, \mathbf{C}|\mathbf{X}) = \ln p(\mathbf{X}|\boldsymbol{\pi}, \boldsymbol{\mu}, \mathbf{C}) \qquad (3.87)$$

$$= \sum_{n=1}^{N} \ln \left( \sum_{k=1}^{K} \pi_k \mathcal{N}(\mathbf{x}_n|\boldsymbol{\mu}_k, \mathbf{C}_k) \right). \qquad (3.88)$$

In the next subsection, we will describe an algorithm for solving the maximum likelihood estimation problem. Meanwhile, let us look at some examples of fitting Gaussian mixture models. With $x$ a one-dimensional variable, a synthetic dataset is generated from mixing three Gaussian distributions, and the result from fitting a three-component (i.e. $K = 3$) Gaussian mixture model to this dataset is shown in Fig. 3.10(a).

In Fig. 3.10(b), $\mathbf{x}$ is a two-dimensional variable describing the daily air pressure and wind speed, and a two-component Gaussian mixture model is fitted to this dataset. The mean position $\boldsymbol{\mu}_k$ for the two Gaussian components are not far enough for the model distribution to show two separate peaks. It is interesting to compare with the $K$-means clustering result from Fig. 2.14(d). The difference between the Gaussian mixture model and $K$-means clustering is that $K$-means clustering did a 'hard' division of the data points into two clusters whereas the mixture model provides a 'soft' probability density distribution to describe the dataset. For more details on using Gaussian mixture models for clustering, see Section 10.2.3.

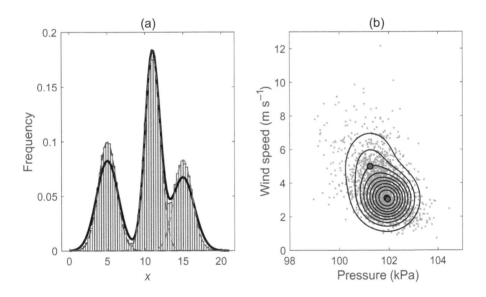

Figure 3.10 (a) Histogram of a synthetic dataset generated from mixing three Gaussian distributions (shown individually by the three dashed curves). Fitting a three-component Gaussian mixture model to the data yields the solid curve with three peaks. (b) A two-component Gaussian mixture model is fitted to the daily pressure and wind speed data (indicated by dots), from Vancouver, BC during 2013–2017. The PDF from the Gaussian mixture model is shown by the contours, with the contour interval being 0.02. The two circles indicate the mean position $\boldsymbol{\mu}_k$ for the two Gaussian components. [Data source: weatherstats.ca based on Environment and Climate Change Canada data.]

### 3.12.1   Expectation–Maximization (EM) Algorithm  B⊗

The maximum likelihood solution is usually obtained by using the iterative expectation–maximization (EM) algorithm (Dempster et al., 1977; Meng and van Dyk, 1997; McLachlan and Krishnan, 2008). Setting the derivative of $\mathcal{L}$ with respect to $\boldsymbol{\mu}_k$ in (3.88) to zero yields

$$-\sum_{n=1}^{N} \frac{\pi_k \, \mathcal{N}(\mathbf{x}_n | \, \boldsymbol{\mu}_k, \mathbf{C}_k)}{\sum_{j=1}^{K} \pi_j \, \mathcal{N}(\mathbf{x}_n | \, \boldsymbol{\mu}_j, \mathbf{C}_j)} \, \mathbf{C}_k(\mathbf{x}_n - \boldsymbol{\mu}_k) = 0. \tag{3.89}$$

Substituting in (3.85) gives

$$\sum_{n=1}^{N} q(z_{nk}) \mathbf{C}_k(\mathbf{x}_n - \boldsymbol{\mu}_k) = 0. \tag{3.90}$$

Left multiplying by $\mathbf{C}_k^{-1}$ and writing

$$Q_k \equiv \sum_{n=1}^{N} q(z_{nk}), \tag{3.91}$$

we get

$$\boldsymbol{\mu}_k = \frac{1}{Q_k} \sum_{n=1}^{N} q(z_{nk}) \mathbf{x}_n. \tag{3.92}$$

Similarly, by setting the derivative of $\mathcal{L}$ with respect to $\mathbf{C}_k$ to zero, one can derive

$$\mathbf{C}_k = \frac{1}{Q_k} \sum_{n=1}^{N} q(z_{nk})(\mathbf{x}_n - \boldsymbol{\mu}_k)(\mathbf{x}_n - \boldsymbol{\mu}_k)^{\mathrm{T}}. \tag{3.93}$$

Finally, to maximize $\mathcal{L}$ with respect to the mixing coefficients $\pi_k$, we need to take into account the constraint that the mixing coefficients sum to one. The constraint can be incorporated by introducing a Lagrange multiplier $\lambda$ when optimizing the quantity

$$\mathcal{L} + \lambda \left( \sum_{k=1}^{K} \pi_k - 1 \right), \tag{3.94}$$

yielding

$$\sum_{n=1}^{N} \frac{\mathcal{N}(\mathbf{x}_n | \, \boldsymbol{\mu}_k, \mathbf{C}_k)}{\sum_{j=1}^{K} \pi_j \, \mathcal{N}(\mathbf{x}_n | \, \boldsymbol{\mu}_j, \mathbf{C}_j)} + \lambda = 0. \tag{3.95}$$

Multiplying through by $\pi_k$ and summing over $k$, we get $\lambda = -N$. Eliminating $\lambda$ from (3.95), multiplying through by $\pi_k$ and invoking (3.85) and (3.91), we get

$$\pi_k = \frac{Q_k}{N}. \tag{3.96}$$

Although the model parameters are given by (3.92), (3.93) and (3.96), we cannot estimate the parameters from them directly as they depend on $q$ in

(3.85), which in turn depends on the model parameters in a complicated way. Thus, we need an iterative algorithm to estimate the parameters. The EM algorithm iterates two steps – (i) compute the posterior probability $q$ in (3.85) and (ii) compute the model parameters using (3.92), (3.93) and (3.96).

---

The steps in the EM algorithm are as follows:

1. Make an initial guess of the parameters $\boldsymbol{\mu}_k$, $\mathbf{C}_k$ and $\pi_k$.

2. **E step**: Compute

$$q(z_{nk}) = \frac{\pi_k \, \mathcal{N}(\mathbf{x}_n | \boldsymbol{\mu}_k, \mathbf{C}_k)}{\sum_{j=1}^{K} \pi_j \, \mathcal{N}(\mathbf{x}_n | \boldsymbol{\mu}_j, \mathbf{C}_j)}, \tag{3.97}$$

$$Q_k = \sum_{n=1}^{N} q(z_{nk}). \tag{3.98}$$

3. **M step**: Re-estimate the parameters

$$\boldsymbol{\mu}_k = \frac{1}{Q_k} \sum_{n=1}^{N} q(z_{nk}) \, \mathbf{x_n}, \tag{3.99}$$

$$\mathbf{C}_k = \frac{1}{Q_k} \sum_{n=1}^{N} q(z_{nk})(\mathbf{x}_n - \boldsymbol{\mu}_k)(\mathbf{x}_n - \boldsymbol{\mu}_k)^{\mathrm{T}}, \tag{3.100}$$

$$\pi_k = \frac{Q_k}{N}. \tag{3.101}$$

4. Compute the log likelihood

$$\mathcal{L} = \sum_{n=1}^{N} \ln \left( \sum_{k=1}^{K} \pi_k \, \mathcal{N}(\mathbf{x}_n | \boldsymbol{\mu}_k, \mathbf{C}_k) \right). \tag{3.102}$$

Check if the convergence criterion for $\mathcal{L}$ or for the parameters has been satisfied. If not, go to step 2 for more iteration.

---

In general, the log likelihood function will have multiple local maxima, and there is no guarantee that the EM algorithm will find the largest of the maxima.

## 3.13   Kernel Density Estimation  B☺

For much of this chapter, we have used various probability distributions governed by a small number of parameters to fit the observed data. This approach is known as *parametric* density estimation. Unfortunately, sometimes the model distribution is a poor match to the true distribution, for example trying to use a unimodal Gaussian distribution to fit a dataset with two peaks.

*Non-parametric* density estimation is an alternative approach which avoids making strong assumptions about the form of the distribution. A very simple example is the histogram (see Section 2.12.1). For a variable $x$, the histogram partitions $x$ into bins, with bin width $\Delta_i$ for the $i$th bin, then counts the number of data points $n_i$ landing in the $i$th bin. To convert the counts $n_i$ to probability density, we define a probability density $p_i$ (uniform within the $i$th bin) as

$$p_i = \frac{n_i}{N\Delta_i}, \tag{3.103}$$

where $N$ is the total number of data points and the requirement $\int p(x)\mathrm{d}x = 1$ is easily seen to be satisfied. Figure 3.11(a) shows a histogram probability density distribution of temperature, which displays two peaks. The disadvantage of the histogram is having sharp edges at the boundaries of the bins, so the probability density is not a continuous function.

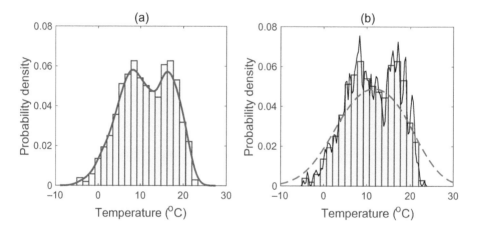

Figure 3.11 (a) Probability density distribution from histogram of daily temperature at Vancouver, BC (2014–2015), and from kernel density estimation (with Gaussian kernel and bandwidth $h = 1.4°\mathrm{C}$) as shown by the smooth curve. (b) Kernel density estimation with $h = 0.2°\mathrm{C}$ (thin solid curve) and $h = 5°\mathrm{C}$ (dashed curve). [Data source: weatherstats.ca based on Environment and Climate Change Canada data.]

*Kernel density estimation* (KDE) is similar to the histogram approach, but uses a smooth kernel function $K$ to yield a smooth probability density estimation for a dataset (Silverman, 1986; Wand and Jones, 1995; Scott, 2015). Like the histogram approach, KDE sums up the contributions from individual data points $x_n$ to construct the PDF. Unlike the histogram, the contribution from a single data point $x_n$ is not a small vertical bar, but a smooth function centred around $x_n$, that is, the PDF from KDE is

$$p(x) = \frac{1}{Nh} \sum_{n=1}^{N} K\left(\frac{x - x_n}{h}\right), \tag{3.104}$$

where $h > 0$ is a parameter known as the bandwidth. A common choice for the kernel function $K$ is the *Gaussian kernel* function

$$G(z) = \frac{1}{\sqrt{2\pi}} \exp\left(-\frac{z^2}{2}\right), \tag{3.105}$$

which is simply the standard Gaussian function from (3.17). Thus,

$$p(\mathbf{x}) = \frac{1}{Nh} \sum_{n=1}^{N} \frac{1}{\sqrt{2\pi}} \exp\left(-\frac{(x - x_n)^2}{2h^2}\right). \tag{3.106}$$

There are many ways to estimate the bandwidth. Silverman's rule of thumb (Silverman, 1986; Sheather, 2004) estimates the bandwidth by

$$\hat{h} = 0.9AN^{-1/5}, \tag{3.107}$$

where $A = \min($sample standard deviation, (sample interquartile range)$/1.34)$. Figure 3.11(a) shows the Gaussian KDE with $h = 1.4°C$ from Silverman's rule of thumb, yielding a smoothed version of the histogram function. However, if $h$ is chosen to be too small, as in Fig. 3.11(b), the resulting PDF appears noisy, and when $h$ is too large, fine structures are smoothed out.

For a multi-dimensional variable $\mathbf{x} \in \mathbb{R}^D$, the KDE PDF is given by

$$p(\mathbf{x}) = \frac{1}{Nh^D} \sum_{n=1}^{N} K\left(\frac{\mathbf{x} - \mathbf{x}_n}{h}\right). \tag{3.108}$$

Using a multivariate Gaussian kernel, that is, $\mathcal{N}(\mathbf{x}|\boldsymbol{\mu}, \mathbf{C})$ from (3.39), with $\boldsymbol{\mu}$ set to the zero vector and $\mathbf{C}$ to the identity matrix, we get

$$p(\mathbf{x}) = \frac{1}{Nh^D} \sum_{n=1}^{N} \frac{1}{(2\pi)^{D/2}} \exp\left(-\frac{\|\mathbf{x} - \mathbf{x}_n\|^2}{2h^2}\right). \tag{3.109}$$

## 3.14   Re-expressing Data  [A]☺

In many situations, it is beneficial to re-express (i.e. transform) the data before one performs any data analysis. We have already encountered standardizing variables in (2.29). Often, the data are very skewed, for example wind speed, precipitation amount and pollutant concentration are all positively skewed (i.e.

skewed to the right) as they cannot have negative values. Many data methods work best when the observed data distribution is approximately Gaussian.
Thus, there is advantage in re-expressing heavily skewed distributions into non-
skewed, near-Gaussian distributions. In addition, many data methods assume a
linear relation between two variables, for example linear regression, so if the observed data display non-linear behaviour, one may want to re-express one of the
variables to give a more linear relation before performing linear data analysis.

The simplest transformations for re-expressing data is via power transformations $(y = x^\lambda)$ and logarithmic transformations $(y = \log x)$. For instance, if
the original data $x > 0$, using the log transform stretches the transformed data
to $-\infty < y < \infty$. However, if $x$ is the precipitation amount, $x$ may contain zero
values and $\log(0)$ is undefined. A common trick is to replace the zero values in
the data record by a very small positive value.

The widely used Box–Cox transformations (Box and D. R. Cox, 1964) are
simply a shifted and scaled version of the above transformations, that is,

$$y^{(\lambda)} = \begin{cases} (x^\lambda - 1)/\lambda & \text{if } \lambda \neq 0, \\ \ln x & \text{if } \lambda = 0, \end{cases} \tag{3.110}$$

where ln is the natural logarithm function and $\lambda$ is commonly estimated from
maximum likelihood. The function $\ln x$ for $\lambda = 0$ can be derived from $(x^\lambda - 1)/\lambda$
by taking the limit $\lambda \to 0$.

Data analysis is then performed on the transformed data $y^{(\lambda)}$, and the result
can be expressed in the original variable by the reverse transform

$$\tilde{x} = \begin{cases} (\lambda y^{(\lambda)} + 1)^{1/\lambda} & \text{if } \lambda \neq 0, \\ \exp(y^{(\lambda)}) & \text{if } \lambda = 0. \end{cases} \tag{3.111}$$

The transforms $y^{(\lambda)}$ are illustrated in Fig. 3.12. If an observed data distribution is skewed to the right, a Box–Cox transform with $\lambda < 1$ will reduce the
excessive right skewness, as seen in Fig. 3.13 for wind speed and non-zero precipitation. In contrast, for the left-skewed relative humidity, a transform with
$\lambda > 1$ is used to reduce the left skewness.

The Box–Cox transform is restricted to $x > 0$. To accommodate non-positive
values of $x$, Box and D. R. Cox (1964) proposed adding an extra shift parameter
$\lambda_2 \geq 0$, thus

$$y^{(\lambda)} = \begin{cases} [(x + \lambda_2)^{\lambda_1} - 1]/\lambda_1 & \text{if } \lambda_1 \neq 0, \\ \ln(x + \lambda_2) & \text{if } \lambda_1 = 0, \end{cases} \tag{3.112}$$

so $y^{(\lambda)}$ is defined for $x > -\lambda_2$. This approach has two disadvantages: (i) the
value of $\lambda_2$ is arbitrary and (ii) one may encounter a new data point in the future
where $x \leq -\lambda_2$, and $y^{(\lambda)}$ becomes undefined. For instance, the lowest value in
a temperature record is $-10°C$. All is well with $y^{(\lambda)}$ by choosing $\lambda_2 = 15°C$,
until one night the temperature drops to $-16°C$.

To overcome the limitations of the Box–Cox transform, Yeo and R. A. Johnson (2000) proposed a one-parameter transform that works for $-\infty < x < \infty$,

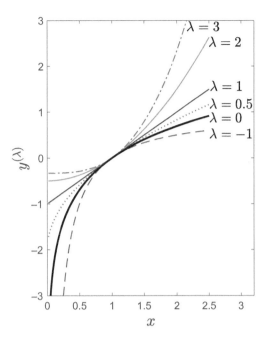

Figure 3.12 Box–Cox transform for various values of the parameter $\lambda$. $\lambda = 1$ gives a straight line, while $\lambda > 1$ have transformed variables more positively (i.e. right) skewed than the original variables, and $\lambda < 1$ have transformed variables more negatively (i.e. left) skewed than the original.

$$y^{(\lambda)} = \begin{cases} [(x+1)^\lambda - 1]/\lambda & \text{if } \lambda \neq 0, x \geq 0, \\ \ln(x+1) & \text{if } \lambda = 0, x \geq 0, \\ -[(-x+1)^{(2-\lambda)} - 1]/(2-\lambda) & \text{if } \lambda \neq 2, x < 0, \\ -\ln(-x+1) & \text{if } \lambda = 2, x < 0. \end{cases} \tag{3.113}$$

Also quite commonly used is a flexible system of distributions, based on three families of transformations, developed by N. L. Johnson (1949), which maps an observed, non-Gaussian variable to one conforming to the standard Gaussian distribution, using up to four parameters (Niermann, 2006).

## 3.15 Student *t*-distribution  B☺

To prepare us for the next chapter on statistical inference, we will study two probability distributions that are widely used in statistical inference, namely the Student *t*-distribution and the chi-squared distribution.

The Student *t*-distribution (or simply the *t*-distribution) is a one-parameter continuous probability distribution developed for estimating the mean of a Gaussian population when the sample size is small and the population standard deviation is unknown. 'Student' was the pseudonym of William Sealy Gosset, who published his paper in 1908,[6] though the distribution had actually been derived earlier by others in the late nineteenth century.

---

[6] Gosset worked for Guinness brewery, which forbade its employees from publishing their work.

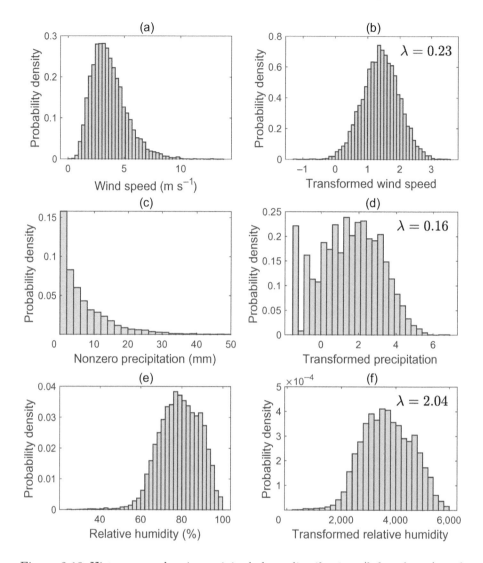

Figure 3.13 Histograms showing original data distribution (left column) and Box–Cox transformed distribution (right column). (a), (c) and (e) are the wind speed, non-zero precipitation and relative humidity in Vancouver, BC during 1993–2017, while (b), (d) and (f) are the corresponding transformed distributions, with $\lambda = 0.23$, $0.16$ and $2.04$, respectively. [Data source: weatherstats.ca based on Environment and Climate Change Canada data.]

The $t$-distribution is used in the Student $t$-test for estimating (i) whether the mean of a population has a value within the range of a null hypothesis, (ii) the statistical significance of the difference between two sample means, (iii) the confidence intervals for the difference between two population means and (iv) the significance and confidence intervals for correlation and linear regression analysis.

Let $x_1, \ldots, x_N$ be independent samples from a Gaussian distribution $\mathcal{N}(\mu, \sigma^2)$, with population mean $\mu$ and population variance $\sigma^2$. The sample mean $\bar{x}$ is given in (3.25), and the sample variance $s^2$ is

$$s^2 = \frac{1}{N-1} \sum_{i=1}^{N} (x_i - \bar{x})^2. \tag{3.114}$$

The factor of $N-1$ in the denominator arose from the fact that there are only $N-1$ degrees of freedom[7] in the calculation, as $\bar{x}$ in (3.25) imposes a constraint on the $x_i$ values in (3.114), thereby reducing the degrees of freedom from $N$ to $N-1$.

As the sample mean $\bar{x}$ is also a Gaussian random variable with population mean $\mu$ and population variance $\sigma^2/N$, the standardized variable

$$z = \frac{\bar{x} - \mu}{\sigma/\sqrt{N}} \tag{3.115}$$

follows a standard Gaussian distribution, that is, $\mathcal{N}(0, 1)$. It can be proved that the random variable

$$t = \frac{\bar{x} - \mu}{s/\sqrt{N}}, \tag{3.116}$$

where $s$ has replaced $\sigma$, follows the Student $t$-distribution with $N-1$ degrees of freedom (Zehna, 1970).

The PDF of the $t$-distribution is given by

$$p(t|\nu) = \frac{\Gamma((\nu+1)/2)}{\sqrt{\nu\pi}\,\Gamma(\nu/2)} \left(1 + \frac{t^2}{\nu}\right)^{-(\nu+1)/2}, \qquad -\infty < t < \infty, \tag{3.117}$$

where $\Gamma$ is the gamma function, and the single parameter $\nu$ is the number of degrees of freedom, with $\nu = N - 1$. The distribution does not depend on $\mu$ nor $\sigma$. The $t$-distribution resembles the standard Gaussian distribution but has heavier tails, as seen in Fig. 3.14. As $\nu \to \infty$, the $t$-distribution approaches the standard Gaussian distribution.

## 3.16  Chi-squared Distribution  B☺

The chi-squared distribution (i.e. $\chi^2$-distribution) with $N$ degrees of freedom is the distribution of a sum of the squares of $N$ independent standard Gaussian random variables. The chi-squared distribution is widely used in statistical

---

[7] The *degrees of freedom* is the number of values in the final calculation of a statistic that are free to vary.

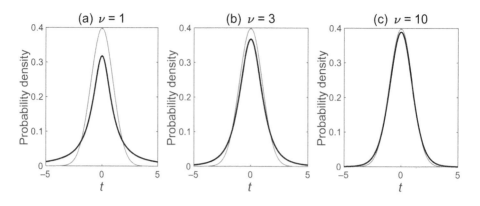

Figure 3.14 The $t$-distribution with various degrees of freedom: (a) $\nu = 1$, (b) $\nu = 3$ and (c) $\nu = 10$. As $\nu$ increases, the $t$-distribution approaches the standard Gaussian distribution $\mathcal{N}(0, 1)$, shown by a light, thin curve.

inference, especially in hypothesis testing and in estimating confidence intervals. The distribution was first published in German in 1876 by Friedrich Robert Helmert and later rediscovered by the English statistician Karl Pearson in 1900.

Let $x_1, \ldots, x_N$ be independent samples from the standard Gaussian distribution $\mathcal{N}(0, 1)$, and let

$$\chi^2 = \sum_{i=1}^{N} x_i^2, \tag{3.118}$$

then $\chi^2$ follows the chi-squared distribution with $N$ degrees of freedom. The probability density of the chi-squared distribution is given by

$$p(z \mid N) = \frac{1}{2^{(N/2)} \Gamma(N/2)} \, z^{(N/2)-1} \, e^{-(z/2)}, \quad \text{for } z > 0, \tag{3.119}$$

and $p(z \mid N) = 0$ for $z \leq 0$, with $\Gamma$ denoting the gamma function. Figure 3.15 illustrates the chi-squared PDF as the degrees of freedom changes. As the chi-squared distribution is the sum of $N$ independent random variables with finite mean and variance, it converges to the Gaussian distribution as $N \to \infty$, according to the central limit theorem.

The chi-squared distribution is a special form of the gamma distribution Ga,

$$p(z \mid N) = \text{Ga}(z \mid N/2, 2), \tag{3.120}$$

with Ga given in (3.59).

It can be shown that the sum of independent chi-squared variables also follows a chi-squared distribution, that is, if $z^{(1)}, \ldots, z^{(k)}$ are independent chi-squared variables with $N_1, \ldots, N_k$ degrees of freedom, respectively, then $z = z^{(1)} + \cdots + z^{(k)}$ is chi-squared distributed with $N_1 + \cdots + N_k$ degrees of freedom.

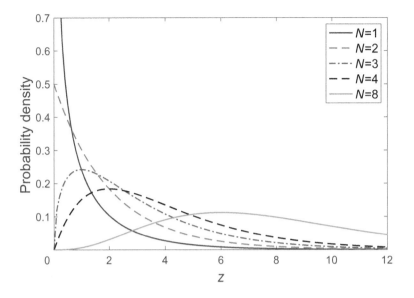

Figure 3.15 The chi-squared distribution with various degrees of freedom $(N)$. The PDF has mean $N$ and variance $2N$. As $N$ increases, the PDF approaches the Gaussian distribution.

# Exercises

### 3.1

Over 100 years, there have been 25 years with at least one category 5 Atlantic hurricane. (a) Over a 10-year period, what is the probability of having at least two such years (i.e. with at least one category 5 hurricane occurring over the year)?

### 3.2

A statistician hired by an insurance company to study hurricane damage in the USA came across the paper by Bove et al. (1998), which found that the annual number of Atlantic hurricanes making landfall on the USA was affected by El Niño (warm) and La Niña (cold) conditions (see Section 9.1.5) in the tropical Pacific Ocean, with the mean annual number being 1.04 hurricanes during warm years, 2.23 during cold years and 1.61 during neural years. The insurance company suffers a net annual loss if three or more hurricanes make landfall in a year. What is the probability of a loss during warm years and during cold years?

### 3.3

For a lake, the number of toxic algal blooms in a year has been recorded over 50 years. The number of blooms in a year being 0, 1, 2, 3, 4 and 5 has occurred in 11, 14, 13, 8, 3 and 1 year(s), respectively. The number of blooms being six or higher has never been observed. Assuming the blooms follow a Poisson distribution, find the probability of having six blooms in a year.

$\boxed{\textbf{3.4}}$

Winters at a village can be classified as either dry, rainy or snowy. From the historical record, a researcher finds that for two consecutive winters, the probability that both are dry is 0.25, while the probability that both are rainy is 0.16. The record is not long enough to show two consecutive snowy winters. Assuming consecutive winters to be independent events, what is the probability of having two consecutive snowy winters?

$\boxed{\textbf{3.5}}$

A fishery manager found that the weight of his sampled sockeye salmon follows the Gaussian distribution with mean 4.2 kg and standard deviation 0.4 kg. (a) What fraction of the salmon has weight above 5.0 kg? (b) Find the weight corresponding to the 95th percentile.

$\boxed{\textbf{3.6}}$

Use a two-component Gaussian mixture model to fit the daily temperature and relative humidity data from Vancouver, BC during 2013–2017. Are there two distinct peaks in the model density distribution? [Data source: weather-stats.ca based on Environment and Climate Change Canada data]

$\boxed{\textbf{3.7}}$

Prove that for the Box–Cox transform, in the limit $\lambda \to 0$, $(x^\lambda - 1)/\lambda$ becomes $\ln x$. [Hint: use the Taylor series expansion for $e^z$].

$\boxed{\textbf{3.8}}$

Prove that the chi-squared distribution has mean $N$ and variance $2N$. [Hint: If $n$ is a positive integer, $\Gamma(n) = (n-1)!$].

# 4

# Statistical Inference

Using data analysis to infer the properties of an underlying probability distribution is called statistical inference. For instance, it is used for deriving estimates of the parameters of the underlying probability distribution and for testing hypotheses. The observed dataset is assumed to have been sampled from a larger population.

## 4.1 Hypothesis Testing $\boxed{A}$ ☺

The most common use of statistical inference is in hypothesis testing. For example, a researcher wants to test the hypothesis that the mean number of daily human deaths in a city is greater during higher air pollution days than during lower pollution days. How would one test the hypothesis?

The basic steps in hypothesis testing consist of:

1. Choose the null hypothesis $H_0$, for example there is no difference in the mean number of deaths between higher pollution days and lower pollution days. The null hypothesis is a 'straw man' set up against the alternative hypothesis, which is the actual hypothesis the researcher is interested in.

2. Choose the alternative hypothesis $H_1$, for example the mean number of deaths is greater during higher pollution days than during lower pollution days.

3. Choose a test statistic. Here, it is the mean number of daily deaths during higher pollution days.

4. Estimate the null distribution, that is, the distribution under $H_0$. Using data from the lower pollution days, one can construct the distribution.

5. If the test statistic occurs too far at the high end tail of the null distribution, one can reject $H_0$ as being too improbable. The critical value for rejecting $H_0$ is chosen to be $\alpha$ (Fig. 4.1(a)).

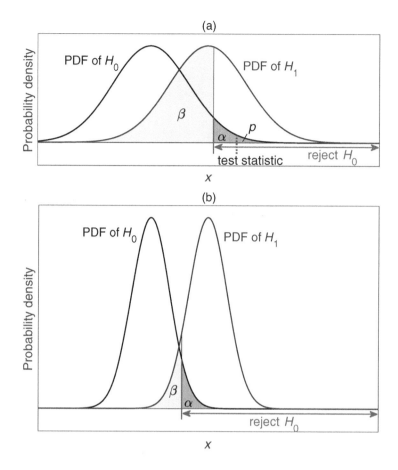

Figure 4.1 (a) Probability density functions of the null hypothesis $H_0$ and the alternative hypothesis $H_1$. The vertical solid line marks the critical value for rejecting $H_0$ at the $\alpha$ level ($\alpha = 0.05$); that is, if the value of the test statistic lies to the right of the critical value, $H_0$ is rejected. The dark shaded region of area $\alpha$ underneath the PDF of $H_0$ indicates that there is a probability of $\alpha$ where $H_0$ is actually true though the test statistic turned up in the region for rejecting $H_0$ (type I error). The value of the test statistic is marked by the vertical dotted line, and $p$ is the dark shaded area to its right. The light shaded region of area $\beta$ underneath the PDF of $H_1$ indicates a probability of $\beta$ for failing to reject $H_0$ when $H_1$ is true (type II error). The *power* of the test is $1 - \beta$, that is, the area under the $H_1$ PDF to the right of the critical value, and is the probability of correctly accepting $H_1$. (b) The test statistic is next estimated using many more data points, so the spread of the PDF is much reduced. For the same $\alpha$ value, the vertical line marking the critical value is shifted to the left; the area $\beta$ is much reduced and the power much enhanced.

The choice of the critical value, also known as the rejection level or *significance level*, is somewhat arbitrary, but $\alpha = 0.05$ is the most common choice, followed by $\alpha = 0.10$ and $0.01$. If the value of the test statistic is more extreme than the critical value, that is, it lies to the right of the critical value in Fig. 4.1(a), then the null hypothesis is rejected at the significance level of $\alpha$. The area underneath the $H_0$ distribution curve to the right of the test statistic is $p$; thus, if $p \leq \alpha$, the null hypothesis is rejected. If the test statistic lies to the left of the critical value, that is, $p > \alpha$, the null hypothesis is not rejected. The $p$-value is obtained from the tail distribution of the $H_0$ distribution curve, that is, the complementary cumulative distribution function (CCDF) $\tilde{F}$ in (2.17), with

$$p = \tilde{F}(t), \tag{4.1}$$

where $t$ is the value of the test statistic.

There are some common misconceptions regarding *significance tests* (von Storch and Zwiers, 1999, p. 74). First, it is incorrect to interpret the $p$-value as the probability that the null hypothesis is true. The correct interpretation of the *p-value* is that it is the probability of observing a value of the test statistic that is as or more extreme than what was observed in the sample, assuming the null hypothesis is true. Second, statisticians are irked by non-statisticians referring to the 5% significance test as the '95% significance test', which seems to be based on incorrectly interpreting $1 - p$ as the probability that the null hypothesis is false.

Another interpretation of $\alpha$ comes from the fact that it is the area under the $H_0$ PDF in the 'reject $H_0$' region (Fig. 4.1(a)); thus, $\alpha$ is the probability of rejecting $H_0$ when $H_0$ is actually true. This type of false positive error is called a Type I error. Under the $H_0$ PDF, to the left of the critical value, is a region of area $1 - \alpha$, which is the probability of correctly accepting $H_0$.

In Fig. 4.1(a), the light shaded region of area $\beta$ under the $H_1$ PDF to the left of the critical value indicates a different type of error. Here, $H_0$ is not rejected, but $H_1$ is true. Thus, $\beta$ is the probability of this false negative error, called a Type II error. Under the $H_1$ PDF to the right of the critical value is a region with area $1 - \beta$, that is, $1 - \beta$ is the probability of correctly accepting $H_1$ and is called the *power* of the test. In Fig. 4.1(a), the area $\beta$, that is, the probability of Type II error, is quite large, hence the power $1 - \beta$ is modest. If we choose a smaller $\alpha$ to reduce the probability of Type I error, we will increase $\beta$, the probability of Type II error, and vice versa. The four possible outcomes of hypothesis testing are summarized in Table 4.1.

Suppose more data are available and the test statistic is estimated with a larger sample. This reduces the spread of the $H_0$ and $H_1$ PDFs, as in Fig. 4.1(b). With the same value for $\alpha$, the critical value is now located more to the left and the area $\beta$ is much decreased when compared to Fig. 4.1(a), thereby greatly increasing $1 - \beta$, the power of the test.

The significance test that has been carried out is a *one-tailed test* (a.k.a. one-sided test), which is used when there is a prior reason (e.g. a physical or biological reason) to expect the $H_1$ PDF to shift to one particular side of the $H_0$ distribution. In our example of air pollution, there is clear medical evidence that

Table 4.1 Possible outcomes of hypothesis testing, with the probability of each outcome given in parentheses.

|            | Accept $H_0$                     | Accept $H_1$                        |
|------------|----------------------------------|-------------------------------------|
| $H_0$ true | True negative                    | False positive, Type I error        |
|            | $(1 - \alpha)$                   | $(\alpha)$                          |
| $H_1$ true | False negative, Type II error    | True positive                       |
|            | $(\beta)$                        | $(1 - \beta)$                       |

air pollution is detrimental to human health (Cohen et al., 2017); thus, the shift in the $H_1$ PDF can only be in the direction of increasing number of deaths. In contrast, if one studies whether precipitation days change the number of deaths in a city relative to non-precipitation days, there is no strong physical/biological reason that the $H_1$ PDF can only be shifted in one direction relative to the $H_0$ PDF – for example, precipitation may increase traffic fatalities but reduce heat stroke fatalities – therefore, a *two-tailed test* (a.k.a. two-sided test) seems more appropriate.

In the two-tailed test, if the test statistic is below the $\alpha/2$ quantile or above the $1 - \alpha/2$ quantile of the $H_0$ cumulative distribution function (CDF), the null hypothesis is rejected. In contrast, for the one-tailed test, the null hypothesis is rejected when the test statistic is below the $\alpha$ quantile of the $H_0$ CDF if the $H_1$ PDF can only be shifted to the left of the $H_0$ distribution. If the $H_1$ PDF can only be shifted to the right, the null hypothesis is rejected if the test statistic is above the $1 - \alpha$ quantile. Thus, rejection of the null hypothesis is harder under the two-tailed test than under the one-tailed test for the same value of $\alpha$. The procedures described here for the one-tailed and two-tailed tests are general – the $H_0$ PDF of course depends on the particular type of test used (e.g. the PDF could be the $t$-distribution, the standard Gaussian $\mathcal{N}(0, 1)$, etc.), as we shall see in the following sections.

## 4.2   Student *t*-test  $\boxed{A}$ ☺

The Student $t$-test, or simply $t$-test, is any statistical hypothesis test where, under the null hypothesis, the test statistic follows a Student $t$-distribution (see Section 3.15).

In many situations, a test statistic would follow a Gaussian distribution if the value of a scaling term in the test statistic were known. However, the scaling term (e.g. the population standard deviation $\sigma$) is often unknown and has to be replaced by an estimate based on the data (e.g. the sample standard deviation $s$), whereby the test statistics follow a $t$-distribution instead of a Gaussian; therefore, a $t$-test is performed.

The $t$-test is widely used for estimating (i) whether the mean of a population has a value within the range of a null hypothesis, (ii) the statistical significance of the difference between two sample means, (iii) the confidence intervals for the difference between two population means and (iv) the significance and confidence intervals for correlation and linear regression analysis.

The Hotelling $T^2$ test is a generalization of the $t$-test to multi-dimensional data (Hotelling, 1931; Mardia et al., 1979; Wilks, 2011).

## 4.2.1 One-Sample $t$-test  [A]☺

The simplest application of the $t$-test is on whether the mean computed from a sample of size $N$ allows one to reject the null hypothesis that the population mean is $\mu_0$. For example, a team of biologists has found the mean weight of a fish species to be $\mu_0$ in a large lake with low nutrient content. From a nearby small lake with higher nutrient content, they found a larger mean weight from a sample of 20 fish. They would like to perform the $t$-test to determine if the null hypothesis can be rejected at the 0.05 level. They decide to use the one-tailed test instead of the two-tailed test, since they know the small lake is more nutrient rich.

The test statistic

$$t = \frac{\bar{x} - \mu_0}{s/\sqrt{N}} \tag{4.2}$$

(where $N$ is the sample size, $\bar{x}$ the sample mean, $s$ the sample standard deviation and $\mu_0$ the population mean) follows the $t$-distribution with $N - 1$ degrees of freedom. For our fish example, assume the biologists have measurements over a large sample size to get an accurate estimate of the population mean $\mu_0 = 2.65$ kg for the large lake. For the small lake, they have $N = 20$, $\bar{x} = 2.91$ kg and $s = 0.73$ kg, giving $t = 1.59$. For the $t$-distribution, the cumulative distribution function (CDF) can be evaluated for a given value of $t$ and $N - 1$ degrees of freedom using almost any statistical package. The $p$-value for the one-sided test (see Fig. 4.1) is given by

$$p = 1 - \text{CDF}(|t|; N - 1). \tag{4.3}$$

With $t = 1.59$ and $N - 1 = 19$, $p = 0.064$; therefore, the null hypothesis cannot be rejected as $p > 0.05$. The biologists decide that they need to collect more data. Now with $N = 40$, and assuming $\bar{x}$ and $s$ are unchanged, the $p$-value is 0.015, and the null hypothesis can be rejected at the 0.05 level.

If the two-tailed test is used instead of the one-tailed test, the $p$-value is doubled to 0.030, and the null hypothesis can still be rejected at the 0.05 level.

## 4.2.2 Independent Two-Sample $t$-test  [A]☺

Another common application of the $t$-test is on whether the difference in the means observed from two independent samples is significant. For example, as mentioned in Section 4.1, the mean number of human deaths in a city could be different between high air pollution days and low pollution days. In the study of climate change, the means of climate variables (e.g. temperature, precipitation, etc.) from climate models representing future climate with increased greenhouse gas concentration are compared with those from models representing current climate.

Assume the two samples have the same population variance $\sigma^2$. If the first sample has $N_a$ data points $x_a$ giving sample mean $\overline{x_a}$ and sample standard deviation $s_a$, and the second sample has $N_b$ data points $x_b$ giving $\overline{x_b}$ and $s_b$, it can be shown that the $t$-statistic for testing whether the means are different is given by (Zehna, 1970; von Storch and Zwiers, 1999)

$$t = \frac{\overline{x_a} - \overline{x_b}}{s_p\sqrt{1/N_a + 1/N_b}}, \tag{4.4}$$

where the pooled standard deviation of the two samples is estimated by $s_p$,

$$s_p = \sqrt{\frac{(N_a - 1)s_a^2 + (N_b - 1)s_b^2}{N_a + N_b - 2}}, \tag{4.5}$$

with the $t$-distribution having $N_a + N_b - 2$ degrees of freedom.

As an example, suppose a researcher wanted to test the hypothesis that during El Niño episodes (when the surface water of the eastern-central equatorial Pacific is anomalously warm), the mean winter (December–February) temperature in Winnipeg, Manitoba, in central Canada, is higher than during non-El Niño episodes. Using the definition of warm episodes (i.e. El Niño) by the Climate Prediction Center, US National Weather Service,[1] the winter seasonal data of Winnipeg, from DJF 1964–1965 to DJF 2017–2018 (obtained from weatherstats.ca based on Environment and Climate Change Canada data) was partitioned into an El Niño set (with $N_a = 19$) and a non-El Niño set (with $N_b = 35$). With sample mean $\overline{x_a} = -13.85°$C and $\overline{x_b} = -15.43°$C, and sample standard deviation $s_a = 2.85°$C and $s_b = 2.75°$C, the (one-tailed) independent two-sample $t$-test gave $t = 2.00$, $p = 0.026$,[2] allowing the null hypothesis to be rejected at the 5% level, thus affirming the far-reaching global effects of the El Niño phenomenon (Hoerling et al., 1997).

There are many situations where the assumption of the two samples having the same population variance $\sigma^2$ is unjustifiable. For instance, in the example of high and low air pollution days, the population variance of the number of deaths in a city is likely to be much greater during high pollution days, as the variance of the pollutant concentration is much larger during high pollution days than during low pollution days.

---

[1] Warm and cold periods occur when the 3-month moving average of the sea surface temperature anomalies in the Niño 3.4 region (5°N–5°S, 120°–170°W) exceeds a threshold of 0.5°C above or 0.5°C below the 30-year normal. Warm and cold episodes (i.e. El Niño and La Niña, respectively) are defined by having the threshold exceeded for a minimum of five consecutive over-lapping seasons, according to the Climate Prediction Center, US National Weather Service.

[2] When there is serial correlation in the data, the sample size $N$ needs to be replaced by the effective sample size $N_{eff}$ in the $t$-test (see Section 4.2.4). With the lag-1 autocorrelation coefficient of 0.121 from (4.9), $N_{a,eff} = 14.9$ and $N_{b,eff} = 27.5$, yielding $p = 0.042$, still allowing the null hypothesis to be rejected at the 5% level.

The *Welch t-test* extended the test to two independent samples with different population variance (Welch, 1947; Ruxton et al., 2006). For this test,

$$t = \frac{\overline{x_a} - \overline{x_b}}{\sqrt{s_a^2/N_a + s_b^2/N_b}}, \tag{4.6}$$

where $t$ follows the usual $t$-distribution but with the number of degrees of freedom approximated by

$$\nu \approx \left(\frac{s_a^2}{N_a} + \frac{s_b^2}{N_b}\right)^2 \left[\frac{(s_a^2/N_a)^2}{N_a - 1} + \frac{(s_b^2/N_b)^2}{N_b - 1}\right]^{-1}. \tag{4.7}$$

## 4.2.3 Dependent *t*-test for Paired Samples  A ☺

In some situations, the data in the two samples are not independent. For instance, suppose one is to determine if the mean of some atmospheric daily data (e.g. temperature, pollutant concentration, etc.) at Los Angeles is different from the mean at San Diego. Since San Diego is only about 180 km from Los Angeles, many of their atmospheric variables are correlated. In such circumstances, it is best to study the difference of the daily data between Los Angeles and San Diego and apply the one-sample $t$-test to the mean of the difference to see if the null hypothesis can be rejected.

Let $x_a$ and $x_b$ be the paired data, that is, in our example, the $i$th data point in $x_a$ was measured on the same day as the $i$th data point in $x_b$, with $i = 1, \ldots, N$. Let the difference variable $x_d = x_a - x_b$. The paired difference $t$-statistic is

$$t = \frac{\overline{x_d} - \mu_0}{s_d/\sqrt{N}}, \tag{4.8}$$

where $\mu_0$ is the population mean of $x_d$, usually taken to be zero under the null hypothesis (i.e. no difference in the population means of $x_a$ and $x_b$), $\overline{x_d}$ and $s_d$ are, respectively, the sample mean and standard deviation of $x_d$, and the $t$-distribution has $N - 1$ degrees of freedom.

If we had mistakenly used the independent two-sample $t$-test, that is, (4.4) and (4.5), on dependent paired samples, the sample standard deviation would have been over-estimated, leading to an under-estimation of the $t$-statistic and an over-estimation of the $p$-value, thereby making us unable to reject the null hypothesis even when the $t$-test for paired samples had rejected the null hypothesis.

## 4.2.4 Serial Correlation  A ☺

The $t$-tests mentioned above all assumed independent data; however, environmental data often display serial correlation. For instance, if the winter weather pattern is typically three rainy days followed by five dry days, then on average there are only two events (a rain event and a dry event) over eight days. If the sample consists of 80 days of data, then there are only about 20 independent events over 80 days, and simply using $N = 80$ in the $t$-tests is incorrect. In Section 2.11.2, we have introduced the lag-$l$ autocorrelation $\rho(l)$ in (2.56), the effective sample size $N_{\text{eff}}$ in (2.57) and $N_{\text{eff}}$ for the auto-regressive process AR(1) in (2.59). The question is, can we still use the $t$-tests on serially correlated data, but with $N$ replaced by $N_{\text{eff}}$ from (2.59)?

For $N_{\text{eff}} \geq 30$, one can indeed simply replace $N$ by $N_{\text{eff}}$ in (4.2) for the one-sample $t$-test, and in (4.8) for the dependent $t$-test for paired samples, as confirmed by Monte Carlo experiments[3] (Zwiers and von Storch, 1995). However, for $N_{\text{eff}} < 30$, the $t$-tests using $N_{\text{eff}}$ did not perform well, with the true null hypothesis rejected far fewer times than expected. For $N_{\text{eff}} < 30$, a better method is a 'table look-up test' with tables generated by Monte Carlo experiments (Zwiers and von Storch, 1995; von Storch and Zwiers, 1999).

For the independent two-sample $t$-tests (Section 4.2.2) with serially correlated data, the lag-1 autocorrelation coefficient $\rho(1)$ can be obtained from a pooled estimate using data from $x_a$ and $x_b$ (Zwiers and von Storch, 1995),

$$\rho(1) = \frac{\sum_{i=1}^{N_a-1}[(x_{a,i} - \overline{x_a})(x_{a,i+1} - \overline{x_a})] + \sum_{i=1}^{N_b-1}[(x_{b,i} - \overline{x_b})(x_{b,i+1} - \overline{x_b})]}{\sum_{i=1}^{N_a}(x_{a,i} - \overline{x_a})^2 + \sum_{i=1}^{N_b}(x_{b,i} - \overline{x_b})^2}.$$

$$(4.9)$$

Using (2.58) and (2.59), the effective sample sizes $N_{a,\text{eff}}$ and $N_{b,\text{eff}}$ can be obtained and used in place of $N_a$ and $N_b$ in the $t$-tests in (4.4) and (4.6). This apporach worked well as confirmed by Monte Carlo experiments when $N_{a,\text{eff}} + N_{b,\text{eff}} \geq 30$. When $N_{a,\text{eff}} + N_{b,\text{eff}} < 30$, again the true null hypothesis was rejected far fewer times than expected, so a 'table look-up test' is preferred for small effective sample size (Zwiers and von Storch, 1995; von Storch and Zwiers, 1999).

To illustrate the difference between using $N$ and $N_{\text{eff}}$, let us use the daily temperature data during the winter of 2017–2018 from Vancouver, BC and Victoria, BC, with Victoria located 93 km SSW of Vancouver. First, perform the independent two-sample $t$-test using $N_a = N_b = 90$ in (4.4) and (4.5). The sample mean temperatures are $\overline{x_a} = 3.88°C$ for Vancouver, $\overline{x_b} = 4.42°C$ for Victoria, with corresponding sample standard deviation $s_a = 2.88°C$ and $s_b = 2.53°C$, yielding $t = -1.35$ and $p = 0.090$ from (4.3) for the one-tailed test and $p = 0.179$ for the two-tailed test for rejecting the null hypothesis at the 5% level. One could justify the use of the one-tailed test since Victoria lies to the south of Vancouver; thus, a higher mean temperature is expected. Both tests failed to reject the null hypothesis at the 5% level as $p > 0.05$.

Next, perform the dependent $t$-test for paired samples, as the two cities are close together and are affected by the same weather systems. With $N = 90$,

---

[3] To obtain numerical solutions for problems which are too complicated to solve analytically, *Monte Carlo methods* use repeated random sampling to obtain numerical results. Stanislaw Ulam in 1946 named this random sampling approach, based on the Monte Carlo Casino in Monaco where his uncle spent much time and money (Metropolis, 1987).

$\overline{x_\mathrm{d}} = -0.55°\mathrm{C}$, $s_\mathrm{d} = 0.98°\mathrm{C}$ and $\mu_0 = 0°\mathrm{C}$, (4.8) gives $t = -5.29$ and (4.3) gives $p = 4.35 \times 10^{-7}$ for the one-tailed test (or $p = 8.69 \times 10^{-7}$ for the two-tailed test); therefore, the null hypothesis is easily rejected at the 5% level.

Next, we try to take into account serial correlation in the dependent $t$-test for paired samples by replacing $N$ by $N_\mathrm{eff}$ from (2.59). With $N_\mathrm{eff} = 37.1$ for the paired data $x_\mathrm{d}$, (4.8) yields $t = -3.40$, corresponding to $p = 8.34 \times 10^{-4}$ for the one-tailed test (or $p = 1.67 \times 10^{-3}$ for the two-tailed test). The $p$-values are not as small as when $N = 90$, but the null hypothesis is still rejected at the 5% level.

In general, when samples $x_\mathrm{a}$ and $x_\mathrm{b}$ are not independent, the dependent $t$-test for paired samples is better than the independent two-sample $t$-test in rejecting the null hypothesis. The independent two-sample $t$-test we performed earlier did not take into account serial correlation. Taking into account serial correlation would further raise the $p$-value, making it even harder to reject the null hypothesis.

## 4.2.5 Significance Test for Correlation ▣☺

After computing the sample Pearson correlation $\rho$ (2.53) for $N$ observations of a pair of variables $(x, y)$, one often wants to estimate whether the correlation between $x$ and $y$ can be considered significantly different from 0, that is, perform a test of the null hypothesis $\hat{\rho} = 0$, where $\hat{\rho}$ is the population correlation. A common approach involves transforming $\rho$ to the variable $t$,

$$t = \rho\sqrt{\frac{N-2}{1-\rho^2}}, \tag{4.10}$$

which in the null case of $\hat{\rho} = 0$ is distributed as the Student's $t$-distribution, with $\nu = N - 2$ degrees of freedom (Bickel and Doksum, 1977, p. 221). One can then perform either a two-sided or a one-sided $t$-test for statistical significance of the correlation.

For example, with $N = 32$ data pairs, $\rho$ was found to be 0.36. Is this correlation significant at the 5% level? In other words, if the true correlation is zero ($\hat{\rho} = 0$), is there a less than 5% chance that we could obtain $\rho \geq 0.36$ for our sample? To answer this, if we choose to perform a two-tailed $t$-test, we need to find the value $t_{0.975}$ in the $t$-distribution, where $t > t_{0.975}$ occurs less than 2.5% of the time and $t < -t_{0.975}$ occurs less than 2.5% of the time (as the $t$-distribution is symmetrical), so altogether $|t| > t_{0.975}$ occurs less than 5% of the time. Using suitable statistical software, we find that with $\nu = 32 - 2 = 30$, $t_{0.975} = 2.04$.

We now need to transform from $t$ back to $\rho$. From (4.10), we get

$$\rho = \frac{t}{\sqrt{N - 2 + t^2}}, \tag{4.11}$$

so substituting in $t_{0.975} = 2.04$ yields $\rho_{0.05} = 0.349$, that is, less than 5% of the sample correlation values exceeds the critical value $\rho_{0.05}$ in magnitude if $\hat{\rho} = 0$.

Thus, our $\rho = 0.36 > \rho_{0.05}$ is significant at the 5% level based on a two-tailed $t$-test. For other problems, the one-tailed $t$-test may be more appropriate, as discussed in Section 4.1.

Figure 4.2 shows how $\rho_{0.05}$ depends on the sample size $N$. A seemingly 'high' correlation of 0.8 may not be statistically significant at the 5% level when $N$ is small, whereas a low correlation of 0.2 can be significant when $N$ is large. It is harder to reject the null hypothesis when using the two-tailed test instead of the one-tailed test. For a given $N$, $\rho_{0.05}$ is higher in the two-tailed test than the one-tailed test, as the two-tailed test uses $t_{0.975}$ while the one-tailed test uses $t_{0.95}$.

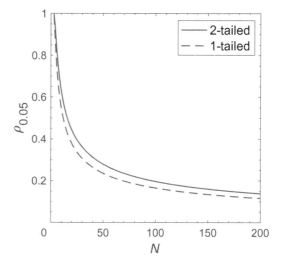

Figure 4.2 The critical correlation value at the 5% level ($\rho_{0.05}$) as a function of the sample size $N$, for the two-tailed test (solid curve) and the one-tailed test (dashed).

For moderately large $N$ ($N \geq 30$), an alternative test involving the Fisher transformation (Section 4.4.2) can be used (Bickel and Doksum, 1977, pp. 221–223).

For non-Gaussian data, the robust Spearman rank correlation (Section 2.11.3) is often used instead of the Pearson correlation. See von Storch and Zwiers (1999, section 8.2.3) on hypothesis testing for the Spearman correlation.

The presence of serial correlation in the data would affect the hypothesis testing, leading to a more frequent rejection of the null hypothesis than expected from the chosen level of significance – for example, the $t$-test at the 5% significance level would actually reject the null hypothesis with greater than 5% probability. Fortunately, the influence of serial correlation on hypothesis testing involving the correlation coefficient is not as strong as on testing involving the mean, at least when correlations are small (von Storch and Zwiers, 1999, p. 149). Ebisuzaki (1997) compared several methods for taking into account serial correlation when performing hypothesis testing on the correlation coefficient.

In some situations, one has *a pair of correlations* and one would like to know if their difference is statistically significant. For example, given three sets of data $x_a$, $x_b$ and $x_c$, one computes the correlation $\rho_{ac}$ between $x_a$ and $x_c$, and

$\rho_{bc}$ between $x_b$ and $x_c$. This situation occurs when one wants to evaluate how well two models making predictions $x_a$ and $x_b$ agree with the observed data $x_c$. Suppose $\rho_{ac} > \rho_{bc}$; is the difference in the correlation statistically significant at some given level? A significance test is given by Steiger (1980). Alternatively, one can estimate confidence intervals (Section 4.4.2) for the difference between a pair of correlations (Zou, 2007).

## 4.3 Non-parametric Alternatives to *t*-test  B☺

*Non-parametric statistics* involves using statistical methods that do not rely on data obeying a particular probability distribution specified by parameters (e.g. the Gaussian distribution specified by its mean and variance parameters).[4] From Section 2.1, we have seen there are several types of data that are not made up of continuous variables, for example ordinal data such as education level (elementary school graduate, high school graduate, some college and college graduate). While the education level categories can be ranked 1, 2, 3 and 4, the separations between the categories are not really integers. Non-parametric statistics has flourished as many data variables do not belong to a distribution such as Gaussian, gamma, and so on, or may contain outliers from errors in observations.

In non-parametric statistics, *distribution-free statistical methods* are used in hypothesis testing, without making assumptions about the underlying probability distributions of the data – in contrast to parametric methods such as the t-test, which assumes Gaussian data. The Gaussian assumption does not work well for many environmental variables, since many of them are bounded at one end (e.g. streamflow, precipitation, wind speed and pollutant concentration are all non-negative) or bounded at both ends (e.g. specific humidity, cloud cover, fractional snow cover, sea ice concentration). In this section, we study non-parametric alternatives to the t-test, with the Wilcoxon–Mann–Whitney test as an alternative to the independent two-sample t-test (Section 4.2.2) and the Wilcoxon signed-rank test as an alternative to the dependent t-test for paired samples (Section 4.2.3). The dependency on distributions is removed as these non-parametric tests are performed on ranked data.

### 4.3.1 Wilcoxon–Mann–Whitney Test  C☺

The Wilcoxon–Mann–Whitney (WMW) test is also called the Mann–Whitney $U$-test or the Wilcoxon rank sum test, as Mann and Whitney (1947) generalized the test introduced earlier by Wilcoxon (1945).[5]

Given two samples containing $N_a$ and $N_b$ data points, the WMW test is used to test if the locations of the two samples are the same. Location is the non-parametric counterpart of the mean, as used in the t-test. The observations

---

[4] A second meaning of non-parametric statistics is that the methods do not assume the structure of a model is fixed – that is, the statistical model is allowed to grow depending on the complexity of the data.

[5] An equivalent test was published in German by Gustav Deuchler in 1914 (Kruskal, 1957).

within each sample are independent and observations from the two samples are also independent. The data from the two samples are combined into a single dataset $D$ with $N = N_a + N_b$ data points. The data in $D$ are ranked, for example $5, -2, 5, 1000$ are assigned the ranks $2.5, 1, 2.5, 4$, where the tied values are given their average rank. With $N$ data points, the sum of the ranks is $N(N + 1)/2$.

The Mann–Whitney $U$-statistic is given by

$$U_a = R_a - \frac{N_a(N_a + 1)}{2}, \tag{4.12}$$

where $R_a$ is the sum of ranks from sample A. As $U_a$ is simply $R_a$ minus the constant $N_a(N_a+1)/2$, there is no real difference between using $R_a$ as originally introduced by Wilcoxon or $U_a$ by Mann and Whitney. If all the data from sample A have ranks $\leq$ ranks of the data from sample B, then $R_a = N_a(N_a + 1)/2$ and $U_a$ takes on its minimum value of $0$. Of course, one can also compute $U$ for sample B, that is,

$$U_b = R_b - \frac{N_b(N_b + 1)}{2}. \tag{4.13}$$

As $R_a + R_b = N(N + 1)/2$,

$$U_a + U_b = \frac{N(N + 1)}{2} - \frac{N_a(N_a + 1)}{2} - \frac{N_b(N_b + 1)}{2}. \tag{4.14}$$

As $N = N_a + N_b$, one can show that

$$U_a + U_b = N_a N_b, \tag{4.15}$$

that is, $U_a$ and $-U_b$ only differ by a constant. When either $U_a$ or $U_b$ takes on its minimum value of $0$, the other $U$-statistic takes on its maximum value of $N_a N_b$.

The null hypothesis $H_0$ is that the two samples have the same distribution. The distribution of the $U$-statistic under $H_0$ can be obtained from combinatorial arguments (Bickel and Doksum, 1977). For moderately large sample sizes ($U_a$, $U_b \gtrsim 20$), one can use the Gaussian approximation. The standardized variable

$$z = \frac{U - \mu_U}{\sigma_U}, \tag{4.16}$$

has mean and standard deviation given by

$$\mu_U = \frac{N_a N_b}{2}, \tag{4.17}$$

$$\sigma_U = \left[ \frac{N_a N_b (N + 1)}{12} \right]^{1/2}, \tag{4.18}$$

and $z$ approximately follows the standard Gaussian distribution $\mathcal{N}(0, 1)$. When the number of tied ranks is not small, $\sigma_U$ should be corrected by

$$\sigma_{\text{corrected}} = \left[ \frac{N_a N_b}{12} \left( (N + 1) - \sum_{j=1}^{g} \frac{t_j^3 - t_j}{N(N - 1)} \right) \right]^{1/2}, \tag{4.19}$$

where $g$ is the number of tied groups and $t_j$ is the size of tied group $j$ (an untied observation is considered to be a tied group of size 1) (Hollander et al., 2014).

The usual alternative hypothesis $H_1$ is that the two samples belong to distributions with different locations (i.e. two distributions with the same shape but shifted from each other). The test is carried out as a two-tailed test as described in Section 4.1, using the standard Gaussian distribution when $U_a$, $U_b \gtrsim 20$. A one-tailed test can be used when one is testing if the location of distribution B is shifted to one particular side of distribution A. Serial correlation effects on the WMW test can be alleviated by pre-whitening the data (Yue and C. Y. Wang, 2002).

The rank sum $R$ or the $U$-statistic characterizes the location of a sample but does not have the simple interpretation of other location statistics such as the mean or the median. As the WMW test is a non-parametric counterpart of the $t$-test, it is common in the literature to report the median of the samples instead of the mean. This might have contributed to a common belief that the WMW test is for comparing the medians of two samples. In reality, this test only compares the rank sum (more precisely, $U$) of two samples, not the median. If the two samples are from distributions of different shapes, it is possible to have similar medians but different rank sums, or vice versa.

When comparing tests, instead of just comparing their power $1 - \beta$ (Fig. 4.1), it is more helpful to compare the sample size needed for a test to reach a certain power with the sample size of an alternative test. The *relative efficiency* is the ratio of the sample sizes from two tests with the same power. The relative efficiency of the WMW test to the $t$-test is $N_t/N_W$, where $N_t$ is the sample size of the $t$-test and $N_W$ the sample size of the WMW test. For Gaussian data, in the limit of large sample size, $N_t/N_W = 0.955$, independent of the significance level $\alpha$ and $\beta$ (Bickel and Doksum, 1977). This means that for Gaussian data, the $t$-test is more efficient than the WMW test as it requires only 95.5% the amount of data to attain the same power. For non-Gaussian data, this ratio can be considerably larger than 1, that is, the WMW test is more efficient.

In Section 4.2.2, we used the independent two-sampled $t$-test to test the hypothesis that during El Niño episodes (when the surface water of the eastern-central equatorial Pacific is warm), the mean winter (December–February) temperature in Winnipeg, Canada, is higher than during non-El Niño episodes. We now use the WMW test on snowpack, measured in snow water equivalent (SWE).[6] SWE is not well described by a Gaussian distribution, since SWE is non-negative, therefore right-skewed.

During the 69-year period of 1950–2018, there were 24 warm (El Niño) episodes and 22 cold (La Niña) episodes in the equatorial Pacific according to the Climate Prediction Center of the US National Weather Service. The maximum winter SWE at Grouse Mountain,[7] a popular ski resort in British

---

[6] *Snow water equivalent* (SWE), a common quantity used by hydrologists and water managers to measure the amount of liquid water contained within a snowpack, is the depth of water that would theoretically result if the whole snowpack is melted instantaneously.

[7] Grouse Mountain (49° 23′ N, 123° 05′ W, 1,100 m elevation) is one of the sites monitored under the BC Snow Survey program by the River Forecast Centre, British Columbia, which

Columbia, yielded an El Niño set (with $N_a = 24$) and a La Niña set (with $N_b = 22$) (Fig. (4.3). With sample median SWE of 1,104 mm during El Niño

Figure 4.3 Histogram of the winter snow water equivalent distribution at Grouse Mountain, BC for all winters, El Niño winters and La Niña winters. [Data source: River Forecast Centre, British Columbia.]

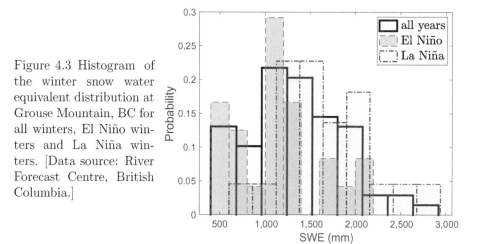

and 1,596 mm during La Niña, the (one-tailed) WMW test gave $p = 0.00024$, easily rejecting the null hypothesis, with the alternative hypothesis being that SWE during El Niño episodes was lower than SWE during La Niña episodes. Using the $t$-test yielded $p = 0.00029$. As for serial correlation in the SWE data, the lag-1 autocorrelation from (4.9) is only 0.026, so the effect of serial correlation is almost negligible, with $N_{a,\text{eff}} = 23$ (versus $N_a = 24$) and $N_{b,\text{eff}} = 21$ (vs $N_b = 22$).

### 4.3.2   Wilcoxon Signed-Rank Test  $\boxed{\text{C}}$ ☺

The Wilcoxon signed-rank test was published together with the Wilcoxon rank sum test in Wilcoxon (1945). The Wilcoxon signed-rank test is commonly used to compare the locations of two dependent samples, serving as a non-parametric substitute for the dependent $t$-test for paired samples (Section 4.2.3). The Wilcoxon signed-rank test can also be used with only one sample, analogous to the one-sample $t$-test (Section 4.2.1).

First, compute the difference between the paired samples $(x_i, y_i)$, $i = 1, \ldots, N_0$, by letting

$$d_i = x_i - y_i, \quad i = 1, \ldots, N_0. \tag{4.20}$$

Zero values of $d_i$ are ignored in the test, so we will assume we have a set of non-zero $d_i$, $i = 1, \ldots, N$. The magnitude of $d_i$, that is, $|d_i|$ is assigned a rank $r_i$. For instance, if $[d_1, d_2, d_3, d_4] = [6, -1, 9, -6]$, then $[r_1, r_2, r_3, r_4] = [2.5, 1, 4, 2.5]$, since 6 and $-6$ are tied in magnitude.

---

performs manual snow surveys up to eight times per year around the beginning of every month from January through June with extra measurements at mid-month in May and June. The maximum winter SWE is obtained by taking the maximum of the measured values during the whole snow season.

Let $R_+$ be the sum of all $r_i$ with $d_i > 0$, and $R_-$ be the sum of all $r_i$ with $d_i < 0$, and $R = R_+ - R_-$. The three statistics, $R_+$, $R_-$ and $R$ can be used interchangeably as they only differ by constants: $R_+ + R_-$ is simply the sum of the ranks of $N$ numbers (i.e. $1 + 2 + \cdots + N$), thus

$$R_+ + R_- = \frac{N(N+1)}{2}, \tag{4.21}$$

$$R = 2R_+ - \frac{N(N+1)}{2} = \frac{N(N+1)}{2} - 2R_-. \tag{4.22}$$

Under the null hypothesis $H_0$, the two distributions are the same; thus, the distribution of the differences is symmetric about zero. For smaller values of $N$, the null distribution of the differences can be found exactly by generating all $2^N$ sign permutations of the ranked differences (Rey and Neuhäuser, 2011). The value of the test statistic is calculated for for each permutation. The proportion of permutations with a value $\geq$ the observed statistic is the $p$-value.

For $N \gtrsim 20$, the standardized variable

$$R'_+ = \frac{R_+ - \mu_R}{\sigma_R} \tag{4.23}$$

approximately obeys the standard Gaussian distribution $\mathcal{N}(0, 1)$, with

$$\mu_R = \frac{N(N+1)}{4}, \tag{4.24}$$

$$\sigma_R = \left[\frac{N(N+1)(2N+1) + T}{24}\right]^{1/2}, \tag{4.25}$$

where $T = 0$ if there are no ties. When there are ties, the correction term

$$T = -\frac{1}{2}\sum_{j=1}^{g}(t_j^3 - t_j), \tag{4.26}$$

where $g$ is the number of tied groups and $t_j$ is the size of tied group $j$ (an untied observation is considered to be a tied group of size 1) (Hollander et al., 2014). A two-tailed test or a one-tailed test can be carried out as described in Section 4.1.

Examples of applications to environmental sciences include Dibike, Gachon, et al. (2008), who used the Wilcoxon signed-rank test to make a quantitative evaluation of the reliability of statistically downscaled climate data from coupled global climate models with respect to observed daily temperature and precipitation in northern Canada. Feng et al. (2013) used the Wilcoxon signed-rank test as a one-sampled test for changes in rainfall seasonality in the tropics.

## 4.4 Confidence Interval $\boxed{A}$ ☺

From observed data, a *confidence interval* (CI) can be computed as an estimate of the range within which the true value of an unknown population parameter may be located. The CI has an associated confidence level (e.g. 95%), such that, if CIs for a specified confidence level are computed repeatedly (assuming an infinite number of independent samples are available), the proportion of intervals that contain the true value of the population parameter equals the confidence level.

While confidence intervals provide a range of potential values of the unknown population parameter, the interval computed from a particular sample may not contain the true value of the parameter. For instance, if the confidence level is 95%, then by computing CIs repeatedly from an infinite number of independent samples, 95% of the CIs will contain the true population parameter, while 5% will not. Under strict frequentist interpretation, it is incorrect to say that a computed 95% CI has a 95% probability of containing the true population parameter. The reason is that once a CI is computed from a sample, the interval either contains the true parameter value or it does not, so there is no 95% probability.

CIs can be related to statistical hypothesis testing. An alternative interpretation of the 95% CI is: The confidence interval represents values for the population parameter for which the difference between the parameter and the observed estimate is not statistically significant at the 5% level (D. R. Cox and Hinkley, 1974, pp. 214, 225, 233). In other words, given the observed statistic, the CI can be obtained by finding the values of the population parameter for which the null hypothesis cannot be rejected. This allows us to construct CIs using the familiar hypothesis testing framework, but with the roles of the population parameter and the observed test statistic reversed — that is, instead of having the population parameter fixed and looking for the critical value of the test statistic, we now have the observed statistic fixed and look for the range of values for the population parameter.

### 4.4.1   Confidence Interval for Population Mean   B☺

Let $x_1, \ldots, x_N$ be a sample of independent and identically distributed (i.i.d.) data from a Gaussian distribution with population mean $\mu$ and variance $\sigma^2$. With $\bar{x}$ denoting the sample mean, the standardized variable

$$z = \frac{\bar{x} - \mu}{\sigma/\sqrt{N}} \tag{4.27}$$

follows the standard Gaussian distribution $\mathcal{N}(0,1)$, as mentioned earlier in (3.115). Choose the confidence level $C$ for the CI, for example, for the 95% CI, $C = 0.95$. Let $\alpha = 1 - C$, so $C = 0.95$ gives $\alpha = 0.05$.

There are two cases: (i) the population standard deviation $\sigma$ is known or (ii) $\sigma$ is not known.

First, consider case (i):

Let $z_L$ and $z_U$ denote, respectively, the $\alpha/2$ quantile and the $(1 - \alpha/2)$ quantile of $\mathcal{N}(0,1)$. With a two-tailed test, within the interval $z_L < z < z_U$, the null hypothesis cannot be rejected at the $\alpha$ level. Substituting (4.27) into $z_L < z < z_U$ gives

$$z_L \frac{\sigma}{\sqrt{N}} < \bar{x} - \mu < z_U \frac{\sigma}{\sqrt{N}}, \tag{4.28}$$

$$-z_U \frac{\sigma}{\sqrt{N}} < \mu - \bar{x} < -z_L \frac{\sigma}{\sqrt{N}}, \tag{4.29}$$

$$\bar{x} - z_U \frac{\sigma}{\sqrt{N}} < \mu < \bar{x} - z_L \frac{\sigma}{\sqrt{N}}. \tag{4.30}$$

Since $\mathcal{N}(0,1)$ is a symmetric distribution, $z_L = -z_U$; hence, the confidence interval for the population mean $\mu$ is given by

$$\bar{x} - z_U \frac{\sigma}{\sqrt{N}} < \mu < \bar{x} + z_U \frac{\sigma}{\sqrt{N}}, \tag{4.31}$$

where we recall $\sigma/\sqrt{N}$ is the standard error of the mean in (3.26). Specifically, for the 95% CI, $\alpha = 0.05$ and $z_U = 1.96$, giving

$$\bar{x} - 1.96 \frac{\sigma}{\sqrt{N}} < \mu < \bar{x} + 1.96 \frac{\sigma}{\sqrt{N}}. \tag{4.32}$$

Next, consider case (ii): As the population standard deviation $\sigma$ is unknown, we need to substitute with the sample standard deviation $s$, and the $z$ is replaced by $t$, which follows the $t$-distribution with $N-1$ degrees of freedom as mentioned in (3.116). Following the steps in case (i), we have the confidence interval of the population mean given by

$$\bar{x} - t_U \frac{s}{\sqrt{N}} < \mu < \bar{x} + t_U \frac{s}{\sqrt{N}}, \tag{4.33}$$

where $t_U$ is the $(1 - \alpha/2)$ quantile of the $t$-distribution with $N - 1$ degrees of freedom.

The difference between cases (i) and (ii) arises from the fact that case (ii) uses the sample standard deviation instead of the population standard deviation, which introduces more uncertainty, so the CI is broader as the $t$-distribution used in (ii) has heavier tails than the standard Gaussian distribution used in (i). As $N$ increases, $s$ approaches $\sigma$ and the $t$-distribution approaches the Gaussian, so the CI in (4.33) approaches that in (4.31).

The formulas for the CI were derived under the assumption that the data were independent. For serially correlated data, the effective sample size $N_{\text{eff}}$ may be much smaller than $N$, that is, using $N$ gives CIs that are too narrow.

Let us again use the daily temperature data $x_a$ and $x_b$ from Vancouver, BC and Victoria, BC, respectively, during the winter of 2017–2018, to illustrate CIs. Using (2.59) gave $N_{\text{eff}} = 9.44$ for Vancouver and 11.8 for Victoria. With $N(= 90)$ replaced by $N_{\text{eff}}$ in (4.33), the 95% CI for the population mean $= [1.74, 6.02]\,°C$ for Vancouver and $[2.79, 6.05]\,°C$ for Victoria. As the two CIs largely overlapped, we need to use a more sensitive test — finding the CI of the *difference* in the daily temperature between Vancouver and Victoria. For the difference variable $x_d = x_a - x_b$, $N_{\text{eff}} = 37.1$, $\bar{x}_d = -0.54°C$ and the CI $= [-0.87, -0.22]\,°C$. The CI now lying entirely below $0°C$ indicates the population mean temperature at Victoria is higher than the population mean at

Vancouver within 95% confidence. Note that $N_{\mathrm{eff}} = 37.1$ for $x_{\mathrm{d}}$ is much larger than the $N_{\mathrm{eff}}$ values of 9.44 for $x_{\mathrm{a}}$ and 11.8 for $x_{\mathrm{b}}$, as the lag-1 autocorrelation $\rho(1) = 0.42$ for $x_{\mathrm{d}}$ is much smaller than $\rho(1) = 0.81$ for $x_{\mathrm{a}}$ and $\rho(1) = 0.77$ for $x_{\mathrm{b}}$.

## 4.4.2   Confidence Interval for Correlation  $\boxed{\text{B}}$☺

As the sample Pearson correlation $\rho$ is bounded within $[-1, 1]$, the Fisher transformation introduces a variable

$$F(\rho) = \frac{1}{2} \ln \left( \frac{1+\rho}{1-\rho} \right) = \mathrm{artanh}(\rho), \tag{4.34}$$

with ln the natural logarithm, artanh the inverse hyperbolic tangent function and $F(\rho) \in (-\infty, \infty)$. The inverse transformation is

$$\rho = \tanh F, \tag{4.35}$$

where tanh is the hyperbolic tangent function.

If $(x, y)$ obeys a bivariate Gaussian distribution with population correlation $\hat{\rho}$ and the data pairs $(x_i, y_i)$ $(i = 1, \ldots, N)$ are i.i.d., then $F$ is approximately Gaussian-distributed for large $N$, with mean

$$F(\hat{\rho}) = \frac{1}{2} \ln \left( \frac{1+\hat{\rho}}{1-\hat{\rho}} \right) \tag{4.36}$$

and standard error SE$= 1/\sqrt{N-3}$  (R. A. Fisher, 1915; R. A. Fisher, 1921; David, 1938; Bickel and Doksum, 1977, pp. 221–223).

The $z$-score for standardizing $F$ is given by

$$z = \frac{F - F(\hat{\rho})}{\mathrm{SE}} = [F(\rho) - F(\hat{\rho})]\sqrt{N-3}, \tag{4.37}$$

with $z$ approximately obeying the standard Gaussian distribution.

Let $z_{\mathrm{L}}$ and $z_{\mathrm{U}}$ denote, respectively, the $\alpha/2$ quantile and the $(1 - \alpha/2)$ quantile of $\mathcal{N}(0, 1)$, and $z_{\mathrm{L}} = -z_{\mathrm{U}}$. The CI for $F(\hat{\rho})$, at confidence level $C = 1 - \alpha$, is given by

$$F(\rho) - z_{\mathrm{U}}/\sqrt{N-3} \; < \; F(\hat{\rho}) \; < \; F(\rho) + z_{\mathrm{U}}/\sqrt{N-3}. \tag{4.38}$$

To transform from $F(\hat{\rho})$ back to $\hat{\rho}$, one applies the tanh function to (4.38), yielding the CI for $\hat{\rho}$,

$$\tanh \left( F(\rho) - z_{\mathrm{U}}/\sqrt{N-3} \right) \; < \; \hat{\rho} \; < \; \tanh \left( F(\rho) + z_{\mathrm{U}}/\sqrt{N-3} \right). \tag{4.39}$$

When $N$ is not large ($N \lesssim 25$), the above CI estimate is inaccurate. David (1938) calculated the CI exactly and provided look-up tables for $N \leq 25$ and $N = 50, 100, 200$ and $400$.

In Section 4.2.5, we have mentioned that one often has *a pair of correlations* and one would like to know if their difference is statistically significant. For instance, with two models making predictions $x_a$ and $x_b$ and given observed data $x_c$, one computes the correlation between model prediction and observation for the two models and would like to test if the difference between the two models is statistically significant. Zou (2007) provides confidence interval estimates for the difference between a pair of correlations.

# 4.5  Goodness-of-Fit Tests  B ☺

From a given dataset $\{x_1, \ldots, x_N\}$, one can construct an empirical probability distribution. One often would like to know if the empirical distribution from the sample agrees with a reference distribution, for example the Gaussian distribution or the gamma distribution, and so on. Goodness-of-fit tests describe how well the empirical distribution fits the reference distribution. For continuous variables, the best known goodness-of-fit test is the *Kolmogorov–Smirnov test* (KS test), named after Andrey Kolmogorov and Nikolai Smirnov.

In some situations, one has two datasets and one would like to compare the two empirical distributions derived separately from the two samples to determine if the two samples came from the same (unknown) underlying distribution. The one-sample goodness-of-fit tests mentioned above are often modified to become two-sample goodness-of-fit tests.

## 4.5.1  One-Sample Goodness-of-Fit Tests  B ☺

In the KS test, the empirical cumulative distribution function $F_N(x)$ derived from $N$ data points is compared with a reference cumulative distribution $F(x)$ via the KS test statistic

$$D_N = \max_x |F_N(x) - F(x)|, \tag{4.40}$$

which measures the maximum distance between the empirical distribution and the reference distribution. The empirical distribution function is given by

$$F_N(x) = \frac{1}{N} \operatorname{count}(x_i \leq x), \tag{4.41}$$

where $\operatorname{count}(x_i \leq x)$ simply counts the number of data points $x_i$ ($i = 1, \ldots, N$) with $x_i \leq x$. Thus, $F_N$ has the limiting values $F_N(-\infty) = 0$ and $F_N(\infty) = 1$. Starting from $F_N(-\infty)$, as one proceeds in the $+x$ direction, $F_N$ increases by a step of $1/N$ whenever it reaches a data point $x_i$. ($F_N$ increases by $m/N$ if $m$ data points have the same value).

As $N \to \infty$, $\sqrt{N}D_N$ obeys the Kolmogorov distribution, for which the cumulative distribution function is given by

$$P(K) = 1 - 2\sum_{j=1}^{\infty}(-1)^{j-1}e^{-2j^2 K^2} \tag{4.42}$$

(Feller, 1948). For the KS test, the null hypothesis is that the two distributions are the same. The null hypothesis can be rejected at level $\alpha$ if

$$D_N \geq K_\alpha/\sqrt{N}, \tag{4.43}$$

where the critical value $K_\alpha$ from the Kolmogorov distribution satisfies $P(K_\alpha) = 1 - \alpha$.

$$K_\alpha = \sqrt{-\frac{1}{2}\ln(\alpha/2)}, \tag{4.44}$$

whereby $K_\alpha = 1.224, 1.358$ and $1.628$ for $\alpha = 0.10, 0.05$ and $0.01$, respectively.

If $N$ is not necessarily large, the approximate formula from Stephens (1974, table 1A) can be used to reject the null hypothesis,

$$D_N \geq K_\alpha/\left(\sqrt{N} + 0.12 + 0.11/\sqrt{N}\right), \tag{4.45}$$

which approaches (4.43) for large $N$. A multivariate KS test for a sample of vectors $\{\mathbf{x}_1, \ldots, \mathbf{x}_N\}$ has been developed by Justel et al. (1997).

While the KS test is widely used, there are other tests with higher power, for example, the *Anderson–Darling* (AD) test (T. W. Anderson and Darling, 1954; Stephens, 1974; Marsaglia, 2004). The AD test belongs to the family of quadratic empirical distribution function statistics, where the test statistic is of the form

$$D^2 = N \int_{-\infty}^{\infty} [F_N(x) - F(x)]^2 \, w(x) \, dF(x), \tag{4.46}$$

with $w(x)$ being a weight function. The AD test chooses

$$w(x) = \frac{1}{F(x)[1 - F(x)]}, \tag{4.47}$$

which increases the weight at the two tails ($F \to 0$ and $F \to 1$), thus making the AD test more sensitive at the distribution tails than the KS test. The AD test statistic is

$$A^2 = N \int_{-\infty}^{\infty} \frac{[F_N(x) - F(x)]^2}{F(x)[1 - F(x)]} \, dF(x). \tag{4.48}$$

With the data arranged in ascending order, that is, $x_1 \leq \cdots \leq x_N$, $A^2$ can be expressed as

$$A^2 = -\sum_{i=1}^{N} \frac{2i-1}{N} \left[\ln\left(F(x_i)\right) + \ln\left(1 - F(x_{N+1-i})\right)\right] - N \tag{4.49}$$

(T. W. Anderson and Darling, 1954). For $N \geq 5$, the null hypothesis can be rejected when $A^2 \geq A_\alpha^2$, where the critical value $A_\alpha^2 = 1.933, 2.492$ and $3.875$ for $\alpha = 0.10, 0.05$ and $0.01$, respectively (Stephens, 1974, table 1A; Stephens, 1986). Most statistical software packages contain the KS and Anderson–Darling tests. Goodness-of-fit tests for discrete data are also available (Stephens, 1986, section 4.17; Arnold and Emerson, 2011).

There is a pitfall in comparing an empirical distribution with a reference distribution, namely, some or all of the parameters for the reference distribution may have been computed using the same sample used to obtain the empirical distribution. For instance, to test if the empirical distribution is Gaussian, one usually obtains the mean and variance parameters of the Gaussian distribution from the same sample. Using the above formulas will have the critical values set too high, thereby failing to reject the null hypothesis for many datasets.

When the reference is a Gaussian distribution and its parameters are obtained from the sample, the corrected critical values of the KS test (computed by a Monte Carlo method) can be obtained from Lilliefors (1967, table 1). Stephens (1974, table 1A) provides approximate formulas for the corrected critical values of the KS test and the Anderson–Darling test. Stephens (1986) provides a large collection of goodness-of-fit tests for various reference distributions, including the Gaussian, exponential, extreme value, Weibull, gamma, logistic, Cauchy and von Mises distributions.

Figure 4.4 shows the empirical cumulative distribution function for two common indices for the El Niño-Southern Oscillation (ENSO), the dominant mode of interannual climate variability in the equatorial Pacific with global implications. The Niño3.4 index is the monthly sea surface temperature anomalies (i.e. deviations from the mean) for the Niño 3.4 region in the central equatorial Pacific (5°S–5°N, 170°W–120°W) (Fig. 9.3). The atmospheric index is the Southern Oscillation index (SOI), defined by the normalized air pressure difference between Tahiti and Darwin. The two indices were standardized for comparison with the standard Gaussian distribution. From inspection, the SOI is much closer to the Gaussian distribution than the Niño3.4 is. The KS test, the Lilliefors test and the Anderson–Darling test all rejected the null hypothesis (with the Gaussian reference) at the 5% level for Niño3.4 but none rejected the null hypothesis for SOI.

Razali and Wah (2011) compared the power of four tests for determining if a sample came from a Gaussian distribution, and found that the Shapiro–Wilk test was generally the best, followed by the Anderson–Darling test, the Lilliefors test and the KS test. Romao et al. (2010) compared the power of 33 goodness-of-fit tests for Gaussianity. Note that the reference distributions $F(x)$ in the KS test and the Anderson–Darling test are general and not restricted to the Gaussian distribution, in contrast to methods such as the Shapiro–Wilk test, which is restricted to Gaussian.

## 4.5.2 Two-Sample Goodness-of-Fit Tests B☺

Kolmogorov first introduced the one-sample goodness-of-fit test in 1933. In 1939, Smirnov extended the one-sample test to the two-sample test for determining if two samples belong to the same (unknown) underlying distribution. Thus, the two-sample Kolmogorov–Smirnov (KS) test is also simply called the Smirnov test. The two-sample KS test has found many applications in environmental sciences. For instance, in assessing climate change, one often would like to know if data (from observations or from a numerical model) during a later

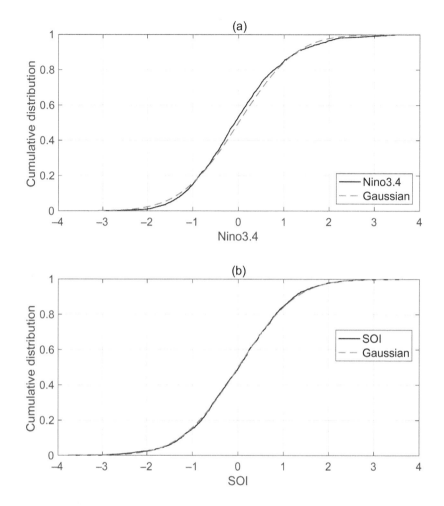

Figure 4.4 Empirical cumulative distribution function (solid curve) for (a) the standardized Niño3.4 index and (b) the standardized SOI index, using monthly data from 1870 to 2017, with the reference distribution, the standard Gaussian, shown by the dashed curve. Close inspection reveals the empirical distribution curves to be non-smooth, as they vary by steps. $D_N$, the KS test statistic, is simply the maximum vertical distance between the solid and dashed curves. [Data source: Climatic Research Unit, University of East Anglia.]

period would still belong to the same underlying distribution as data from an earlier period.

Given a first sample containing $N$ data points and a second sample with $M$ points, the two-sample KS statistic is

$$D_{N,M} = \max_x |F_{1,N}(x) - F_{2,M}(x)|, \tag{4.50}$$

where $F_{1,N}$ and $F_{2,M}$ are the empirical cumulative distribution functions from sample 1 and sample 2, respectively. In the limit of large $N$ and $M$, the null hypothesis (i.e. the two samples came from the same underlying distribution) can be rejected at level $\alpha$ if

$$D_{N,M} \geq K_\alpha \sqrt{\frac{N+M}{NM}} \, , \tag{4.51}$$

where $K_\alpha$ is the same as that in (4.44) for the one-sample KS test. The 148-year SOI record was divided into two equal halves, and the two-sample KS test was performed on the samples from the two halves (Fig. 4.5). The null hypothesis was rejected at the 5% level, suggesting that the underlying distribution for the SOI had changed over time.

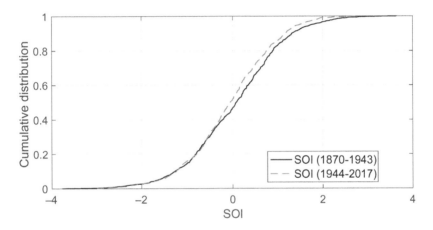

Figure 4.5 The empirical distribution functions for SOI during 1870–1943 and during 1944–2017. The $D_{N,M}$ test statistic from the two-sample KS test is the maximum vertical distance between the two curves. [Data source: Climatic Research Unit, University of East Anglia.]

The two-sample KS test has been applied to assess the impact of climate change by comparing climate model simulations for an earlier period with a future period. The KS test was used to evaluate the statistical significance of the changes found between the two periods for wind energy (Nolan et al., 2012) and for air temperature (Argueso et al., 2014). Another application of the KS test was to evaluate how well climate model simulation data agreed with observed data, for example in Californian streamflow (Maurer et al., 2010).

The two-sample Anderson–Darling test was compared with the two-sample KS test in Engmann and Cousineau (2011). The greater sensitivity of the AD test for the tails of distributions could be valuable in climate change assessment, as climate change often affects the tails of the distribution (i.e. extreme events) more strongly than the central part of the distribution.

# 4.6    Test of Variances  $\boxed{\text{B}}$☺

The $t$-test in Section 4.2 was used to test the mean (i.e. first moment) of samples. Tests on variances (second moment) of samples are also used in many situations. The classical statistical test on variances is the $F$-test.

An $F$-test is any statistical test where the test statistic obeys an $F$ distribution under the null hypothesis. It is commonly used to test for the null hypothesis that two Gaussian populations have the same variance. For instance, in climate change studies, researchers want to detect changes not only in the climate mean, but also in the climate variability – for example, have storms or El Niño events become more intense or more frequent? In such studies, one has to test whether the climate variability (from observations or model simulations) has changed significantly between an earlier period and a later or future period (Giorgi, Bi, et al., 2004). The $F$-test is also widely used to test whether any of the independent variables in a multiple linear regression model are significant.

Given two independent samples $x_1, \ldots, x_N$ and $y_1, \ldots, y_M$, both consisting of i.i.d. random variables. With sample variances $s_x^2$ and $s_y^2$, the $F$ statistic, defined by the ratio of the two variances,

$$F = \frac{s_x^2}{s_y^2}, \tag{4.52}$$

obeys the $F$ distribution under the null hypothesis, with degrees of freedom $\nu_1 = N - 1$ and $\nu_2 = M - 1$. The probability density function $p(F; \nu_1, \nu_2)$ and the cumulative distribution function $P(F; \nu_1, \nu_2)$ of the $F$ distribution are illustrated in Fig. 4.6. The $F$ distribution (also called Snedecor's $F$ distribution or Fisher–Snedecor distribution) was first obtained by G.W. Snedecor, who named it '$F$' in honour of Sir Ronald A. Fisher, as Fisher had earlier obtained the distribution of $\ln F$ (Rahman, 1968, p. 377). The mathematical form of the $F$ distribution can be found in most standard statistics textbooks, for example Rahman (1968) and Zehna (1970).

The null hypothesis $H_0$ is that the two population variances are equal, that is, $\sigma_x^2 = \sigma_y^2$. The test is whether the null hypothesis can be rejected at the $\alpha$ level. There are three possible choices for an alternative hypothesis $H_1$.

(1) $H_1$: $\sigma_x^2 < \sigma_y^2$. This requires a lower one-tailed test, that is, the null hypothesis is rejected when the test statistic $F = s_x^2/s_y^2$ is below the $\alpha$ quantile of the CDF.

(2) $H_1$: $\sigma_x^2 > \sigma_y^2$. This requires an upper one-tailed test, that is, $H_0$ is rejected when $F$ is above the $1 - \alpha$ quantile.

(3) $H_1$: $\sigma_x^2 \neq \sigma_y^2$. This requires a two-tailed test, that is, $H_0$ is rejected when $F$ is below the $\alpha/2$ quantile or above the $1 - \alpha/2$ quantile.

The $F$-test can be inaccurate if the data are non-Gaussian. Accounting for serial correlation in the data has also been proposed for the $F$-test (X. L. L. Wang, 2008; Y. X. Sun, 2013). The $F$-test is not very powerful, so sometimes the significance level is set at 0.10 instead of 0.05 (von Storch and Zwiers, 1999, section 6.7.3).

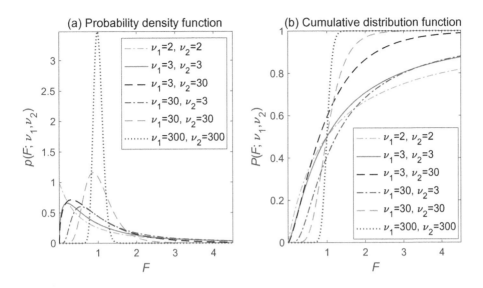

Figure 4.6 (a) PDF and (b) CDF of the $F$ distribution with $\nu_1$ and $\nu_2$ degrees of freedom.

## 4.7 Mann–Kendall Trend Test  B☺

Climate change studies have generated increasing interest in testing for trends in time series data. The simplest way to estimate a linear trend is to use linear regression (Section 5.1) to estimate the slope from the data. A non-parametric test for linear trends was proposed by (Mann, 1945) using the Kendall rank correlation (i.e. Kendall's tau) (Section 2.11.4).

Given a sample containing $x_1, \ldots, x_N$ measurements (in the order in which they were collected over time), the Mann–Kendall test statistic is defined by

$$S = \sum_{i=1}^{N-1} \sum_{j=i+1}^{N} \operatorname{sgn}(x_j - x_i), \qquad (4.53)$$

with the sign function

$$\operatorname{sgn}(x) \equiv \begin{cases} 1, & \text{if } x > 0, \\ 0, & \text{if } x = 0, \\ -1, & \text{if } x < 0. \end{cases} \qquad (4.54)$$

A positive trend in the data will be associated with $S > 0$ and a negative trend with $S < 0$.

The null hypothesis $H_0$ is that the data $x_1, \ldots, x_N$ are a sample of $N$ i.i.d. random variables (i.e. no trend present), while the alternative hypothesis $H_1$ is that a monotonic trend is present. Under $H_0$, $S$ has mean $\mu = 0$ and variance

$$\sigma^2 = \frac{N(N-1)(2N+5) - \sum_{j=1}^{g} t_j(t_j - 1)(2t_j + 5)}{18}, \qquad (4.55)$$

where the second term is only present when there are ties in the data, with $g$ being the number of tied groups and $t_j$ the size of tied group $j$. For $N \geq 10$, the null distribution can be approximated by the standard Gaussian $\mathcal{N}(0, 1)$ provided the following $z$-transformation is used:

$$z = \begin{cases} (S-1)/\sigma, & \text{if } S > 0, \\ 0, & \text{if } S = 0, \\ (S+1)/\sigma, & \text{if } S < 0, \end{cases} \qquad (4.56)$$

where adjustments of $\pm 1$ were made for continuity correction. Either a two-tailed or a one-tailed significance test can be performed as in Section 4.1. For $N \leq 10$, the distribution values can be found in R. O. Gilbert (1987, p. 272, table A18).

Next, apply the Mann–Kendall trend test to the maximum winter snow water equivalent observed at Grouse Mountain, BC. The two-tailed trend test gave an insignificant $p$-value of 0.328 for the 81 years of data from 1938 to 2018 (Fig. 4.7(a)). For Sinclair Pass further east, the trend test gave $p = 0.024$, for a negative trend significant at the 5% level (Fig. 4.7(b)). Of the seven stations in the BC Snow Survey program with manual snow survey measurements dating back to 1938, the trend test found no significant trends for five of the stations (Grouse Mountain, McColluch, Summerland Reservoir, Glacier and Ferguson) and significant negative trends at the 5% level for two stations (Sinclair Pass and Nelson).

When given time series showing a seasonal cycle, one can separate the data into individual seasons or months. For instance, with monthly data, the test statistics can be computed separately for each month, that is, $S_1$ for January data, ..., $S_m$, ..., $S_{12}$ for December data, with the final test statistic $S'$ being the sum of the 12 monthly $S_m$ statistics (Hirsch et al., 1982).

Serial correlation effects can be corrected by using an effective sample size $N_{\text{eff}}$. After removing the linear trend from the data, the lag-1 autocorrelation is computed and $N_{\text{eff}}$ can be approximated for the lag-1 autoregressive process by (2.58) (Yue and C. Y. Wang, 2004). The Mann–Kendall test is then performed with $N_{\text{eff}}$ replacing $N$ in (4.55). More general analyses on serial correlation effects in the Mann–Kendall test are given by Hamed and Rao (1998) and Hamed (2009).

## 4.8   Bootstrapping  $\boxed{\text{A}}$☺

Upon computing a statistic $\theta$ from a dataset, one would like to assess the uncertainty in the sample statistic, for example by estimating the confidence interval (CI) or testing for statistical significance. If the statistic is simple (e.g. the mean) and the data follow the Gaussian distribution, one can use the techniques in the earlier parts of this chapter (e.g. Sections 4.2 and 4.4). For complicated $\theta$ and/or non-Gaussian data, *resampling methods* such as bootstrapping (Efron and Tibshirani, 1993; Davison and Hinkley, 1997; Lahiri, 2003) are commonly used.

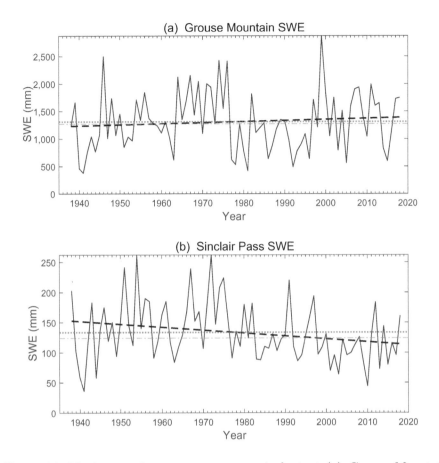

Figure 4.7 Maximum winter snow water equivalent at (a) Grouse Mountain and (b) Sinclair Pass, BC from 1938 to 2018, with linear trend (dashed line), mean (dotted) and median (dot-dashed). Sinclair Pass $(50° 40' N, 117° 58' W,$ 1,370 m elevation) is located just west of the Canadian Rockies, while Grouse Mountain $(49° 23' N, 123° 05' W,$ 1,100 m elevation) is located near the west coast (hence the much large winter SWE values). [Data source: River Forecast Centre, British Columbia.]

   To compute the CI, one needs a large number of samples, so one can compute $\theta$ from each sample and see how $\theta$ is distributed. Unfortunately, our dataset contains only one sample, so how does one get more samples? If one has access to the population, one can obtain more samples from the population. However, in most situations, one does not have the option of getting more samples from the population. In bootstrapping, one takes the single given sample to be the 'population' and takes samples repeatedly from the 'population'.

Suppose the given sample has five data points $\{x_1, x_2, x_3, x_4, x_5\}$. Bootstrapping samples are obtained by randomly choosing data points (with replacement) from this 'population' of five points, where 'with replacement' means a data point can be chosen more than once in a bootstrap sample. For instance, the first bootstrap sample can be $\{x_3, x_4, x_4, x_1, x_5\}$, the second bootstrap sample can be $\{x_5, x_1, x_5, x_4, x_2\}$, and the resampling continues until there are a total of $B$ bootstrap samples. Each bootstrap sample has the same number of elements as the original sample, and a data point from the original sample can appear more than once or not at all in the bootstrap sample. For instance, in the first bootstrap sample $\{x_3, x_4, x_4, x_1, x_5\}$, the point $x_2$ is missing while $x_4$ appears twice. Given a dataset $\mathcal{X} = \{x_1, \ldots, x_N\}$, what is the average percentage of data in $\mathcal{X}$ chosen in a bootstrap sample? The percentage turns out to be approximately 63.2% when $N$ is large (see Exercise 4.7).

The statistic $\theta$ is computed $B$ times using the bootstrap samples, yielding $B$ values of $\theta^*$ (with the asterisk indicating a statistic from a bootstrap sample). Ideally, one would like B to be as large as possible, so $B = 1{,}000$ or $10{,}000$ is commonly used. However, Efron, Rogosa, et al. (2015) pointed out that randomness in the bootstrap standard error is usually negligible for $B > 100$ (where 'negligible' means small relative to the randomness caused by variations in the original dataset) and even $B = 25$ often gives satisfactory results, which is useful to know if computing $\theta$ a large number of times is very expensive.

Bootstrapping was first introduced by Efron (1979) and became very popular with the increase of computing power. The main advantage of bootstrap lies in the fact that it is a very simple method for deriving estimates of standard errors and CIs for (complicated) statistics of the distribution. The bootstrap CI is asymptotically more accurate than the standard CI obtained from sample standard error assuming Gaussianity (DiCiccio and Efron, 1996).

From our original sample, we obtained the sample statistic $\theta_\mathrm{s}$ and from the bootstrap samples, $B$ values of $\theta^*$. We want to estimate the CI, where the population statistic or parameter $\theta_\mathrm{p}$ lies within the CI with a probability of $1 - \alpha$.

The simplest and most intuitive (but not the most accurate) way to construct the CI is by the *percentile method*, where all the values of $\theta^*$ obtained from the bootstrap samples are arranged in ascending order, that is,

$$\theta^*_{(1)} \leq \theta^*_{(2)} \leq \cdots \leq \theta^*_{(B)}. \tag{4.57}$$

If $B = 10{,}000$, then the 95% CI is simply taken to be

$$\mathrm{CI} = [\theta^*_{(250)}, \ \theta^*_{(9750)}], \tag{4.58}$$

that is, one takes the $\alpha/2$ and the $(1 - \alpha/2)$ quantiles of the $\theta^*$ values from the bootstrap samples, with $\alpha = 0.05$ for the 95% CI. The problem with this method is that there is no theoretical justification for the assumption that the quantiles of $\theta^*$ provide a good approximation for the CI of $\theta_\mathrm{p}$; thus, the use of this method is not advisable.

We need to know how $\theta_s$ fluctuates around the true value $\theta_p$, that is, the distribution of

$$\delta = \theta_s - \theta_p. \tag{4.59}$$

The bootstrap samples provide a way to estimate the distribution of $\delta$ by computing

$$\delta^* = \theta^* - \theta_s \tag{4.60}$$

and using the distribution of $\delta^*$ as an approximation for the distribution of $\delta$. From the $\alpha/2$ and the $(1 - \alpha/2)$ quantiles of the $\theta^*$ distribution, we get

$$[\delta^*_{\alpha/2}, \, \delta^*_{1-\alpha/2}] = [\theta^*_{\alpha/2} - \theta_s, \, \theta^*_{1-\alpha/2} - \theta_s]. \tag{4.61}$$

From (4.59), we have $\theta_p = \theta_s - \delta$, and equating the quantiles of $\delta^*$ with those of $\delta$, we obtain the CI for $\theta_p$ as

$$\text{CI} = [\theta_s - \delta^*_{1-\alpha/2}, \, \theta_s - \delta^*_{\alpha/2}]. \tag{4.62}$$

Substituting in (4.61) gives

$$\text{CI} = [2\theta_s - \theta^*_{1-\alpha/2}, \, 2\theta_s - \theta^*_{\alpha/2}]. \tag{4.63}$$

This is known as the *basic method* for computing bootstrap CI (Davison and Hinkley, 1997, pp. 27–29).

There are many other more sophisticated methods for estimating CI from bootstrapping (DiCiccio and Efron, 1996; Davison and Hinkley, 1997; Puth et al., 2015). The *bias-corrected and accelerated* (BCa) *method* developed by Efron (1987) is probably the most commonly recommended for its accuracy (Puth et al., 2015) and is described in Efron and Tibshirani (1993), Davison and Hinkley (1997, pp. 203–211) and Wilks (2011, pp. 175–177).

As an example, let us examine the maximum winter snow water equivalent (SWE) at Glacier, BC ($51° 15'$ N, $117° 29'$ W, 1,250 m elevation) for the period 1950–2018. The histogram of the SWE shows the distribution to be right-skewed relative to a Gaussian (Fig. 4.8(a)). The SWE sample mean is $\bar{x} = 726.9$ mm. We want an estimate of the CI for the sample mean. Assuming Gaussianity, the standard error of the mean from (3.26) is 18.2 mm, and the 95% CI from (4.33) is $[690.6, 763.3]$ mm.

Figure 4.8(b) shows the distribution of the sample mean for 10,000 bootstrap samples. The distribution is still slightly right-skewed. The BCa method gives $\text{CI} = [694.3, 766.2]$ mm, which is shifted to the right of the CI of $[690.6, 763.3]$ mm from the Gaussian standard error method mentioned above, and the sample mean (solid vertical line) is no longer located at the centre of the CI from BCa.

The distribution of the sample standard deviation (Fig. 4.8(c)) shows the bootstrap distribution to be also right-skewed. The difference between the statistic from the original sample (solid vertical line) and the median of the bootstrap statistic (dashed line), and the differences between the CI from the BCa method and the CIs from the basic method and the percentile method are all larger in Fig. 4.8(c) than in Fig. 4.8(b).

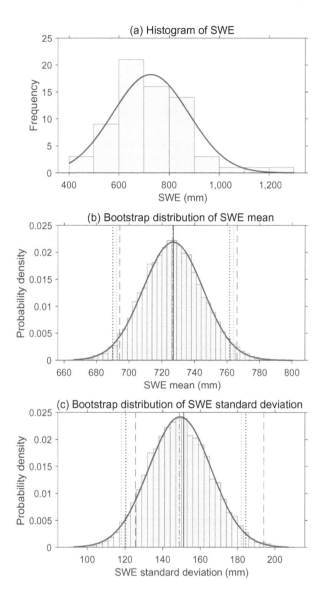

Figure 4.8 (a) Histogram of the maximum winter SWE at Glacier, BC. Distribution of the SWE (b) sample mean and (c) sample standard deviation from
10,000 bootstrap samples. The fitted Gaussian curve is also shown. The vertical lines show the statistic from the original sample (solid), the median of the
bootstrap statistic (dot-dashed), the 95% CI from the BCa method (dashed),
the basic method (darkly dotted) and the percentile method (lightly dotted).
[Data source: River Forecast Centre, British Columbia.]

The CI for the median and many other statistics (e.g. the 0.90 quantile) are hard to estimate theoretically, but are easily computed using bootstrapping. Bootstrap CI can also be computed easily for two or multi-sample statistics. For instance, for the ratio of the variance between sample A (with $N_A$ data points) and sample B (with $N_B$ points), bootstrap sampling would choose $N_A$ points from sample A with replacement and $N_B$ points from sample B with replacement. For two-sample statistics such as the correlation between two variables $x$ and $y$, bootstrap sampling would randomly choose pairs of data $(x_i, y_i)$ $(i = 1, \ldots, N)$ with replacement. Bootstrapping can also be used for hypothesis testing, for example comparing the means of two independent samples (Efron and Tibshirani, 1993).

In our SWE example, the lag-1 autocorrelation is only 0.046, so the effects of serial correlation can be ignored. However, for many time series, the effects of serial correlation should not be ignored. To retain serial information, *block bootstrapping* is used, that is, perform bootstrap sampling by choosing blocks of data instead of individual data points. Given a dataset with $N$ data points, *moving block bootstrapping* (Künsch, 1989) splits the data into $N - L + 1$ overlapping blocks of length $L$, that is, block 1 consists of $\{x_1, x_2, \ldots, x_L\}$, block 2, $\{x_2, x_3, \ldots, x_{L+1}\}$, $\ldots$, and block $N - L + 1$, $\{x_{N-L+1}, \ldots, x_N\}$. The bootstrap sampling here simply randomly chooses a block of data with replacement for a total of $N/L$ times, so the bootstrap sample contains $N$ data points.

What value should one use for the block length $L$? Researchers often use $L \approx N/N_{\text{eff}}$, with $N_{\text{eff}}$ being the effective sample size (Section 2.11.2). Actually, the optimal $L$ increases with the autocorrelation in the data and with the sample size $N$; thus, $L \approx N/N_{\text{eff}}$ is much smaller than the optimal $L$ when $N$ is large (see Exercise 4.8). When the statistic is the mean or the variance, the optimal $L$ is of the form $AN^{1/3}$, where $A$ depends on the autocorrelation function (Bühlmann and Künsch, 1999; Politis and White, 2004; Patton et al., 2009). For other statistics, $L$ can be of the order $N^{1/4}$ or $N^{1/5}$ (P. Hall et al., 1995). If the statistic is the mean, and the data follows a first order autoregressive process, Wilks (1997) provides an implicit equation for $L$,

$$L = (N - L + 1)^{(2/3)(1 - N_{\text{eff}}/N)}, \tag{4.64}$$

with $N_{\text{eff}}$ given by (2.58).

## 4.9 Field Significance  [B]☺

Environmental scientists often work with data fields, that is, observations from multiple stations or data from a spatial grid. Instead of testing for statistical significance at one station as in the previous sections of this chapter, we now perform testing at all the individual stations, then assess whether there is statistical significance for the field as a whole. For instance, one may be interested if climate change has led to a significant trend in the streamflow over a river basin where there are many stations along the river and its tributaries. From the large number of significance tests at individual stations, how does one conclude if there is statistical significance over the whole basin? In other words, if there are $M$ stations in a field, and $k$ of them indicate (local) statistical significance at the 5% level, how large does $k$ have to be in order to have field significance at the 5% level? The naive answer, namely $k \geq 0.05M$, is unfortunately incorrect as pointed out by Livezey and W. Y. Chen (1983).

As an analogy, suppose at each station, we have a box containing 19 black marbles and 1 white marble. The probability of randomly picking a white marble is $\alpha = 1/20$. At each of the $M$ stations, we randomly pick a marble and count the number of white marbles chosen. The probability of having $k$ white marbles picked from the $M$ trials is given by the binomial distribution (3.5), that is, $\mathrm{Bin}(k | M, \alpha)$.

Suppose the field has $M = 40$ stations. Table 4.2 gives the probability distribution of getting $k$ white marbles. The probability of having no white marbles is 0.129, one white marble is 0.271, etc. The cumulative distribution function (CDF) is also given, where $\mathrm{CDF}(k = 1)$ means the probability of having either $k = 0$ or 1 is 0.399. The complementary cumulative distribution function (CCDF) is $1 - \mathrm{CDF}$, for example $\mathrm{CCDF}(k = 1)$ gives the probability of having $k \geq 2$ as being 0.601. To have the CCDF $\leq 0.05$, we need to go as far as $\mathrm{CCDF}(k = 4) = 0.048$, that is, we need five or more white marbles to reject the null hypothesis. Note 5 out of 40 is 12.5%, much higher than the 5% from the naive answer. In other words, if we have performed significance tests at the 5% level at the $M$ local stations, we need at least 12.5% of the local tests being significant before we can reject the null hypothesis at the 5% level for the field.

Table 4.2 Probability distribution function (PDF), cumulative distribution function (CDF) and complementary cumulative distribution function (CCDF) for various values of $k$, with the PDF being $\mathrm{Bin}(k | 40, 0.05)$.

| $k$ | 0 | 1 | 2 | 3 | 4 | 5 | 6 |
|---|---|---|---|---|---|---|---|
| PDF | 0.129 | 0.271 | 0.278 | 0.185 | 0.090 | 0.034 | 0.010 |
| CDF | 0.129 | 0.399 | 0.677 | 0.862 | 0.952 | 0.986 | 0.997 |
| CCDF | 0.871 | 0.601 | 0.323 | 0.138 | 0.048 | 0.014 | 0.003 |

So far, we have assumed there is no spatial correlation between neighbouring stations. Suppose we are analysing data from a spatial grid with $M = 1000$. If we assume the stations are independent, the binomial distribution $\mathrm{Bin}(k | 1000, 0.05)$ has CCDF $\leq 0.05$ first occurring at $\mathrm{CCDF}(k = 63) = 0.038$. In other words, we need at least 63 out of 1000, that is, 6.3% of the local tests being significant before we can reject the null hypothesis at the 5% level for the field. Note 6.3% is not as far off from the naive answer of 5% as before. Indeed, the naive answer gets better as $M$ becomes large (Livezey and W. Y. Chen, 1983, figure 3).

Now, the bad news – the binomial distribution $\mathrm{Bin}(k | 1000, 0.05)$ is not appropriate when there is spatial correlation between the stations. It is quite common to encounter atmospheric patterns with much larger spatial scale than the distance between neighbouring stations or grids. For instance, we can have 1,000 grid points, but only 40 effective degrees of freedom (Livezey and W. Y. Chen, 1983). An idealized way to visualize this is to assume the field of 1,000 grid points is divided into 40 regions of 25 points each. Within each region, the strong spatial correlation ensures all grid points to have the same local significance test result. There are effectively only 40 independent local significance tests and the PDF is simply $\mathrm{Bin}(k|\,40, 0.05)$, as given in Table 4.2. Thus, we are back to requiring at least 12.5% of the local tests being significant before we can reject the null hypothesis for the field, much higher than the 6.3% obtained by ignoring spatial correlation. Early researchers often erroneously concluded spatial patterns to be significant at the 5% level when they found the percentage of grid points being locally significant exceeded 5% (Livezey and W. Y. Chen, 1983).

For tests of field significance, a Monte Carlo approach is most commonly used (Livezey and W. Y. Chen, 1983; Lettenmaier et al., 1994; Shabbar, Bonsal, et al., 1997; Douglas et al., 2000; Renard et al., 2008), though Bretherton, Widmann, et al. (1999) proposed ways to estimate the effective number of spatial degrees of freedom in a time-varying field without the use of a Monte Carlo method. The key idea in the Monte Carlo approach is to generate a distribution under the null hypothesis, taking care to preserve the spatial correlation among stations.

Among Monte Carlo methods, the bootstrap method (Section 4.8) is commonly used (Douglas et al., 2000; Renard et al., 2008): Let $x_i^{(j)}$ denote the $i$th measurement for the variable $x$ at station $j$ $(i = 1, \ldots, N,\ j = 1, \ldots, M)$.

(a) Choose a bootstrap sample over the index $i$, that is, randomly pick an index value from $i = 1, \ldots, N$, with replacement, for a total of $N$ times. For instance, $N = 5$ and $M = 2$, and the first bootstrap sample chooses $i = 4, 1, 4, 5, 3$, yielding the data $x_4^{(1)}, x_1^{(1)}, x_4^{(1)}, x_5^{(1)}, x_1^{(1)}$ for station 1 and $x_4^{(2)}, x_1^{(2)}, x_4^{(2)}, x_5^{(2)}, x_1^{(2)}$ for station 2. Although the sequence has been randomly shuffled from the bootstrap sampling, the spatial correlation is unchanged since all stations have had their sequence shuffled the same way. Repeat until a total of $B$ bootstrap samples have been generated.

If the statistical test involves two variables $x$ and $y$, for example testing for significance in the correlation between $x^{(j)}$ and $y^{(j)}$, $(j = 1, \ldots, M)$, then the bootstrap sample also contains a different random sampling of $i = 1, \ldots, N$, with replacement, for $y^{(j)}$. For instance, the first bootstrap sample can contain:
$$[x_4^{(1)}, x_1^{(1)}, x_4^{(1)}, x_5^{(1)}, x_1^{(1)}],\ [x_4^{(2)}, x_1^{(2)}, x_4^{(2)}, x_5^{(2)}, x_1^{(2)}],$$
$$[y_2^{(1)}, y_5^{(1)}, y_5^{(1)}, y_1^{(1)}, y_3^{(1)}]\ \text{and}\ [y_2^{(2)}, y_5^{(2)}, y_5^{(2)}, y_1^{(2)}, y_3^{(2)}].$$

(b) Compute the statistical test $t_b^{(j)}$ at each station $j$ for each bootstrap sample $(b = 1, \ldots, B)$.

(c) Let $k_b$ be the total number of test results that are significant at level $\alpha$ over the $M$ stations, for the $b$th bootstrap sample.

(d) From $k_b$ $(b = 1, \ldots, B)$, generate an empirical CDF.

(e) For field significance at level $\alpha_f$, the critical value $k_{\text{critical}}$ is simply the $1 - \alpha_f$ quantile of the empirical CDF. (A common choice is to have $\alpha_f = \alpha$).

(f) If $k$ is the number of test results that are significant at level $\alpha$ over the $M$ stations for the original data, field significance at level $\alpha_f$ requires $k > k_{\text{critical}}$.

Bootstrap sampling produces an approximation of the distribution under the null hypothesis. For instance, if we are testing for a linear trend in $x$, the random shuffling of the sequence by bootstrap sampling eliminates any trend in the data while preserving the spatial correlation among stations. Similarly, if we are testing for significant correlation between $x$ and $y$, the different random shuffling used for $x$ and $y$ eliminates any correlation between $x$ and $y$ but retains the spatial correlation among stations for $x$ and for $y$.

If serial correlation is a concern, one may want to use a block bootstrapping approach, for example the moving block bootstrapping method, where one selects a block of data (e.g. a year of data) instead of an individual time point repeatedly (Section 4.8).

One disadvantage of the above method is that $k_b$ only counts the number of test results which are significant at level $\alpha$ over the $M$ stations, that is, $p \leq \alpha$ where $p$ from (4.1) is the $p$-value from the test. It does not distinguish between a station where $p \ll \alpha$ from one where $p$ is only slight less than $\alpha$. The *false discovery rate* (FDR) method (Benjamini and Hochberg, 1995; Ventura et al., 2004; Wilks, 2006; Wilks, 2011) uses the magnitude of $p$ more fully and is expected to be a more sensitive test. However, comparison between the bootstrap method and the FDR method shows the FDR method to be slightly more accurate when the spatial correlation is weak and overly conservative (i.e. failing to detect field significance) when spatial correlation is strong (Renard et al., 2008, table 1).

# Exercises

4.1

A fisheries scientist is studying if the arsenic concentration in farmed salmon is higher or lower than in wild salmon. For 100 farmed salmon, he measured the mean arsenic level $\overline{x_a} = 0.77$ mg/kg of wet weight, and for 20 wild salmon, $\overline{x_b} = 0.68$ mg/kg. The sample standard deviation $s_a = 0.28$ mg/kg and $s_b = 0.27$ mg/kg. (a) Can the null hypothesis be rejected for the two means at the 5% level using a two-tailed $t$-test? (b) If not, would measuring 40 additional wild salmon reject the null hypothesis (assuming $\overline{x_b}$ and $s_b$ are unchanged with the additional data)? (c) If instead of adding 40 wild salmon, the scientists added 40 farmed salmon, would that reject the null hypothesis (assuming $\overline{x_a}$ and $s_a$ unchanged)?

**4.2**

From the given dataset, with the Nino index = 1 indicating El Niño conditions and 0 indicating non-El Niño conditions in the equatorial Pacific: (a)
apply the independent two-sampled $t$-test (at the 5% level) to the mean winter (December–February) temperature in Winnipeg, MB, Canada to test the
hypothesis that the mean winter temperature during El Niño is warmer than
during non-El Niño conditions; (b) determine the 95% confidence intervals for
the population mean of the winter temperature during El Niño and during non-
El Niño conditions and (c) apply the Welch $t$-test (for two independent samples
with different population variance) and see how the result differs from the $t$-
test. [Data source: definition of El Niño from Climate Prediction Center, US
National Weather Service and winter seasonal temperature data of Winnipeg
from weatherstats.ca based on Environment and Climate Change Canada data.]

**4.3**

A farmer is trying to test if adding a certain chemical to the soil enhances the
fruit production of his apple trees. For his six trees, the fruit yield in the year
before adding the chemical and the year with the added chemical are recorded
(see Table 4.3). (a) Does the chemical significantly enhance the yield at the 0.05
level? (b) If not, how many apple trees would he need to test in order to show
the result is significant (assuming the mean and the standard deviation of the
yield remain the same)?

Table 4.3 Fruit yield in bushels.

| Tree | 1 | 2 | 3 | 4 | 5 | 6 |
|---|---|---|---|---|---|---|
| yield (no chemical) | 9.1 | 15.6 | 11.5 | 18.2 | 8.7 | 12.9 |
| yield (with chemical) | 10.8 | 15.1 | 13.1 | 17.5 | 11.2 | 13.6 |

**4.4**

A dataset contains $N$ observations of a pair of variables $x$ and $y$.
(a) If $N = 6$ and the sample Pearson correlation $\rho = 0.80$, can the null hypothesis be rejected at the 5% level using the two-tailed $t$-test?
(b) If $N = 1,000$ and $\rho = -0.080$, can the null hypothesis be rejected at the 5%
level?
(c) Before you compute the 95% confidence interval (CI) for the population
correlation $\hat{\rho}$, would you expect the CI to include the value 0, based on your
answer in (b)?
(d) Compute the 95% CI for $\hat{\rho}$ in (b).

**4.5**

During the 69-year period of 1950–2018, there were 24 warm (El Niño)
episodes and 22 cold (La Niña) episodes in the equatorial Pacific. From the
given Grouse Mountain, BC, maximum annual snow water equivalent (SWE)
data and the list of years of El Niño and La Niña, use the Mann–Whitney–

Wilcoxon test (a) to test SWE during El Niño and during non-El Niño (i.e. all data except data during El Niño), and (b) to test SWE during non-La Niña and during La Niña. Which one has the stronger effect on Grouse Mountain SWE – El Niño or La Niña? [Data source: definition of El Niño and La Niña from Climate Prediction Center, US National Weather Service and SWE data from River Forecast Centre, British Columbia.]

**4.6**

Given the monthly sea surface temperature anomalies for the Niño 3.4 region in the central equatorial Pacific during 1870–2017 (Fig. 9.3), divide the data record into two equal halves. Plot the empirical cumulative distribution function obtained from each half. Apply the two-sample Kolmogorov–Smirnov test at the 5% level to check if data from the first half and data from the second half belong to the same underlying distribution. Repeat using the Niño 1+2 anomalies. [Data source: Climatic Research Unit, University of East Anglia]

**4.7**

For bootstrap resampling applied to a dataset with $N$ observations, derive an expression for the average fraction of data in the original dataset drawn in a bootstrap sample. Show that this fraction $\approx 0.632$ for large $N$.

**4.8**

Estimate the bootstrap block length $L$ by solving the implicit equation (4.64) for (a) $N = 100$, (b) $N = 1,000$ and (c) $N = 10,000$, with (i) $N/N_{\text{eff}} = 2$ and (ii) $N/N_{\text{eff}} = 20$. Round off $L$ to nearest integer.

**4.9**

Assuming no spatial correlation between stations in a field containing $M$ stations, and $k$ stations are found to be locally statistically significant, find the critical value $k_{\text{critical}}$ for field significance at the 5% level when $M = 3, 10, 30, 100, 300, 1,000, 3,000$ and $10,000$. Use a semi-log plot to show $k_{\text{critical}}/M \times 100\%$ as a function of $M$. Also draw a horizontal line indicating the 5% level.

# 5

# Linear Regression

Originally introduced by Galton (1985), linear regression is a linear approach to modelling the relationship between a scalar response variable (also known as the dependent variable or output variable) and one or more predictor variables (a.k.a. independent variables or input variables). If there is only one predictor variable, one has simple linear regression. With more than one predictor variable, the problem is called *multiple* linear regression. One must not confuse this problem with *multivariate* linear regression, where multiple correlated dependent variables are predicted.

## 5.1  Simple Linear Regression  $\boxed{\text{A}}$ ☺

For now, consider *simple linear regression* (SLR), where there is only one independent variable $x$ and one dependent variable $y$, and the dataset contains $N$ pairs of $(x, y)$ measurements. The relation is

$$y_i = \hat{y}_i + \epsilon_i = a_0 + a_1 x_i + \epsilon_i, \qquad i = 1, \ldots, N, \tag{5.1}$$

where $a_0$ and $a_1$ are the regression parameters or coefficients, $\hat{y}_i$ is the $y_i$ predicted or described by the linear regression relation and $\epsilon_i$ is the error or the residual unaccounted for by the regression, with $\epsilon_i$ usually assumed to obey the Gaussian distribution with zero mean (Fig. 5.1).

As regression is commonly used as a prediction tool (i.e. given $x$, use the regression relation to predict $y$), $x$ is referred to as the *predictor* and $y$ as the *response variable* (or just 'response' for brevity).[1] The error

$$\epsilon_i = y_i - \hat{y}_i = y_i - a_0 - a_1 x_i.^2 \tag{5.2}$$

---

[1] The predictor variable is also called the regressor, covariate, explanatory variable, independent variable, input variable and *feature* in ML. The response variable is also called the regressand, dependent variable, output variable and *predictand*. Curiously, the term 'predictand', widely used in the literature for atmospheric science, oceanography, and so on, is not well known in the statistical and ML communities.

[2] In ML and other non-linear regression models, the error is usually defined with the opposite sign, that is, $\hat{y}_i - y_i$, instead of $y_i - \hat{y}_i$, which is the convention used in linear regression.

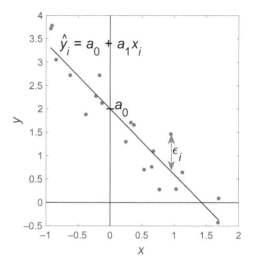

Figure 5.1 Illustrating linear regression. A straight line $\hat{y}_i = a_0 + a_1 x_i$ is fitted to the data, where the parameters $a_0$ and $a_1$ are determined from minimizing the sum of the square of the error $\epsilon_i$, which is the vertical distance between the $i$th data point and the line. The slope of the line is given by $a_1$ and the $y$-intercept by $a_0$.

Linear regression minimizes the sum of squared errors (SSE),

$$\text{SSE} = \sum_{i=1}^{N} \epsilon_i{}^2 = \sum_{i=1}^{N} (y_i - a_0 - a_1 x_i)^2, \tag{5.3}$$

to yield the best estimate of the parameters $a_0$ and $a_1$. Because SSE is minimized, this method is also referred to as the *linear least squares* or *ordinary least squares* method ('linear' because the model (5.1) depends on the parameters $a_0$ and $a_1$ linearly).

To minimize the SSE, we first partially differentiate the SSE in (5.3) with respect to $a_0$ and $a_1$,

$$\frac{\partial \sum_i (y_i - a_0 - a_1 x_i)^2}{\partial a_0} = -2 \sum_{i=1}^{N} (y_i - a_0 - a_1 x_i), \tag{5.4}$$

$$\frac{\partial \sum_i (y_i - a_0 - a_1 x_i)^2}{\partial a_1} = -2 \sum_{i=1}^{N} (y_i - a_0 - a_1 x_i) x_i, \tag{5.5}$$

then set the derivatives to zero, yielding the *normal equations*,

$$\sum_{i=1}^{N} (y_i - a_0 - a_1 x_i) = 0, \tag{5.6}$$

$$\sum_{i=1}^{N} (y_i - a_0 - a_1 x_i) x_i = 0, \tag{5.7}$$

from which we will solve for the optimal estimates $\hat{a}_0$ and $\hat{a}_1$.

From (5.6), we have

$$\hat{a}_0 = \frac{1}{N} \sum_i y_i - \frac{\hat{a}_1}{N} \sum_i x_i, \quad \text{i.e. } \hat{a}_0 = \overline{y} - \hat{a}_1 \overline{x}, \tag{5.8}$$

where $\bar{x}$ and $\bar{y}$ are the sample means of $x$ and $y$, respectively. Substituting (5.8) into (5.7) gives

$$\sum_i x_i y_i - \frac{1}{N}\left(\sum_i x_i\right)\left(\sum_i y_i\right) + \frac{\hat{a}_1}{N}\left(\sum_i x_i\right)\left(\sum_i x_i\right) - \hat{a}_1 \sum_i x_i x_i = 0.$$

$$(5.9)$$

Thus,

$$\hat{a}_1 = \frac{\sum_i x_i y_i - N\bar{x}\,\bar{y}}{\sum_i x_i^2 - N\bar{x}\,\bar{x}}.$$

$$(5.10)$$

Equations (5.8) and (5.10) provide the optimal values $\hat{a}_0$ and $\hat{a}_1$ for minimizing the SSE, thereby yielding the best straight line fit to the data in the $x$-$y$ plane. The parameter $\hat{a}_1$ gives the slope of the regression line, while $\hat{a}_0$ gives the $y$-intercept.

Since regression and correlation are two approaches to extract linear relations between two variables, one would expect the two to be related. Equation (5.10) can be rewritten as

$$\hat{a}_1 = \frac{\sum_i (x_i - \bar{x})(y_i - \bar{y})}{\sum_i (x_i - \bar{x})^2}$$

$$(5.11)$$

(see Exercise 5.1). Comparing with the expression for the Pearson sample correlation, (2.53), we see that

$$\hat{a}_1 = \rho_{xy}\frac{s_y}{s_x},$$

$$(5.12)$$

that is, the slope of the regression line is the correlation coefficient $\rho_{xy}$ times the ratio of the sample standard deviation of $y$ to that of $x$ (Section 2.5).

If $s_x$ and $s_y$ are held constant, increasing $\rho_{xy}$ in (5.12) will increase $\hat{a}_1$, the slope of the regression line. If $s_x$ and $\rho_{xy}$ are held constant, increasing $s_y$ will increase the magnitude of $\hat{a}_1$. If $s_y$ and $\rho_{xy}$ are held constant, increasing $s_x$ will decrease the magnitude of $\hat{a}_1$.

## 5.1.1   Partition of Sums of Squares   A☺

The total variation in $y$ is described by the sum of squares (total) (SST),

$$\text{SST} = \sum_{i=1}^{N}(y_i - \bar{y})^2.$$

$$(5.13)$$

The variance is simply the SST divided by the number of degrees of freedom, $N - 1$. SST can be partitioned,

$$\text{SST} = \text{SSR} + \text{SSE},$$

$$(5.14)$$

where

$$\text{SSR} = \sum_{i=1}^{N} (\hat{y}_i - \bar{y})^2, \tag{5.15}$$

$$\text{SSE} = \sum_{i=1}^{N} (y_i - \hat{y}_i)^2 = \sum_{i=1}^{N} \epsilon_i^2. \tag{5.16}$$

Thus, the total sum of squares (SST) can be partitioned into two: the first part is that accounted for by the regression relation, that is, the sum of squares due to regression (SSR), and the second part is the sum of squared errors (SSE). The proof of (5.14) is given in Section (5.2.2), which is for the more general multiple linear regression.

There is a simple relation between the ratio SSE/SST and the Pearson correlation:

$$\text{SSE} = \sum_{i=1}^{N} (y_i - \hat{y}_i)^2 = \sum_{i=1}^{N} \left[ y_i - (a_0 + a_1 x_i) \right]^2. \tag{5.17}$$

Substituting in (5.8), we have

$$\text{SSE} = \sum_i (y_i - a_1 x_i - \bar{y} + a_1 \bar{x}) = \sum_i [(y_i - \bar{y}) - a_1 (x_i - \bar{x})]^2$$

$$= S_{yy} - 2 a_1 S_{xy} + a_1^2 S_{xx}, \tag{5.18}$$

where

$$S_{yy} = \sum_i (y_i - \bar{y})^2, \quad S_{xy} = \sum_i (x_i - \bar{x})(y_i - \bar{y}), \quad S_{xx} = \sum_i (x_i - \bar{x})^2. \tag{5.19}$$

From (5.11), $a_1 = S_{xy}/S_{xx}$, thus

$$\text{SSE} = S_{yy} - a_1 S_{xy} = S_{yy} \left( 1 - \frac{S_{xy}^2}{S_{xx} S_{yy}} \right) = S_{yy}(1 - \rho_{xy}^2), \tag{5.20}$$

where

$$\rho_{xy} = \frac{S_{xy}}{\sqrt{S_{xx} S_{yy}}} \tag{5.21}$$

is the Pearson sample correlation from (2.53). With SST $= S_{yy}$, we get

$$\frac{\text{SSE}}{\text{SST}} = 1 - \rho_{xy}^2. \tag{5.22}$$

Since $\hat{y}$ is a linear function of $x$, $\rho_{xy} = \rho_{\hat{y}y} = \rho_{y\hat{y}}$,

$$\frac{\text{SSE}}{\text{SST}} = 1 - \rho_{y\hat{y}}^2, \tag{5.23}$$

where $\rho_{y\hat{y}}$ is the correlation between the dependent variable $y$ and $\hat{y}$, its predicted value from the regression relation.

Since SSE/SST is the fraction of SST not accounted for by the regression relation, that means $1 - \rho_{y\hat{y}}^2$ can be interpreted as the fraction of SST not accounted for by the regression relation. For example, if the correlation $\rho_{y\hat{y}} = 0.5$, then $1 - \rho_{y\hat{y}}^2 = 0.75$, meaning that 75% of the variability of $y$ is not accounted for by the regression relation. On the other hand, from (5.14), the fraction of SST accounted for by the regression relation is

$$\frac{\text{SSR}}{\text{SST}} = \rho_{xy}^2 = \rho_{y\hat{y}}^2. \tag{5.24}$$

We will learn more about the properties of simple linear regression by studying multiple linear regression (MLR) in the next section. Since MLR has multiple predictors, SLR is just a special case of MLR limited to a single predictor.

## 5.1.2   Confidence Interval for Regression Parameters  B☺

Since the regression parameters $\hat{a}_0$ and $\hat{a}_1$ were derived from data containing noise, $\hat{a}_0$ and $\hat{a}_1$ are themselves random variables, and we would like to estimate confidence intervals for them. It will be shown below that the variance of the slope parameter $\hat{a}_1$ is given by

$$\text{var}(\hat{a}_1) = \frac{\sigma^2}{\sum_{i=1}^{N}(x_i - \overline{x})^2}, \tag{5.25}$$

with $\sigma^2 = \text{var}(\epsilon_i)$ for all $i$, while the variance of the intercept parameter $\hat{a}_0$ is given by

$$\text{var}(\hat{a}_0) = \sigma^2 \left[ \frac{1}{N} + \frac{\overline{x}^2}{\sum_{i=1}^{N}(x_i - \overline{x})^2} \right] = \sigma^2 \left[ \frac{\sum_{i=1}^{N} x_i^2}{N \sum_{i=1}^{N}(x_i - \overline{x})^2} \right]. \tag{5.26}$$

As the population noise variance $\sigma^2$ is not usually known, the sample noise variance $s^2$ is used instead in the above formulas, where

$$s^2 = \frac{1}{N-2} \sum_{i=1}^{N}(y_i - \hat{a}_0 - \hat{a}_1 x_i)^2, \tag{5.27}$$

as there are $N - 2$ degrees of freedom (due to the regression relation having two parameters).

If the error is Gaussian, the $(1 - \alpha) \times 100\%$ confidence interval for the slope and intercept parameters are given respectively by

$$\text{CI for } \hat{a}_1 = \hat{a}_1 \pm \frac{t(1 - \tfrac{1}{2}\alpha \,|\, N - 2)\, s}{[\sum_i (x_i - \overline{x})^2]^{1/2}}, \tag{5.28}$$

$$\text{CI for } \hat{a}_0 = \hat{a}_0 \pm t(1 - \tfrac{1}{2}\alpha \,|\, N - 2)\, s \left[ \frac{\sum_i x_i^2}{N \sum_i (x_i - \overline{x})^2} \right]^{1/2}, \tag{5.29}$$

where $t(1 - \frac{1}{2}\alpha \,|\, N - 2)$ is the $(1 - \frac{1}{2}\alpha)$ quantile of the t-distribution (Section 3.15) with $N - 2$ degrees of freedom.

The reader less keen on mathematical derivation can skip the reminder of this subsection as we derive (5.25) and (5.26). From (5.11) and (5.1), the variance of $\hat{a}_1$ is given by

$$
\begin{aligned}
\mathrm{var}(\hat{a}_1) &= \mathrm{var}\left( \frac{\sum_i (x_i - \bar{x})(y_i - \bar{y})}{\sum_i (x_i - \bar{x})^2} \right) \\
&= \mathrm{var}\left( \frac{\sum_i (x_i - \bar{x})(a_0 + a_1 x_i + \epsilon_i - \bar{y})}{\sum_i (x_i - \bar{x})^2} \right) \\
&= \mathrm{var}\left( \frac{\sum_i (x_i - \bar{x})\epsilon_i}{\sum_i (x_i - \bar{x})^2} \right),
\end{aligned}
\tag{5.30}
$$

since only $\epsilon_i$ is a random variable, the rest are treated as constants inside $\mathrm{var}(\ldots)$. Invoking the identity

$$
\mathrm{var}(c_1 z_1 + c_2 z_2) = c_1^2 \, \mathrm{var}(z_1) + c_2^2 \, \mathrm{var}(z_2),
\tag{5.31}
$$

with $c_1$, $c_2$ being constants and $\mathrm{cov}(z_1, z_2) = 0$, we have

$$
\mathrm{var}(\hat{a}_1) = \frac{\sum_i (x_i - \bar{x})^2 \, \mathrm{var}(\epsilon_i)}{[\sum_i (x_i - \bar{x})^2]^2} = \frac{\sigma^2 \sum_i (x_i - \bar{x})^2}{[\sum_i (x_i - \bar{x})^2]^2},
\tag{5.32}
$$

as $\mathrm{var}(\epsilon_i) = \sigma^2$ for all $i$ and $\mathrm{cov}(\epsilon_i, \epsilon_j) = 0$ for $i \neq j$. Thus, we obtain (5.25) for the variance of the slope parameter $\hat{a}_1$.

From (5.8), the variance of $\hat{a}_0$ is given by

$$
\begin{aligned}
\mathrm{var}(\hat{a}_0) &= \mathrm{var}(\bar{y} - \hat{a}_1 \bar{x}) \\
&= \mathrm{var}(\bar{y}) + \bar{x}^2 \, \mathrm{var}(\hat{a}_1) - 2\bar{x} \, \mathrm{cov}(\bar{y}, \hat{a}_1).
\end{aligned}
\tag{5.33}
$$

The final covariance term can be shown to vanish (Bickel and Doksum, 1977, p. 272; Draper and H. Smith, 1981, p. 28) and the first term can be written as

$$
\mathrm{var}(\bar{y}) = \mathrm{var}\left( \frac{1}{N} \sum_{i=1}^{N} y_i \right) = \mathrm{var}\left( \frac{1}{N} \sum_{i=1}^{N} (a_0 + a_1 x_i + \epsilon_i) \right) = \mathrm{var}\left( \frac{1}{N} \sum_{i=1}^{N} \epsilon_i \right),
\tag{5.34}
$$

as only $\epsilon_i$ is a random variable. Thus,

$$
\mathrm{var}(\bar{y}) = \frac{1}{N^2} \sum_{i=1}^{N} \mathrm{var}(\epsilon_i) = \frac{\sigma^2}{N}.
\tag{5.35}
$$

Substituting (5.25) into the second term of (5.33), we obtain (5.26) for the variance of the intercept parameter $\hat{a}_0$.

### 5.1.3 Confidence Interval and Prediction Interval for the Response Variable ☐☺

Using the optimal estimates $\hat{a}_0$ and $\hat{a}_1$, we substitute (5.8) into the regression relation (5.1) to get

$$\hat{y} = \bar{y} + \hat{a}_1(x - \bar{x}).\qquad(5.36)$$

Since the regression relation has been built from $N$ observations, we can drop the subscript $i$ ($i = 1, \ldots, N$), as the regression relation (5.36) can be used to predict a value $\hat{y}$ from any given $x$. Since the original observations contained random noise, $\bar{y}$, $\hat{a}_1$ and $\hat{y}$ are random variables. We want to estimate the variance and the confidence interval of $\hat{y}$, the response variable predicted by the regression relation.

The variance of $\hat{y}$,

$$\mathrm{var}(\hat{y}) = \mathrm{var}(\bar{y}) + (x - \bar{x})^2 \, \mathrm{var}(\hat{a}_1),\qquad(5.37)$$

follows from (5.31), with $x$ and $\bar{x}$ treated as constants, and from $\mathrm{cov}(\bar{y}, \hat{a}_1) = 0$ (Draper and H. Smith, 1981, p. 28). Substituting in (5.35) and (5.25), and using the sample variance $s^2$ instead of $\sigma^2$, we have

$$\mathrm{var}(\hat{y}) = s^2 \left[ \frac{1}{N} + \frac{(x - \bar{x})^2}{\sum_i (x_i - \bar{x})^2} \right].\qquad(5.38)$$

If the error is Gaussian, the $(1 - \alpha) \times 100\%$ confidence interval for $\hat{y}$ is given by

$$\text{CI for } \hat{y} = \hat{y} \pm t(1 - \tfrac{1}{2}\alpha \,|\, N - 2)\, s \left[ \frac{1}{N} + \frac{(x - \bar{x})^2}{\sum_i (x_i - \bar{x})^2} \right]^{1/2},\qquad(5.39)$$

where $t(1 - \tfrac{1}{2}\alpha \,|\, N - 2)$ is the $(1 - \tfrac{1}{2}\alpha)$ quantile of the $t$-distribution (Section 3.15) with $N - 2$ degrees of freedom.

Individual $y$ values will also contain noise, that is, $y = \hat{y} + \epsilon$, so

$$\mathrm{var}(y) = s^2 \left[ 1 + \frac{1}{N} + \frac{(x - \bar{x})^2}{\sum_i (x_i - \bar{x})^2} \right].\qquad(5.40)$$

The $(1 - \alpha) \times 100\%$ prediction interval for $y$ is given by

$$\text{PI for } y = \hat{y} \pm t(1 - \tfrac{1}{2}\alpha \,|\, N - 2)\, s \left[ 1 + \frac{1}{N} + \frac{(x - \bar{x})^2}{\sum_i (x_i - \bar{x})^2} \right]^{1/2}.\qquad(5.41)$$

PI gives the spread of individual $y$ values, while CI gives the uncertainty in the conditional mean of $y$ given $x$ (as predicted by the regression relation). Thus, PI is usually much wider than CI. PI is useful for indicating the spread of future $y$ values when given future $x$ values.

As an example, consider SLR with daily weather data from Vancouver, BC, using visibility as the response variable and relative humidity as the predictor.[3] In Fig. 5.2, as $x$ moves further away from the mean position $\bar{x}$, the confidence interval gets wider. This behaviour results from the $(x - \bar{x})^2$ term in (5.39).

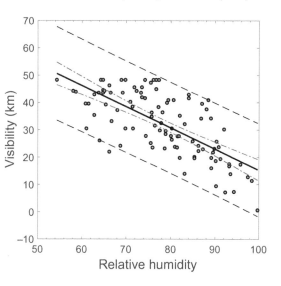

Figure 5.2      Simple linear regression with visibility as the response variable and relative humidity as predictor. The 95% predictor intervals are indicated by the dashed lines and the 95% confidence intervals by the dot-dashed lines. [Data source:          weatherstats.ca based on Environment and Climate   Change   Canada data.]

## 5.1.4   Serial Correlation   [B] ☺

In the SLR example in Fig. 5.2, the data were selected at least 20 days apart, thereby avoiding positive serial correlation (i.e. positive autocorrelation) in the weather data. For many datasets, ignoring positive serial correlation can lead to overestimation in the degrees of freedom, hence underestimation in the confidence intervals.

A common test for serial correlation in the regression error or residual $\epsilon$ is the *Durbin–Watson statistic* $d$ (Durbin and Watson, 1950; Durbin and Watson, 1951; Durbin and Watson, 1971; von Storch and Zwiers, 1999),

$$d = \frac{\sum_{i=1}^{N-1}(\epsilon_{i+1} - \epsilon_i)^2}{\sum_{i=1}^{N} \epsilon_i^2} , \qquad (5.42)$$

where the numerator is the sum of the squares of the difference between consecutive regression errors and the denominator is the SSE, with $0 \leq d \leq 4$ (see Exercise [5.4]). Positive serial correlation gives $d < 2$, while negative serial correlation gives $d > 2$.

The Durbin–Watson statistic is used to test the null hypothesis, that is, no serial correlation in the regression errors. Most statistical packages provide

---

[3] From the record containing 25 years of daily data (1993–2017), every 20th data point was chosen (i.e. using 457 out of the 9,131 available data points). From the 457 data points, 100 points were randomly selected for the SLR example.

two-tailed, right-tailed and left-tailed tests. A small $p$-value indicates significant serial correlation in the regression errors.

When there is significant positive serial correlation in the regression errors, the effective sample size $N_{\text{eff}}$ can be much smaller than the sample size $N$. Assuming an auto-regressive (AR) process of order 1, for large $N$, $N_{\text{eff}}$ can be approximated by (2.58) and (2.59). Replacing $N$ by $N_{\text{eff}}$ when estimating the sample variance $s^2$ in (5.27) gives

$$s^2_{\text{eff}} = \frac{1}{N_{\text{eff}} - 2} \sum_{i=1}^{N} (y_i - \hat{a}_0 - \hat{a}_1 x_i)^2. \tag{5.43}$$

In general, the denominator term is $N_{\text{eff}} - M$, where $M$ is the number of regression parameters. When estimating the CIs for regression parameters $\hat{a}_1$ and $\hat{a}_0$ in (5.28) and (5.29), $s_{\text{eff}}$ should be used instead of $s$ (Sallenger et al., 2012).

Similarly, for $\hat{y}$, the variance $\text{var}(\hat{y})$ in (5.38) should be adjusted by using $s_{\text{eff}}$ instead of $s$, that is,

$$\text{var}(\hat{y}) = s^2_{\text{eff}} \left[ \frac{1}{N} + \frac{(x - \bar{x})^2}{\sum_i (x_i - \bar{x})^2} \right]. \tag{5.44}$$

(Wilks, 2011, p. 232). For individual $y$ values, $\text{var}(y)$ in (5.40) should be adjusted by

$$\text{var}(y) = s^2 + s^2_{\text{eff}} \left[ \frac{1}{N} + \frac{(x - \bar{x})^2}{\sum_i (x_i - \bar{x})^2} \right]. \tag{5.45}$$

# 5.2  Multiple Linear Regression  $\boxed{\text{A}}$☺

Often one encounters situations where there are multiple predictors $x_j$, $(j = 1, \dots , m)$ for the response variable $y$. This type of multiple linear regression (MLR) has the form

$$y_i = a_0 + \sum_{j=1}^{m} x_{ij} a_j + \epsilon_i, \qquad i = 1, \dots , N, \tag{5.46}$$

with $N$ the sample size.[4]

In vector form,

$$\mathbf{y} = \mathbf{X}\mathbf{a} + \boldsymbol{\epsilon}, \tag{5.47}$$

where

$$\mathbf{y} = \begin{bmatrix} y_1 \\ \vdots \\ y_N \end{bmatrix}, \quad \mathbf{a} = \begin{bmatrix} a_0 \\ \vdots \\ a_m \end{bmatrix}, \quad \boldsymbol{\epsilon} = \begin{bmatrix} \epsilon_1 \\ \vdots \\ \epsilon_N \end{bmatrix}, \tag{5.48}$$

---

[4] The regression problem is linear because $y$ is a linear function of $a_j$. Therefore, $y$ can be a non-linear function of the predictor variables $x_j$ and the problem is still linear regression, for example, $y = a_0 + a_1 x_1 + a_2 x_2 + a_3 x_1^2 + a_4 x_1 x_2 + a_5 x_2^2$.

and the *design matrix* $\mathbf{X}$ is given by

$$\mathbf{X} = \begin{bmatrix} 1 & x_{11} & \cdots & x_{1m} \\ \vdots & \vdots & \ddots & \vdots \\ 1 & x_{N1} & \cdots & x_{Nm} \end{bmatrix}. \tag{5.49}$$

The sum of squared errors is given by

$$\mathrm{SSE} = \epsilon^{\mathrm{T}} \epsilon = (\mathbf{y} - \mathbf{X}\mathbf{a})^{\mathrm{T}} (\mathbf{y} - \mathbf{X}\mathbf{a}), \tag{5.50}$$

where the superscript T denotes the transpose. To minimize SSE with respect to $\mathbf{a}$, we compute the partial derivatives of SSE and set them to zero, that is,

$$\begin{bmatrix} \partial\mathrm{SSE}/\partial a_0 \\ \vdots \\ \partial\mathrm{SSE}/\partial a_m \end{bmatrix} = \begin{bmatrix} 0 \\ \vdots \\ 0 \end{bmatrix}, \tag{5.51}$$

yielding the *normal equations*,

$$\mathbf{X}^{\mathrm{T}}(\mathbf{y} - \mathbf{X}\mathbf{a}) = \mathbf{0}. \tag{5.52}$$

The normal equations (5.6) and (5.7) in the previous section on simple linear regression are from the special case $m = 1$ in (5.51).

With

$$\mathbf{X}^{\mathrm{T}}\mathbf{X}\,\mathbf{a} = \mathbf{X}^{\mathrm{T}}\mathbf{y}, \tag{5.53}$$

left multiplying the equation by the inverse matrix $(\mathbf{X}^{\mathrm{T}}\mathbf{X})^{-1}$ gives

$$(\mathbf{X}^{\mathrm{T}}\mathbf{X})^{-1}(\mathbf{X}^{\mathrm{T}}\mathbf{X})\,\mathbf{a} = (\mathbf{X}^{\mathrm{T}}\mathbf{X})^{-1}\mathbf{X}^{\mathrm{T}}\mathbf{y}, \tag{5.54}$$

yielding the optimal parameters (i.e. regression coefficients)

$$\hat{\mathbf{a}} = (\mathbf{X}^{\mathrm{T}}\mathbf{X})^{-1}\,\mathbf{X}^{\mathrm{T}}\mathbf{y} = \mathbf{X}^{+}\mathbf{y}, \tag{5.55}$$

where the matrix

$$\mathbf{X}^{+} \equiv (\mathbf{X}^{\mathrm{T}}\mathbf{X})^{-1}\,\mathbf{X}^{\mathrm{T}} \tag{5.56}$$

is called the *Moore–Penrose inverse* of the matrix $\mathbf{X}$. The estimator in (5.55) is called the *ordinary least squares* (OLS) estimator. The value of $\mathbf{y}$ predicted by the regression relation is

$$\hat{\mathbf{y}} = \mathbf{X}\hat{\mathbf{a}} = \mathbf{X}\mathbf{X}^{+}\mathbf{y}. \tag{5.57}$$

In some special situations, it is appropriate to force the regression line to pass through the origin, for example the response is proportional to the predictors. To perform MLR without the intercept term $a_0$, simply delete $a_0$ from $\mathbf{a}$ in (5.48) and delete the first column of '1' from the $\mathbf{X}$ matrix in (5.49).

While one can solve for the regression parameters using (5.55) directly, this approach is not usually recommended by numerical analysts, since computing the inverse of a matrix can have poor numerically stability (Forsythe et al., 1977).[5] Instead, the regression parameters are obtained from the linear least squares (a.k.a. ordinary least squares) solution of

$$\mathbf{Xa} = \mathbf{y}, \tag{5.58}$$

using either the QR decomposition method or the singular value decomposition (SVD) method (Hansen et al., 2013, chapter 5), which are both numerically more stable than using (5.55). In terms of operation count,[6] when $N \gg m$, both QR and SVD involve $O(2m^2N)$ multiplications, where $O(.)$ means 'of the order of'. Hence, if we double the sample size $N$, the computer time for MLR will approximately double, but if we double the number of predictors $m$, the computer time will quadruple.

## 5.2.1 Gauss–Markov Theorem  B☺

The ordinary least squares (OLS) estimator in (5.55) has some nice properties. The *Gauss–Markov theorem* (Mardia et al., 1979, section 6.6.1; Hastie, Tibshirani, et al., 2009, section 3.2.2) states that:
Assuming

- the noise has zero mean, $\mathrm{E}[\epsilon_i] = 0$,

- the covariance $\mathrm{cov}(\epsilon_i, \epsilon_j) = 0$ for all $i \neq j$,

- the variance $\mathrm{var}(\epsilon_i) = \sigma^2 < \infty$, for all $i$,

then the OLS estimator is the *best linear unbiased estimator* (BLUE), where unbiased means $\mathrm{E}[\hat{\mathbf{a}}] = \mathbf{a}$, 'best' means the variance of $\hat{\mathbf{a}}$ is the smallest among all linear unbiased estimators, and 'linear' means the estimator is a linear function of $\mathbf{y}$, as $\hat{\mathbf{a}}$ is a linear function of $\mathbf{y}$ in (5.55).

If the assumption $\mathrm{cov}(\epsilon_i, \epsilon_j) = 0$ for all $i \neq j$ is not satisfied, the Mahalanobis distance should be used instead of the Euclidean distance in the SSE in (5.50), giving rise to the generalized least squares (GLS) method described in Section 5.10.

---

[5] In numerical linear algebra, numerical stability means a small error in the input data leads to a small error in the final result (i.e. the regression parameters in our MLR problem). Having poor numerical stability means a small error in the input data can lead to a large error in the final result.

[6] Operation count involves counting the number of computer operations (additions and multiplications) in an algorithm. Usually, only multiplications are counted as multiplications are slower than additions (especially in early computers). For instance, multiplying an $m \times n$ matrix by an $n \times k$ matrix involves a total of $mnk$ multiplications.

## 5.2.2   Partition of Sums of Squares   $\boxed{\text{B}}$☺

Readers less keen on mathematical details can skip the rest of this paragraph, where we derive (5.14). In vector form, (5.13) is

$$\text{SST} = \|\mathbf{y} - \overline{\mathbf{y}}\|^2 = \|\mathbf{y} - \hat{\mathbf{y}} + \hat{\mathbf{y}} - \overline{\mathbf{y}}\|^2$$
$$= \|\mathbf{y} - \hat{\mathbf{y}}\|^2 + \|\hat{\mathbf{y}} - \overline{\mathbf{y}}\|^2 + 2(\mathbf{y} - \hat{\mathbf{y}})^{\text{T}}(\hat{\mathbf{y}} - \overline{\mathbf{y}})$$
$$= \|\boldsymbol{\epsilon}\|^2 + \|\hat{\mathbf{y}} - \overline{\mathbf{y}}\|^2 + 2\boldsymbol{\epsilon}^{\text{T}}(\hat{\mathbf{y}} - \overline{\mathbf{y}}). \tag{5.59}$$

Substituting in (5.15), (5.50) and (5.57), we get

$$\text{SST} = \text{SSE} + \text{SSR} + 2\boldsymbol{\epsilon}^{\text{T}}(\mathbf{X}\hat{\mathbf{a}} - \overline{\mathbf{y}})$$
$$= \text{SSE} + \text{SSR} + 2(\boldsymbol{\epsilon}^{\text{T}}\mathbf{X})\hat{\mathbf{a}} - 2\boldsymbol{\epsilon}^{\text{T}}\overline{\mathbf{y}}. \tag{5.60}$$

The term $2(\boldsymbol{\epsilon}^{\text{T}}\mathbf{X})\hat{\mathbf{a}}$ vanishes because

$$\boldsymbol{\epsilon}^{\text{T}}\mathbf{X} = (\mathbf{y} - \hat{\mathbf{y}})^{\text{T}}\mathbf{X} = \mathbf{y}^{\text{T}}\mathbf{X} - \hat{\mathbf{y}}^{\text{T}}\mathbf{X} = \mathbf{y}^{\text{T}}\mathbf{X} - \mathbf{X}^{\text{T}}\hat{\mathbf{y}}$$
$$= \mathbf{y}^{\text{T}}\mathbf{X} - \mathbf{X}^{\text{T}}(\mathbf{X}\mathbf{X}^{+}\mathbf{y}) = \mathbf{y}^{\text{T}}\mathbf{X} - \mathbf{X}^{\text{T}}\mathbf{X}(\mathbf{X}^{\text{T}}\mathbf{X})^{-1}\mathbf{X}^{\text{T}}\mathbf{y}$$
$$= \mathbf{y}^{\text{T}}\mathbf{X} - \mathbf{X}^{\text{T}}\mathbf{y} = 0, \tag{5.61}$$

where we have invoked (5.57) and (5.56). The final term in (5.60) also vanishes since

$$\boldsymbol{\epsilon}^{\text{T}}\overline{\mathbf{y}} = \sum_i \epsilon_i \overline{y} = \overline{y} \sum_i \epsilon_i = 0, \tag{5.62}$$

as $\sum_i \epsilon_i = 0$.

How well the regression fitted the data can be characterized by the *coefficient of determination $R^2$*,

$$R^2 = \frac{\text{SSR}}{\text{SST}} = 1 - \frac{\text{SSE}}{\text{SST}}, \tag{5.63}$$

where $R^2$ approaches 1 when the fit is very good. $R$ is called the *multiple correlation coefficient*, as it can be shown that it equals $\rho_{y\hat{y}}$, the Pearson correlation between the dependent variable $y$ and $\hat{y}$ from the regression relation (see Exercise $\boxed{5.5}$) The term 'multiple' refers to the use of multiple predictors in the regression relation.

## 5.2.3   Standardized Predictors   $\boxed{\text{A}}$☺

One problem with MLR is that the predictor variables may have very different magnitudes. For instance, if the first predictor $x_1$ is of order $10^3$ and the second predictor $x_2$ is of order $10^{-2}$, and if the two have comparable influence on $y$, that is, $a_1 x_1$ is of similar magnitude as $a_2 x_2$, then $|a_2| \sim 10^5 |a_1|$, that is, the magnitude of $a_j$ does not indicate the relative importance of the predictor $x_j$. One option is to standardize each predictor variable as in (2.29), so all the predictor variables have zero mean and unit standard deviation. The standardized predictors are then used in the MLR.

There are several advantages in standardizing the predictor variables before performing MLR:

- (a) The numerical stability of the MLR solution may be improved (Belsley et al., 1980, appendix 3B).

- (b) The magnitude of a regression parameter $a_j$ indicates the relative importance of a predictor $x_j$.

- (c) The $y$-intercept $a_0$ gives the value of $y$ when the predictors are set to zero. Some variables are meaningless if set to zero (e.g. the observed surface air pressure at a station may lie in the range [98, 104] hPa). When predictors are standardized, $a_0$ gives the value of $y$ when the predictors are all at their mean value (i.e. 0), which is much more meaningful.

- (d) Sometimes one may want to model quadratic behaviour, for example by having $x_2 = x_1^2$. If $x_1$ lies in the range [98, 104] hPa, without standardizing $x_1$, $x_2$ would behave more like a linear function of $x_1$ than a quadratic.

- (e) To avoid overfitting (i.e. fitting to noise in the data), methods such as ridge regression (Section 5.7) puts a penalty on the use of large $|a_j|$. It is essential to have standardized predictors when using this type of approach.

Let us illustrate MLR with daily weather data from Vancouver, BC, using visibility (in km) as the response variable and relative humidity, pressure and air temperature as the three predictors. With 25 years of daily data (1993–2017), we chose every 20th data point for the MLR, that is, using 457 out of the 9,131 available data points. With standardized predictors, the regression parameters $\hat{a}_0, \hat{a}_1, \hat{a}_2$ and $\hat{a}_3$ are 31.26, $-6.59, 1.63$ and 1.19, respectively, and the coefficient of determination $R^2 = 0.45$. The magnitudes of $\hat{a}_1, \hat{a}_2$ and $\hat{a}_3$ indicate relative humidity to be the most important predictor followed by pressure then temperature.

If we use only two predictors, namely relative humidity and pressure, the regression parameters $\hat{a}_0, \hat{a}_1$ and $\hat{a}_2$ are 31.26, $-7.03$ and 1.54, respectively, with $R^2 = 0.44$ (Fig. 5.3).

## 5.2.4 Analysis of Variance (ANOVA)   B☺

Analysis of variance (ANOVA) is a statistical method in which the variability in a dataset is divided into distinct components. The term ANOVA was first introduced in R. A. Fisher (1918). Most statistical packages can provide ANOVA on the MLR results. For our MLR example on visibility, the ANOVA statistics are given in Table 5.1.

The ANOVA table has a row for each predictor (excluding the constant term):

- SumSq gives $\text{SSR}_j$, the SSR accounted for by the individual predictor $x_j$, that is,

$$\text{SSR}_j = \text{SSR}_{\text{excl}.j} - \text{SSR}, \qquad (5.64)$$

where $\text{SSR}_{\text{excl}.j}$ is the SSR from the regression using all predictors but excluding the predictor $x_j$.

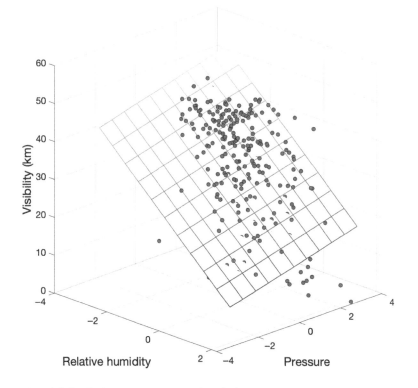

Figure 5.3 Multiple linear regression for daily weather variables in Vancouver, BC, where visibility is the response variable $y$ and relative humility and pressure are the two standardized predictors. The MLR predicted values $\hat{y}_i$ lie on a two-dimensional plane as indicated by the grid, with the observed $y_i$ values indicated by the circles. The grid is tilted downward as relative humidity increases and upward as pressure increases, as expected from the regression parameters $\hat{a}_1 = -7.03$ and $\hat{a}_2 = 1.54$. The vertical distance between a data point $(\mathbf{x}_i, y_i)$ and its projected value $(\mathbf{x}_i, \hat{y}_i)$ on the plane is the error $\epsilon_i$. When there are $m$ predictors in the regression relation, $(\mathbf{x}_i, \hat{y}_i)$ lies on an $m$-dimensional hyperplane. [Data source: weatherstats.ca based on Environment and Climate Change Canada data.]

- DF (degrees of freedom) is 1 for each term in the regression model and $N - M$ for the error term, where $M$ is the number of regression parameters (including $a_0$) in the model.

- MeanSq (mean square or mean sum of squares) is defined by MeanSq = SumSq/DF, with the bottom row giving the mean squared error (MSE) as being 67.3.

- The statistic $F = $ MeanSq/MSE is used to test the null hypothesis that the corresponding regression parameter is zero. When the null hypothesis

Table 5.1 Analysis of variance (ANOVA) in the multiple linear regression with visibility as the response variable and standardized relative humidity ($x_1$) and standardized pressure ($x_2$) as the predictors. The five columns are (1) SumSq, the sum of squares accounted for by the term, (2) DF, the degrees of freedom, (3) MeanSq, the mean square, (4) the $F$ statistic and (5) the $p$-value.

|       | SumSq | DF  | MeanSq | F     | $p$-value            |
|-------|-------|-----|--------|-------|----------------------|
| $x_1$ | 22423 | 1   | 22423  | 333.1 | $3.26 \times 10^{-56}$ |
| $x_2$ | 1074  | 1   | 1074   | 16.0  | $7.54 \times 10^{-5}$  |
| Error | 30562 | 454 | 67.3   |       |                      |

is true, $F$ follows the $F$ distribution (Section 4.6), with degrees of freedom $\nu_1 = 1$ and $\nu_2 = N - M = 454$ in this example.

- The corresponding $p$-values for the $F$ statistic for both predictors are very small, much below 0.05, meaning the predictors are both significant. If we reduce the sample size from 457 to 100 by randomly picking 100 data points from the 457 points, the $p$-value for the two predictors becomes $6.15 \times 10^{-16}$ and 0.123, respectively, that is, the second predictor is not significant at the 5% level. The $p$-value can be helpful for rejecting insignificant predictors, thereby reducing the number of predictors for the regression relation.

The process of choosing good predictors for MLR can be quite unstable, as pointed out by (Breiman, 2001b, p. 206). Suppose there are 10 predictor variables and there is correlation among them. With one subsample, the best predictors chosen for MLR may be predictors 2, 3 and 6, while for another subsample, the best predictors maybe 1, 4 and 9.

## 5.2.5 Confidence and Prediction Intervals  B☺

The symmetric covariance matrix of the regression errors is

$$
\mathbf{V}(\boldsymbol{\epsilon}) \;\equiv\; \mathrm{E}(\boldsymbol{\epsilon}\boldsymbol{\epsilon}^{\mathrm{T}}) =
\begin{bmatrix}
\mathrm{var}(\epsilon_1) & \mathrm{cov}(\epsilon_1, \epsilon_2) & \cdots & \mathrm{cov}(\epsilon_1, \epsilon_N) \\
\mathrm{cov}(\epsilon_2, \epsilon_1) & \mathrm{var}(\epsilon_2) & \cdots & \mathrm{cov}(\epsilon_2, \epsilon_N) \\
\vdots & \vdots & \ddots & \vdots \\
\mathrm{cov}(\epsilon_N, \epsilon_1) & \mathrm{cov}(\epsilon_N, \epsilon_2) & \cdots & \mathrm{var}(\epsilon_N)
\end{bmatrix}
$$

$$
= \; \sigma^2 \mathbf{I}, \tag{5.65}
$$

where $\mathrm{var}(\epsilon_i) = \sigma^2$ for all $i = 1, \ldots, N$ and $\mathrm{cov}(\epsilon_i, \epsilon_k) = 0$ for $i \neq k$ and $\mathbf{I}$ is the identity matrix.

The regression model (5.47) can be viewed as a random vector $\mathbf{y}$ with mean $\mathrm{E}(\mathbf{y}) = \mathbf{X}\mathbf{a}$. The covariance matrix for $\mathbf{y}$ is

$$
\mathbf{V}(\mathbf{y}) = \mathrm{E}\left([\mathbf{y} - \mathrm{E}(\mathbf{y})][\mathbf{y} - \mathrm{E}(\mathbf{y})]^{\mathrm{T}}\right) = \mathrm{E}(\boldsymbol{\epsilon}\boldsymbol{\epsilon}^{\mathrm{T}}) = \sigma^2 \mathbf{I}. \tag{5.66}
$$

For the covariance matrix of the regression parameters $\hat{\mathbf{a}}$, (5.55) gives

$$\mathbf{V}(\hat{\mathbf{a}}) = \mathbf{V}(\mathbf{X}^+\mathbf{y}) = \mathbf{X}^+ \mathbf{V}(\mathbf{y}) \mathbf{X}^{+\mathrm{T}}, \tag{5.67}$$

as $\mathbf{X}^+$ is treated as a matrix of constants when computing the covariance matrix. Thus,

$$\begin{aligned}
\mathbf{V}(\hat{\mathbf{a}}) &= \mathbf{X}^+ \sigma^2 \mathbf{I} \mathbf{X}^{+\mathrm{T}} = \sigma^2 \mathbf{X}^+ \mathbf{X}^{+\mathrm{T}} \\
&= \sigma^2 (\mathbf{X}^{\mathrm{T}}\mathbf{X})^{-1} \mathbf{X}^{\mathrm{T}} [(\mathbf{X}^{\mathrm{T}}\mathbf{X})^{-1} \mathbf{X}^{\mathrm{T}}]^{\mathrm{T}} \\
&= \sigma^2 (\mathbf{X}^{\mathrm{T}}\mathbf{X})^{-1} \mathbf{X}^{\mathrm{T}} \mathbf{X} (\mathbf{X}^{\mathrm{T}}\mathbf{X})^{-1},
\end{aligned} \tag{5.68}$$

as $(\mathbf{X}^{\mathrm{T}}\mathbf{X})^{-1}$ is symmetric. Therefore,

$$\mathbf{V}(\hat{\mathbf{a}}) = \sigma^2 (\mathbf{X}^{\mathrm{T}}\mathbf{X})^{-1}. \tag{5.69}$$

As $\sigma^2$ is generally not known, the sample noise variance $s^2$ is used instead in (5.69), where

$$s^2 = \frac{1}{N - M} \, \boldsymbol{\epsilon}^{\mathrm{T}} \boldsymbol{\epsilon}, \tag{5.70}$$

with $M$ the number of regression parameters.

If the error is Gaussian, the $(1 - \alpha) \times 100\%$ confidence interval for the regression parameter $\hat{a}_j$ $(j = 0, \ldots, m)$ is given by

$$\text{CI for } \hat{a}_j = \hat{a}_j \pm t(1 - \tfrac{1}{2}\alpha \,|\, N - M) \, s\sqrt{c_{j'j'}}, \tag{5.71}$$

where $t(1 - \tfrac{1}{2}\alpha \,|\, N - M)$ is the $(1 - \tfrac{1}{2}\alpha)$ quantile of the $t$-distribution (Section 3.15) with $N - M$ degrees of freedom $(M = m + 1)$ and $c_{j'j'}$ $(j' = j + 1)$ is the $j'$th diagonal element of $(\mathbf{X}^{\mathrm{T}}\mathbf{X})^{-1}$.[7,8]

From (5.57), the variance of the regression output $\hat{y}_i$ is given by

$$\text{var}(\hat{y}_i) = \text{var}(\mathbf{x}_i \, \hat{\mathbf{a}}), \tag{5.72}$$

where $\mathbf{x}_i^{\mathrm{T}}$ is the $i$th row of the design matrix $\mathbf{X}$ in (5.49). Thus,

$$\text{var}(\hat{y}_i) = \mathbf{x}_i^{\mathrm{T}} \mathbf{V}(\hat{\mathbf{a}}) \, \mathbf{x}_i = \sigma^2 \mathbf{x}_i^{\mathrm{T}} (\mathbf{X}^{\mathrm{T}}\mathbf{X})^{-1} \mathbf{x}_i, \tag{5.73}$$

upon substituting in (5.69).

Assuming Gaussian error and using $s$ in place of $\sigma$, the $(1 - \alpha) \times 100\%$ confidence interval for $\hat{y}$ is given by

$$\text{CI for } \hat{y}_i = \hat{y}_i \pm t(1 - \tfrac{1}{2}\alpha \,|\, N - M) \, s\sqrt{\mathbf{x}_i^{\mathrm{T}} (\mathbf{X}^{\mathrm{T}}\mathbf{X})^{-1} \mathbf{x}_i}. \tag{5.74}$$

---

[7] The index $j' = j+1$ is necessary since $j = 0, \ldots, m$ (as $a_j$ represents $a_0, \ldots, a_m$), whereas the matrix $(\mathbf{X}^{\mathrm{T}}\mathbf{X})^{-1}$ has row and column indices running from $j' = 1, \ldots, M$, $(M = m + 1)$.

[8] The CIs in (5.71) would lead to a rectangular *joint confidence region* in the multi-dimensional regression parameter space, which is not realistic. The joint confidence region is an ellipsoid even if the regression parameters are independent. If there is correlation between the regression parameters, the axes of the ellipsoid will not be aligned with the $a_j$ axes. See von Storch and Zwiers (1999, section 8.4.7) for details.

With $y_i = \hat{y}_i + \epsilon_i$, $\mathrm{var}(y_i)$ is given by

$$\mathrm{var}(y_i) = \sigma^2 \left[1 + \mathbf{x}_i^{\mathrm{T}} \left(\mathbf{X}^{\mathrm{T}}\mathbf{X}\right)^{-1} \mathbf{x}_i\right]. \tag{5.75}$$

Dropping the subscript $i$ from $\mathbf{x}_i$ and $y_i$, the prediction interval for $y$ is given by

$$\text{PI for } y = \hat{y} \pm t(1 - \tfrac{1}{2}\alpha \,|\, N - M)\, s\sqrt{1 + \mathbf{x}^{\mathrm{T}} \left(\mathbf{X}^{\mathrm{T}}\mathbf{X}\right)^{-1} \mathbf{x}}. \tag{5.76}$$

Thus, given a future $\mathbf{x}$, one can obtain a PI for the future $y$ value.

## 5.3 Multivariate Linear Regression  B☺

In the previous section on multiple linear regression (MLR), the response is a scalar variable $y$. When the response is an $l$-dimensional vector $\mathbf{y}$, the method is called *multivariate linear regression* or *general linear model*.

In multiple linear regression (5.46),

$$y_i = a_0 + \sum_{j=1}^{m} x_{ij}\, a_j + \epsilon_i, \qquad i = 1, \ldots, N,$$

with $N$ the sample size. In multivariate linear regression,

$$y_{ik} = a_{0k} + \sum_{j=1}^{m} x_{ij}\, a_{jk} + \epsilon_{ik}, \qquad i = 1, \ldots, N, \quad k = 1, \ldots, l. \tag{5.77}$$

Equation (5.77) can be expressed in vector form (Mardia et al., 1979),

$$\mathbf{Y} = \mathbf{X}\mathbf{A} + \mathbf{E}, \tag{5.78}$$

where

$$\mathbf{Y} = \begin{bmatrix} y_{11} & \cdots & y_{1l} \\ \vdots & \ddots & \vdots \\ y_{N1} & \cdots & y_{Nl} \end{bmatrix}, \quad \mathbf{X} = \begin{bmatrix} 1 & x_{11} & \cdots & x_{1m} \\ \vdots & \vdots & \ddots & \vdots \\ 1 & x_{N1} & \cdots & x_{Nm} \end{bmatrix}, \tag{5.79}$$

$$\mathbf{A} = \begin{bmatrix} a_{01} & \cdots & a_{0l} \\ \vdots & \ddots & \vdots \\ a_{m1} & \cdots & a_{ml} \end{bmatrix} \quad \text{and} \quad \mathbf{E} = \begin{bmatrix} \epsilon_{11} & \cdots & \epsilon_{1l} \\ \vdots & \ddots & \vdots \\ \epsilon_{N1} & \cdots & \epsilon_{Nl} \end{bmatrix}, \tag{5.80}$$

where $\mathbf{Y}$ is the response matrix, $\mathbf{X}$ the design matrix, $\mathbf{A}$ the matrix of regression parameters or coefficients and $\mathbf{E}$ the error or residual matrix.

Similar to the derivation of (5.50)–(5.55), we obtain the solution for the regression parameters

$$\hat{\mathbf{A}} = \mathbf{X}^{+}\mathbf{Y}, \tag{5.81}$$

with the Moore–Penrose inverse

$$\mathbf{X}^{+} \equiv \left(\mathbf{X}^{\mathrm{T}}\mathbf{X}\right)^{-1} \mathbf{X}^{\mathrm{T}}. \tag{5.82}$$

The value of $\mathbf{Y}$ predicted by the regression relation is

$$\hat{\mathbf{Y}} = \mathbf{X}\hat{\mathbf{A}} = \mathbf{X}\mathbf{X}^+\mathbf{Y}. \tag{5.83}$$

The solutions for MLR, that is, (5.55) and (5.57), is simply a special case of (5.81) and (5.83), where the $l$-dimensional vector $\mathbf{y}$ is reduced to a scalar $y$ (by setting $l = 1$). When given a vector $\mathbf{y}$, instead of using multivariate linear regression, we can simply solve its components $y_1, \ldots, y_l$ individually using the MLR solution. In other words, if we solve the individual components one at a time using MLR, we get the same result from using multivariate linear regression.

It has been pointed out that if the components of $\mathbf{y}$ are correlated, then using multivariate linear regression or using MLR for individual components of $\mathbf{y}$ fails to utilize the information contained in the correlation between the components. Methods have been developed to utilize the correlation between the components of $\mathbf{y}$ (Breiman and Friedman, 1997; Hastie, Tibshirani, et al., 2009, section 3.7). In Section 9.4, canonical correlation analysis is shown to also utilize the correlation information among the response variables.

## 5.4   Online Learning with Linear Regression $\boxed{\text{C}}$☹

In ML, online machine learning or *online learning* is an approach used when data become available in a sequential order. The model is updated repeatedly as new data become available, as opposed to the traditional *batch learning* approach, which generates the model by learning from the entire training dataset at once. Once the new data have been used by an online learning method to update the model, the data can be discarded as they are not needed for future model updates.

When given a gigantic dataset for which one lacks the computer resources to train the batch model over the entire dataset, a 'divide and conquer' approach is used, which trains manageable portions of the dataset repeatedly by an online learning method.

In the age of the Internet and Big Data, it is increasingly common to encounter situations where copious amounts of new data arrive continually. If a multivariate linear regression model has been built, one would like to update the model as new data arrive. Using the batch algorithm (5.81) to update the model would be very inefficient since the model needs to be retrained with all available data instead of only the new data. Fortunately, the *recursive least squares* algorithm (Chong and Zak, 2013) allows the multivariate linear regression model to be updated efficiently using only the new data.

From Section 5.3, the multivariate linear regression problem is given by (5.78), (5.79) and (5.80), with $\hat{\mathbf{A}}$, the solution for the matrix of regression parameters, given by (5.81) and (5.82), that is,

$$\hat{\mathbf{A}} = \mathbf{X}^+\mathbf{Y} = (\mathbf{X}^{\mathrm{T}}\mathbf{X})^{-1}\mathbf{X}^{\mathrm{T}}\mathbf{Y}, \tag{5.84}$$

$$\hat{\mathbf{A}} = \mathbf{K}^{-1}\mathbf{X}^{\mathsf{T}}\mathbf{Y}, \tag{5.85}$$

with

$$\mathbf{K} \equiv \mathbf{X}^{\mathsf{T}}\mathbf{X}, \tag{5.86}$$

being the covariance matrix and $\mathbf{K}^{-1}$ the inverse covariance matrix. The value of $\mathbf{Y}$ predicted by the regression relation is

$$\hat{\mathbf{Y}} = \mathbf{X}\hat{\mathbf{A}} = \mathbf{X}\mathbf{K}^{-1}\mathbf{X}^{\mathsf{T}}\mathbf{Y}. \tag{5.87}$$

To derive the recursive least squares method, we assume the dataset with $N$ observations is composed of two parts, with $N_0$ and $N_1$ observations, respectively. The response matrix $\mathbf{Y}$ and the design matrix $\mathbf{X}$ can be partitioned into

$$\mathbf{Y} = \begin{bmatrix} \mathbf{Y}_0 \\ \mathbf{Y}_1 \end{bmatrix} \quad \text{and} \quad \mathbf{X} = \begin{bmatrix} \mathbf{X}_0 \\ \mathbf{X}_1 \end{bmatrix}, \tag{5.88}$$

where $\mathbf{Y}_0$ and $\mathbf{Y}_1$ are the first $N_0$ rows and final $N_1$ rows of $\mathbf{Y}$, respectively, while similarly, $\mathbf{X}_0$ and $\mathbf{X}_1$ are the first $N_0$ and final $N_1$ rows of $\mathbf{X}$.

If initially given only the first $N_0$ data points, the solution is

$$\hat{\mathbf{A}}^{(0)} = \mathbf{K}_0^{-1}\mathbf{X}_0^{\mathsf{T}}\mathbf{Y}_0, \tag{5.89}$$

where

$$\mathbf{K}_0 \equiv \mathbf{X}_0^{\mathsf{T}}\mathbf{X}_0, \tag{5.90}$$

Next, $\mathbf{X}_1$ and $\mathbf{Y}_1$ become available, so (5.85) gives the next estimate of $\hat{\mathbf{A}}$ to be

$$\hat{\mathbf{A}}^{(1)} = \mathbf{K}_1^{-1} \begin{bmatrix} \mathbf{X}_0 \\ \mathbf{X}_1 \end{bmatrix}^{\mathsf{T}} \begin{bmatrix} \mathbf{Y}_0 \\ \mathbf{Y}_1 \end{bmatrix}, \tag{5.91}$$

where

$$\mathbf{K}_1 = \begin{bmatrix} \mathbf{X}_0 \\ \mathbf{X}_1 \end{bmatrix}^{\mathsf{T}} \begin{bmatrix} \mathbf{X}_0 \\ \mathbf{X}_1 \end{bmatrix}. \tag{5.92}$$

For online learning, we need to express $\hat{\mathbf{A}}^{(1)}$ as a function of $\hat{\mathbf{A}}^{(0)}$, $\mathbf{X}_1$, $\mathbf{Y}_1$ and $\mathbf{K}_1$, but not a function of the earlier dataset $\{\mathbf{X}_0, \mathbf{Y}_0\}$. First, $\mathbf{K}_1$ can be expressed as

$$\mathbf{K}_1 = [\mathbf{X}_0^{\mathsf{T}} \; \mathbf{X}_1^{\mathsf{T}}] \begin{bmatrix} \mathbf{X}_0 \\ \mathbf{X}_1 \end{bmatrix} \tag{5.93}$$

$$= \mathbf{K}_0 + \mathbf{X}_1^{\mathsf{T}}\mathbf{X}_1. \tag{5.94}$$

Next, in (5.91),

$$\begin{bmatrix} \mathbf{X}_0 \\ \mathbf{X}_1 \end{bmatrix}^{\mathsf{T}} \begin{bmatrix} \mathbf{Y}_0 \\ \mathbf{Y}_1 \end{bmatrix} = \mathbf{X}_0^{\mathsf{T}}\mathbf{Y}_0 + \mathbf{X}_1^{\mathsf{T}}\mathbf{Y}_1$$

$$= (\mathbf{K}_0\mathbf{K}_0^{-1})\mathbf{X}_0^{\mathsf{T}}\mathbf{Y}_0 + \mathbf{X}_1^{\mathsf{T}}\mathbf{Y}_1$$

$$= \mathbf{K}_0\hat{\mathbf{A}}^{(0)} + \mathbf{X}_1^{\mathsf{T}}\mathbf{Y}_1$$

$$= (\mathbf{K}_1 - \mathbf{X}_1^{\mathsf{T}}\mathbf{X}_1)\,\hat{\mathbf{A}}^{(0)} + \mathbf{X}_1^{\mathsf{T}}\mathbf{Y}_1, \tag{5.95}$$

upon invoking (5.89) and (5.94). Substituting (5.95) into (5.91), we get

$$\hat{\mathbf{A}}^{(1)} = \mathbf{K}_1^{-1} \left( \mathbf{K}_1 \hat{\mathbf{A}}^{(0)} - \mathbf{X}_1^T \mathbf{X}_1 \hat{\mathbf{A}}^{(0)} + \mathbf{X}_1^T \mathbf{Y}_1 \right)$$

$$= \hat{\mathbf{A}}^{(0)} + \mathbf{K}_1^{-1} \mathbf{X}_1^T \left( \mathbf{Y}_1 - \mathbf{X}_1 \hat{\mathbf{A}}^{(0)} \right). \tag{5.96}$$

We can now generalize the iteration of (5.94) and (5.96) to step $k$ ($k \geq 1$):

$$\mathbf{K}_{k+1} = \mathbf{K}_k + \mathbf{X}_{k+1}^T \mathbf{X}_{k+1}, \tag{5.97}$$

$$\hat{\mathbf{A}}^{(k+1)} = \hat{\mathbf{A}}^{(k)} + \mathbf{K}_{k+1}^{-1} \mathbf{X}_{k+1}^T \left( \mathbf{Y}_{k+1} - \mathbf{X}_{k+1} \hat{\mathbf{A}}^{(k)} \right). \tag{5.98}$$

While these iterative equations allow us to update our regression model as new data $\{\mathbf{X}_{k+1}, \mathbf{Y}_{k+1}\}$ become available, they are not the most efficient. The reason is that updating $\hat{\mathbf{A}}^{(k+1)}$ in (5.98) requires $\mathbf{K}_{k+1}^{-1}$, and obtaining $\mathbf{K}_{k+1}^{-1}$ by computing the inverse of $\mathbf{K}_{k+1}$ is not very efficient. Therefore, instead of updating $\mathbf{K}_{k+1}$ using (5.97), we will try to derive an equation for updating $\mathbf{K}_{k+1}^{-1}$ iteratively.

From (5.97),

$$\mathbf{K}_{k+1}^{-1} = \left( \mathbf{K}_k + \mathbf{X}_{k+1}^T \mathbf{X}_{k+1} \right)^{-1}. \tag{5.99}$$

Using the Woodbury matrix identity,[9] we have

$$\mathbf{K}_{k+1}^{-1} = \mathbf{K}_k^{-1} - \mathbf{K}_k^{-1} \mathbf{X}_{k+1}^T \left( \mathbf{I} + \mathbf{X}_{k+1} \mathbf{K}_k^{-1} \mathbf{X}_{k+1}^T \right)^{-1} \mathbf{X}_{k+1} \mathbf{K}_k^{-1}, \tag{5.100}$$

with $\mathbf{I}$ being the identity matrix.

Letting

$$\mathbf{P}_{k+1} \equiv \mathbf{K}_{k+1}^{-1}, \tag{5.101}$$

the equations for iteratively updating the inverse covariance matrix $\mathbf{P}_{k+1}$ and the regression parameter matrix $\hat{\mathbf{A}}^{(k+1)}$ are:

$$\mathbf{P}_{k+1} = \mathbf{P}_k - \mathbf{P}_k \mathbf{X}_{k+1}^T \left( \mathbf{I} + \mathbf{X}_{k+1} \mathbf{P}_k \mathbf{X}_{k+1}^T \right)^{-1} \mathbf{X}_{k+1} \mathbf{P}_k, \tag{5.102}$$

$$\hat{\mathbf{A}}^{(k+1)} = \hat{\mathbf{A}}^{(k)} + \mathbf{P}_{k+1} \mathbf{X}_{k+1}^T \left( \mathbf{Y}_{k+1} - \mathbf{X}_{k+1} \hat{\mathbf{A}}^{(k)} \right). \tag{5.103}$$

## 5.5    Circular and Categorical Data  $\boxed{\text{A}}$☺

In regression problems, one sometimes encounters predictors that are circular variables or categorical variables. Circular data can be used to indicate direction

---

[9] Let $\mathbf{K}$ be a non-singular matrix and $\mathbf{U}$ and $\mathbf{V}$ be matrices such that $\mathbf{I} + \mathbf{V}\mathbf{K}^{-1}\mathbf{U}$ is non-singular. The *Woodbury matrix identity* says that $\mathbf{K} + \mathbf{U}\mathbf{V}$ is non-singular and

$$(\mathbf{K} + \mathbf{U}\mathbf{V})^{-1} = \mathbf{K}^{-1} - (\mathbf{K}^{-1}\mathbf{U})(\mathbf{I} + \mathbf{V}\mathbf{K}^{-1}\mathbf{U})^{-1}(\mathbf{V}\mathbf{K}^{-1}).$$

The identity can be proved by showing that the right hand side of the above equation right multiplied by $(\mathbf{K} + \mathbf{U}\mathbf{V})$ equals the identity matrix $\mathbf{I}$.

(e.g. wind direction) or cyclical time (e.g. hour of day). If one directly inputs wind direction data into a regression model, the regression model interprets $360°$ as being very different from $0°$; thus, it cannot correctly use the circular data in the regression.

Categorical variables can be used to separate the data into groups or categories. For instance, satellite data of the Earth's surface may be categorized into observations of water, land, snow or ice. If one naively sets 'water' = 1, 'land' = 2, 'snow' = 3 and 'ice' = 4, and inputs the data into a regression model, the model interprets ice as being three times as different from water as land is from water, that is, one has unintentionally supplied nonsense information to the model.

Circular data can be used in correlation and regression studies (N. I. Fisher, 1995; A. Lee, 2010). For simple LR having a circular variable $\theta$ of period $T$ as predictor, the model output can be written as

$$\hat{y}_i = a_0 + b_0 \cos\left(2\pi(\theta_i - \theta_0)/T\right), \quad i = 1, \dots, N, \tag{5.104}$$

with $N$ observations and regression parameters $a_0$, $b_0$ and $\theta_0$ (N. I. Fisher, 1995, section 6.2.1). This can be rewritten as

$$\hat{y}_i = a_0 + a \cos(2\pi\theta_i/T) + b \sin(2\pi\theta_i/T), \tag{5.105}$$

with regression parameters $a_0$, $a$ and $b$. This is the form we can easily apply the linear regression solution (5.57), as we can view $\cos(2\pi\theta_i/T)$ and $\sin(2\pi\theta_i/T)$ as simply two predictors $x_1$ and $x_2$ in MLR.

In the latitude–longitude $(\phi, \lambda)$ coordinate system, if the area of study is global ($-90° \leq \phi \leq 90°$ and $-180° \leq \lambda \leq 180°$), the longitude $\lambda$ (but not the latitude $\phi$) must be treated as a circular variable.

In general, for regression problems, any circular variable with period $T$ can be input into the regression model as *two* predictor variables $\cos(2\pi\theta_i/T)$ and $\sin(2\pi\theta_i/T)$. This applies to not only linear regression models but also neural network models, and so on. It is not uncommon to encounter published papers using only one of $\cos(2\pi\theta_i/T)$ or $\sin(2\pi\theta_i/T)$, which is not exactly right, as two sinusoidals are needed to specify the phase of a circular variable. For instance, if one wants to input the seasonal cycle but only a cosine term is included as predictor, then one can indicate winter conditions (with $\cos 0 = 1$) and summer ($\cos \pi = -1$), but spring ($\cos(\pi/2) = 0$) and autumn ($\cos(3\pi/2) = 0$) conditions cannot be distinguished by the model as the cosine term has the same zero value for both.

For categorical data, the standard approach is to use *one-hot encoding* (a.k.a. one-of-$C$ encoding). For our example of satellite observations of the Earth's surface being grouped into water, land, snow or ice, we would introduce four binary variables. The first binary variable is 1 if the surface is water and 0 otherwise, the second variable is 1 if land and 0 otherwise, and so on. Writing the four binary variables together as a vector, $[1\,0\,0\,0]$ indicates water, $[0\,1\,0\,0]$

land, [0 0 1 0] snow and [0 0 0 1] represents ice. For the general situation of $C$ categories, one would introduce $C$ binary variables, where the variables take on the value 1 for their particular category and 0 for all other categories. The $C$ binary variables are inputs into the regression model (again, regression models can include neural network models, etc.).

## 5.6  Predictor Selection  $\boxed{C}$ ☺

A major problem with multiple linear regression (MLR) is that often a large number of predictors are available, but only a fraction of them are actually significant. If all predictors are used in building an MLR model, one often *overfits* the data, that is, too many parameters are used in the model so that one is fitting not only to the signal but also to the noise in the data. While the fit to the data may appear very impressive from the small SSE, such overfitted MLR models generally perform poorly when used to make predictions based on new predictor data. Automatic procedures, such as stepwise multiple regression (Draper and H. Smith, 1981), have been devised to eliminate insignificant predictors, thereby avoiding an overfitted MLR model.

Predictor selection (i.e. feature selection in ML) is an issue that divides the statistics community from the ML community. Statisticians like models that are more interpretable; thus, using predictor selection to delete some predictors certainly helps to make the model simpler and more interpretable. However, there are disadvantages in using predictor selection. Suppose among a number of predictors, predictors $x_3$ and $x_7$ are well correlated. For a particular sample, the predictor selection method chose $x_3$ and deleted $x_7$. For a different sample, the method chose $x_7$ and deleted $x_3$. Thus, predictor selection is not always reproducible due to sampling noise. The other problem with discarding predictors is that one is discarding information – since $x_7$ is not the same variable as $x_3$, it is providing some information that is not available from $x_3$. The ML community does not consider discarding information a good practice; instead, this community focuses on developing good *regularization*[10] techniques to prevent overfitting.

In statistics, *stepwise regression* (or stepwise multiple regression) (Draper and H. Smith, 1981; von Storch and Zwiers, 1999) has been very widely used for predictor selection. We will first describe the method, then discuss why it has been strongly criticized in recent times. The initial step involves fitting a constant $y = a_0$ to the data. Next, find the predictor (among $m$ predictors) that gives the best regression (based on the coefficient of determination $R^2$) that passes a significance test (e.g. the $F$ test), so the regression relation now contains the constant $a_0$ and one predictor. In the next step, look among the remaining $m - 1$ predictors for the best predictor based on $R^2$ that passes the significance test, and add that predictor to the regression equation, so there are now two predictors used. The algorithm then checks if some of the predictors

---

[10] Regularization refers to a variety of techniques used in models to solve ill-posed problems or to prevent overfitting.

can be deleted, as sometimes predictors added at an earlier stage are no longer significant after the addition of new predictors. This iterative process of adding and deleting predictors is repeated until no further changes can be made to the regression equation.

The use of this very popular method has been seriously challenged (Derksen and Keselman, 1992; Flom and Cassell, 2007; Harrell, 2015):

> Stepwise variable selection has been a very popular technique for many years, but if this procedure had just been proposed as a statistical method, it would most likely be rejected because it violates every principle of statistical estimation and hypothesis testing. [(Harrell, 2015, p. 67)]

Harrell (2015, p. 68) compiled a long list of problems with stepwise regression, his main points being:

- The method gives $R^2$ values that are biased to the high side.

- The $F$ and $\chi^2$ tests quoted next to each predictor variable on output do not have the claimed distributions.

- The $F$ test is intended to test a pre-specified hypothesis, not to be used repeatedly as done in the method, whereby one could obtain many 'significant' results by chance.

- The standard errors of the regression parameters are biased low.

- The $p$-values are too small.

- The absolute value of the regression parameters are biased high.

- When dealing with well correlated predictors (i.e. collinearity), the method selects predictors arbitrarily.

Various improvements to stepwise regression have been proposed for predictor selection, but none can resolve all the above problems. Unlike predictor selection methods that set the regression parameters to zero for the predictors they rejected, *shrinkage methods* only shrink the regression parameters towards zero.

> Variable selection does not compete well with shrinkage methods that simultaneously model all potential predictors. [Harrell (2015, p. 69)]

Thus, shrinkage methods often offer a better alternative to the unreliable predictor selection methods. In the next two sections, we will study two common shrinkage methods, namely ridge regression (Section 5.7) and lasso (Section 5.8).

Let us compare MLR and stepwise regression (plus ridge regression and lasso) when working with datasets with relatively small sample size so there is a risk of overfitting. From the daily weather data from Vancouver, BC, visibility is chosen as the response variable and relative humidity, pressure, air temperature and wind speed as the four predictors. With 25 years of daily data (1993–2017), we chose every 10th data point for the MLR, that is, using 914 out of the 9,131 available data points. In Experiment A, for each trial, $N = 100$ data points were chosen randomly as training data (from the pool of 914 data points), with the data not chosen for training used later as test data. MLR, stepwise regression, ridge regression and lasso were trained and their prediction RMSE on the test data were recorded.[11] The trials were repeated 1,000 times. Difference in the RMSE with respect to the MLR model for each trial is shown in the boxplot (Fig. 5.4) for stepwise regression, ridge regression and lasso. In Experiment B, training sample size was reduced to $N = 30$. In Experiment C, with $N = 30$, the third and fourth predictors were replaced by random numbers from a standard Gaussian distribution. Stepwise regression outperformed MLR in Experiment C and underperformed MLR in Experiments A and B. Stepwise regression also underperformed the other two shrinkage methods in Experiments A and B.

Let us examine which predictors were selected by stepwise regression. With four predictors, there are 16 possible model architectures depending on whether a predictor is included or not in the regression. Using 1 and 0 to indicate on/off for a predictor, for example, [1 0 0 1] means predictors $x_1$ and $x_4$ are used, but $x_2$ and $x_3$ are rejected, the frequency distribution of the model architectures selected by stepwise regression is shown in Fig. 5.5(a). For experiment A, with $N = 100$ training data, stepwise regression gave a very confused picture on which model to use, as five models have been chosen with frequency $\geq 10\%$, namely model [1 0 0 1] (27%), [1 1 0 0] (22%), [1 1 1 0] (18%), [1 0 0 0] (12%) and [1 0 1 1] (10%). Thus, from one sample to another, stepwise regression can give completely inconsistent recommendations on which predictors to select. In contrast, the shrinkage method lasso chose model [1 1 1 1] (i.e. using all four predictors) with a frequency of 70% and no other models were chosen with frequency $\geq 10\%$ (Fig. 5.5(b)). In the two experiments with $N = 30$, stepwise regression chose the single-predictor model [1 0 0 0] with a frequency of 49% in Experiment B and 65% in Experiment C.

In summary, stepwise regression, with its propensity to reject predictors, performed better than MLR only when the dataset contained many irrelevant predictors and the sample size was relatively small (Experiment C). However, with multiple relevant predictors in Experiments A and B, stepwise regression tended to reject many relevant predictors, and the set of selected predictors varied from one trial to another (i.e. the set of selected predictors was very sensitive to sampling noise), resulting in worse prediction RMSE than MLR and the other two shrinkage methods.

---

[11] When comparing the performance of various models, the performance scores (e.g. RMSE, correlation, etc.) need to be computed on test data not used in model training. Comparing the performance scores on training data is very misleading since models with more adjustable parameters will have better performance scores on the training data.

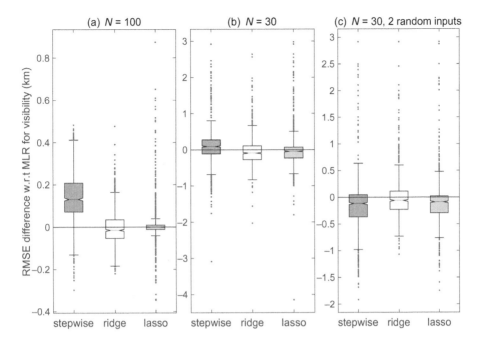

Figure 5.4 Boxplot showing the RMSE difference with respect to the MLR model, that is, the RMSE of stepwise regression, ridge regression and lasso minus the RMSE of the MLR model, for 1,000 trials. Visibility in Vancouver, BC is the response variable, while humidity, pressure, air temperature and wind speed are the four predictors. The sample size of the training data was (a) $N = 100$, (b) $N = 30$ and (c) $N = 30$. In (c), the third and fourth predictors were replaced by random numbers from a standard Gaussian distribution. The RMSE was computed on test data, that is, data not chosen for model training. A positive RMSE difference means the model is underperforming the MLR. See Section 2.12.3 for an explanation of boxplots. [Data source: weatherstats.ca based on Environment and Climate Change Canada data.]

## 5.7 Ridge Regression B☺

In Section 5.6, we noted that predictor selection using a common method such as stepwise regression does not select predictors reliably. Such methods set the regression parameters to zero for the predictors they rejected. A better alternative is to shrink the magnitude of the regression parameters towards zero by using shrinkage methods.

Among shrinkage methods, *regularized least squares* is a family of methods for solving the least-squares problem using regularization to constrain the solution, for example to avoid overfitting. Among regularized least squares methods, the most common regularization method used is the *Tikhonov regularization*,

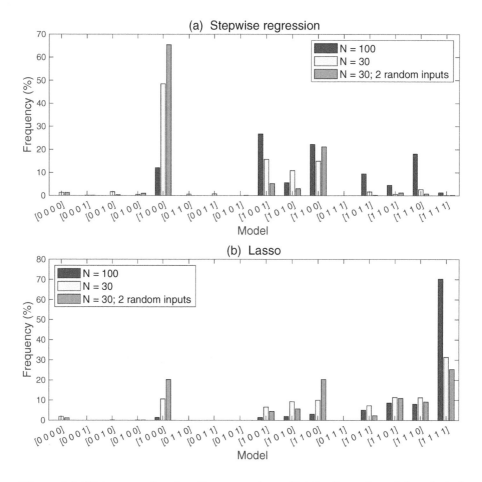

Figure 5.5 Histogram showing the percentage distribution of models selected by (a) stepwise regression and (b) lasso from 1,000 trials with visibility as the response variable. With four predictors, there are 16 possible model architectures, for example, model [1 0 0 0] indicates only the first predictor was used and [1 1 1 1] indicates that all four predictors were used.

named after Andrey Nikolayevich Tikhonov, who first published on the subject in 1943. This regularization when applied to multiple or multivariate linear regression is called *ridge regression* (Hoerl and Kennard, 1970).

To estimate the regression parameters in MLR, the SSE was minimized in (5.50). In ridge regression, the *objective function* $J$ to be minimized is not the SSE, but SSE plus a regularization term. In scalar form (with $N$ denoting the sample size and $m$ the number of predictors),

$$J = \sum_{i=1}^{N} \left( y_i - a_0 - \sum_{j=1}^{m} x_{ij}\, a_j \right)^2 + \lambda \sum_{j=1}^{m} a_j^2, \tag{5.106}$$

where the first term on the right hand side is the SSE and the second term is the regularization term, with $\lambda$ the regularization parameter ($\lambda \geq 0$). When $J$ is being minimized, the regularization term suppresses the magnitude of the weight or regression parameters $a_j$; hence, the term is also called a *weight penalty* or *weight decay* term.[12] Note there is no suppression on the magnitude of $a_0$ since one does not want to push the $y$-intercept $a_0$ towards 0.

For the weight penalty term to work evenly on the predictors, the predictors need to have similar magnitude, so standardizing all predictors is considered an essential initial step before performing ridge regression. With standardized predictors, taking the mean of (5.46) gives

$$a_0 = \bar{y}, \tag{5.107}$$

with $\bar{y}$ being the mean of $y$. We can centre the response variable by subtracting $\bar{y}$ from $y$. For the centred $y$, $a_0$ is no longer present in the objective function (5.106), which can now be expressed in vector form,

$$J = (\mathbf{y} - \mathbf{Xa})^{\mathrm{T}}(\mathbf{y} - \mathbf{Xa}) + \lambda \mathbf{a}^{\mathrm{T}}\mathbf{a}, \tag{5.108}$$

where

$$\mathbf{a} = \begin{bmatrix} a_1 \\ \vdots \\ a_m \end{bmatrix}, \quad \mathbf{X} = \begin{bmatrix} x_{11} & \cdots & x_{1m} \\ \vdots & \ddots & \vdots \\ x_{N1} & \cdots & x_{Nm} \end{bmatrix}. \tag{5.109}$$

Similar to the derivation of (5.51)–(5.55) for MLR, one can easily show that the ridge regression solution is

$$\hat{\mathbf{a}} = (\mathbf{X}^{\mathrm{T}}\mathbf{X} + \lambda \mathbf{I})^{-1} \mathbf{X}^{\mathrm{T}}\mathbf{y}, \tag{5.110}$$

with $\mathbf{I}$ being the $m \times m$ identity matrix. When $\lambda$ is zero, this solution simply reduces to the MLR solution. The new term $\lambda \mathbf{I}$ adds a positive constant to the diagonal of $\mathbf{X}^{\mathrm{T}}\mathbf{X}$.

If the sample covariance matrix $\mathbf{X}^{\mathrm{T}}\mathbf{X}$ is not of full rank, the matrix is singular and the matrix inverse does not exist. Even when the matrix is only near-singular (e.g. due to collinearity, i.e. strong correlation between some predictors), small errors in the elements of $\mathbf{X}$ lead to large changes in the computed $(\mathbf{X}^{\mathrm{T}}\mathbf{X})^{-1}$. $\mathbf{X}^{\mathrm{T}}\mathbf{X}$ is said to be *ill-conditioned*[13] and the regression parameters

---

[12] 'Weight decay' is more commonly used in ML, while 'weight penalty' is more common in statistics.

[13] A matrix $\mathbf{A}$ is said to be ill-conditioned or poorly conditioned if its *condition number $C$*, defined as the ratio of the largest singular value of $\mathbf{A}$ to the smallest singular value of $\mathbf{A}$, satisfies $C \gg 1$. A large $C$ is a warning that the inverse of $A$ cannot be computed accurately.

cannot be estimated accurately. Adding $\lambda\mathbf{I}$ improves the numerical stability when computing the inverse of $\mathbf{X}^T\mathbf{X}$. The original main purpose of ridge regression was to improve numerical stability, with Hoerl and Kennard (1970) first naming the method 'ridge regression' as it has a ridge added along the diagonal of the sample covariance matrix.

The optimal value for $\lambda$ is commonly estimated by *cross-validation*. In $K$-fold cross-validation, the data record with $N$ points is divided into $K$ (approximately) equal segments. Start with an initial guess for $\lambda$. The first segment, designated as *validation data*, is segregated and the remaining $K-1$ segments are used to train the model. The trained model is used to make predictions on the validation data segment. Next, the second segment is designated validation data and segregated, and the first segment plus the third to final segments are pooled together as training data. Again the model is trained and used to predict on the segregated validation segment. This process is repeated until the final segment is designated validation data and the first of $K-1$ segments are used for model training. A performance score, for example the RMSE, is computed from the model predictions on all $K$ segments of validation data. This process is iterated for different values of $\lambda$, and the $\lambda$ with the best performance score over the validation data is chosen as the optimal $\lambda$. The advantage of $K$-fold cross-validation is that all data are used for both training and validation, with each data point used for validation exactly once. Using a smaller $K$ means fewer models have to be trained, resulting in lower computational cost, but the resulting smaller training dataset gives less accurate models. A common choice for $K$ is 10; for lower computational cost, $K = 5$ is also common.

The values of $\lambda$ used in the search for the optimal $\lambda$ tend to be spaced logarithmically, for example from $10^{-1}$ to $10^{-5}$. A very coarse grid search could use, for example, $\lambda = 10^{-1}$, $10^{-2}$, $10^{-3}$, $10^{-4}$ and $10^{-5}$. A less coarse grid search could generate the grid values by letting

$$\lambda = \lambda_0(10^{-1/2})^i, \quad (i = i_{min}, \ldots, i_{max}). \tag{5.111}$$

For instance, with the choice $\lambda_0 = 1$, $i_{min} = 2$ and $i_{max} = 10$, one gets the values:
$10^{-1}$, $3.16\times10^{-2}$, $10^{-2}$, $3.16\times10^{-3}$, $10^{-3}$, $3.16\times10^{-4}$, $10^{-4}$, $3.16\times10^{-5}$ and $10^{-5}$.
One can use $\lambda = \lambda_0(10^{-1/3})^i$ for a finer grid or $\lambda_0(10^{-1/4})^i$ for an even finer grid. The grid search can also be done twice – a coarse grid search to find the approximate location of the optimal $\lambda$, then a fine grid search in that vicinity to pinpoint the optimal $\lambda$.

An alternative to minimizing the objective function (5.106) is to have the weight penalty expressed as an explicit constraint, that is,

$$J = \sum_{i=1}^{N}\left(y_i - a_0 - \sum_{j=1}^{m}x_{ij}a_j\right)^2, \quad \text{subject to} \sum_{j=1}^{m}a_j^2 \le b. \tag{5.112}$$

From the method of Lagrange multipliers (see Appendix B), an explicitly constrained optimization problem such as (5.112) can be shown to be equivalent to an unconstrained optimization problem containing a Lagrange multiplier term (5.106), where the exact relation between the constraint bound $b$ and the Lagrange multiplier $\lambda$ is data dependent. From the explicit constraint, it is clear that the regression parameters are confined to within a hypersphere of radius $b$, thereby preventing overfitting from using large $|a_j|$.

In Fig. 5.4, ridge regression outperformed MLR in all three experiments. For $N > 300$, the advantage of ridge regression over MLR in the prediction RMSE disappears (not shown) since overfitting is no longer a problem for MLR when the sample size greatly exceeds the number of regression parameters, while the weight penalty term introduces a bias in the solution.

Regularization via weight penalty or weight decay is very commonly used in neural networks and other non-linear regression/classification models as these non-linear models tend to have a large number of model parameters, so overfitting is a much more serious issue than in linear models (Section 8.4.1).

## 5.8  **Lasso**  $\boxed{\text{C}}$☺

Lasso (least absolute shrinkage and selection operator) performs multiple linear regression with both weight penalty and predictor selection, thereby enhancing prediction accuracy and model interpretability. Although originally introduced in geophysics (Santosa and Symes, 1986), lasso was independently rediscovered and popularized by Tibshirani (1996), who named the method 'lasso'.

As in ridge regression, predictors are usually standardized first prior to performing lasso. Unlike ridge regression, which uses the $L_2$ norm to constrain the weights $a_j$ in (5.112), lasso uses the $L_1$ norm in the constraint,[14] that is,

$$J = \sum_{i=1}^{N}\left(y_i - a_0 - \sum_{j=1}^{m} x_{ij}\, a_j\right)^2, \quad \text{subject to} \sum_{j=1}^{m} |a_j| \le b. \tag{5.113}$$

The alternative Lagrangian form converts the constraint to a weight penalty term in the objective function (P. E. Gill et al., 1981, chapter 5), that is,

$$J = \sum_{i=1}^{N}\left(y_i - a_0 - \sum_{j=1}^{m} x_{ij}\, a_j\right)^2 + \lambda \sum_{j=1}^{m} |a_j|, \tag{5.114}$$

where $\lambda$ is a Lagrange multiplier. The problem is solved using quadratic programming, where efficient algorithms allow lasso to be comparable to ridge regression in terms of computational cost. As in ridge regression, cross-validation is used to find the optimal $\lambda$ value.

An interesting difference between lasso and ridge regression is that while ridge regression can make some regression parameters or weights small, it does not set them to zero, whereas lasso sets some of the small weights to zero,

---

[14] In general, for a real number $p$, the $L_p$ norm or $p$-norm of an $m$-dimensional vector $\mathbf{x}$ is defined by $\|\mathbf{x}\|_p \equiv (|x_1|^p + \cdots + |x_m|^p)^{1/p}$. For $p = 2$, we get back the familiar $L_2$ or Euclidean norm, $\|\mathbf{x}\|_2 = \sum_{j=1}^{m} x_j^2$. For $p = 1$, the $L_1$ norm is given by $\|\mathbf{x}\|_1 = \sum_{j=1}^{m} |x_j|$.

which is equivalent to rejecting the corresponding predictors, that is, performing predictor selection. This difference is illustrated in Fig. 5.6. For lasso, the solution of (5.113) is the lowest value of the SSE that satisfies the constraint, with the constraint boundary $(|a_1| + |a_2|) \leq b$ given by the diamond-shaped region. For ridge regression, the constraint boundary in (5.112) is a circular region given by $(a_1^2 + a_2^2) \leq b$. Since the lasso constraint region has corners, the interception point given by the small square in Fig. 5.6(a) has $a_1 = 0$. With the circular region in Fig. 5.6(b), the interception point gives non-zero values for $a_1$ and $a_2$. When there are more predictors, the diamond constraint region in lasso becomes a hypercube, with many more opportunities for weights $a_j$ to be zero than in ridge regression with its hyperspherical constraint region, that is, lasso finds a solution with greater *sparsity*.

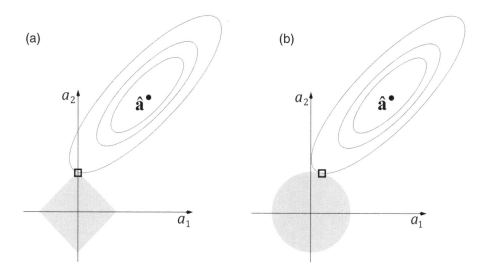

Figure 5.6 Schematic diagram illustrating the SSE contours (ellipses) and the constraint region for (a) lasso (diamond region) and (b) ridge regression (circular region) in a two-dimensional regression parameter space $a_1$-$a_2$. The solution is indicated by the small square, marking where the lowest value of the SSE function intercepts the constraint region. The dot in the centre of the ellipses marks â, where the minimum of the SSE occurs. [Adapted from Hastie, Tibshirani, et al. (2009, figure 3.11), which was based on Tibshirani (1996, figure 2).]

In Fig. 5.4, lasso showed less spread (in terms of the height of the boxes or the distance between the upper and lower whiskers) than stepwise regression and ridge regression. In Fig. 5.5, Lasso tended to retain more predictors than stepwise regression.

## 5.9  Quantile Regression  $\boxed{\text{C}}$ ☺

For linear regression, a more statistical interpretation is that the model predicted response $\hat{y} = E(y|\mathbf{x})$, that is, the predicted response by the linear regression model is the conditional mean of the response variable for given value(s) of the predictor $\mathbf{x}$ (Abraham and Ledolter, 1983, section 2.1). In contrast, quantile regression (Koenker and Bassett, 1978; Koenker and Hallock, 2001) aims at estimating either the conditional median or other quantiles of the response variable for given values of the predictor. The median, a more robust statistic than the mean, is simply the 50th percentile or the 0.50 quantile.

In linear regression, the objective is to minimize the sum of squared errors (SSE) (5.16), that is, the objective function is of the form

$$J = \sum_{i=1}^{N} L(y_i - \hat{y}_i), \qquad (5.115)$$

where $N$ is the sample size, $y$ the observed response, $\hat{y}$ the model response and the loss function $L$ for error $\epsilon$ is

$$L(\epsilon) = \epsilon^2, \qquad (5.116)$$

with $\epsilon = y - \hat{y}$. By using absolute errors instead of squared errors in the loss, that is,

$$L(\epsilon) = |\epsilon| \qquad (5.117)$$

(Fig. 5.7(a)), the optimization yields the conditional median instead of the conditional mean for $\hat{y}$. Since large errors have much less impact on the objective function if the absolute error loss is used instead of the quadratic loss (i.e. squared errors), the solution using absolute error loss is much more resistant to outliers in the training data. The proof that minimizing the (i) squared errors and (ii) absolute errors yields an estimate of the conditional (i) mean and (ii) median, respectively, is given for the more general non-linear regression problem in Section 8.2.

For the $q_\alpha$ quantile (see Section 2.9), the loss $L_\alpha$ is given by

$$L_\alpha(\epsilon) = \begin{cases} \alpha\epsilon & \text{if } \epsilon \geq 0, \\ (\alpha - 1)\epsilon & \text{if } \epsilon < 0. \end{cases} \qquad (5.118)$$

The loss function for $q_{0.95}$ and $q_{0.10}$ are shown in Fig. 5.7(b). Setting $\alpha = 0.5$, the loss function reduces to the absolute value function multiplied by $\frac{1}{2}$.

The quantile regression equation is similar to that in MLR with $m$ predictors, that is,

$$\hat{y}_i = a_0 + \sum_{j=1}^{m} x_{ij} a_j, \qquad i = 1, \dots, N. \qquad (5.119)$$

The objective function to be minimized is

$$J_\alpha = \sum_{i=1}^{N} L_\alpha(y_i - \hat{y}_i), \qquad (5.120)$$

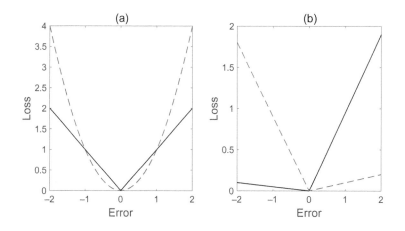

Figure 5.7 The loss $L$ as a function of the error $\epsilon$ using (a) absolute errors (solid curve) for finding the conditional median and squared errors (dashed curve) for the conditional mean, and (b) for finding the conditional 0.95 quantile (solid) and the conditional 0.10 quantile (dashed).

so that $\hat{y}$ gives the estimate of the conditional $\alpha$ quantile. The minimization problem can be reformulated as a linear programming problem (Koenker and Bassett, 1978, appendix), where the simplex algorithm is commonly used (P. E. Gill et al., 1991, chapter 8).

Figure 5.8 shows quantile regression with sea level air pressure as the predictor and wind speed as the response variable, both from Vancouver, BC, where 914 data points were used (by choosing every 10th data point) from 25 years of daily data (1993–2017). The regression lines for the 0.95 quantile, the median and the 0.05 quantile are shown. The steeper slope of the 0.95 quantile indicates a stronger dependence between wind speed and pressure for the 0.95 quantile than for the median and the 0.05 quantile.

One known problem is that quantile regression lines for different $\alpha$ values may cross one another. Constrained versions of linear and non-linear quantile regression have been developed to avoid the crossing of quantile lines or curves (Bondell et al., 2010; Cannon, 2011b; Cannon, 2018).

## 5.10   Generalized Least Squares  [C] ☺

In Section 5.2, we have used linear least squares to solve the multiple linear regression problem,

$$\mathbf{y} = \mathbf{X}\mathbf{a} + \boldsymbol{\epsilon}, \tag{5.121}$$

with $\mathbf{y}$, $\mathbf{a}$, $\boldsymbol{\epsilon}$ and $\mathbf{X}$ defined in (5.48) and (5.49). The residual or noise $\boldsymbol{\epsilon}$ is usually assumed to be Gaussian with zero mean. In *ordinary least squares* (OLS), the residual is assumed to satisfy $\text{cov}(\epsilon_i, \epsilon_j) = 0$ for $i \neq j$. By minimizing the SSE,

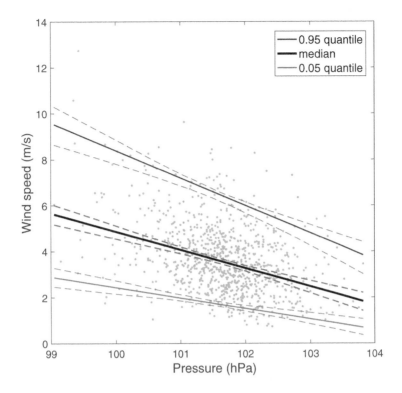

Figure 5.8 Quantile regression with wind speed as response and pressure as predictor, with data from Vancouver, BC. Bootstrap resampling was used to estimate the 95% confidence intervals (dashed lines) around the regression lines. [Data source: weatherstats.ca based on Environment and Climate Change Canada data.]

$$\text{SSE} = \epsilon^{\mathrm{T}}\epsilon = (\mathbf{y} - \mathbf{Xa})^{\mathrm{T}}(\mathbf{y} - \mathbf{Xa}), \qquad (5.122)$$

the optimal parameters (i.e. regression coefficients) were found to be

$$\hat{\mathbf{a}} = (\mathbf{X}^{\mathrm{T}}\mathbf{X})^{-1}\mathbf{X}^{\mathrm{T}}\mathbf{y}. \qquad (5.123)$$

In *generalized least squares* (GLS), the assumption of $\text{cov}(\epsilon_i, \epsilon_j) = 0$ for $i \neq j$ is dropped. For instance, the residual can have serial correlation or heteroscedasticity.[15] Instead of minimizing the SSE using the Euclidean distance,

---

[15] A random variable is *heteroscedastic* if there are sub-populations with different variability than others. For instance, a heteroscedastic time series may display a change in the variance or skewness with time.

we use the Mahalanobis distance (Section 2.13), that is, instead of minimizing (5.122), we minimize

$$\epsilon^T C^{-1} \epsilon = (y - Xa)^T C^{-1} (y - Xa), \qquad (5.124)$$

where $C$, the covariance matrix of the residual, has elements $C_{ij} = \text{cov}(\epsilon_i, \epsilon_j)$. It can be shown that the optimal parameters are given by

$$\hat{a} = (X^T C^{-1} X)^{-1} X^T C^{-1} y. \qquad (5.125)$$

When $C = I$, (5.125) reduces to the OLS solution of (5.123).

### 5.10.1   Optimal Fingerprinting in Climate Change  $\boxed{C}$ ☺

GLS has been applied to detect the expected 'fingerprints' of climate change (K. Hasselmann, 1993; G. Hegerl and Zwiers, 2011). *Optimal fingerprinting* methods evaluate how well climate changes found in observed data ($y$) can be explained by the fingerprints, that is, patterns of response ($X$) to external forcings found in climate model simulations. Separate external forcings, for example solar, volcanic, greenhouse gas and aerosol, generate the individual columns of $X$, with the coefficients in $\hat{a}$ indicating the importance of the contribution from the individual forcing signals. The residual $\epsilon$ is the internal variability of the climate system. $C$ is usually computed from $\epsilon$ found in the fluctuations of the unforced climate model runs (M. R. Allen and Tett, 1999). The role of $C^{-1}$ in (5.125) can be interpreted as allowing the optimal fingerprinting method to maximize the squared signal-to-noise ratio for each individual signal component (K. Hasselmann, 1993).

There have been further extensions beyond GLS used in climate change detection. For instance, in GLS, only $y$ has a noise term $\epsilon$ in (5.121). In *total least squares* (TLS), the predictors also have noise terms (M. R. Allen and P. A. Stott, 2003), that is,

$$y = \sum_{j=1}^{m} (x_j - \epsilon_j) a_j + \epsilon_0, \qquad (5.126)$$

where $\epsilon_0$ is the noise in $y$ and $\epsilon_j$ is the noise in predictor $x_j$ ($j = 1, \ldots, m$), with $x_j$ being the $j$th column vector in $X$. DelSole, Trenary, et al. (2019) discussed the various methods for estimating confidence intervals in optimal fingerprinting.

## Exercises

$\boxed{5.1}$

From (5.11), derive (5.10).

5.2

The Old Faithful geyser in Yellowstone National Park, Wyoming, USA, erupts intermittently. From the given dataset, study the behaviour of the geyser using simple linear regression with the eruption duration (in minutes) as predictor and the wait time (in minutes) till the next eruption as the response variable. [Data source: R manual, based on Härdle (1991, table 3)]

5.3

*Teleconnections* refer to the climate variability links between non-contiguous regions on the Earth's surface. For instance, the El Niño-Southern Oscillation (ENSO), the dominant mode of interannual variability in the tropical Pacific [commonly measured by the Niño 3.4 region (see Fig. 9.3) sea surface temperature index] can extend its influence to the extratropics by atmospheric teleconnection. Some of the best known teleconnections are the Pacific-North American (PNA) pattern, the North Atlantic Oscillation (NAO) and the Arctic Oscillation (AO) (Nigam and Baxter, 2015).

Consider the given dataset containing the winter air temperature (averaged over December to February) for Vancouver, BC on the west coast of Canada and for Toronto, Ontario, in eastern Canada, and four predictors (the Niño3.4 index and the PNA, NAO and AO teleconnection indices, all averaged over the same three months). (a) Find the Pearson correlation between each pair of predictor variables. (b) Find the lag-1 autocorrelation of each predictor variable. (c) Find the lag-1 autocorrelation of the Vancouver winter temperature and (d) use simple linear regression to study the relation between the response variable (Vancouver temperature) and each of the four predictor variables individually. (e) Repeat (c) and (d) using the Toronto winter temperature instead. [Data source: The temperatures at the Vancouver and Toronto international airports were obtained from the Government of Canada climate.weather.gc.ca website, while the four predictors were from Climate Prediction Center, US National Weather Service]

5.4

For the Durbin–Watson statistic $d$ in (5.42), prove that (a) $d = 0$ if $\epsilon_i = \epsilon_{i+1}$, (b) $d = 4$ if $\epsilon_i = -\epsilon_{i+1}$ and (c) $d = 2$ if $\text{cov}(\epsilon_i, \epsilon_{i+1}) = 0$. Assume large sample sizes in (b) and (c).

5.5

Show that for multiple linear regression, the square of the Pearson correlation between the dependent variable $y$ and $\hat{y}$ predicted by the regression relation, that is, $\rho^2_{y\hat{y}}$, is equal to the coefficient of determination $R^2 = \text{SSR/SST}$. [Hint: $\text{cov}(y, \hat{y}) = \text{cov}(\hat{y} + \epsilon, \hat{y})$].

5.6

Using the dataset from Exercise (5.3), (a) study the relation between the Vancouver winter temperature and the four predictors (Niño3.4, PNA, NAO and AO) using multiple linear regression (MLR) and ANOVA. (b) Repeat by using the Toronto winter temperature instead. What happens if you drop AO from the predictors in the MLR for the Toronto temperature?

5.7

From the given dataset containing wind direction ($\theta$) (in degrees) and ozone concentration at a weather station in Milwaukee in 1975, perform linear regression with ozone concentration as the response variable. (a) Try different predictors in the following four models runs: (i) $\theta$, (ii) $\cos(2\pi\theta/360)$ and $\sin(2\pi\theta/360)$, (iii) $\cos(2\pi\theta/360)$ and (iv) $\sin(2\pi\theta/360)$. (b) Add $60°$ to $\theta$ and repeat the four runs. Explain if there is any difference in the RMSE between the runs in (b) and the corresponding ones in (a). [Data source: R. Johnson and Wehrly (1977, table 1)]

5.8

In this exercise, we learn how validation data help to prevent overfitting from using too many predictors. Consider the given dataset containing the winter snow water equivalent (SWE) for Grouse Mountain, BC, Canada (the winter SWE being the maximum observed SWE from January to June) and four predictors (the Niño3.4 index and the PNA, NAO and AO teleconnection indices, averaged December to February) (see Exercise 5.3). Use the first 45 data points for training and the remaining 23 points for validation. Compute the RMSE over the training data and the RMSE over the validation data for (a) the MLR model predicting SWE using all four predictors. (b) Drop the weakest predictor based on the $p$-values for individual predictors and repeat MLR using three predictors. (c) Again drop the weakest predictor and repeat using two predictors and (d) drop the weaker predictor and repeat using one predictor. Which MLR model would you pick based on the best validation RMSE? Would you have picked this model if you only looked at the $p$-values of individual predictors based on the training data?

5.9

In this exercise, we study the effect of collinearity among predictors and the advantage of ridge regression. Consider the given dataset containing the winter snow water equivalent (SWE) for Grouse Mountain, BC as used in Exercise 5.8. Use only the first two predictors, that is, the Niño3.4 and PNA indices. Standardize the two predictors and the SWE. Next, construct a collinear predictor from the second predictor by adding a small noise term, namely $0.001\,\mathcal{N}(0,1)$, with $\mathcal{N}(0,1)$ being the standard Gaussian distribution, that is, the second and third predictors are very strongly correlated. Use the first 45 data points for training and the remaining 23 points for validation. Compute the RMSE over the training data and the RMSE over the validation data for the MLR model predicting SWE from the three predictors. Next perform ridge regression with the weight penalty parameter being small, $\lambda = 10^{-5}$, and being relatively large, $\lambda = 0.01$. Repeat the MLR and the two ridge regression model runs 100 times using different random noise in the third predictor. From the 100 trials, compute and compare the following statistics from the MLR and from the two ridge regression models: standard deviation for the regression parameters $\hat{a}_0, \hat{a}_1, \hat{a}_2$ and $\hat{a}_3$, the mean and standard deviation for the RMSE over training data and for the RMSE over validation data.

# 6

# Neural Networks

The human brain is an organ of marvel – with a massive network of about $10^{11}$ interconnecting neural cells called neurons, it performs highly parallel computing. The brain is exceedingly robust and fault tolerant. After all, we still recognize a friend whom we have not seen in a decade, though many of our brain cells have since died and our friend has become fat and bald. A neuron is only a very simple processor, and its 'clock speed'[1] is actually quite slow, in the order of 100 Hz, about ten million times slower than that of a modern computer; yet, the human brain beats the computer on many tasks involving vision, motor control, common sense, and so on. Hence, the power of the brain lies not in its clock speed, nor in the computing power of a neuron, but in its massive network structure. What computational capability is offered by a massive network of interconnected neurons is a question that has greatly intrigued scientists, leading to the development of the field of *neural networks* (NN).

Of course, there are medical researchers who are interested in how real neural networks function. However, there are far more scientists/engineers from all disciplines who are interested in *artificial* neural networks (ANN), that is, how to borrow ideas from biological neural network structures to develop better techniques in computing, artificial intelligence, data analysis, modelling and prediction. Neural networks have become dominant in ML, and their use has spread from science to engineering, commerce and countless other fields.

There are two main types of learning problems, *supervised* and *unsupervised* learning (Section 1.4). In supervised learning, predictor data $\{\mathbf{x}_1, \ldots, \mathbf{x}_N\}$ and response data $\{\mathbf{y}_1, \ldots, \mathbf{y}_N\}$ from $N$ observations are supplied to the model, which produces outputs $\{\hat{\mathbf{y}}_1, \ldots, \hat{\mathbf{y}}_N\}$. The model learning is 'supervised' in the sense that the model output is guided towards the response data, usually by minimizing an *objective function* (also called a *loss function*, *cost function* or, less commonly, *error function*). Regression and classification problems involve supervised learning. In contrast, for unsupervised learning, only input data are provided, and the model discovers the natural patterns or structure in the input data. Principal component analysis (Chapter 9) and clustering (Chapter 13)

---

[1] For computers, clock speed is the rate at which a processor can complete a processing cycle. Modern computers have clock speeds measured in GHz ($10^9$ hertz).

involve unsupervised learning. Neural networks are more commonly used in supervised learning, but are also used in unsupervised learning.

There are many types of neural networks, some mainly of interest to the biomedical researcher, others with broad applications. The most widely used type of NN is the *feed-forward* neural network, where the signal in the model only proceeds forward from the inputs through any intermediate layers to the outputs without any feedback. The feed-forward NN is represented by the multi-layer perceptron (MLP) model (Section 6.3). The next two sections outline some of the historical developments leading to the MLP model. NN can be applied to regression problems or classification problems. In this chapter, from Section 6.3 onward, the focus is on regression problems, as we will turn to classification problems in later chapters (e.g. Chapter 12). Examples of the widespread use of the MLP model in the environmental sciences have been provided in many books, including Abrahart, Kneale, et al. (2004) (on hydrology), S. E. Haupt, Pasini, et al. (2009), Hsieh (2009), Blackwell and F. W. Chen (2009) (on atmospheric remote sensing), Lakshmanan (2012) (on geospatial images) and Krasnopolsky (2013). Since 2012, NN has undergone explosive growth with the advent of deep NN models (i.e. models with many layers of hidden nodes or hidden neurons), which are presented separately in Chapter 15 on deep learning.

Traditionally, the MLP model is trained using non-linear optimization, a technically somewhat complicated subject that is presented in a separate chapter (Chapter 7). Furthermore, MLP with its numerous adjustable parameters is prone to overfitting; therefore, Chapter 8 focuses on how to build models which are neither overfitting nor underfitting. Fortunately, it is now known that non-linear optimization can be avoided altogether in randomized NN models such as the extreme learning machine (ELM) model, which only requires linear least squares optimization, the same as that used in multiple linear regression. Thus, Section 6.4 on ELM provides a shortcut to running simple MLP models (i.e. with only one layer of hidden nodes) without non-linear optimization. Radial basis function NN (Section 6.5) is another common NN model that can be solved using only linear least squares optimization. NN models can also be used to model conditional probability distributions (Section 6.6) and non-linear quantile regression (Section 6.7). The reader who is keen on moving quickly to solving MLP models using non-linear optimization can skip Sections 6.4 – 6.7 and jump to Chapter 7.

# 6.1   McCulloch and Pitts Model   $\boxed{\text{B}}$☺

From neurobiology, it is known that a nerve cell, called a neuron, receives stimuli (signals) from its neighbours and may become activated (Fig. 6.1). The earliest NN model of significance is the McCulloch and Pitts (1943) model. Their model neuron is a binary threshold unit, that is, it receives a weighed sum of its inputs from other units, and outputs either 1 or 0 depending on whether the sum exceeds a threshold. This simulates a neuron receiving signals from its neighbouring neurons, and depending on whether the strength of the overall

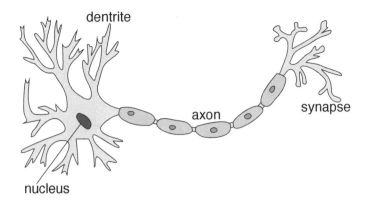

dentrite

axon

synapse

nucleus

Figure 6.1 In a neuron, dendrites receive signals from other neurons. If the total stimulus exceeds some threshold, the neuron becomes activated, firing a signal down its axon to the synapses at its end. Synapses are sites where neurotransmitting chemicals are released into the space between neurons so the signal can be picked up by neighbouring neurons. [Image source: Quasar Jarosz at English Wikipedia.]

stimulus exceeds the threshold, the neuron either becomes activated (outputs the value 1) or remains at rest (value 0). For a neuron, if $x_i$ denotes the input signal from the $i$th neighbour, which is weighed by a *weight parameter* $w_i$, the output of the neuron $y$ is given by

$$y = H\left(\sum_i w_i x_i + b\right),\tag{6.1}$$

where $b$ is called an *offset* or *bias* parameter, and $H$ is the Heaviside step function,

$$H(z) = \begin{cases} 1 & \text{if } z \geq 0 \\ 0 & \text{if } z < 0. \end{cases}\tag{6.2}$$

By adjusting $b$, the threshold level for the firing of the neuron can be changed. McCulloch and Pitts proved that networks made up of such neurons are capable of performing any computation a digital computer can, though there is no provision that such an NN computer is necessarily faster or easier. In the McCulloch and Pitts model, the neurons are very similar to conventional logical gates; however, there is no algorithm provided for finding the appropriate weight and offset parameters for a particular problem.

## 6.2  Perceptrons  $\boxed{\text{A}}$ ☺

The next major advance is the *perceptron* model of Rosenblatt (1958) and Rosenblatt (1962) (and similar work by Widrow and Hoff (1960)). The perceptron model consists of an input layer of neurons connected to an output layer of neurons (Fig. 6.2).

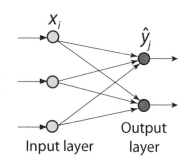

Figure 6.2 The perceptron model consists of a layer of input neurons connected directly to a layer of output neurons.

Neurons are also referred to as *nodes* or *units* in the NN literature. The key advance is the introduction of a *learning algorithm*, which finds the weight and offset parameters of the network for a particular problem. An output neuron

$$\hat{y}_j = f\left(\sum_i w_{ji} x_i + b_j\right), \qquad (6.3)$$

where $x_i$ denotes an input, $f$ a specified transfer function known as an *activation function*, $w_{ji}$ the weight parameter connecting the $i$th input neuron to the $j$th output neuron and $b_j$ the offset or bias parameter of the $j$th output neuron. Note that $-b_j$ is also called the *threshold* parameter by some authors.

A more compact notation eliminates the distinction between weight and offset parameters by expressing $\sum_i w_i x_i + b$ as $\sum_i w_i x_i + w_0 = \sum_i w_i x_i + w_0 1$, that is, $b$ can be regarded as simply the weight $w_0$ for an extra constant input $x_0 = 1$, and (6.3) can be written as

$$\hat{y}_j = f\left(\sum_i w_{ji} x_i\right), \qquad (6.4)$$

with the summation of $i$ starting from $i = 0$. This convenient notation incorporating the offsets into the weights will often be used in this book to simplify equations.

Given data of the inputs and outputs, one can train the network so that the model output values $\hat{y}_j$ derived from the inputs using (6.3) are as close as possible to the data $y_j$ (also called the *target*), by finding the optimal weight and offset parameters in (6.3) – that is, try to fit the model to the data by adjusting the parameters. With the parameters known, (6.3) gives the output variable $\hat{y}_j$ as a function of the input variables. A detailed explanation of the training process will be given later.

The step function was originally used for the activation function, but it has been replaced by other functions. A commonly used activation function is the *logistic sigmoidal function*, or simply the *logistic function*, where 'sigmoidal' means an S-shaped function:

$$f(x) = \frac{1}{1 + e^{-x}}. \tag{6.5}$$

This function has an asymptotic value of 0 when $x \to -\infty$, and rises smoothly as $x$ increases, approaching the asymptotic value of 1 as $x \to +\infty$ (Fig. 6.3(a)).

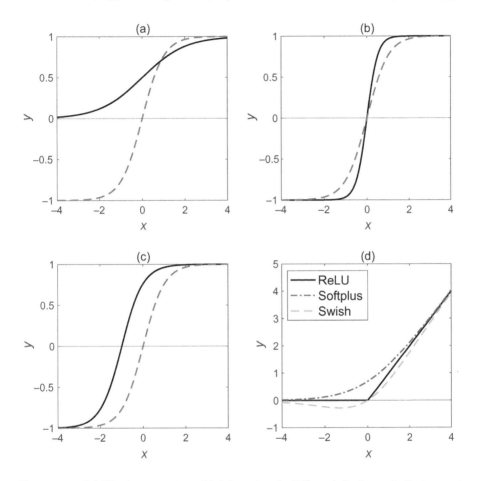

Figure 6.3 (a) The logistic sigmoidal function (solid) and the hyperbolic tangent function tanh (dashed). (b) Effect of the weight $w$ in $f(wx)$ as seen by comparing $f(2x)$ (solid) and $f(x)$ (dashed), with tanh used for $f$. (c) Effect of the offset parameter $b$ in $f(wx + b)$ by comparing $f(x + 1)$ (solid) and $f(x)$ (dashed). (d) Three unbounded activation functions: the rectified linear unit (ReLU) function $f(x) = \max(0, x)$, the softplus function $f(x) = \log(1 + e^x)$ and the swish function given in (15.2).

From a biological perspective, the shape of the logistic function resembles the activity of a neuron. Suppose the 'activity' of a neuron is 0 if it is at rest, or 1 if it is activated. A neuron is subjected to stimulus from its neighbouring neurons, each with activity $x_i$, transmitting a stimulus of $w_i x_i$ to this neuron. With the summed stimulus from all its neighbours being $\sum_i w_i x_i$, whether this neuron is activated depends on whether the summed stimulus exceeds a threshold $c$ or not, that is, the activity of this neuron is 1 if $\sum_i w_i x_i \geq c$, but is 0 otherwise. This behaviour is captured by the Heaviside step function $H\left(\sum_i w_i x_i - c\right)$. To have a differentiable function with a smooth transition between 0 and 1 instead of an abrupt step, one can use the logistic function $f$ in (6.5), that is, $f\left(\sum_i w_i x_i + b\right)$, where $b = -c$ is the offset parameter. An alternative is the hyperbolic tangent function (tanh), which, unlike the logistic function, has a mean value of 0 (Fig. 6.3(a)).

The role of the weight and offset parameters can be readily seen in the univariate form $f(wx + b)$, where the tanh function is used for $f$ and a larger magnitude for $w$ gives $f$ a steeper transition from $-1$ to 1 (Fig. 6.3(b)). (In the limit of $|w| \to +\infty$, $f$ approaches a step function). Increasing $b$ slides the whole curve to the left along the $x$-axis (Fig. 6.3(c)).

Unbounded activation functions, such as the rectified linear unit (ReLU), the softplus function and the swish function (Fig. 6.3(d)), are used in deep neural networks (i.e. NN with many layers of neurons) (see Chapter 15). A caveat is that activation functions designed to perform well in deep NN models may underperform tanh in shallow NN models (Goodfellow, Bengio, et al., 2016, p. 219).

## 6.2.1   Limitation of Perceptrons   [B]☺

The advent of the perceptron model led to great excitement; however, the serious limitations of the perceptron model were soon recognized (Minsky and Papert, 1969). Simple examples are provided by the use of perceptrons to model the Boolean logical operators AND and XOR (the exclusive OR). For $y = x_1.\text{AND}.x_2$, $y$ is TRUE only when both $x_1$ and $x_2$ are TRUE. For $y = x_1.\text{XOR}.x_2$, $y$ is TRUE only when exactly one of $x_1$ or $x_2$ is TRUE. Let 0 denote FALSE and 1 denote TRUE, then Fig. 6.4 shows the simple perceptron model that represents $y = x_1.\text{AND}.x_2$, mapping from $(x_1, x_2)$ to $y$ in the following manner:

$$(0,0) \to 0,$$
$$(0,1) \to 0,$$
$$(1,0) \to 0,$$
$$(1,1) \to 1.$$

However, a perceptron model for $y = x_1.\text{XOR}.x_2$ does not exist! One cannot find a perceptron model which will map:

Figure 6.4 The perceptron model for computing $y = x_1.\text{AND}.x_2$. The activation function $f$ used is the Heaviside step function $H$ in (6.2). The threshold $-b = 1.5$ is exceeded by $w_1 x_1 + w_2 x_2$ in $H(w_1 x_1 + w_2 x_2 + b)$ only when both inputs have the value 1.

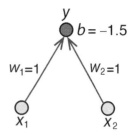

$$(0,0) \to 0,$$
$$(0,1) \to 1,$$
$$(1,0) \to 1,$$
$$(1,1) \to 0.$$

The difference between the two problems is shown in Fig. 6.5. The AND problem is *linearly separable*, that is, the input data can be classified correctly with a linear (i.e. hyperplanar) decision boundary, whereas the XOR problem is not linearly separable. It is easy to see why the perceptron model is limited to a linearly separable problem. If the activation function $f(z)$ has a decision boundary at $z = c$, (6.3) implies that the decision boundary for the $j$th output neuron is given by $\sum_i w_{ji} x_i + b_j = c$, which is the equation of a straight line in the input $\mathbf{x}$-space.

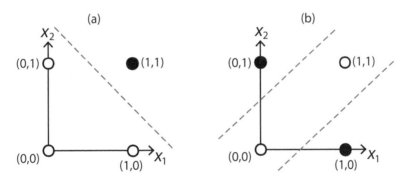

Figure 6.5 The classification of the input data $(x_1, x_2)$ by the Boolean logical operator (a) AND and (b) XOR (exclusive OR). In (a), the decision boundary separating the TRUE domain (black circle) from the FALSE domain (white circles) can be represented by a straight (dashed) line, hence the problem is linearly separable, whereas in (b), two lines are needed, rendering the problem not linearly separable.

For an input $\mathbf{x}$-space with dimension $d = 2$, there are 16 possible Boolean functions (among them AND and XOR), and 14 of the 16 are linearly separable. When $d = 3$, 104 out of 256 Boolean functions are linearly separable. When

$d = 4$, the fraction of Boolean functions that are linearly separable drops further – only 1,882 out of 65,536 are linearly separable (Rojas, 1996). As $d$ gets large, the set of linearly separable functions forms a very tiny subset of the total set (Bishop, 1995). Interest in NN research waned following the realization that the perceptron model is restricted to linearly separable problems.

## 6.3   Multi-layer Perceptrons (MLP)   $\boxed{\text{A}}$ ☺

When the limitations of the perceptron model (Section 6.2) were realized, it was felt that the model might have greater power if additional '*hidden*' layers of neurons were placed between the input layer and the output layer. Unfortunately, there was then no algorithm that would solve for the parameters of the *multi-layer perceptrons* (MLP). Revival of interest in NN did not occur until the mid-1980s – largely through the highly influential paper by Rumelhart et al. (1986a), which rediscovered the *back-propagation algorithm* to solve the MLP problem, as the algorithm had actually been derived by earlier researchers (Schmidhuber, 2015, section 5.5).

Another name for MLP is *feed-forward neural network* (FFNN). The name FFNN came about because signal flow in the network only proceeds forward from the input to the output, unlike in the recurrent neural network (RNN), where the signal flow can loop around the network. FFNN, or MLP, has been the most widely used model in ML. It has been extensively applied to the environmental sciences, with numerous examples provided in several books (S. E. Haupt, Pasini, et al., 2009; Hsieh, 2009; Krasnopolsky, 2013).

Figure 6.6 The multi-layer perceptron (MLP) or feed-forward neural network (FFNN) model with one 'hidden layer' of neurons or nodes sandwiched between the input layer and the output layer. There are $m_1$ nodes in the input layer, $m_2$ nodes in the hidden layer and $m_3$ nodes in the output layer.

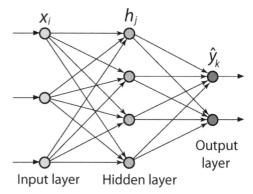

Figure 6.6 illustrates an MLP with one hidden layer. The inputs $x_i$ ($i = 1, \ldots, m_1$) are mapped to the hidden layer of neurons $h_j$ ($j = 1, \ldots, m_2$) by

$$h_j = f\left(\sum_{i=1}^{m_1} w_{ji} x_i + b_j\right), \tag{6.6}$$

and then on to the outputs $\hat{y}_k$ $(k = 1, \ldots, m_3)$,

$$\hat{y}_k = g\left(\sum_{j=1}^{m_2} \tilde{w}_{kj} h_j + \tilde{b}_k\right), \tag{6.7}$$

where $f$ and $g$ are activation functions, $w_{ji}$ and $\tilde{w}_{kj}$ weight parameter matrices, and $b_j$ and $\tilde{b}_k$ are offset or bias parameters.[2] In the network shown in Fig. 6.6, there are three input nodes $(m_1 = 3)$, four hidden nodes $(m_2 = 4)$ and two output nodes $(m_3 = 2)$. From inputs to outputs, there are two layers of mapping functions, namely (6.6) and (6.7); hence, this model is called a two-layer MLP or FFNN model. Additional hidden layers can be added, that is, an $L$-layer MLP model contains $L - 1$ hidden layers.

To train the NN to learn from a dataset, we need to minimize the *objective function $J$* (also known as the *loss function, cost function* or *error function*), defined here to be one half the *mean squared error* (MSE) between the model output $\hat{y}_k$ and the given response data (a.k.a. target data) $y_k$, that is,

$$J = \frac{1}{N} \sum_{n=1}^{N} \left[\frac{1}{2} \sum_{k=1}^{m_3} (\hat{y}_{kn} - y_{kn})^2\right], \tag{6.8}$$

where the additional subscript $n$ indicates the $n$th observation, with $n = 1, \ldots, N$. In NN literature, a conventional scale factor of $\frac{1}{2}$ is usually included, though it and the factor $\frac{1}{N}$ can be omitted or replaced by any positive constant (in other parts of this book, the constant factors $\frac{1}{2}$ and $\frac{1}{N}$ may be dropped from the objective function). An optimization algorithm is needed to find the weight and offset parameter values that minimize the objective function, that is, the MSE between the model output and the target data. The MSE is the most common form used for the objective function in non-linear regression problems, as minimizing the MSE is equivalent to maximizing the likelihood function assuming Gaussian error distribution (see Section 8.1). For non-linear classification (Chapter 11), the negative log-likelihood (i.e. cross-entropy) objective function is commonly used instead of the MSE.

The standard approach to finding the minimum of $J$ is to use a non-linear optimization algorithm.[3] Unfortunately, non-linear optimization problems generally have multiple local minima in the objective function, that is, starting with different initial weights, the optimization algorithm often converges to different local minima (Section 7.1). Details on how to perform the non-linear optimization will be presented in Chapter 7, as there are many choices of optimization algorithms. The problem of multiple local minima is best handled by using an ensemble approach, that is, build an ensemble of MLP NN models instead of a single model (Section 8.7).

---

[2] In ML literature, the offset or bias parameters are often simply referred to as 'bias', although they are completely unrelated to the statistical bias of a model or an estimator. This book avoids ambiguity by never referring to an offset or bias parameter as 'bias'.

[3] It is possible to avoid non-linear optimization, see Sections 6.4 and 6.5, where linear optimization is used.

To illustrate the use of an ensemble, consider a synthetic dataset consisting of the signal $y_{\text{signal}} = \cos(x-1)$ and 121 values of $x$ spaced equally in the interval $[-6, 6]$. Gaussian noise with a quarter of the standard deviation of $y_{\text{signal}}$ was added to give the target data $y$. An ensemble of 25 members was built using an MLP model with a single hidden layer containing 4 hidden nodes.[4] Figure 6.7(a) compares the ensemble average and the true signal over the interval $[-8, 8]$, that is, there is some extrapolation beyond the training domain of $[-6, 6]$. In Fig. 6.7(b), individual ensemble members are shown by thin curves. Two ensemble members corresponding to shallow local minima solutions failed to fit the data well, for example, the dashed curve is basically constant for $x < 0$ and the dot-dashed curve is mainly constant in the interval $-4 < x < 6$. These solutions mistook the signal oscillations as noise to be ignored. Obviously, one should use an ensemble instead of a single MLP model (the rigorous proof is given in Section 8.7.2). Some of the ensemble members also gave large amplitude extrapolations. How extrapolation can insidiously ruin the skills of MLP and other non-linear regression models is discussed in Section 16.9.

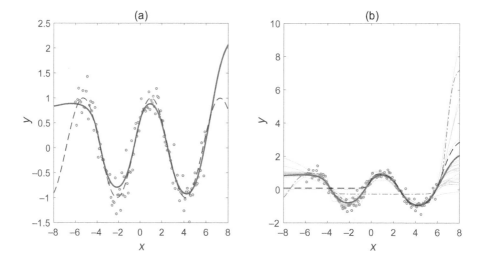

Figure 6.7 (a) The ensemble average (solid curve) of an MLP ensemble with 25 members, the true signal (dashed) and the training data (circles). (b) The 25 individual ensemble members are shown by thin curves in addition to the ensemble average (solid curve), with two of the individual members (corresponding to shallow local minima in the objective function) highlighted by the dashed curve and the dot-dash curve. Different vertical scales are used in (a) and (b).

---

[4] Some technical aspects that the reader can skip for now (as these topics are covered in the next two chapters): The optimal number of hidden nodes for the MLP ensemble was determined by a five-fold *cross-validation* procedure (Section 8.5). To avoid overfitting, each ensemble member was trained with 15% of the available training data reserved for *early stopping* (Section 8.4.2). Finally, using the optimal number of hidden nodes, the MLP ensemble was again trained without cross-validation.

In general, non-linear optimization is difficult, and convergence can be drastically slowed or numerically inaccurate if the input variables are poorly scaled. For instance, if an input variable has a mean exceeding its standard deviation, it is strongly recommended that the mean be subtracted from the variable to remove the systematic bias. Having input variables with very different variances also negatively impacts the non-linear optimization. Thus, it is strongly recommended that input data be *standardized* before applying the NN model, that is, each variable is transformed by (2.29), so each standardized variable will have zero mean and unit standard deviation.

As well as the logistic function, another commonly used sigmoidal activation function is the *hyperbolic tangent function* (Fig. 6.3),

$$f(x) = \tanh x = \frac{e^x - e^{-x}}{e^x + e^{-x}}. \tag{6.9}$$

The tanh function can be viewed as a scaled version of the logistic function (6.5), as the two are related by:

$$\tanh(x) = 2 \operatorname{logistic}(2x) - 1. \tag{6.10}$$

While the range of the logistic function is $(0, 1)$, the range of the tanh function is $(-1, 1)$. Thus, the output of the logistic function has a positive systematic bias that is not present in the output of the tanh function. It has been found empirically that a network using logistic activation functions in the hidden layer(s) tends to converge slower during model training than a corresponding network using tanh activation functions (LeCun, Kanter, et al., 1991). Thus, the tanh activation function is to be preferred over the logistic activation function in the hidden layer(s). For deep NN, that is, NN with many layers of hidden nodes, model error over test data was found to be much larger when using the logistic activation function rather than the tanh function (Glorot and Bengio, 2010). Furthermore, in deep NN models, unbounded activation functions such as ReLU Fig. 6.3(d) have replaced the bounded activation functions (e.g. tanh and logistic) (see Chapter 15).

When an output variable is bounded within the interval $(a, b)$, we can rescale the variable to lie within $(0, 1)$ or $(-1, 1)$ and use the logistic or the tanh function for the output activation function $g$. When the output is not restricted to a bounded interval, the identity activation function is commonly used for $g$, that is, the output is simply a linear combination of the hidden neurons in the layer before,

$$\hat{y}_k = \sum_{j=1}^{m_2} \tilde{w}_{kj} h_j + \tilde{b}_k. \tag{6.11}$$

For the one-hidden-layer NN, this means the output is just a linear combination of activation functions $f$, that is,

$$\hat{y}_k = \sum_{j=1}^{m_2} \tilde{w}_{kj} f\left(\sum_{i=1}^{m_1} w_{ji} x_i + b_j\right) + \tilde{b}_k. \tag{6.12}$$

In vector notation, this can be rewritten as

$$\hat{\mathbf{y}} = \sum_{j=1}^{m_2} \tilde{\mathbf{w}}_j \, f\left(\mathbf{w}_j \cdot \mathbf{x} + b_j\right) + \tilde{\mathbf{b}}. \tag{6.13}$$

There is some confusion in the literature on how to count the number of layers. The most common convention is to count the number of layers with adjustable weights/parameters. Some researchers count the total number of layers of neurons (which is not as relevant since the complexity of the MLP is governed by the layers with weights), and will refer to the one-hidden-layer MLP in Fig. 6.6 as a three-layer MLP. In this book, we will follow the standard terminology and refer to the MLP in Fig. 6.6 as a two-layer MLP or a one-hidden-layer MLP. A useful shorthand notation for describing the number of inputs, hidden and output neurons is the $m_1$-$m_2$-$m_3$ notation, where a 3-4-2 network denotes a one-hidden-layer MLP with three input, four hidden and two output nodes.

The total number of (weight and offset) parameters in an $m_1$-$m_2$-$m_3$ network is

$$N_{\mathrm{p}} = (m_1 + 1)\, m_2 + (m_2 + 1)\, m_3, \tag{6.14}$$

of which $m_1 m_2 + m_2 m_3 = m_2(m_1 + m_3)$ are weight parameters and $m_2 + m_3$ are offset parameters. In multiple linear regression problems with $m$ predictors and one response variable, that is, $y = a_0 + a_1 x_1 + \cdots + a_m x_m$, there are $m+1$ parameters. For corresponding non-linear regression with an $m$-$m_2$-1 MLP network, there will be $N_{\mathrm{p}} = (m + 1)m_2 + (m_2 + 1)$ parameters, usually greatly exceeding the number of parameters in the multiple linear regression model. Furthermore, the parameters of an NN model are, in general, very difficult to interpret, unlike the parameters in a multiple linear regression model, which have straightforward interpretations.

Incidentally, in a one-hidden-layer MLP, if the activation functions at both the hidden and output layers are linear then it is easily shown that the outputs are simply linear combinations of the inputs. The hidden layer is, therefore, redundant and can be deleted altogether. In this case, the MLP model simply reduces to multiple linear regression. This demonstrates that the presence of a non-linear activation function at the hidden layer is essential for the MLP model to have non-linear modelling capability.

While there can be exceptions (Weigend and Gershenfeld, 1994), MLP are normally used with $N_{\mathrm{p}} < N$, $N$ being the number of observations. Ideally, one would like to have $N_{\mathrm{p}} \ll N$, but in many environmental problems this is unattainable. In most climate problems, decent climate data have been available only after World War II. Another problem is that the number of predictors can be very large, although there can be strong correlations among predictors. In this case, principal component analysis (PCA) (Section 9.1) is commonly used for *dimension reduction*, that is, PCA is applied to the predictor variables, and the leading few principal components (PCs) are extracted then standardized to serve as inputs to the MLP network, to greatly reduce the number of input variables and, therefore, $N_{\mathrm{p}}$. With the advent of convolutional neural networks

(Chapter 15) in recent years, there is no longer a need to reduce predictors using PCA.

If there are multiple outputs, one has two choices: build a single NN with multiple outputs or build multiple NN models, each with a single output. If the output variables are correlated among themselves, then the single NN approach often leads to higher skills, since training separate networks for individual outputs does not take into account the relations between the output variables. An example is found in Krasnopolsky, Gemmill, et al. (1999), where the microwaves coming from the ocean surface were influenced by four variables – the surface wind speed, columnar water vapour, columnar liquid water and sea surface temperature. An MLP model with these four outputs and input from satellite microwave channels retrieves wind speed better than the single output model with only wind speed. On the other hand, if the output variables are uncorrelated (e.g. the outputs are principal components), then training separate networks often leads to slightly higher skills, as this approach focuses the single-output NN (with fewer parameters than the multiple-output NN) on one output variable without distraction from the other uncorrelated output variables.

## 6.3.1 Comparison with Polynomials B☺

For modelling a non-linear regression relation such as $\hat{y} = f(\mathbf{x})$, why can one not use the familiar Taylor series expansion instead of MLP? The Taylor expansion is of the form

$$\hat{y} = a_0 + \sum_{i_1=1}^{m} a_{i_1} x_{i_1} + \sum_{i_1=1}^{m} \sum_{i_2=1}^{m} a_{i_1 i_2} x_{i_1} x_{i_2} + \sum_{i_1=1}^{m} \sum_{i_2=1}^{m} \sum_{i_3=1}^{m} a_{i_1 i_2 i_3} x_{i_1} x_{i_2} x_{i_3} + \cdots ,$$

(6.15)

where there are $m$ input variables. In practice, the series is truncated, and only terms up to order $k$ are kept, that is, $\hat{y}$ is approximated by a $k$th order polynomial and can be written more compactly as

$$\hat{y} = \sum_{i=1}^{L} c_i \phi_i + c_0,$$

(6.16)

where $\phi_i$ represents the terms $x_{i_1}$, $x_{i_1} x_{i_2}$, ..., $c_i$ the corresponding parameters, and there are $L$ terms in the summation. $L$ grows with $m$ and $k$ at the order of $L \sim m^k$ (Bishop, 1995, exercise 1.8); thus, there are about $m^k$ parameters. Even with moderate $k$, $L \sim m^k$ means the number of model parameters rises at an unacceptable rate as $m$ increases. Barron (1993) showed that for the $m$-$m_2$-1 MLP model, the summed squared error is of order $O(1/m_2)$ (independent of $m$). In contrast, for the polynomial approximation, the summed squared error is at least $O(1/L^{2/m}) = O(1/m^{2k/m})$, yielding an unacceptably slow rate of convergence for large $m$. The polynomial model suffers from the *curse of dimensionality*, (Section 1.5), that is, the method is unusable at high dimensions.

Another major disadvantage with polynomials is that power functions $x^p$ can grow very aggressively when $x$ lies outside the training domain. For example, suppose $x$ is within $[-1, 1]$ during training, but a new test data point has $x = 1.2$. The power function $x^p$ takes on the value 1.44 (for $p = 2$), 1.73 ($p = 3$), 2.07 ($p = 4$), 2.49 ($p = 5$) and 6.19 ($p = 10$). In contrast, the sigmoidal-shaped activation functions in the MLP model are bounded. Thus, when a test value of $x$ lies outside the training domain, the polynomial model will tend to produce a spike in its predicted $\hat{y}$ value, whereas the MLP prediction is well behaved.

## 6.3.2  Hidden Neurons  [A]☺

So far, we have described only the MLP model with one hidden layer, which is sufficient for many problems. Let us extend to MLP models with more than one hidden layer, where each hidden layer receives input from the preceding layer of neurons, similar to (6.6). Prior to the advent of deep NN with many hidden layers in the twenty-first century (see Chapter 15), the activation function is usually a sigmoidal-shaped function (e.g. tanh). How many layers of hidden neurons does one need? And how many neurons in each hidden layer?

Studies such as Cybenko (1989), Hornik et al. (1989) and Hornik (1991) have shown that, given enough hidden neurons in an MLP with one hidden layer, the network can approximate arbitrarily well any continuous function $y = f(\mathbf{x})$. Thus, even though the original perceptron is of limited power, by adding one hidden layer, the MLP has become a successful *universal approximator*. There is, however, no guidance as to how many hidden neurons are needed. Experience tells us that a complicated continuous function $f$ with many bends needs many hidden neurons, while an $f$ with few bends needs fewer hidden neurons.

Intuitively, it is not hard to understand why a one-hidden-layer MLP with enough hidden neurons can approximate any continuous function. For instance, the function can be approximated by a sum of sinusoidal functions under Fourier decomposition. One can in turn approximate a sinusoidal curve by a series of small steps, that is, a sum of Heaviside step functions (or sigmoidal-shaped functions) (Bishop, 1995, section 4.3; Duda et al., 2001, section 6.2). Thus, the function can be approximated by a linear combination of step functions or sigmoidal-shaped functions, which is exactly the architecture of the one-hidden-layer MLP. Alternatively, one can think of the continuous function as being composed of localized bumps, each of which can be approximated by a sum of sigmoidal functions. Of course, the sinusoidal and bump constructions are only conceptual aids; the actual NN does not try deliberately to sum sigmoidal functions to yield sinusoidals or bumps.

Prior to the advent of deep NN models, most MLP models were run with a single hidden layer. In real-world applications, one may encounter some very complicated non-linear relations where a very large number of hidden neurons is needed if a single hidden layer is used, whereas if two or more hidden layers are used, a more modest number of hidden neurons suffices and gives greater accuracy. Using a large number of hidden layers was not practical until the development of deep NN models about a decade or so ago, which is a separate topic for Chapter 15.

Hidden neurons have been somewhat of a mystery to new users of NN models. They are intermediate variables needed to carry out the computation from the inputs to the outputs and are generally not easy to interpret. If the hidden neurons are few, then they might be viewed as a low-dimensional phase space describing the state of the system (Hsieh and B. Tang, 1998). For most NN applications, it is not worth spending time to find interpretations for the large number of hidden neurons.

### 6.3.3 Monotonic Multi-layer Perceptron Model  Ⓒ☺

In many situations, the relation between a predictor and the response variable can be monotonic,[5] as dictated by the physics, biology or mathematics. For instance, wind power increases monotonically with wind speed, and seasonal water availability increases monotonically with snowpack. Incorporating monotonic relation between predictor(s) and response in MLP models has been developed by H. Zhang and Z. Zhang (1999); Lang (2005); Cannon (2011b) and Cannon (2018).

The standard MLP with one or more hidden layers is a universal approximator, that is, being able to approximate any continuous function to arbitrary accuracy (Section 6.3.2). Imposing monotonic relations puts constraints on the model weights, so that the monotonic MLP (MonMLP) model needs two or more hidden layers for it to be a universal approximator (Lang, 2005). In practice, a one-hidden-layer MonMLP model, though not a universal approximator, is adequate for modelling simple functional relationships (Cannon, 2018).

Consider a MonMLP model with two hidden layers using tanh as the activation function (Lang, 2005). The first layer of hidden nodes $h_j^{(1)}$,

$$h_j^{(1)} = \tanh\left(\sum_i w_{ji}^{(1)} x_i + b_j^{(1)}\right), \tag{6.17}$$

with $x_i$ being the inputs and $w_{ji}^{(1)}$ and $b_j^{(1)}$, adjustable parameters, is followed by the second layer of hidden nodes $h_j^{(2)}$,

$$h_k^{(2)} = \tanh\left(\sum_j w_{kj}^{(2)} h_j^{(1)} + b_k^{(2)}\right), \tag{6.18}$$

which is then followed by the output

$$\hat{y} = \sum_k w_k^{(3)} h_k^{(2)} + b^{(3)}. \tag{6.19}$$

The MonMLP model imposes

$$w^{(l)} \geq 0, \quad \text{(for all } l \geq 2\text{)}, \tag{6.20}$$

---

[5] For all $x_1$ and $x_2$ with $x_1 \leq x_2$, a function $f$ is said to be monotonically increasing if $f(x_1) \leq f(x_2)$, and monotonically decreasing if $f(x_1) \geq f(x_2)$.

(this requirement holding even when there are more than two hidden layers in the network). For the first layer of weights, restrictions are imposed only when monotonic relations with certain predictors $x_m$ are desired, that is,

$$w_{jm}^{(1)} \begin{cases} \geq 0, & \text{for monotonically increasing relation with } x_m \\ \leq 0, & \text{for monotonically decreasing relation with } x_m, \end{cases} \tag{6.21}$$

for all $j$. The proof for why such a network allows monotonic relations for certain predictors and non-monotonic relations for other predictors is given by Lang (2005).

To implement the restrictions on the weights in (6.21), we can let

$$w_{jm}^{(1)} = \begin{cases} \exp(\tilde{w}_{jm}^{(1)}), & \text{for monotonically increasing relation with } x_m \\ -\exp(\tilde{w}_{jm}^{(1)}), & \text{for monotonically decreasing relation with } x_m, \end{cases} \tag{6.22}$$

with $-\infty < \tilde{w}_{jm} < \infty,$[6] and for (6.20),

$$w^{(l)} = \exp(\tilde{w}^{(l)}), \quad (\text{for all } l \geq 2). \tag{6.23}$$

In practice, there is no need for MonMLP to model monotonically decreasing relation with $x_m$, since one can replace a predictor $x_m$ by $-x_m$, and the relation switches to a monotonically increasing relation. Thus, the predictors can be separated into two groups, a group with monotonically increasing relation with the response variable and a second group without monotonicity constraints. The first group will be denoted by $x_{m \in M}$ and the second group $x_{i \in I}$, where $M$ is the set of indices for predictors with a monotonically increasing relation with the response and $I$ is the set for predictors without monotonicity constraints.

Thus, the first layer of hidden nodes $h_j^{(1)}$ is given by

$$h_j^{(1)} = \tanh\left(\sum_{m \in M} \exp(w_{jm}^{(1)}) x_m + \sum_{i \in I} w_{ji}^{(1)} x_i + b_j^{(1)}\right), \tag{6.24}$$

where, for brevity, the tilde has been dropped from $\tilde{w}_{jm}^{(1)}$. The second layer of hidden nodes $h_j^{(2)}$ is given by

$$h_k^{(2)} = \tanh\left(\sum_j \exp(w_{kj}^{(2)}) h_j^{(1)} + b_k^{(2)}\right), \tag{6.25}$$

and the output by

$$\hat{y} = \sum_k \exp(w_k^{(3)}) h_k^{(2)} + b^{(3)}. \tag{6.26}$$

One area of application of MonMLP is in quantile regression. One can view $q_\alpha$, the quantile $q$ at level $\alpha$ (see Section 2.9), as a function $q(\alpha)$, which must be a

---

[6] Technically speaking, the function $\exp(z)$, being $> 0$, can only get arbitrarily close to zero, with $-\infty < z < \infty$.

monotonically increasing function, that is, $q(\alpha_1) \leq q(\alpha_2)$ if $\alpha_1 \leq \alpha_2$. Otherwise, the quantiles from a model can cross, for example $q_{0.95}(x)$ somehow ended up lower than $q_{0.90}(x)$ for some value of the predictor $x$. Thus, MonMLP has been used in quantile regression neural network models (see Section 6.7) to prevent crossing quantiles (Cannon, 2011b; Cannon, 2018). MonMLP has been applied to the modelling of intensity-duration-frequency curves for precipitation (used in the design of civil infrastructure such as culverts, storm sewers, dams and bridges) (Cannon, 2018) and to the forecasting of seasonal water availability (Fleming and Goodbody, 2019).

## 6.4 Extreme Learning Machines $\boxed{\text{A}}$ ☺

From (6.13), an MLP model with one hidden layer can be expressed as

$$\hat{\mathbf{y}} = \sum_{j=1}^{m_2} \mathbf{a}_j h\left(\mathbf{w}_j \cdot \mathbf{x} + b_j\right) + \mathbf{a}_0, \tag{6.27}$$

where $\mathbf{x}$ is the input vector of dimension $m_1$ and $\hat{\mathbf{y}}$ the output vector of dimension $m_3$, $\mathbf{w}_j$ and $b_j$ are, respectively, the weight and offset parameters in the hidden layer (with $m_2$ hidden nodes), $\mathbf{a}_j$ and $\mathbf{a}_0$ are the weight and offset parameters in the output layer and $h$ is a non-linear activation function. The output $\hat{\mathbf{y}}$ is linearly related to the parameters $\mathbf{a}_j$ and $\mathbf{a}_0$, but non-linearly to $\mathbf{w}_j$ and $b_j$ (due to $h$ being non-linear); thus, non-linear optimization is needed in MLP to find the model parameters that minimize the mean squared error (MSE) between the model output $\hat{\mathbf{y}}$ and the target data $\mathbf{y}$. Non-linear optimization by the gradient descent type of methods is computationally expensive and converges to multiple local minima. In addition, as non-linear models can easily overfit (i.e. fit to the noise in the data), some form of regularization (e.g. weight penalty) is usually added to MLP models (Section 8.4).

The key idea in a *randomized neural network* approach (Broomhead and Lowe, 1988; Schmidt et al., 1992; Pao et al., 1994) is that not all the parameters in a neural network model need to be solved – some parameters can be randomly chosen and held constant. If $\mathbf{w}_j$ and $b_j$ are randomly chosen constants in the model, and the only adjustable parameters are $\mathbf{a}_j$ and $\mathbf{a}_0$, then these parameters can be found using only linear optimization, as $\hat{\mathbf{y}}$ is linearly related to $\mathbf{a}_j$ and $\mathbf{a}_0$ in (6.27). Among the various randomized NN models, the version that is most commonly used is called *extreme learning machine* (ELM) (G.-B. Huang, Zhu, et al., 2006; G.-B. Huang, D. H. Wang, et al., 2011; G. Huang, G.-B. Huang, et al., 2015).[7] Cao and Z. Lin (2015) conducted a survey on the use of ELM on high dimensional and large data applications.

---

[7] The name 'extreme learning machine' was criticized as being unnecessary (L. P. Wang and C. R. Wan, 2008; G.-B. Huang, 2008), as the ELM is basically the same model as the one proposed earlier by Schmidt et al. (1992) but without the $\mathbf{a}_0$ parameter and is also related to the models of Broomhead and Lowe (1988) and Pao et al. (1994).

The ELM model has an NN structure similar to an MLP with one hidden layer, that is,

$$\hat{\mathbf{y}}^{(n)} = \sum_{j=1}^{L} \mathbf{a}_j h\left(\mathbf{w}_j \cdot \mathbf{x}^{(n)} + b_j\right), \quad (n = 1, \ldots, N), \tag{6.28}$$

where $N$ is the number of observations and, for notational brevity, $L \equiv m_2$ (the number of hidden nodes) and $m \equiv m_3$ (the dimension of $\hat{\mathbf{y}}$). The offset or bias parameter $\mathbf{a}_0$ in (6.27) has been omitted as it is not necessary for the ELM model (G.-B. Huang, 2014). In matrix notation, with $\hat{\mathbf{y}}_n \equiv \hat{\mathbf{y}}^{(n)}$ and $\mathbf{x}_n \equiv \mathbf{x}^{(n)}$ $(n = 1, \ldots, N)$, (6.28) is simply

$$\hat{\mathbf{Y}} = \mathbf{H}\mathbf{A}, \tag{6.29}$$

where the hidden layer output matrix $\mathbf{H}$ of dimension $N \times L$ is

$$\mathbf{H} = \begin{bmatrix} h(\mathbf{w}_1 \cdot \mathbf{x}_1 + b_1) & \cdots & h(\mathbf{w}_L \cdot \mathbf{x}_1 + b_L) \\ \vdots & \ddots & \vdots \\ h(\mathbf{w}_1 \cdot \mathbf{x}_N + b_1) & \cdots & h(\mathbf{w}_L \cdot \mathbf{x}_N + b_L) \end{bmatrix} \tag{6.30}$$

and the output matrix $\hat{\mathbf{Y}}$ of dimension $N \times m$ and the parameter matrix $\mathbf{A}$ of dimension $L \times m$ are given by

$$\hat{\mathbf{Y}} = \begin{bmatrix} \hat{\mathbf{y}}_1^{\mathrm{T}} \\ \vdots \\ \hat{\mathbf{y}}_N^{\mathrm{T}} \end{bmatrix} \quad \text{and} \quad \mathbf{A} = \begin{bmatrix} \mathbf{a}_1^{\mathrm{T}} \\ \vdots \\ \mathbf{a}_L^{\mathrm{T}} \end{bmatrix}. \tag{6.31}$$

Unlike MLP, ELM only treats $\mathbf{a}_j$ as adjustable parameters, while $\mathbf{w}_j$ and $b_j$ are treated as randomly assigned constants, so the $\mathbf{H}$ matrix is known, while $\mathbf{A}$ contains the unknown parameters. ELM finds the least-squares solution of the *linear* system (G.-B. Huang, D. H. Wang, et al., 2011),

$$\mathbf{Y} = \hat{\mathbf{Y}} + \mathbf{E} = \mathbf{H}\mathbf{A} + \mathbf{E}, \tag{6.32}$$

where the target data matrix $\mathbf{Y}$ of dimension $N \times m$ and the error matrix $\mathbf{E}$ of dimension $N \times m$ are given by

$$\mathbf{Y} = \begin{bmatrix} \mathbf{y}_1^{\mathrm{T}} \\ \vdots \\ \mathbf{y}_N^{\mathrm{T}} \end{bmatrix} \quad \text{and} \quad \mathbf{E} = \begin{bmatrix} \epsilon_{11} & \cdots & \epsilon_{1m} \\ \vdots & \ddots & \vdots \\ \epsilon_{N1} & \cdots & \epsilon_{Nm} \end{bmatrix}. \tag{6.33}$$

Equation (6.32) is of the same mathematical form as for the multivariate linear regression problem in (5.78), except that the $\mathbf{X}$ matrix has been replaced by the $\mathbf{H}$ matrix. The linear system for $\mathbf{A}$ is solved via the Moore–Penrose inverse $\mathbf{H}^+$ as in (5.81), that is,

$$\hat{\mathbf{A}} = \mathbf{H}^+\mathbf{Y}, \text{ with } \mathbf{H}^+ = \mathbf{P}\mathbf{H}^{\mathrm{T}} \text{ and } \mathbf{P} = (\mathbf{H}^{\mathrm{T}}\mathbf{H})^{-1}, \tag{6.34}$$

where $\mathbf{P}$ is the inverse covariance matrix of the hidden layer outputs. The model output is

$$\hat{\mathbf{Y}} = \mathbf{H}\hat{\mathbf{A}} = \mathbf{H}\mathbf{H}^{+}\mathbf{Y}. \qquad (6.35)$$

For ELM to work well, the random parameters $\mathbf{w}_j$ and $b_j$ need to be chosen within an appropriate range (Parviainen et al., 2010; A. R. Lima et al., 2015). For instance, using a small range for $\mathbf{w}_j$ would limit the ELM to nearly linear models. The original ELM MATLAB code (April 2004) from G.-B. Huang's website (https://personal.ntu.edu.sg/egbhuang/elm_random_hidden_nodes.html) expects input training data to be scaled to the interval $[-1, 1]$ and chooses the random values of $\mathbf{w}_j$ uniformly from the interval $(-1, 1)$ and $b_j$ uniformly from $(0, 1)$.[8]

The ELM is almost always run as an ensemble of models with different random values chosen for $\mathbf{w}_j$ and $b_j$. An ensemble has typically 20–100 ensemble members, and for regression problems, the final model output is the average of the individual member outputs. G. B. Huang, Zhu, et al. (2006) considered $L$, the number of hidden nodes, as the only parameter that needs to be tuned for the ELM, since having too small or too large a value for $L$ leads to underfitting or overfitting, respectively (Fig. 6.8(a)). The optimal $L$ is usually determined by trying various values for $L$ and selecting the best $L$ based on model performance (e.g. the RMSE) on some *validation data*[9] (Fig. 6.8(b)). The parameter $c$ in (6.36) can also be tuned on validation data to improve ELM performance for some problems (Parviainen et al., 2010; A. R. Lima et al., 2015).

Since the parameters $\mathbf{w}_j$ and $b_j$ are not tuned in ELM, ELM is typically run with considerably more hidden nodes than MLP (A. R. Lima et al., 2015, table 4). For ELM, the number of adjustable parameters in $\mathbf{A}$ is $Lm$, or just $L$ when there is only one output variable. For comparison, an MLP with $m_1$ inputs, $L$ hidden nodes and 1 output has $(m_1 + 1)L + (L + 1) = m_1 L + 2L + 1$ adjustable parameters.

---

[8] When using the ELM for non-linear regression on nine environmental datasets, A. R. Lima et al. (2015) found that using these default intervals was not optimal. With input variables in the training data standardized to zero mean and unit variance (which is less sensitive to outliers than scaling them to the interval $[-1, 1]$), A. R. Lima et al. (2015) used uniformly distributed random values from the interval $[-r, r]$ for the elements of $\mathbf{w}_j$, where

$$r = c m_1^{-0.5}, \qquad (6.36)$$

with $m_1$ the input dimension, and $c = 1$ for the tanh activation function and $c = 2$ for the logistic function, due to the relation between the two functions in (6.10). For the uniformly distributed offset parameter $b_j$, the interval of $[-c, c]$ was recommended (with $c = 1$ for the tanh activation function and $c = 2$ for the logistic function).

[9] Validation data are data segregated from training data. As they have not been used in training the model, they provide independent data for checking the performance of the trained model. Validation was mentioned in Section 5.7 for determining the optimal ridge regression parameter. The crucial concept of model validation is studied in depth in Chapter 8, especially in Section 8.5 on cross-validation.

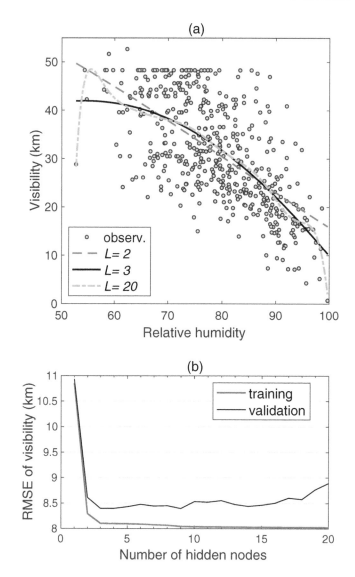

Figure 6.8 The relative humidity–visibility data from Fig. 5.2 (but with 457 instead of 100 data points chosen) are used to train an ensemble of 100 ELM models (using the logistic activation function in the hidden layer), with the ensemble averaged output in (a) shown for $L = 2$, 3 and 20 hidden nodes, with underfitting seen when $L = 2$ and overfitting when $L = 20$. (b) With five-fold cross-validation (see Section 8.5), the RMSE of the validation data (dashed curve) bottoms at $L = 3$ hidden nodes. RMSE of the training data (solid curve) keeps on decreasing as $L$ increases, as the model fits closer to the noisy data. [Data source: weatherstats.ca based on Environment and Climate Change Canada data.]

## 6.4.1 Online Learning  $\boxed{\text{B}}$☺

Thus, the MLP model requiring non-linear optimization has been simplified in ELM to just solving a linear least squares problem as in multivariate linear regression. So far, the ELM model is set up for batch learning. To update a model when new data become available, a batch model has the disadvantage in that it needs to be retrained with the entire data record instead of just the new data.

In contrast to batch learning, *online learning* is much less expensive computationally when new data become available continually, as it updates a model using only the new data. In Section 5.4, we found that it was straightforward to update a multivariate linear regression model using a recursive least squares algorithm as new data become available. Similarly, an online learning version of the ELM model can be used to update ELM inexpensively using recursive linear least squares as new data arrive continually.

The *online sequential extreme learning machine* (OSELM) model (N.-Y. Liang et al., 2006; Lan et al., 2009) has two phases: (i) an initialization phase, where the batch version of ELM in (6.34) is trained with some initial data, then (ii) the online sequential learning phase, where the model is frequently updated with the newly available data. In the initialization phase, the initial training dataset $\{\mathbf{X}_0, \mathbf{Y}_0\} = \{\mathbf{x}^{(n)}, \mathbf{y}^{(n)}\}_{n=1}^{N_0}$ is used to generate $\mathbf{H}_0$, $\mathbf{A}_0$ and $\mathbf{P}_0$ using (6.30) and (6.34), with $N_0$ being the number of data points available during the initialization phase.

During step $k + 1$ in the online sequential learning phase ($k = 0, 1, \ldots$), the $(k + 1)$th chunk of new data $\{\mathbf{X}_{k+1}, \mathbf{Y}_{k+1}\}$ (with $N_{k+1}$ observations) becomes available, and $\mathbf{H}_{k+1}$ is calculated using only the new data. $N_{k+1}$ can be as small as 1, that is, the model can be updated whenever a single observation becomes available. From N.-Y. Liang et al. (2006), analogous to our results for online linear regression, that is, (5.102) and (5.103), we obtain the updated model,

$$\mathbf{P}_{k+1} = \mathbf{P}_k - \mathbf{P}_k\mathbf{H}_{k+1}^{\mathrm{T}}(\mathbf{I} + \mathbf{H}_{k+1}\mathbf{P}_k\mathbf{H}_{k+1}^{\mathrm{T}})^{-1}\mathbf{H}_{k+1}\mathbf{P}_k, \tag{6.37}$$

$$\hat{\mathbf{A}}_{k+1} = \hat{\mathbf{A}}_k + \mathbf{P}_{k+1}\mathbf{H}_{k+1}^{\mathrm{T}}(\mathbf{Y}_{k+1} - \mathbf{H}_{k+1}\hat{\mathbf{A}}_k), \tag{6.38}$$

with $\mathbf{I}$ being the identity matrix.

In summary, ELM has some notable advantages over MLP. Linear optimization in ELM is much simpler than non-linear optimization in MLP. ELM does not need regularization (such as weight penalty) typically used in MLP. Over nine environmental regression problems, A. R. Lima et al. (2015, tables 2–4) found that ELM performed with similar accuracy to MLP and was usually computationally cheaper (except when $L$ is very large in ELM). With new data becoming available continually, ELM can be updated inexpensively by OSELM. In environmental sciences, OSELM has been successfully applied to streamflow forecasting (A. R. Lima et al., 2016; Yadav et al., 2016) and to air quality forecasting (H. Peng et al., 2017).

A limitation of OSELM is that $L$, the number of hidden nodes, is fixed as online learning proceeds. As online data accumulate, information on longer time-scale variability becomes available. The fixed $L$ (determined by the relatively small amount of initial training data) will be too small to model the longer term variability, and the model becomes sub-optimal from underfitting. There is a need for algorithms allowing model complexity (i.e. the number of hidden nodes) to change during the online learning phase. The variable complexity OS-ELM (VC-OSELM) model (A. R. Lima, Hsieh, et al., 2017) dynamically adds or removes hidden nodes in the OSELM as needed, so the model complexity can be self-adapted.

## 6.4.2 Random Vector Functional Link  Ⓒ☺

ELM can be regarded as a special case of a randomized NN model called the random vector functional link (RVFL) model, proposed earlier by Pao et al. (1994). RVFL is very similar to ELM, but it allows direct links between the inputs and the outputs (Fig. 6.9), that is, (6.28) for the ELM is changed to

$$\hat{\mathbf{y}}^{(n)} = \sum_{j=1}^{L} \mathbf{a}_j h\left(\mathbf{w}_j \cdot \mathbf{x}^{(n)} + b_j\right) + \sum_{j=1}^{d} \tilde{\mathbf{a}}_j \, x_{nj}, \tag{6.39}$$

where the first term is the same as that in ELM and the second term allows direct linear mapping from the inputs $x_{nj}$ to the outputs (there being $d$ input variables and $n = 1, \ldots, N$ observations). The direct link architecture is an example of *skip connection*, as the input signal is allowed to bypass the hidden layer in Fig. 6.9. Skip connection architectures later became very common in deep NN models, such as Resnet (Section 15.2.4) and U-net (Section 15.3.1).

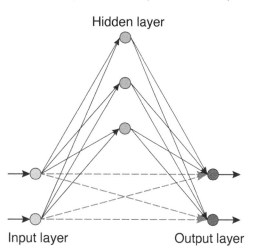

Figure 6.9 The random vector functional link (RVFL) model is similar in architecture to an MLP model except direct linear mapping from the input layer to the output layer is allowed (dashed arrows). In this example, the RVFL model has two inputs, three hidden nodes and two outputs.

Letting $a_{L+j} \equiv \tilde{a}_j$ $(j = 1, \ldots, d)$ and redefining the index $j$ in the second summation, we obtain

$$\hat{\mathbf{y}}^{(n)} = \sum_{j=1}^{L} \mathbf{a}_j h\left(\mathbf{w}_j \cdot \mathbf{x}^{(n)} + b_j\right) + \sum_{j=L+1}^{L+d} \mathbf{a}_j \, x_{n,j-L}, \tag{6.40}$$

so that in matrix notation, with $\hat{\mathbf{y}}_n \equiv \hat{\mathbf{y}}^{(n)}$ and $\mathbf{x}_n \equiv \mathbf{x}^{(n)}$ $(n = 1, \ldots, N)$, (6.40) is again of the form

$$\hat{\mathbf{Y}} = \mathbf{HA}, \tag{6.41}$$

where the hidden layer output matrix $\mathbf{H}$ of dimension $N \times (L + d)$ is

$$\mathbf{H} = \begin{bmatrix} h(\mathbf{w}_1 \cdot \mathbf{x}_1 + b_1) & \cdots & h(\mathbf{w}_L \cdot \mathbf{x}_1 + b_L) & x_{11} & \cdots & x_{1d} \\ \vdots & \ddots & \vdots & \vdots & \ddots & \vdots \\ h(\mathbf{w}_1 \cdot \mathbf{x}_N + b_1) & \cdots & h(\mathbf{w}_L \cdot \mathbf{x}_N + b_L) & x_{N1} & \cdots & x_{Nd} \end{bmatrix}, \tag{6.42}$$

and the output matrix $\hat{\mathbf{Y}}$ of dimension $N \times m$ and the parameter matrix $\mathbf{A}$ of dimension $(L + d) \times m$ are given by

$$\hat{\mathbf{Y}} = \begin{bmatrix} \hat{\mathbf{y}}_1^{\mathrm{T}} \\ \vdots \\ \hat{\mathbf{y}}_N^{\mathrm{T}} \end{bmatrix} \quad \text{and} \quad \mathbf{A} = \begin{bmatrix} \mathbf{a}_1^{\mathrm{T}} \\ \vdots \\ \mathbf{a}_{L+d}^{\mathrm{T}} \end{bmatrix}. \tag{6.43}$$

The solution for the RVFL model is of the same form as the ELM solution, as given by (6.34) and (6.35). The online learning version of RVFL is also analogous to that for the ELM.

With direct link between inputs and outputs, RVFL makes it easier to model the linear relation between the inputs and the outputs, freeing the weights $\mathbf{w}_j$ and $b_j$ to concentrate on modelling the non-linear part of the relation. In tests conducted over 16 classification problems, Y. S. Zhang et al. (2019) found RVFL having better accuracy but requiring more CPU time for training than ELM.

## 6.5  Radial Basis Functions  B☺

While sigmoidal-shaped activation functions (e.g. the hyperbolic tangent function) are widely used in feed-forward NN, *radial basis functions* (RBF), typically Gaussian-shaped functions, introduced by Broomhead and Lowe (1988), are also commonly used.

Radial basis function methods originated in the problem of *exact interpolation*, where every input vector is required to be mapped exactly to the corresponding target vector (Powell, 1987). First, consider a one-dimensional target space. The output of the mapping $f$ is a linear combination of basis functions $g$

$$f(\mathbf{x}) = \sum_{j=1}^{k} w_j\, g(\|\mathbf{x} - \mathbf{c}_j\|, \sigma), \tag{6.44}$$

where each basis function is specified by its centre $\mathbf{c}_j$ and a width parameter $\sigma$. The name 'radial basis function' arises because the function $g$, depending on the distance $\|\mathbf{x} - \mathbf{c}_j\|$, is radially symmetric about the centre $\mathbf{c}_j$. In the case of exact interpolation, if there are $N$ observations, then there are $k = N$ basis functions to allow for the exact interpolation, and each $\mathbf{c}_j$ corresponds to one of the input data vectors. There is a number of choices for the basis functions, the most common being the Gaussian function

$$g(r, \sigma) = \exp\left(-\frac{r^2}{2\sigma^2}\right). \tag{6.45}$$

In NN applications, exact interpolation is undesirable, as it would mean an exact fit to noisy data. The remedy is simply to choose $k < N$, that is, use fewer (often far fewer) basis functions than the number of observations. This prevents an exact fit but allows a smooth interpolation of the noisy data. By adjusting the number of basis functions used, one can obtain the desired level of closeness of fit. The mapping is now

$$f(\mathbf{x}) = \sum_{j=1}^{k} w_j\, g(\|\mathbf{x} - \mathbf{c}_j\|, \sigma_j) + w_0, \tag{6.46}$$

where (i) the centres $\mathbf{c}_j$ are no longer given by the input data vectors but are determined during training, (ii) instead of using a uniform $\sigma$ for all the basis functions, each basis function has its own width $\sigma_j$, determined from training, and (iii) an offset parameter $w_0$ has been added.

If the output is multivariate, the mapping simply generalizes to

$$f_i(\mathbf{x}) = \sum_{j=1}^{k} w_{ji}\, g(\|\mathbf{x} - \mathbf{c}_j\|, \sigma_j) + w_{0i}, \tag{6.47}$$

for the $i$th output variable.

Moody and Darken (1989) introduced *normalized* (a.k.a. *renormalized*) radial basis functions, where the RBF $g(\|\mathbf{x} - \mathbf{c}_j\|, \sigma_j)$ in (6.46) is replaced by the normalized version

$$\frac{g(\|\mathbf{x} - \mathbf{c}_j\|, \sigma_j)}{\sum_{l=1}^{k} g(\|\mathbf{x} - \mathbf{c}_l\|, \sigma_l)}. \tag{6.48}$$

Using RBF can lead to holes, that is, regions where the basis functions all give little support (Fig. 6.10); thus, normalized RBF is also widely used as it provides a more efficient basis.

While RBF neural networks can be trained like MLP networks by back-propagation (termed adaptive RBFs), RBFs are most commonly used in a non-adaptive manner, that is, the training is performed in two separate stages, where the first stage uses unsupervised learning to find the centres and widths of the RBFs, followed by supervised learning via linear least squares to minimize the MSE between the network output and the target data, similar to the linear least squares approach in extreme learning machines (ELM) (Section 6.4).

Let us describe the procedure of the non-adaptive RBF. In the literature, there are many different ways to set up the RBF basis functions (Schwenker et al., 2001). First choose $k$, the number of RBFs to be used. To find the

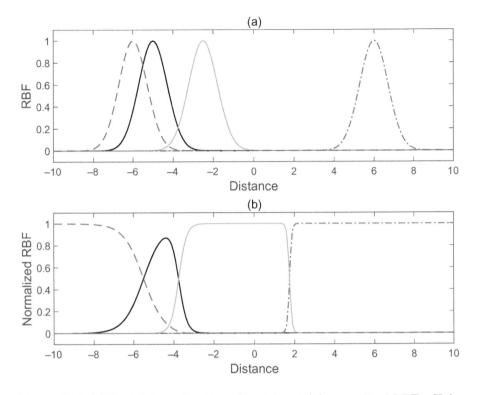

Figure 6.10 (a) Radial basis functions (RBFs) and (b) normalized RBFs. Holes are present in (a), where RBFs with fixed width $\sigma$ are used. This problem is avoided in (b) with the normalized RBFs.

centres $\mathbf{c}_j$ of the RBFs, one commonly uses *K-means clustering* (Section 2.16) or self-organizing maps (SOMs) (Section 10.4).

The width parameters $\sigma_j$ are often estimated in a somewhat ad hoc manner. For the $j$th centre $\mathbf{c}_j$, compute the distance $r_{jl}$ between $\mathbf{c}_j$ and its $l$th closest neighbouring centre. Average the distances over the $m$ closest neighbouring centres, that is, let

$$r_j = \frac{1}{m} \sum_{l=1}^{m} r_{jl}. \tag{6.49}$$

Let $\sigma_j = \alpha r_j$, where $\alpha$ is a heuristic scale factor allowing one to choose wider or narrower basis functions. Fortunately, the RBF model is not very sensitive to the width parameter, so a default value of $\alpha = 1$ is likely to work for most problems (Hertz et al., 1991, p. 249).

With the basis functions $g(\|\mathbf{x} - \mathbf{c}_j\|, \sigma_j)$ now determined, the only task left is to find the weights $w_{ji}$ in the equation

$$f_i(\mathbf{x}) = \sum_{j=0}^{k} w_{ji}\, g_j(\mathbf{x}), \tag{6.50}$$

which is the same as (6.47), with $g_j(\mathbf{x}) = g(\|\mathbf{x}-\mathbf{c}_j\|, \sigma_j)$ $(j = 1,\ldots,k)$, $g_0(\mathbf{x}) = 1$, and the summation starting from $j = 0$ to incorporate the offset parameter $w_{0i}$. The network output $f_i(\mathbf{x})$ is to approximate the target data $y_i(\mathbf{x})$ by minimizing the MSE, which is simply a linear least squares problem.

Given $N$ observations, the linear least squares problem in matrix notation is

$$\mathbf{Y} = \hat{\mathbf{Y}} + \mathbf{E} = \mathbf{GW} + \mathbf{E}, \tag{6.51}$$

where $\mathbf{Y}$ is the target data matrix, $\hat{\mathbf{Y}}$ the model output and $\mathbf{E}$ the error or residual in the least squares fit, with the matrix elements $(\mathbf{Y})_{ni} = y_i(\mathbf{x}^{(n)})$, $(n = 1,\ldots,N)$, $(\hat{\mathbf{Y}})_{ni} = f_i(\mathbf{x}^{(n)})$, $(\mathbf{G})_{nj} = g_j(\mathbf{x}^{(n)})$ and $(\mathbf{W})_{ji} = w_{ji}$.

The linear least squares solution via the Moore–Penrose inverse $\mathbf{G}^+$ is analogous to that in (6.34), that is,

$$\mathbf{W} = \mathbf{G}^+\mathbf{Y} = (\mathbf{G}^{\mathrm{T}}\mathbf{G})^{-1}\mathbf{G}^{\mathrm{T}}\mathbf{Y}. \tag{6.52}$$

Overfitting occurs when $k$, the number of basis functions, is too large, while underfitting occurs when $k$ is too small. To choose the optimal $k$, a common approach is to run the RBF model over a range of values for $k$, then select the best $k$ based on the model performance (e.g. the RMSE) over some validation data.

In summary, the RBF NN is most commonly trained in two distinct stages. The first stage involves finding the centres and widths of radial basis functions by unsupervised learning of the input data (with no consideration of the output target data). The second stage involves finding the best linear least squares fit to the target data (supervised learning). In contrast, in the multi-layer perceptron (MLP) NN approach, all weights are trained together under supervised learning. The supervised learning in MLP involves non-linear optimizations that usually have multiple minima in the objective function, whereas the supervised learning in RBF involves only linear least squares optimization, hence no multiple minima in the objective function – a main advantage of the RBF over the MLP approach. However, that the basis functions are computed in the RBF NN approach without considering the target data can be a major drawback, especially when the input dimension is large. The reason is that many of the input variables may have significant variance but have no influence on fitting the output to the target data, yet these irrelevant inputs introduce a large number of basis functions. The second stage training may then involve solving a very large, ill-conditioned matrix problem. To alleviate this problem, one can add a regularization term with a regularization parameter $\lambda$ (a.k.a. weight penalty or weight decay parameter) as in ridge regression (5.110), thereby replacing the solution (6.52) with

$$\mathbf{W} = (\mathbf{G}^{\mathrm{T}}\mathbf{G} + \lambda\mathbf{I})^{-1}\,\mathbf{G}^{\mathrm{T}}\mathbf{Y}, \tag{6.53}$$

where the best value for $\lambda$ can be chosen based on model performance on some validation data from repeated runs with various $\lambda$ values.

One can summarize the architectures of the three types of models, the two-layer MLP, the ELM and the non-adaptive RBF as follows: The weights in the first layer of mapping (from inputs to hidden layer) and the weights in the second layer (from hidden layer to outputs) are treated differently in the three types of models. In the two-layer MLP, the weights in both layers are optimized under supervised learning, thereby requiring non-linear optimization. In the ELM, the weights in the first layer are randomly chosen and held constant, then the weights in the second layer are optimized under supervised learning, requiring only linear least squares optimization. In RBF, the first layer of weights is derived from unsupervised learning (e.g. $K$-means clustering), then the weights in the second layer are optimized under supervised learning, also requiring only linear least squares optimization. After the non-adaptive RBF has been trained, its accuracy can be further improved by an optional supervised learning phase, where all its parameters ($\mathbf{c}_j$, $\sigma_j$ and $w_{ji}$) are further adjusted using non-linear optimization (Schwenker et al., 2001).

## 6.6 Modelling Conditional Distributions $\boxed{\text{B}}$☺

So far, the NN models have been used for non-linear regression, that is, the output $y$ is related to the inputs $\mathbf{x}$ by a non-linear function, $y = f(\mathbf{x})$. In many applications, one is less interested in a single predicted value for $y$ given by $f(\mathbf{x})$ than in $p(y|\mathbf{x})$, a conditional probability distribution of $y$ given $\mathbf{x}$. With $p(y|\mathbf{x})$, one can easily obtain a single predicted value for $y$ by taking the mean, the median or the mode (i.e. the location of the peak) of the distribution $p(y|\mathbf{x})$. In addition, the distribution provides an estimate of the uncertainty in the predicted value for $y$. For example, if the forecasted air temperature is $27°C$ on two separate days, but the $95\%$ confidence intervals are $[25, 29]°C$ and $[24, 35]°C$, managers of utility companies would find the two CIs much more informative than the predicted mean of $27°C$.

Suppose we have selected an appropriate distribution function, which is governed by some parameters $\boldsymbol{\theta}$. For example, the gamma distribution (Section 3.8) is governed by two parameters ($\boldsymbol{\theta} = [a, b]^T$):

$$\text{Ga}(y \,|\, a, b) = \frac{(y/b)^{a-1} \exp(-y/b)}{b \, \Gamma(a)}, \qquad 0 \le y < \infty, \qquad (6.54)$$

where $a > 0$, $b > 0$ and $\Gamma$ is the gamma function.

In general, the parameters of the given distribution are to be functions of the inputs $\mathbf{x}$, that is, $\boldsymbol{\theta} = \boldsymbol{\theta}(\mathbf{x})$. The conditional distribution $p(y|\mathbf{x})$ is now replaced by $p(y|\boldsymbol{\theta}(\mathbf{x}))$. The functions $\boldsymbol{\theta}(\mathbf{x})$ can be approximated by an NN (e.g. an MLP, ELM or an RBF) model, that is, inputs of the NN model are $\mathbf{x}$ while the outputs are $\boldsymbol{\theta}$. The NN used to model the parameters of a conditional probability density distribution is called a *conditional density network* (CDN) model. To train the NN model, we need an objective function.

To obtain an objective function, we turn to the principle of *maximum likelihood* (Section 3.5). If we have a probability distribution $p(\mathbf{y}|\boldsymbol{\theta})$, and we have a dataset $D$ containing the observed values $\mathbf{y}^{(n)}$, $n = 1, \ldots, N$, then the parameters $\boldsymbol{\theta}$ can be found by maximizing the likelihood function $p(D|\boldsymbol{\theta})$, that is, the parameters $\boldsymbol{\theta}$ should be chosen so that the likelihood of observing $D$ is maximized. Note that $p(D|\boldsymbol{\theta})$ is a function of $\boldsymbol{\theta}$ as $D$ is known, and the output $\mathbf{y}$ can be multivariate.

Since the observed data have $N$ observations, and if we assume independent observations so we can multiply their probabilities together, the likelihood function is then

$$L = p(D|\boldsymbol{\theta}) = \prod_{n=1}^{N} p(\mathbf{y}^{(n)}|\boldsymbol{\theta}^{(n)}) = \prod_{n=1}^{N} p(\mathbf{y}^{(n)}|\boldsymbol{\theta}(\mathbf{x}^{(n)})), \tag{6.55}$$

where $\boldsymbol{\theta}(\mathbf{x}^{(n)})$ are determined by the weights $\mathbf{w}$ (including all weight and offset parameters) of the NN model. Thus,

$$L = \prod_{n=1}^{N} p(\mathbf{y}^{(n)}|\mathbf{w}, \mathbf{x}^{(n)}). \tag{6.56}$$

Mathematically, maximizing the likelihood function is equivalent to minimizing the negative logarithm of the likelihood function, since the logarithm is a monotonically increasing function. Thus, we choose the objective function to be

$$J = -\ln L = -\sum_{n=1}^{N} \ln p(\mathbf{y}^{(n)}|\mathbf{w}, \mathbf{x}^{(n)}), \tag{6.57}$$

where we have converted the (natural) logarithm of a product of $N$ terms to a sum of $N$ logarithmic terms. Since $\mathbf{x}^{(n)}$ and $\mathbf{y}^{(n)}$ are known from the given data, the unknowns $\mathbf{w}$ are optimized to attain the minimum $J$. Once the weights of the NN model are solved, then for any input $\mathbf{x}$, the NN model outputs $\boldsymbol{\theta}(\mathbf{x})$, and we obtain the conditional distribution $p(\mathbf{y}|\mathbf{x})$ from $p(\mathbf{y}|\boldsymbol{\theta}(\mathbf{x}))$. An $\mathcal{R}$ package for building CDN models, with many choices for the type of distribution, is described in Cannon (2012).

In our example, where $p(y|\boldsymbol{\theta})$ is the gamma distribution (6.54), we have to ensure that the outputs of the NN model satisfy the restriction that both parameters ($a$ and $b$) of the gamma distribution have to be positive. This can be accommodated easily by letting

$$a = \exp(z_1), \quad b = \exp(z_2), \tag{6.58}$$

where $z_1$ and $z_2$ are the NN model outputs, with $-\infty < z_1 < \infty$ and $-\infty < z_2 < \infty$.

For extreme value analysis, the generalized extreme value (GEV) distribution has been presented in Section 3.11. A CDN model has been developed where an MLP NN model is used to predict the parameters of the GEV distribution (Cannon, 2010; Cannon, 2011a).

## 6.6.1  Mixture Density Network  B⊖

The disadvantage of specifying a parametric form for the conditional distribution
is that even adjusting the parameters may not lead to a good fit to the observed
data. One way to produce an extremely flexible distribution function is to use
a mixture (i.e. a weighted sum) of simple distribution functions to produce a
*mixture model*:

$$p(\mathbf{y}|\mathbf{x}) = \sum_{k=1}^{K} a_k(\mathbf{x})\phi_k(\mathbf{y}|\mathbf{x}), \tag{6.59}$$

where $K$ is the number of components (also called *kernels*) in the mixture, $a_k(\mathbf{x})$
is the (non-negative) *mixing coefficient* and $\phi_k(\mathbf{y}|\mathbf{x})$ the conditional distribution
from the $k$th kernel. There are many choices for the kernel distribution function
$\phi$, the most popular choice being the Gaussian function

$$\phi_k(\mathbf{y}|\mathbf{x}) = \frac{1}{(2\pi)^{m/2}\,\sigma_k^m(\mathbf{x})} \exp\left(-\frac{\|\mathbf{y} - \boldsymbol{\mu}_k(\mathbf{x})\|^2}{2\sigma_k^2(\mathbf{x})}\right), \tag{6.60}$$

where the Gaussian kernel function is centred at $\boldsymbol{\mu}_k(\mathbf{x})$ with variance $\sigma_k^2(\mathbf{x})$,
and $m$ is the dimension of the output vector $\mathbf{y}$. With large enough $K$, and with
properly chosen $\boldsymbol{\mu}_k(\mathbf{x})$ and $\sigma_k(\mathbf{x})$, $p(\mathbf{y}|\mathbf{x})$ of any form can be approximated to
arbitrary accuracy by the Gaussian mixture model.

In *mixture density network* (MDN) models (Bishop, 1994; Nabney, 2002;
Bishop, 2006), neural network models (e.g. MLP) can be used to approximate
the parameters of the Gaussian mixture model, namely $\boldsymbol{\mu}_k(\mathbf{x})$, $\sigma_k(\mathbf{x})$ and $a_k(\mathbf{x})$.
There are a total of $m \times K$ parameters in $\boldsymbol{\mu}_k(\mathbf{x})$, and $K$ parameters in each of
$\sigma_k(\mathbf{x})$ and $a_k(\mathbf{x})$, hence a total of $(m + 2)K$ parameters. Let $\mathbf{z}$ represent the
$(m + 2)K$ outputs of the NN model. Since there are constraints on $\sigma_k(\mathbf{x})$
and $a_k(\mathbf{x})$, they cannot simply be the direct outputs from the NN model. As
$\sigma_k(\mathbf{x}) > 0$, we need to represent them as

$$\sigma_k = \exp\left(z_k^{(\sigma)}\right), \tag{6.61}$$

where $z_k^{(\sigma)}$ are the NN model outputs related to the $\sigma$ parameters.

From the normalization condition

$$\int p(\mathbf{y}|\mathbf{x})\mathrm{d}\mathbf{y} = 1, \tag{6.62}$$

we obtain, via (6.59) and (6.60), the constraints

$$\sum_{k=1}^{K} a_k(\mathbf{x}) = 1, \quad 0 \le a_k(\mathbf{x}) \le 1. \tag{6.63}$$

To satisfy these constraints, $a_k(\mathbf{x})$ is related to the NN output $z_k^{(a)}$ by a *softmax*
*function*, that is,

$$a_k = \frac{\exp\left(z_k^{(a)}\right)}{\sum_{k'=1}^{K} \exp\left(z_{k'}^{(a)}\right)}. \tag{6.64}$$

As the softmax function is widely used in NN models involved in classification problems, it is discussed in detail in Section 12.7. Since there are no constraints on $\boldsymbol{\mu}_k(\mathbf{x})$, they can simply be the NN model outputs directly, that is,

$$\mu_{jk} = z_{jk}^{(\mu)}. \tag{6.65}$$

The objective function for the NN model is again obtained via the likelihood as in (6.57), with

$$J = -\sum_n \ln \left( \sum_{k=1}^{K} a_k(\mathbf{x}^{(n)}) \phi_k(\mathbf{y}^{(n)} | \mathbf{x}^{(n)}) \right). \tag{6.66}$$

Once the NN model weights $\mathbf{w}$ are solved from minimizing $J$, we obtain the mixture model parameters $\boldsymbol{\mu}_k(\mathbf{x})$, $\sigma_k(\mathbf{x})$ and $a_k(\mathbf{x})$ from the NN model outputs, hence the conditional distribution $p(\mathbf{y}|\mathbf{x})$ via (6.59) and (6.60).

To get a specific $\mathbf{y}$ value given $\mathbf{x}$, we calculate the mean of the conditional distribution using (6.59) and (6.60), yielding

$$\mathrm{E}[\mathbf{y}|\mathbf{x}] = \int \mathbf{y}\, p(\mathbf{y}|\mathbf{x}) \mathrm{d}\mathbf{y} = \sum_k a_k(\mathbf{x}) \int \mathbf{y}\, \phi_k(\mathbf{y}|\mathbf{x}) \mathrm{d}\mathbf{y} = \sum_k a_k(\mathbf{x})\boldsymbol{\mu}_k(\mathbf{x}). \tag{6.67}$$

Using (6.59), (6.60) and (6.67), one can show that the variance of the conditional distribution about the conditional mean is given by (Takada, 2018, p. 31)

$$s^2(\mathbf{x}) = \mathrm{E}\left[ \left\| \mathbf{y} - \mathrm{E}[\mathbf{y}|\mathbf{x}] \right\|^2 \Big| \mathbf{x} \right]$$

$$= \sum_k a_k(\mathbf{x}) \left[ m\sigma_k(\mathbf{x})^2 + \left\| \boldsymbol{\mu}_k(\mathbf{x}) - \sum_{k=1}^{K} a_k(\mathbf{x})\boldsymbol{\mu}_k(\mathbf{x}) \right\|^2 \right]. \tag{6.68}$$

Thus, the Gaussian mixture model not only gives the conditional distribution $p(\mathbf{y}|\mathbf{x})$ but also provides, for a given $\mathbf{x}$, a specific estimate for $\mathbf{y}$ and a measure of its uncertainty via the conditional mean (6.67) and variance (6.68).

For an example, let us apply the MDN model to the same dataset used in Fig. 6.8(a), where relative humidity is the predictor and visibility is the response. For an MDN with $K = 2$ (i.e. a mixture model containing two Gaussians), the MLP neural network model used to model the MDN has one input, two hidden nodes and six outputs (from which we obtain $a_1(x)$, $a_2(x)$, $\mu_1(x)$, $\mu_2(x)$, $\sigma_1(x)$ and $\sigma_2(x)$). From the six outputs, we can construct the conditional density $p(y|x)$ from (6.59) and (6.60) and the mean of the conditional density from (6.67). Figure 6.11(a) displays $p(y|x)$ by contours and the conditional mean by a solid curve. Clearly, $p(y|x)$ provides more useful information than just the conditional mean. For instance, for relative humidity $\approx 70$, $p(y|x)$ displays two distinct peaks (as supported by the observations), whereas the conditional mean curve runs through the valley between the two peaks.

The simple mixture model involves mixing two Gaussian functions (with centres $\mu_1$ and $\mu_2$ and standard deviations $\sigma_1$ and $\sigma_2$), where mixing coefficients

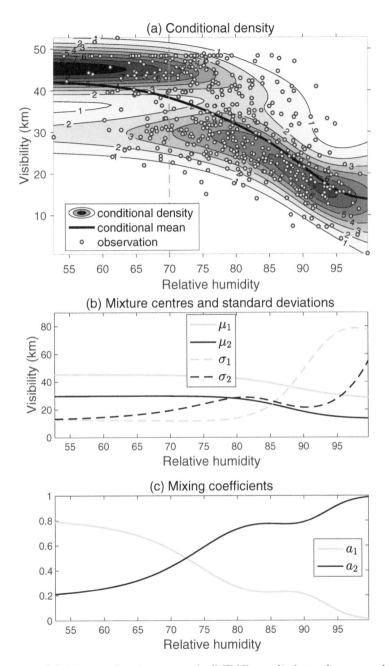

Figure 6.11 (a) Mixing density network (MDN) applied to the same dataset used in Fig. 6.8(a), with contours showing the conditional density $p(y|x)$ and the solid curve the conditional mean. The contour labels need to be multiplied by 100 to give the value for $p(y|x)$. Along the vertical dashed line at relative humidity $= 70$, the conditional density shows two peaks (at visibility around 30 km and 45 km). (b) The centres $\mu_1$ and $\mu_2$ and the standard deviations $\sigma_1$ and $\sigma_2$ for the two Gaussian functions in the mixture model and (c) the mixing coefficients $a_1$ and $a_2$.

$a_1$ and $a_2$ determine how much weight to give to each of the two Gaussians. Figure 6.11(b) shows that at low relative humidity, the centres $\mu_1$ and $\mu_2$ are located at 45 km and 30 km visibility, respectively. The centres shift to lower visibility values as relative humidity increases, while the widths of the Gaussians ($\sigma_1$ and $\sigma_2$) increase at high relative humidity. The first Gaussian has a larger mixing coefficient $a_1$ than the second Gaussian at low relative humidity, but $a_1$ drops to zero at high relative humidity, reducing $p(y|x)$ to a single Gaussian.

## 6.7   Quantile Regression   $\boxed{\text{C}}$ ☺

Quantiles (Section 2.9) are an effective way to present probabilistic information, and we have seen quantile regression as an extension of multiple linear regression to quantiles (Section 5.9). *Quantile regression neural network* (QRNN) models have been developed to model non-linear regression relations between the predictor variables and the quantile response variables (J. W. Taylor, 2000; Cannon, 2011b; Q. F. Xu et al., 2017; Cannon, 2018).[10]

Earlier QRNN models suffer from quantile crossing, for example, the 0.8 quantile $q_{0.8}(x)$ lies below the 0.7 quantile $q_{0.7}(x)$ for high values of $x$ (relative humidity) in Fig. 6.12(a). Cannon (2018) resolved the quantile crossing problem by imposing monotonic constraint on the quantiles using the monotonic multi-layer perceptron (MonMLP) model (Section 6.3.3). The resulting model, the *monotone composite quantile regression neural network* (MCQRNN), solves for $K$ quantiles ($q_{\alpha_1} < \cdots < q_{\alpha_k} < \cdots < q_{\alpha_K}$) altogether (as 'composite' means all the quantiles are solved together instead of one by one).

Given the predictors $x_i$ ($i = 1, \ldots, I$) and the response $y$ for $n = 1, \ldots, N$ observations, the objective function, similar to that in quantile linear regression (Section 5.9), is

$$J = \frac{1}{KN} \sum_{k=1}^{K} \sum_{n=1}^{N} L_{\alpha_k}\Big(y(n) - \hat{y}_{\alpha_k}(n)\Big), \qquad (6.69)$$

where the model output $\hat{y}_{\alpha_k}$ is simply the estimate for the quantile $q_{\alpha_k}$, and the loss function

$$L_\alpha(\epsilon) = \begin{cases} \alpha\,\epsilon & \text{if } \epsilon \geq 0, \\ (\alpha - 1)\,\epsilon & \text{if } \epsilon < 0. \end{cases} \qquad (6.70)$$

That the derivative of this function is undefined at $\epsilon = 0$ presents an obstacle for using gradient-descent non-linear optimization algorithms (Chapter 7) in the MonMLP model. Instead, a smoothed approximation of the loss function is used, that is,

$$\tilde{L}_\alpha(\epsilon) = \begin{cases} \alpha\,h(\epsilon) & \text{if } \epsilon \geq 0, \\ (1 - \alpha)\,h(\epsilon) & \text{if } \epsilon < 0, \end{cases} \qquad (6.71)$$

---

[10] In Section 6.6.1, the mixture density network models the conditional distribution $p(\mathbf{y}|\mathbf{x})$ using a sum of Gaussian functions. While one can compute the quantiles from $p(\mathbf{y}|\mathbf{x})$, this approach using a sum of Gaussians is not likely to be very accurate for the extreme quantiles if the true distribution has a non-Gaussian tail distribution.

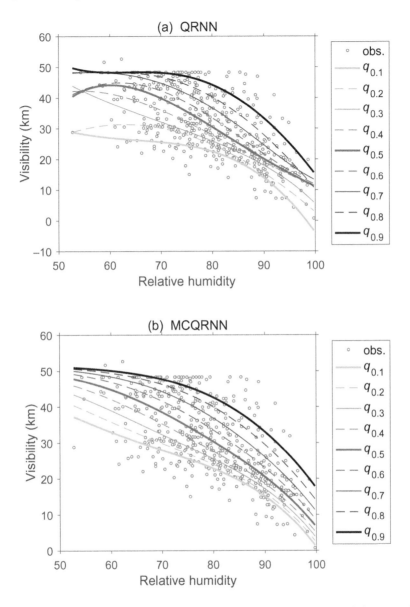

Figure 6.12 Non-linear quantile regression by neural network models as applied to the same dataset as in Fig. 6.11, using (a) the basic QRNN model and (b) the MCQRNN model. The observations (circles) and the 0.1, 0.2, ..., 0.9 quantiles are shown. Crossing of quantile curves is seen in (a) but not in (b).

where $h$ is the Huber function,[11]

$$h(\epsilon) = \begin{cases} \dfrac{\epsilon^2}{2\delta} & \text{if } 0 \le |\epsilon| \le \delta, \\[2mm] |\epsilon| - \dfrac{\delta}{2}, & \text{otherwise,} \end{cases} \tag{6.72}$$

___
[11] There is more than one way to define the Huber function; for example, it can be multiplied

as illustrated in Fig. 6.13. When $|\epsilon|$ is not larger than $\delta$, $h$ behaves like the

Figure 6.13 The Huber error
function $h(\epsilon)$ (with parameter
$\delta = 1$) (solid curve) plotted
versus the error $\epsilon$. The squared
error function (dashed) and
the absolute error function
(dot-dashed) are also shown
for comparison. The vertical
dashed lines at $\epsilon = \pm\delta = \pm1$
indicate where the Huber func-
tion changes from behaving like
the squared error function to
behaving like the absolute error
function, which is less sensitive
to outliers.

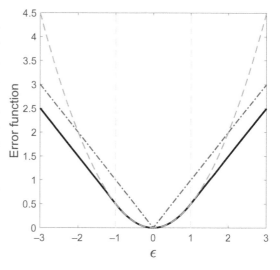

squared error, while for larger $\epsilon$, $h$ behaves like the absolute error. In MCQRNN,
$\delta$ is chosen to be small, so $h$ is very similar to the absolute value function but
is differentiable at the origin, hence gradient-descent non-linear optimization
algorithms can be used.

To prevent crossing quantile curves, the quantile levels $\alpha_1 < \cdots < \alpha_k < \cdots <
\alpha_K$ are added as an extra predictor, with monotonic relation to the quantile
response variable $\hat{y}_{\alpha_k}$ (i.e. $q_{\alpha_k}$) imposed. The original input data matrix $\mathbf{X}$ and
output vector $\mathbf{y}$ are given by

$$\mathbf{X} = \begin{bmatrix} x_{11} & \cdots & x_{1I} \\ \vdots & \ddots & \vdots \\ x_{N1} & \cdots & x_{NI} \end{bmatrix}, \quad \mathbf{y} = \begin{bmatrix} y_1 \\ \vdots \\ y_N \end{bmatrix}. \tag{6.73}$$

To add $\alpha_k$ as an extra predictor in MCQRNN, one needs to make $K$ copies of
$\mathbf{X}$ and $\mathbf{y}$ to make a stacked input matrix and a stacked output matrix, that is,

---

by a positive scalar. A common version is to define it as $h$ in (6.72) but multiplied by $\delta$ (Hastie,
Tibshirani, et al., 2009, equation (12.38)).

$$\mathbf{X}^{(s)} = \begin{bmatrix} \alpha_1 & x_{11} & \cdots & x_{1I} \\ \vdots & \vdots & \ddots & \vdots \\ \alpha_1 & x_{N1} & \cdots & x_{NI} \\ \alpha_2 & x_{11} & \cdots & x_{1I} \\ \vdots & \vdots & \ddots & \vdots \\ \alpha_2 & x_{N1} & \cdots & x_{NI} \\ \vdots & \vdots & \vdots & \vdots \\ \alpha_K & x_{11} & \cdots & x_{1I} \\ \vdots & \vdots & \ddots & \vdots \\ \alpha_K & x_{N1} & \cdots & x_{NI} \end{bmatrix}, \quad \mathbf{y}^{(s)} = \begin{bmatrix} y_1 \\ \vdots \\ y_N \\ y_1 \\ \vdots \\ y_N \\ \vdots \\ y_1 \\ \vdots \\ y_N \end{bmatrix}, \quad (6.74)$$

are used as the input and output data for training the MCQRNN model. A weight penalty term as used in ridge regression (Section 5.7) can be added to avoid overfitting (see Section 8.4.1). The monotonic constraint can also be imposed on (a subset of) the predictors $x_i$ if desirable.

In Fig. 6.12, the QRNN and MCQRNN models were applied to the same relative humidity–visibility dataset used previously in Figs. 6.8 and 6.12. The QRNN model estimated each quantile individually, whereas the MCQRNN estimated all nine quantiles in a single optimization with monotonic constraint, thereby preventing the quantile curves from crossing as seen in the QRNN results. A weight penalty term was used in the objective function for both models, with the weight penalty parameter $\lambda = 0.001$ (Section 8.4.1).

# 6.8 Historical Development of NN in Environmental Science $\boxed{\text{B}}$☺

In Section 1.2, it was noted that ML methods were adopted in various branches of ES at different speeds depending on three factors: (i) whether existing physical-based models are working well, (ii) whether the branch is data rich (i.e. having large effective sample sizes) and (iii) whether non-linearity is needed to model the data. For example: (i) Hydrology adopted NN and other ML methods more readily than meteorology, as physical-based models have more difficulty in forecasting streamflow than weather. (ii) Oceanography and climate science are hampered by being relatively data poor among the environmental sciences. (iii) Averaging daily data to form seasonal or annual data reduces the non-linear relations in the data (Fig. 1.3), thereby hampering the use of non-linear ML models in climate science.

In this section, to provide some historical context, we list some of the early papers introducing NN models into the various branches of ES.

### 6.8.1   Remote Sensing   B☺

By deploying a *radiometer* (i.e. an instrument measuring radiation at certain frequency bands) on a satellite, *remote sensing* allows us to indirectly measure an astonishing number of variables in the air, land and water (Lillesand et al., 2015; Emery and Camps, 2017; Elachi and van Zyl, 2021), such as temperature, precipitation, wind, clouds, sea ice, snow, sea level displacements, vegetation on land and chlorophyll concentration in the ocean, and so on. There are two types of remote sensing – *passive* sensing, where only a receiver is used, and *active* sensing, where both a transmitter and a receiver are used. Active sensing, which operates similarly to radar, requires far more power to operate than passive sensing, as the transmitter has to emit a powerful enough signal beam to Earth below so that the return signal can be detected on the satellite. However, active sensing has good control of the signal to be detected, since the signal is generated originally from the transmitter. One requirement of active remote sensing is that the transmitted radiation must be able to reflect strongly from the Earth's surface back to the receiver; thus, only microwaves have been used in active sensing, as other radiation (e.g. visible and infrared) are too strongly absorbed at the Earth's surface. In contrast to active sensing, passive sensing simply receives whatever radiation is coming from the Earth. As the radiation passes through the atmosphere, it is scattered, absorbed and re-emitted by the clouds, aerosols and molecules in the atmosphere.

Let $\mathbf{v}$ denote some variables of interest on Earth and $\mathbf{s}$, measurements made by the satellite. For instance, $\mathbf{s}$ can be the measurements made at several frequency bands by the radiometer, while $\mathbf{v}$ can be the chlorophyll and sediment concentrations in the surface water. The goal is to find a function $\mathbf{f}$ (a retrieval algorithm), where

$$\mathbf{v} = \mathbf{f}(\mathbf{s}) \tag{6.75}$$

allows the retrieval of $\mathbf{v}$ from the satellite measurements. This retrieval process is not always straightforward, since two different $\mathbf{v}$ can yield the same $\mathbf{s}$ (e.g. snow on the ground and clouds may both appear white as seen by the satellite), so the problem can be ill-posed. In fact, this is an inverse problem to a forward problem (which is well-posed):

$$\mathbf{s} = \mathbf{F}(\mathbf{v}), \tag{6.76}$$

where the variables on earth cause a certain $\mathbf{s}$ to be measured by the satellite's radiometer. One common way to resolve the ill-posedness is to use more frequency bands (e.g. snow and clouds are both white in the visible band, but in the thermal infrared (IR) band they can be separated by their different temperatures). NN and other ML methods for non-linear regression and classification have been widely used to solve the retrieval problem (6.75) for many types of satellite data (Atkinson and Tatnall, 1997; Krasnopolsky, 2007; Blackwell and F. W. Chen, 2009; Maxwell et al., 2018).

Let us proceed through the various frequency bands – visible, infrared and microwave:

**Visible light**

*Land cover*

In vegetative land cover applications, Benediktsson et al. (1990) used an MLP NN to classify Landsat satellite images into nine forest classes and water, using four channels from Landsat and three pieces of topographic information (elevation, slope and aspect) as input. Performance comparison between MLP NN and other classifiers in vegetation classification using satellite data were undertaken by C. Huang et al. (2002) and Foody and Mathur (2004).

Gopal and Woodcock (1996) used MLP NN to estimate the fraction of dead coniferous trees in an area. Forests are burnt in forest fires, then regenerated. Using data from four channels (two visible, one short-wave IR and one near-IR) of the SPOT satellite's vegetation sensor, Fraser and Z. Li (2002) used MLP to estimate the post-fire regeneration age of a forest area.

In urban land cover applications, Del Frate, Pacifici, et al. (2007) and Pacifici et al. (2007) applied MLP NN classifiers to very high-resolution (about 1m resolution) satellite images to monitor changes in urban land cover.

*Ocean colour*

In biologically productive regions, the ocean waters are more greenish due to the presence of phytoplankton. From satellite images of the ocean in the visible band, one can estimate the chlorophyll concentrations in the surface water, hence the phytoplankton concentration, which is key to the ocean's biological productivity.

Keiner and Yan (1998) developed an MLP NN model to retrieve the chlorophyll and suspended sediment concentrations in estuarine water by using the three visible frequency channels of the Landsat Thematic Mapper. Schiller and Doerffer (1999) used an MLP model to estimate the concentrations of phytoplankton pigment, suspended sediments, coloured dissolved organic matter and aerosol over turbid coastal waters using data from the Medium Resolution Imaging Spectrometer (MERIS). Gross, Thiria, and Frouin (1999) and Gross, Thiria, Frouin, and Greg (2000) developed an MLP model to estimate the phytoplankton pigment concentration, where the inputs were the five visible frequency channels from the Sea-viewing Wide Field-of-view Sensor (SeaWiFS). NN methods have also been used to extract information about absorbing aerosols from SeaWiFS data (Jamet et al., 2005; Brajard, Jamet, et al., 2006).

**Infrared sensing**

*Clouds*

J. Lee et al. (1990) used MLP NN models to classify cloud types (cumulus, stratocumulus and cirrus) from a single visible channel of Landsat satellite imagery. However, most studies of remote sensing of clouds tend to rely more on the infrared (IR) channels from instruments such as the Advanced Very High Resolution Radiometer (AVHRR). Bankert (1994) used probabilistic NN

to classify cloud types (cirrus, cirrocumulus, cirrostratus, altostratus, nimbo-
stratus, stratocumulus, stratus, cumulus, cumulonimbus and clear) in satellite
images over maritime regions observed by the AVHRR instrument, with the
method extended to a much larger database of clouds over both land and ocean
in Tag et al. (2000).

In other remote sensing applications, undetected clouds are a source of con-
tamination. For instance, when measuring sea surface temperature from satellite radiometers like the AVHRR, the presence of clouds can introduce large
errors in the temperature. Using the thermal infrared and visible channels in
the AVHRR, Yhann and Simpson (1995) successfully detected clouds (including
subpixel clouds and cirrus clouds) with MLP NN classifiers. Miller and Emery
(1997) also used an MLP NN classifier model to detect clouds in AVHRR images, where clouds were classified into eight types in addition to the cloudless
condition.

### *Precipitation*

From satellite images of clouds, one can estimate the precipitation rate. The
PERSIANN (Precipitation Estimation from Remotely Sensed Information using
Artificial Neural Networks) system is an automated system for retrieving rainfall
from the Geostationary Operational Environmental Satellites (GOES) longwave
infrared images (GOES-IR) at a resolution of $0.25° \times 0.25°$ every half-hour (K. L.
Hsu, Gao, et al., 1997; Sorooshian et al., 2000), with many subsequent upgrades
to the PERSIANN system (Nguyen et al., 2018). Precipitation can also be
estimated from Doppler weather radar, using MLP NN (Teschl et al., 2007).

### *Snow*

Snow in mountainous regions is a major source of water supply and an
important piece of climate information. Simpson and McIntire (2001) classified
clear land, cloud and areal extent of snow in IR satellite images of mid-latitude
regions using a feed-forward MLP NN and, alternatively, a recurrent NN for
classifying sequences of images.

### Passive microwave sensing

In satellite remote sensing, the shorter is the wavelength of the radiation, the
finer the spatial resolution of the image. Thus microwaves, with their relatively
long wavelengths, are at a disadvantage compared to visible light and infrared
sensing in terms of spatial resolution. However, microwaves from the Earth's
surface generally can reach the satellite even with moderately cloudy skies,
whereas visible light and infrared radiation cannot. Although microwaves can
pass through a cloudy atmosphere, they interact with the water vapour and
the liquid water in the atmosphere, thus making the inversion problem more
complicated.

Stogryn et al. (1994) applied an MLP NN to retrieve surface *wind speed* from
several channels of the Special Sensor Microwave Imager (SSM/I), a passive
microwave sensor. Two NN models were developed, one for clear sky and one

for cloudy sky conditions. Krasnopolsky, Breaker, et al. (1995) showed that a single NN with the same architecture can generate the same accuracy as the two NNs for clear and cloudy skies.

Since the microwaves coming from the ocean surface were influenced by the surface wind speed, the columnar water vapour, the columnar liquid water and the sea surface temperature (SST), Krasnopolsky, Gemmill, et al. (1999) retrieved these four variables altogether, that is, the NN had four outputs. NN models have also been developed to monitor *snow* characteristics on a global scale using SSM/I data (Cordisco et al., 2006).

### Active microwave sensing

#### *Altimeter*

The *altimeter* emits a pulsed radar beam vertically downward from the satellite. The beam is strongly reflected at the sea surface, and the reflected pulse is measured by a receiver in the altimeter. From the travel time of the pulse, the distance between the satellite and the sea level can be estimated. If the satellite orbital position is known, then this distance measured by the altimeter gives the sea level displacements. If the sea surface is smooth as a mirror, the reflected pulse will be sharp, whereas if the surface is wavy, the reflected pulse will be stretched out in time, since the part of the pulse hitting the wave crest will reflect before the part hitting the wave trough. Hence, from the sharpness of the reflected pulse, the surface wave condition can be estimated. Furthermore, the surface wave condition is strongly related to the local wind condition, hence retrieval of surface *wind speed* is possible from the altimetry data. MLP NN was used in the retrieval of surface wind speed from altimeter data (Glazman and Greysukh, 1993b; Glazman and Greysukh, 1993a).

#### *Scatterometer*

Compared to surface wind speed retrieval, it is vastly more difficult to retrieve surface *wind direction* from satellite data. The *scatterometer* is designed to retrieve both wind speed and direction through the use of multiple radar beams. The inversion problem of retrieving the wind speed and wind direction has been extensively studied (Badran et al., 1991; Thiria et al., 1993; Mejia et al., 1998; K. S. Chen et al., 1999; Cornford et al., 1999). In Richaume et al. (2000), two MLP NN models were used, the first to retrieve the wind speed and the second to retrieve the wind azimuth.

#### *Synthetic aperture radar*

For active remote sensing, one is limited to microwaves since other radiation (e.g. visible, IR) cannot reflect strongly from the Earth's surface back to the satellite. The spatial resolution of microwave images is very coarse due to the long wavelength used. To obtain high-resolution microwave images (with resolution down to ∼1 m), the *synthetic aperture radar* (SAR) has been developed.

For a radar operating at a given frequency, the larger the size of the receiver antenna, the finer is the observed spatial resolution. Since the satellite is flying, if an object on the ground is illuminated by the radar beam at time $t_1$ and continues to be illuminated till time $t_2$, then all the reflected data collected by the satellite between $t_1$ and $t_2$ can be combined as though the data have been collected by a very large antenna. From SAR images, MLP NN has been used to classify land-cover (Dobson et al., 1995), detect oil spill on the ocean surface (Del Frate, Petrocchi, et al., 2000) and retrieve wind speeds globally at about 30 m resolution (Horstmann et al., 2003).

## 6.8.2   Hydrology   B ☺

When precipitation falls on the land surface, a portion of the water soaks into the ground. This portion, known as *infiltration*, is determined by the permeability of the soils/rocks and the surface vegetation. If the land is unable to absorb all the water, water flows on the surface as *surface runoff*, eventually reaching streams and lakes. The land area which contributes surface runoff to a stream or lake is called the *watershed*, which can be a few hectares in size to thousands of square kilometres. Streamflow is fed by surface runoff, interflow (the flow below the surface but above groundwater) and groundwater.

The relation between streamflow and precipitation is highly complex, since water from precipitation is affected by the type of soil and vegetation in the watershed, or is stored as snow or ice, before it eventually feeds into the streamflow. Because of this complexity, physical-based (a.k.a. 'conceptual') models, which try to model the physical mechanisms of the hydrological processes, are not very skillful in forecasting streamflow from precipitation data. MLP NN models adapted by hydrologists to predict streamflow (Crespo and Mora, 1993; Karunanithi et al., 1994; K. L. Hsu, Gupta, et al., 1995; Minns and M. J. Hall, 1996) quickly became popular, as they are much simpler to develop than the physical-based models, while offering better prediction skills.

*General circulation models* (GCM), a.k.a. *global climate models*, are the main tools for studying global climate and climate change. Using physics-based numerical models, Manabe and Bryan (1969) developed the first GCM by coupling an atmospheric model to an ocean model. As outputs from GCM are rather coarse in spatial resolution, Cannon and Whitfield (2002) used NN to *downscale* (Section 16.11) the GCM output to yield five-day averaged streamflow estimates.

Snowpack on mountains are important for storing winter precipitation and releasing the water during spring/summer. NN models allow retrieval of snow water equivalent (SWE) and snow depth from microwave satellite data (Chang and L. Tsang, 1992; Tedesco et al., 2004).

As a reflection of the wide acceptance of ML methods in hydrology, there have been numerous review papers published on the subject (Maier and Dandy, 2000; Dawson and Wilby, 2001; Abrahart, Anctil, et al., 2012; Zounemat-Kermani et al., 2020; Sit et al., 2020).

## 6.8.3 Atmospheric Science [B]☺

**Severe weather**

For *tornado* forecasting, Marzban and Stumpf (1996) used an MLP NN classifier, with input from Doppler weather radar data, to forecast 'tornado' or 'no tornado' at a lead time of 20 minutes. Their work has been extended to prediction of damaging wind (defined as the existence of either or both of the following conditions: tornado, wind gust $\geq 25$ m s$^{-1}$) (Marzban and Stumpf, 1998). Baik and Hwang (1998) used NN for *tropical cyclone* intensity prediction. Marzban and Witt (2001) used an NN classifier to predict the size of severe *hail*, for three classes of hailstone size (coin, golf ball and baseball size) (Section 16.1).

**Air quality**

The pollutants most commonly investigated in air quality studies include ozone,[12] nitrogen oxides (denoted by the symbol NO$_x$ to include nitric oxide (NO) and nitrogen dioxide (NO$_2$)), sulfur oxides (SO$_x$, especially sulfur dioxide SO$_2$) and particulate matter (PM) from smoke and dust (e.g. PM$_{10}$ and PM$_{2.5}$ are the fraction of suspended particles with diameter $\leq 10$ $\mu$m and $\leq 2.5$ $\mu$m, respectively).

Yi and Prybutok (1996) applied MLP NN to forecast the daily maximum ozone concentration in an industrialized urban area in Texas. Other NN models for forecasting ozone concentration include Comrie (1997), Gardner and Dorling (2000) and Cannon and Lord (2000).

For other pollutants, SO$_2$ forecasts by MLP NN models were developed by Boznar et al. (1993), NO$_x$ by Gardner and Dorling (1999) and NO$_2$ by Kolehmainen et al. (2001).

For particulate matter, MLP models were used to forecast PM$_{2.5}$ (Pérez et al., 2000) and both PM$_{10}$ and PM$_{2.5}$ (McKendry, 2002).

## 6.8.4 Oceanography [B]☺

**Temperature**

Due to the ocean's vast heat storage capacity, upper ocean temperature and heat content anomalies have a major influence on global climate variability. The best known large-scale interannual variability in *sea surface temperature* (SST) is the El Niño-Southern Oscillation (ENSO), a coupled ocean–atmosphere interaction involving the oceanic phenomenon El Niño in the tropical Pacific, and the associated atmospheric phenomenon, the Southern Oscillation (Section 9.1.5). The coupled interaction results in anomalously warm SST in the eastern equatorial Pacific during El Niño episodes, and cool SST in the central equatorial Pacific during La Niña episodes. Grieger and Latif (1994) used an MLP NN to

---

[12] Ground-level ozone is a harmful pollutant, in contrast to the beneficial ozone in the stratosphere, which forms a protective shield against harmful ultraviolet radiation from the sun.

reconstruct the ENSO attractor in a state space spanned by the first four PCA of the combined SST, zonal wind stress and upper ocean heat content data.

For forecasting the tropical Pacific seasonal SST anomalies, Tangang et al. (1997) used MLP NN with wind stress as predictors. Sea level pressure (SLP) replaced wind stress as predictors in the later NN models of Tangang et al. (1998), Tangang, B. Tang, et al. (1998), B. Tang, Hsieh, et al. (2000), and Yuval (2000). Yuval (2001) used bootstrap resampling of the data (Section 8.7.1) to generate an ensemble of NN models and used the ensemble spread to estimate the forecast uncertainty. Thus far, the SST forecasts have been restricted to specific regions in the tropical Pacific, for example, the Niño3.4 region (Fig. 9.3). Wu, Hsieh, and B. Tang (2006) extended the NN model to forecast over the whole tropical Pacific.

Using NN, Ali et al. (2004) estimated the ocean's *subsurface temperature* structure from surface conditions (SST, sea surface height, wind stress, net radiation and net heat flux) at an Arabian Sea mooring.

**Sea level height**

For sea level height, MLP NN models have been developed to forecast the non-tidal component at Galveston Bay, Texas, using the two wind components, barometric pressure, and the previously observed water level anomaly as predictors (D. T. Cox et al., 2002). NN have also been used to forecast the water level in coastal inlets along the South Shore of Long Island, New York (W. Huang et al., 2003) and the sea level changes during storm surges along the Polish coast (Sztobryn, 2003).

# Exercises

### 6.1

Let $x$, $y$ and $z$ be binary variables with the value of 0 or 1, and $z = f(x, y)$. We have encountered in Section 6.2 two special cases of $f$, namely the AND logical function and the XOR (i.e. the exclusive OR) function. There are a total of 16 such possible $f(x, y)$ logical functions. Which of these 16 functions cannot be represented by a perceptron model with two input neurons connected directly to an output neuron by a step function?

### 6.2

Show that for a multi-layer perceptron (MLP) NN with one hidden layer, if the activation function for the hidden layer is linear, then the NN reduces to one without any hidden layer.

### 6.3

Consider an MLP NN model containing $m_1$ inputs, one hidden layer with $m_2$ neurons and $m_3$ outputs. (a) If the activation function in the hidden layer is the tanh function, what happens if we flip the signs of all the weights (including offset parameters) feeding into a particular hidden neuron and also flip the signs of the weights leading out of that neuron? For a given set of weights for the NN, how many equivalent sets of weights are there due to the 'sign-flip' symmetries? (b) Furthermore, the weights associated with one hidden neuron can be interchanged with those of another hidden neuron without affecting the NN outputs. Therefore, show that there are a total of $m_2! \, 2^{m_2}$ equivalent sets of weights from the 'sign-flip' and the 'interchange' symmetries.

**6.4**

For the logistic sigmoidal activation function $f(x)$ in (6.5), show that its derivative $f'(x)$ can be expressed in terms of $f(x)$. Also show that for the activation function $\tanh(x)$, its derivative can also be expressed in terms of the tanh function.

**6.5**

Build a non-linear regression model for predicting the winter snow water equivalent (SWE) for the dataset provided on the book website (same dataset as used in Exercise 5.8), with the Niño 3.4 and the PNA climate indices as the two predictors. Use only the first 40 years of data to build the model, reserving the remaining 28 years as independent test data. Compute the RMSE and Pearson correlation between the observed and predicted SWE values over the test data. Compare these performance scores with those from multiple linear regression.

**6.6**

Build a non-linear regression model for predicting the daily precipitation amount (only for days with positive precipitation) for the training dataset provided on the book website, with sea-level pressure, 700-hPa specific humidity and 500-hPa geopotential height as the three predictors (Cannon, 2011b). Use predictors from the separate test dataset to predict the response variable. Compute the RMSE and Pearson correlation between the observed and predicted response variable over the test data. Compare these performance scores with those from multiple linear regression. [Note: the given response variable is actually the fourth root of the prediction amounts, as the fourth root transform makes the precipitation distribution more Gaussian, and the predictors have been standardized.]

# 7

# Non-linear Optimization

As mentioned in the previous chapter, the multi-layer perceptron (MLP) or feed-forward neural network (FFNN) model generally requires non-linear optimization, whereas multiple linear regression (MLR) only requires linear optimization (i.e. linear least squares). The few exceptions of neural network models requiring only linear optimization occur when the MLP has only one hidden layer and the first layer of weights are chosen randomly, as in extreme learning machines (ELM) (Section 6.4), in random vector functional link (RVFL) models (Section 6.4.2) or in radial basis function (RBF) models (Section 6.5), where the radial basis functions are determined first by unsupervised learning. Thus, an understanding of non-linear optimization is essential when using general NN models and especially deep NN models (i.e. NN models with many hidden layers).

Optimization methods can be divided into two broad types: (A) methods that require the objective function $J$ to be differentiable, so information provided by the gradient $\nabla J$ can be used to find the descent direction towards a minimum of $J$ and (B) methods that do not require a differentiable objective function. Type (A) methods covered in this chapter include gradient descent (or steepest descent) (Section 7.5), stochastic gradient descent (Section 7.6), conjugate gradient (Section 7.7), quasi-Newton (Section 7.8) and Levenberg–Marquardt (Section 7.9). Type (B) methods covered include hill climbing (Section 7.11) and evolutionary algorithms (Section 7.10), of which genetic algorithm (Section 7.12) and differential evolution (Section 7.13) are examined.

## 7.1 Extrema and Saddle Points $\boxed{\text{A}}$ ☺

To appreciate the difference between linear optimization and non-linear optimization, consider the MLR model output

$$\hat{y} = w_0 + \sum_{l=1}^{L} w_l f_l, \tag{7.1}$$

where $f_l = f_l(x_1, \ldots, x_m)$ are functions of the predictor variables $x_1, \ldots, x_m$. For instance, the polynomial fit is a special case with $f_l$ consisting of $x_1$, $x_2$, $x_1^2$, $x_1 x_2$, $x_2^2$, and so on. Although the response variable $\hat{y}$ can be non-linearly related to the predictor variables $x_1, \ldots, x_m$ (as $f_l$ is in general a non-linear function), $\hat{y}$ is a linear function of the parameters or weights $\{w_l\}$. It follows that the objective function

$$J = \sum (\hat{y} - y)^2 \tag{7.2}$$

(with $y$ the target data and the summation over all observations) is a quadratic function of the weights $\{w_l\}$. This means the objective function $J(w_0, \ldots, w_L)$ is a parabolic surface, which has a single minimum, the global minimum.

In contrast, when $y$ is a non-linear function of $\{w_l\}$, the objective function surface is in general filled with numerous hills and valleys, that is, there are usually many local minima besides the global minimum (if there are symmetries among the weights, there can even be multiple global minima). Thus, non-linear optimization involves finding a global minimum among many local minima. The difficulty faced by the optimization algorithm is similar to that encountered by a robot rover sent to explore the rugged surface of a distant planet. The rover can easily fall into a hole or a steep valley and be unable to escape from it, thereby never reaching its final objective, the global minimum. Thus, non-linear optimization is vastly more tricky than linear optimization, with no guarantee that the algorithm actually finds the global minimum, as it may become trapped by a local minimum or a saddle point.

For the simple example in Fig. 7.1, $J$ is a function of a single scalar variable $w$. The multiple minima and maxima (all referred to as *extrema*) have zero gradient, that is, $J' = 0$, with $J'$ denoting the first derivative of $J$. For the second derivative of $J$, $J'' > 0$ (positive curvature) occurs at a minimum, whereas $J'' < 0$ (negative curvature), at a maximum. If $J' = 0$ and $J'' = 0$, then the point is either a saddle point or it lies in a flat region. Thus, the second derivative is a useful test to determine if a point with zero gradient is a minimum or a maximum, or neither if the second derivative also vanishes.

Not all local minima are bad. In Fig. 7.1, the shallow local minima A and F are bad (see Fig. 6.7(b) for examples of bad local minima solutions in MLP), but the deeper local minima B and D are good, as their $J$ values are not far from that found at the global minimum C. Often the non-linear models are trained multiple times with different random initial weights, so the optimization algorithm settles into different local minima. The model from a single training run is called an ensemble member or a 'committee member' in ML. These multiple runs of the model together form an ensemble or a committee. The multiple ensemble members would yield multiple predictions, so an ensemble average is usually computed by taking the mean of the multiple predictions if $\hat{y}$ is a continuous variable. If $\hat{y}$ is a discrete variable, usually a vote is taken among the multiple predictions, that is, select the most common outcome chosen by the ensemble members. See Section 8.7 for a detailed explanation of ensemble methods.

If **w** is more than one-dimensional, the second partial derivatives of $J$ are placed in a matrix **H**, called the *Hessian matrix*, with elements

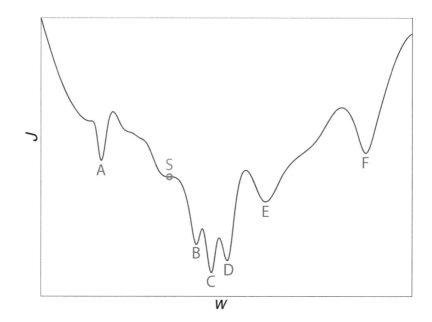

Figure 7.1 A schematic diagram illustrating an objective function $J(w)$. There are multiple local minima labelled A, B, D, E and F, and a global minimum labelled C. Local minima B and D, being close to C, are likely to give good solutions, whereas the shallow minima A and F, poor solutions. The point labelled S is a saddle point – the gradient (that is, slope) of the curve is zero but the point is neither a maximum nor a minimum.

$$(\mathbf{H})_{ij} \equiv \frac{\partial^2 J}{\partial w_i \partial w_j} . \tag{7.3}$$

If the second partial derivatives are continuous, the order of the partial differentiation can be interchanged, that is, $(\mathbf{H})_{ij} = (\mathbf{H})_{ji}$, and the Hessian matrix is symmetric. The objective functions used in NN models usually have a symmetric Hessian almost everywhere.

A saddle point is better illustrated in a two-dimensional $\mathbf{w}$ space (Fig. 7.2), where it is shown as having a local minimum in the $w_1$ dimension but a local maximum in the $w_2$ dimension. In general, the direction of largest positive curvature need not lie along either the $w_1$ or the $w_2$ directions. A real, symmetric Hessian matrix can be decomposed into a set of real eigenvalues and a corresponding set of eigenvectors forming an orthogonal basis. An eigenvalue gives the second derivative in the direction of the corresponding eigenvector. Thus, the eigenvector corresponding to the largest positive eigenvalue gives the direction of largest positive curvature.

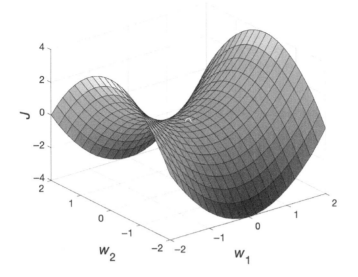

Figure 7.2 A saddle point on the surface $J = w_1^2 - w_2^2$, with the saddle point at $(0,0)$ marked by a semi-circle. At $(0,0)$, $J$ concaves up (i.e. positive curvature) along the $w_1$ dimension and concaves down (negative curvature) along the $w_2$ dimension.

In a high-dimensional $\mathbf{w}$ space, convergence to saddle points can be a more serious problem than convergence to local minima. The reason is that at a point of zero gradient, the chances that all the directions of the eigenvectors have positive curvature (in order for the point to be a local minimum) becomes increasingly improbable as the number of dimensions of $\mathbf{w}$ increases. If even one of the directions does not have positive curvature, the point is a saddle point. For Gaussian processes, the number of saddle points to local minima increases exponentially with the dimensionality of $\mathbf{w}$, and an optimization algorithm that can rapidly escape from saddle points has been developed (Dauphin et al., 2014).

## 7.2 Gradient Vector in Optimization $\boxed{\text{A}} \odot$

In essence, with MLP models, one needs to minimize the objective function $J$ with respect to $\mathbf{w}$ (with components $w_i$, representing all the weight and offset/bias parameters), that is, find the optimal weights that will minimize $J$. The gradient vector $\nabla J$, with components $\partial J / \partial w_i$, plays a vital role in the optimization process. It is common to solve the minimization problem using an iterative procedure. Suppose the current approximation of the solution is $\mathbf{w}_0$. A Taylor series expansion of $J(\mathbf{w})$ around $\mathbf{w}_0$ yields

$$J(\mathbf{w}) = J(\mathbf{w}_0) + (\mathbf{w} - \mathbf{w}_0)^{\mathrm{T}} \nabla J(\mathbf{w}_0) + \frac{1}{2}(\mathbf{w} - \mathbf{w}_0)^{\mathrm{T}} \mathbf{H}_0 (\mathbf{w} - \mathbf{w}_0) + \ldots, \quad (7.4)$$

where $\nabla J(\mathbf{w}_0)$ denotes the gradient vector evaluated at the point $\mathbf{w}_0$, and the Hessian matrix $\mathbf{H}_0$ contains the second order derivatives of $J$ in (7.3) evaluated at $\mathbf{w}_0$.

Applying the gradient operator to (7.4), with the gradient of the constant term $J(\mathbf{w}_0)$ being zero, we obtain

$$\nabla J(\mathbf{w}) = \nabla J(\mathbf{w}_0) + \mathbf{H}_0 (\mathbf{w} - \mathbf{w}_0) + \ldots \quad (7.5)$$

Next, let us derive an iterative scheme for finding the optimal $\mathbf{w}$, with $\mathbf{w}_0$ being the current approximation of the optimal solution. At the optimal $\mathbf{w}$, $\nabla J(\mathbf{w}) = 0$, and (7.5), with higher order terms ignored, yields

$$\mathbf{H}_0 (\mathbf{w} - \mathbf{w}_0) = -\nabla J(\mathbf{w}_0). \quad (7.6)$$

Left multiplication by $\mathbf{H}_0^{-1}$ gives

$$\mathbf{w} = \mathbf{w}_0 - \mathbf{H}_0^{-1} \nabla J(\mathbf{w}_0), \quad (7.7)$$

which provides a way to estimate the optimal $\mathbf{w}$ from our current estimate $\mathbf{w}_0$. This suggests the following iterative scheme for proceeding from step $k$ to step $k + 1$:

$$\mathbf{w}_{k+1} = \mathbf{w}_k - \mathbf{H}_k^{-1} \nabla J(\mathbf{w}_k). \quad (7.8)$$

This is known as *Newton's method*.[1]

It is worth noting that if $\mathbf{w}_0$ is a minimum of $J(\mathbf{w})$, then the gradient of $J$ at $\mathbf{w}_0$ is zero, that is, $\nabla J(\mathbf{w}_0) = 0$. The second term in (7.4) vanishes, and ignoring the higher order terms, (7.4) reduces to

$$J(\mathbf{w}) \approx J(\mathbf{w}_0) + \frac{1}{2}(\mathbf{w} - \mathbf{w}_0)^{\mathrm{T}} \mathbf{H}_0 (\mathbf{w} - \mathbf{w}_0). \quad (7.10)$$

As the first term on the right hand side is a constant and the second term is a quadratic, this is an equation describing a parabolic surface. Thus, near a minimum, assuming the Hessian matrix is non-zero, the objective function has an approximately parabolic surface.

In the multi-dimensional case, if $\mathbf{w}$ is of dimension $L$, then the Hessian matrix $\mathbf{H}_k$ is of dimension $L \times L$. Computing $\mathbf{H}_k^{-1}$, the inverse of an $L \times L$

---

[1] In the one-dimensional case, (7.8) reduces to the familiar form

$$w_{k+1} = w_k - \frac{J'(w_k)}{J''(w_k)}, \quad (7.9)$$

for finding a root of $J'(w) = 0$, where the prime and double prime denote the first and second derivatives, respectively.

matrix, may be computationally too costly. Simplification is needed, resulting in quasi-Newton methods.

In essence, the optimization procedure involves two parts: (i) estimate the gradient vector $\nabla J$ by *back-propagation* and (ii) supply the $\nabla J$ information to a gradient descent optimization algorithm, such as a quasi-Newton algorithm or a stochastic gradient descent algorithm. In Section 7.3, we describe back-propagation, then in subsequent sections, we study some of the optimization algorithms using the information provided by $\nabla J$. The algorithms are divided into first-order methods (e.g. gradient descent and stochastic gradient descent), using only the first-order derivatives of $J$, and second-order methods (e.q. quasi-Newton and Levenberg–Marquardt), using the second order derivatives.

# 7.3   Back-Propagation   $\boxed{\text{B}}$☹

To find the optimal weights that would minimize the objective function $J$ of an MLP NN model, one needs to know $\nabla J$, the gradient of $J$ with respect to the weights. The *back-propagation* algorithm gives $\nabla J$ through the backward propagation of the model errors (i.e. $\hat{y} - y$). In fact, the MLP problem could not be solved until the introduction of the back-propagation algorithm, as popularized by Rumelhart et al. (1986a). The basic back-propagation algorithm had actually been independently discovered/rediscovered many times before (Schmidhuber, 2015, section 5.5), as far back as 1960 in control theory (Kelley, 1960).

The original back-propagation algorithm is composed of two parts: the first part computes the gradient of $J$ by backward propagation of the model errors, while the second part descends along the gradient towards the minimum of $J$. This descent method is called *gradient descent* or *steepest descent*, and is notoriously inefficient, thus rendering the original back-propagation algorithm a very slow method. Nowadays, the term 'back-propagation' is used somewhat ambiguously – it could mean the original back-propagation algorithm or, more likely, it could mean using only the first part involving the backward error propagation to compute the gradient of $J$, to be followed by a more efficient descent algorithm, such as the *stochastic gradient descent* algorithm.

An analogy is a motor boat, composed of two parts, namely the boat and the motor mounted at the stern. After some years, the motor appears underpowered, so the owner removes the motor and mounts a more powerful motor. Here, back-propagation is the boat and steepest descent is the original motor, which has long since been replaced.

While the original back-propagation algorithm (using steepest descent) is not recommended for actual computation because of its slowness, it is presented in this section because of its historical significance and, more importantly, because it is the easiest algorithm to understand. In later sections, replacements for the obsolete steepest descent algorithm are presented.

In (6.8), the objective function $J$ is evaluated over all observations. It is convenient to define the objective function $J^{(n)}$ associated with the $n$th observation or training pattern, that is,

$$J^{(n)} = \frac{1}{2} \sum_k \left[ \hat{y}_k^{(n)} - y_k^{(n)} \right]^2. \tag{7.11}$$

In the following derivation, we will be dealing with $J^{(n)}$, though for brevity we will simply use the symbol $J$. In batch training, the overall $J$ in (6.8) is simply the mean of $J^{(n)}$ over all $n = 1, \ldots, N$ observations.

In the two-layer network described by (6.6) and (6.7), letting $w_{j0} \equiv b_j$, $\tilde{w}_{k0} \equiv \tilde{b}_k$, and $x_0 = 1 = h_0$ allows the equations to be expressed more compactly as

$$h_j = f\left( \sum_i w_{ji} x_i \right), \tag{7.12}$$

$$\hat{y}_k = g\left( \sum_j \tilde{w}_{kj} h_j \right), \tag{7.13}$$

where the summation indices start from 0 to include the offset terms. We can introduce an even more compact notation by putting all the $w_{ji}$ and $\tilde{w}_{kj}$ weights into a single weight vector $\mathbf{w}$. In back-propagation, the weights are first assigned random initial values and then adjusted according to the gradient of the objective function, that is,

$$\Delta \mathbf{w} = -\eta \, \nabla J. \tag{7.14}$$

The weights are adjusted by $\Delta \mathbf{w}$, proportional to the negative gradient vector $-\nabla J$ by the scale factor $\eta$, called the *learning rate* in ML. In component form, we have

$$\Delta w_{ji} = -\eta \, \frac{\partial J}{\partial w_{ji}}, \quad \text{and} \quad \Delta \tilde{w}_{kj} = -\eta \, \frac{\partial J}{\partial \tilde{w}_{kj}}. \tag{7.15}$$

Note that $\eta$ determines the size of the step taken by the optimization algorithm as it descends along the direction of the objective function gradient. At the current step $l$, the weights are adjusted for the next step $l+1$ by

$$\mathbf{w}(l+1) = \mathbf{w}(l) + \Delta \mathbf{w}(l). \tag{7.16}$$

Let us introduce the symbols

$$s_j = \sum_i w_{ji} x_i, \quad \text{and} \quad \tilde{s}_k = \sum_j \tilde{w}_{kj} h_j, \tag{7.17}$$

hence (7.12) and (7.13) become

$$h_j = f(s_j), \quad \text{and} \quad \hat{y}_k = g(\tilde{s}_k). \tag{7.18}$$

The objective function gradients in (7.15) can be solved by applying the chain rule in differentiation, for example

$$\frac{\partial J}{\partial \tilde{w}_{kj}} = \frac{\partial J}{\partial \tilde{s}_k} \frac{\partial \tilde{s}_k}{\partial \tilde{w}_{kj}} \equiv -\tilde{\delta}_k \frac{\partial \tilde{s}_k}{\partial \tilde{w}_{kj}}, \tag{7.19}$$

where $\tilde{\delta}_k$ is called the *sensitivity* of the $k$th output neuron. Further application of the chain rule yields

$$\tilde{\delta}_k \equiv -\frac{\partial J}{\partial \tilde{s}_k} = -\frac{\partial J}{\partial \hat{y}_k}\frac{\partial \hat{y}_k}{\partial \tilde{s}_k} = (y_k - \hat{y}_k)\, g'(\tilde{s}_k), \qquad (7.20)$$

where (7.11) has been differentiated, and $g'$ is the derivative of $g$ from (7.18).

From (7.17), we have

$$\frac{\partial \tilde{s}_k}{\partial \tilde{w}_{kj}} = h_j. \qquad (7.21)$$

Substituting this and (7.20) into (7.19) and (7.15), we obtain the weight update or learning rule for the weights connecting the hidden and output layers:

$$\Delta \tilde{w}_{kj} = \eta\, \tilde{\delta}_k h_j = \eta\, (y_k - \hat{y}_k)\, g'(\tilde{s}_k) h_j. \qquad (7.22)$$

If linear activation functions are used at the output neurons, that is, $g$ is the identity map, then $g' = 1$.

Similarly, to obtain the learning rule for the weights connecting the input layer to the hidden layer, we use the chain rule repeatedly to get:

$$\frac{\partial J}{\partial w_{ji}} = \sum_k \frac{\partial J}{\partial \hat{y}_k}\frac{\partial \hat{y}_k}{\partial \tilde{s}_k}\frac{\partial \tilde{s}_k}{\partial h_j}\frac{\partial h_j}{\partial s_j}\frac{\partial s_j}{\partial w_{ji}} = -\sum_k \tilde{\delta}_k \tilde{w}_{kj} f'(s_j) x_i. \qquad (7.23)$$

Thus, the learning rule for the weights connecting the input layer to the hidden layer is

$$\Delta w_{ji} = \eta \sum_k \tilde{\delta}_k \tilde{w}_{kj} f'(s_j) x_i. \qquad (7.24)$$

This can be expressed in a similar form as (7.22), that is,

$$\Delta w_{ji} = \eta\, \delta_j x_i, \quad \text{with} \quad \delta_j = \left(\sum_k \tilde{\delta}_k \tilde{w}_{kj}\right) f'(s_j). \qquad (7.25)$$

Equations (7.22) and (7.25) give the back-propagation algorithm. The model output error $(\hat{y}_k - y_k)$ is propagated backwards by (7.22) to update the weights connecting the hidden to output layers, and then further back by (7.25) to update the weights connecting the input to hidden layers.

In general, the weights are randomly initialized at the start of the optimization process. The inputs are mapped forward by the network, and the output model error is obtained. The error is then back-propagated to update the weights. This process of mapping the inputs forward and then back-propagating the error is iterated until the objective function satisfies some convergence criterion. The derivation of the back-propagation algorithm here is for a network with only one hidden layer. The algorithm can be extended in a straightforward manner to a network with two or more hidden layers by repeated use of the chain rule. For a more general derivation of back-propagation, see Goodfellow, Bengio, et al. (2016, section 6.5).

## 7.4   Training Protocol   $\boxed{\text{A}}$☺

There are two main training protocols, *sequential training* and *batch training*.
In sequential training, each pattern (i.e. example or observation) is presented to
the network and the weights are updated. The next pattern is then presented,
until all patterns in the training dataset have been presented once to the network
– called an *epoch*. The same patterns are then presented repeatedly for many
epochs until the objective function convergence criterion is satisfied. A variant
of sequential training is *stochastic training*, where a pattern is randomly selected
from the training dataset, presented to the network for weight updating, and
the process is repeated. For very large patterns presenting storage problems,
*online training* can be used, where a pattern is presented to the network and
the weights updated repeatedly, before moving on to the next pattern. As each
pattern is presented only once, there is no need to store all the patterns on the
computer.

In batch training, all patterns in the training dataset are presented to the
network before the weights are updated. This process is repeated for many
epochs. The objective function $J$ is the mean of $J^{(n)}$ over all $n = 1, \ldots, N$
observations, as defined in (6.8). However, if the training data are redundant,
that is, a pattern may be presented several times in the dataset, then stochastic
training can be more efficient than batch training.

The objective function convergence criterion is a subtle issue. Many MLP
applications do not train until convergence to the global minimum – this can
be a great shock to researchers trained in classical optimization! The reason is
that data contain both signal and noise. Given enough hidden neurons, an MLP
can have enough weights to fit the training data to arbitrary accuracy, which
means it is also fitting to the noise in the data, an undesirable condition known
as *overfitting*. When one obtains an overfitted solution, it will not perform well
when making new predictions (Figs. 1.4 and 1.5).

In other words, one is interested not in using an NN to fit a given dataset
to arbitrary accuracy, but in using the NN to *learn* the underlying relationship in the given data, that is, be able to generalize from a given dataset,
so that the extracted relationship even fits new data not used in training the
model. Various methods have been devised to improve learning, as presented in
Chapter 8.

To prevent overfitting, the dataset is often divided into two parts, one for
training, the other for *validation*. As the number of training epochs increases,
the objective function evaluated over the training data decreases. However, the
objective function evaluated over the validation data will drop but eventually increase as training epochs increase (Fig. 7.3), indicating that the training dataset
is already overfitted. When the objective function evaluated over the validation
data reaches a minimum, it gives a useful signal that this is the appropriate time
to stop the training, as additional training epochs only contribute to overfitting.
This very common approach to avoid overfitting is called *early stopping*. What
fraction of the available data one should reserve for validation is examined in
Section 8.4.2.

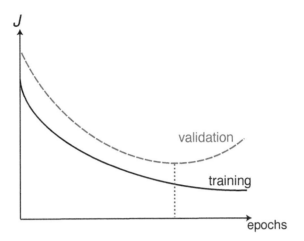

Figure 7.3 Schematic diagram illustrating the behaviour of the objective function $J$ as the number of training epochs increases. Evaluated over the training data, the objective function (solid curve) decreases with increasing number of epochs; however, evaluated over an independent set of validation data, the objective function (dashed curve) initially drops but eventually rises with increasing number of epochs, indicating that overfitting has occurred when a large number of training epochs is used. The minimum in the objective function evaluated over the validation data (as marked by the vertical dotted line) indicates when training should be stopped to avoid overfitting.

Similarly, the objective function evaluated over the training data generally decreases as the number of hidden neurons increases. Again, the objective function evaluated over a validation set will drop initially but eventually increase due to overfitting from excessive number of hidden neurons. Thus, the minimum of the objective function over the validation data can also give an indication on how many hidden neurons to use.

## 7.5 Gradient Descent Method $\boxed{\text{A}}$ ☺

We have already encountered the *gradient descent* or *steepest descent* method in the back-propagation algorithm, where according to (7.14), the weights are updated at the $k$th iterative step by

$$\mathbf{w}_{k+1} = \mathbf{w}_k - \eta \nabla J(\mathbf{w}_k), \tag{7.26}$$

with $\eta$ being the learning rate. Clearly (7.26) is a major simplification of Newton's method (7.8), with the learning rate $\eta$, a scalar, replacing $\mathbf{H}^{-1}$, the inverse of the Hessian matrix. One also tries to reach the optimal $\mathbf{w}$ by descending along the negative gradient of $J$ in (7.26), hence the name gradient descent or steepest descent, as the negative gradient gives the direction of steepest descent. An analogy is a hiker trying to descend in thick fog from a mountain to the bottom of a valley by taking the steepest descending path. One might be tempted to think that following the direction of steepest descent should allow the hiker to reach the bottom most efficiently; alas, this approach is surprisingly inefficient, as we shall see.

The learning rate $\eta$ can be either a fixed constant or calculated by a line minimization algorithm. In the former case, one simply takes a step of fixed size along the direction of the negative gradient of $J$. In the latter, one proceeds along the negative gradient of $J$ until one reaches the minimum of $J$ along that direction (Fig. 7.4).

Figure 7.4 The gradient descent approach starts from the weights $\mathbf{w}_k$ estimated at step $k$ of an iterative optimization process. The descent path $\mathbf{d}_k$ is chosen along the negative gradient of the objective function $J$, which is the steepest descent direction. Note that $\mathbf{d}_k$ is perpendicular to the $J$ contour where $\mathbf{w}_k$ lies. The descent along $\mathbf{d}_k$ proceeds until it is tangential to another contour at $\mathbf{w}_{k+1}$, which is the minimum of $J$ along the $\mathbf{d}_k$ direction, thereby giving the optimal step size $\eta$ in the descend along $\mathbf{d}_k$. At $\mathbf{w}_{k+1}$, the direction of steepest descent is given by $-\nabla J(\mathbf{w}_{k+1})$. The process is iterated.

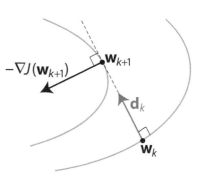

More precisely, suppose at step $k$, we have estimated weights $\mathbf{w}_k$. We then descend along the negative gradient of the objective function at $\mathbf{w}_k$, that is, go along the direction

$$\mathbf{d}_k = -\nabla J(\mathbf{w}_k). \tag{7.27}$$

Proceed along $\mathbf{d}_k$, with our path described by $\mathbf{w}_k + \eta\,\mathbf{d}_k$, until we reach the minimum of $J$ along this direction. Going further along this direction would mean we would actually be ascending rather than descending, so we should stop at this minimum of $J$ along $\mathbf{d}_k$, which occurs at

$$\frac{\partial}{\partial \eta} J(\mathbf{w}_k + \eta\,\mathbf{d}_k) = 0, \tag{7.28}$$

thereby yielding the optimal step size $\eta$ (Fig. 7.4). Performing the partial differentiation by $\eta$ in (7.28) gives

$$\mathbf{d}_k^{\mathrm{T}} \nabla J(\mathbf{w}_k + \eta\,\mathbf{d}_k) = 0. \tag{7.29}$$

With

$$\mathbf{w}_{k+1} = \mathbf{w}_k + \eta\,\mathbf{d}_k, \tag{7.30}$$

we can rewrite the above equation as

$$\mathbf{d}_k^{\mathrm{T}} \nabla J(\mathbf{w}_{k+1}) = 0. \tag{7.31}$$

As this means the dot product of the vectors $\mathbf{d}_k$ and $\nabla J(\mathbf{w}_{k+1})$ is zero, the two vectors are perpendicular, that is,

$$\mathbf{d}_k \perp \nabla J(\mathbf{w}_{k+1}). \tag{7.32}$$

At $\mathbf{w}_{k+1}$, the new direction of steepest descent is $\mathbf{d}_{k+1} = -\nabla J(\mathbf{w}_{k+1})$, hence (7.31) gives

$$\mathbf{d}_k^{\mathrm{T}} \mathbf{d}_{k+1} = 0, \quad \text{that is,} \quad \mathbf{d}_k \perp \mathbf{d}_{k+1}. \tag{7.33}$$

As the new direction $\mathbf{d}_{k+1}$ is orthogonal to the previous direction $\mathbf{d}_k$, this results in an inefficient zigzag path of descent (Fig. 7.5(a)).

The other alternative of using fixed learning rate is also inefficient, as a small learning rate results in taking too many small steps (Fig. 7.5(b)), while a large learning rate size results in an even more severely zigzagged path of descent (Fig. 7.5(c)) and can be numerically unstable if the step size is too large. A common approach is to use a 'fixed' learning rate but to gradually shrink the learning rate as the optimization progresses. For instance, one can start with a learning rate of $\eta = \eta_0$ for the first $k_0$ steps, and then have $\eta$ decaying by $O(1/k)$ as $k$ increases, that is,

$$\eta = \frac{\eta_0 \, k_0}{\max(k, k_0)} \tag{7.34}$$

(Bengio, 2012).

Another way to reduce the zigzag in the gradient descent scheme is to add '*momentum*' to the descent direction (Rumelhart et al., 1986b), that is,

$$\mathbf{d}_k = -\nabla J(\mathbf{w}_k) + \mu \, \mathbf{d}_{k-1}, \tag{7.35}$$

with $\mu$ the momentum parameter. Here, the momentum or memory of $\mathbf{d}_{k-1}$ prevents the new direction $\mathbf{d}_k$ from being orthogonal to $\mathbf{d}_{k-1}$, thereby reducing the zigzag (Fig. 7.5(d)). The momentum parameter can be chosen by the user or chosen automatically in some algorithms, such as the conjugate gradient method (Section 7.7).

## 7.6 Stochastic Gradient Descent $\boxed{\text{A}}$ ☺

Stochastic methods have been extensively developed for search and optimization problems (Spall, 2003). For NN models, the *stochastic gradient descent* (SGD) method has become probably the most commonly used optimization approach in recent years, especially since the advent of deep neural networks and very large datasets. It is a stochastic approximation of the gradient descent method, as it replaces the gradient computed using the entire dataset by an approximation using only a randomly selected (small) subset of the data. In contrast, the traditional gradient descent under batch training requires the entire dataset

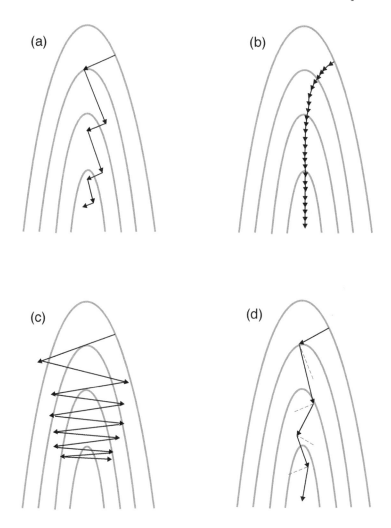

Figure 7.5 The gradient descent method with (a) line minimization (i.e. optimal step size $\eta$), (b) a fixed step size that is too small, (c) a fixed step size that is too large and (d) momentum, which reduces the zigzag behaviour during descent. The direction of steepest descent is indicated by dashed lines in (d). [Follows Masters (1995, chapter 1).]

be available in the computer memory, which can be infeasible for very large datasets.

To illustrate this approach, let us start with the objective function $J$ from (6.8) but with only a single output variable $\hat{y}$ for simplicity, that is,

$$J = \frac{1}{2N} \sum_{n=1}^{N} (\hat{y}_n - y_n)^2, \qquad (7.36)$$

where $y$ is the observed response variable and the summation is over all available data $n = 1, \ldots, N$ for traditional batch training. In SGD, the summation is only over a subset $B$, called a *mini-batch*, containing $N_B$ randomly selected data points, that is,

$$J = \frac{1}{2N_B} \sum_{n \in B} (\hat{y}_n - y_n)^2 . \tag{7.37}$$

Typically, $N_B$ ranges from one to a few hundred. $\nabla J$ is then estimated by back-propagation using only $N_B$ data points.

The advantage of having $\nabla J$ computed using very few data points means this step is cheap computationally, so one can afford to iterate many more steps during the gradient descent. Using $\nabla J$ estimated from a small number of data points adds noise to the optimization process. Ironically, this injection of noise helps to prevent the optimization algorithm from settling quickly into shallow local minima.

Using a very small value of $N_B$ (e.g. $N_B = 1$) has the disadvantage of injecting too much noise in the estimation of $\nabla J$ and making many iterative steps, which can be computationally expensive. Instead, $N_B$ is usually chosen based on the particular computer's architecture for efficient computing. For instance, $N_B$ is often chosen to be a power of two, which fits the memory requirements of the GPU or CPU hardware (e.g. $N_B = 32, 64, 128, 256$, etc.), with $N_B = 32$ being a good default value (Bengio, 2012). In summary, mini-batch SGD can have several advantages over the traditional batch training of gradient descent: (a) being less likely to converge to shallow local minima, (b) having no large memory requirement during computation and (c) may be faster computationally.

Interestingly, data models requiring linear optimization, such as MLR or kernel methods (Chapter 12), can also benefit from SGD when the enormous size of the dataset makes even linear optimization computationally unaffordable under batch learning (Goodfellow, Bengio, et al., 2016, section 5.9). SGD is also naturally adapted for online learning, that is, training/updating the model as new data arrive continually.

SGD is usually run with the learning rate $\eta$ shrinking gradually as the optimization progresses, as in (7.34). SGD generally runs well without momentum, but for some problems having momentum as in (7.35) helps (Bengio, 2012).

In the last decade, a number of methods have been developed to adapt the learning rate to individual model weights/parameters, such as AdaGrad, RMSProp and Adam (Goodfellow, Bengio, et al., 2016).

## 7.7 Conjugate Gradient Method $\boxed{\text{C}}$☹

The idea of adding momentum in the optimization algorithm to avoid inefficient zigzagging during gradient descent has been introduced in (7.35) and Fig. 7.5. The momentum parameter $\mu$ can be specified by the user or determined automatically, for example in the conjugate gradient method. The linear conjugate

gradient method was developed by Hestenes and Stiefel (1952) and extended
to non-linear problems by Fletcher and Reeves (1964). Assume that at step $k$,
(7.29) is satisfied. Now we want to find the next direction $\mathbf{d}_{k+1}$, which preserves
what was achieved in (7.29). In (7.29), the gradient of $J$ in the direction of $\mathbf{d}_k$
has been made 0; now starting from $\mathbf{w}_{k+1}$, we want to find the new direction
$\mathbf{d}_{k+1}$, such that the gradient of $J$ in the direction of $\mathbf{d}_k$ remains 0 (to lowest
order), as we travel along $\mathbf{d}_{k+1}$, that is,

$$\mathbf{d}_k^{\mathrm{T}} \, \nabla J(\mathbf{w}_{k+1} + \eta \, \mathbf{d}_{k+1}) = 0. \tag{7.38}$$

Using (7.5), we can write

$$\nabla J(\mathbf{w}_{k+1} + \eta \, \mathbf{d}_{k+1}) \approx \nabla J(\mathbf{w}_{k+1}) + \mathbf{H} \eta \, \mathbf{d}_{k+1}. \tag{7.39}$$

Thus, (7.38) becomes

$$0 = \mathbf{d}_k^{\mathrm{T}} \, \nabla J(\mathbf{w}_{k+1} + \eta \, \mathbf{d}_{k+1}) \approx \mathbf{d}_k^{\mathrm{T}} \, \nabla J(\mathbf{w}_{k+1}) + \eta \, \mathbf{d}_k^{\mathrm{T}} \, \mathbf{H} \mathbf{d}_{k+1}. \tag{7.40}$$

As the first term drops out according to (7.31), we obtain (to lowest order) the
conjugate direction property

$$\mathbf{d}_k^{\mathrm{T}} \, \mathbf{H} \, \mathbf{d}_{k+1} = 0, \tag{7.41}$$

where $\mathbf{d}_{k+1}$ is said to be *conjugate* to $\mathbf{d}_k$.

Next, we try to obtain an estimate for the momentum parameter $\mu$. Let

$$\mathbf{g}_k \equiv \nabla J(\mathbf{w}_k). \tag{7.42}$$

Substituting (7.35) for $\mathbf{d}_{k+1}$ in (7.41) yields

$$\mathbf{d}_k^{\mathrm{T}} \, \mathbf{H} \, (-\mathbf{g}_{k+1} + \mu \, \mathbf{d}_k) = 0. \tag{7.43}$$

Thus,

$$\mu \, \mathbf{d}_k^{\mathrm{T}} \, \mathbf{H} \, \mathbf{d}_k = \mathbf{d}_k^{\mathrm{T}} \, \mathbf{H} \, \mathbf{g}_{k+1} = \mathbf{g}_{k+1}^{\mathrm{T}} \, \mathbf{H} \, \mathbf{d}_k, \tag{7.44}$$

as $\mathbf{H}^{\mathrm{T}} = \mathbf{H}$. While $\mu$ can be calculated from this equation, it involves the
Hessian matrix, which is in general not known and is computationally costly to
estimate.

Ignoring higher order terms and following (7.5), we have

$$\mathbf{g}_{k+1} = \mathbf{g}_k + \mathbf{H}(\mathbf{w}_{k+1} - \mathbf{w}_k) = \mathbf{g}_k + \eta \, \mathbf{H} \mathbf{d}_k, \tag{7.45}$$

upon substituting in (7.30). Thus,

$$\mathbf{H} \mathbf{d}_k = (\mathbf{g}_{k+1} - \mathbf{g}_k)/\eta. \tag{7.46}$$

Substituting this into (7.44) gives an estimate for the momentum parameter,

$$\mu = \frac{\mathbf{g}_{k+1}^{\mathrm{T}}(\mathbf{g}_{k+1} - \mathbf{g}_k)}{\mathbf{d}_k^{\mathrm{T}}(\mathbf{g}_{k+1} - \mathbf{g}_k)}. \tag{7.47}$$

This way to estimate $\mu$ is called the *Hestenes–Stiefel* method.

A far more commonly used conjugate gradient algorithm is the *Polak–Ribiere* method. First, write (7.31) as

$$\mathbf{d}_k^{\mathrm{T}}\,\mathbf{g}_{k+1} = 0. \tag{7.48}$$

From (7.35),

$$\mathbf{d}_k^{\mathrm{T}}\,\mathbf{g}_k = -\mathbf{g}_k^{\mathrm{T}}\,\mathbf{g}_k + \mu\,\mathbf{d}_{k-1}^{\mathrm{T}}\,\mathbf{g}_k = -\mathbf{g}_k^{\mathrm{T}}\,\mathbf{g}_k, \tag{7.49}$$

where (7.48) has been invoked. Substituting (7.48) and (7.49) into the denominator of (7.47) gives

$$\mu = \frac{\mathbf{g}_{k+1}^{\mathrm{T}}(\mathbf{g}_{k+1} - \mathbf{g}_k)}{\mathbf{g}_k^{\mathrm{T}}\mathbf{g}_k}, \tag{7.50}$$

which is the Polak–Ribiere method (Polak and Ribiere, 1969; Polak, 1971).

Another commonly used algorithm is the *Fletcher–Reeves* method (Fletcher and Reeves, 1964), with

$$\mu = \frac{\mathbf{g}_{k+1}^{\mathrm{T}}\mathbf{g}_{k+1}}{\mathbf{g}_k^{\mathrm{T}}\mathbf{g}_k}, \tag{7.51}$$

which follows from the Polak–Ribiere version, as it can be shown that $\mathbf{g}_{k+1}^{\mathrm{T}}\mathbf{g}_k = 0$ to lowest order (Bishop, 1995). Since the objective function is not exactly a quadratic, the higher order terms cause the two versions to differ from each other. Which version is better depends on the particular problem, though the Polak–Ribiere version is generally considered to have the better performance (Luenberger, 1984; Haykin, 1999).

We still have to find the optimal step size $\eta$ along the search direction $\mathbf{d}_k$, that is, the minimum of the objective function $J$ along $\mathbf{d}_k$ has to be located by finding the $\eta$ that minimizes $J(\mathbf{w}_k + \eta\,\mathbf{d}_k)$. For notational simplicity, we will write $J(\mathbf{w}_k + \eta\,\mathbf{d}_k)$ as $J(\eta)$. To avoid dealing with costly Hessian matrices, a *line search* algorithm is commonly used. The basic line search procedure is as follows:

(1) Find three points $a$, $b$ and $c$ along the search direction such that $J(a) > J(b)$ and $J(b) < J(c)$. As the objective function is continuous, this then guarantees a minimum has been *bracketed* within the interval $(a, c)$.

(2) Fit a parabolic curve to pass through $J(a)$, $J(b)$ and $J(c)$ (Fig. 7.6). The minimum of the parabola is at $\eta = d$.

(3) Next, choose the three points among $a$, $b$, $c$ and $d$ with the three lowest values of $J$. Repeat (2) until convergence to the minimum. More sophisticated algorithms include the widely used Brent's method (Brent, 1973; Press et al., 1986).

Figure 7.6 Using line search to find
the minimum of the function $J(\eta)$.
First, three points $a$, $b$ and $c$ are found
with $J(a) > J(b)$ and $J(b) < J(c)$, so
that the minimum is bracketed within
the interval $(a, c)$. Next a parabola is
fitted to pass through the three points
(dashed curve). The minimum of the
parabola is at $\eta = d$. Next the three
points among $a$, $b$, $c$ and $d$ with the
three lowest values of $J$ are selected,
and a new parabola is fitted to the
three selected points, with the process iterated until convergence to the
minimum of $J$.

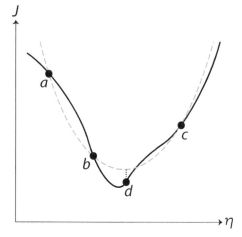

Conjugate gradient is very widely used for linear optimization problems
where the objective function $J$ is a quadratic function of the weights. For
nonlinear optimization problems, $J$ is not a quadratic function, so the conjugate direction condition (7.41) no longer implies (7.38) due to the presence of
higher order terms, that is, the approximation that gradient of $J$ in the direction of $\mathbf{d}_k$ is almost 0 as we travel along $\mathbf{d}_{k+1}$ is no longer accurate. Thus,
the conjugate gradient method is less widely used for non-linear problems than
for linear problems (Murphy, 2012, p. 249). In practice, the method works fine
for non-quadratic $J$ provided once every few steps $\mu$ is set to zero in (7.35)
so the algorithm does a steepest descent along the $-\nabla J$ direction for one step
(Goodfellow, Bengio, et al., 2016, p. 306). Although conjugate gradient was
developed for batch learning, it has been found to run efficiently for NN models
under mini-batch learning (i.e. using only a small subset of data to compute
the gradient) and can be superior to SGD in many aspects, especially in high-
dimensional problems (i.e. problems with a large number of weights) (Q. V. Le
et al., 2011).

## 7.8   Quasi-Newton Methods  $\boxed{\text{C}}$☺

Earlier in this chapter we encountered Newton's method as (7.8), which can be
expressed as

$$\mathbf{w}_{k+1} = \mathbf{w}_k - \mathbf{G}_k \mathbf{g}_k, \qquad (7.52)$$

with $\mathbf{g}_k (\equiv \nabla J(\mathbf{w}_k))$, the gradient of the objective function, and $\mathbf{G}_k (\equiv \mathbf{H}_k^{-1})$,
the inverse of the Hessian matrix. As this form was derived by ignoring terms
above the quadratic in the objective function, this form is highly effective near
a minimum, where the objective function has generally a parabolic surface, but
may not be effective further away from the minimum because of the higher

order terms in the objective function. Hence, a simple modification of Newton's method is to have

$$\mathbf{w}_{k+1} = \mathbf{w}_k - \eta_k \mathbf{G}_k \mathbf{g}_k, \tag{7.53}$$

for some scalar step size $\eta_k$.

Newton's method is extremely expensive for higher-dimensional problems, since at each iteration the inverse of the Hessian matrix has to be computed. *Quasi-Newton methods* (Nocedal and Wright, 2006; Luenberger and Y. Ye, 2016) try to reduce computational costs by making simpler estimates of $\mathbf{G}_k$. Quasi-Newton methods are also related to gradient and conjugate gradient methods. In fact, the simplest approximation, that is, replacing $\mathbf{G}_k$ by the identity matrix $\mathbf{I}$ in (7.53), yields the gradient descent method. The commonly used quasi-Newton methods also preserve the conjugate direction property (7.41) as in conjugate gradient methods. Quasi-Newton methods are second-order methods as they try to approximate the second order derivatives in the Hessian matrix $\mathbf{H}$, whereas gradient descent and SGD are first-order methods using only first-order derivative information from $\nabla J$. The conjugate gradient methods (Section 7.7) are somewhere between the first order and second order methods as they used $\mathbf{H}$ in their derivation but at the end, cleverly avoided $\mathbf{H}$.

The first successful quasi-Newton method is the DFP (*Davidon–Fletcher–Powell*) method (Davidon, 1959; Fletcher and Powell, 1963; Luenberger and Y. Ye, 2016). The procedure is to start at step $k = 0$, with any $\mathbf{w}_0$ and any symmetric positive definite matrix $\mathbf{G}_0$. Then, iterate the following steps:

(1) Set $\mathbf{d}_k = -\mathbf{G}_k \mathbf{g}_k$.

(2) Minimize $J(\mathbf{w}_k + \eta_k \mathbf{d}_k)$ with respect to $\eta_k \geq 0$. One then computes $\mathbf{w}_{k+1}$, $\mathbf{p}_k \equiv \eta_k \mathbf{d}_k$ and $\mathbf{g}_{k+1}$.

(3) Set

$$\mathbf{q}_k = \mathbf{g_{k+1}} - \mathbf{g_k}, \text{ and}$$

$$\mathbf{G}_{k+1} = \mathbf{G}_k + \frac{\mathbf{p}_k \mathbf{p}_k^{\mathrm{T}}}{\mathbf{p}_k^{\mathrm{T}} \mathbf{q}_k} - \frac{\mathbf{G}_k \mathbf{q}_k \mathbf{q}_k^{\mathrm{T}} \mathbf{G}_k}{\mathbf{q}_k^{\mathrm{T}} \mathbf{G}_k \mathbf{q}_k}. \tag{7.54}$$

(4) Update $k$, then return to (1) if needed.

The conjugate direction property (7.41) is preserved in the DFP method (Luenberger and Y. Ye, 2016, section 10.3). If one chooses the initial approximation $\mathbf{G}_0 = \mathbf{I}$, the DFP method becomes the conjugate gradient method.

The most popular quasi-Newton method is the *BGFS* (Broyden–Fletcher–Goldfarb–Shanno) method (Broyden, 1970; Fletcher, 1970; Goldfarb, 1970; Shanno, 1970). Instead of (7.54), the update for the BFGS method (Polak, 1971; Luenberger and Y. Ye, 2016, section 10.4) is

$$\mathbf{G}_{k+1}^{\mathrm{BFGS}} = \mathbf{G}_{k+1}^{\mathrm{DFP}} + \mathbf{v}_k \mathbf{v}_k^{\mathrm{T}}, \tag{7.55}$$

where $\mathbf{G}_{k+1}^{\mathrm{DFP}}$ is given by (7.54) and

$$\mathbf{v}_k = (\mathbf{q}_k^{\mathrm{T}} \mathbf{G}_k \mathbf{q}_k)^{1/2} \left( \frac{\mathbf{p}_k}{\mathbf{p}_k^{\mathrm{T}} \mathbf{q}_k} - \frac{\mathbf{G}_k \mathbf{q}_k}{\mathbf{q}_k^{\mathrm{T}} \mathbf{G}_k \mathbf{q}_k} \right). \tag{7.56}$$

The conjugate direction property (7.41) is also preserved in the BFGS method.

In summary, both conjugate gradient and quasi-Newton methods avoid using the inverse of the Hessian matrix. However, quasi-Newton methods do try to approximate the inverse Hessian, while preserving the conjugate direction property of the conjugate gradient methods. Thus, quasi-Newton methods can be regarded as a further extension of the conjugate gradient methods by incorporating an approximation of the inverse Hessian to simulate Newton's method, which leads to faster convergence than the conjugate gradient methods. Another advantage is that the line search, which is of critical importance in the conjugate gradient methods, need not be performed to high accuracy in the quasi-Newton methods, as its role is not as critical. The major disadvantage of the quasi-Newton methods is the large storage associated with carrying the $L \times L$ matrix $\mathbf{G}_k$ (as $\mathbf{w}$ has $L$ elements), that is, the memory requirement for the quasi-Newton methods is of order $O(L^2)$ versus $O(L)$ for the conjugate gradient methods. For large $L$, the memory requirement of $O(L^2)$ may become prohibitive, and the conjugate gradient method has the advantage.

To reduce the large memory requirement, Shanno (1978) proposed *limited memory quasi-Newton methods*. In the limited memory version of the BFGS method (L-BFGS), after $\mathbf{G}_k$ is iterated for $m$ steps ($m$ often $< 10$), the process is restarted from $\mathbf{G}_0$. The storage requirement is moderate since the inverse Hessian requires only storing $m$ pairs of the vectors $\mathbf{p}_k$ and $\mathbf{q}_k$ (Luenberger and Y. Ye, 2016), thereby reducing the memory requirement to $O(L)$. Q. V. Le et al. (2011) found L-BFGS to perform well against the SGD method and well against the conjugate gradient method for problems with $< 10^4$ parameters or weights.

## 7.9  Non-linear Least Squares Methods  [C] ☺

So far, the optimization methods presented have not been limited to an objective function of a specific form. In many situations, the objective function or error function $\mathcal{E}$ involves a sum of squares, that is,

$$\mathcal{E} = \frac{1}{2} \sum_n (\epsilon^{(n)})^2 = \frac{1}{2} \|\boldsymbol{\epsilon}\|^2, \tag{7.57}$$

where we have used $\mathcal{E}$ instead of $J$ to avoid confusion with the Jacobian matrix $\mathbf{J}$ later, $\epsilon^{(n)}$ is the error associated with the $n$th observation or pattern and $\boldsymbol{\epsilon}$ is a vector with elements $\epsilon^{(n)}$ (and the conventional scale factor $\frac{1}{2}$ has been added to avoid the appearance of a factor of 2 in the derivatives). We now derive optimization methods specially designed to deal with such non-linear least squares problems.

Suppose at the $k$th step, we are at the point $\mathbf{w}_k$ in weight space, and we take the next step to $\mathbf{w}_{k+1}$. To first order, the Taylor expansion of the function $\epsilon(\mathbf{w})$ about $\mathbf{w}_k$ allows us to write

$$\epsilon(\mathbf{w}_{k+1}) = \epsilon(\mathbf{w}_k) + \mathbf{J}_k\,(\mathbf{w}_{k+1} - \mathbf{w}_k), \tag{7.58}$$

where $\mathbf{J}_k$ is the Jacobian matrix $\mathbf{J}$ at step $k$, with the elements of $\mathbf{J}$ given by

$$(\mathbf{J})_{ni} = \frac{\partial \epsilon^{(n)}}{\partial w_i}. \tag{7.59}$$

Substituting (7.58) into (7.57), we have

$$\begin{aligned}
\frac{1}{2}\|\epsilon(\mathbf{w}_{k+1})\|^2 &= \frac{1}{2}\|\epsilon(\mathbf{w}_k) + \mathbf{J}_k\,(\mathbf{w}_{k+1} - \mathbf{w}_k)\|^2 \\
&= \frac{1}{2}\|\epsilon(\mathbf{w}_k)\|^2 + \epsilon^{\mathrm{T}}(\mathbf{w}_k)\,\mathbf{J}_k\,(\mathbf{w}_{k+1} - \mathbf{w}_k) \\
&\quad + \frac{1}{2}(\mathbf{w}_{k+1} - \mathbf{w}_k)^{\mathrm{T}}\,\mathbf{J}_k^{\mathrm{T}}\mathbf{J}_k\,(\mathbf{w}_{k+1} - \mathbf{w}_k).
\end{aligned} \tag{7.60}$$

To find the optimal $\mathbf{w}_{k+1}$, differentiate the right hand side of the above equation by $\mathbf{w}_{k+1}$ and set the result to zero, yielding

$$\begin{aligned}
\epsilon^{\mathrm{T}}(\mathbf{w}_k)\,\mathbf{J}_k + \mathbf{J}_k^{\mathrm{T}}\mathbf{J}_k\,(\mathbf{w}_{k+1} - \mathbf{w}_k) &= 0, \\
\mathbf{J}_k^{\mathrm{T}}\mathbf{J}_k\,(\mathbf{w}_{k+1} - \mathbf{w}_k) &= -\mathbf{J}_k^{\mathrm{T}}\,\epsilon(\mathbf{w}_k).
\end{aligned} \tag{7.61}$$

Solving for $\mathbf{w}_{k+1}$, we get

$$\mathbf{w}_{k+1} = \mathbf{w}_k - (\mathbf{J}_k^{\mathrm{T}}\mathbf{J}_k)^{-1}\mathbf{J}_k^{\mathrm{T}}\,\epsilon(\mathbf{w}_k), \tag{7.62}$$

which is known as the *Gauss–Newton method*. Equation (7.62) resembles the normal equations (5.55) encountered previously in the multiple linear regression problem. In fact, one can regard the Gauss–Newton method as replacing a non-linear least squares problem with a sequence of linear least squares problems (7.62).

For the sum-of-squares error function (7.57), the gradient can be written as

$$\nabla \mathcal{E} = \mathbf{J}^{\mathrm{T}}\epsilon(\mathbf{w}), \tag{7.63}$$

while the Hessian matrix has elements

$$(\mathbf{H})_{ij} = \frac{\partial^2 \mathcal{E}}{\partial w_i \partial w_j} = \sum_n \left\{ \frac{\partial \epsilon^{(n)}}{\partial w_i}\frac{\partial \epsilon^{(n)}}{\partial w_j} + \epsilon^{(n)}\frac{\partial^2 \epsilon^{(n)}}{\partial w_i \partial w_j} \right\}. \tag{7.64}$$

If the error function depends on the weights linearly, then the second derivatives, that is, the second term in (7.64), vanish. Even when the error function is not a linear function of the weights, we will ignore the second order terms in (7.64) and approximate the Hessian by

$$\mathbf{H} = \mathbf{J}^{\mathrm{T}}\mathbf{J}. \tag{7.65}$$

By (7.63) and (7.65), we can regard (7.62) as an approximation of Newton's method (7.8), with the inverse of the Hessian matrix being approximated by $(\mathbf{J}^T\mathbf{J})^{-1}$.

While (7.62) can be used repeatedly to reach the minimum of the error function, the pitfall is that the step size may become too large, so the first order Taylor approximation (7.58) becomes inaccurate, and the Gauss–Newton method may not converge at all. To correct this problem, the *Levenberg–Marquardt* method (Levenberg, 1944; Marquardt, 1963) adds a penalty term to the error function (7.60), that is,

$$\frac{1}{2}\|\boldsymbol{\epsilon}(\mathbf{w}_{k+1})\|^2 = \frac{1}{2}\|\boldsymbol{\epsilon}(\mathbf{w}_k) + \mathbf{J}_k\left(\mathbf{w}_{k+1} - \mathbf{w}_k\right)\|^2 + \lambda\|\mathbf{w}_{k+1} - \mathbf{w}_k\|^2, \qquad (7.66)$$

where large step size is penalized by the last term, with a larger parameter $\lambda$ tending to give smaller step size. Again minimizing this penalized error function with respect to $\mathbf{w}_{k+1}$ yields

$$\mathbf{w}_{k+1} = \mathbf{w}_k - (\mathbf{J}_k^T\mathbf{J}_k + \lambda\mathbf{I})^{-1}\mathbf{J}_k^T\,\boldsymbol{\epsilon}(\mathbf{w}_k), \qquad (7.67)$$

where $\mathbf{I}$ is the identity matrix. For small $\lambda$, this Levenberg–Marquardt formula reduces to the Gauss–Newton method, while for large $\lambda$, this reduces to the gradient descent method. While the Gauss–Newton method converges very quickly near a minimum, the gradient descent method is robust even when far from a minimum. A common practice is to change $\lambda$ during the optimization procedure. Start with some arbitrary value of $\lambda$, say $\lambda = 0.1$. If the error function decreases after taking a step by (7.67), reduce $\lambda$ by a factor of 10, and the process is repeated. If the error function increases after taking a step, discard the new $\mathbf{w}_{k+1}$, increase $\lambda$ by a factor of 10, and repeat step (7.67). The whole process is repeated until convergence. In essence, the Levenberg–Marquardt method improves on the robustness of the Gauss–Newton method by switching to gradient descent when far from a minimum and then switching back to the Gauss–Newton method for fast convergence when close to a minimum.

Note that if $\mathbf{w}$ is of dimension $L$, the Hessian $\mathbf{H}$ is of dimension $L \times L$, while the Jacobian $\mathbf{J}$ is of dimension $N \times L$, where $N$ is the number of observations or patterns. Since $N$ is likely to be considerably larger than $L$, the Jacobian matrix may require even more memory storage than the Hessian matrix. Thus, the storage of the Jacobian in Gauss–Newton and Levenberg–Marquardt methods renders them most demanding on memory, surpassing even the quasi-Newton methods, which require the storage of an approximate Hessian matrix. A limited memory Levenberg–Marquardt method has been developed, allowing the method to run efficiently on datasets containing a large number of observations (Wilamowski and H. Yu, 2010).

# 7.10  Evolutionary Algorithms  $\boxed{\text{B}}$☺

All the optimization methods presented so far (except SGD) belong to the class of methods known as *deterministic optimization*, in that each step of the optimization process is determined by explicit formulas. SGD introduces stochasticity by randomly selecting a small subset of data to estimate the gradient of the objective function. While deterministic optimization methods tend to converge to a minimum efficiently, they often converge to a nearby local minimum. To find a global minimum, one usually has to introduce some stochastic element into the search. A simple way is to repeat the optimization process many times, each starting from different random initial weights. These multiple runs will likely converge to different minima, and one hopes that the lowest minimum among them is the desired global minimum. Of course, there is no guarantee that the global minimum has been found. Nevertheless, by using a large enough number of runs and broadly distributed random initial weights, the global minimum can usually be found by such an approach. From Section 7.1, it has been pointed out that in many ML problems, it is not essential that one finds the global minimum, as there are usually multiple local minima that give good solutions, and for continuous model output, taking an ensemble average of the output from multiple runs usually gives reasonable predictions.

Unlike deterministic optimization, *stochastic optimization* methods repeatedly introduce randomness during the search process to avoid getting trapped in a local minimum. Evolution of living organisms is a prime example of stochastic optimization. Starting from simple organic molecules, Nature evolves increasingly complex and superior organisms by stochastic optimization where stochasticity comes from genetic mutation and crossover of DNA during sexual reproduction.

In computer science, *evolutionary computation* uses the computer to study evolution or simulate evolutionary processes. In particular, *evolutionary algorithms* (EA) use concepts from evolution to solve optimization problems (Simon, 2013; Eiben and J. E. Smith, 2015). Genetic algorithm (GA) is the most popular among the EAs. Other EAs include particle swarm optimization, ant colony optimization, differential evolution, and so on. Other stochastic optimization algorithms such as hill climbing and simulated annealing are often also grouped under EA.

Another reason for the rise of EA is that not all optimization problems have an objective function $J$ that allows easy estimation of its gradient so deterministic methods can use the gradient information for fast convergence to a local minimum. For instance, for the MLP model with one hidden layer, $N_h$, the number of hidden neurons or nodes, is an adjustable parameter – usually called a *hyperparameter*. In ML, a hyperparameter is a parameter that is set before the model training begins and is held constant throughout the model training. In MLP, $N_h$ is chosen before the model weights are optimized via back-propagation. Since $N_h$ is an integer, it is impossible to compute the gradient of $J$ with respect to $N_h$. How to find optimal hyperparameters is examined in Section 8.6.

Instead of minimizing an objective function $J$ (representing MSE, cost, loss, etc.) as done in deterministic optimization, most EA methods maximize a 'fitness' function $f$, since the main principle of evolution is 'survival of the fittest', so maximizing fitness is the objective. The two views are easily reconciled by regarding $f$ as being $-J$, so maximizing $f$ is equivalent to minimizing $J$.

In the following sections, we start with the simplest method, hill climbing (Section 7.11), then proceed to the most popular method, genetic algorithm (Section 7.12) and finally to a newer method, differential evolution (Section 7.13), probably the best EA algorithm for optimizing in continuous space.

# 7.11   Hill Climbing   ☐C☺

Hill climbing is a very simple method for finding a local maximum of the fitness function $f(\mathbf{w})$. The $L$-dimensional weight or parameter vector $\mathbf{w}$ contains elements that can be discrete or continuous variables. There are several variants of hill climbing, of which the most basic is the *steepest ascent* hill climbing algorithm. It involves looking at all directions and proceeding in the direction that gives the largest increase in fitness.

Another variant is the *random mutation* hill climbing algorithm (Simon, 2013, section 2.6). This variant randomly selects a dimension $l \in \{1, \ldots, L\}$ to perturb and as soon as a higher fitness is found, $\mathbf{w}^{(k)}$ is replaced by the new perturbed value. This type of search is called *greedy* – a greedy algorithm is any algorithm that follows the locally optimal choice at each stage of the optimization. The steepest ascent variant is non-greedy as it conservatively evaluates all $L$ perturbed versions of $\mathbf{w}^{(k)}$ before making an update for $\mathbf{w}^{(k)}$, whereas the random mutation variant is greedy as it jumps to a perturbed value with higher fitness without bothering to check out other dimensions. Among four variants of hill climbing algorithms tested, Simon (2013, table 2.1) found random mutation to be the most efficient.

The random mutation hill climbing algorithm is as follows:

[1] Start with an initial guess $\mathbf{w}^{(0)}$. Set counter $k = 0$.

[2] Evaluate the fitness function $f(\mathbf{w}^{(k)})$.

[3] Mutation: Randomly select a dimension $l \in \{1, \ldots, L\}$ and randomly perturb the $l$th dimension of $\mathbf{w}^{(k)}$ to get $\mathbf{w}^{(k)}_{\text{pert}}$.

[4] Evaluate $f(\mathbf{w}^{(k)}_{\text{pert}})$.

[5] If $f(\mathbf{w}_{\text{pert}}) > f(\mathbf{w}^{(k)})$,
    set $k = k + 1$, $\mathbf{w}^{(k)} = \mathbf{w}_{\text{pert}}$ and $f(\mathbf{w}^{(k)}) = f(\mathbf{w}_{\text{pert}})$.

[6] Iterate by going to step [3] unless some convergence criterion is met.

Hill climbing packages usually have some options for the user on ways to perturb $\mathbf{w}^{(k)}$ during the mutation step. As hill climbing tends to converge to local maxima, it is common to make multiple runs with different initial guess value $\mathbf{w}^{(0)}$ and choose the result from the run attaining the highest fitness value.

In continuous spaces, ridges in $f(\mathbf{w})$ can render hill climbing very inefficient. For instance, suppose the maximum is located on a narrow ridge, which lies diagonally in the $w_1$-$w_2$ space. Hill climbing cannot advance along the ridge in this diagonal direction as it can only advance in either the $w_1$ or the $w_2$ direction, hence it is forced to take a zig-zag path to the maximum. If the ridge is steep and narrow, the zig-zag steps have to be very small, hence the algorithm is inefficient. In contrast, gradient ascent (i.e. gradient descent applied to $-f$) can travel along the diagonal direction, so will be much more efficient compared to hill climbing. The advantage of hill climbing is that it does not require gradient information, which may not be available for some problems.

# 7.12  Genetic Algorithm  $\boxed{\text{C}}$ ☺

Among EA methods, *genetic algorithms* (GA) were inspired by biological evolution where crossover of genes from parents and genetic mutations result in a stochastic process that can lead to superior descendants after many generations (R. L. Haupt and S. E. Haupt, 2004; Simon, 2013). A history of the development of GA is given in Simon (2013, section 3.3).

The weight vector $\mathbf{w}$ of a model can be treated as a long strand of DNA, and an ensemble of solutions is treated like a population of organisms. A part of the weight vector of one solution can be exchanged with a part from another solution to form a new solution, analogous to the *crossover* of DNA material from two parents. For instance, two parents have weight vectors $\mathbf{w}$ and $\mathbf{w}'$. A random position is chosen (in this example just before the third weight parameter) for an incision, and the second part of $\mathbf{w}'$ is connected to the first part of $\mathbf{w}$ and vice versa in the offspring, that is,

$$[w_1, w_2, w_3, w_4, w_5, w_6] \qquad\qquad [w_1', w_2', w_3, w_4, w_5, w_6]$$
$$- \text{ crossover } \rightarrow \qquad\qquad\qquad (7.68)$$
$$[w_1', w_2', w_3', w_4', w_5', w_6'] \qquad\qquad [w_1, w_2, w_3', w_4', w_5', w_6'].$$

Genetic *mutation* can be simulated by randomly perturbing one of the weights $w_l$ in the weight vector $\mathbf{w}$, that is, randomly choose an $l$ and replace $w_l$ by $w_l + \epsilon$ for some random $\epsilon$. These two processes introduce many new offspring, but only the relatively fit offspring have a high probability of surviving to reproduce. With the 'survival of the fittest' principle pruning the offspring, successive generations eventually converge towards the global optimum.

One must specify a *fitness function* $f$ to evaluate the fitness of the individuals in a population. If for the $i$th individual in the population, its fitness is $f(i)$, then a fitness probability $P(i)$ can be defined as

$$P(i) = \frac{f(i)}{\sum_{i=1}^{N_\mathrm{p}} f(i)}, \qquad\qquad (7.69)$$

where $N_\mathrm{p}$ is the total number of individuals in a population. Individuals with high $P$ will be given greater chances to reproduce, while those with low $P$ will be given greater chances to die off.

There are many variants of GA – the following describes the basic structure of GA:

[1] Choose the population size ($N_p$) and the number of generations ($N_g$). Initialize the weight vectors of the population. Repeat the following steps $N_g$ times or until some convergence criterion is met.

[2] Calculate the fitness function $f$ and the fitness probability $P$ for each individual in the population.

[3] Select a given number of individuals from the population, where the chance of an individual getting selected is given by its fitness probability $P$.

[4] Duplicate the weight vectors of these individuals, then apply either the crossover or the mutation operation on the various duplicated weight vectors to produce new offspring.

[5] To keep the population size constant, individuals with poor fitness are selected (based on the probability $1 - P$) to die off and are replaced by the new offspring (the fittest individual is never chosen to die).

Finally, after $N_g$ generations, the individual with the greatest fitness is chosen as the solution. To monitor the evolutionary progress over successive generations, one can check the average fitness of the population, simply by averaging the fitness $f$ over all individuals in the population.

In general, the population size need not be huge. For low-dimensional problems ($L < 5$), $N_p = 30$–$50$ is often adequate. For higher-dimensional problems, $N_p$ is usually in the hundreds. The fittest individuals are the 'elite' and are always retained in the next generation. The elite can consist of the single fittest individual, or, for example, the top five per cent fittest individuals. Usually crossover is performed much more often than mutation, for example 4–10 times more frequently. For example, for a GA model with a population of 100, the next generation could consist of 5 elites, 80 crossover offspring and 15 mutation offspring. A GA model using too high a mutation rate behaves like random search, while using too low a mutation rate will likely converge prematurely to a local minimum. The optimal ratio of the number of crossover offspring to the number of mutation offspring is unfortunately problem dependent.

Historically, GA started with having binary variables in $\mathbf{w}$. It was later extended to work with integers and real numbers in $\mathbf{w}$.

GA can be used to perform the non-linear optimization in NN problems, where the fitness function can, for instance, be the negative of the mean squared error. However, deterministic optimization using gradient descent methods would converge much quicker than stochastic optimization methods such as GA.

In summary, there are three advantages with GA:

(i) In problems where the fitness function cannot be expressed in closed analytic form, gradient descent methods cannot be used effectively, whereas GA works well.

(ii) When there are many local optima in the fitness function, gradient descent methods can be trapped too easily.

(iii) GA can utilize parallel processors much more readily than gradient descent algorithms, since in GA different individuals in a population can be computed simultaneously on different processors.

## 7.13   Differential Evolution   $\boxed{\text{C}}$☺

*Differential evolution* (DE) is an evolutionary algorithm but is not biologically motivated. It was originally designed for optimization search in continuous space and first appeared in a technical report by R. Storm and K. Price in 1995, then published later in Storn and K.Price (1997) and K. V. Price et al. (2005). It attracted attention after strong performance in many EA optimization contests starting in the mid-1990s (Das and Suganthan, 2011; Simon, 2013; Das, Mullick, et al., 2016) and in tests against other EA algorithms on multiple benchmarks (Vesterstrøm and Thomsen, 2004). Mutation and crossover in DE are consolidated and simpler with fewer hyperparameters than in GA.

Candidate solutions in the search space undergo *differential mutation*, that is, a candidate solution $\mathbf{w}^{(1)}$ in $\mathbb{R}^L$ is mutated by

$$\tilde{\mathbf{w}}^{(1)} = \mathbf{w}^{(1)} + \mathbf{p}, \tag{7.70}$$

where the perturbation vector $\mathbf{p}$ is the scaled difference between two other randomly chosen candidates $\mathbf{w}^{(2)}$ and $\mathbf{w}^{(3)}$, that is,

$$\mathbf{p} = s\left(\mathbf{w}^{(2)} - \mathbf{w}^{(3)}\right), \tag{7.71}$$

where $s > 0$ is a scale factor or stepsize parameter (Fig. 7.7). Despite the name 'differential', there is no differentiation involved and the method works well for non-differentiable fitness functions.

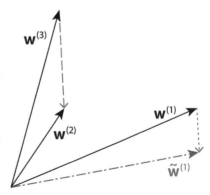

Figure 7.7 Constructing the mutant vector in DE. From three randomly chosen vectors $\mathbf{w}^{(1)}$, $\mathbf{w}^{(2)}$ and $\mathbf{w}^{(3)}$ from the population, the difference vector $\mathbf{w}^{(2)} - \mathbf{w}^{(3)}$ (dashed) is constructed. A scaled version of this difference vector (dotted) is added to $\mathbf{w}^{(1)}$ to give the mutant vector $\tilde{\mathbf{w}}^{(1)}$ (dot-dashed).

The user chooses three parameters: the population size $N_{\mathrm{p}}$, the crossover rate $c$ and the stepsize $s$. Simon (2013, figure 12.2) recommends the ranges $c \in [0.1, 1]$ and $s \in [0.4, 0.9]$.

The basic DE algorithm for maximizing the fitness function $f(\mathbf{w})$ is as follows:

[1] Initialize the candidate solutions $\{\mathbf{w}_i\}$ for $i = 1, \ldots, N_{\mathrm{p}}$.

[2] For each candidate $\mathbf{w}_i$:

    [a] Set $\tilde{\mathbf{w}}_i = \mathbf{w}_i$.

    [b] Mutation step:

        Excluding $\mathbf{w}_i$, randomly choose three different candidates $\mathbf{w}^{(1)}$, $\mathbf{w}^{(2)}$ and $\mathbf{w}^{(3)}$ from the population.

        Compute the mutant vector $\tilde{\mathbf{w}}^{(1)}$,

$$\tilde{\mathbf{w}}^{(1)} = \mathbf{w}^{(1)} + s\left(\mathbf{w}^{(2)} - \mathbf{w}^{(3)}\right). \tag{7.72}$$

    [c] Crossover step: [Randomly crossover elements of $\tilde{\mathbf{w}}^{(1)}$ into $\tilde{\mathbf{w}}_i$]
        Choose a random integer $l_{\mathrm{r}} \in \{1, \ldots, L\}$.

        For each dimension $l$ ($l = 1, \ldots, L$):

        [i] Choose a uniformly distributed random number $r_l \in [0, 1]$.

        [ii] If ($r_l < c$) or ($l = l_{\mathrm{r}}$):

            Set $\tilde{w}_{il} = \tilde{w}_l^{(1)}$, where $\tilde{w}_{il}$ and $\tilde{w}_l^{(1)}$ are the $l$th element of $\tilde{\mathbf{w}}_i$ and $\tilde{\mathbf{w}}^{(1)}$, respectively.

    [d] If $f(\mathbf{w}_i) < f(\tilde{\mathbf{w}}_i)$, set $\mathbf{w}_i = \tilde{\mathbf{w}}_i$.

[3] Iterate for another generation by going to step [2], unless some stopping criterion is met.

The solution is the candidate with the highest fitness $f$ in the final population. In step [ii], crossover occurs in the $l$th dimension when the random number $r_l < c$, the crossover rate. However, there is a chance that the condition $r_l < c$ is not satisfied for all $L$ dimensions; thus, we ensure at least one crossover occurs at $l = l_{\mathrm{r}}$ (a randomly chosen dimension), so $\tilde{\mathbf{w}}_i$ differs from $\mathbf{w}_i$ in at least one dimension.

As an example, consider using DE to find the minimum of the function

$$J(w_1, w_2) = -\cos\left[0.5\left(w_1^2 + w_2^2\right)\right] \exp\left(-0.2\sqrt{w_1^2 + w_2^2}\right), \tag{7.73}$$

which has a global minimum at the origin with $J = -1$ and concentric rings of local minima and local maxima (Fig. 7.8). DE was run with the fitness function $f = -J$, $N_{\mathrm{p}} = 50$, $c = 0.5$ and $s = 0.7$. After 10 iterations, about half the candidates are well on their way to converging towards the global minimum, with the remaining candidates sitting on concentric rings of local minima (Fig. 7.8(c)). After 20 iterations, all 50 candidates are close to the global minimum (Fig. 7.8(d)).

There are several variants of DE. One variant replaces $\mathbf{w}^{(1)}$ in (7.72) by $\mathbf{w}_{\mathrm{best}}$, the fittest individual in the population. This speeds up the convergence

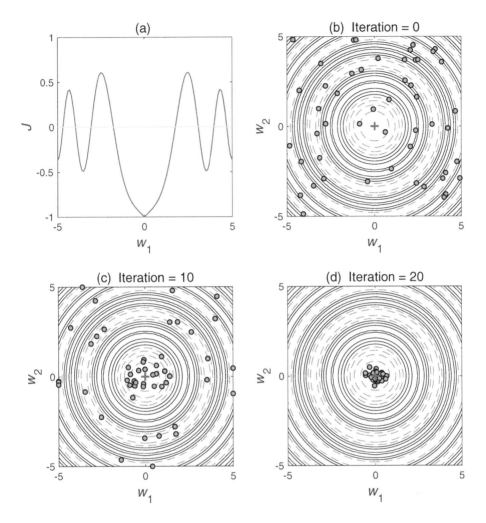

Figure 7.8 (a) The objective function $J$ along the $w_1$ axis. Position of the 50 candidate solutions in the $w_1$-$w_2$ space: (b) at the initial setup and after using the DE algorithm to perform (c) 10 iterations and (d) 20 iterations. If a candidate is perturbed to move beyond the boundary of the interval $[-5, 5]$ in any dimension, it is repositioned to sit right on the boundary. Negative contour values of $J$ are indicated by dashed lines and non-negative contours by solid lines, with the global minimum at $(0, 0)$ marked by the cross.

but at the expense of doing broader exploration of the search space. Another variant is to introduce more variety in the mutation by adding an additional difference vector, $\mathbf{w}^{(4)} - \mathbf{w}^{(5)}$, when computing the mutant vector, that is,

$$\tilde{\mathbf{w}}^{(1)} = \mathbf{w}^{(1)} + s\left(\mathbf{w}^{(2)} - \mathbf{w}^{(3)} + \mathbf{w}^{(4)} - \mathbf{w}^{(5)}\right). \tag{7.74}$$

Originally designed to work in continuous search space, DE has been extended to work in discrete space or mixed discrete–continuous space (Das and Suganthan, 2011; Simon, 2013, chapter 12).

## Exercises

7.1

Find the minimum of the function

$$J(x, y) = -10 + (x + 1)^2 + y^2 + 10 \cos(1.5\,xy) + 5 \sin(xy), \qquad (7.75)$$

using (a) stochastic optimization (e.g. differential evolution or genetic algorithm) and (b) deterministic optimization (e.g. BFGS quasi-Newton or conjugate gradient). Use a population of 50 in (a) and 50 runs with different random initial weights in (b). Comment on the issue of local minima and convergence.

7.2

Find the minimum of the function

$$J(x, y) = -0.5 \exp\{-0.01[x^2 + (y - 1)^2]\} + |\sin 2x|, \qquad (7.76)$$

using (a) stochastic optimization (e.g. differential evolution or genetic algorithm) and (b) deterministic optimization (e.g. BFGS quasi-Newton or conjugate gradient). Use a population of 50 in (a) and 50 runs with different random initial weights in (b). Comment on the issue of local minima and convergence.

7.3

(a) Use a stochastic optimization code (e.g. differential evolution or genetic algorithm) to fit the one-hidden-layer MLP ANN model in (6.13) to the given dataset $\{x, y\}$ from the book website by minimizing the mean squared error. (b) Repeat with a deterministic optimization code (e.g. BFGS quasi-Newton or conjugate gradient). Use (i) one hidden node and (ii) two hidden nodes. Use a population of 50 in (a) and 50 runs with different random initial weights in (b). How reproducible are the results if you initialize the random number generator differently?

# 8

# Learning and Generalization

Data generally contain both signal and noise. Like Ulysses sailing between Scylla and Charybdis, a researcher must steer a careful course between using a model with too little flexibility to model the underlying signal adequately (*underfitting*) and using a model with too much flexibility, which readily fits to the noise (*overfitting*) (Fig. 1.4). Instead of just using a model to find the closest fit to the data, one needs a wiser objective – namely, the model must be able to correctly *learn* the underlying signal or relation in the data and to *generalize*, that is, be able to work well with data the model has not encountered previously. Overfitting is the most common cause of spurious skills reported by ML/statistical models in the literature. Overfitting often infiltrates problems in an inconspicuous manner – anyone learning to use data models almost invariably step into this pitfall on some occasions.

This chapter has four main parts. The first part covers objective functions and errors (Sections 8.1–8.3). The second part covers various regularization techniques to avoid overfitting (Sections 8.4–8.9). Section 8.7 on the ensemble approach is especially important for dealing with the multiple local minima problem in NN models. The third part covers the Bayesian approach to model selection and model averaging (Sections 8.10–8.12). The fourth part covers the recent development of interpretable ML or interpretable AI, that is, how to make a black box model less opaque (Section 8.13).

## 8.1 Mean Squared Error and Maximum Likelihood $\boxed{\text{A}}$ ☺

For many statistical and ML models, for example linear regression and multi-layer perceptron (MLP) ANN models, minimizing the objective function $J$ in-

volves minimizing the *mean squared error* (MSE)[1] between the model outputs
$\hat{\mathbf{y}}$ and the observed target data $\mathbf{y}$ (with components $y_k$), that is,

$$J = \frac{1}{N} \sum_{n=1}^{N} \left\{ \frac{1}{2} \sum_{k=1}^{M} \left[ \hat{y}_k^{(n)} - y_k^{(n)} \right]^2 \right\}, \tag{8.1}$$

where there are $k = 1, \ldots, M$ output variables $\hat{y}_k$, $n = 1, \ldots, N$ observations
and the constant $1/2$ is optional. While minimizing the MSE is quite intuitive,
it can be derived from the broader principle of *maximum likelihood* under the
assumption of Gaussian noise distribution.

Recall from (2.10), for independent events $y_1$ and $y_2$, the probability of both
occurring is $P(y_1, y_2) = P(y_1)P(y_2)$. If we assume that the multivariate target
data $y_k$ are independent random variables, then the conditional distribution of
$\mathbf{y}$ given predictors $\mathbf{x}$ is simply a product of the conditional distribution of the
individual components $y_k$, that is,

$$p(\mathbf{y}|\mathbf{x}) = \prod_{k=1}^{M} p(y_k|\mathbf{x}). \tag{8.2}$$

The target data $y_k$ are made up of noise $\epsilon_k$ plus an underlying signal (which we
are trying to simulate by a model with parameters or weights $\mathbf{w}$ and outputs
$\hat{y}_k$), that is,

$$y_k = \hat{y}_k(\mathbf{x}; \mathbf{w}) + \epsilon_k. \tag{8.3}$$

We assume that the noise $\epsilon_k$ obeys a Gaussian distribution with zero mean and
standard deviation $\sigma$, with $\sigma$ independent of $k$ and $\mathbf{x}$, that is,

$$p(\boldsymbol{\epsilon}) = \prod_{k=1}^{M} p(\epsilon_k) = \frac{1}{(2\pi)^{M/2}\sigma^M} \exp\left( -\frac{\sum_k \epsilon_k^2}{2\sigma^2} \right). \tag{8.4}$$

Since $\hat{y}_k(\mathbf{x}; \mathbf{w})$ is a deterministic signal, that is, given $\mathbf{x}$, the distribution of $y_k$ in
(8.3) arises solely from the distribution of $\epsilon_k$, we substitute $p(\epsilon_k)$ for $p(y_k|\mathbf{x})$ in
(8.2). Writing $\epsilon_k^2$ as $(\hat{y}_k(\mathbf{x}; \mathbf{w}) - y_k)^2$, we get the conditional density distribution

$$p(\mathbf{y}|\mathbf{x}; \mathbf{w}) = \frac{1}{(2\pi)^{M/2}\sigma^M} \exp\left[ -\frac{\sum_k (\hat{y}_k(\mathbf{x}; \mathbf{w}) - y_k)^2}{2\sigma^2} \right]. \tag{8.5}$$

The principle of *maximum likelihood* says that if we have a conditional dis-
tribution $p(\mathbf{y}|\mathbf{x}; \mathbf{w})$, and we have observed values $\mathbf{y}$ given by the dataset $D$ and
$\mathbf{x}$ by the dataset $X$, then the parameters $\mathbf{w}$ can be found by maximizing the
likelihood function $p(D|X; \mathbf{w})$, that is, the parameters $\mathbf{w}$ should be chosen so
that the likelihood of observing $D$ given $X$ is maximized. Note that $p(D|X; \mathbf{w})$
is a function of $\mathbf{w}$ only, since $D$ and $X$ are known.

---

[1] While the MSE objective function is most commonly used in regression problems, the
cross entropy objective function is more commonly used in classification problems (Section
12.6).

The datasets $X$ and $D$ contain the observations $\mathbf{x}^{(n)}$ and $\mathbf{y}^{(n)}$, with $n = 1, \ldots, N$. The likelihood function $L$ is then

$$L = p(D|X; \mathbf{w}) = \prod_{n=1}^{N} p\left(\mathbf{y}^{(n)}|\mathbf{x}^{(n)}; \mathbf{w}\right). \tag{8.6}$$

Instead of maximizing the likelihood function, it is more convenient to minimize the negative log of the likelihood, as the logarithm function is a monotonic function. From (8.5) and (8.6), we end up minimizing the following objective function with respect to $\mathbf{w}$:

$$\tilde{J} = -\ln L = \frac{1}{2\sigma^2} \sum_{n=1}^{N} \sum_{k=1}^{M} \left[\hat{y}_k(\mathbf{x}^{(n)}; \mathbf{w}) - y_k^{(n)}\right]^2$$

$$+ NM \ln \sigma + \frac{NM}{2} \ln(2\pi). \tag{8.7}$$

Since the last two terms are independent of $\mathbf{w}$, they are irrelevant to the minimization process and can be omitted. Other than a constant multiplicative factor, the remaining term in $\tilde{J}$ is the same as the MSE objective function $J$ in (8.1). Thus, minimizing MSE is equivalent to maximizing likelihood with the noise distribution assumed to be Gaussian.

## 8.2 Objective Functions and Robustness $\boxed{\text{A}}$☺

In this section, we examine where the model outputs converge to, under the MSE objective function (8.1) in the limit of infinite sample size $N$ with a flexible enough model (Bishop, 1995). However, the MSE is not the only way to incorporate information about the error between the model output $\hat{y}_k$ and the target data $y_k$ into $J$. We could minimize the *mean absolute error* (MAE) instead of the MSE, that is, define

$$J = \frac{1}{N} \sum_{n=1}^{N} \sum_{k=1}^{M} \left|\hat{y}_k^{(n)} - y_k^{(n)}\right|. \tag{8.8}$$

Any data point $y_k$ lying far from the mean of the distribution of $y_k$ would exert far more influence in determining the solution under the MSE objective function than in the MAE objective function. We will show that, unlike the MAE, the MSE objective function is not resistant to *outliers* (i.e. data points lying far away from the mean, which might have resulted from defective measurements or from exceptional events).

Let us first study the MSE objective function (8.1). With infinite $N$, the sum over the $N$ observations in the objective function can be replaced by integrals, that is,

$$J = \frac{1}{2} \sum_{k} \iint \left[\hat{y}_k(\mathbf{x}; \mathbf{w}) - y_k\right]^2 p(y_k, \mathbf{x}) \, \mathrm{d}y_k \, \mathrm{d}\mathbf{x}, \tag{8.9}$$

where $\mathbf{x}$ and $\mathbf{w}$ are the model inputs and model parameters, respectively, and $p(y_k, \mathbf{x})$ is a joint probability distribution. Since

$$p(y_k, \mathbf{x}) = p(y_k|\mathbf{x})\, p(\mathbf{x}), \tag{8.10}$$

where $p(\mathbf{x})$ is the probability density of the input data and $p(y_k|\mathbf{x})$ is the probability density of the target data conditional on the inputs, we have

$$J = \frac{1}{2}\sum_k \iint \left[\hat{y}_k(\mathbf{x}; \mathbf{w}) - y_k\right]^2 p(y_k|\mathbf{x})\, p(\mathbf{x})\, \mathrm{d}y_k\, \mathrm{d}\mathbf{x}. \tag{8.11}$$

Next we introduce the following conditional averages of the target data:

$$\langle y_k|\mathbf{x}\rangle = \int y_k\, p(y_k|\mathbf{x})\, \mathrm{d}y_k, \tag{8.12}$$

$$\langle y_k^2|\mathbf{x}\rangle = \int y_k^2\, p(y_k|\mathbf{x})\, \mathrm{d}y_k, \tag{8.13}$$

so we can write

$$[\hat{y}_k - y_k]^2 = [\hat{y}_k - \langle y_k|\mathbf{x}\rangle + \langle y_k|\mathbf{x}\rangle - y_k]^2 \tag{8.14}$$
$$= [\hat{y}_k - \langle y_k|\mathbf{x}\rangle]^2 + 2[\hat{y}_k - \langle y_k|\mathbf{x}\rangle][\langle y_k|\mathbf{x}\rangle - y_k]$$
$$+ [\langle y_k|\mathbf{x}\rangle - y_k]^2. \tag{8.15}$$

Upon substituting (8.15) into (8.11), we note that the second term of (8.15) vanishes from the integration over $y_k$ and from (8.12), that is,

$$J = \frac{1}{2}\sum_k \left\{ \int [\hat{y}_k(\mathbf{x}; \mathbf{w}) - \langle y_k|\mathbf{x}\rangle]^2\, p(\mathbf{x})\mathrm{d}\mathbf{x} + \int [\langle y_k|\mathbf{x}\rangle - y_k]^2\, p(\mathbf{x})\mathrm{d}\mathbf{x} \right\}. \tag{8.16}$$

As

$$[\langle y_k|\mathbf{x}\rangle - y_k]^2 = y_k^2 - 2\, y_k\langle y_k|\mathbf{x}\rangle + \langle y_k|\mathbf{x}\rangle^2, \tag{8.17}$$

invoking (8.13) and (8.12) gives

$$J = \frac{1}{2}\sum_k \left\{ \int [\hat{y}_k(\mathbf{x}; \mathbf{w}) - \langle y_k|\mathbf{x}\rangle]^2\, p(\mathbf{x})\mathrm{d}\mathbf{x} + \int [\langle y_k^2|\mathbf{x}\rangle - \langle y_k|\mathbf{x}\rangle^2]\, p(\mathbf{x})\mathrm{d}\mathbf{x} \right\}. \tag{8.18}$$

The second term does not depend on the model output $\hat{y}_k$; thus, it is independent of the model weights $\mathbf{w}$. Therefore, during the search for the optimal weights to minimize $J$, the second term in (8.18) can be ignored. In the first term of (8.18), the integrand cannot be negative, so the minimum of $J$ occurs when this first term vanishes, that is,

$$\hat{y}_k(\mathbf{x}; \mathbf{w}_{\mathrm{opt}}) = \langle y_k|\mathbf{x}\rangle, \tag{8.19}$$

where $\mathbf{w}_{\mathrm{opt}}$ denotes the weights at the minimum of $J$. This is an important result as it shows that the model output is simply the conditional mean of the target data. Thus, in the limit of an infinite number of observations in the

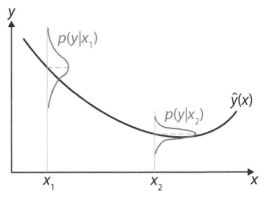

Figure 8.1 The model output $\hat{y}$ is the conditional mean (dashed line) of the target data $y$, with the conditional probability distribution $p(y|x)$ illustrated at $x_1$ and at $x_2$.

dataset, and with the use of a flexible enough model, the model output $\hat{y}_k$ for a given input $\mathbf{x}$ is the conditional mean of the target data at $\mathbf{x}$, as illustrated in Fig. 8.1.

Furthermore, the derivation of this result is quite general, as it does not require the model mapping $\hat{y}_k(\mathbf{x}; \mathbf{w})$ to be restricted to a particular class of models (e.g. the model can be an MLP NN model or a linear regression model, etc.). This result also shows that in non-linear regression problems, in the limit of infinite sample size, overfitting cannot occur, as the model output converges to the conditional mean of the target data. In practice, in the absence of outliers, overfitting ceases to be a problem when the number of independent observations is much larger than the number of model parameters.

Next, we turn to the MAE objective function (8.8). Under infinite $N$, (8.8) becomes

$$J = \sum_k \iint |\hat{y}_k(\mathbf{x}; \mathbf{w}) - y_k| \; p(y_k|\mathbf{x}) \, p(\mathbf{x}) \, dy_k \, d\mathbf{x}. \tag{8.20}$$

This can be rewritten as

$$J = \sum_k \int \tilde{J}_k(\mathbf{x}) \, p(\mathbf{x}) \, d\mathbf{x}, \tag{8.21}$$

where

$$\tilde{J}_k(\mathbf{x}) \equiv \int |\hat{y}_k(\mathbf{x}; \mathbf{w}) - y_k| \; p(y_k|\mathbf{x}) \, dy_k. \tag{8.22}$$

$\tilde{J}_k(\mathbf{x}) \geq 0$, since the integrand of (8.22) is non-negative. Also $J$ in (8.21) is minimized when $\tilde{J}_k(\mathbf{x})$ is minimized. To minimize $\tilde{J}_k(\mathbf{x})$ with respect to the model output $\hat{y}_k$, we set

$$\frac{\partial \tilde{J}_k}{\partial \hat{y}_k} = \int \text{sgn}(\hat{y}_k(\mathbf{x}; \mathbf{w}) - y_k) \; p(y_k|\mathbf{x}) \, dy_k = 0, \tag{8.23}$$

where taking the derivative of the absolute value function in (8.22) led to the sign function $\text{sgn}(z)$, which gives the value $-1$, $0$ or $+1$ if $z < 0$, $z = 0$ or $z > 1$, respectively. For this integral to vanish, the equivalent condition is

$$\int_{-\infty}^{\hat{y}_k} p(y_k|\mathbf{x}) \, \mathrm{d}y_k - \int_{\hat{y}_k}^{\infty} p(y_k|\mathbf{x}) \, \mathrm{d}y_k = 0, \qquad (8.24)$$

which means that $\hat{y}_k(\mathbf{x}; \mathbf{w})$ has to be the conditional *median*, so that the conditional density integrated to the left of $\hat{y}_k$ equals that integrated to the right of $\hat{y}_k$. In statistics, it is well known that the median is robust to outliers whereas the mean is not. For instance, the mean price of a house in a small town can be raised dramatically by the sale of a single palatial home in a given month, while the median, which is the price where there are equal numbers of sales above and below this price, remains stable. Thus, in the presence of outliers, the MSE objective function can produce solutions that are strongly influenced by outliers, whereas the MAE objective function can largely eliminate this undesirable property. However, a disadvantage of the MAE objective function is that it is less sensitive than the MSE objective function, so it may not fit the data as closely.

The Huber function of Huber (1964), seen earlier in Fig. 6.13, is defined by

$$h(\epsilon) = \begin{cases} \dfrac{\epsilon^2}{2\delta} & \text{if } 0 \le |\epsilon| \le \delta, \\[2mm] |\epsilon| - \dfrac{\delta}{2}, & \text{otherwise.} \end{cases} \qquad (8.25)$$

The Huber error function, with $\epsilon = \hat{y}_k - y_k$, combines the desirable properties of the MSE and the MAE functions. When the error $\epsilon$ has magnitude $|\epsilon| \le \delta$, $h$ behaves similarly to the MSE, as $h$ equals $\epsilon^2$ divided by the constant factor $2\delta$. For errors with $|\epsilon| > \delta$, $h$ behaves similarly to the MAE (as $h$ equals $|\epsilon|$ minus a constant $\delta/2$); hence, it is also resistant to outliers. Thus, the Huber error function has the more sensitive nature of the MSE when $|\epsilon|$ is small and the more robust nature of the MAE when $|\epsilon|$ is large.

## 8.3   Variance and Bias Errors  $\boxed{\text{A}}$ ☺

It is important to distinguish between two types of error when fitting a model to a dataset – namely variance error and bias error. To simplify the discussion, we will assume that the model output is a single variable $\hat{y} = f(\mathbf{x})$. The true relation is $y_{\mathrm{T}} = f_{\mathrm{T}}(\mathbf{x})$. The model was trained over a dataset $D$. Let $\mathcal{E}[\cdot]$ denote the expectation or ensemble average over all datasets $D$. Note that $\mathcal{E}[\cdot]$ is not the expectation $\mathrm{E}[\cdot]$ over $\mathbf{x}$ (see Section 2.4), so $\mathcal{E}[\hat{y}] \equiv \bar{y}$ is still a function of $\mathbf{x}$. Thus, the error of $y$ is

$$\begin{aligned} \mathcal{E}[(\hat{y} - y_{\mathrm{T}})^2] &= \mathcal{E}[(\hat{y} - \bar{y} + \bar{y} - y_{\mathrm{T}})^2] \\ &= \mathcal{E}[(\hat{y} - \bar{y})^2] + \mathcal{E}[(\bar{y} - y_{\mathrm{T}})^2] + 2\mathcal{E}[(\hat{y} - \bar{y})(\bar{y} - y_{\mathrm{T}})]. \end{aligned} \qquad (8.26)$$

In the second term, $(\bar{y} - y_{\mathrm{T}})^2$ is unaffected by the $\mathcal{E}$ operator. In the third term, $(\bar{y} - y_{\mathrm{T}})$ is unaffected by the $\mathcal{E}$ operator, so it can be pulled in front of $\mathcal{E}$, giving

$$\mathcal{E}[(\hat{y} - y_{\mathrm{T}})^2] = \mathcal{E}[(\hat{y} - \bar{y})^2] + (\bar{y} - y_{\mathrm{T}})^2 + 2(\bar{y} - y_{\mathrm{T}}) \, \mathcal{E}[\hat{y} - \bar{y}]. \qquad (8.27)$$

Since $\mathcal{E}[\hat{y} - \bar{y}] = \mathcal{E}[\hat{y}] - \bar{y} = 0$, we end up with

$$\mathcal{E}[(\hat{y} - y_{\mathrm{T}})^2] = \mathcal{E}[(\hat{y} - \bar{y})^2] + (\bar{y} - y_{\mathrm{T}})^2. \tag{8.28}$$

The first term, $\mathcal{E}[(\hat{y} - \bar{y})^2]$, is the *variance error*, as it measures the departure of the model output $\hat{y}$ from its expectation $\bar{y}$. The second term, $(\bar{y} - y_{\mathrm{T}})^2$, is the *bias error*, as it measures the departure of $\bar{y}$ from the true value $y_{\mathrm{T}}$. The variance error tells us how much the $\hat{y}$ estimated from a given dataset $D$ can be expected to fluctuate about $\bar{y}$, the expectation over all datasets $D$.

If one uses a model with few adjustable parameters, then the model may have trouble fitting to the true underlying relation accurately, resulting in a large bias error, as illustrated by the linear model fit in Fig. 1.4(a). In this underfitting situation, the variance error is small, since the model is not flexible enough to fit to the particular noise pattern in the data. In contrast, if one uses a model with many adjustable parameters, the model will overfit to the noise closely, resulting in a large variance error (Fig. 1.4(d)), but the bias error will be small. The art of machine learning/statistics hinges on a balanced trade-off between variance error and bias error — with underfitting giving a relatively large bias error to variance error ratio and overfitting giving a relatively small bias error to variance error.

## 8.4 Regularization  $\boxed{\text{A}}$☺

To avoid having the model underfitting the data, one would add more adjustable weights or parameters to the model; however, this could lead to overfitting, as discussed in Section 8.3. *Regularization* is a variety of techniques used to enable a model to neither underfit nor overfit the data, so the model learns properly from the data. The model error over test data (i.e. data not used in training the model) is called the *generalization error*, as opposed to the model *training error* found over the training data. Regularization techniques aim to reduce the model generalization error but not the training error.

### 8.4.1 Weight Penalty  $\boxed{\text{A}}$☺

To prevent overfitting, a very common approach in NN and other ML models is via *regularization* of the objective function, that is, by adding *weight penalty* (commonly called *weight decay* in ML) to the objective function. The MSE objective function (6.8) becomes

$$J = \frac{1}{N}\sum_{n=1}^{N}\left\{ \frac{1}{2}\sum_{k}\left[ \hat{y}_k^{(n)} - y_k^{(n)} \right]^2 \right\} + \frac{\lambda}{2}\sum_{j=1}^{N_{\mathrm{w}}} w_j^2, \tag{8.29}$$

where the second term is the weight penalty term, $\lambda \geq 0$ is the *weight penalty parameter* (a.k.a. *weight decay parameter* or *regularization parameter*) and $w_j$ represents all the weight parameters – but excluding the offset or bias parameters in NN models (Bishop, 2006, p. 258). Sometimes the constant factor of $\frac{1}{2}$ is omitted from this second term. If we omit the first term, (8.29) describes $J$ as being on an $N_w$-dimensional parabolic surface. Note that $\lambda$ is also referred to as a *hyperparameter* as it exerts control over the weight and offset parameters – during non-linear optimization of the objective function, $\lambda$ is held constant while the other parameters (weights) are being searched for their optimal values. With a positive $\lambda$, the selection of larger $|w_j|$ during optimization would increase the value of $J$, so larger $|w_j|$ values are penalized. Choosing a larger $\lambda$ will increase the suppression on the magnitude of the weights during the search for the optimal weights. We have seen this type of regularization used in linear regression problems under the name ridge regression (Section 5.7).

Figure 8.2 illustrates the difference between the optimal weights $\hat{\mathbf{w}}$ found without using weight penalty and the optimal $\mathbf{w}_p$ found with weight penalty in the objective function $J$. Both $w_{p1}$ and $w_{p2}$ from $\mathbf{w}_p$ have smaller magnitude than $\hat{w}_1$ and $\hat{w}_2$, respectively. As $J$ changes much more slowly (i.e. less sensitive) in the $w_1$ direction than in the $w_2$ direction (solid curves in Fig. 8.2), the weight penalty reduces the magnitude of $\hat{w}_1$ much more strongly than $\hat{w}_2$, that is, $|w_{p1}| \ll |\hat{w}_1|$ and $|w_{p2}| < |\hat{w}_2|$. Thus, weight penalty reduces the magnitude of the less sensitive weights more than the sensitive weights.

Figure 8.2 The optimal solution $\hat{\mathbf{w}}$ when there is no weight penalty in $J$ and the optimal solution $\mathbf{w}_p$ when there is weight penalty. The solid contours show $J$ (with only the MSE term), with a minimum at $\hat{\mathbf{w}}$, while the dashed contours show the parabolic contribution from the weight penalty term. Thus, $\mathbf{w}_p$ is the minimum resulting from adding the weight penalty term to the MSE term.

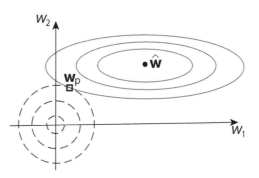

For sigmoidal activation functions such as tanh, the hyperbolic tangent function commonly used in MLP NN models, the effect of weight penalty can be illustrated as follows: For $|wx| \ll 1$, the leading term of the Taylor expansion of $\tanh(wx)$ gives

$$y = \tanh(wx) \approx wx, \tag{8.30}$$

that is, the non-linear activation function tanh is approximated by a linear activation function when the weight $w$ is penalized to have small magnitude and the magnitude of $x$ is O(1). Hence, using a relatively large $\lambda$ to penalize weights would diminish the non-linear modelling capability of the model, thereby alleviating overfitting.

With the weight penalty term in (8.29), it is essential that the input variables have been scaled to similar magnitudes. The reason is that if, for example, the first input variable is much larger in magnitude than the second, then the weights multiplied to the second input variable will have to be much larger in magnitude than those multiplied to the first variable, in order for the second input to exert comparable influence on the output. However, the same weight penalty parameter $\lambda$ acts on both sets of weights, thereby disallowing large weights for the second input, thus reducing the effectiveness of the second input variable. Similarly, if the model output has multiple real variables, the target data for the different variables should be scaled to similar magnitude.

Therefore, when dealing with real unbounded variables, it is common to *standardize* the data first, that is, use (2.29) to transform each variable to having zero mean and unit standard deviation. As a cautionary tale, a graduate student of mine was once trying to compare an MLP NN model and linear regression (LR). No matter how hard he tried, he was only able to show that LR was outperforming NN on his test data. Only when he standardized the input variables was he finally able to show that his NN outperformed LR!

What value should one choose for the weight penalty parameter? A common way to select $\lambda$ is by *validation*, which was mentioned in Section 5.7 on ridge regression and is discussed in detail in Section 8.5 on cross-validation. The basic idea is to divide the available data into two subsets, namely, training data and validation data. Models are trained using the training data for a number of specified $\lambda$ values. Model performance over the validation data is then used to select the optimal $\lambda$ value. The model error (e.g. the MSE) over the training data generally drops as $\lambda$ drops (Fig. 8.3), as decreasing $\lambda$ gives the model more freedom to fit the training data closely; however, the model error over the

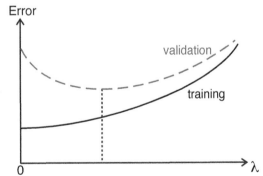

Figure 8.3 Illustrating the model error (e.g. the MSE) for the training data (solid curve) and for the validation data (dashed curve) as a function of the weight penalty parameter $\lambda$. The minimum in the dashed curve gives the optimal $\lambda$ value (as marked by the dotted line).

validation data eventually rises for small enough $\lambda$, as the excessively flexible model has begun overfitting the training data. The values of $\lambda$ used in the search for the optimal $\lambda$ tend to be spaced logarithmically, for example in a very coarse grid search, a loop repeatedly makes model runs using $10^{-5}$, $10^{-4}$, $10^{-3}$, $10^{-2}$ and $10^{-1}$ for the weight penalty parameter, and the run with the smallest validation error gives the optimal $\lambda$ value.

As an example, consider an MLP NN model with a single hidden layer, where there are $m_1$ inputs, $m_2$ hidden neurons and $m_3$ outputs. One often assumes that $m_2$ is chosen large enough so that the model has enough flexibility to accurately capture the underlying relation in the dataset and that validation gives the optimal $\lambda$ that prevents overfitting. In practice, one may not know what $m_2$ value to use. Hence, instead of a single loop of model runs using a number of specified $\lambda$ values, one may also want a second loop using a number of specified $m_2$ values. The run with the smallest validation error gives the best $\lambda$ and $m_2$ values. Thus, validation is used here to determine two model hyperparameters, namely the weight penalty and the number of hidden neurons.

An alternative to the weight penalty approach is the maximum norm constraint approach, which has become popular since the advent of deep neural networks (Section 8.9).

## 8.4.2   Early Stopping  Ⓐ☺

In Section 7.4, we mentioned the *early stopping* method in non-linear optimization, where a portion of the dataset is set aside for validation, and the remainder for training. As the number of training epochs increases, the objective function evaluated over the training data decreases, but the objective function evaluated over the validation data often decreases to a minimum then increases due to overfitting if the model has a relatively large number of adjustable parameters or weights (Fig. 7.3). The minimum gives an indication as to when to stop the training, as additional training epochs only contribute to overfitting. After $N_e$, the optimal number of training epochs, has been determined, one often trains the model again using all training and validation data until the number of epochs reaches $N_e$. Due to its simplicity, early stopping is commonly used as a regularization technique to avoid overfitting, either alone or in conjunction with other regularization techniques (such as weight penalty).

In early stopping, $N_e$ can be considered a hyperparameter. Usually, to determine the optimal value of a hyperparameter by validation, multiple runs of the model using different values of the hyperparameter are needed. In comparison, early stopping is unusually efficient in that only a single run is needed to determine $N_e$. There is a little extra memory needed in that the model parameters/weights have to be stored, so that after the validation error is seen to be in an increasing trend, one can go back and recover the model at the $N_e$th epoch.

What fraction of the data should one reserve for validation? Since more data for validation means fewer data for training the model, there is clearly an optimal fraction for validation. Using more than the optimal fraction for validation results in the training process generalizing less reliably from the smaller training dataset, while using a smaller fraction than the optimal could lead to overfitting.

A theory to estimate this optimal fraction was provided by Amari et al. (1996), assuming a large number of observations $N$. Of the $N$ observations, $fN$ are used for validation and $(1-f)N$ for training. If the number of adjustable parameters in the model is $N_p$, then in the case of $N < 30\,N_p$, the optimal value for $f$ is

$$f_{\text{opt}} = \frac{\sqrt{2N_{\text{p}} - 1} - 1}{2(N_{\text{p}} - 1)}, \tag{8.31}$$

$$\text{hence } f_{\text{opt}} \approx \frac{1}{\sqrt{2N_{\text{p}}}}, \quad \text{for large } N_{\text{p}}. \tag{8.32}$$

For example, if $N_{\text{p}} = 100$, (8.32) gives $f_{\text{opt}} \approx 0.0707$, that is, only 7% of the data should be set aside for validation, with the remaining 93% used for training. If $N_{\text{p}} = 1000$, $f_{\text{opt}} \approx 0.0224$, that is, only 2% of the data are needed for validation.

For the case $N > 30\,N_{\text{p}}$, Amari *et al.* (1996) showed that there is negligible difference between using the optimal fraction for validation and not using early stopping at all (i.e. using the whole dataset for training till convergence). The reason is that in this case there are so many data relative to the number of model parameters that overfitting is not a problem. In fact, using early stopping with $f > f_{\text{opt}}$ leads to poorer results than not using early stopping, as found in numerical experiments by Amari *et al.* (1996). Thus, when $N > 30\,N_{\text{p}}$, overfitting is not a problem, and reserving considerable validation data for early stopping is a poor approach.

As an example, consider the simple polynomial relation

$$y_{\text{signal}} = -x^4 - x^3 + x^2 + x + 1, \tag{8.33}$$

with the random variable $x$ uniformly distributed in the interval $[-1, 1]$. The $y$ data were generated by adding Gaussian noise to $y_{\text{signal}}$, where the Gaussian noise had the same standard deviation as $y_{\text{signal}}$. With 100 training data points generated, an MLP NN model with one hidden layer containing four hidden nodes was used to fit the $(x, y)$ training data. Early stopping was implemented by randomly selecting 17% of the training data for validation, as recommended by (8.31). The NN model was run 25 times with different random initial weights to yield an ensemble with 25 members (Fig. 8.4(a)). Although a couple of ensemble members show sharp kinks at $x \approx 0.5$ and $0.8$, the ensemble average has a smooth curve (see Section 8.7 on ensemble methods). The NN model was then run 25 times without early stopping (and without data reserved for validation) (Fig. 8.4(b)). This time, not only individual members show kinks: even the ensemble average is twisted. In addition, there is much less diversity in the ensemble members in panel (b) than in panel (a). In (a), the early stopping procedure randomly selected 83% of the data for training and 17% for validation, that is, ensemble members would not all use the same data points for training, thereby introducing more diversity in the ensemble members, in contrast to panel (b), where all ensemble members used the same training data.

# 8.5   Cross-Validation   [A] ☺

When there are plentiful data, reserving some data for validation poses no problem — for example, for time series, one can simply use the earlier data for

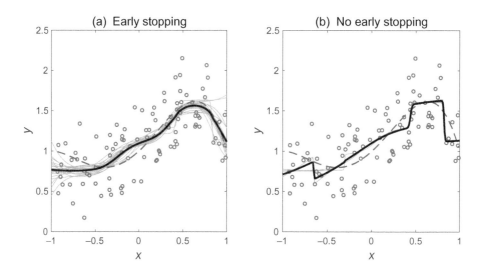

Figure 8.4 (a) The output from 25 runs of an MLP NN model trained using early stopping (thin lines), the ensemble average of the output from the 25 runs (thick line), the signal (dashed line) and the data (circles). (b) Repeat of (a) but without using early stopping during the training of the NN model.

training and the later data for validation. Unfortunately, data are often not plentiful, and one can ill-afford setting aside a fair amount of data for validation, since this means fewer observations for model training. On the other hand, if one reserves few observations for validation so that one can use as many observations for training as possible, the validation error estimate may be very unreliable. Cross-validation is a technique that allows the entire dataset to be used for validation.

Given a data record, $K$-fold *cross-validation* involves dividing the record into $K$ (approximately equal) segments. One segment is reserved as 'validation data', while the other $K - 1$ segments are 'training data' used for model training. This process is repeated $K$ times, so that each segment of the data record has been used as validation data (Fig. 8.5). Thus, a validation performance score (e.g. the RMSE or other performance score) is computed using every data point in the data record. A number of model runs is made using different values for the hyperparameters (e.g. weight penalty, number of adjustable parameters or weights, number of training epochs, etc.). Based on the best validation performance score over the whole data record, one can obtain the optimal hyperparameters. Using the optimal hyperparameters, the model is usually trained again using all available data as training data to obtain the final model.

For example, suppose the data record is 50 years long. In five-fold cross-validation, the record is divided into five segments, that is, years 1–10, 11–20,

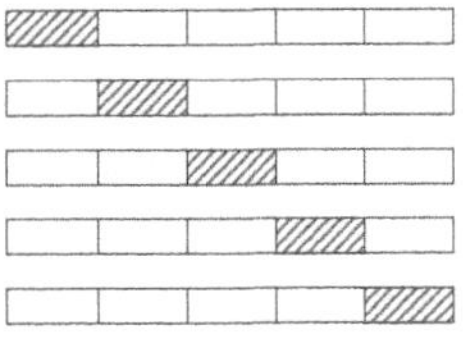

Figure 8.5 In $K$-fold cross-validation (here $K = 5$), the computational loop begins by withholding the first data segment as validation data (striped pattern), using only the remaining data segments for training the model. The trained model is then used to predict the data over the validation segment. Next, the second segment is withheld as validation data and the other segments used as training data, and so on, until the final segment is used as validation data.

31–40, 41–50. First, we reserve years 1–10 for validation, and train the model using data from years 11–50. Next, we reserve years 11-20 for validation, and train using data from years 1–10 and 21–50. This is repeated until the final segment of 41–50 is reserved for validation, with training done using data from years 1–40.

If one has more computing resources, one can try 10-fold cross-validation, where the 50-year record is divided into ten 5-year segments. If one has even more computing resources, one can try 50-fold cross-validation, with the record divided into fifty 1-year segments. At the extreme, one arrives at the *leave-one-out* cross-validation, where the validation segment consists of a single observation. For instance, if the 50-year record contains monthly values, then there are a total of 600 monthly observations, and a 600-fold cross-validation is the same as the leave-one-out approach. Ideally one would want a small validation segment, so the training dataset is not much smaller than the original dataset.

There are several complications when applying cross-validation to environmental data (Elsner and Schmertmann, 1994). With time series data, the neighbouring observations in the data record are often not independent of each other due to serial correlation. For example, if the dataset has serial correlation over a timescale of two years, then reserving one year of data for independent validation would make little sense, since it is correlated with neighbouring observations already used for training. The validation segments need to be longer than the timescale in which there is serial correlation. Even then, at the boundary of a validation segment, there is still correlation between the data immediately to one side that are used for training and those to the other side used for validation. Thus, under cross-validation, serial correlation can lead to an underestimation of the model error over the validation data, especially when using small validation segments, with the effect being worst for leave-one-out cross-validation. Leave-one-out validation is sometimes mistakenly assumed to be the best way to perform $K$-fold cross-validation, since it is computationally the most intensive, when in fact it is the one most susceptible to spurious skills from serial correlation.

To illustrate this effect, imagine we are using leave-one-out cross-validation, and the year 2005 happens to be selected for validation. Data from 2004 and 2006 will be used for training the model, so any information on the interannual/interdecadal climate variability will simply leak from the 2004 and 2006 training data to the validation data in 2005, giving spurious skill. This information leakage effect is less serious if we are cross-validating a longer segment of data, say 2000–2010. One way to alleviate the leakage in leave-one-out cross-validation is to leave a gap between the validation data and the training data. For instance, when validating the 2005 data, the training data consist of all data excluding those from 2003 to 2007.

Because validation data are used for model selection, that is, for choosing the optimal number of adjustable parameters or weights, weight penalty, and so on, the model error over the validation data cannot be considered an accurate model prediction or forecast error, since the validation data have already been used in deriving the model. To assess the model prediction error accurately, the model error needs to be calculated over independent data not used in model training or model selection. Thus, the data record needs to be divided into training data, validation data and *test data* (a.k.a. *verification data*) for measuring the true model prediction or forecast error. One then has to do a *double cross-validation* (a.k.a. *nested cross-validation*), which can be quite expensive computationally. Again consider the example of a 50-year data record, where we want to do a 10-fold cross-testing. We first reserve years 1–5 for test data, and use years 6–50 for training and validation. We then implement a $K$-fold cross-validation over the data from years 6–50 to select the model with the optimal hyperparameters, which we use to make predictions over the test data. Next, we reserve years 6–10 for testing and perform a $K$-fold cross-validation over the data from years 1–5 and 11–50 to select the optimal model hyperparameters (which can be different from the values found in the previous cross-validation). The process is repeated until the model error is computed over test data covering the entire data record (Fig. 8.6).

To alleviate serial correlation across the boundary between the test data and the training data, Fig. 8.7 shows the modified double cross-validation scheme from Zeng et al. (2011). Further discussion on spurious skill in forecast models is pursued in Section 16.8.

*Generalized cross-validation* (GCV) is an extension of the cross-validation method (Golub et al., 1979) and has been applied to MLP NN models (Yuval, 2000) to determine the weight penalty parameter automatically.

## 8.6   Hyperparameter Tuning   $\boxed{\text{A}}$☺

Prior to playing a guitar, the guitarist needs to tune his instrument for the best sound. Similarly, in ML, one needs to tune the hyperparameters for the best model. The hyperparameters can be divided into two categories: (a) optimizer hyperparameters from the optimization algorithm (e.g. learning rate, momentum parameter, mini-batch size, etc.) and (b) model hyperparameters associated with the structure or architecture of the model (e.g. number of hidden nodes, weight penalty, etc.)

Figure 8.6 Double cross-validation involving an outer cross-validation loop CV1 and an inner loop CV2. In CV1, the data record is divided into a number of segments, and the first segment is withheld as test data (horizontally striped), while the remaining segments are used for model training. In CV2, consider only the data for model training: a standard $K$-fold cross-validation loop (here $K = 5$) is used where data are withheld for validation (diagonally striped) to determine the optimal model hyperparameters. CV1 continues by withholding the second segment as test data, and so on, until the final segment is used for test data.

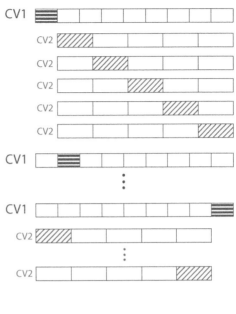

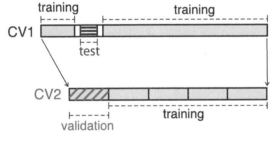

Figure 8.7 A modified double cross-validation scheme to alleviate serial correlation across the boundary between the test data and the training data. In the outer loop (CV1), the training data are shown in grey and the test data are horizontally striped. The data segments (in white) bridging the training data and the test data are not used, to avoid serial correlation leaking information from the training data to the adjacent test data. The test data segment is moved repeatedly from the start of the data record to the end in this cross-validation loop, so forecast performance is tested over the whole record. Meanwhile, in the inner loop (CV2), the training data from CV1 are assembled and divided into training and validation (diagonally striped) data segments, which are rotated under $K$-fold cross-validation to determine the optimal hyperparameters. The model with the optimal hyperparameters is used to predict over the test data segment in CV1. [Adapted from Zeng et al. (2011, figure A.1).]

For category (a), most optimization algorithms provide reasonable default values for their hyperparameters, or are capable of automatically adapting some of the hyperparameters to the given dataset (i.e. the algorithm automatically finds appropriate hyperparameter values from the given data). The learning rate (Section 7.5) is probably the most important among the optimizer hyperparameters – if one has computer resources to tune only one optimizer hyperparameter, this would be the one (Goodfellow, Bengio, et al., 2016, section 11.4.1). The learning rate can be tuned by monitoring the *training* error (i.e. using only training data). The training error tends to decrease gradually as the learning rate increases but then increases very steeply when the learning rate gets too large. One would select the learning rate just before the training error starts to rise. Tuning other hyperparameters would require having validation data in addition to the training data.

When there are only three or fewer hyperparameters to tune, the common approach is to use a *grid search*. For example, a coarse grid search for the weight penalty and the number of hidden nodes make runs using the following values for the weight penalty parameter, $10^{-5}$, $10^{-4}$, $10^{-3}$, $10^{-2}$ and $10^{-1}$, and the following values for the number of hidden nodes, 2, 5, 10, 20, 50, 100 and 200. Note the search values are spaced approximately in equal intervals on a logarithmic scale. The model is run repeatedly in two nested loops, where the outer loops varies the first hyperparameter and the inner loop varies the second hyperparameter, and the run with the smallest *validation* error gives the optimal hyperparameters. Sometimes the search range is off, for example if the smallest validation error occurs when the number of hidden nodes is 200, that is, right at the end of our search range, we will need to expand the search range further. Often, one would start with a coarse grid search, then zero in with a fine grid search. For instance, the validation error is found to be lowest at 10 hidden nodes. The two neighbouring search values used were 5 and 20 hidden nodes, so our fine grid search can be, for example, 6, 7, 8, 9, 10, 12, 14, 16, 18 hidden nodes.

Computationally, grid search becomes very costly as the number of hyperparameters increases. If there are $m$ hyperparameters, and each is searched over a grid of $n$ values, the $m$ nested loops would require $n^m$ model runs, that is, the cost increases exponentially with the number of hyperparameters.

When the number of hyperparameters is not small, *random search* has been shown to be much cheaper computationally than grid search (Bergstra and Bengio, 2012). In random search, random values of a hyperparameter are selected from a uniform distribution over the logarithm of the hyperparameter. Figure 8.8 illustrates a situation where the objective function is sensitive to one hyperparameter but not to another. The random search is seen to locate the best value of the important hyperparameter well while the grid search fails to do so. The grid search is wasteful of computing resources when the objective function is not sensitive to some of the hyperparameters.

Using Bayes theorem (Section 2.14), MacKay (1992a) and MacKay (1992b) introduced a *Bayesian neural network* (BNN) approach that gives an estimate of the optimal weight penalty parameter(s), without the need for validation data.

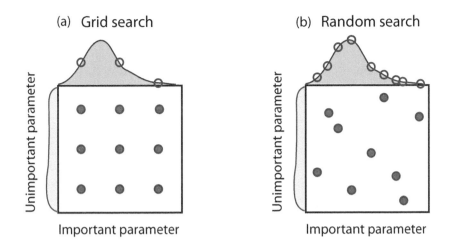

Figure 8.8 Illustrating (a) grid search and (b) random search for hyperparameters. Of the two hyperparameters, one is important and the other unimportant in influencing the objective function. The peaked curve drawn on top of the square illustrates the effectiveness of the important hyperparameter on improving the objective function, while the relatively flat curve on the left side of the square illustrates the ineffectiveness of the unimportant hyperparameter. Both (a) and (b) have nine circles within the square representing nine model runs using different values for the hyperparameters. The peak of the curve for the important hyperparameter was well located by the nine circles projected onto the curve from the nine runs in (b). In contrast, the projected circles from the nine runs in (a) failed to locate the peak of the curve due to multiple projected circles landing on the same spot. [Follows Bergstra and Bengio (2012, figure 1).]

BNN was once very popular (Neal, 1996; Foresee and Hagan, 1997; Nabney, 2002; Bishop, 2006, section 5.7), but the interest in *Bayesian hyperparameter optimization* has waned since the advent of deep NN (Goodfellow, Bengio, et al., 2016, section 11.4.5), though the Bayesian approach still provides an alternative to random search in hyperparameter tuning (Snoek, Rippel, et al., 2015).

## 8.7   Ensemble Methods   A☺

In weather forecasting, it is now standard practice to run a numerical weather prediction model multiple times from slightly perturbed initial conditions, giving an ensemble of model runs (T. Palmer, 2019). The rationale is that the atmospheric models are very unstable to small perturbations in the initial conditions, that is, a tiny error in the initial conditions would lead to a vastly different forecast a couple of weeks later. This behaviour was first noted by Lorenz (1963), which led to the discovery of the 'chaos' phenomenon. From this ensemble of model runs, the averaged forecast over the individual ensemble members is usually issued as the forecast, while the spread of the ensemble members provides information on the uncertainty of the forecast. For climate, the Intergovernmental Panel on Climate Change (IPCC) assessment reports are based on an ensemble of many climate models developed by various institutions around the world (Knutti, Masson, et al., 2013; Eyring et al., 2016).

In ML, one often trains a number of models for a variety of reasons, for example, to deal with the multiple minima in the objective function in MLP NN models, to experiment varying the number of model weights, and so on. One can check the model performance over some validation data and simply select the best performer. However, model skill is dependent on the noise in the validation data, that is, if a different validation dataset is used, a different model may be selected as the best performer. For this reason, it is common to retain multiple models to form an *ensemble* of models and use the ensemble average of their outputs as the desired output. For discrete/categorical output variables (i.e. classification problems), voting is used instead of averaging, that is, the discrete outcome most widely selected by the ensemble members is chosen as the output of the ensemble. In ML jargon, an ensemble of models is often called a *committee*, reflecting the fact that classification problems are far more common than regression problems in the ML literature, so voting is more commonly used than averaging in ensemble methods.

How can one generate an ensemble of models from only one model? With MLP NN models, just using different initial random weights for optimization will usually result in different models, as there are multiple local minima. Having greater diversity among the models in an ensemble is desirable, since greater diversity means weaker correlation between the model errors from different ensemble members, so the errors from different members will tend to cancel each other (Fig. 8.4). In Section 8.7.1, bootstrap resampling is used to generate multiple training datasets to increase the diversity among the ensemble members.

In Section 8.7.2, we proved a very important property – that the expected error of the ensemble average is less than or equal to the average expected error of the individual members in the ensemble. Hence *MLP NN models are almost always run as an ensemble* – partly because of the presence of multiple local minima in the objective function but also because, in practice, an ensemble of NN models almost always outperforms a single NN model.[2]

In Sections 8.7.3 and 8.7.4, the ensemble members need not be weighted equally in forming the ensemble average.

The topic of ensemble methods continues beyond this chapter. In Section 14.2 on *random forests*, the ensemble method is applied to weak learners, that is, relatively simple models with less than stellar performance. By using a large ensemble, weak leaners can be competitive against strong learners. Analogously, in Nature, army ants can overwhelm much stronger opponents such as spiders

---

[2] A caveat: The ML literature is full of papers claiming to have discovered novel models capable of outperforming the good old MLP neural network model. However, in most cases, the outperformance is bogus since the MLP benchmark was run as a single model, not as an ensemble of models, which would have given better performance.

and lizards. The large ensemble can be further enhanced using unequal weights in *boosting* (Section 14.3).

## 8.7.1 Bagging  A ☺

A clever way to generate multiple models from a single model is through *bagging* (abbreviated from Bootstrap AGGregatING) (Breiman, 1996a), developed from the idea of *bootstrapping* (Efron, 1979; Efron and Tibshirani, 1993) in statistics. Under bootstrap resampling, data are drawn randomly from a dataset to form a new training dataset, which is to have the same number of data points as the original dataset (see Section 4.8). A data point in the original dataset can be drawn more than once into a training dataset. On average, 63.2% of the original data is drawn, while 36.8% is not drawn into a training dataset. This is repeated until a large number of training datasets are generated by this bootstrap procedure. During the random draws, predictor and response data pairs are of course drawn together as a pair. In the case of serially correlated data, data segments with length not less than the time interval over which serial correlation exists are drawn instead of individual data points. In the bagging approach, one model can be built from each training set obtained from bootstrap resampling, so from the multiple training sets, an ensemble of models is obtained.

Incidentally, the data not selected during the bootstrap resampling, called *out-of-bag* (OOB) data, are not wasted, as they can be used as validation data. These validation data can be used for model selection. In MLP NN training, these validation data can be used in the *early stopping* approach to avoid overfitting (Section 8.4.2).

For MLP NN models, one can generate an ensemble of models just by training the model multiple times on the same dataset with different initial random weights, as there are usually a large number of multiple minima for the multiple runs to converge to, thereby yielding an ensemble of distinct models. If bagging is used, the ensemble members will be trained using different bootstrap resampled datasets as well as different random initial weights. The members would, therefore, tend to have greater diversity than if the members were all trained using the same original dataset (but with different initial weights). However, there may not be an improvement in the prediction performance using bagging, as tested for classification problems by Maclin and Opitz (1997).

From the bagging ensemble members, confidence intervals can be easily estimated, while prediction intervals can also be estimated with more effort (Heskes, 1997; Carney et al., 1999).

## 8.7.2 Error of Ensemble  B ☹

We next compare the error of the ensemble average to the average error of the individual models in the ensemble. Let $y_T(\mathbf{x})$ denote the true relation, $\hat{y}_m(\mathbf{x})$ the $m$th model relation in an ensemble of $M$ models and $\hat{y}^{(M)}(\mathbf{x})$ the ensemble average. For brevity, we will drop the ˆ and write the latter two as $y_m(\mathbf{x})$ and $y^{(M)}(\mathbf{x})$. The expected mean squared error of the ensemble average is

$$\mathrm{E}\Big[\big(y^{(M)} - y_\mathrm{T}\big)^2\Big] = \mathrm{E}\Bigg[\Big(\frac{1}{M}\sum_{m=1}^{M} y_m - y_\mathrm{T}\Big)^2\Bigg]$$

$$= \mathrm{E}\Bigg[\Big(\frac{1}{M}\sum_{m}(y_m - y_\mathrm{T})\Big)^2\Bigg] = \frac{1}{M^2}\mathrm{E}\Bigg[\Big(\sum_{m}\epsilon_m\Big)^2\Bigg], \quad (8.34)$$

where $\epsilon_m \equiv y_m - y_\mathrm{T}$ is the error of the $m$th model. From the Cauchy inequality, we have

$$\Big(\sum_{m=1}^{M}\epsilon_m\Big)^2 \leq M\sum_{m=1}^{M}\epsilon_m^2, \quad (8.35)$$

hence

$$\mathrm{E}\Big[\big(y^{(M)} - y_\mathrm{T}\big)^2\Big] \leq \frac{1}{M}\sum_{m=1}^{M}\mathrm{E}\big[\epsilon_m^2\big]. \quad (8.36)$$

This proves that the expected error of the ensemble average is less than or equal to the average expected error of the individual models in the ensemble, thereby providing the rationale for using ensemble averages instead of individual models.[3] Note this is a general result, as it applies to an ensemble of dynamical models (e.g. general circulation models) as well as an ensemble of empirical models (e.g. ML or statistical models), or even to a mixture of completely different dynamical and empirical models.

Next we restrict consideration to a single model for generating the ensemble members, for example by training the model with various bootstrap resampled datasets or performing non-linear optimization with random initial weights. We now repeat the variance and bias error calculation of Section 8.3 for an ensemble of models. Again, let $\mathcal{E}[\cdot]$ denote the expectation or ensemble average over all datasets $D$ or over all random initial weights (as distinct from $\mathrm{E}[\cdot]$, the expectation over $\mathbf{x}$). Since all members of the ensemble were generated from a single model, we have for all the $m = 1, \ldots, M$ members,

$$\mathcal{E}[y_m] = \mathcal{E}[y] \equiv \bar{y}, \quad (8.37)$$

where $\bar{y} = \bar{y}(\mathbf{x})$. The expected squared error of the ensemble average $y^{(M)}$ is

$$\mathcal{E}\big[(y^{(M)} - y_\mathrm{T})^2\big] = \mathcal{E}\big[(y^{(M)} - \bar{y} + \bar{y} - y_\mathrm{T})^2\big]$$

$$= \mathcal{E}\big[(y^{(M)} - \bar{y})^2\big] + \mathcal{E}\big[(\bar{y} - y_\mathrm{T})^2\big] + 2\mathcal{E}\big[(y^{(M)} - \bar{y})(\bar{y} - y_\mathrm{T})\big]$$

$$= \mathcal{E}\big[(y^{(M)} - \bar{y})^2\big] + (\bar{y} - y_\mathrm{T})^2 + 2(\bar{y} - y_\mathrm{T})\mathcal{E}\big[y^{(M)} - \bar{y}\big]$$

$$= \mathcal{E}\big[(y^{(M)} - \bar{y})^2\big] + (\bar{y} - y_\mathrm{T})^2, \quad (8.38)$$

---

[3] This result when applied to social systems says that the average error made by a group of individuals is less than or equal to the average error made by an individual.

as $\mathcal{E}\left[y^{(M)} - \bar{y}\right] = 0$. The first term, $\mathcal{E}[(y^{(M)} - \bar{y})^2]$, is the variance error, while the second term, $(\bar{y} - y_T)^2$, is the bias error. Note that the variance error depends on $M$, the ensemble size, whereas the bias error does not.

Let us examine the variance error:

$$\mathcal{E}\left[\left(y^{(M)} - \bar{y}\right)^2\right] = \mathcal{E}\left[\left(\frac{1}{M}\sum_{m=1}^{M} y_m - \bar{y}\right)^2\right] = \mathcal{E}\left[\left(\frac{1}{M}\sum_{m=1}^{M}(y_m - \bar{y})\right)^2\right]$$

$$= \frac{1}{M^2}\mathcal{E}\left[\left(\sum_m \delta_m\right)^2\right], \tag{8.39}$$

where $\delta_m \equiv y_m - \bar{y}$. If the errors of the ensemble members are uncorrelated, that is, $\mathcal{E}\left[\delta_m \delta_n\right] = 0$ if $m \neq n$, then

$$\frac{1}{M^2}\mathcal{E}\left[\left(\sum_m \delta_m\right)^2\right] = \frac{1}{M^2}\mathcal{E}\left[\sum_m \delta_m^2\right] = \frac{1}{M}\mathcal{E}\left[\delta^2\right], \tag{8.40}$$

as $\mathcal{E}[\delta_m^2] = \mathcal{E}[\delta^2]$, for all $m$. Thus, if the member errors are uncorrelated, the variance error of the ensemble average is

$$\mathcal{E}\left[(y^{(M)} - \bar{y})^2\right] = \frac{1}{M}\mathcal{E}\left[\delta^2\right], \tag{8.41}$$

that is,

$$\mathcal{E}\left[(y^{(M)} - \bar{y})^2\right] = \frac{1}{M}\mathcal{E}\left[(y - \bar{y})^2\right], \tag{8.42}$$

where the right hand side is simply the variance error of a single member divided by $M$. Thus, the variance error of the ensemble average $\rightarrow 0$ as $M \rightarrow \infty$. Of course, the decrease in the variance error of the ensemble average will not be as rapid as $M^{-1}$ if the errors of the members are correlated. If the errors of the members are well correlated, there is little cancellation of the errors in the ensemble averaging. In contrast, if the errors of the members are uncorrelated, the errors tend to cancel each other in the ensemble averaging, leading to the variance error of the ensemble average decreasing at $O(M^{-1})$ as $M$ increases, as indicated in (8.42).

As mentioned before, in bagging, ensemble members trained using different bootstrap resampled datasets tend to have greater diversity than ensemble members all trained using the same original dataset (but from different initial weights). This greater diversity means weaker covariance between the errors from different ensemble members, that is, smaller $\left|\mathcal{E}[\delta_m \delta_n]\right|$, hence closer to the ideal limit of $\mathcal{E}\left[\delta_m \delta_n\right] = 0$ $(m \neq n)$, where the variance error of the ensemble average decreases readily with an increase in the ensemble size M as given by (8.42). This may explain why the bagging approach is heavily used in constructing ensemble models.

In summary, for the ensemble average, the variance error can be decreased by increasing the ensemble size $M$, but the bias error is unchanged. This suggests that one should use models with small bias errors and then rely on the ensemble averaging to reduce the variance error. In other words, one would prefer using models that overfit slightly to models that underfit, as ensemble averaging can alleviate the overfitting. In a hydrological study, Cannon and Whitfield (2002, figures 2 and 3) compared the performance of a single MLP NN model using the early stopping method and that of an ensemble of MLP models by bagging (without early stopping). They found that the ensemble approach performed better than the single model; hence, the ensemble method can be an effective way to control overfitting.

In general, when using models with multiple local minima (e.g. MLP NN models), it is not a good idea to predict using a single model. The single model can easily be turned into an ensemble of models using the bagging approach, and the ensemble model almost invariably outperforms the original single model. The ensemble can also give a distribution for estimating prediction intervals.

For regularization purposes, ensemble size $M$ typically ranges around 20–30 for MLP NN models. For large, deep learning NN models, one may have to do with $M$ around 3–10, due to the high computational costs of running large deep learning models, for example, Szegedy et al. (2015) used $M = 6$ in their GoogLeNet model, which won in the Large Scale Visual Recognition Challenge 2014 (ILSVRC2014). In contrast, when the ensemble method is applied to weak learners (i.e. relatively simple models), much larger ensemble size is used, for example, for random forests, $M \geq 100$ is commonly used.

### 8.7.3   Unequal Weighting of Ensemble Members   C☺

So far, all the ensemble members $y_m$ $(m = 1, \ldots, M)$ are equally weighted when forming the ensemble average. A question arises if some members are smarter than others: Is there any advantage in using unequal weights in the ensemble average? The more general ensemble averaging is of the form

$$y^{(M)} = \sum_{m=1}^{M} a_m \, y_m, \tag{8.43}$$

where the weights $a_m$ if all set to $1/M$ reduces to the equal-weighted ensemble average seen earlier in Section 8.7.2.

If the ensemble models are all generated from a single model, for example a dynamical model run with slightly different initial conditions, or a MLP NN model trained with random initial weights or trained on different bootstrap resampled data, there is little justification for unequal weights.

There are, however, many situations where the ensemble members have distinctly different structures. For instance, one can have an ensemble of dynamic climate models built by different institutions, one is combining several very different empirical models (e.g. MLP, support vector machines, random forests, etc.) or the ensemble contains both dynamical models and empirical models. The Intergovernmental Panel on Climate Change (IPCC) assessment reports are based on multiple global climate models contributed by many countries around the world (Knutti, Masson, et al., 2013; Eyring et al., 2016), so various methods have been developed to obtain *multi-model ensemble* averages (Derome et al., 2001; Casanova and Ahrens, 2009; DelSole, Xiaosong Yang, et al., 2013; Knutti, Sedlacek, et al., 2017).

The simplest estimate for $a_m$ is to have $a_m$ proportional to the reciprocal of the mean squared error from the training data for the $m$th member, that is,

$$a_m \propto \left[ \frac{1}{N} \sum_{n=1}^{N} \left( y_m^{(n)} - y^{(n)} \right)^2 \right]^{-1}, \qquad (8.44)$$

where $y^{(n)}$ is the $n$th observation. Thus, ensemble members with larger MSE are given smaller weight when forming the ensemble average. A normalization factor ensuring the weights sum to one,

$$\sum_{m=1}^{M} a_m = 1, \qquad (8.45)$$

is added, giving

$$a_m = \frac{\left[ \sum_{n=1}^{N} \left( y_m^{(n)} - y^{(n)} \right)^2 \right]^{-1}}{\sum_{i=1}^{M} \left[ \sum_{n=1}^{N} \left( y_i^{(n)} - y^{(n)} \right)^2 \right]^{-1}}. \qquad (8.46)$$

In Casanova and Ahrens (2009), $a_m$ was computed at each spatial grid point over the globe.

While this simple scheme takes into account the difference in MSE among the ensemble members, it does not take into account the covariance between the errors of two different ensemble members, seen to be important in Section 8.7.2. To incorporate error covariance in the ensemble averaging, one can compute the error covariance matrix $\mathbf{C}$, with elements

$$C_{ij} = \frac{1}{N} \sum_{n=1}^{N} \left( y_i^{(n)} - y^{(n)} \right) \left( y_j^{(n)} - y^{(n)} \right), \qquad (8.47)$$

where $y_i$ and $y_j$ are the outputs from two ensemble member models. Bishop (1995, section 9.6) gives the derivation for the $a_m$, with

$$a_m = \frac{\sum_{j=1}^{M} (\mathbf{C}^{-1})_{mj}}{\sum_{i=1}^{M} \sum_{j=1}^{M} (\mathbf{C}^{-1})_{ij}}. \qquad (8.48)$$

Equation (8.46) is a special case of (8.48) when all off-diagonal elements of $\mathbf{C}$ are zero. Bishop (1995) mentions that the constraint (8.45) does not prevent $a_m$ from taking on negative values (see Exercise [8.3]), so having the additional constraint

$$a_m \geq 0 \qquad (8.49)$$

can be helpful, but the optimization needed to obtain $a_m$ becomes more complicated and requires linear programming.

More sophisticated ways to combine ensemble members are stacking (Section 8.7.4) from ML and Bayesian model averaging (Section 8.12) using Bayesian statistics.

## 8.7.4   Stacking  Ⓒ☺

*Stacking* or *stacked generalization* (Wolpert, 1992; Breiman, 1996b) is an ensemble method where a new model learns how to best combine the output from a number of existing models. Originally, stacking involves combining the various models linearly, as in (8.43). Based on cross-validation, stacking provides an alternative to the Bayesian approach used in Bayesian model averaging (BMA) (Section 8.12). Comparing BMA and stacking, B. Clarke (2003) found that stacking (Section 8.7.4) has better robustness properties than BMA in most settings.

The existing models are called level-0 models and the new model, the level-1 model, is trained to best combine the output from the level-0 models (Witten et al., 2011, section 8.7). The level-0 models can be various ML/statistical models, dynamical models or a combination of both. There are $m = 1, \ldots, M$ level-0 models, using inputs $\mathbf{x}$ to provide outputs $y_m$. These $y_m$ become the $M$ inputs to the level-1 model, which gives the output $y^{(M)}$. If one solves for the optimal weights $a_m$ in (8.43) as a linear regression problem using training data, overfitting could occur.

To avoid overfitting, the data used to train the level-1 model should be independent from the data used to train the level-0 models. The usual procedure is to use cross-validation, either $K$-fold cross-validation or leave-one-out cross-validation (Section 8.5). The level-0 models are trained using data excluding the validation data. The trained models are then used to predict $y_m$ using $\mathbf{x}$ from the validation data. Their output are then used to train the the level-1 model. Thus, the level-1 model is trained using data not used in training the level-0 models. The rotation of the validation data segments under cross-validation allows all data to be used for training the level-1 model. After the level-1 model has been trained, the level-0 models are retrained using all available training data (i.e. without reserving data for validation). The stacked model is now ready to make predictions using new data.

Since the level-0 models have already done the hard work modelling the data, the level-1 model needs not be as complex as the level-0 models. Simple linear models such as (8.43) can work well as the level-1 model for combining the outputs from complex non-linear level-0 models. Non-linear models such as MLP NN can also be used as the level-1 model, so (8.43) is replaced by

$$y^{(M)} = f(y_1, \ldots, y_M), \qquad (8.50)$$

where $f$ is a non-linear function.

Although stacking is mainly used to combine multiple models, it is possible to use it with only a single level-0 model (Wolpert, 1992). In this situation, the level-1 model corrects the predictions made by the level-0 model.

Krasnopolsky (2007) used an ensemble of 10 MLP NN models to emulate sea level height anomalies using state variables from a dynamical ocean model as input. The outputs of the 10 level-0 ensemble members were then non-linearly combined by a level-1 MLP NN model. The level-1 model was trained with the same training and validation datasets as the level-0 models. The non-linear ensemble averaging by the level-1 NN model is compared with simple averaging (Fig. 8.9). The simple ensemble average has smaller model error standard

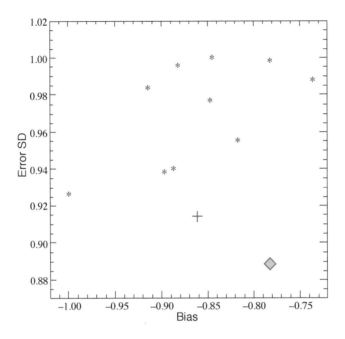

Figure 8.9 Scatter plot of model bias versus standard deviation (SD) of the model error for the 10 individual ensemble members (asterisks), the simple ensemble average (cross) and the non-linear ensemble average by NN (diamond). [Adapted from (Krasnopolsky, 2007, figure 7).]

deviation than all 10 individuals, but its bias is just the average of the bias of the individuals. In contrast, the non-linear NN ensemble method results in even smaller error standard deviation plus a reduction in the bias. The level-1 and level-0 models appeared to have been trained using the same data, as overfitting may not be a problem with the large sample size (536,259) used.

## 8.8  Dropout  $\boxed{\text{B}}$ ☺

In Section 8.7.1, bootstrap resampling was used to generate multiple training datasets to increase the diversity among the ensemble members. For large expensive models (e.g. deep neural networks), running an ensemble with 20–30 members can be prohibitively costly. Fortunately, a low-cost alternative has been developed. *Dropout* is a relatively new regularization technique, first proposed by Hinton, Srivastava, et al. (2012), which randomly deletes input nodes with probability $1 - P_i$ and deletes hidden nodes with probability $1 - P_h$ during model training (Srivastava et al., 2014; Goodfellow, Bengio, et al., 2016).

Typically $P_h = 0.5$ (i.e. 50% probability that a hidden node is retained in the model) and $P_i$ is usually larger, typically $P_i \approx 0.8$ (i.e. 80% probability an input node is retained).

Figure 8.10 illustrates how a large number of NN models of different architecture can be generated by random dropout of inputs and hidden nodes. If there is a total of $M$ inputs and hidden nodes, each of them having two choices (either being included or excluded from the model), then the random dropout procedure allows a total of $2^M$ possible model architectures from the original model. Usually, mini-batch gradient descent is used during model training. The model randomly chooses one of the $2^M$ possible model architectures for one mini-batch, then chooses another architecture for the next mini-batch as the training proceeds. In addition, different bootstrap samples are used, that is, a mini-batch of training sample is generated by randomly picking data (with replacement) from the training set. As many hidden nodes are dropped during model training, one typically compensates by using a larger model. For instance, if 50% of the hidden nodes are dropped, one can compensate by using a model with twice as many hidden nodes.

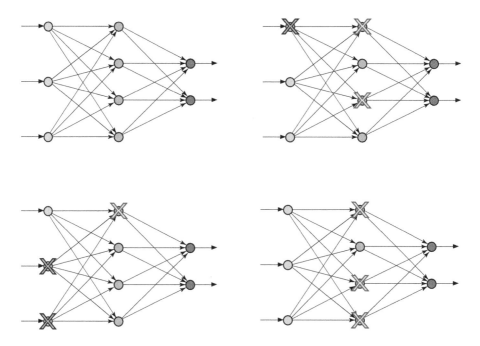

Figure 8.10 The dropout method applied to a neural network model. The original model is shown in the top left quadrant. The other three are versions of the original model but with various input and hidden nodes deleted (as marked by the crosses) during model training. The output nodes are always retained.

After the model has finished training, it is used for predictions. Dropout is not applied when the model is used for predictions, that is, all input and hidden nodes are kept in the prediction model. To compensate for the fact that there are more weights being used during prediction than during training, the weights are multiplied by a scale factor of $P_h$; for example, if 50% of the weights were dropped during training, then the weights are multiplied by a factor of 0.5 when making predictions using the trained model. This weight scaling rule is exact if the NN model has linear activation functions. For a general NN model, this scaling rule is only heuristic, but it has been found to work well in practice (Goodfellow, Bengio, et al., 2016, section 7.12). For deep NN, dropout is usually used in conjunction with the maximum norm constraint on weights (Section 8.9), together with large learning rates.

When the amount of training data available is truly plentiful, there is little danger of overfitting, so dropout is not necessary. Nevertheless, applying dropout with a small dropout rate of 2% can still lead to improved results (Aggarwal, 2018, p. 190).

## 8.9 Maximum Norm Constraint  B☺

In Section 5.7 on ridge regression, we learned that weight penalty and explicit constraint of the weight norm are closely related methods for regularizing linear regression models. The advent of deep neural networks has led to the use of *maximum norm constraint* (a.k.a. max-norm constraint) on weights together with the dropout method (Section 8.8). During the optimization of the objective function $J$, the max-norm constraint is applied to impose an upper bound on the weight norm, that is,

$$\sum_j w_j^2 \le c^2, \quad \text{or} \quad \left(\sum_j w_j^2\right)^{1/2} \le c, \tag{8.51}$$

where the upper bound $c$ is a tunable hyperparameter and $w_j$ the model weights. Typical values of $c$ used are 2, 3 or 4 (Srivastava et al., 2014; Aggarwal, 2018, p. 189). When the weight norm exceeds $c$, the weight vector is rescaled so its norm does not exceed $c$. The max-norm constraint can be applied after each update of the weights during optimization. One can use a single $c$ value for all the weights, a different value for one or more layers of weights or a different value for each hidden node (Goodfellow, Bengio, et al., 2016, section 7.2).

In the older weight penalty or weight decay approach, a term $(\lambda/2)\sum_j w_j^2$ is added to the objective function $J$ in (8.29). The weight penalty pulls the weights towards a smaller norm but does not guarantee the weight norm to be below some bound, which is guaranteed if the max-norm constraint is used instead. On the other hand, the lack of a pull on weights towards small values under the max-norm constraint allows the optimization algorithm to explore a broader weight space, thereby avoiding being trapped by local minima corresponding to small weights under weight penalty.

Using a large learning rate in gradient descent allows a wider and faster search of the weight space, but large weights may lead to the algorithm increasing instead of decreasing the training error, thereby leading to large oscillations or even numerical overflow. The max-norm constraint can prevent this as the weight norm is bounded. Thus, pairing large learning rate with max-norm constraint in the optimization algorithm allows rapid exploration of the weight space (Hinton, Srivastava, et al., 2012).

## 8.10   Bayesian Model Selection  B☺

When there is a choice of models (e.g. different number of adjustable weights or parameters, different weight penalty, etc.), one commonly uses validation or cross-validation to select the optimal model based on best performance over validation data. Cross-validation is computationally quite expensive, since a $K$-fold cross-validation requires training each model $K$ times over the $K$ folds of the dataset (Section 8.5). An alternative is to use *Bayesian model selection*, which does not use validation data but instead estimates the uncertainty in the choice of model from the training data. From Bayes theorem (Section 2.14), the posterior distribution over the models is given by

$$P(M_i|D) = \frac{p(D|M_i)\, P(M_i)}{\sum_i p(D|M_i)\, P(M_i)}, \tag{8.52}$$

where $D$ is the observed dataset, $M_i$ $(i = 1, \ldots, L)$ are the $L$ candidate models and $P(M_i)$ is the prior distribution. The model with the highest $P(M_i|D)$ is preferred under Bayesian model selection.

One can ignore the normalization factor in the denominator and focus on

$$P(M_i|D) \propto p(D|M_i)\, P(M_i). \tag{8.53}$$

If the models are assumed to have equal prior $P(M_i)$, then one only needs to focus on $p(D|M_i)$, called the *marginal likelihood* or *model evidence*, since this term will determine the relative size of $P(M_i|D)$, that is, how strongly a particular model $M_i$ is preferred by the observed data $D$. *Bayes factor* refers to the ratio of two model evidences, for example $p(D|M_i)/p(D|M_j)$.

In Bayesian model selection, one chooses the model with the largest $P(M_i|D)$, which amounts to choosing the model with the largest $p(D|M_i)$ under the assumption of equal prior $P(M_i)$. Figure 8.11 illustrates $p(D|M_i)$ for three different models – a simple model $M_1$, an intermediate model $M_2$ and a complex model $M_3$. While $M_3$ can fit a broad range of data $D$, the normalization of $p(D)$ causes $p(D)$ for $M_3$ to lie below that for $M_2$ for the observed data $D_0$. Thus, Bayesian model selection chooses model $M_2$, thereby avoiding underfitting from $M_1$ and overfitting from $M_3$.

If each model $M_i$ is governed by a set of parameters/hyperparameters $\boldsymbol{\theta}_i$, the marginal likelihood

$$p(D|M_i) = \int p(D|\boldsymbol{\theta}_i, M_i)\, p(\boldsymbol{\theta}_i|M_i)\, \mathrm{d}\boldsymbol{\theta}_i. \tag{8.54}$$

Figure 8.11 Schematic diagram of
model selection based on the highest model evidence, with $p(D)$
shown for three models – a simple
model $M_1$, an intermediate model
$M_2$ and a complex model $M_3$. The
simple model can only fit a narrow
range of observed data $D$, while
the complex model can fit a broad
range. For the observed data $D_0$,
the highest $p(D)$ (along the vertical dashed line) is found for model
$M_2$. [Follows Bishop (2006, figure 3.13).]

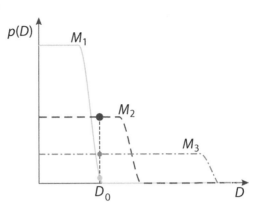

Evaluating the integral can be difficult, so approximations are often made.

## 8.11   Information Criterion   B☺

A common application of Bayesian model selection is to choose a model with
optimal complexity, that is, an optimal number of adjustable model parameters
so that the model is neither overfitting nor underfitting. Various simplifying assumptions are made to avoid computing the difficult integral in (8.54). Schwarz
(1978) proposed the Bayesian information criterion (BIC) (a.k.a. Schwarz information criterion). Other approaches led to the Akaike information criterion
(AIC) (Akaike, 1974) (Section 8.11.2) and other types of information criteria.
Some researchers prefer BIC over AIC since BIC is Bayesian. Their belief that
AIC is non-Bayesian is, in fact, incorrect, as AIC can be derived using a Bayesian
approach with a 'savvy' prior on models instead of the equal prior for all models used in BIC (Burnham and D. R. Anderson, 2004, section 4). Alternatively,
both AIC and BIC can be derived from a non-Bayesian (frequentist) approach
(Burnham and D. R. Anderson, 2004, section 5). BIC tends to select models
with fewer parameters than AIC, so the trade-off is that there is a risk of underfitting with BIC and overfitting with AIC. Overall, AIC or AICc (AIC with
a correction for finite sample size) seems to perform better than BIC (Burnham
and D. R. Anderson, 2002; Burnham and D. R. Anderson, 2004; Vrieze, 2012),
though some authors prefer BIC (von Storch and Zwiers, 1999, section 12.2.11).

### 8.11.1   Bayesian Information Criterion   C☺

For candidate models $M_1, \ldots, M_L$, each model $M_i$ has likelihood $l_i(D|\boldsymbol{\theta}_i)$ [$\equiv$
$p(D|\boldsymbol{\theta}, M_i)$ in (8.54)], characterized by the parameter vector $\boldsymbol{\theta}_i$ containing $k_i$
elements. The marginal likelihood in (8.54) can be rewritten as

$$p_i(D) = \int l_i(D|\boldsymbol{\theta}_i)\,\pi_i(\boldsymbol{\theta}_i)\,\mathrm{d}\boldsymbol{\theta}_i, \tag{8.55}$$

where $p_i(D) \equiv p(D|M_i)$ is the marginal likelihood, that is, the likelihood of the $i$th model, and $\pi_i(\boldsymbol{\theta}_i) \equiv p(\boldsymbol{\theta}|M_i)$ in (8.54), with $\pi_i(\boldsymbol{\theta}_i)$ being the prior distribution for $\boldsymbol{\theta}$ under model $M_i$.

To render the integral tractable, the Bayesian information criterion (BIC) of Schwarz (1978) applied the Laplace approximation on $l_i(D|\boldsymbol{\theta}_i)$. The Laplace approximation is based on the fact that for large sample size $N$, the integrand in (8.55) is concentrated in a neighbour of the mode (i.e. highest peak) of the $l_i(D|\boldsymbol{\theta}_i)$ distribution, with the mode located at $\hat{\boldsymbol{\theta}}_i$. The behaviour of $l_i(D|\boldsymbol{\theta}_i)$ around $\hat{\boldsymbol{\theta}}_i$ is approximated by the Taylor series expansion, with second order terms retained (first order terms vanished since $l_i(D|\hat{\boldsymbol{\theta}}_i)$ is a maximum) (Konishi and Kitagawa, 2008, section 9.1.2).

With

$$\mathrm{BIC} \equiv -2 \log p_i(D), \tag{8.56}$$

retaining only the leading terms and dropping subscript $i$ for brevity, one gets

$$\mathrm{BIC} \approx -2 \log l(D|\hat{\boldsymbol{\theta}}) + k \log N, \tag{8.57}$$

where log denotes the natural logarithm and details of the derivation can be found in Konishi and Kitagawa (2008, section 9.1) or Burnham and D. R. Anderson (2002, section 6.4.1). Among the $L$ models, the one with the lowest BIC is preferred. When striving for the lowest BIC, the first term on the right hand side of (8.57) tries to maximize the log likelihood (log $l$), while the second term penalizes $k$, the number of model parameters. In other words, BIC tries to choose a model with high log likelihood but without using too many parameters.

In Section 8.1, we have seen the connection between maximizing likelihood and minimizing mean squared error (MSE). For linear models, assuming Gaussian errors, the maximized log likelihood can be rewritten (Burnham and D. R. Anderson, 2002, p. 12) as

$$\log l(D|\hat{\boldsymbol{\theta}}) = -\frac{1}{2}N \log(\hat{\sigma}^2) - \frac{N}{2} \log(2\pi) - \frac{N}{2}, \tag{8.58}$$

where

$$\hat{\sigma}^2 = \frac{1}{(N-k)} \sum_{n=1}^{N} (\hat{y}_n - y_n)^2. \tag{8.59}$$

Assuming the sample size $N$ is much larger than $k$ (the number of model parameters), $N - k \approx N$, so $\hat{\sigma}^2$ is essentially the minimized MSE. As the final two terms in (8.58) are additive constants that do not affect likelihood-based inference, they are often omitted, leaving

$$\log l(D|\hat{\boldsymbol{\theta}}) = -\frac{1}{2}N \log(\hat{\sigma}^2), \tag{8.60}$$

and turning (8.57) into

$$\text{BIC} = N \log(\hat{\sigma}^2) + k \log N. \tag{8.61}$$

BIC has two main limitations:
(a) The approximation is only valid for sample size $N$ much larger than $k$, the number of model parameters.
(b) Model selection by BIC does not work well in high-dimensional problems (Giraud, 2015, p. 50).

## 8.11.2 Akaike Information Criterion $\boxed{\text{C}}$ ☺

While the Akaike information criterion (AIC) can be derived using a Bayesian approach with a 'savvy' prior on models instead of the equal prior for all models used in BIC (Burnham and D. R. Anderson, 2004, section 4), AIC was originally developed from information theory by Akaike (1973) and Akaike (1974). AIC was derived as an estimate of the expected relative Kullback–Leibler (KL) divergence (Section 2.17.3) between a candidate model and the true model, where a smaller divergence means the closer is the candidate to the truth. For derivation of the AIC, see Burnham and D. R. Anderson (2002, section 2.2) or Konishi and Kitagawa (2008, section 3.4). The AIC is given by

$$\text{AIC} = -2 \log l(D|\hat{\boldsymbol{\theta}}_i) + 2k, \tag{8.62}$$

which is similar to the BIC in form, but the penalty term is $2k$ instead of $k \log N$ in (8.57). For large sample size $N$, $k \log N$ in BIC is larger than $2k$ in AIC, hence BIC may select a model with smaller $k$, that is, fewer parameters, than AIC.

Again, assuming Gaussian errors, (8.58) with the two additive constants dropped allows AIC to be rewritten as

$$\text{AIC} = N \log(\hat{\sigma}^2) + 2k. \tag{8.63}$$

Let us illustrate AIC and BIC on a simple test problem with four different noise levels. Let the signal be a fifth order polynomial (with six parameters),

$$y_{\text{signal}} = 0.1\,x^5 - 0.2\,x^4 - 0.3\,x^3 + 0.4\,x^2 + x + 1, \tag{8.64}$$

with the random variable $x$ uniformly distributed in $[-2, 2]$. The $y$ data were composed of $y_{\text{signal}}$ plus Gaussian noise having standard deviation equal to (a) 0.25, (b) 0.5, (c) 1 and (d) 1.5 times the standard deviation of $y_{\text{signal}}$. With a training sample size of $N = 200$, a polynomial linear regression model was fitted to the data where the order of the polynomial (1.4) varied from 1 to 10. In Fig. 8.12, the AIC and BIC values are plotted as a function of the number of parameters. For the lowest noise level (a), both AIC and BIC selected the correct model with six parameters. For the next noise level (b), neither picked the correct model, with AIC picking seven parameters and BIC five parameters. For (c), AIC correctly picked six parameters, while BIC picked five parameters.

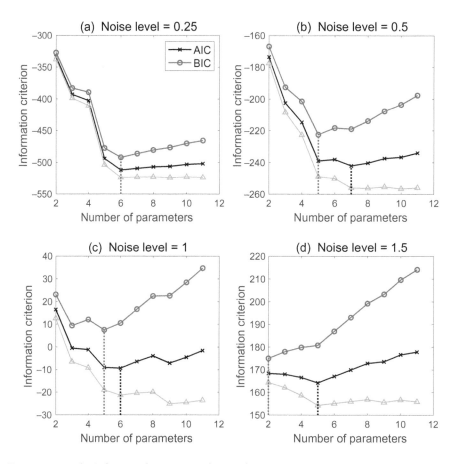

Figure 8.12 AIC (crosses) and BIC (circles) values for polynomial linear regression models of order 1 to 10 (i.e. with corresponding number of parameters from 2 to 11). The signal $y_{\text{signal}}$ was generated by a fifth order polynomial (with six parameters). Gaussian noise with standard deviation equal to (a) 0.25, (b) 0.5, (c) 1 and (d) 1.5 times the standard deviation of $y_{\text{signal}}$ was added to the signal. Model is selected based on lowest AIC or BIC (as indicated by the vertical dotted lines). In (a), both AIC and BIC selected the correct model with six parameters. The first term in AIC and BIC, that is, $N \log(\hat{\sigma}^2)$, is also shown (triangles).

For the highest noise level (d), AIC chose five parameters, while BIC chose only two parameters.

Next, 10,000 runs were made with different random noise, and the probability of selecting a model with $k$ parameters is plotted in Fig. 8.13. At the lowest noise level (a), BIC slightly outperformed AIC, as AIC had a slightly higher probability of selecting models with more than six parameters. At the next noise level (b), BIC slightly underperformed AIC as BIC had almost 40% probability selecting a model with five parameters. For (c), the mode for AIC and BIC occurred at five and three parameters, respectively, while for the highest noise level (d), the mode for AIC and BIC occurred at five and two parameters, respectively. This example is consistent with other studies that found that BIC tends to select models which underfit (Burnham and D. R. Anderson, 2002; Burnham and D. R. Anderson, 2004; Vrieze, 2012).

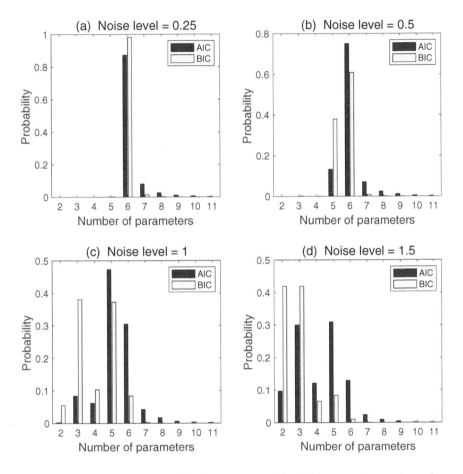

Figure 8.13 The probability of selecting a model with a certain number of parameters using AIC and BIC for four different noise levels, as estimated from 10,000 runs with different random noise. The true model had six parameters.

AIC was derived assuming large $N$. For small sample size, AIC can select models with too many parameters. $AIC_c$ is AIC corrected for small sample sizes, with

$$\text{AIC}_c = -2\log l(D|\hat{\boldsymbol{\theta}}_i) + 2k\left(\frac{N}{N-k-1}\right), \tag{8.65}$$

where the penalty term is multiplied by the correction factor $N/(N-k-1)$ (Burnham and D. R. Anderson, 2002, section 2.4). This can be written as

$$\text{AIC}_c = -2\log l(D|\hat{\boldsymbol{\theta}}_i) + 2k + \frac{2k(k+1)}{N-k-1}. \tag{8.66}$$

Substituting in (8.62), $\text{AIC}_c$ can be expressed in terms of AIC, that is,

$$\text{AIC}_c = \text{AIC} + \frac{2k(k+1)}{N-k-1}. \tag{8.67}$$

Burnham and D. R. Anderson (2002, p. 66) recommend using $\text{AIC}_c$ instead of AIC when $N/k_{\max} < 40$, where $k_{\max}$ is the maximum $k$ found in all the candidate models. As $N$ becomes large, the second term in (8.67) becomes negligible and $\text{AIC}_c$ becomes indistinguishable from AIC.

In summary, AIC tries to select a model by minimizing the Kullback–Leibler divergence between the true distribution and a candidate model, whereas BIC tries to select a model that maximizes the posterior model probability.

> From a practical perspective, AIC and BIC might be distinguished as follows. AIC could be advocated when the primary goal of the modeling application is predictive; that is, to build a model that will effectively predict new outcomes. BIC could be advocated when the primary goal of the modeling application is descriptive; that is, to build a model that will feature the most meaningful factors influencing the outcome, based on an assessment of relative importance. As the sample size grows, predictive accuracy improves as subtle effects are admitted to the model. AIC will increasingly favor the inclusion of such effects; BIC will not. (Cavanaugh and Neath, 2019).

## 8.12   Bayesian Model Averaging  C☺

Given models $M_1, \ldots, M_L$ and observed data $D$, in Bayesian model selection, we select only one model based on the highest $P(M_i|D)$ in (8.52) and use that model to make predictions. Alternatively, we can use *Bayesian model averaging* (BMA) to utilize all the $M_1, \ldots, M_L$ models in a weighted ensemble average when making predictions (Raftery, Madigan, et al., 1997; Hoeting et al., 1999; Wasserman, 2000).

If $y$ is the quantity to be predicted, then the probability density function of $y$ is given by

$$p(y) = \sum_{i=1}^{L} p(y\,|M_i, D)\,P(M_i|D), \tag{8.68}$$

(Hoeting et al., 1999). Here, $p(y \mid D)$ is a weighted sum over the individual model's posterior distribution $p(y \mid M_i, D)$, with $P(M_i \mid D)$ providing the weighting factor for each model.

Probabilistic weather forecasting is mainly based on ensembles where dynamical models are run multiple times with slightly different initial conditions or model physics, as proposed originally by Epstein (1969) and Leith (1974). Instead of simply averaging the predictions of the individual ensemble members, Raftery, Gneiting, et al. (2005) proposed using BMA to combine the predictions of the ensemble members $M_i$ weighted by $P(M_i \mid D)$.

With each model $M_i$ providing a biased-corrected forecast $f_i$, the BMA predictive model is given by

$$p(y \mid f_1, \ldots, f_L) = \sum_{i=1}^{L} w_i \, g_i(y \mid f_i), \tag{8.69}$$

where the weighting factor $w_i$ is the posterior probability of model $M_i$ being the best as based on the performance of $M_i$ during the training period, with $\sum_i w_i = 1$. The conditional distribution $g_i(y \mid f_i)$ is assumed to be a Gaussian distribution with variance $\sigma^2$ centred on a linear function $a_i + b_i f_i$, that is,

$$g_i(y \mid f_i) = \mathcal{N}(a_i + b_i f_i, \, \sigma^2). \tag{8.70}$$

The log likelihood function (Section 3.5) is

$$\log l(w_1, \ldots, w_L, \sigma^2) = \sum_{s,t} \log \left( \sum_{i=1}^{L} w_i \, g_i(y_{st} \mid f_{ist}) \right), \tag{8.71}$$

where the summation is over the $s$ and $t$ (space and time) indices. The EM algorithm (Section 3.12.1) is used to find the maximum likelihood solution, yielding estimates for $w_1, \ldots, w_L$ and $\sigma^2$. When working with non-Gaussian variables (e.g. streamflow), Box–Cox transformation (Section 3.14) is first applied to render the variable closer to being Gaussian (Najafi et al., 2011).

The BMA predictive mean is given by the conditional expectation

$$\mathrm{E}[y \mid f_1, \ldots, f_L] = \sum_{i=1}^{L} w_i \, (a_i + b_i f_i), \tag{8.72}$$

while the BMA predictive variance is

$$\mathrm{var}(y_{st} \mid f_{1st}, \ldots, f_{Lst}) = \sum_{i=1}^{L} w_i \left[ (a_i + b_i f_{ist}) - \sum_{i=1}^{L} w_i \, (a_i + b_i f_{ist}) \right]^2 + \sigma^2 \tag{8.73}$$

(Raftery, Gneiting, et al., 2005). The variance is composed of two parts, the first being the between-model variance and the second being the within-model variance $\sigma^2$.

The application of BMA has been extended from weather forecasting to climate models and climate change predictions (S. K. Min and Hense, 2006; S.-K. Min et al., 2007; Miao et al., 2014; Najafi et al., 2011) and to El Ninõ-Southern Oscillation (ENSO) forecasts (H. Zhang, Chu, et al., 2019). BMA is the Bayesian alternative to the methods presented in Sections 8.7.3 and 8.7.4 for unequal weighting of ensemble members. Comparing BMA and stacking, the results from B. Clarke (2003) suggest that stacking (Section 8.7.4) has better robustness properties than BMA. There does not appear to be any study comparing the performance of stacking (using linear or non-linear level-1 models) versus that of BMA in the environmental sciences.

# 8.13  Interpretable ML  $\boxed{\text{C}}$ ☺

For users familiar with multiple linear regression, a major disappointment with NN models is that there is nothing equivalent to the familiar regression parameters in MLR, which tells us about the effectiveness of individual predictor variables in the regression model. To determine which predictors are important in an NN model (or other ML models), after the model has been trained, one can select an individual predictor, randomly reshuffle the order of its data[4] and see how the model error changes. Randomly shuffling an important predictor will lead to large increase in the model error, while shuffling an unimportant predictor will show little change in the error. By randomly shuffling the predictors one at a time, the relative importance of individual predictors is revealed (Gagne, S. E. Haupt, et al., 2019, figure 7).

In recent years, the ML community has placed more effort in developing *interpretable ML*, or *interpretable AI* (Murdoch et al., 2019; Escalante et al., 2018; Masis, 2021). Environmental scientists have also been trying to make ML less opaque, with reviews of recent interpretable ML techniques given by McGovern, Lagerquist, et al. (2019) and Ebert-Uphoff and Hilburn (2020).

*Backward optimization* (Gagne, S. E. Haupt, et al., 2019; McGovern, Lagerquist, et al. 2019; Toms et al., 2020) is a technique used to reveal the input pattern most closely associated with a particular model output. For example, after developing an NN model using meteorological fields as input to predict the binary output 'storm' or 'no storm' some time into the future, one can use backward optimization to find the pattern in the input field most closely associated with the output 'storm'. The backward optimization method is explained in Toms et al. (2020, figure 2) and used to find the sea surface temperature anomaly pattern over the world ocean most closely associated with warmer temperatures on the west coast of North America as predicted by a two-hidden layer MLP NN model.

---

[4] Randomly reshuffling the data is preferred to setting all values of a predictor to zero. To see this, consider a non-linear system where predictor $x_2$ is important only if predictor $|x_1| > 0.5$ (assuming both predictors have been standardized). Setting $x_1 \equiv 0$ would never allow the importance of $x_2$ to be manifested.

*Layerwise relevance propagation* (LRP) is a method for identifying the relevance of each input to the output of an NN model (Toms et al., 2020, figure 3). It finds the input features important for a network's output for a single input sample or image. For example, given a single input image, the resulting output from LRP is a heatmap with the same dimensions as the original image, revealing the regions of the image most important for generating the NN model output for that particular image. LRP can be applied to samples or images not used in the NN model training, that is, it can analyse any sample of the same dimensions as those used to train the NN model. Examples of LRP applications are given by Toms et al. (2020); Barnes et al. (2020), and Ebert-Uphoff and Hilburn (2020).

# Exercises

8.1

Let $y_{signal} = \sin(2\pi x)$ be the signal, with the random variable $x$ uniformly distributed in the interval $[-1, 1]$. The $y$ data are generated by adding Gaussian noise to $y_{signal}$, where the Gaussian noise has the same standard deviation as $y_{signal}$. Generate 200 training data points and 1,000 test data points. Fit an ensemble of MLP NN models with one hidden layer to the $(x, y)$ training data, using some form of regularization (e.g. early stopping by reserving 15% of the training data for validation and training with 85% of the training data). Vary the number of hidden neurons (at 2, 3, 4, 5, 6, 7 and 8) and the ensemble size (at 25, 50 and 100). From the test data, determine (i) the average RMSE of an individual model's prediction of $y$ and (ii) the RMSE of the ensemble average prediction of $y$.
(a) Are the values in (ii) smaller than the corresponding values in (i)?
(b) Do the ensemble average using 100 ensemble members outperform that using 25 ensemble members?
(c) Repeat with the Gaussian noise at 0.5 and 2 times the standard deviation of $y_{signal}$. Do your answers in (a) and (b) still hold at the lower and higher noise levels?

8.2

Let $y_{signal} = \sin(2\pi x)$ be the signal, with the random variable $x$ uniformly distributed in the interval $[-1, 1]$. The $y$ data are generated by adding Gaussian noise to $y_{signal}$, where the Gaussian noise has the same standard deviation as $y_{signal}$. Generate 200 training data points and 1,000 test data points. Fit an ensemble of MLP NN models with one hidden layer to the $(x, y)$ training data, using some form of regularization (e.g. early stopping by reserving 15% of the training data for validation and training with 85% of the training data). Vary the number of hidden neurons (at 2, 3, 4, 5, 6, 7 and 8) and the ensemble size (at 25, 50 and 100). Train the ensemble members with random initial weights using (i) the training data, then repeat using (ii) bootstrap resampled training data (i.e. bagging).

From the test data, compute:
(1) the standard deviation of (individual member output minus the ensemble average output), (2) the average RMSE of an individual model's prediction of $y$ and (3) the RMSE of the ensemble average prediction of $y$.
(a) Does bagging give a larger standard deviation in (1) than without bagging?
(b) Does the average RMSE of an individual model's prediction in (2) improve or worsen with bagging? Why?
(c) Does the RMSE of the ensemble average prediction in (3) improve or worsen with bagging?

### 8.3

Given the error covariance matrix $\mathbf{C}$ for ensemble members in (8.47), find the optimal weights $a_m$ for the ensemble members $m = 1, \ldots, 4$ and explain the reasons for the weight distribution (especially if there is negative weight) for the following three cases:
(a)

$$\mathbf{C} = \begin{bmatrix} 4 & 0 & 0 & 0 \\ 0 & 2 & 0 & 0 \\ 0 & 0 & 3 & 0 \\ 0 & 0 & 0 & 1 \end{bmatrix}. \tag{8.74}$$

(b)

$$\mathbf{C} = \begin{bmatrix} 1 & 0 & 0.9 & 0 \\ 0 & 1 & 0 & 0 \\ 0.9 & 0 & 1 & 0 \\ 0 & 0 & 0 & 1 \end{bmatrix}. \tag{8.75}$$

(c)

$$\mathbf{C} = \begin{bmatrix} 1 & -0.8 & 0.8 & 0 \\ -0.8 & 1 & -0.5 & 0 \\ 0.8 & -0.5 & 1 & 0 \\ 0 & 0 & 0 & 1 \end{bmatrix}. \tag{8.76}$$

### 8.4

Consider the given dataset containing the winter snow water equivalent (SWE) for Grouse Mountain, BC, Canada (the winter SWE being the maximum observed SWE from January to June) and four predictors (the Niño3.4 index and the PNA, NAO and AO teleconnection indices, averaged December to February). [In Exercise 5.8, validation was used to choose predictors from this dataset]. (a) Use all 68 data points to train four models with different number of predictors: Model 1 uses all four predictors, Model 2 uses Niño3.4, PNA and NAO, Model 3 uses Niño3.4 and PNA and Model 4 uses only PNA as predictor. Use information criteria AIC and BIC for model selection. Are the results consistent with what was found in Exercise 5.8? (b) Repeat, but use only the first 53 data points for training.

# 9

# Principal Components and Canonical Correlation

As one often encounters datasets with more than a few variables, multivariate statistical techniques have been commonly used to extract the information contained in these datasets. In the environmental sciences, examples of multivariate datasets are ubiquitous – the air temperatures recorded by all the weather stations around the globe, the satellite images composed of numerous pixels, the gridded output from a global climate model, and so on. The number of variables or time series from these datasets ranges from thousands to millions. Without a mastery of multivariate techniques, one would be overwhelmed by these gigantic datasets. In this chapter, we review several multivariate statistical techniques – namely principal component analysis and its many variants, canonical correlation analysis and the related maximum covariance analysis. These methods, using standard matrix techniques such as singular value decomposition, are relatively easy to use but suffer from being *linear* methods, a limitation which has been removed in recent decades by various non-linear ML methods, for example, autoencoders (a.k.a. auto-associative neural networks) (Section 10.5), convolutional neural networks (Section 15.2) and U-nets (Section 15.3.1).

## 9.1 Principal Component Analysis (PCA)  [A] ☺

Datasets containing more than a few variables are often hard to interpret. *Principal component analysis* (PCA) is a method for reducing the dimensionality of such datasets, thereby improving interpretability. By linearly combining the variables of the dataset, PCA generates new uncorrelated variables that maximize the amount of variance captured. PCA has various names in various fields – in meteorology and oceanography, it is often called *empirical orthogonal function* (EOF) analysis.

This classical method dates back to Pearson (1901) and Hotelling (1933). Books on PCA include Preisendorfer (1988) and Jolliffe (2002). Over the years, as datasets gradually grew in size, with an increasing number of variables, this classical method has been gaining wider usage and novel applications. A caveat on the limitations of PCA is provided by Monahan, J. C. Fyfe, et al. (2009). Recent review papers include Hannachi et al. (2007) and Jolliffe and Cadima (2016).

### 9.1.1 Geometric Approach to PCA  A ☺

Consider a dataset containing variables $x_1, \ldots, x_m$, where each variable has been sampled $n$ times. In many situations, the $m$ variables are time series each containing $n$ observations in time. For instance, one may have a dataset containing the monthly air temperature measured at $m$ stations over $n$ months. If $m$ is a large number, we would like to capture the essence of $x_1, \ldots, x_m$ by a smaller set of variables $z_1, \ldots, z_k$ (i.e. $k < m$; and hopefully $k \ll m$, for truly large $m$). We first begin with a geometric approach, which is more intuitive than the standard eigenvector approach to PCA.

Let us start with only two variables, $x_1$ and $x_2$, as illustrated in Fig. 9.1. Clearly the bulk of the variance is along the $z_1$-axis. If $r_i$ is the distance between

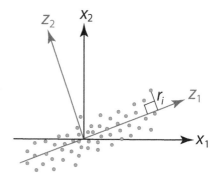

Figure 9.1 The PCA problem formulated as a minimization of the sum of $r_i^2$, where $r_i$ is the shortest distance from the $i$th data point to the first PCA axis $z_1$.

the $i$th data point and the $z_1$-axis, then the optimal $z_1$-axis is found by minimizing $\sum_{i=1}^{n} r_i^2$. This type of geometric approach to PCA was first proposed by Pearson (1901). Note that PCA treats all variables equally, whereas linear regression divides variables into independent and dependent variables (cf. Fig. 9.1 and Fig. 5.1); hence, the straight line described by $z_1$ is in general different from a regression line.

In 3-D, $z_1$ is the best 1-D line fit to the data, while $z_1$ and $z_2$ span a 2-D plane giving the best plane fit to the data. In general, with an $m$-dimensional dataset, we want to find the $k$-dimensional hyperplane giving the best fit.

### 9.1.2 Eigenvector Approach to PCA  A ☺

The more systematic eigenvector approach to PCA is due to Hotelling (1933). Here again with the 2-D example, a data point is transformed from its old coordinates $(x_1, x_2)$ to new coordinates $(z_1, z_2)$ via a rotation of the coordinate system (Fig. 9.2):

$$z_1 = x_1 \cos\theta + x_2 \sin\theta,$$
$$z_2 = -x_1 \sin\theta + x_2 \cos\theta. \tag{9.1}$$

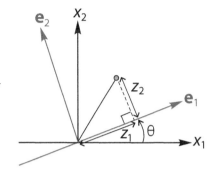

Figure 9.2 Rotation of coordinate axes by an angle $\theta$ in a two-dimensional space.

In the general $m$-dimensional problem, we want to introduce new coordinates

$$z_j = \sum_{l=1}^{m} e_{jl}\, x_l, \qquad j = 1, \ldots, m. \tag{9.2}$$

The objective is to find

$$\mathbf{e}_1 = [e_{11}, \ldots, e_{1m}]^{\mathrm{T}}, \tag{9.3}$$

which maximizes var($z_1$), where $z_1$ is the projected distance from a data point onto the $\mathbf{e}_1$ direction (Fig. 9.2). In other words, find the coordinate transformation such that the variance of the dataset along the direction of $\mathbf{e}_1$ is maximal. With

$$\mathbf{x} = [x_1, \ldots, x_m]^{\mathrm{T}}, \tag{9.4}$$

the projected distance of the data point $\mathbf{x}$ onto the vector $\mathbf{e}_1$ is given by the dot product $\mathbf{x} \cdot \mathbf{e}_1$, that is,

$$\mathbf{e}_1^{\mathrm{T}}\mathbf{x} = \sum_{l=1}^{m} e_{1l}\, x_l = z_1, \tag{9.5}$$

where (9.2) has been invoked.

The variance of $z_1$ is given by

$$\operatorname{var}(z_1) = \mathrm{E}\left[(z_1 - \bar{z}_1)(z_1 - \bar{z}_1)\right] = \mathrm{E}\left[\mathbf{e}_1^{\mathrm{T}}(\mathbf{x} - \bar{\mathbf{x}})(\mathbf{x} - \bar{\mathbf{x}})^{\mathrm{T}}\mathbf{e}_1\right], \tag{9.6}$$

where we have used the vector identity $\mathbf{a}^{\mathrm{T}}\mathbf{b} = \mathbf{b}^{\mathrm{T}}\mathbf{a}$ and the bar above a variable denotes the mean of that variable. Thus,

$$\operatorname{var}(z_1) = \mathbf{e}_1^{\mathrm{T}}\,\mathrm{E}\left[(\mathbf{x} - \bar{\mathbf{x}})(\mathbf{x} - \bar{\mathbf{x}})^{\mathrm{T}}\right]\mathbf{e}_1 = \mathbf{e}_1^{\mathrm{T}}\mathbf{C}\,\mathbf{e}_1, \tag{9.7}$$

where the *covariance matrix* $\mathbf{C}$ [see (2.33)] is given by

$$\mathbf{C} = \mathrm{E}\left[(\mathbf{x} - \bar{\mathbf{x}})(\mathbf{x} - \bar{\mathbf{x}})^{\mathrm{T}}\right]. \tag{9.8}$$

Clearly, the larger is the vector norm $\|\mathbf{e}_1\|$, the larger $\mathrm{var}(z_1)$ will be. Therefore, we need to place a constraint on $\|\mathbf{e}_1\|$ while we try to maximize $\mathrm{var}(z_1)$. Let us impose a normalization constraint $\|\mathbf{e}_1\| = 1$, that is,

$$\mathbf{e}_1^{\mathrm{T}} \mathbf{e}_1 = 1. \tag{9.9}$$

Thus, our optimization problem is to find $\mathbf{e}_1$ that maximizes $\mathbf{e}_1^{\mathrm{T}} \mathbf{C} \mathbf{e}_1$, subject to the constraint

$$\mathbf{e}_1^{\mathrm{T}} \mathbf{e}_1 - 1 = 0. \tag{9.10}$$

The method of Lagrange multipliers is commonly used to tackle optimization under constraints (see Appendix B). Instead of finding stationary points of $\mathbf{e}_1^{\mathrm{T}} \mathbf{C} \mathbf{e}_1$, we search for the stationary points of the Lagrange function $L$,

$$L = \mathbf{e}_1^{\mathrm{T}} \mathbf{C} \mathbf{e}_1 - \lambda \, (\mathbf{e}_1^{\mathrm{T}} \mathbf{e}_1 - 1), \tag{9.11}$$

where $\lambda$ is a Lagrange multiplier. Differentiating $L$ by the elements of $\mathbf{e}_1$, and setting the derivatives to zero, we obtain

$$\mathbf{C} \mathbf{e}_1 - \lambda \mathbf{e}_1 = 0, \tag{9.12}$$

which says that $\lambda$ is an eigenvalue of the covariance matrix $\mathbf{C}$, with $\mathbf{e}_1$ the eigenvector. Multiplying this equation by $\mathbf{e}_1^{\mathrm{T}}$ on the left, then invoking (9.9) and (9.7), and writing $\lambda$ as $\lambda_1$, we obtain

$$\lambda_1 = \mathbf{e}_1^{\mathrm{T}} \mathbf{C} \mathbf{e}_1 = \mathrm{var}(z_1). \tag{9.13}$$

Since $\mathbf{e}_1^{\mathrm{T}} \mathbf{C} \mathbf{e}_1$ is maximized, so are $\mathrm{var}(z_1)$ and $\lambda_1$. Thus, by solving the eigen problem for the matrix $\mathbf{C}$, we obtain the eigenvector $\mathbf{e}_1$ and the eigenvalue $\lambda_1$. The new coordinate $z_1$, called the first *principal component* (PC), is easily found from (9.5), and its variance is given by $\lambda_1$. Together, the first eigenvector and the first PC make up the first *mode* of PCA.

Next, we want to find the second mode. Our task is to find $\mathbf{e}_2$ that maximizes $\mathrm{var}(z_2) = \mathbf{e}_2^{\mathrm{T}} \mathbf{C} \mathbf{e}_2$, subject to the constraint $\mathbf{e}_2^{\mathrm{T}} \mathbf{e}_2 = 1$, and the constraint that $z_2$ be uncorrelated with $z_1$, that is, the covariance between $z_2$ and $z_1$ be zero,

$$\mathrm{cov}(z_1, z_2) = 0. \tag{9.14}$$

As $\mathbf{C} = \mathbf{C}^{\mathrm{T}}$, we can write

$$\begin{aligned} 0 = \mathrm{cov}(z_1, z_2) &= \mathrm{cov}(\mathbf{e}_1^{\mathrm{T}} \mathbf{x}, \mathbf{e}_2^{\mathrm{T}} \mathbf{x}) = \mathrm{E}[\mathbf{e}_1^{\mathrm{T}} (\mathbf{x} - \overline{\mathbf{x}}) \, \mathbf{e}_2^{\mathrm{T}} (\mathbf{x} - \overline{\mathbf{x}})] \\ &= \mathrm{E}[\mathbf{e}_1^{\mathrm{T}} (\mathbf{x} - \overline{\mathbf{x}})(\mathbf{x} - \overline{\mathbf{x}})^{\mathrm{T}} \mathbf{e}_2] = \mathbf{e}_1^{\mathrm{T}} \mathrm{E}[(\mathbf{x} - \overline{\mathbf{x}})(\mathbf{x} - \overline{\mathbf{x}})^{\mathrm{T}}] \mathbf{e}_2 \\ &= \mathbf{e}_1^{\mathrm{T}} \mathbf{C} \mathbf{e}_2 = \mathbf{e}_2^{\mathrm{T}} \mathbf{C} \mathbf{e}_1 = \mathbf{e}_2^{\mathrm{T}} \lambda_1 \mathbf{e}_1 = \lambda_1 \mathbf{e}_2^{\mathrm{T}} \mathbf{e}_1 = \lambda_1 \mathbf{e}_1^{\mathrm{T}} \mathbf{e}_2 . \end{aligned} \tag{9.15}$$

Hence, the orthogonality condition

$$\mathbf{e}_2^{\mathrm{T}} \mathbf{e}_1 = 0 \tag{9.16}$$

can be used as a constraint in place of (9.14).

Upon introducing another Lagrange multiplier $\gamma$, we want to find an $\mathbf{e}_2$ that gives a stationary point of the Lagrange function $L$, with

$$L = \mathbf{e}_2^{\mathrm{T}} \mathbf{C} \mathbf{e}_2 - \lambda \left( \mathbf{e}_2^{\mathrm{T}} \mathbf{e}_2 - 1 \right) - \gamma \mathbf{e}_2^{\mathrm{T}} \mathbf{e}_1. \tag{9.17}$$

This equation is analogous to 9.11 but with the $\gamma$ term adding in the orthogonality constraint. Differentiating $L$ by the elements of $\mathbf{e}_2$, and setting the derivatives to zero, we obtain

$$\mathbf{C} \mathbf{e}_2 - \lambda \mathbf{e}_2 - \gamma \mathbf{e}_1 = 0. \tag{9.18}$$

Left multiplying this equation by $\mathbf{e}_1^{\mathrm{T}}$ yields

$$\mathbf{e}_1^{\mathrm{T}} \mathbf{C} \mathbf{e}_2 - \lambda \mathbf{e}_1^{\mathrm{T}} \mathbf{e}_2 - \gamma \mathbf{e}_1^{\mathrm{T}} \mathbf{e}_1 = 0. \tag{9.19}$$

On the left hand side, the first two terms are both zero from (9.15), while the third term is simply $\gamma$, so $\gamma = 0$. With $\lambda$ written as $\lambda_2$, (9.18) becomes

$$\mathbf{C} \mathbf{e}_2 - \lambda_2 \mathbf{e}_2 = 0. \tag{9.20}$$

Once again $\lambda_2$ is an eigenvalue of $\mathbf{C}$, with $\mathbf{e}_2$ the eigenvector. As

$$\lambda_2 = \mathbf{e}_2^{\mathrm{T}} \mathbf{C} \mathbf{e}_2 = \mathrm{var}(z_2), \tag{9.21}$$

which is maximized, this $\lambda_2$ is as large as possible with $\lambda_2 < \lambda_1$ (the case $\lambda_2 = \lambda_1$ is degenerate and will be discussed later). Hence, $\lambda_2$ is the second largest eigenvalue of $\mathbf{C}$, with $\lambda_2 = \mathrm{var}(z_2)$. This process can be repeated for $z_3$, $z_4, \ldots$

How do we reconcile the geometric approach of the previous subsection and the present eigenvector approach? First, let us subtract the mean $\overline{\mathbf{x}}$ from $\mathbf{x}$, so the transformed data are centred around the origin with $\overline{\mathbf{x}} = 0$. In the geometric approach, we minimize the distance between the data points and the new axis (Fig. 9.1). If the unit vector $\mathbf{e}_1$ gives the direction of the new axis, then the projection of a data point (described by the vector $\mathbf{x}$) onto $\mathbf{e}_1$ is $(\mathbf{e}_1^{\mathrm{T}} \mathbf{x}) \mathbf{e}_1$. The component of $\mathbf{x}$ normal to $\mathbf{e}_1$ is $\mathbf{x} - (\mathbf{e}_1^{\mathrm{T}} \mathbf{x}) \mathbf{e}_1$. Thus, minimizing the distance between the data points and the new axis amounts to minimizing

$$\epsilon = \mathrm{E}\left[ \|\mathbf{x} - (\mathbf{e}_1^{\mathrm{T}} \mathbf{x}) \mathbf{e}_1\|^2 \right]. \tag{9.22}$$

Since $\mathbf{e}_1^{\mathrm{T}} \mathbf{x}$ is a scalar, (9.22) can be rewritten as follows:

$$\begin{aligned}
\epsilon &= \mathrm{E}\left[ \left[ \mathbf{x} - (\mathbf{e}_1^{\mathrm{T}} \mathbf{x}) \mathbf{e}_1 \right]^{\mathrm{T}} \left[ \mathbf{x} - (\mathbf{e}_1^{\mathrm{T}} \mathbf{x}) \mathbf{e}_1 \right] \right] \\
&= \mathrm{E}\left[ \mathbf{x}^{\mathrm{T}} \mathbf{x} - 2(\mathbf{e}_1^{\mathrm{T}} \mathbf{x})(\mathbf{e}_1^{\mathrm{T}} \mathbf{x}) + (\mathbf{e}_1^{\mathrm{T}} \mathbf{x}) \mathbf{e}_1^{\mathrm{T}} (\mathbf{e}_1^{\mathrm{T}} \mathbf{x}) \mathbf{e}_1) \right] \\
&= \mathrm{E}\left[ \mathbf{x}^{\mathrm{T}} \mathbf{x} - (\mathbf{e}_1^{\mathrm{T}} \mathbf{x})(\mathbf{e}_1^{\mathrm{T}} \mathbf{x}) \right] = \mathrm{E}\left[ \|\mathbf{x}\|^2 \right] - \mathrm{E}\left[ (\mathbf{e}_1^{\mathrm{T}} \mathbf{x})^2 \right] \\
&= \mathrm{var}(\mathbf{x}) - \mathrm{var}(\mathbf{e}_1^{\mathrm{T}} \mathbf{x}) = \mathrm{var}(\mathbf{x}) - \mathrm{var}(z_1), \tag{9.23}
\end{aligned}$$

where $\text{var}(\mathbf{x}) = \text{E}[\|\mathbf{x}\|^2]$, as $\overline{\mathbf{x}} = 0$. Since $\text{var}(\mathbf{x})$ is constant, minimizing $\epsilon$ is equivalent to maximizing $\text{var}(z_1)$. Thus, the geometric approach of minimizing the distance between the data points and the new axis is equivalent to the eigenvector approach in finding the largest eigenvalue $\lambda_1$, that is, maximizing $\text{var}(z_1)$.

So far, $\mathbf{C}$ is the data covariance matrix, but one could instead use the data *correlation* matrix, which is equivalent to performing PCA using standardized variables. This approach allows an easy interpretation for the elements of the eigenvectors under the Hotelling scaling of eigenvectors (see Section 9.1.6).

In *combined PCA*, where two or more variables with different units are combined into one large data matrix for PCA – for example, finding the PCA modes of the combined sea surface temperature data and the sea level pressure data – then one needs to standardize the variables, so that $\mathbf{C}$ is the correlation matrix.

### 9.1.3   Real and Complex Data   Ⓒ☺

In general, for $\mathbf{x}$ real,
$$\mathbf{C} \equiv \text{E}\big[(\mathbf{x} - \overline{\mathbf{x}})(\mathbf{x} - \overline{\mathbf{x}})^{\text{T}}\big], \tag{9.24}$$
implies that $\mathbf{C}^{\text{T}} = \mathbf{C}$, that is, $\mathbf{C}$ is a real, symmetric matrix. A *positive semi-definite matrix* $\mathbf{A}$ is defined by the property that for any vector $\mathbf{v} \neq \mathbf{0}$, it follows that $\mathbf{v}^{\text{T}}\mathbf{A}\mathbf{v} \geq 0$ (Strang, 2005). From the definition of $\mathbf{C}$ (9.24), it is clear that $\mathbf{v}^{\text{T}}\mathbf{C}\mathbf{v} \geq 0$ is satisfied. Thus, $\mathbf{C}$ is a real, symmetric, positive semi-definite matrix.

If $\mathbf{x}$ is complex, then
$$\mathbf{C} \equiv \text{E}\big[(\mathbf{x} - \overline{\mathbf{x}})(\mathbf{x} - \overline{\mathbf{x}})^{\text{T}*}\big], \tag{9.25}$$
with complex conjugation denoted by the superscript asterisk. As $\mathbf{C}^{\text{T}*} = \mathbf{C}$, $\mathbf{C}$ is a *Hermitian matrix*. It is also a positive semi-definite matrix. Theorems on Hermitian, positive semi-definite matrices (Strang, 2005) tell us that $\mathbf{C}$ has real eigenvalues
$$\lambda_1 \geq \lambda_2 \geq \cdots \geq \lambda_m \geq 0, \qquad \sum_{j=1}^{m} \lambda_j = \text{var}(\mathbf{x}), \tag{9.26}$$
with corresponding orthonormal eigenvectors, $\mathbf{e}_1, \ldots, \mathbf{e}_m$, and that the $k$ eigenvectors corresponding to $\lambda_1, \ldots, \lambda_k$ minimize
$$\epsilon_k = \text{E}\bigg[\Big\|(\mathbf{x} - \overline{\mathbf{x}}) - \sum_{j=1}^{k} \big(\mathbf{e}_j^{\text{T}}(\mathbf{x} - \overline{\mathbf{x}})\big)\mathbf{e}_j\Big\|^2\bigg], \tag{9.27}$$
which can be expressed as
$$\epsilon_k = \text{var}(\mathbf{x}) - \sum_{j=1}^{k} \lambda_j. \tag{9.28}$$

This is an extension of (9.23) to $k$ modes, with $\lambda_j$ being the variance of the $j$th PC.

## 9.1.4 Orthogonality Relations  A☺

Thus, PCA amounts to finding the eigenvectors and eigenvalues of **C**. The orthonormal eigenvectors then provide a basis, that is, the data **x** can be expanded in terms of the eigenvectors $\mathbf{e}_j$:

$$\mathbf{x} - \overline{\mathbf{x}} = \sum_{j=1}^{m} a_j(t)\, \mathbf{e}_j, \tag{9.29}$$

where $a_j(t)$ are the expansion coefficients. When the variables in **x** are time series, the index $t$ is simply the time, and $a_j(t)$ are also time series. To obtain $a_j(t)$, left multiply the above equation by $\mathbf{e}_i^{\mathrm{T}}$,

$$\mathbf{e}_i^{\mathrm{T}}(\mathbf{x} - \overline{\mathbf{x}}) = \sum_{j=1}^{m} a_j(t)\, \mathbf{e}_i^{\mathrm{T}} \mathbf{e}_j = a_i(t), \tag{9.30}$$

upon invoking the orthonormal property of the eigenvectors,

$$\mathbf{e}_i^{\mathrm{T}} \mathbf{e}_j = \delta_{ij}, \tag{9.31}$$

with $\delta_{ij}$ the Kronecker delta function. Rewriting the index $i$ as $j$, we get

$$a_j(t) = \mathbf{e}_j^{\mathrm{T}}(\mathbf{x} - \overline{\mathbf{x}}), \tag{9.32}$$

that is, $a_j(t)$ is obtained by projecting the data anomaly vector $\mathbf{x} - \overline{\mathbf{x}}$ onto the eigenvector $\mathbf{e}_j$, as the right hand side of this equation is simply a dot product between the two vectors. Except for having the mean $\overline{\mathbf{x}}$ removed prior to performing PCA, the expansion coefficient $a_j(t)$ in (9.32) is the same as $z_j$ defined earlier in (9.2).

The nomenclature varies considerably in the literature: $a_j$ are called principal components, scores, temporal coefficients and amplitudes, while the eigenvectors $\mathbf{e}_j$ are also referred to as principal vectors, loadings, spatial patterns and EOFs (Empirical Orthogonal Functions). In this book, we prefer calling $a_j$ *principal components* (PCs), $\mathbf{e}_j$ eigenvectors or EOFs (and their elements $e_{ji}$ *loadings*), and $j$ the mode number.

Note that for time series data, $a_j$ is a function of time while $\mathbf{e}_j$ is a function of space, hence the names temporal coefficients and spatial patterns describe them well. However, in many cases, the dataset may not consist of time series. For instance, the dataset could be plankton collected from various oceanographic stations – $t$ then becomes the index for a station, while 'space' here could represent the various plankton species, and the data $\mathbf{x}(t) = [x_1(t), \ldots, x_m(t)]^{\mathrm{T}}$ could be the amount of species $1, \ldots, m$ found at station $t$. Another example comes from the multi-channel satellite image data, where images of the Earth's surface have been collected at several frequency channels. Here $t$ becomes the location index for a pixel in an image, and 'space' indicates the various frequency channels.

There are two important properties of PCAs. The expansion $\sum_{j=1}^{k} a_j(t)\mathbf{e}_j(\mathbf{x})$, with $k \leq m$, explains more of the variance of the data than any other linear combination $\sum_{j=1}^{k} b_j(t)\mathbf{f}_j(\mathbf{x})$. Thus, PCA provides the most efficient way to compress data linearly, using $k$ eigenvectors $\mathbf{e}_j$ and corresponding PCs $a_j$.

The second important property is that the PCs in the set $\{a_j\}$ are uncorrelated. We can write

$$a_j(t) = \mathbf{e}_j^{\mathrm{T}}(\mathbf{x} - \overline{\mathbf{x}}) = (\mathbf{x} - \overline{\mathbf{x}})^{\mathrm{T}}\mathbf{e}_j. \qquad (9.33)$$

For $i \neq j$,

$$\begin{aligned}
\mathrm{cov}(a_i, a_j) &= \mathrm{E}\big[\mathbf{e}_i^{\mathrm{T}}(\mathbf{x} - \overline{\mathbf{x}})(\mathbf{x} - \overline{\mathbf{x}})^{\mathrm{T}}\mathbf{e}_j\big] = \mathbf{e}_i^{\mathrm{T}}\,\mathrm{E}\big[(\mathbf{x} - \overline{\mathbf{x}})(\mathbf{x} - \overline{\mathbf{x}})^{\mathrm{T}}\big]\mathbf{e}_j \\
&= \mathbf{e}_i^{\mathrm{T}}\mathbf{C}\,\mathbf{e}_j = \mathbf{e}_i^{\mathrm{T}}\lambda_j\,\mathbf{e}_j = \lambda_j\mathbf{e}_i^{\mathrm{T}}\mathbf{e}_j = 0, \qquad (9.34)
\end{aligned}$$

implying zero correlation between $a_i(t)$ and $a_j(t)$. Thus, PCA extracts the uncorrelated modes of variability of the data field. Note that no correlation between $a_i(t)$ and $a_j(t)$ only means no linear relation between the two: there may still be non-linear relations between them, which can be extracted by non-linear PCA methods (Chapter 10).

When applying PCA to lat–lon gridded data over the globe, one should take into account the decrease in the area of a grid cell with latitude, as longitudinal lines converge towards the poles and the east–west grid distance shrinks in proportion to $\cos\phi$, with $\phi$ being the latitude. By scaling the variance from each grid cell by the area of the cell (which is proportional to $\cos\phi$), one can avoid having the anomalies in the higher latitudes overweighted in the PCA (North et al., 1982). This scaling of the variance can be accomplished simply by multiplying the anomaly $x_l - \overline{x}_l$ at the $l$th grid cell by the factor $(\cos\phi)^{1/2}$ for that cell.

### 9.1.5  PCA of the Tropical Pacific Climate Variability  Ⓐ☺

Let us illustrate the PCA technique with data from the tropical Pacific, a region renowned for the *El Niño* phenomenon (Neelin et al., 1998; McPhaden et al., 2006; A. J. Clarke, 2008; Sarachik and Cane, 2010; Timmermann et al., 2018). Every 2–10 years, a sudden warming of the coastal waters occurs off Peru. As the maximum warming occurs around Christmas, the local fishermen called this warming 'El Niño' (the Child in Spanish), after the Christ child. During normal times, the easterly equatorial winds drive surface waters offshore, forcing the cool, nutrient-rich, sub-surface waters to upwell, thereby replenishing the nutrients in the surface waters, hence the rich biological productivity and fisheries in the Peruvian waters (off the west coast of South America). During an El Niño, upwelling suddenly stops and the Peruvian fisheries crash. A major El Niño can bring a maximum anomalous warming of 5°C or more to the surface waters off Peru, where the term 'anomalous' means deviation from the climatological seasonal cycle. Sometimes the opposite of an El Niño develops, that is, anomalously cool waters appear in the equatorial Pacific, and this has been named '*La Niña*' (the girl in Spanish). Unlike El Niño episodes, which were documented as far back as 1726, La Niña episodes were not noticed until the last few decades, because its cool sea surface temperature (SST) anomalies are located much further offshore than the El Niño warm anomalies, and La Niña does not harm the Peruvian fisheries. The SST anomalies averaged over some regions (Niño 1+2, Niño 3, Niño 3.4, Niño 4, etc.) of the equatorial Pacific (Fig. 9.3) are shown in Fig. 9.4, where El Niño warm episodes and La Niña cool episodes can easily be seen. For instance, the very strong El Niño episodes of 1982–1983, 1997–1998 and 2015–2016 are manifested as sharp peaks and the strong La Niña of 1973–1974 and 1988–1989 as deep troughs in the Niño 3.4 SST anomaly time series, which is widely used as an index for the occurrence of El Niño/La Niña episodes.

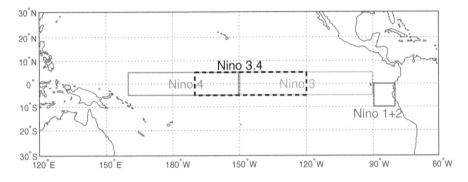

Figure 9.3 Regions of interest in the equatorial Pacific for sea surface temperature anomalies associated with the El Niño/La Niña phenomenon: Niño 1+2 (0°–10°S, 80°W–90°W), Niño 3 (5°S–5°N, 150°W–90°W), and Niño 4 (5°S–5°N, 160°E–150°W). Niño 3.4 (5°S-5°N, 170°W-120°W, marked by dashed box) straddles Niño 3 and Niño 4. SST anomalies averaged over each of these regions are used as climate indices.

Let us perform PCA on the monthly tropical Pacific SST from NOAA (Reynolds and T. M. Smith, 1994; T. M. Smith et al., 1996) for the period January 1950 to August 2000 (with the original 2° by 2° resolution data combined into 4° by 4° gridded data). The SST at each grid point had its climatological seasonal cycle removed to obtain the SST anomaly, then the time series was smoothed by a three-month moving average filter (Section 11.6.1). The SST anomaly field has two spatial dimensions, but can easily be rearranged into the form of $\mathbf{x}(t)$ for the analysis with PCA. The first six PCA modes account for 51.8%, 10.1%, 7.3%, 4.3%, 3.5% and 3.1%,[1] respectively, of the total SST variance. The spatial patterns (i.e. the eigenvectors) for the first three modes are shown in Fig. 9.5, and the associated PC time series in Fig. 9.6.

All three modes have their most intense variability located close to the equator (Fig. 9.5). Only until the fourth mode and beyond (not shown), do we find modes where their most intense variability occurs off equator. It turns out that

---

[1] The percentage of variance accounted for by mode $k$ is $\lambda_k / \left( \sum_{j=1}^{m} \lambda_j \right) \times 100\%$.

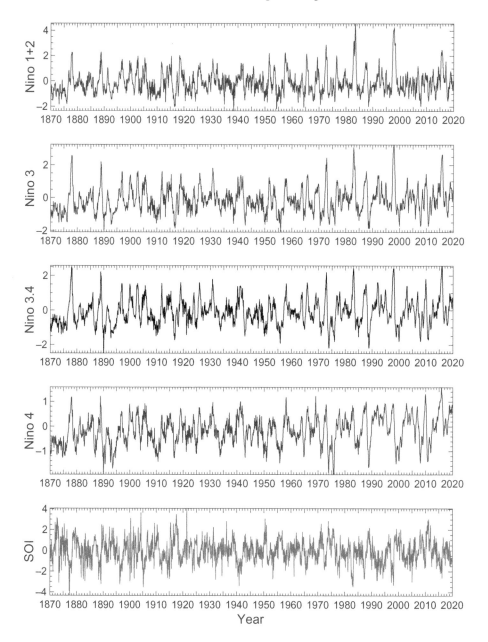

Figure 9.4 The monthly SST anomalies in Niño 1+2, Niño 3, Niño 3.4 and Niño 4 (in °C), and the monthly Southern Oscillation Index, SOI. During El Niño episodes, the SST rises in Niño 3 and Niño 3.4 (and less consistently in Niño 1+2), while the SOI drops. The reverse occurs during a La Niña episode. The grid mark for a year marks the January of that year. [Data source: Climate Research Unit, University of East Anglia.]

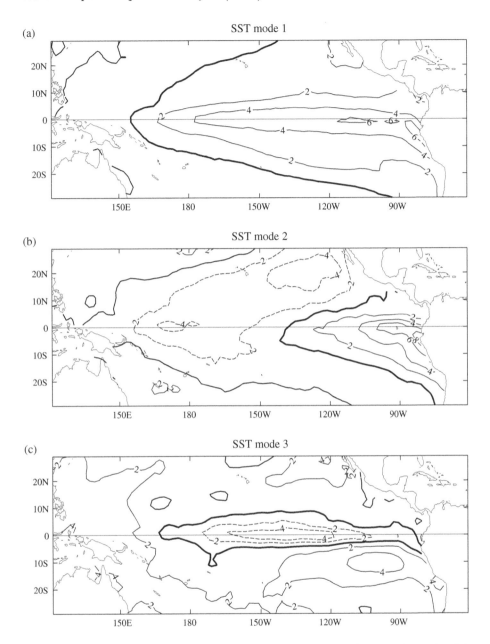

Figure 9.5 The spatial patterns (i.e. eigenvectors or EOFs) of PCA modes (a) 1, (b) 2 and (c) 3 for the SST anomalies. Positive contours are indicated by the solid curves, negative contours by dashed curves and the zero contour by the thick solid curve. The contour unit is 0.01°C. The eigenvectors have been normalized to unit norm. [Reproduced from Hsieh (2001b, figure 7).]

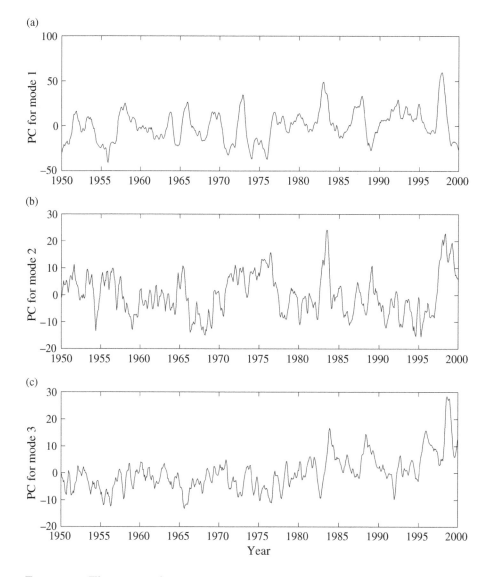

Figure 9.6 The principal component time series for the SST anomaly modes (a) 1, (b) 2 and (c) 3. [Source: Hsieh (2009)]

the first three modes are all related to the El Niño/La Niña phenomenon. That it takes at least three modes to represent this phenomenon accurately is a result of the limitations of a linear approach like PCA – later we will see how a single non-linear PCA mode by neural network modelling can accurately represent the El Niño/La Niña oscillation (Section 10.6).

Mode 1 (Fig. 9.5(a)) shows the largest SST anomalies occurring in the eastern and central equatorial Pacific. The mode 1 PC (Fig. 9.6) closely resembles the Niño 3.4 index (Fig. 9.4). The SST anomalies associated with mode 1 at a given time are the product of the PC $a_1(t)$ and the spatial EOF pattern $\mathbf{e}_1$ in (9.29). Thus, the large positive value of PC mode 1 during 1997–1998 would have mode 1 contributing positive SST anomalies in the eastern-central equatorial Pacific, while the negative value of PC mode 1 during 1988–1989 would have mode 1 contributing negative SST anomalies in the same region.

Mode 2 (Fig. 9.5(b)) has, along the equator, positive anomalies near the east and negative anomalies further west. Its PC (Fig. 9.6(b)) shows positive values during both El Niño and La Niña episodes. Mode 3 (Fig. 9.5(c)) shows the largest anomaly occurring in the central equatorial Pacific, and its PC (Fig. 9.6(c)) shows a rising trend after the mid-1970s.

Since the late nineteenth century, it has been known that the normal high air pressure (at sea level) in the eastern equatorial Pacific and the low pressure in the western equatorial Pacific and Indian Ocean may undergo a seesaw oscillation once every 2–10 years. The '*Southern Oscillation*' (termed by Sir Gilbert Walker in the 1920s) is the east–west seesaw oscillation in the sea level pressure (SLP) centred in the equatorial Pacific. The SLP of Tahiti (in the eastern equatorial Pacific) minus that of Darwin (in the western equatorial Pacific/Indian Ocean domain) is commonly called the *Southern Oscillation Index* (SOI). Clearly the SOI is negatively correlated with the Niño 3.4 SST index (Fig. 9.4), that is, when an El Niño occurs, the SLP of the eastern equatorial Pacific drops relative to the SLP of the western equatorial Pacific/Indian Ocean domain. By the mid-1960s, El Niño, the oceanographic phenomenon, has been found to be strongly connected to the Southern Oscillation, the atmospheric phenomenon, and the coupled phenomenon named the *El Niño-Southern Oscillation* (ENSO) (Neelin et al., 1998; McPhaden et al., 2006; A. J. Clarke, 2008; Sarachik and Cane, 2010; Timmermann et al., 2018). As ENSO exerts a strong influence on global climate variability, ENSO has spurred the development of seasonal climate prediction (Goddard et al., 2001).

Let us also consider the tropical Pacific monthly SLP data from COADS (Comprehensive Ocean-Atmosphere Data Set) (Woodruff et al., 1987) during January 1950 to August 2000. The 2° by 2° resolution data were combined into 10° longitude by 4° latitude gridded data, with climatological seasonal cycle removed (to give the SLP anomalies), and the anomalies smoothed by a three-month moving average. PCA of the data resulted in the first six modes accounting for 29.5%, 16.5%, 4.8%, 3.4%, 2.5% and 2.2%, respectively, of the total variance. The first three spatial modes (Fig. 9.7) and their associated PCs (Fig. 9.8) are also shown. The first mode describes the east–west seesaw in the SLP anomalies associated with the Southern Oscillation (Fig. 9.7(a)).

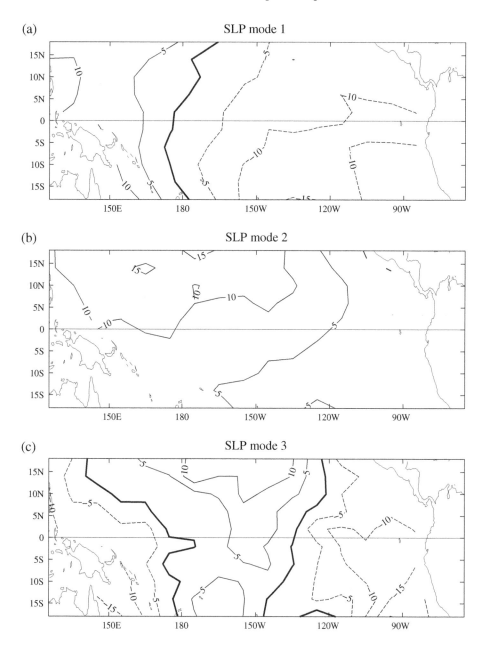

Figure 9.7 The spatial patterns of PCA modes (a) 1, (b) 2 and (c) 3 for the SLP anomalies. The contour unit is 0.01 hPa. Positive contours are indicated by the solid curves, negative contours by dashed curves and the zero contour by the thick solid curve. [Source: Hsieh (2009)]

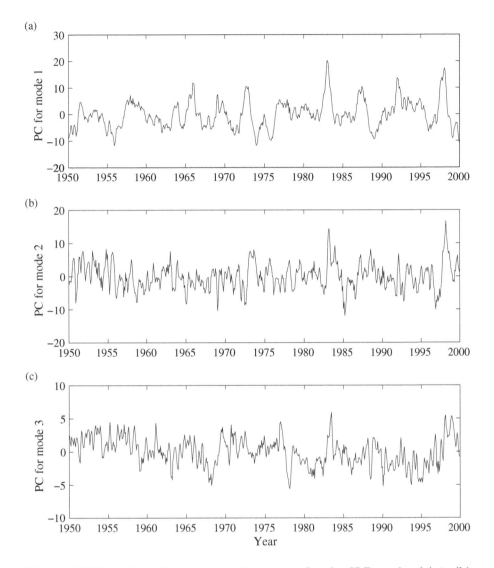

Figure 9.8 The principal component time series for the SLP modes (a) 1, (b) 2 and (c) 3. PC for mode 1 is strongly correlated with the SST mode 1 PC in Fig. 9.6(a). [Source: Hsieh (2009)]

## 9.1.6  Scaling the PCs and Eigenvectors  [A]☺

There are various options for scaling the PCs $\{a_j(t)\}$ and the eigenvectors $\{\mathbf{e}_j\}$. One can introduce an arbitrary scale factor $\alpha$,

$$a'_j = \frac{1}{\alpha}\, a_j, \qquad \mathbf{e}'_j = \alpha\, \mathbf{e}_j, \tag{9.35}$$

so that (9.29) becomes

$$\mathbf{x} - \bar{\mathbf{x}} = \sum_{j=1}^{m} a'_j \mathbf{e}'_j. \tag{9.36}$$

Thus, $a_j(t)$ and $\mathbf{e}_j$ are defined only up to an arbitrary scale factor. With $\alpha = -1$, one reverses the sign of both $a_j(t)$ and $\mathbf{e}_j$, which is often done to make them more interpretable.

Our choice for the scaling has so far been

$$\mathbf{e}_i^{\mathrm{T}} \mathbf{e}_j = \delta_{ij}, \tag{9.37}$$

which was the choice of Lorenz (1956), who called his eigenvectors $\{\mathbf{e}_i\}$ with unit norm 'empirical orthogonal functions' (EOF). Under this choice of scaling, the variance of $\mathbf{x}$ is contained in $\{a_j(t)\}$:

$$\mathrm{E}\big[a_j^2\big] = \mathrm{E}\big[\mathbf{e}_j^{\mathrm{T}}(\mathbf{x} - \bar{\mathbf{x}})(\mathbf{x} - \bar{\mathbf{x}})^{\mathrm{T}} \mathbf{e}_j\big] = \mathbf{e}_j^{\mathrm{T}} \mathrm{E}\big[(\mathbf{x} - \bar{\mathbf{x}})(\mathbf{x} - \bar{\mathbf{x}})^{\mathrm{T}}\big] \mathbf{e}_j$$
$$= \mathbf{e}_j^{\mathrm{T}} \mathbf{C}\, \mathbf{e}_j = \mathbf{e}_j^{\mathrm{T}} \lambda_j\, \mathbf{e}_j = \lambda_j \mathbf{e}_j^{\mathrm{T}} \mathbf{e}_j = \lambda_j, \tag{9.38}$$

hence

$$\mathrm{var}(\mathbf{x}) = \sum_{j=1}^{m} \lambda_j = \sum_{j=1}^{m} \mathrm{E}\big[a_j^2\big]. \tag{9.39}$$

Another common choice is Hotelling's original scaling

$$a'_j = \frac{1}{\sqrt{\lambda_j}}\, a_j, \qquad \mathbf{e}'_j = \sqrt{\lambda_j}\, \mathbf{e}_j, \tag{9.40}$$

from which

$$\|\mathbf{e}'_j\|^2 = \lambda_j, \tag{9.41}$$

$$\mathrm{var}(\mathbf{x}) = \sum_{j=1}^{m} \lambda_j = \sum_{j=1}^{m} \|\mathbf{e}'_j\|^2, \tag{9.42}$$

$$\mathrm{cov}(a'_i, a'_j) = \delta_{ij}. \tag{9.43}$$

The variance $\mathrm{var}(\mathbf{x})$ is now contained in $\{\mathbf{e}'_j\}$ instead of in $\{a_j\}$, as seen in (9.39). In sum, regardless of the arbitrary scale factor, the PCA eigenvectors are orthogonal and the PCs are uncorrelated.

If PCA is performed using the data correlation matrix instead of the data covariance matrix, which is equivalent to performing PCA on the standardized variables $\tilde{x}_l$, that is, $x_l$ with mean removed and division by standard deviation, then one can show that, under Hotelling's scaling, the Pearson correlation

$$\rho\big(a'_j(t), \tilde{x}_l(t)\big) = e'_{jl}, \tag{9.44}$$

the $l$th element of $\mathbf{e}'_j$ (Jolliffe, 2002, p. 25). Hence the $l$th element of $\mathbf{e}'_j$ conveniently provides the correlation between the PC $a'_j$ and the standardized variable $\tilde{x}_l$, which is a reason why Hotelling's scaling (9.40) is also widely used.

### 9.1.7 Degeneracy of Eigenvalues $\boxed{\text{A}}$☺

A degenerate case arises when $\lambda_i = \lambda_j$, $(i \neq j)$. When two eigenvalues are equal, their eigenspace is 2-D, that is, a plane in which any two orthogonal vectors can be chosen as the eigenvectors – hence the eigenvectors are not unique. If $l$ eigenvalues are equal, $l$ non-unique orthogonal vectors can be chosen in the $l$-dimensional eigenspace.

A simple example of degeneracy is illustrated by a propagating plane wave in the 2-D $x$-$y$ plane,

$$h(x, y, t) = A \cos(kx - \omega t), \tag{9.45}$$

which can be expressed in terms of two standing waves:

$$h = A \cos(kx) \cos(\omega t) + A \sin(kx) \sin(\omega t). \tag{9.46}$$

If we perform PCA on $h(x, y, t)$, we get two modes with equal eigenvalues. To see this, note that in the $x$-$y$ plane, $\cos(kx)$ and $\sin(kx)$ are orthogonal, while $\cos(\omega t)$ and $\sin(\omega t)$ are uncorrelated, so (9.46) satisfies the properties of PCA modes in that the eigenvectors are orthogonal and the PCs are uncorrelated. Equation (9.46) is a PCA decomposition, with the two modes having the same amplitude $A$, hence the eigenvalues $\lambda_1 = \lambda_2$, and the case is degenerate. Thus, propagating waves in the data lead to degeneracy in the eigenvalues. If one finds eigenvalues of very similar magnitudes from a PCA analysis, that implies near degeneracy and there may be propagating waves in the data. In reality, noise in the data usually precludes $\lambda_1 = \lambda_2$ exactly. Nevertheless, when $\lambda_1 \approx \lambda_2$, the near degeneracy causes the eigenvectors to be rather poorly defined (i.e. very sensitive to noise in the data) (North et al., 1982).

### 9.1.8 A Smaller Covariance Matrix $\boxed{\text{A}}$☺

Let the data matrix be

$$\mathbf{X} = \begin{bmatrix} x_{11} & \cdots & x_{1m} \\ \vdots & \ddots & \vdots \\ x_{n1} & \cdots & x_{nm} \end{bmatrix}, \tag{9.47}$$

where $m$ is the number of spatial points and $n$ the number of time points. Assuming

$$\frac{1}{n} \sum_{i=1}^{n} x_{ij} = 0, \tag{9.48}$$

that is, the temporal mean has been subtracted from the data, then the data covariance matrix

$$\mathbf{C} = \frac{1}{n}\mathbf{X}^{\mathrm{T}}\mathbf{X} \tag{9.49}$$

is an $m \times m$ matrix. The theory of singular value decomposition (SVD) (see Section 9.1.11) tells us that the non-zero eigenvalues of $\mathbf{X}^{\mathrm{T}}\mathbf{X}$ (an $m \times m$ matrix) are exactly the non-zero eigenvalues of $\mathbf{XX}^{\mathrm{T}}$ (an $n \times n$ matrix).

In most problems, the size of the two matrices can be very different. For instance, for global $5° \times 5°$ monthly sea level pressure data collected over 70 years, the total number of spatial grid points is $m = 2{,}592$, while the number of time points is $n = 840$. Obviously, it will be much easier to solve the eigen problem for the $840 \times 840$ matrix than that for the $2{,}592 \times 2{,}592$ matrix.

Hence, when $n < m$, considerable computational savings can be made by first finding the eigenvalues $\{\lambda_j\}$ and eigenvectors $\{\mathbf{v}_j\}$ for the alternative smaller covariance matrix

$$\mathbf{C}' = \frac{1}{n}\mathbf{XX}^{\mathrm{T}}, \tag{9.50}$$

that is, solve the eigen problem

$$\left(\frac{1}{n}\mathbf{XX}^{\mathrm{T}}\right)\mathbf{v}_j = \lambda_j\mathbf{v}_j. \tag{9.51}$$

Left multiplying the equation by $\mathbf{X}^{\mathrm{T}}$ yields

$$\mathbf{X}^{\mathrm{T}}\frac{1}{n}\mathbf{XX}^{\mathrm{T}}\mathbf{v}_j = \mathbf{X}^{\mathrm{T}}\lambda_j\mathbf{v}_j,$$

$$\left(\frac{1}{n}\mathbf{X}^{\mathrm{T}}\mathbf{X}\right)(\mathbf{X}^{\mathrm{T}}\mathbf{v}_j) = \lambda_j(\mathbf{X}^{\mathrm{T}}\mathbf{v}_j). \tag{9.52}$$

This equation is easily seen to be of the form

$$\mathbf{C}\,\mathbf{e}_j = \lambda_j\mathbf{e}_j, \tag{9.53}$$

with $\mathbf{C}$ given by (9.49) and

$$\mathbf{e}_j = \mathbf{X}^{\mathrm{T}}\mathbf{v}_j, \tag{9.54}$$

which means $\mathbf{e}_j$ is an eigenvector for $\mathbf{C}$ corresponding to the eigenvalue $\lambda_j$. In summary, solving the eigen problem for the smaller matrix $\mathbf{C}'$ yields the eigenvalues $\{\lambda_j\}$ and eigenvectors $\{\mathbf{v}_j\}$. The eigenvectors $\{\mathbf{e}_j\}$ for the bigger matrix $\mathbf{C}$ are then easily obtained from (9.54). This trick is often used to allow large computational savings when performing PCA.

## 9.1.9   How Many Modes to Retain   B ☺

One of the most common uses of PCA is for *dimensionality reduction* – that is, the higher PCA modes, which basically contain noise, are rejected to reduce the dimension of the dataset. How does one decide how many modes to retain? There are some 'rules of thumb'. One of the simplest approaches is to plot the variance from each mode (i.e. eigenvalues $\lambda_j$) as a function of the mode number $j$. Usually, from the graph (called a *scree graph*), as the mode number increases, one can see the initial steep decline in the variance with mode number eventually changing somewhat abruptly to a more gradual decline, which is regarded as the noise region. One can then retain the first $k$ modes just before reaching the region of gradual decline.

Alternatively, a very simple test is the Kaiser test, where modes with eigenvalues $\lambda$ less than the mean value $\overline{\lambda}$ are ignored. Jolliffe (1972) found that a cut-off level of $\overline{\lambda}$ was too high in practice, and suggested using $0.7\overline{\lambda}$ instead (Jolliffe, 2002, section 6.1.2).

Computationally more involved is the Monte Carlo test or 'Rule N' (Preisendorfer, 1988, section 5d), which involves setting up random data matrices $\mathbf{R}_l$ ($l = 1, \ldots, L$) of the same size as the data matrix $\mathbf{X}$. The random elements are Gaussian distributed, with the variance of the random data matching the variance of the actual data. Principal component analysis is performed on each of the random matrices, yielding eigenvalues $\{\lambda_j^{(l)}\}$. For each $j$, one examines the distribution of the $L$ values of $\lambda_j^{(l)}$ and finds the 95th percentile $\lambda_{0.95}(j)$, which is exceeded by only 5% of the $\lambda_j^{(l)}$ values. Modes with $\lambda_j$ that fail to rise above the $\lambda_{0.95}(j)$ level are then rejected.

One problem with this approach is that if the first mode accounts for a very large percentage of the variance, the small amount of remaining variance will not be able to allow $\lambda_2$ to exceed $\lambda_{0.95}(2)$. Instead, one should redo the Monte Carlo runs with variance of the random Gaussian data matching the observed variance minus the variance of the first mode (Jolliffe, 2002, section 6.1.7). If $\lambda_2$ is above the new $\lambda_{0.95}(1)$ level, mode 2 is accepted. One would then repeat the Monte Carlo runs for mode 3 by having the variance of the random data matching the original variance minus the variance from mode 1 and mode 2, and check if $\lambda_3$ is above the new $\lambda_{0.95}(1)$ level.

A weakness of this approach is the use of random Gaussian data, which is not appropriate if the observed data are non-Gaussian (e.g. precipitation amount). One can instead perform the Monte Carlo runs using bootstrap resampled data (Section 4.8) from the observed data.

If the data have strong serial correlation, the dimension of $\mathbf{R}_l$ should be reduced, with the effective sample size $n_{\text{eff}}$ (Section 2.11.2) replacing the sample size $n$.

Since the Monte Carlo method performs PCA on $L$ matrices and $L$ is typically about 100–1000, it can be costly for large data matrices. Hence, asymptotic methods based on the central limit theorem are often used in the case of large data matrices (Mardia et al., 1979, pp. 230–237; Preisendorfer, 1988, pp. 204–206). Other rules for deciding how many modes to retain are given in Jolliffe (2002, chapter 6).

## 9.1.10   Temporal and Spatial Mean Removal  B☺

Given a data matrix $\mathbf{X}$ as in (9.47), what type of mean are we trying to remove from the data? So far, we have removed the temporal mean, that is, the average of the $j$th column, from each datum $x_{ij}$. We could instead have removed the spatial mean, that is, the average of the $i$th row, from each datum $x_{ij}$.

Which type of mean should be removed is very dependent on the type of data one has. For most applications, one removes the temporal mean. However, for satellite-sensed sea surface temperature data, the precision is much better than the accuracy. Also, the subsequent satellite image may be collected by a different satellite, which would have different systematic errors. Therefore, it may be more appropriate to subtract the spatial mean of an image from each pixel.

It is also possible to remove both the temporal and spatial means, by subtracting the average of the $j$th column and then the average of the $i$th row from each datum $x_{ij}$.

## 9.1.11   Singular Value Decomposition  B☹

Instead of solving the eigen problem of the data covariance matrix $\mathbf{C}$, a computationally more efficient way to perform PCA is via *singular value decomposition* (SVD) of $\mathbf{X}^{\mathrm{T}}$, the transpose of the $n \times m$ data matrix $\mathbf{X}$ given in (9.47) (Kelly, 1988). Without loss of generality, we can assume $m \geq n$, then the SVD theorem (Strang, 2005) says that

$$
\mathbf{X}^{\mathrm{T}} = \mathbf{ESF}^{\mathrm{T}} = 
\begin{array}{c}
\overset{\mathbf{E}}{\left[\begin{array}{c|c}
\underset{m \times n}{\mathbf{E}'} & 0 \\
\end{array}\right]}
\end{array}
\begin{array}{c}
\overset{\mathbf{S}}{\left[\begin{array}{c}
\underset{n \times n}{\mathbf{S}'} \\ \hline 0 \\
\end{array}\right]}
\end{array}
\begin{array}{c}
\overset{\mathbf{F}^{\mathrm{T}}}{\left[\phantom{xx}\right]}_{n \times n}
\end{array}.
\tag{9.55}
$$

The $m \times m$ matrix $\mathbf{E}$ contains an $m \times n$ sub-matrix $\mathbf{E}'$ – and if $m > n$, some zero column vectors. The $m \times n$ matrix $\mathbf{S}$ contains the diagonal $n \times n$ sub-matrix $\mathbf{S}'$, and possibly some zero row vectors. $\mathbf{F}^{\mathrm{T}}$ is an $n \times n$ matrix. [If $m < n$, one can apply the above decomposition to $\mathbf{X}$ instead of $\mathbf{X}^{\mathrm{T}}$ and interchange the indices $m$ and $n$ in (9.55)].

$\mathbf{E}$ and $\mathbf{F}$ are *orthonormal matrices*, that is, they satisfy

$$
\mathbf{E}^{\mathrm{T}}\mathbf{E} = \mathbf{I}, \qquad \mathbf{F}^{\mathrm{T}}\mathbf{F} = \mathbf{I},
\tag{9.56}
$$

where $\mathbf{I}$ is the identity matrix. The leftmost $n$ columns of $\mathbf{E}$ contain the $n$ *left singular vectors*, and the columns of $\mathbf{F}$ the $n$ *right singular vectors*, while the diagonal elements of $\mathbf{S}'$ are the *singular values*.

The covariance matrix $\mathbf{C}$ can be rewritten as

$$
\mathbf{C} = \frac{1}{n}\mathbf{X}^{\mathrm{T}}\mathbf{X} = \frac{1}{n}\mathbf{ESS}^{\mathrm{T}}\mathbf{E}^{\mathrm{T}},
\tag{9.57}
$$

where (9.55) and (9.56) have been invoked. The matrix

$$\mathbf{SS}^{\mathrm{T}} \equiv \Lambda, \tag{9.58}$$

is diagonal and zero everywhere, except in the upper left $n \times n$ corner, containing $\mathbf{S}'^2$.

Right multiplying (9.57) by $n\mathbf{E}$ gives

$$n\mathbf{CE} = \mathbf{E}\Lambda, \tag{9.59}$$

where (9.58) and (9.56) have been invoked, and $\Lambda$ contains the eigenvalues for the matrix $n\mathbf{C}$. Instead of solving the eigen problem (9.59), we use SVD to get $\mathbf{E}$ from $\mathbf{X}^{\mathrm{T}}$ by (9.55). Equation (9.59) implies that there are only $n$ eigenvalues in $\Lambda$ from $\mathbf{S}'^2$, and the eigenvalues = (singular values)$^2$. As (9.59) and (9.53) are equivalent except for the constant $n$, the eigenvalues in $\Lambda$ are simply $n\lambda_j$, with $\lambda_j$ the eigenvalues from (9.53).

Similarly, for the other covariance matrix

$$\mathbf{C}' = \frac{1}{n}\mathbf{XX}^{\mathrm{T}}, \tag{9.60}$$

we can rewrite it as

$$\mathbf{C}' = \frac{1}{n}\mathbf{FS}'^2\mathbf{F}^{\mathrm{T}}, \tag{9.61}$$

and, ultimately,

$$n\mathbf{C}'\mathbf{F} = \mathbf{FS}'^2 = \mathbf{F}\Lambda. \tag{9.62}$$

Hence, the eigen problem (9.62) has the same eigenvalues as (9.59).

The PCA decomposition (assuming $\bar{\mathbf{x}} = 0$)

$$\mathbf{x}(t) = \sum_{j=1}^{m} \mathbf{e}_j a_j(t), \tag{9.63}$$

where $\mathbf{x}$ is a column vector with $m$ elements, is equivalent to the matrix form

$$\mathbf{X} = \mathbf{AE}^{\mathrm{T}} = \sum_{j=1}^{m} \mathbf{a}_j \mathbf{e}_j^{\mathrm{T}}, \tag{9.64}$$

where $\mathbf{X}$ is an $n \times m$ matrix, the eigenvector $\mathbf{e}_j$ is the $j$th column in the matrix $\mathbf{E}$ and the PC $a_j(t)$ is the vector $\mathbf{a}_j$, the $j$th column in the matrix $\mathbf{A}$. From (9.64) and the transpose of (9.55), we obtain

$$\mathbf{A} = \mathbf{FS}^{\mathrm{T}}. \tag{9.65}$$

Thus, by the SVD (9.55), we obtain the eigenvectors $\mathbf{e}_j$ from $\mathbf{E}$, and the PCs $a_j(t)$ from $\mathbf{A}$ in (9.65). We can also right multiply (9.64) by $\mathbf{E}$, and invoke (9.56) to get

$$\mathbf{A} = \mathbf{XE}. \tag{9.66}$$

There are many computing packages (e.g. MATLAB) with standard codes for performing SVD. Kelly (1988) pointed out that the SVD approach to PCA is at least twice as fast as the eigen approach, as SVD requires $O(mn^2)$ operations to compute, while the eigen approach requires $O(mn^2)$ to compute the smaller of $\mathbf{C}$ or $\mathbf{C}'$, then $O(n^3)$ to solve the eigen problem, and then $O(mn^2)$ to get the PCs.

## 9.1.12 Missing Data ⓒ☺

Missing data produce gaps in data records. Estimating values for the missing data is called *imputation*. If the gaps are small, one can interpolate the missing values using neighbouring data. If the gaps are not small, then instead of

$$\mathbf{C} = \frac{1}{n}\mathbf{X}^{\mathrm{T}}\mathbf{X}, \tag{9.67}$$

(assuming that the means have been removed from the data), one computes

$$c_{kl} = \frac{1}{n'}\sum_i{}' x_{ki}\, x_{il}, \tag{9.68}$$

where the prime denotes that the summation is only over $i$ with neither $x_{ki}$ nor $x_{il}$ missing – with a total of $n'$ terms in the summation. The eigenvectors $\mathbf{e}_j$ can then be obtained from this new covariance matrix. The principal components $a_j$ in (9.33) cannot be computed from

$$a_j(t_l) = \sum_i e_{ji}\, x_{il}, \tag{9.69}$$

as some values of $x_{il}$ are missing. Instead, $a_j$ is estimated (von Storch and Zwiers, 1999, section 13.2.8) as a least squares solution to minimizing $\mathrm{E}[\|\mathbf{x} - \sum a_j\mathbf{e}_j\|^2]$, that is,

$$a_j(t_l) = \frac{\sum_i' e_{ji}\, x_{il}}{\sum_i' |e_{ji}|^2}, \tag{9.70}$$

where for a given value of $l$, the superscript prime means that the summations are only over $i$ for which $x_{il}$ is not missing.

PCA is also used to supply missing data. With climate data, one often finds the earlier parts of the data records to be sparse. Suppose the data record can be divided into two parts: $\mathbf{X}$, which contains no missing values and $\tilde{\mathbf{X}}$, which contains missing values. From (9.64), PCA applied to $\mathbf{X}$ yields $\mathbf{E}$, which contains the eigenvectors $\mathbf{e}_j$. The PCs for $\tilde{\mathbf{X}}$ are then computed from (9.70)

$$\tilde{a}_j(t_l) = \frac{\sum_i' e_{ji}\, \tilde{x}_{il}}{\sum_i' |e_{ji}|^2}. \tag{9.71}$$

The missing values in $\tilde{\mathbf{X}}$ are filled in $\tilde{\mathbf{X}}'$, where

$$\tilde{\mathbf{X}}' = \tilde{\mathbf{A}}\mathbf{E}^{\mathrm{T}}, \tag{9.72}$$

where the $j$th column of $\tilde{\mathbf{A}}$ is given by $\tilde{a}_j(t_l)$. More sophisticated interpolation of missing data by PCA is described by A. Kaplan et al. (2000) and Jolliffe (2002, section 13.6). Missing data imputation has also been performed using non-linear ML methods such as NN and support vector machines (Richman, Trafalis, et al., 2009) and more recently using deep NN (i.e. deep learning) models (Section 15.3).

## 9.2   Rotated PCA   B☺

In PCA, the linear mode that accounts for the most variance of the dataset is sought. However, as illustrated in Fig. 9.9, the resulting eigenvectors may not

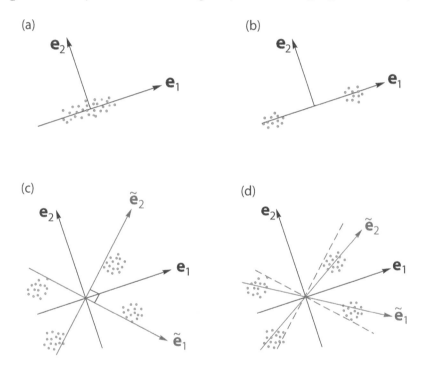

Figure 9.9 The case of PCA applied to a dataset composed of (a) a single cluster, (b) two clusters, (c) three and (d) four clusters. In (c), an *orthonormal rotation* has yielded rotated eigenvectors $\tilde{\mathbf{e}}_j$, $(j = 1, 2)$, which pass much closer to the data clusters than the unrotated eigenvectors $\mathbf{e}_j$. In (d), an *oblique rotation* is used instead of an orthonormal rotation to spear through the data clusters, while the dashed lines indicate the orthonormally rotated eigenvectors. Eigenvectors that failed to approach any data clusters generally bear little resemblance to physical states. [Follows Preisendorfer (1988, figure 7.3).]

align closely to local data clusters, so the eigenvectors may not represent actual physical states well. Rotated PCA (RPCA) methods can rotate the PCA eigenvectors so that they point closer to the local clusters of data points. Thus, the rotated eigenvectors may bear greater resemblance to actual physical states than the unrotated eigenvectors (Section 9.2.1). There is an alternative approach to RPCA, where rotation can be used to find more focused or localized patterns in the loadings (i.e. eigenvectors) (Section 9.2.2). Rotated PCA, also called rotated EOF analysis, is a more general but also more subjective technique than PCA.

Let us quickly review the rotation of vectors and matrices. Given a matrix $\mathbf{P}$ composed of the column vectors $\mathbf{p}_1, \ldots, \mathbf{p}_m$, and a matrix $\mathbf{Q}$ containing the column vectors $\mathbf{q}_1, \ldots, \mathbf{q}_m$, $\mathbf{P}$ can be transformed into $\mathbf{Q}$ by $\mathbf{Q} = \mathbf{PR}$, that is,

$$q_{il} = \sum_{j=1}^{m} p_{ij} r_{jl}, \tag{9.73}$$

where $\mathbf{R}$ is an $m \times m$ rotation matrix with elements $r_{jl}$. When $\mathbf{R}$ is orthonormal, that is,

$$\mathbf{R}^{\mathrm{T}} \mathbf{R} = \mathbf{I}, \tag{9.74}$$

the rotation is called an *orthonormal rotation*. Clearly,

$$\mathbf{R}^{-1} = \mathbf{R}^{\mathrm{T}}, \tag{9.75}$$

for an orthonormal rotation.

Earlier in this chapter, we already encountered a simple orthonormal rotation when rotating the coordinate axes by an angle $\theta$ in Fig. 9.2. Equation (9.1) can be rewritten as

$$\mathbf{z} = \mathbf{R}^{\mathrm{T}} \mathbf{x}, \text{ or } \mathbf{z}^{\mathrm{T}} = \mathbf{x}^{\mathrm{T}} \mathbf{R}, \tag{9.76}$$

where

$$\mathbf{z} = \begin{bmatrix} z_1 \\ z_2 \end{bmatrix}, \quad \mathbf{R}^{\mathrm{T}} = \begin{bmatrix} \cos\theta & \sin\theta \\ -\sin\theta & \cos\theta \end{bmatrix}, \text{ and } \mathbf{x} = \begin{bmatrix} x_1 \\ x_2 \end{bmatrix}. \tag{9.77}$$

The reader can easily verify that the rotation matrix $\mathbf{R}$ indeed satisfies the orthonormal condition (9.74).

Given the data matrix $\mathbf{X}$ in (9.64), we can rewrite it as follows:

$$\mathbf{X} = \mathbf{A}\mathbf{E}^{\mathrm{T}} = \mathbf{A}\mathbf{R}\mathbf{R}^{-1}\mathbf{E}^{\mathrm{T}} = \tilde{\mathbf{A}}\tilde{\mathbf{E}}^{\mathrm{T}}, \tag{9.78}$$

with

$$\tilde{\mathbf{A}} = \mathbf{A}\mathbf{R}, \tag{9.79}$$

and

$$\tilde{\mathbf{E}}^{\mathrm{T}} = \mathbf{R}^{-1}\mathbf{E}^{\mathrm{T}}. \tag{9.80}$$

If $\mathbf{R}$ is orthonormal, substituting (9.75) into (9.80) and taking the transpose gives

$$\tilde{\mathbf{E}} = \mathbf{E}\mathbf{R}. \tag{9.81}$$

Note that $\mathbf{E}$ has been rotated into $\tilde{\mathbf{E}}$, and $\mathbf{A}$ into $\tilde{\mathbf{A}}$.

To see the orthogonality properties of the rotated eigenvectors, we note that

$$\tilde{\mathbf{E}}^\mathrm{T}\tilde{\mathbf{E}} = \mathbf{R}^\mathrm{T}\mathbf{E}^\mathrm{T}\mathbf{E}\mathbf{R} = \mathbf{R}^\mathrm{T}\mathbf{D}\mathbf{R}, \tag{9.82}$$

where the diagonal matrix $\mathbf{D}$ is

$$\mathbf{D} = \mathrm{diag}(\mathbf{e}_1^\mathrm{T}\mathbf{e}_1, \ldots, \mathbf{e}_m^\mathrm{T}\mathbf{e}_m). \tag{9.83}$$

If $\mathbf{e}_j^\mathrm{T}\mathbf{e}_j = 1$, for all $j$, then $\mathbf{D} = \mathbf{I}$ and (9.82) reduces to

$$\tilde{\mathbf{E}}^\mathrm{T}\tilde{\mathbf{E}} = \mathbf{R}^\mathrm{T}\mathbf{R} = \mathbf{I}, \tag{9.84}$$

which means the $\{\tilde{\mathbf{e}}_j\}$ are orthonormal. Hence, the rotated eigenvectors $\{\tilde{\mathbf{e}}_j\}$ are orthonormal only if the original eigenvectors $\{\mathbf{e}_j\}$ are orthonormal. If $\{\mathbf{e}_j\}$ are orthogonal but not orthonormal, then $\{\tilde{\mathbf{e}}_j\}$ are in general not orthogonal.

From (9.34), the PCs $\{a_j(t_l)\}$ are uncorrelated, that is, their covariance matrix is diagonal,

$$\mathbf{C}_{\mathbf{AA}} = \mathrm{diag}(\alpha_1^2, \ldots, \alpha_m^2), \tag{9.85}$$

where the diagonal elements are given by

$$\mathbf{a}_j^\mathrm{T}\mathbf{a}_j = \alpha_j^2. \tag{9.86}$$

With the rotated PCs, their covariance matrix is

$$\begin{aligned}
\mathbf{C}_{\tilde{\mathbf{A}}\tilde{\mathbf{A}}} &= \frac{1}{n}\tilde{\mathbf{A}}^\mathrm{T}\tilde{\mathbf{A}} = \frac{1}{n}\mathbf{R}^\mathrm{T}\mathbf{A}^\mathrm{T}\mathbf{A}\mathbf{R} \\
&= \mathbf{R}^\mathrm{T}\frac{1}{n}\mathbf{A}^\mathrm{T}\mathbf{A}\,\mathbf{R} = \mathbf{R}^\mathrm{T}\,\mathbf{C}_{\mathbf{AA}}\,\mathbf{R}.
\end{aligned} \tag{9.87}$$

For orthonormal rotations, if $\mathbf{C}_{\mathbf{AA}} = \mathbf{I}$, that is, $\mathbf{a}_j^\mathrm{T}\mathbf{a}_j = 1$, for all $j$,

$$\mathbf{C}_{\tilde{\mathbf{A}}\tilde{\mathbf{A}}} = \mathbf{R}^\mathrm{T}\mathbf{I}\mathbf{R} = \mathbf{I}, \tag{9.88}$$

that is, $\{\tilde{\mathbf{a}}_j\}$ are uncorrelated.

To summarize, there are two cases:

Case a: If we choose $\mathbf{e}_j^\mathrm{T}\mathbf{e}_j = 1$ for all $j$, then we cannot have $\mathbf{a}_j^\mathrm{T}\mathbf{a}_j = 1$ for all $j$. This implies that $\{\tilde{\mathbf{a}}_j\}$ are not uncorrelated, but $\{\tilde{\mathbf{e}}_j\}$ are orthonormal.

Case b: If we choose $\mathbf{a}_j^\mathrm{T}\mathbf{a}_j = 1$ for all $j$, then we cannot have $\mathbf{e}_j^\mathrm{T}\mathbf{e}_j = 1$ for all $j$. This implies that $\{\tilde{\mathbf{a}}_j\}$ are uncorrelated, but $\{\tilde{\mathbf{e}}_j\}$ are not orthonormal.

Thus, there is a notable difference between PCA and RPCA: PCA can have both $\{\mathbf{e}_j\}$ orthonormal and $\{\mathbf{a}_j\}$ uncorrelated, but RPCA can only possess one of these two properties.

In general, out of a total of $m$ PCA modes, only the $k$ leading ones are selected for rotation, while the higher modes are discarded as noise. As there are many possible criteria for rotation, there are many RPCA schemes – Richman (1986) listed 5 orthogonal and 14 oblique rotation schemes. The *varimax* scheme proposed by Kaiser (1958) is the most popular among orthogonal rotation schemes, though it underperforms some of the oblique rotation schemes (e.g. promax) in tests performed by Richman (1986, tables IX–XII).

From (9.64), the symmetry between $\mathbf{A}$ and $\mathbf{E}$ in the PCA decomposition $\mathbf{X} = \mathbf{A}\mathbf{E}^{\mathrm{T}}$ allows dual approaches to rotation: (i) one can rotate the eigenvectors $\{\mathbf{e}_j\}$, that is, the column vectors in $\mathbf{E}$. As the $\{\mathbf{e}_j\}$ vectors form a frame, this is called E-frame rotation by Preisendorfer (1988, section 7d), a.k.a. T-mode decomposition (Richman, 1986). Alternatively, (ii) one can perform A-frame rotation (a.k.a. S-mode decomposition), that is, rotate the principal component vectors $\{\mathbf{a}_j\}$, the column vectors in $\mathbf{A}$ (Preisendorfer, 1988, section 7g). We will study the former rotation in Section 9.2.1 and the latter in Section 9.2.2.

## 9.2.1   E-frame Rotation  [B]☺

For illustration, suppose that only the first two eigenvectors are chosen for rotation. The data are first projected onto the two PCA eigenvectors $\mathbf{e}_j$ ($j = 1, 2$) to get the first two PCs

$$a_j(t_l) = \sum_{i=1}^{m} e_{ji} x_{il}. \tag{9.89}$$

With the rotated eigenvectors $\tilde{\mathbf{e}}_j$, the rotated PCs are

$$\tilde{a}_j(t_l) = \sum_{i=1}^{m} \tilde{e}_{ji} x_{il}. \tag{9.90}$$

A common objective in rotation is to make $\tilde{a}_j^2(t_l)$ either as large as possible or as close to zero as possible, that is, to maximize the *variance* of the square of the rotated PCs. Figure 9.10(a) illustrates a rotation that has yielded $|\tilde{a}_1| < |a_1|$ and $|a_2| < |\tilde{a}_2|$, that is, instead of intermediate magnitudes for $a_1, a_2$, the rotated PCs have either larger or smaller magnitudes. Geometrically, this means the rotated axes (i.e. the eigenvectors) point closer to actual data points than the unrotated axes; hence, the rotated eigenvectors have closer resemblance to observed states than the unrotated ones (see e.g. Fig. 9.9(c)). In the event that the rotated vector $\tilde{\mathbf{e}}_2$ actually passes through the data point in Fig. 9.10(a), then $|\tilde{a}_1|$ is zero, while $|\tilde{a}_2|$ assumes its largest possible value.

The varimax criterion is to find the rotation that maximizes

$$f(\tilde{\mathbf{A}}) = \sum_{j=1}^{k} \mathrm{var}(\tilde{a}_j^2), \tag{9.91}$$

that is,

$$f(\tilde{\mathbf{A}}) = \sum_{j=1}^{k} \left\{ \frac{1}{n} \sum_{l=1}^{n} \left[ \tilde{a}_j^2(t_l) \right]^2 - \left[ \frac{1}{n} \sum_{l=1}^{n} \tilde{a}_j^2(t_l) \right]^2 \right\}, \tag{9.92}$$

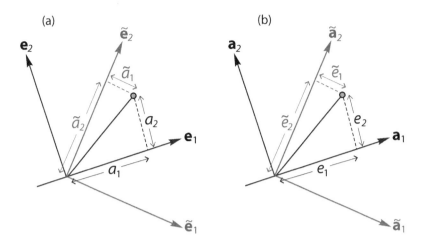

Figure 9.10 (a) In E-frame rotation, the rotated eigenvector $\tilde{\mathbf{e}}_2$ points much closer to the direction of the data point (small circle) than the original eigenvectors $\mathbf{e}_1$ and $\mathbf{e}_2$, with the coordinates of the data point in the original unrotated system being $(a_1, a_2)$ and, in the rotated system, $(\tilde{a}_1, \tilde{a}_2)$. (b) In A-frame rotation, the rotated PC vector $\tilde{\mathbf{a}}_2$ points much closer to the direction of the data point than the original vectors $\mathbf{a}_1$ and $\mathbf{a}_2$. The roles of $\mathbf{a}$ and $\mathbf{e}$ have been reversed between (a) and (b).

where only the first $k$ modes are retained for rotation. Kaiser (1958) found an iterative algorithm for finding the rotation matrix $\mathbf{R}$ (see also Preisendorfer, 1988, pp. 273–277).

If $\{\tilde{\mathbf{e}}_j\}$ are not required to be orthogonal, one can concentrate on the eigenvectors individually, so as to best 'spear' through local data clusters (Fig. 9.9(d)), as done in oblique rotations.

Let us again use the tropical Pacific SST to illustrate the RPCA modes. The PC1–PC2 values from PCA are shown as dots in a scatter plot (Fig. 9.11), where the cool La Niña states lie in the upper left corner, and the warm El Niño states in the upper right corner. The first PCA eigenvector would lie along the horizontal axis, and the second PCA along the vertical axis, neither of which would come close to the El Niño nor the La Niña states.

Using the varimax criterion of (9.92), a rotation is performed on the first three PCA eigenvectors. The direction of the first RPCA eigenvector, indicated by the dot-dashed line, spears through the cluster of El Niño states in the upper right corner, thereby yielding a more accurate description of the SST anomalies during El Niño (Fig. 9.12(a)) than the first PCA mode (Fig. 9.5(a)), which did not fully represent the intense warming of Peruvian waters during El Niño. In terms of variance explained, the first RPCA mode explained only 91.7% as much variance as the first PCA mode. The direction of the second RPCA eigenvector

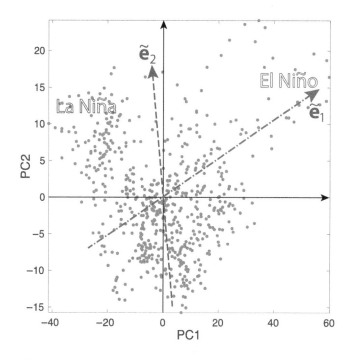

Figure 9.11 PCA with and without rotation for the tropical Pacific SST anomalies. In the PC1–PC2 plane of the scatter plot, where the monthly data are
shown as dots, the cool La Niña states lie in the upper left corner, while the
warm El Niño states lie in the upper right corner. The first PCA eigenvector would lie along the horizontal direction and the second eigenvector along
the vertical direction. A varimax rotation is performed on the first three PCA
eigenvectors. The direction of the first RPCA eigenvector $\tilde{e}_1$ (dot-dashed line)
spears through the cluster of El Niño states in the upper right corner, thereby
yielding a more accurate description of the SST anomalies during an El Niño.
The direction of the second RPCA eigenvector $\tilde{e}_2$ (dashed line) is orthogonal
to the first RPCA eigenvector (though not discernible in this two-dimensional
projection of three-dimensional vectors). Note the axes have different scales for
clarity. [Adapted from Hsieh (2001b, figure 8).]

(the dashed line in Fig. 9.11) did not improve much on the second PCA mode,
with the RPCA spatial pattern shown in Fig. 9.12(b) (cf. Fig. 9.5(b)).

   The dichotomy between PCA and RPCA methods arises because it is generally impossible to have a linear solution simultaneously (a) explaining maximum
global variance of the data and (b) approaching local data clusters. Thus, PCA
excels in (a), while RPCA excels in (b). However, with non-linear PCA using
neural networks or other methods (Chapter 10), both objectives (a) and (b) can
be attained together, thereby unifying the PCA and RPCA approaches.

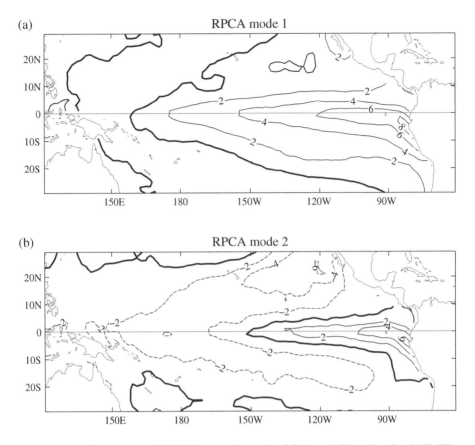

Figure 9.12 The varimax RPCA spatial modes (a) 1 and (b) 2 for the SST. The contour unit is 0.01°C. More intense SST anomalies are found in the eastern equatorial waters off Peru (i.e. just off the west coast of South America) in the RPCA mode 1 than in the PCA mode 1. [Adapted from Hsieh (2001b, figure 9).]

### 9.2.2 A-frame Rotation B☺

In Section 9.2.1 on E-frame rotation, rotating the eigenvectors $\{\mathbf{e}_j\}$ under the varimax criterion (9.92) maximizes the variance of the rotated squared PCs. From the symmetry between $\{\mathbf{e}_j\}$ and $\{\mathbf{a}_j\}$ in (9.64), we will now explore A-frame rotation, that is, rotating the principal component vectors $\{\mathbf{a}_j\}$ (Fig. 9.10(b)) – this alternative is actually much more commonly used than E-frame rotation as it is relatively closer to factor analysis.

*Factor analysis* (FA) is a popular statistical method related to but different from PCA (Gorsuch, 1983). Originally developed by Charles Spearman in 1904, FA is widely used in psychology and other fields, but in the environmental sciences it is nowhere as popular as PCA and RPCA. This is likely due to the fact that FA is computationally more expensive than PCA. This is not an issue today, but in the early days of limited computing power, typical atmospheric datasets with orders of magnitude more variables than datasets in psychology probably led to PCA but not FA being widely used in atmospheric and related sciences. Rotation is widely used in FA, and RPCA has borrowed the rotation techniques from FA. Rotating the relatively small number of leading principal component vectors $\{\mathbf{a}_j\}$ in RPCA is a less costly approach than applying FA directly to a dataset containing many variables.

In contrast to (9.92) under E-frame rotation, A-frame rotation applies the varimax criterion to maximize the variance of the rotated squared loadings $\tilde{e}_{ji}^2$ for the first $k$ modes, that is, maximize

$$f(\tilde{\mathbf{E}}) = \sum_{j=1}^{k} \left\{ \frac{1}{m} \sum_{i=1}^{m} [\tilde{e}_{ji}^2]^2 - \left[ \frac{1}{m} \sum_{i=1}^{m} \tilde{e}_{ji}^2 \right]^2 \right\}. \qquad (9.93)$$

Maximizing the variance of the rotated squared loadings means that the squared loadings are made as large as possible or as close to zero as possible (Fig. 9.10(b)), so many of the loadings are essentially set to zero. If, for instance, we have $m$ variables covering a spatial grid, the rotated loadings with many essentially zero would yield patterns that have more localized anomalies than in the unrotated patterns.

The value of A-frame rotated PCA is well illustrated in the study of atmospheric teleconnections. *Teleconnections* refers to climate anomalies related at large distances (typically thousands of kilometers). The most prominent atmospheric teleconnection is the Southern Oscillation, the seesaw oscillation in the sea-level pressure between Tahiti and Darwin, Australia, which is part of the El Niño-Southern Oscillation (ENSO) phenomenon. The atmospheric response to a heating anomaly in the tropical ocean was first studied with an ideal dynamical model by A. E. Gill (1980). The Tropical Ocean Global Atmosphere program (TOGA) was a 10-year study (1985–1994) under the World Climate Research Programme (WCRP) to understand how the tropical ocean affects the global atmosphere (Trenberth et al., 1998). Several atmospheric teleconnection patterns outside the tropics were revealed by applying PCA and RPCA techniques to observed data (Horel, 1981; Wallace and Gutzler, 1981).

In Wallace and Gutzler (1981), PCA was performed on the monthly 500 hPa geopotential height[2] anomalies in the Northern Hemisphere for the winter months of December, January and February during 15 winters (1962–1963 to 1976–1977), with the EOF spatial patterns for modes 1 and 2 shown in Figs. 9.13(a) and (b), respectively. In contrast, Horel (1981) performed rotated PCA on the same data (using A-frame varimax rotation on the first 19 $\mathbf{a}_j$ vectors), with the corresponding modes 1 and 2 shown in Figs. 9.13(c) and (d), respec-

---

[2] Geopotential height in the atmosphere is approximately the height above sea level of a given pressure level (with the pressure specified in units of hectopascal (hPa) or, equivalently, in non-SI units of millibar (mb)). For instance, if a station reports the 500 hPa height as 5,400 m, it means that the the atmospheric pressure equals 500 hPa at a height of approximately 5,400 m above sea level at that station.

tively. The non-rotated modes are sprawled out over much of the domain, while the rotated modes are indeed more focused and confined to a part of the domain.

The first rotated EOF (Fig. 9.13(c)) shows the Western Pacific pattern (Wallace and Gutzler, 1981, figure 24) – a north–south seesaw oscillation in the western North Pacific corresponding to changes in the Aleutian low and the jet stream. The second rotated EOF (Fig. 9.13(d)) shows the Pacific-North American (PNA) pattern (Wallace and Gutzler, 1981, figure 16) – a string of four alternating L-H-L-H anomaly centres stretching from the tropics to the northern North Pacific then across the North American continent. The percentage of variance accounted for is 15.9% for PCA mode 1 versus 12.9% for RPCA mode 1, and 14.3% for PCA mode 2 versus 8.7% for RPCA mode 2.

Further study of teleconnections from the 700 hPa geopotential height anomalies (for all months of the year during 1950–1984) using RPCA (under A-frame rotation) was carried out by Barnston and Livezey (1987). RPCA was also performed on the surface air temperature over continental USA and on the North Pacific SST by J. E. Walsh and Richman (1981).

Several real-time *daily* climate indices derived from PCA or RPCA are available from the website of the Climate Prediction Center, National Weather Service, USA,
www.cpc.ncep.noaa.gov/products/precip/CWlink/daily_ao_index/
teleconnections.shtml:

- Arctic Oscillation (AO) index is the first PC of the 1,000 hPa geopotential height anomalies poleward of $20°N$.

- Antarctic Oscillation (AAO) index is the first PC of the 700 hPa height anomalies poleward of $20°S$.

- North Atlantic Oscillation (NAO) index is from RPCA of the 500 hPa height anomalies poleward of $20°N$.

- Pacific-North American (PNA) pattern index is from RPCA of the 500 hPa height anomalies poleward of $20°N$.

Monthly climate indices for 10 teleconnection patterns found in the 500 hPa height anomalies poleward of $20°N$ using RPCA, as in Barnston and Livezey (1987), are also available from the website of the Climate Prediction Center, National Weather Service, USA,
www.cpc.ncep.noaa.gov/data/teledoc/telecontents.shtml

The ten indices are:

- North Atlantic Oscillation (NAO),
- East Atlantic Pattern (EA),
- East Atlantic/Western Russia pattern (EATL/WRUS),
- Scandinavia pattern (SCAND),
- Polar/Eurasia pattern,

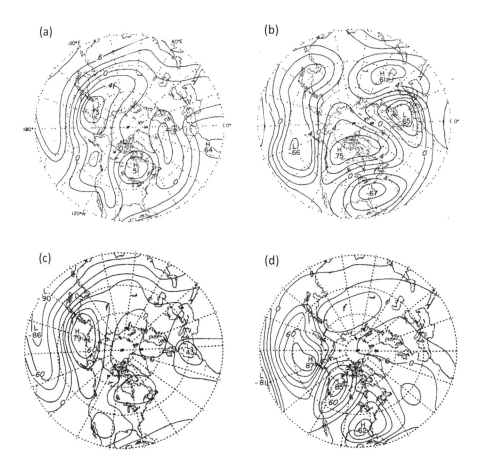

Figure 9.13 Eigenvectors (i.e. EOFs) from PCA on the winter 500 hPa geopotential height anomalies showing the loadings for (a) mode 1 and (b) mode 2. [Reproduced from Wallace and Gutzler (1981, figure 27), ©American Meteorological Society. Used with permission.] Eigenvectors from rotated PCA for (c) mode 1 and (d) mode 2. [Reproduced from Horel (1981, figure 2), ©American Meteorological Society. Used with permission.] The loading at any location is the correlation between the PC and the local 500 hPa height anomaly, as in (9.44). The contour interval is 0.2. The sign of an eigenvector is arbitrary, that is, the entire loading pattern can be multiplied by $-1$.

- West Pacific pattern (WP),
- East Pacific-North Pacific pattern (EP-NP),
- Pacific/North American pattern (PNA),
- Tropical/Northern Hemisphere pattern (TNH), exists during December–February,

- Pacific Transition pattern (PT), exists during August–September.

### 9.2.3   Advantages and Disadvantages of Rotation  Ⓐ☺

The two ways of performing rotation in the last two subsections can be seen as working with either the data matrix $\mathbf{X}$ or its transpose $\mathbf{X}^T$. In PCA, using the transpose of the data matrix does not change the results (but can be exploited to save considerable computational time by working with the smaller data covariance matrix, as seen in Section 9.1.8). Taking the transpose of $\mathbf{X} = \mathbf{AE}^T$ gives $\mathbf{X}^T = \mathbf{EA}^T$, that is, the role of the PCs and the loadings have been interchanged. In E-frame rotation, the eigenvectors in $\mathbf{E}$ are rotated to maximize the variance of the PCs if using the varimax criterion. In A-frame rotation, the PC vectors in $\mathbf{A}$ are rotated to maximize the variance of the loadings. Richman (1986, section 6) reported that applying S-mode decomposition, that is, A-frame rotation, yielded loading patterns with far fewer anomaly centres than observed in typical 700 mb height anomaly maps of the Northern Hemisphere, whereas applying T-mode decomposition, that is, E-frame rotation, yielded loading patterns in good agreement with commonly observed patterns. From idealized data with known properties, Compagnucci and Richman (2008) found that A-frame rotation results can be interpreted as teleconnection patterns and E-frame results as flow patterns.

To summarize, we list the disadvantages of the PCA and those of the RPCA (Richman, 1986; Jolliffe, 1987; Richman, 1987; Huth and Beranová, 2021). There are four main disadvantages with PCA:

(i) Domain shape dependence – often the PCA spatial modes (i.e. eigenvectors) are related more to the spatial harmonics (a.k.a. *Buell patterns*) rather than to physical states, a consequence being that the loading patterns tend to sprawl over the whole spatial domain. Huth and Beranová (2021) argued that using PCA to find atmospheric teleconnection patterns would often result in spurious patterns.

(ii) Subdomain instability – if the domain is divided into two parts, then the PCA mode 1 spatial patterns for the subdomains may not be similar to the spatial mode calculated for the whole domain, as illustrated in Fig. 9.14.

(iii) Degeneracy – if $\lambda_i \approx \lambda_j$, the near degeneracy of eigenvalues means that the eigenvectors $\mathbf{e}_i$ and $\mathbf{e}_j$ cannot be estimated accurately by PCA. Rotation applied to the ill-defined PCA modes can help in their interpretation (Jolliffe, 1989).

(iv) Neglect of regional correlated patterns – small regional correlated patterns tend to be ignored by PCA, as PCA spatial modes tend to be related to the dominant spatial harmonics.

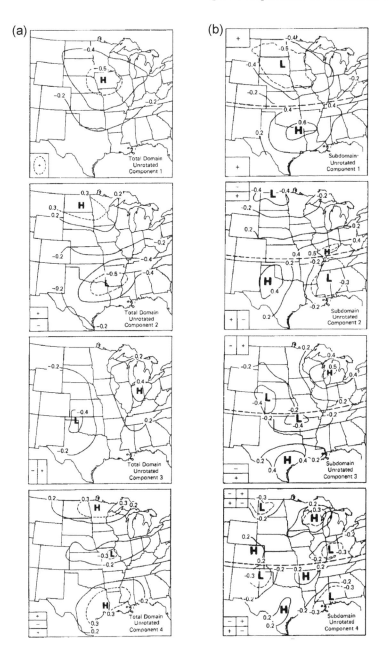

Figure 9.14 First four PCA spatial modes of three-day precipitation (May–August) over central USA. (a) Left panels show the four modes computed for the whole domain. (b) Right panels show the modes computed separately for the northern and southern halves (as separated by the dashed line). Insets show the basic harmonic patterns found by the modes. [Reproduced from Richman (1986, figure 2).]

RPCA improves on all (i)–(iv) above.

There are also three disadvantages with RPCA:

(a) Many possible choices for the rotation criterion – Richman (1986) listed 19 types of rotation scheme. Furthermore, the rotational can be performed on the A-frame or the E-frame, yielding different results.

(b) Dependence on $k$, the number of PCA modes chosen for rotation – if the first $k$ PCA modes are selected for rotation, changing $k$ can lead to large changes in the RPCAs. For instance, in RPCA, if one first chooses $k = 3$, then one chooses $k = 4$, the first three RPCAs are changed. In contrast, in PCA, if one first chooses $k = 3$, then $k = 4$, the first three PCAs are unchanged.

(c) Dependence on how the PCA eigenvectors and PCs have been scaled before rotation is performed – RPCA gives different results if a different scaling convention (Section 9.1.6) has been used in the PCA.

Since PCA maximizes the variance explained (Jolliffe, 2002, property A1, pp. 11–13), it is tempting to conclude that the variance explained by the first $k$ orthonormally rotated PCA modes is $\leq$ the variance of the first $k$ PCA modes[3] ($k \geq 2$ since rotation has to involve at least two PCA modes). To understand why this conclusion is actually incorrect (M. B. Richman, pers. comm.), consider $m \geq 2$ variables and we choose to rotate the first two PCA modes. Let $\lambda_1$ and $\lambda_2$ denote the variance of PCA mode 1 and mode 2, respectively, and $\tilde{\lambda}_1$ and $\tilde{\lambda}_2$, those of RPCA modes 1 and 2. We have $\tilde{\lambda}_1 \leq \lambda_1$, as the variance is maximized by PCA. The two PCA modes span a 2-D space that is also spanned by the two RPCA modes, so the variance contained in the two RPCA modes is exactly the same as the variance contained in the two PCA modes, that is, $\tilde{\lambda}_1 + \tilde{\lambda}_2 = \lambda_1 + \lambda_2$. As $\tilde{\lambda}_1 \leq \lambda_1$, this implies $\tilde{\lambda}_2 \geq \lambda_2$.

On the issue of which RPCA modes to retain, the methods in Section 9.1.9 for deciding how many PCA modes to retain are not applicable for RPCA. Instead, Richman (1986) proposed using congruence[4] to decide if a particular RPCA mode is worth retaining or not.

## 9.3 PCA for Two-Dimensional Vectors $\boxed{C}$☺

When one has vector variables, for example wind velocity or ocean current $(u, v)$, there are several options for performing PCA. (a) One can simply apply PCA

---

[3] The author drew this incorrect conclusion in Hsieh (2009, p. 46).

[4] *Congruence* $g$ is of the same form as the Pearson correlation coefficient in (2.53) but without subtracting the mean values $\bar{x}$ and $\bar{y}$, that is, $g = \left(\sum_i x_i y_i\right) / \left[\left(\sum_i x_i^2\right)\left(\sum_i y_i^2\right)\right]^{1/2}$.

to the $u$ field and to the $v$ field separately. (b) One can do a *combined PCA*, that is, treat the $v$ variables as though they are extra $u$ variables, so the data matrix becomes

$$\mathbf{X} = \begin{bmatrix} u_{11} & \cdots & u_{1m} & v_{11} & \cdots & v_{1m} \\ \vdots & \ddots & \vdots & \vdots & \ddots & \vdots \\ u_{n1} & \cdots & u_{nm} & v_{n1} & \cdots & v_{nm} \end{bmatrix}, \tag{9.94}$$

where $m$ is the number of spatial points or observational stations and $n$ is the number of observations in time. As $u$ and $v$ are components of a vector field, their relation is not fully taken into account using either option (a) or option (b). Options (a) and (b) can easily be generalized to work with a vector field having more than two dimensions.

If the vector field is two-dimensional, then one has option (c) as well, namely one can combine $u$ and $v$ into a complex variable and perform a complex PCA (Hardy, 1977; Hardy and Walton, 1978). This is the best option for taking into account the fact that $u$ and $v$ are components of a vector field.

Defining the complex velocity $w$ by

$$w = u + iv, \tag{9.95}$$

we have the complex data matrix

$$\mathbf{X} = \begin{bmatrix} w_{11} & \cdots & w_{1m} \\ \vdots & \ddots & \vdots \\ w_{n1} & \cdots & w_{nm} \end{bmatrix}. \tag{9.96}$$

Applying PCA to $\mathbf{X}$ allows the complex data matrix to be expressed as

$$\mathbf{X} = \sum_{j=1}^{m} \mathbf{a}_j \mathbf{e}_j^{*\mathrm{T}}, \tag{9.97}$$

where the superscript $*$ denotes the complex conjugate. Since the covariance matrix is Hermitian and positive semi-definite (see Section 9.1.3), the eigenvalues of $\mathbf{C}$ are real and non-negative, though $\mathbf{e}_j$ and $\mathbf{a}_j$ are in general complex.

If we write the $l$th component of $\mathbf{a}_j$ as

$$a_{lj} = |a_{lj}| \, \mathrm{e}^{i\theta_{lj}}, \tag{9.98}$$

where $|a_{lj}|$ is the magnitude and $\theta_{lj}$ the argument or phase of the complex number, then

$$X_{li} = \sum_{j=1}^{m} |a_{lj}| \, \mathrm{e}^{i\theta_{lj}} \, e_{ji}^{*}. \tag{9.99}$$

One can interpret $\mathrm{e}^{i\theta_{lj}} e_{ji}^{*}$ as each complex element of $\mathbf{e}_j^{*\mathrm{T}}$ being rotated by the same angle $\theta_{lj}$ at the $l$th instance in time. Similarly, the magnitude of each element of $\mathbf{e}_j^{*\mathrm{T}}$ is amplified by the same factor $|a_{lj}|$.

For an idealized example, suppose complex PCA has been performed on wind velocity data collected at four stations. The first mode gives $\mathbf{e}_1^{*\mathrm{T}}$, revealing a clockwise circulation pattern in Fig. 9.15(a). Assume at the first time instance ($l = 1$), the first PC $a_{l1} = 1$. At the next time instance $l = 2$, the first PC has changed to $a_{l1} = 1.5\,\mathrm{e}^{-\mathrm{i}\pi/2}$. From (9.99), the wind velocity contributed by mode 1 is given by $1.5\,\mathrm{e}^{-\mathrm{i}\pi/2}\,\mathbf{e}_{1i}^*$, that is, all the wind vectors are rotated clockwise by an angle of $\pi/2$ radians (90°), with the magnitude of the wind vector enhanced by a factor of 1.5, yielding the strong wind convergence in Fig. 9.15(b). At time instance $l = 3$, $a_{l1} = \mathrm{e}^{-\mathrm{i}\pi}$, giving a wind pattern that is identical to that at $l = 1$ but with the wind direction reversed, yielding a counterclockwise circulation (Fig. 9.15(c)). At $l = 4$, $a_{l1} = 0.5\,\mathrm{e}^{-\mathrm{i}3\pi/2}$, giving a weak wind divergence (Fig. 9.15(d)).

Figure 9.15 A complex PCA mode representing a two-dimensional velocity field: (a) At the first time instance ($l=1$), the PC$=1$ and the eigenvector gives a clockwise circulation pattern. (b) At $l=2$, the PC is $1.5\,\mathrm{e}^{-\mathrm{i}\pi/2}$ and the mode gives a strong convergent flow pattern. (c) At $l=3$, the PC is $\mathrm{e}^{-\mathrm{i}\pi}$, with the mode giving a counterclockwise circulation pattern. (d) At $l=4$, the PC is $0.5\,\mathrm{e}^{-\mathrm{i}3\pi/2}$, with the mode giving a weak divergent flow pattern.

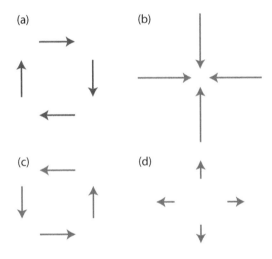

When PCA is applied to real variables, the real $\mathbf{e}_j$ and $\mathbf{a}_j$ can both be multiplied by $-1$. When PCA is applied to complex variables, an arbitrary phase $\phi_j$ can be attached to the complex $\mathbf{a}_j$ and $\mathbf{e}_j$, as follows

$$\mathbf{X} = \sum_{j=1}^{m} \left(\mathbf{a}_j \mathrm{e}^{\mathrm{i}\phi_j}\right)\left(\mathrm{e}^{-\mathrm{i}\phi_j}\mathbf{e}_j^{*\mathrm{T}}\right). \tag{9.100}$$

Often, the arbitrary phase is chosen to give an easier interpretation of the mode. For instance, for the tropical Pacific wind field, $\phi_j$ can be chosen such that $\mathbf{a}_j \mathrm{e}^{\mathrm{i}\phi_j}$ lies mainly along the real axis (Legler, 1983). In the ocean, dynamical theory predicts the near-surface wind-driven current to spiral and diminish with depth, in the shape of an 'Ekman spiral'. This fascinating spiral shape in the current was detected by using the complex PCA (Stacey et al., 1986).

# 9.4 Canonical Correlation Analysis (CCA) B☺

Given a set of variables $\{x_j\}$, PCA finds the linear modes accounting for the maximum amount of variance in the dataset. When there are two sets of variables $\{x_i\}$ and $\{y_j\}$, *canonical correlation analysis* (CCA), first introduced by Hotelling (1936), finds the modes of maximum correlation between $\{x_i\}$ and $\{y_j\}$, rendering CCA a standard tool for discovering linear relations between two fields, that is, CCA can be viewed as a 'double-barrelled PCA'. CCA is a generalization of the Pearson correlation between two variables $x$ and $y$ to two sets of variables $\{x_i\}$ and $\{y_j\}$. A variant of the CCA method finds the modes of maximum *covariance* between $\{x_i\}$ and $\{y_j\}$ (see Section 9.5) – this variant is called the *maximum covariance analysis* (MCA) by von Storch and Zwiers (1999).

In PCA, one finds a linear combination of the $x_j$ variables, that is, $\mathbf{e}_1^T\mathbf{x}$ (assuming $\bar{\mathbf{x}} = 0$), which has the largest variance (subject to $\|\mathbf{e}_1\| = 1$). Next, one finds $\mathbf{e}_2^T\mathbf{x}$ with the largest variance, but with $\mathbf{e}_2^T\mathbf{x}$ uncorrelated with $\mathbf{e}_1^T\mathbf{x}$, and similarly for the higher modes.

In CCA, one finds $\mathbf{f}_1$ and $\mathbf{g}_1$, so that the correlation between $\mathbf{f}_1^T\mathbf{x}$ and $\mathbf{g}_1^T\mathbf{y}$ is maximized. Next, find $\mathbf{f}_2$ and $\mathbf{g}_2$ so that the correlation between $\mathbf{f}_2^T\mathbf{x}$ and $\mathbf{g}_2^T\mathbf{y}$ is maximized, with $\mathbf{f}_2^T\mathbf{x}$ and $\mathbf{g}_2^T\mathbf{y}$ uncorrelated with both $\mathbf{f}_1^T\mathbf{x}$ and $\mathbf{g}_1^T\mathbf{y}$, and so forth for the higher modes.

## 9.4.1 CCA Theory B☺

Consider two data matrices

$$\mathbf{X} = (x_{li}), \qquad l = 1, \ldots, n, \quad i = 1, \ldots m_x, \tag{9.101}$$

and

$$\mathbf{Y} = (y_{lj}), \qquad l = 1, \ldots, n, \quad j = 1, \ldots, m_y, \tag{9.102}$$

that is, $\mathbf{X}$ and $\mathbf{Y}$ need not to have the same number of columns (i.e. variables), but need to have the same the number of rows (observations). Let $\mathbf{x}(t_l)$ and $\mathbf{y}(t_l)$ be the $l$th column vectors in $\mathbf{X}^T$ and $\mathbf{Y}^T$. Assume the data have been centred, that is, the mean values $\bar{\mathbf{x}} = \bar{\mathbf{y}} = 0$. Let

$$u = \mathbf{f}^T\mathbf{x}, \quad v = \mathbf{g}^T\mathbf{y}. \tag{9.103}$$

The Pearson correlation

$$\rho = \frac{\text{cov}(u, v)}{\sqrt{\text{var}(u)\,\text{var}(v)}} = \frac{\text{cov}(\mathbf{f}^T\mathbf{x}, \mathbf{g}^T\mathbf{y})}{\sqrt{\text{var}(u)\,\text{var}(v)}} = \frac{\mathbf{f}^T\text{cov}(\mathbf{x}, \mathbf{y})\mathbf{g}}{\sqrt{\text{var}(\mathbf{f}^T\mathbf{x})\,\text{var}(\mathbf{g}^T\mathbf{y})}}, \tag{9.104}$$

where we have invoked

$$\begin{aligned}
\text{cov}(\mathbf{f}^T\mathbf{x}, \mathbf{g}^T\mathbf{y}) &= \text{E}\big[\mathbf{f}^T\mathbf{x}(\mathbf{g}^T\mathbf{y})^T\big] = \text{E}\big[\mathbf{f}^T\mathbf{x}\mathbf{y}^T\mathbf{g}\big] \\
&= \mathbf{f}^T\text{E}\big[\mathbf{x}\mathbf{y}^T\big]\mathbf{g} = \mathbf{f}^T\text{cov}(\mathbf{x}, \mathbf{y})\,\mathbf{g}.
\end{aligned} \tag{9.105}$$

We want $u$ and $v$, the two *canonical variates* or *canonical correlation coordinates*, to have maximum correlation between them, that is, $\mathbf{f}$ and $\mathbf{g}$ are chosen to maximize $\rho$. We are of course free to normalize $\mathbf{f}$ and $\mathbf{g}$ as we like, because if $\mathbf{f}$ and $\mathbf{g}$ maximize $\rho$, so will $a\mathbf{f}$ and $b\mathbf{g}$, for any positive $a$ and $b$. We choose the normalization condition

$$\mathrm{var}(\mathbf{f}^{\mathrm{T}}\mathbf{x}) = 1 = \mathrm{var}(\mathbf{g}^{\mathrm{T}}\mathbf{y}). \tag{9.106}$$

Since

$$\mathrm{var}(\mathbf{f}^{\mathrm{T}}\mathbf{x}) = \mathrm{cov}(\mathbf{f}^{\mathrm{T}}\mathbf{x}, \mathbf{f}^{\mathrm{T}}\mathbf{x}) = \mathbf{f}^{\mathrm{T}}\mathrm{cov}(\mathbf{x},\mathbf{x})\,\mathbf{f} \equiv \mathbf{f}^{\mathrm{T}}\mathbf{C}_{xx}\mathbf{f}, \tag{9.107}$$

and similarly,

$$\mathrm{var}(\mathbf{g}^{\mathrm{T}}\mathbf{y}) = \mathbf{g}^{\mathrm{T}}\mathbf{C}_{yy}\mathbf{g}, \tag{9.108}$$

(9.106) implies

$$\mathbf{f}^{\mathrm{T}}\mathbf{C}_{xx}\mathbf{f} = 1, \quad \mathbf{g}^{\mathrm{T}}\mathbf{C}_{yy}\mathbf{g} = 1. \tag{9.109}$$

Under (9.106), (9.104) reduces to

$$\rho = \mathbf{f}^{\mathrm{T}}\mathbf{C}_{xy}\mathbf{g}, \tag{9.110}$$

where $\mathbf{C}_{xy} = \mathrm{cov}(\mathbf{x},\mathbf{y})$.

The problem is to maximize (9.110) subject to constraints (9.109). We will again use the method of Lagrange multipliers (Appendix B), where we incorporate the constraints into the Lagrange function $L$,

$$L = \mathbf{f}^{\mathrm{T}}\mathbf{C}_{xy}\mathbf{g} + \alpha\left(\mathbf{f}^{\mathrm{T}}\mathbf{C}_{xx}\mathbf{f} - 1\right) + \beta\left(\mathbf{g}^{\mathrm{T}}\mathbf{C}_{yy}\mathbf{g} - 1\right), \tag{9.111}$$

where $\alpha$ and $\beta$ are the unknown Lagrange multipliers. To find the stationary points of $L$, we need

$$\frac{\partial L}{\partial \mathbf{f}} = \mathbf{C}_{xy}\mathbf{g} + 2\alpha\mathbf{C}_{xx}\mathbf{f} = 0 \tag{9.112}$$

and

$$\frac{\partial L}{\partial \mathbf{g}} = \mathbf{C}_{xy}^{\mathrm{T}}\mathbf{f} + 2\beta\mathbf{C}_{yy}\mathbf{g} = 0. \tag{9.113}$$

From $\mathbf{C}_{xx}^{-1}$(9.112) and $\mathbf{C}_{yy}^{-1}$(9.113), we obtain, respectively,

$$\mathbf{C}_{xx}^{-1}\mathbf{C}_{xy}\mathbf{g} = -2\alpha\mathbf{f} \tag{9.114}$$

and

$$\mathbf{C}_{yy}^{-1}\mathbf{C}_{xy}^{\mathrm{T}}\mathbf{f} = -2\beta\mathbf{g}. \tag{9.115}$$

Substituting (9.115) into (9.114) to eliminate $\mathbf{g}$ yields

$$\mathbf{C}_{xx}^{-1}\mathbf{C}_{xy}\mathbf{C}_{yy}^{-1}\mathbf{C}_{xy}^{\mathrm{T}}\,\mathbf{f} \equiv \mathbf{M}_f\mathbf{f} = \lambda\mathbf{f}, \tag{9.116}$$

with $\lambda = 4\alpha\beta$. Similarly, substituting (9.114) into (9.115) to eliminate $\mathbf{f}$ gives

$$\mathbf{C}_{yy}^{-1}\mathbf{C}_{xy}^{\mathrm{T}}\mathbf{C}_{xx}^{-1}\mathbf{C}_{xy}\,\mathbf{g} \equiv \mathbf{M}_g\mathbf{g} = \lambda\mathbf{g}. \tag{9.117}$$

Both these equations can be viewed as eigenvalue equations, with $\mathbf{M}_f$ and $\mathbf{M}_g$ sharing the same non-zero eigenvalues $\lambda$. As $\mathbf{M}_f$ and $\mathbf{M}_g$ are known from the data, $\mathbf{f}$ can be found by solving the eigenvalue problem (9.116). Then, $\beta\mathbf{g}$ can be obtained from (9.115). Since $\beta$ is unknown, the magnitude of $\mathbf{g}$ is unknown, and the normalization conditions (9.109) are used to determine the magnitude of $\mathbf{g}$ and $\mathbf{f}$. Alternatively, one can use (9.117) to solve for $\mathbf{g}$ first, then obtain $\mathbf{f}$ from (9.114) and the normalization condition (9.109). The matrix $\mathbf{M}_f$ is of dimension $m_x \times m_x$, while $\mathbf{M}_g$ is $m_y \times m_y$, so one usually picks the smaller of the two to solve the eigenvalue problem.

From (9.110),

$$\rho^2 = \mathbf{f}^\mathrm{T}\mathbf{C}_{xy}\mathbf{g}\,\mathbf{g}^\mathrm{T}\mathbf{C}_{xy}^\mathrm{T}\mathbf{f} = 4\alpha\beta\left(\mathbf{f}^\mathrm{T}\mathbf{C}_{xx}\mathbf{f}\right)\left(\mathbf{g}^\mathrm{T}\mathbf{C}_{yy}\mathbf{g}\right), \tag{9.118}$$

where (9.112) and (9.113) have been invoked. From (9.109), (9.118) reduces to

$$\rho^2 = 4\alpha\beta = \lambda. \tag{9.119}$$

The eigenvalue problems (9.116) and (9.117) yield $m$ number of $\lambda$s, with $m = \min(m_x, m_y)$. Assuming the $\lambda$s to be all distinct and non-zero, we have for each $\lambda_j$ ($j = 1, \dots, m$) a pair of eigenvectors, $\mathbf{f}_j$ and $\mathbf{g}_j$, and a pair of canonical variates, $u_j$ and $v_j$, with correlation $\rho_j = \sqrt{\lambda_j}$ between the two. It can also be shown that

$$\mathrm{cov}(u_j, u_k) = \mathrm{cov}(v_j, v_k) = \delta_{jk}, \quad \text{and} \quad \mathrm{cov}(u_j, v_k) = 0 \quad \text{if} \quad j \neq k \tag{9.120}$$

(Mardia et al., 1979, section 10.2.1).

Let us write the forward mappings from the variables $\mathbf{x}(t)$ and $\mathbf{y}(t)$ to the canonical variates $\mathbf{u}(t) = [u_1(t), \dots, u_m(t)]^\mathrm{T}$ and $\mathbf{v}(t) = [v_1(t), \dots, v_m(t)]^\mathrm{T}$ as

$$\mathbf{u} = \left[\mathbf{f}_1^\mathrm{T}\mathbf{x}, \dots, \mathbf{f}_m^\mathrm{T}\mathbf{x}\right]^\mathrm{T} = \mathcal{F}^\mathrm{T}\mathbf{x}, \quad \mathbf{v} = \mathcal{G}^\mathrm{T}\mathbf{y}, \tag{9.121}$$

where the columns of the $\mathcal{F}$ matrix are the vectors $\mathbf{f}_1, \dots, \mathbf{f}_m$ and, similarly, the columns of $\mathcal{G}$ are $\mathbf{g}_1, \dots, \mathbf{g}_m$.

Next, we need to find the inverse mapping from $\mathbf{u} = [u_1, \dots, u_m]^\mathrm{T}$ and $\mathbf{v} = [v_1, \dots, v_m]^\mathrm{T}$ to the original variables $\mathbf{x}$ and $\mathbf{y}$. Let

$$\mathbf{x} = \mathbf{F}\mathbf{u}, \quad \mathbf{y} = \mathbf{G}\mathbf{v}. \tag{9.122}$$

We note that

$$\mathrm{cov}(\mathbf{x}, \mathbf{u}) = \mathrm{cov}\left(\mathbf{x}, \mathcal{F}^\mathrm{T}\mathbf{x}\right) = \mathrm{E}\left[\mathbf{x}(\mathcal{F}^\mathrm{T}\mathbf{x})^\mathrm{T}\right] = \mathrm{E}\left[\mathbf{x}\mathbf{x}^\mathrm{T}\mathcal{F}\right] = \mathbf{C}_{xx}\mathcal{F} \tag{9.123}$$

and

$$\mathrm{cov}(\mathbf{x}, \mathbf{u}) = \mathrm{cov}(\mathbf{F}\mathbf{u}, \mathbf{u}) = \mathbf{F}\,\mathrm{cov}(\mathbf{u}, \mathbf{u}) = \mathbf{F}. \tag{9.124}$$

Equations (9.123) and (9.124) imply

$$\mathbf{F} = \mathbf{C}_{xx}\mathcal{F}. \tag{9.125}$$

Similarly,

$$G = C_{yy}\mathcal{G}. \tag{9.126}$$

Thus, the inverse mappings $\mathbf{F}$ and $\mathbf{G}$ (from the canonical variates to $\mathbf{x}$ and $\mathbf{y}$) can be calculated from the forward mappings $\mathcal{F}$ and $\mathcal{G}$. The matrix $\mathbf{F}$ is composed of column vectors $\mathbf{F}_j$, and $\mathbf{G}$, of column vectors $\mathbf{G}_j$. Note that $\mathbf{F}_j$ and $\mathbf{G}_j$ are the *canonical correlation patterns* associated with $u_j$ and $v_j$, the canonical variates. In general, orthogonality of vectors within a set is not satisfied by any of the four sets $\{\mathbf{F}_j\}$, $\{\mathbf{G}_j\}$, $\{\mathbf{f}_j\}$ and $\{\mathbf{g}_j\}$. Figure 9.16 schematically illustrates the canonical correlation patterns.

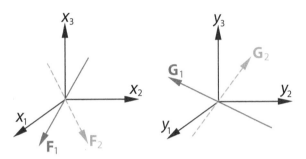

Figure 9.16 Illustrating the CCA solution in the $\mathbf{x}$ and $\mathbf{y}$ spaces. The vectors $\mathbf{F}_1$ and $\mathbf{G}_1$ are the canonical correlation patterns for mode 1, $u_1(t)$ is the amplitude of the fluctuation along $\mathbf{F}_1$ and $v_1(t)$ is the amplitude along $\mathbf{G}_1$. The vectors $\mathbf{F}_1$ and $\mathbf{G}_1$ have been chosen so that the correlation between $u_1$ and $v_1$ is maximized. Next, $\mathbf{F}_2$ and $\mathbf{G}_2$ are found, with $u_2(t)$ the amplitude of the fluctuation along $\mathbf{F}_2$ and $v_2(t)$ the amplitude along $\mathbf{G}_2$. The correlation between $u_2$ and $v_2$ is again maximized, but with $\text{cov}(u_1, u_2) = \text{cov}(v_1, v_2) = \text{cov}(u_1, v_2) = \text{cov}(v_1, u_2) = 0$. In general, $\mathbf{F}_2$ is not orthogonal to $\mathbf{F}_1$, and $\mathbf{G}_2$ is not orthogonal to $\mathbf{G}_1$. Unlike PCA, $\mathbf{F}_1$ and $\mathbf{G}_1$ need not be oriented in the direction of maximum variance. Solving for $\mathbf{F}_1$ and $\mathbf{G}_1$ is analogous to performing rotated PCA in the $\mathbf{x}$ and $\mathbf{y}$ spaces separately, with the rotations determined from maximizing the correlation between $u_1$ and $v_1$. [Source: Hsieh (2009)]

After the CCA model has been built, it can be used to predict $\mathbf{y}$ from a new $\mathbf{x}$. As the canonical variates $u_j$ and $v_j$ have correlation $\rho_j = \sqrt{\lambda_j}$, it can be shown that one can predict $v_j$ from $u_j$ by

$$\hat{v}_j = \rho_j u_j, \tag{9.127}$$

(see Exercise 9.6) . The predicted $\hat{\mathbf{y}}$ is given by

$$\hat{\mathbf{y}} = \mathbf{G}\hat{\mathbf{v}}, \tag{9.128}$$

where $\hat{\mathbf{v}} = [\hat{v}_1, \ldots, \hat{v}_k]$, with the first $k$ modes used for prediction.

## 9.4.2   Pre-filter with PCA   B☺

When $\mathbf{x}$ and $\mathbf{y}$ contain many variables, it is common to use PCA to pre-filter the data to reduce the dimensions of the datasets, that is, apply PCA to $\mathbf{x}$ and $\mathbf{y}$ separately, extract the leading PCs, then apply CCA to the leading PCs of $\mathbf{x}$ and $\mathbf{y}$.

Using Hotelling's scaling for PCA in (9.40), we express the PCA expansions as

$$\mathbf{x} = \sum_j a_j' \mathbf{e}_j', \quad \mathbf{y} = \sum_j a_j'' \mathbf{e}_j''. \tag{9.129}$$

CCA is then applied to

$$\tilde{\mathbf{x}} = \left[ a_1', \dots, a_{k_x}' \right]^{\mathrm{T}}, \quad \tilde{\mathbf{y}} = \left[ a_1'', \dots, a_{k_y}'' \right]^{\mathrm{T}}, \tag{9.130}$$

where only the first $k_x$ and $k_y$ modes are used. The main reason for using the PCA pre-filtering is that when the number of variables is not small relative to the sample size, the CCA method may become unstable (Bretherton, C. Smith, et al., 1992). The reason is that in the relatively high-dimensional $\mathbf{x}$ and $\mathbf{y}$ spaces, among the many dimensions and using correlations calculated with relatively small samples, CCA can often find directions of high correlation but with little variance, thereby extracting a spurious leading CCA mode, as illustrated in Fig. 9.17. This problem can be avoided by pre-filtering using PCA, as this avoids applying CCA directly to high-dimensional input spaces (Barnett and Preisendorfer, 1987).

With Hotelling's scaling,

$$\operatorname{cov}(a_j', a_k') = \delta_{jk}, \quad \operatorname{cov}(a_j'', a_k'') = \delta_{jk}, \tag{9.131}$$

and (9.130), we have

$$\mathbf{C}_{\tilde{x}\tilde{x}} = \mathbf{C}_{\tilde{y}\tilde{y}} = \mathbf{I}. \tag{9.132}$$

Equations (9.116) and (9.117) simplify to

$$\mathbf{C}_{\tilde{x}\tilde{y}} \mathbf{C}_{\tilde{x}\tilde{y}}^{\mathrm{T}} \mathbf{f} \equiv \mathbf{M}_f \mathbf{f} = \lambda \mathbf{f}, \tag{9.133}$$

$$\mathbf{C}_{\tilde{x}\tilde{y}}^{\mathrm{T}} \mathbf{C}_{\tilde{x}\tilde{y}} \mathbf{g} \equiv \mathbf{M}_g \mathbf{g} = \lambda \mathbf{g}. \tag{9.134}$$

As $\mathbf{M}_f$ and $\mathbf{M}_g$ are positive semi-definite symmetric matrices, the eigenvectors $\{\mathbf{f}_j\}$ $\{\mathbf{g}_j\}$ are now sets of orthogonal vectors. Equations (9.125) and (9.126) simplify to

$$\mathbf{F} = \mathcal{F}, \quad \mathbf{G} = \mathcal{G}. \tag{9.135}$$

Thus, $\{\mathbf{F}_j\}$ and $\{\mathbf{G}_j\}$ are also two sets of orthogonal vectors and are identical to $\{\mathbf{f}_j\}$ and $\{\mathbf{g}_j\}$, respectively. Because of these nice properties, pre-filtering by PCA (with the Hotelling scaling) is recommended when $\mathbf{x}$ and $\mathbf{y}$ have many variables (relative to the sample size). However, the orthogonality holds only in the reduced dimensional spaces, $\tilde{\mathbf{x}}$ and $\tilde{\mathbf{y}}$. If transformed into the original space $\mathbf{x}$ and $\mathbf{y}$, $\{\mathbf{F}_j\}$ and $\{\mathbf{G}_j\}$ are in general not two sets of orthogonal vectors.

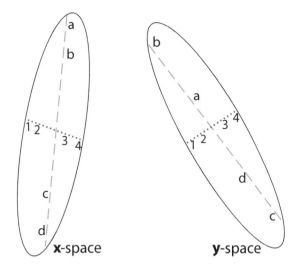

Figure 9.17 Illustrating how CCA may end up extracting a spurious leading mode when working with relatively high-dimensional input spaces. With the ellipses denoting the data clouds in the two input spaces, the dotted lines, illustrate directions with little variance but by chance with high correlation (as illustrated by the perfect order in which the data points 1, 2, 3 and 4 are arranged in the **x** and **y** spaces). Since CCA finds the correlation of the data points along the dotted lines to be higher than that along the dashed lines (where the data points a, b, c and d in the **x**-space are ordered as b, a, d and c in the **y**-space), the dotted lines are chosen as the first CCA mode. Maximum covariance analysis (MCA), which looks for modes of maximum covariance instead of maximum correlation, would select the longer dashed lines over the shorter dotted lines, since the lengths of the lines do count in the covariance but not in the correlation; thus, MCA is stable even without pre-filtering by PCA. [Source: Hsieh (2009)]

As an example, for the tropical Pacific, CCA was performed on the first six PCs of the monthly sea level pressure (SLP) anomaly field and the first six PCs of the SST anomaly field (1950–2000). Figure 9.18 shows the mode 1 CCA manifesting clearly the Southern Oscillation pattern in the SLP and the El Niño/La Niña pattern in the SST.

The canonical variates $u$ and $v$ (not shown) fluctuate with time, both attaining high values during El Niño, low values during La Niña and neutral values around zero during normal conditions.

Using CCA, Barnett and Preisendorfer (1987) studied the monthly and seasonal forecast skills for the surface air temperature over continental USA from global sea level pressure and sea surface temperature. In seasonal climate prediction, CCA has been used for forecasting ENSO (Barnston and Ropelewski, 1992)

Figure 9.18 The CCA mode 1 for (a) the SLP anomalies and (b) the SST anomalies of the tropical Pacific. As $u_1(t)$ and $v_1(t)$ fluctuate together from one extreme to the other as time progresses, the SLP and SST anomaly fields oscillate as standing wave patterns, changing between an El Niño state and a La Niña state. The pattern in (a) is scaled by $\tilde{u}_1 = [\max(u_1) - \min(u_1)]/2$ and (b) by $\tilde{v}_1 = [\max(v_1) - \min(v_1)]/2$. Contour interval is 0.5 hPa in (a) and 0.5°C in (b). [Reproduced from Hsieh (2001a, figure 6), ©American Meteorological Society. Used with permission.]

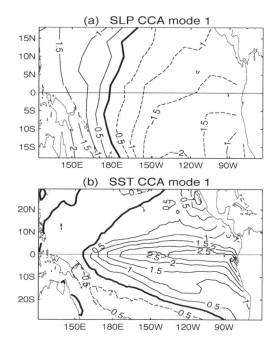

and the surface air temperature and total precipitation over Canada (Shabbar and Barnston, 1996). Air pollution and meteorological data in Athens, Greece have been studied using PCA and CCA (Statheropoulos et al., 1998). In hydrology, CCA has been applied to regional flood frequency analysis by studying the relation between two set of variables represented by watershed characteristics and by flood peaks (Ouarda et al., 2001). For phytoplankton and cyanobacteria in Canadian lakes, Levine and Schindler (1999) used CCA to relate one set of potentially interrelated variables (e.g. phytoplankton composition) to another set of variables (e.g. physicochemical parameters).

Non-linear generalization of CCA by neural network methods is presented in Section 10.8.

## 9.5  Maximum Covariance Analysis  $\boxed{\text{B}}$ ☺

Instead of maximizing the correlation as in CCA, one can maximize the covariance between two datasets. This method is also called 'singular value decomposition' (SVD) by some researchers. However, von Storch and Zwiers (1999, section 14.1.7) proposed the name *maximum covariance analysis* (MCA) as being more appropriate. The reason is that the name SVD is already used to denote a matrix technique (Section 9.1.11), so there is potential confusion in using it to denote a statistical technique. In this book, we will follow the suggestion of von Storch and Zwiers and refer to the statistical method as MCA, using the name SVD only for the matrix technique.

MCA is identical to CCA except that it maximizes the covariance instead of the correlation. As mentioned in the previous subsection, CCA can be unstable when working with a relatively large number of variables, in that directions with high correlation but negligible variance may be selected by CCA as leading modes, hence the recommended pre-filtering of data by PCA before applying CCA to the leading PCs. By using covariance instead of correlation, MCA does not have the unstable nature of CCA (Fig. 9.17) and does not need pre-filtering by PCA. Bretherton, C. Smith, et al. (1992) compared MCA and CCA.

In MCA, one simply performs SVD (see Section 9.1.11) on the data covariance matrix $\mathbf{C}_{xy}$,

$$\mathbf{C}_{xy} = \mathbf{U}\mathbf{S}\mathbf{V}^{\mathrm{T}}, \tag{9.136}$$

where the matrix $\mathbf{U}$ contains the left singular vectors $\mathbf{f}_i$, $\mathbf{V}$ the right singular vectors $\mathbf{g}_i$ and $\mathbf{S}$ the singular values. Maximum covariance between $u_i$ and $v_i$ is attained (Bretherton, C. Smith, et al., 1992) with

$$u_i = \mathbf{f}_i^{\mathrm{T}}\mathbf{x}, \quad v_i = \mathbf{g}_i^{\mathrm{T}}\mathbf{y}. \tag{9.137}$$

The inverse transform is given by

$$\mathbf{x} = \sum_i u_i\mathbf{f}_i, \quad \mathbf{y} = \sum_i v_i\mathbf{g}_i. \tag{9.138}$$

For most applications, MCA yields rather similar results to CCA (with PCA pre-filtering) (Bretherton, C. Smith, et al., 1992; Wallace, C. Smith, et al., 1992).

The matrix technique SVD can also be used to solve the CCA problem: instead of solving the eigenequations (9.116) and (9.117), simply perform SVD on $(\mathbf{C}_{xx}^{-1/2}\mathbf{C}_{xy}\mathbf{C}_{yy}^{-1/2})$. When the data have been prefiltered by PCA, then instead of solving eigenequations (9.133) and (9.134), simply perform SVD on $\mathbf{C}_{\tilde{x}\tilde{y}}$ (Bretherton, C. Smith, et al., 1992). Conversely, eigenequations can be used instead of SVD to solve the MCA problem (von Storch and Zwiers, 1999, section 14.1.7).

## Exercises

9.1

Suppose principal component analysis (PCA) allows one to write the data matrix $\mathbf{X} = \mathbf{A}\mathbf{E}^{\mathrm{T}}$, where

$$\mathbf{A} = \begin{bmatrix} 1 & 2 \\ 3 & -1 \\ -2 & 1 \\ -1 & -3 \end{bmatrix}, \quad \mathbf{E}^{\mathrm{T}} = \begin{bmatrix} 0.7071 & 0.7071 \\ -0.7071 & 0.7071 \end{bmatrix},$$

with $\mathbf{E}$ the matrix of eigenvectors and $\mathbf{A}$ the principal components. Write down the $2\times2$ rotation matrix $\mathbf{R}$ for rotating eigenvectors anticlockwise by $30°$. Derive the rotated matrix of eigenvectors and the rotated principal components.

9.2

In principal component analysis (PCA) under the Hotelling (1933) scaling (Section 9.1.6), one has $\text{cov}(a_i', a_j') = \delta_{ij}$, where $a_i'$ is the $i$th principal component. Suppose each variable $x_l'(t)$ in a dataset has been standardized (i.e. mean removed and divided by its standard deviation). Prove that upon applying PCA (with Hotelling scaling) to this standardized dataset, the $l$th element of the $j$th eigenvector $(e_{jl}')$ is simply the correlation between $a_j'(t)$ and $x_l'(t)$, that is, $\rho(a_j'(t), x_l'(t)) = e_{jl}'$.

9.3

Consider the example of a two-dimensional vector field in Fig. 9.15. If one treats the $v$ components as extra $u$ variables as in (9.94) and performs combined PCA, can one represent the behaviour in Fig. 9.15 by a single combined PCA mode?

9.4

Analyse the dataset from the book website using principal component analysis (PCA). The dataset contains four time series $x_1, x_2, x_3$ and $x_4$.

9.5

From the given dataset containing observations of chlorophyll concentration (chl), dissolved oxygen (dox), suspended particulate matter (spm), salinity (sal) and water temperature (temp) in San Francisco Bay, standardize the five variables and perform (a) principal component analysis (PCA). Compare the PCA eigenvectors with those from (b) E-frame varimax rotated PCA (RPCA) and (c) A-frame varimax RPCA for the first $k$ modes, using four different $k$ values ($k = 2, 3, 4$ and $5$). Also compare the percentage of variance explained by the leading $k$ modes in (a), (b) and (c).

9.6

In CCA, as the canonical variate $v_j$ is uncorrelated with the canonical variate $u_k$ for $j \neq k$, show by using simple linear regression that one can predict $v_j$ from $u_j$ by letting

$$\hat{v}_j = \rho_j u_j, \tag{9.139}$$

where $\rho_j$ is the correlation between $v_j$ and $u_j$ found from CCA.

9.7

Using the data file from the book website, perform canonical correlation analysis between the two groups of time series, $x_1, x_2, x_3$ and $y_1, y_2, y_3$.

# 10

# Unsupervised Learning

In Section 1.4, we discussed the two main types of learning – supervised and unsupervised learning. Given some training data $\{\mathbf{x}_i, \mathbf{y}_i\}$, $(i = 1, ..., N)$, supervised learning tries to find a mapping from the input variables $\mathbf{x}$ to the output variables $\mathbf{y}$. Linear regression in Chapter 5 and MLP neural networks in Section 6.3 are examples of supervised learning. In contrast to supervised learning, unsupervised learning has only input data $\mathbf{x}$ to work with – the goal here is to find structure within the $\mathbf{x}$ data. Principal component analysis in Section 9.1 is an example of unsupervised learning.

In this chapter, we will pursue the topic of unsupervised learning further. Clustering or cluster analysis (Section 10.1) has two main approaches, non-hierarchical (Section 10.2) and hierarchical (Section 10.3). Self-organizing maps (SOM) (Section 10.4), a form of discrete non-linear principal component analysis inspired by neural networks, is also used for clustering. Autoencoders or auto-associative neural networks (Section 10.5), built from MLP NN models, are used for non-linear principal component analysis (NLPCA) (Section 10.6), with other techniques for non-linear dimensionality reduction briefly mentioned in Section 10.7. Similarly, non-linear canonical correlation analysis (NLCCA) has been developed from MLP NN models (Section 10.8).

## 10.1 Clustering ☐B☺

*Clustering* or *cluster analysis* groups objects (i.e. data points) into clusters, so that objects within a cluster are more similar to each other than to those in another cluster. Clustering methods (R. Xu and Wunsch, 2005; R. Xu and Wunsch, 2008) can be divided into *hierarchical* and *non-hierarchical* methods. In non-hierarchical clustering, such as $K$-means clustering (Section 2.16), the relationship between clusters is not determined. In contrast, hierarchical clustering repeatedly links the pair of closest clusters until every data object is included in the hierarchy, which structurally resembles an upside-down tree. While hierarchical clustering gives more information about the relationship between clusters, it is rather sensitive to noise and tends to underperform non-hierarchical clustering methods (Gong and Richman, 1995; M. Bao and Wallace, 2015).

Non-hierarchical clustering methods can be further divided into two categories, hard clustering and soft clustering methods. Under *hard clustering*, an object either belongs to a cluster or it does not, that is, its membership in a cluster is indicated by a binary number, 1 or 0. Under *soft clustering* (fuzzy clustering) an object can belong to multiple clusters with a different probability for each cluster.

Clustering methods have been used in many areas of environmental science: Wolker (1987) applied hierarchical clustering to sea level pressure, surface wind, cloudiness and sea surface temperature to study the Southern Oscillation phenomenon over the tropical Atlantic, eastern Pacific and Indian Oceans. R. G. Fovell and M. Y. C. Fovell (1993) used hierarchical clustering to study the climate zones over the conterminous United States. J. M. Davis et al. (1998) used $K$-means and hierarchical clustering to study the relation between meteorological conditions and air quality (ozone) in Houston, Texas. Whitfield and Cannon (2000) studied variations in climate and hydrology in Canada by applying $K$-means clustering to find regions with similar decadal variations in temperature, precipitation and streamflow. Pfeffer et al. (2003) used $K$-means clustering on vegetation observations and topographic attributes to map alpine vegetation. Torrecilla et al. (2011) used hierarchical clustering on phytoplankton pigment data and spectra of the absorption coefficient and remote-sensing reflectance to discriminate different phytoplankton assemblages in open ocean environments. An important caveat is that different clustering methods can choose different numbers of clusters, as encountered when using cluster analysis to determine the low-frequency atmospheric circulation regimes (Christiansen, 2007a).

## 10.1.1 Distance Measure B☺

A *distance* (a.k.a. *dissimilarity*) measure is needed to determine how close two objects are, and there are many choices. The most common one is the Euclidean distance, where for two data points $\mathbf{x}_i$ and $\mathbf{x}_j$ (both being $m$-dimensional vectors), the distance

$$d_{ij} = \left[ \sum_{l=1}^{m} \left( x_{il} - x_{jl} \right)^2 \right]^{1/2}, \tag{10.1}$$

with $x_{il}$ and $x_{jl}$ ($l = 1, \ldots, m$) denoting the elements of $\mathbf{x}_i$ and $\mathbf{x}_j$, respectively. Since computing the square root is expensive, the squared Euclidean distance often replaces the Euclidean distance in the distance measure, that is,

$$d_{ij} = \sum_{l=1}^{m} \left( x_{il} - x_{jl} \right)^2. \tag{10.2}$$

For $K$-means clustering, there is no difference between using Euclidean distance and squared Euclidean distance, so squared distance is used as it is computationally faster. For other clustering methods, the two distance measures may yield different outcomes. If $\mathbf{x}$ contains variables with different units (e.g. temperature and specific humidity), one should standardize the variables before performing clustering, otherwise the distance measure will be dominated by the variable(s) with large standard deviation.

More generally, the Minkowski distance

$$d_{ij} = \left[ \sum_{l=1}^{m} \left| x_{il} - x_{jl} \right|^{\lambda} \right]^{1/\lambda}, \tag{10.3}$$

which reduces to the Euclidean distance when $\lambda = 2$. When $\lambda = 1$, the Minkowski distance reduces to the city-block distance (a.k.a. Manhattan distance),

$$d_{ij} = \sum_{l=1}^{m} \left| x_{il} - x_{jl} \right|. \tag{10.4}$$

If there are $N$ data points, $d_{ij}$ form the elements of an $N \times N$ distance or dissimilarity matrix $D$.

For ordinal variables, for example low, medium, high, it is common to code the values as real numbers, such as 1, 2, 3, so one can use one of the above distance measures.

For categorial variables, such as red, green, blue, one commonly uses the Hamming distance

$$d_{ij} = \sum_{l=1}^{m} I\left( x_{il} \neq x_{jl} \right), \tag{10.5}$$

where

$$I\left( x_{il} \neq x_{jl} \right) = \begin{cases} 1 & \text{if } x_{il} \neq x_{jl} \\ 0 & \text{if } x_{il} = x_{jl}, \end{cases} \tag{10.6}$$

that is, one simply counts the number of non-matching elements between $\mathbf{x}_i$ and $\mathbf{x}_j$.

## 10.1.2  Model Evaluation  B☺

The most common question in clustering is: What is the optimal value for $K$, the number of clusters in the model? In other words, after running the clustering model a number of times with different values of $K$, how does one pick the best model? Unlike supervised learning, the lack of target data in unsupervised learning makes it difficult to evaluate model performance. There are two types of model evaluation – internal evaluation and external evaluation. In *internal evaluation*, the data used for evaluation are the same data used for training the model. In *external evaluation*, the model is evaluated against a benchmark or gold standard, for example a set of objects with correct labels assigned by human experts; hence, this approach is also called supervised evaluation in contrast to the unsupervised approach used in internal evaluation.

Typical internal evaluation criteria favour models attaining high intra-cluster similarity (i.e. high similarity between objects within a cluster) and low inter-cluster similarity. However, good scores on an internal evaluation criterion do not necessarily guarantee good performance on new data. Common internal evaluation measures include the Calinski–Harabasz index (Calinski and Harabasz, 1974), the Davies–Bouldin index (D. L. Davies and Bouldin, 1979), the silhouette coefficient (Rousseeuw, 1987) and the Gap statistic (Tibshirani et al., 2001). A comparison of 30 internal evaluation criteria by Milligan and Cooper (1985) found that the Calinski–Harabasz index outperformed the others (including the Davies–Bouldin index, but the newer silhouette and gap indices were not included in the comparison). In the comparison made in Tibshirani et al. (2001, table 1), the gap method generally had the best performance, followed by the Calinski–Harabasz method, both ahead of the silhouette method. Charrad et al. (2014) compared 30 internal evaluation criteria available in NbClust, the R package for clustering.

With $N$ data points (objects) and $K$ clusters, the Calinski–Harabasz index is defined by a variance ratio

$$I_{\mathrm{CH}} = \frac{\mathrm{SS_B}/(K-1)}{\mathrm{SS_W}/(N-K)}, \tag{10.7}$$

where $\mathrm{SS_B}/(K-1)$ is the overall between-cluster variance and $\mathrm{SS_W}/(N-K)$ is the overall within-cluster variance. The between-cluster sum of squares, $\mathrm{SS_B}$, is given by

$$\mathrm{SS_B} = \sum_{k=1}^{K} n_k \|\mathbf{c}_k - \mathbf{c}\|^2, \tag{10.8}$$

where the $k$th cluster containing $n_k$ data points has mean position $\mathbf{c}_k$, and $\mathbf{c}$ is the mean of all sample data, with $\|\dots\|$ denoting the Euclidean distance norm. The within-cluster sum of squares, $\mathrm{SS_W}$, is given by

$$\mathrm{SS_W} = \sum_{k=1}^{K} \sum_{\mathbf{x} \in C_k} \|\mathbf{x} - \mathbf{c}_k\|^2, \tag{10.9}$$

where $\mathbf{x}$ is a data point in the $k$th cluster $C_k$. As the objective is to find clusters with large $\mathrm{SS_B}$ and small $\mathrm{SS_W}$, one would search for the model with the largest $I_{\mathrm{CH}}$, that is, from a number of model runs with different $K$ values, the optimal number of clusters is given by the model with the largest $I_{\mathrm{CH}}$.

In the *gap statistic* approach (Tibshirani et al., 2001) to internal evaluation, the sum of pairwise distances for all data points in a cluster $C_k$ is given by

$$D_k = \sum_{i,j \in C_k} d_{ij}, \tag{10.10}$$

with the squared Euclidean distance commonly used for the distance measure $d_{ij}$, and the pooled within-cluster sum of squares around the cluster mean is given by

$$W_K = \sum_{k=1}^{K} \frac{1}{2n_k} D_k \tag{10.11}$$

(where the factor of 2 compensates for the double counting in $D_k$). As $W_K$ is a measure of the compactness of the clustering, a small value of $W_K$ is desirable. The gap statistic is simply the difference between $\log(W_K)$ and the expected value from a null reference distribution, that is, a distribution with no obvious clustering,

$$\text{Gap}(K) = \text{E}_N^*[\log(W_K)] - \log(W_K), \qquad (10.12)$$

where $\text{E}_N^*$ denotes the expectation under a sample of size $N$ from the reference distribution. $\text{E}_N^*[\log(W_K)]$ is estimated by Monte Carlo sampling from the null distribution a total of $B$ times, where the null distribution can be the uniform distribution over the range of the observed data. From the $B$ Monte Carlo samples, we obtain $B$ values of $\log(W_K^*)$, and averaging them gives an estimate for $\text{E}_N^*[\log(W_K)]$. Let $\text{sd}_K$ denote the standard deviation computed from the $B$ values of $\log(W_K^*)$. A standard error $s_K$ is defined by

$$s_K = \text{sd}_K \sqrt{1 + (1/B)}. \qquad (10.13)$$

Finally, the optimal $K$ value is chosen to be the smallest $K$ satisfying

$$\text{Gap}(K) \geq \text{Gap}(K + 1) - s_{K+1}, \qquad (10.14)$$

that is, we proceed from small to large $K$, stopping at the first $K$ value that has $\text{Gap}(K)$ exceeding $\text{Gap}(K + 1)$ within a tolerance of one standard error. The gap approach is computationally expensive because the clustering method needs to be applied to each of the $B$ Monte Carlo samples.

Dudoit and Fridlyand (2002) modified the gap statistic so the optimal $K$ value is chosen to be the smallest $K$ satisfying

$$\text{Gap}(K) \geq \text{Gap}_{\max} - s_{\max}, \qquad (10.15)$$

where $\text{Gap}_{\max}$ is the global maximum of Gap found among all the $K$ values tested, and $s_{\max}$ is the standard error at $\text{Gap}_{\max}$. Many statistical packages offer more than one choice for the gap statistic.

Figure 10.1 illustrates $K$-means clustering with $K = 2$, 3, 4 and 5 clusters applied to the daily air pressure and temperature data at Vancouver, BC, Canada during 2013–2017, seen earlier in Fig. 2.14. The pressure and temperature data have very different units, so it is necessary to perform clustering on the standardized data. The gap statistic and the Calinski–Harabasz index were used to determine the optimal number of clusters (the gap statistic can even be used for testing $K = 1$, whereas the Calinski–Harabasz index only works for $K \geq 2$). Both the Dudoit and Fridlyand (2002) version of Gap in (10.15) and the Calinski–Harabasz index found $K = 3$ to be optimal. However, the original Gap statistic in (10.14) would choose $K = 1$ according to Fig. 10.1(e), that is, no splitting of the original dataset. The three clusters make physical sense: in the Pacific Northwest region of North America, summer tends to be sunny while winter has numerous weather systems passing through, so it makes sense to have one cluster representing summer and two clusters representing winter. In winter, during high pressure days, the clear skies increase outgoing long-wave radiation, leading to colder temperatures than during low pressure days where the clouds reduce the outgoing long-wave radiation, as manifested in the two lower centroids in Fig. 10.1(b).

Next, consider *external evaluation*, where the cluster model output is evaluated against a benchmark or gold standard, that is, a dataset with correct labels (R. Xu and Wunsch, 2008, section 10.2). Without a benchmark, one must forego external evaluation and can only rely on internal evaluation.

Let $C = \{C_1, \ldots, C_K\}$ be the set of clusters from the model and $G = \{G_1, \ldots, G_{K'}\}$ be the clusters from the benchmark. For a pair of data points $x_i$ and $x_j$, there are four possible cases:

- Case 1: $x_i$ and $x_j$ belong to the same cluster in $C$ and to the same cluster in $G$.

- Case 2: $x_i$ and $x_j$ belong to the same cluster in $C$ but to different clusters in $G$.

- Case 3: $x_i$ and $x_j$ belong to the different clusters in $C$ but to the same cluster in $G$.

- Case 4: $x_i$ and $x_j$ belong to the different clusters in $C$ and to different clusters in $G$.

Let $a$, $b$, $c$ and $d$ denote the number of pairs of points belonging to case 1, 2, 3 and 4, respectively. The total number of pairs of points is $M = a + b + c + d$, with $M$ given by the binomial coefficient

$$M = \binom{N}{2} = \frac{N(N-1)}{2}. \tag{10.16}$$

Below are some of the common indices used to evaluate the cluster model relative to the benchmark:
The *Rand index* (Rand, 1971),

$$R = \frac{a+d}{a+b+c+d} = \frac{a+d}{M}, \tag{10.17}$$

where $a + d$ gives the number of pairs of data points where the cluster model is in agreement with the benchmark, and $0 \le R \le 1$. For large datasets with many clusters, $a$ can be dwarfed by $d$. When $d \gg a$, $R$ becomes insensitive to $a$. Since one would rather have the index being insensitive to $d$ than being insensitive to $a$, it may be desirable to delete $d$ from the $R$ index, resulting in the *Jaccard index*,

$$J = \frac{a}{a+b+c}. \tag{10.18}$$

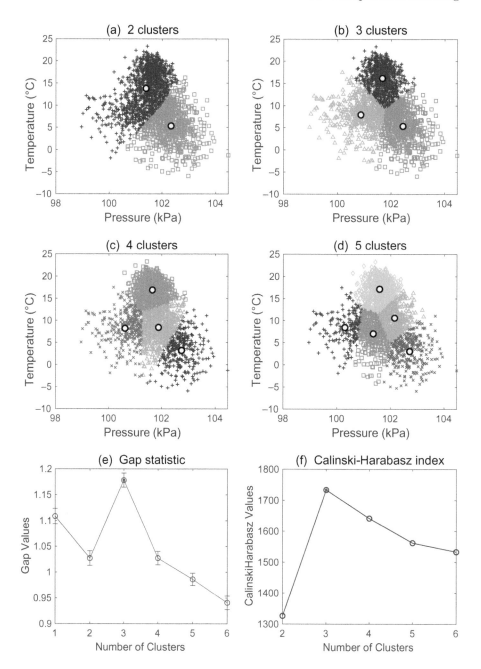

Figure 10.1  $K$-means clustering of the daily air pressure and temperature data at Vancouver, BC, Canada, using (a) 2, (b) 3, (c) 4 and (d) 5 clusters, with their centroids marked by circles. (e) The gap statistic from (10.15) and (f) the Calinski–Harabasz index, with both choosing $K = 3$ clusters as optimal.

Another index that ignores $d$ is the *Fowlkes and Mallows index* (Fowlkes and Mallows, 1983):

$$I_{\mathrm{FM}} = \sqrt{\frac{a}{a+b}\frac{a}{a+c}}. \tag{10.19}$$

Any of the above indices, denoted generically by $I$, can be normalized with respect to a reference distribution (Jain and Dubes, 1988), where the normalized index

$$I' = \frac{I - \mathrm{E}^*(\mathrm{I})}{\max(I) - \mathrm{E}^*(\mathrm{I})}, \tag{10.20}$$

with $\mathrm{E}^*(\mathrm{I})$ being the expected value of $I$ under the reference distribution and $\max(I)$, the maximum possible value of $I$. Using this approach, $R'$, the adjusted Rand index with respect to a null reference distribution (Hubert and Arabie, 1985) is given by

$$R' = \frac{R - \mathrm{E}^*(R)}{1 - \mathrm{E}^*(R)}. \tag{10.21}$$

$R'$ removes the random effect from $R$, but $R'$ is no longer bounded below by 0 and can become negative if $R$ is less than the expected value $E^*(R)$.

In the following sections, we will look at various clustering methods. Section 10.2 covers non-hierarchical clustering methods such as $K$-means clustering, nucleated agglomerative clustering and Gaussian mixture model clustering, where the latter gives the options of hard and soft clustering. Section 10.3 covers hierarchical clustering, including different choices of linkages and Ward's method.

## 10.2 Non-hierarchical Clustering $\boxed{\text{B}}$☺

### 10.2.1 $K$-means Clustering $\boxed{\text{B}}$☺

In Section 2.16, we have already presented the popular $K$-means clustering method, so here we will provide more details. The objective of $K$-means clustering is to group the $N$ data points $(\mathbf{x}_1, \mathbf{x}_2, \ldots, \mathbf{x}_N \in \mathbb{R}^m)$ into $K$ sets or clusters $(C_1, C_2, \ldots, C_K)$, which minimize the within-cluster sum of squares (i.e. squared Euclidean distance),

$$\sum_{k=1}^{K} \sum_{\mathbf{x} \in C_k} \|\mathbf{x} - \mathbf{c}_k\|^2, \tag{10.22}$$

where $\mathbf{c}_k$ is the centroid or centre of the cluster $C_k$, computed by taking the mean of all the data points belonging to the cluster.

The standard algorithm to solve $K$-means clustering is Lloyd's algorithm (first appearing in an unpublished report by Stuart Lloyd of Bell Labs in 1957) and is described in Section 2.16. In terms of computational cost, the operation count for Lloyd's algorithm is $O(NKm)$ for one iteration (R. Xu and Wunsch, 2005, table II). Since $K$ and $m$ are usually much smaller than $N$, this is considered $O(N)$, which is relatively fast among clustering methods [e.g. hierarchical clustering is of $O(N^2)$]. The time complexity for Lloyd's algorithm is $O(NKmT)$, where $T$ is the number of iterations. Faster algorithms for solving $K$-means clustering are listed in R. Xu and Wunsch (2008, p. 69).

The main disadvantage of $K$-means is that it often converges to a local minimum instead of the global minimum. To alleviate this problem, one can run the algorithm multiple times with different seeds for the random number generator and choose the model with the smallest within-cluster sum of squares. Other approaches to avoiding poor local minima include using hill climbing or some type of evolutionary algorithm (e.g. genetic algorithm) (Sections 7.10–7.12). The optimal $K$ is determined by running the cluster model multiple times with different $K$ values and using either internal or external evaluation to choose the best model (Section 10.1.2).

## 10.2.2   Nucleated Agglomerative Clustering  Ⓒ☺

Nucleated agglomerative clustering has been used to analyse environmental data since the mid-1980s (Gong and Richman, 1995). The method starts with $K$-means clustering with a specified initial $K_{\max}$. The number of clusters are then decreased one at a time by merging until the number of clusters reaches a specified final $K_{\min}$.

The basic steps are:

- Step 1: Run $K$-means clustering using $K_{\max}$ clusters.

- Step 2: Reduce the number of clusters by one using Ward's method. The pair of clusters chosen to be merged is the pair that minimizes the increase in the overall within-cluster sum of squares (10.9) from the merging. The increase in the within-cluster sum of squares by merging clusters $C_k$ and $C_j$ can be shown to be

$$\frac{n_k n_j}{n_k + n_j} \sum_{l=1}^{m} \left( c_{kl} - c_{jl} \right)^2, \tag{10.23}$$

  where the cluster $C_k$ has $n_k$ members and centroid $\mathbf{c}_k$ (with elements $c_{kl}$) and the cluster $C_j$ has $n_j$ members and centroid $\mathbf{c}_j$ (with elements $c_{jl}$).

- Step 3.  Run $K$-means clustering initialized by the reduced clusters in Step 2.

- Step 4: Iterate Steps 2 and 3 until $K = K_{\min}$.

This method provides a hierarchy of cluster solutions as $K$ varies between $K_{\max}$ and $K_{\min}$. In a comparison of 23 clustering methods using precipitation data, Gong and Richman (1995) found that nucleated agglomerative clustering performed best among hard-clustering methods.

## 10.2.3  Gaussian Mixture Model  Ⓒ☺

In Section 3.12, we studied the Gaussian mixture model, and it can be used as a clustering method. It gives the options of hard clustering and soft clustering, where a data point can belong to multiple clusters with a different probability for each cluster.

In the Gaussian mixture model, the probability distribution $p(\mathbf{x})$ is modelled by a sum of $K$ Gaussian functions,

$$p(\mathbf{x}) = \sum_{k=1}^{K} \pi_k \, \mathcal{N}(\mathbf{x}|\,\boldsymbol{\mu}_k, \boldsymbol{\Sigma}_k), \qquad (10.24)$$

where $\mathcal{N}$ is the multivariate Gaussian distribution (3.39) with mean $\boldsymbol{\mu}_k$ and covariance $\boldsymbol{\Sigma}_k$ for the $k$th component, and $\pi_k$ is the mixing coefficient. For $p(\mathbf{x})$ to be a valid probability density, the mixing coefficients must satisfy

$$0 \le \pi_k \le 1, \qquad (10.25)$$

$$\sum_{k=1}^{K} \pi_k = 1. \qquad (10.26)$$

Given the $\mathbf{x}$ data, the model parameters $\pi_k$, $\boldsymbol{\mu}_k$ and $\boldsymbol{\Sigma}_k$ can be found from using the expectation-maximization (EM) algorithm (Section 3.12.1).

Each Gaussian function can be considered a cluster with centroid $\boldsymbol{\mu}_k$. For hard clustering, place the point $\mathbf{x}$ in the $j$th cluster where the maximum value of $\pi_k \mathcal{N}(\mathbf{x}|\,\boldsymbol{\mu}_k, \boldsymbol{\Sigma}_k)$ in (10.24) occurs at $k = j$. For soft clustering, the probability of $\mathbf{x}$ belonging to the $j$th cluster is

$$\frac{\pi_j \, \mathcal{N}(\mathbf{x}|\,\boldsymbol{\mu}_j, \boldsymbol{\Sigma}_j)}{\sum_{k=1}^{K} \pi_k \, \mathcal{N}(\mathbf{x}|\,\boldsymbol{\mu}_k, \boldsymbol{\Sigma}_k)}. \qquad (10.27)$$

Another advantage is that the Gaussian clusters can be ellipsoidal in shape instead of being restricted to spherical clusters as in $K$-means clustering using squared Euclidean distance (Fig. 10.2).

## 10.3  Hierarchical Clustering  Ⓒ☺

Hierarchical clustering (HC) methods not only group objects (i.e. data points) into clusters but also arrange the clusters into a hierarchical structure, with the shape of an inverted tree. There are two approaches to hierarchical clustering, *agglomerative* HC (a bottom-up approach to building the inverted tree structure) and *divisive* HC (a top-down approach). In agglomerative HC, clusters are merged; that is, starting with small clusters each containing a single data point, two closest clusters are merged and the process is repeated until reaching a single cluster containing all the data points. In divisive HC, one starts with one big cluster containing all the data points, then proceeds to repeatedly divide a cluster into two smaller clusters. The agglomerative approach is computationally much faster than the divisive approach, so we will only study the former.

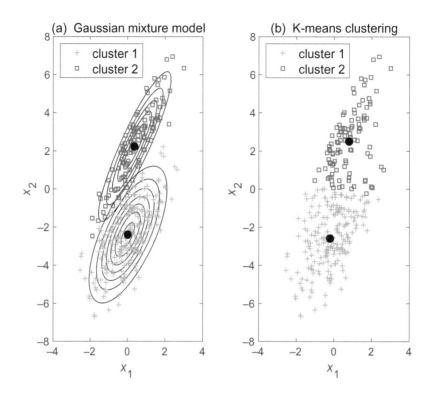

Figure 10.2 2-D data generated from two elliptical Gaussian distributions, with 100 data points in the upper group and 200 in the lower group. (a) Gaussian mixture model clustering of the 300 data points, with $K = 2$, solid circles indicating the centroids and the contours showing the two separate mixture components in (10.24). (b) $K$-means clustering failing to extract two elliptical-shaped clusters.

There are several choices for the *linkage* measure, used in deciding which two clusters are close enough to justify merging them in agglomerative HC. The linkage measure determines the 'distance' between two clusters. All pairs of clusters have their distance determined by the linkage measure, and the pair with the smallest linkage value is merged. Among the most commonly used linkage measures are:

- Maximum distance or complete linkage defines the distance between two clusters $A$ and $B$ as being the maximum distance between a pair of data points, with one point belonging to $A$ and the other to $B$,

$$d(A, B) = \max_{\mathbf{x}_i \in A, \, \mathbf{x}_j \in B} d_{ij}, \qquad (10.28)$$

where $d_{ij}$ is a distance measure for two data points $\mathbf{x}_i$ and $\mathbf{x}_j$ (see Section 10.1.1 for the various choices for $d_{ij}$).

- Minimum distance or single linkage defines the distance between two clusters $A$ and $B$ as being the minimum distance between a pair of data points, with one point in $A$ and the other in $B$,

$$d(A, B) = \min_{\mathbf{x}_i \in A, \mathbf{x}_j \in B} d_{ij}. \tag{10.29}$$

Although not uncommon due to its simplicity, this method can yield chains of small clusters of little practical use (Gong and Richman, 1995, figure 6).

- Average linkage defines the distance between clusters $A$ and $B$ as the average distance between all possible pairs of points, with one point in $A$ and the other in $B$,

$$d(A, B) = \frac{1}{n_A n_B} \sum_{i=1}^{n_A} \sum_{j=1}^{n_B} d_{ij}, \tag{10.30}$$

with $n_A$ and $n_B$ being the number of points in cluster $A$ and $B$, respectively.

Another popular choice is *Ward's method*[1] (Ward, 1963). At each step, this approach finds the pair of clusters that, when merged, would lead to the minimum increase in the total within-cluster sum of squared Euclidean distance $SS_W$, as defined in (10.9). In a comparison of 23 clustering methods on precipitation data, Gong and Richman (1995) found Ward's method to perform best among the hierarchical methods but that hierarchical methods underperformed non-hierarchical methods.

The hierarchical inverted-tree structure can be plotted in a dendrogram. A dendrogram consists of many '⊓'-shaped lines connecting two clusters in a hierarchical tree (Fig. 10.3(a)). Along the vertical axis is plotted the distance measure of two clusters (or in the case of Ward's methods, the $SS_W$). The length of the shorter vertical line in ⊓ represents the distance between the two clusters to be merged. By moving the cut-off level represented by the horizontal dashed line in the dendrogram, one can vary the number of clusters retained. Here the cut-off is set at three clusters, with the clusters shown in Fig. 10.3(b). The optimal number of clusters is $K = 3$ as chosen by the Calinski–Harabasz index, but is just $K = 1$ with the gap statistic, for both variants of Gap in (10.14) and (10.15). Comparing with the $K$-means result in Fig. 10.1(b), the three clusters obtained by Ward's method (Fig. 10.3(b)) are more jagged in appearance. The maximum Calinski–Harabasz index at $K = 3$ is 1,396 for Ward's method and 1,734 for $K$-means. This index as defined in (10.7) is the ratio of the overall between-cluster variance to the overall within-cluster variance; thus, a lower ratio for Ward's method reflects a poorer clustering result compared with $K$-means. Furthermore, the gap statistic ended up choosing $K = 1$ because the standard error bar is much wider for Ward's method (Fig. 10.3(c)) than for $K$-means (Fig. 10.1(e)), which reflects the poor robustness of HC methods.

---

[1] Also called the minimum variance method, though it is actually the within-cluster sum of squared Euclidean distance that is minimized.

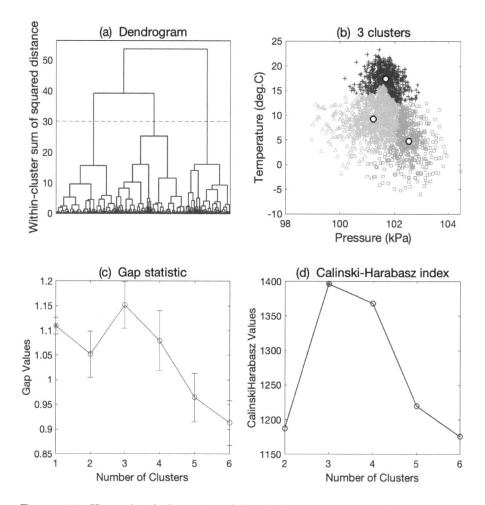

Figure 10.3 Hierarchical clustering of the daily air pressure and temperature data at Vancouver, BC, Canada using Ward's method on standardized data. (a) Dendrogram, where the cut-off level (horizontal dashed line) is set to three clusters, (b) the three clusters, (c) gap statistic and (d) Calinski–Harabasz index.

The common criticism for classical HC algorithms is that they lack robustness and are, hence, sensitive to noise and outliers. Once an object is assigned to a cluster, it will not be considered again, which means that HC algorithms are not capable of correcting possible previous misclassification. (R. Xu and Wunsch, 2005, p. 651).

In environmental science applications, classical HC methods have been found to underperform other clustering methods (Gong and Richman, 1995; M. Bao and Wallace, 2015). Furthermore, in terms of computational speed, with sample size $N$, classical HC algorithms are $O(N^2)$ versus $O(N)$ for $K$-means, hence HC is too slow for large datasets. More recently, HC algorithms have been developed to work efficiently with large $N$. For instance, the BIRCH HC method is $O(N)$ and has improved robustness (T. Zhang et al., 1996). There does not seem to be an authoritative comparison between the new HC methods and the non-hierarchical methods.

## 10.4 Self-Organizing Map $\boxed{\text{C}}$☺

The *self-organizing map* (SOM) method, introduced by Kohonen (1982), approximates a dataset in multi-dimensional space by a flexible grid (typically of one or two dimensions) of cluster centres (Kohonen, 2001). Widely used for clustering, SOM can also be regarded as a discrete version of non-linear principal component analysis (Section 10.6) (Cherkassky and Mulier, 2007, section 6.3).

As with many neural network models, self-organizing maps have a biological background. In neurobiology, it is known that many structures in the cortex of the brain are 2-D or 1-D. However, even the perception of colour involves three types of light receptor. Besides colour, human vision also processes information about the shape, size, texture, position and movement of an object. So the question naturally arises as to how 2-D networks of neurons can process higher dimensional signals in the brain.

Among various possible 2-D grids, rectangular and hexagonal grids are most commonly used in SOM. Though the rectangular grid was introduced first, the hexagonal grid with six neighbouring grid points is more flexible, hence is now more popular. For a two-dimensional grid, the grid points or units (a.k.a. nodes or neurons) are indexed by $\mathbf{i}_j = (l, m)$, where $l$ and $m$ take on integer values, with $l = 1, \ldots, L$, $m = 1, \ldots, M$, and $j = 1, \ldots, LM$ (if a one-dimensional grid is desired, simply set $M = 1$).

To initialize the training process, principal component analysis (PCA) (Section 9.1) is usually performed on the dataset, and the grid $\mathbf{i}_j$ is mapped to $\mathbf{z}_j(0)$ lying on the plane spanned by the two leading PCA eigenvectors. As training proceeds, the initial flat 2-D surface of $\mathbf{z}_j(0)$ is bent to fit the data. The original SOM was trained in a flow-through manner (i.e. observations are presented one at a time during training), though algorithms for batch training are now also available. In flow-through training, there are two steps to be iterated, starting with $n = 1$:

- Step (i): At the $n$th iteration, randomly select an observation $\mathbf{x}(n)$ from the data space and find among the points $\mathbf{z}_j(n-1)$ the one with the shortest (Euclidean) distance to $\mathbf{x}(n)$. Call this closest neighbour $\mathbf{z}_k$, with the corresponding unit $\mathbf{i}_k$ called the *best matching unit* (BMU).

- Step (ii): Let

$$\mathbf{z}_j(n) = \mathbf{z}_j(n-1) + \eta\, h\big(\|\mathbf{i}_j - \mathbf{i}_k\|^2\big)\,[\mathbf{x}(n) - \mathbf{z}_j(n-1)], \qquad (10.31)$$

where $\eta$ is the learning rate parameter and $h$ is a neighbourhood or kernel function. The neighbourhood function gives more weight to the grid points

$\mathbf{i}_j$ near $\mathbf{i}_k$ than those far away, an example being a Gaussian drop-off with distance.

Note that the distances between neighbours are computed for the fixed grid points $(\mathbf{i}_j = (l, m))$, not for their corresponding positions $\mathbf{z}_j$ in the data space. Typically, as $n$ increases, the learning rate $\eta$ is decreased gradually from the initial value of 1 towards 0, while the width of the neighbourhood function is also gradually narrowed (Cherkassky and Mulier, 2007, section 6.3.3).

While SOM has been commonly used as a clustering tool, it should be pointed out that it may underperform simpler techniques such as $K$-means clustering. Balakrishnan et al. (1994) found that $K$-means clustering had fewer points misclassified compared with SOM, and the classification accuracy of SOM worsened as the number of clusters in the data increased. Mingoti and J. O. Lima (2006) tested SOM against $K$-means and other clustering methods over a large number of datasets and found that SOM did not perform well in almost all cases. On the other hand, some studies found SOM outperforming $K$-means clustering (Bacao et al., 2005; G. F. Lin and L. H. Chen, 2006), so whether SOM as a clustering tool is better than $K$-means may be problem dependent. Thus, the value of SOM lies primarily in its role as discrete non-linear PCA, rather than as a clustering algorithm. In other words, SOM is most useful when one suspects there is an underlying low-dimensional (2-D or 1-D) manifold structure in the data. Some units in a 2-D grid in SOM may even be bypassed by having no data points assigned (as seen in Fig. 10.5), whereas in $K$-means clustering a cluster cannot be empty.

As an example, consider the famous Lorenz 'butterfly'-shaped attractor from chaos theory (Lorenz, 1963). Describing idealized atmospheric convection, the Lorenz system is governed by three (non-dimensionalized) differential equations:

$$\dot{x} = -ax + ay, \quad \dot{y} = -xz + bx - y, \quad \dot{z} = xy - cz, \qquad (10.32)$$

where the overhead dot denotes a time derivative, and $a, b$ and $c$ are three parameters of the dynamical system. A chaotic system is generated by choosing $a = 10, b = 28$ and $c = 8/3$. The Lorenz data are fitted by a 2-D SOM and a 1-D SOM in Fig. 10.4. The one-dimensional fit resembles a discrete version of the non-linear PCA solution (Monahan, 2000) found using auto-associative NN (i.e. autoencoder) (Section 10.6).

A propagating wave can also be represented in a two-dimensional SOM. Y. Liu, Weisberg, and Mooers (2006) illustrated a sinusoidal wave travelling to the right by a $3 \times 4$ SOM (Fig. 10.5). As time progresses, the BMU rotates counterclockwise around the SOM (i.e. the patterns of the $3 \times 4$ SOM are manifested sequentially as time progresses), producing a travelling wave pattern. The counterclockwise movement around the SOM means that the two nodes at the centre of the SOM are not excited, hence their associated patterns (patterns (5) and (8) in the figure) are spurious.

Using different random numbers during initialization, SOM can converge to different solutions. Two quantitative measures of mapping quality are commonly used to choose the best model solution: average *quantization error* (QE)

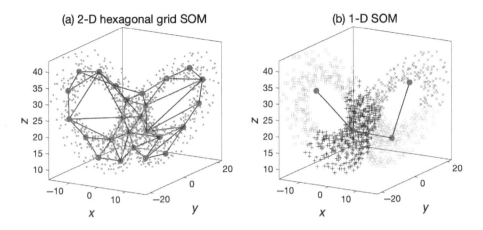

Figure 10.4 (a) A two-dimensional self-organizing map (SOM) where a $5 \times 5$ hexagonal mesh is fitted to the Lorenz (1963) attractor data (dots) and (b) a one-dimensional SOM with four units, dividing the data points into four clusters.

and *topographic error* (TE) (Kohonen, 2001, p. 161; Kiviluoto, 1996). The measure QE is the average distance between each data point $\mathbf{x}$ and $\mathbf{z}_k$ of its BMU. The TE value gives the fraction of data points for which the first BMU and the second BMU are not neighbouring units. Smaller QE and TE values indicate better mapping quality. By increasing the number of units, QE can be further decreased; however, TE will eventually rise, indicating that one is using an excessive number of units. Thus, QE and TE are helpful for finding the appropriate number of units in SOM.

It is worth mentioning that the *generative topographic mapping* (GTM) model (Bishop et al., 1998) is a probabilistic alternative to SOM, analogous to the Gaussian mixture model (Section 10.2.3) being a probabilistic alternative to $K$-means clustering. In GTM, a node $\mathbf{i}_j$ in a regular grid is mapped to a corresponding point $\mathbf{z}_j$ in the data space as in SOM, except that in GTM each $\mathbf{z}_j$ is the centre of a Gaussian density function. Hence, GTM is basically a Gaussian mixture alternative of SOM.

## 10.4.1  Applications of SOM  Ⓒ☺

SOM has been used in many studies in the environmental sciences, as given in the review by Y. Liu and Weisberg (2011). Cavazos (1999) applied a $2 \times 2$ SOM to cluster the winter daily precipitation in northeastern Mexico and southeastern Texas, while Cavazos (2000) used SOM to study the wintertime precipitation in the Balkans. Hewitson and Crane (2002) applied SOM to identify the January sea level pressure (SLP) anomaly patterns in northeastern USA.

The SOM method has been widely used to classify satellite data, including ocean colour (Yacoub et al., 2001; El Hourany et al., 2019), sea surface tem-

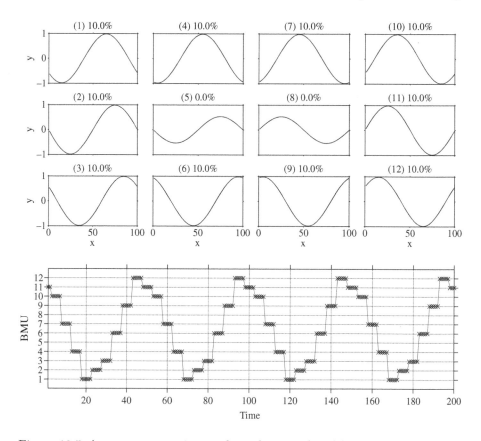

Figure 10.5 A wave propagating to the right is analysed by a $3 \times 4$ SOM. The frequency of occurrence of each SOM pattern is given by the percentage on top of each panel. As time progresses, the best matching unit (BMU) rotates counterclockwise around the $3 \times 4$ SOM patterns, where the SOM patterns (5) and (8) are bypassed (as indicated by their frequency of occurrence of 0.0%). [Reproduced from Y. Liu, Weisberg, and Mooers (2006, figure 1).]

perature (SST) (Richardson et al., 2003), sea level height (Hardman-Mountford et al., 2003), scatterometer winds (Richardson et al., 2003) and aerosol type and optical thickness (Niang et al., 2006). Villmann et al. (2003) applied SOM not only to clustering low-dimensional spectral data from the LANDSAT thematic mapper but also to high-dimensional hyperspectral AVIRIS (Airborne Visible-Near Infrared Imaging Spectrometer) data where there are about 200 frequency bands. A 2-D SOM with a mesh of $40 \times 40$ was applied to AVIRIS data to classify the land surface geology.

SOM was used to study ocean currents on the West Florida Shelf (Y. G. Liu and Weisberg, 2005; Y. G. Liu, Weisberg, and Shay, 2007). N. C. Johnson (2013) applied a 1-D SOM on tropical Pacific SST anomalies to identify the various flavours of El Niño and La Niña. Finnis et al. (2009a) and Finnis et al. (2009b) applied SOM to daily mean SLP anomaly data from reanalysis and general circulation models in a study of synoptic hydroclimatology of Arctic watersheds. In their cluster analysis of the Northern Hemisphere wintertime 500-hPa flow regimes, M. Bao and Wallace (2015) found SOM to outperform Ward's method.

For evaluating the performance of 22 IPCC global climate models, SOM was used to classify the characteristic daily patterns of sea level pressure over a region in North America (Radić and G. K. C. Clarke, 2011). SOM was also used to study future changes in autumn atmospheric river events in British Columbia, Canada, as projected by five global climate models (Radić, Cannon, et al., 2015).

## 10.5   Autoencoder   $\boxed{\text{A}}$☺

An *autoencoder* (a.k.a. *auto-associative neural network*) is a neural network model that uses unsupervised learning to *code* (i.e. extract essential information from) the input data. The model architecture is a feedforward multi-layer perceptron model with multiple hidden layers. The most important hidden layer is called the *code* or *bottleneck* layer as it contains the extracted code from the input (Fig. 10.6). While a standard MLP for non-linear regression has input data $\mathbf{x}$ and output target data $\mathbf{y}$, the autoencoder uses $\mathbf{x}$ for both the input and target data (hence the term auto-associative NN). The network can be divided into two parts, an *encoder* mapping from the input $\mathbf{x}$ to the code $\mathbf{u}$, that is,

$$\mathbf{u} = \mathbf{f}_{\text{encoder}}(\mathbf{x}), \tag{10.33}$$

and a *decoder* mapping from the code to the output, that is,

$$\mathbf{x}' = \mathbf{f}_{\text{decoder}}(\mathbf{u}). \tag{10.34}$$

At first thought, training the model output to resemble the input data seems to be a useless exercise. However, a bottleneck with a small number of nodes or neurons greatly reduces the underlying dimension of the data, similar to using principal component analysis (PCA) to reduce the dimension of a dataset. The autoencoders are trained much like the standard MLP model, that is, the model weights are adjusted to minimize an objective function $J$,

$$J = \langle \|\mathbf{x} - \mathbf{x}'\|^2 \rangle + \Omega, \tag{10.35}$$

where the first term is the MSE (mean squared error) of the model output, with $\langle \ldots \rangle$ denoting the average over all training data, and the second term $\Omega$ is a regularization term added to improve model learning.

There are many applications for autoencoders (Goodfellow, Bengio, et al., 2016, chapter 14), among them:

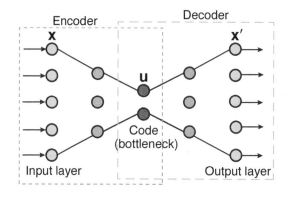

Figure 10.6 The autoencoder or auto-associative neural network is a feedfor-ward MLP NN model, mapping from the input layer to the output layer while passing through several hidden layers, including a bottleneck layer called the 'code', given by the bottleneck nodes **u**. To reduce clutter, the connecting ar-rows linking the nodes in the network are not drawn. The model output **x**′ is trained towards the target data **x**, the same as the input data **x**. The first part of the network, called the *encoder*, maps from the input layer to the bot-tleneck layer, while the second part, the *decoder*, maps from the bottleneck to the output layer.

- *Dimensionality reduction*: PCA has been widely used for dimensionality re-duction (Section 9.1.9). Autoencoders serve as a non-linear generalization of PCA, allowing efficient dimensionality reduction of a dataset (Kramer, 1991). The non-linear mapping of high-dimensional data to a low-dimensional man-ifold often allows a clearer interpretation of the underlying structure of the data.

- *Denoising*: When training the denoising autoencoder, the target data are still **x**, but some form of noise is added to the input data (or a fraction of the input is randomly set to zero) when training the autoencoder. This makes it harder for the autoencoder to overfit. Once the autoencoder has been trained, the model output **x**′ is a denoised version of the noisy input data (Vincent et al., 2010; Goodfellow, Bengio, et al., 2016, sections 14.2.2 and 14.5). The denoising autoencoder has been used in a model for thunderstorm predictions (Kamangir et al., 2020).

- *Variational autoencoder*: Unlike the traditional autoencoder, where the en-coder maps $\mathbf{x} \to \mathbf{u}$, the variational autoencoder has the encoder mapping $\mathbf{x} \to p(\mathbf{u}|\mathbf{x})$, that is, instead of mapping to a single value **u** in the code or bottleneck, the encoder now gives the conditional distribution $p(\mathbf{u}|\mathbf{x})$ (Kingma and Welling, 2014). By randomly sampling from this distribution, one can use the decoder to generate many new patterns **x**′.

While the early autoencoders (Kramer, 1991) have only one hidden layer between the input layer and the code layer and again one hidden layer between the code layer and the output layer (Fig. 10.6), the advent of deep learning has led to deep autoencoders with more hidden layers in the encoder and in the decoder.

The advantages of increased depth in the autoencoder architecture are (Goodfellow, Bengio, et al., 2016, section 14.3):

- Depth can exponentially reduce the computational cost of representing some functions.

- Depth can exponentially decrease the amount of training data needed to learn some functions.

Furthermore, Hinton and Salakhutdinov (2006) found experimentally that deep autoencoders yield better compression compared with shallow or linear autoencoders.

In recent years, a sparsity regularization term $\Omega_{\text{sparse}}$ is often added to the objective function $J$ in (10.35) (Goodfellow, Bengio, et al., 2016, section 14.2.1; Arpit et al., 2016). For instance, a common form used is

$$\Omega_{\text{sparse}} = \lambda_{\text{sparse}} \sum_i |u_i|, \qquad (10.36)$$

where $\lambda_{\text{sparse}}$ is a hyperparameter and the summation is over all hidden nodes $u_i$ in the code layer. The $L_1$ norm in (10.36) has previously been used to find sparse solutions with the lasso method (Fig. 5.6). The code layer in a *sparse autoencoder* does not need to have a low-dimensional bottleneck structure. With a large enough $\lambda_{\text{sparse}}$, most of the $u_i$ can be turned off ($u_i = 0$), effectively reducing the dimension of the code layer. Other choices for the sparsity regularization term are given in Arpit et al. (2016).

With the advent of deep learning, the convolutional neural network (CNN) has become widely used for analysing image data (Section 15.2). The encoder–decoder architecture has been adapted to work with CNN in the U-net method (Section 15.3.1), where high-resolution image data are mapped to a lower-dimensional space before being mapped back to a high-resolution image.

In Section 10.6, we illustrate the simplest application of an autoencoder by extracting a non-linear PCA mode. In Section 10.7, we also briefly mention other techniques for non-linear dimensionality reduction.

# 10.6 Non-linear Principal Component Analysis B ☺

In Chapter 6, we saw ML methods non-linearly generalizing the linear regression method. In this section, we will explore using autoencoders (i.e. auto-associative neural networks) to non-linearly generalize principal component analysis (PCA).

Figure 10.7 illustrates the difference between linear regression, PCA, non-linear regression and non-linear PCA.

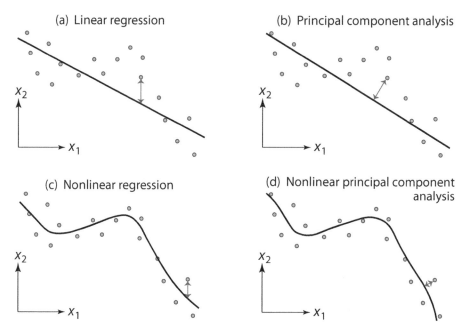

Figure 10.7 Schematic comparison of methods: (a) The linear regression line minimizes the mean squared error (MSE) in the response variable $x_2$, with the error of a particular data point indicated by the double-headed arrow. (b) Principal component analysis (PCA) minimizes the MSE in all variables, with the double-headed arrow perpendicular to the line. (c) Non-linear regression methods produce a curve minimizing the MSE in the response variable. (d) Non-linear PCA methods use a curve that minimizes the MSE of all variables. In both (c) and (d), the smoothness of the curve can be varied by the method. [Follows Hastie and Stuetzle (1989, figure 1).]

Principal component analysis can be performed using neural network (NN) methods (Oja, 1982; Sanger, 1989). However, far more interesting is the non-linear generalization of PCA, where the straight line (or hyperplane) in PCA is replaced by a curve (or manifold) that minimizes the mean squared error (MSE) (Fig. 10.7(d)).

Kramer (1991) proposed using an autoencoder NN model for non-linear PCA (NLPCA). For simplicity, consider an autoencoder where the bottleneck layer has only a single node (Fig. 10.8). The encoder and the decoder part of the network each has only one hidden layer of nodes, that is, $\mathbf{h}^{(x)}$ and $\mathbf{h}^{(u)}$, respectively. The encoder part maps from the inputs $\mathbf{x}$ through the hidden layer $\mathbf{h}^{(x)}$ to the bottleneck node $u$, that is, the encoder performs a non-linear mapping $u = f(\mathbf{x})$, where $u$ will turn out to be the non-linear principal component (NLPC).

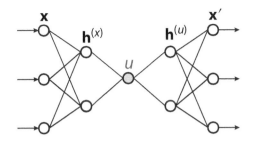

Figure 10.8 An autoencoder for NLPCA, with a feedforward NN architecture. The input layer $\mathbf{x}$ is followed by the encoding layer $\mathbf{h}^{(x)}$, then the code or bottleneck layer (with a single node $u$ for simplicity), the decoding layer $\mathbf{h}^{(u)}$ and finally the output layer $\mathbf{x}'$. Squeezing the input information through a bottleneck accomplishes dimensionality reduction, with $u$ giving the non-linear principal component (NLPC). [Adapted from Hsieh (2001b).]

The decoder part inversely maps from $u$ back to the original higher-dimensional $\mathbf{x}$-space, after passing through the hidden layer $\mathbf{h}^{(u)}$, with the objective that the outputs $\mathbf{x}' = \mathbf{g}(u)$ be as close as possible to the inputs $\mathbf{x}$. Thus, $\mathbf{g}(u)$ non-linearly generates a curve in the $\mathbf{x}$-space, a one-dimensional approximation of the original data. To minimize the MSE (mean squared error) of this approximation, the objective function $J = \langle \|\mathbf{x} - \mathbf{x}'\|^2 \rangle$ (where $\langle \dots \rangle$ denotes taking the average over all training data) is minimized to solve for the weight and offset parameters of the NN. We will add a regularization term $\Omega$ to $J$ later for improved learning.

Let us write down the four layers of mapping involved in the NLPCA model in Fig. 10.8. The activation function $f_1$ maps from $\mathbf{x}$, the input vector of length $m$, to the first hidden layer (the encoding layer) $\mathbf{h}^{(x)}$, a vector of length $l$,[2] with elements

$$h_k^{(x)} = f_1\Big( \big(\mathbf{W}^{(x)}\mathbf{x} + \mathbf{b}^{(x)}\big)_k \Big), \quad k = 1, \dots, l, \tag{10.37}$$

where $\mathbf{W}^{(x)}$ is an $l \times m$ weight matrix, $\mathbf{b}^{(x)}$, a vector of length $l$ containing the offset parameters. Similarly, a second activation function $f_2$ maps from the encoding layer to the bottleneck layer containing a single node $u$,

$$u = f_2\big( \mathbf{w}^{(x)} \cdot \mathbf{h}^{(x)} + \tilde{b}^{(x)} \big). \tag{10.38}$$

Next, an activation function $f_3$ maps from $u$ to the final hidden layer (the decoding layer) $\mathbf{h}^{(u)}$,

$$h_k^{(u)} = f_3\Big( \big(\mathbf{w}^{(u)} u + \mathbf{b}^{(u)}\big)_k \Big), \quad k = 1, \dots, l. \tag{10.39}$$

---

[2] There is no restriction on $l$ relative to $m$, for example, $l$ can be $> m$ if needed.

Then $f_4$ maps from $\mathbf{h}^{(u)}$ to the output $\mathbf{x}'$, with

$$x'_i = f_4\Big(\big(\mathbf{W}^{(u)}\mathbf{h}^{(u)} + \tilde{\mathbf{b}}^{(u)}\big)_i\Big), \quad i = 1, \ldots, m. \tag{10.40}$$

The objective function $J = \langle \|\mathbf{x} - \mathbf{x}'\|^2 \rangle$ is minimized by finding the optimal values of $\mathbf{W}^{(x)}$, $\mathbf{b}^{(x)}$, $\mathbf{w}^{(x)}$, $\tilde{b}^{(x)}$, $\mathbf{w}^{(u)}$, $\mathbf{b}^{(u)}$, $\mathbf{W}^{(u)}$ and $\tilde{\mathbf{b}}^{(u)}$. The MSE between the NN output $\mathbf{x}'$ and the original data $\mathbf{x}$ is thus minimized. The activation functions $f_1$ and $f_3$ are generally non-linear (e.g. the hyperbolic tangent function), while $f_2$ and $f_4$ are usually taken to be the identity function, which simplifies (10.38) and (10.40) to

$$u = \mathbf{w}^{(x)} \cdot \mathbf{h}^{(x)} + \tilde{b}^{(x)}, \tag{10.41}$$

$$x'_i = \big(\mathbf{W}^{(u)}\mathbf{h}^{(u)} + \tilde{\mathbf{b}}^{(u)}\big)_i. \tag{10.42}$$

For non-linear optimization to work well, appropriate scaling of the input variables is needed. One can use standardized variables. However, if the input variables are themselves the leading principal components, then standardization would exaggerate the influence of the higher PCA modes. In this situation, it would be better to scale each variable by subtracting its mean and dividing by the standard deviation of the first PC.

For the regularization term $\Omega$, besides weight penalty, we can also add the normalization conditions $\langle u \rangle = 0$ and $\langle u^2 \rangle = 1$, so the NLPC $u$ has approximately zero mean and unit standard deviation. The regularized objective function $J$ takes the form

$$J = \langle \|\mathbf{x} - \mathbf{x}'\|^2 \rangle + \lambda\Big[ \langle u \rangle^2 + \big(\langle u^2 \rangle - 1\big)^2 \Big]$$

$$+ P\sum_k \Big[ \sum_i \big(W^{(x)}_{ki}\big)^2 + \big(w^{(x)}_k\big)^2 + \big(w^{(u)}_k\big)^2 + \sum_i \big(W^{(u)}_{ik}\big)^2 \Big], \tag{10.43}$$

where $\lambda$ is a hyperparameter controlling the normalization terms for $u$, and $P$ is a weight penalty hyperparameter, with a larger $P$ yielding a less flexible curve for the NLPCA solution. In practice, the weight penalty term can be simplified to only including $\big(W^{(x)}_{ki}\big)^2$, that is, the weights in the encoding layer, and for most suitably scaled problems, taking $\lambda = 1$ suffices (Hsieh, 2001b, 2004).

The total number of (weight and offset) parameters used by the NLPCA is $2ml + 4l + m + 1$, though the number of effectively free parameters is less due to the weight penalty and the constraints on $\langle u \rangle$ and $\langle u^2 \rangle$.

The choice of $l$, the number of hidden nodes in each of the encoding and decoding layers, is important, as too large a value leads to overfitting and too small a value, underfitting. If $f_4$ is the identity function, and $l = 1$, then (10.42) implies that all $x'_i$ are linearly related to a single hidden node, hence there can only be a linear relation between the $x'_i$ variables. Thus, for non-linear solutions, we need to look at $l \geq 2$.

The percentage of the variance explained by the NLPCA mode is given by

$$\Big(1 - \frac{\langle \|\mathbf{x} - \mathbf{x}'\|^2 \rangle}{\langle \|\mathbf{x} - \overline{\mathbf{x}}\|^2 \rangle}\Big) \times 100\%, \tag{10.44}$$

with $\bar{\mathbf{x}}$ being the mean of $\mathbf{x}$.

In effect, the linear relation ($u = \mathbf{e}\cdot\mathbf{x}$) in PCA is now generalized to $u = f(\mathbf{x})$, where $f$ can be any non-linear continuous function representable by an MLP NN mapping from the input layer to the bottleneck, and $\langle \|\mathbf{x}-\mathbf{g}(u)\|^2 \rangle$ is minimized. That the classical PCA is indeed a linear version of this NLPCA can be readily seen by replacing all the activation functions with the identity function, thereby removing the non-linear modelling capability of the NLPCA. Then the forward map to $u$ involves only a linear combination of the original variables as in PCA.

The residual, $\mathbf{x} - \mathbf{g}(u)$, can be input into the same network to extract the second NLPCA mode, and so on for higher modes. However, it is more common to use a bottleneck layer with $k > 1$ nodes to extract a $k$-dimensional manifold than to extract one-dimensional curves repeatedly. For example, if one is interested in extracting the first two modes, the first approach obtains the representation $\mathbf{g}_1(u_1) + \mathbf{g}_2(u_2)$ for the model output, while the latter obtains the more general functional form $\mathbf{g}(u_1, u_2)$, hence should be more effective in dimensionality reduction. However, the disadvantage with the second approach is that $u_1$ and $u_2$ are not ordered hierarchically in the way that PC1 accounts for more variance than PC2 in PCA.

Scholz and Vigario (2002) proposed a hierarchical approach to NLPCA: a good algorithmic implementation of the *hierarchical* PCA should fulfill two important properties – 'scalability' and 'stability'. Scalability means that the first $k$ components or modes explain as much variance as possible in a $k$-dimensional subspace of the data space. Stability means that the $i$th component of a $k$-dimensional solution is identical to the $i$th component of a $k'$-dimensional solution ($k \neq k'$).

PCA satisfies both scalability and stability. NLPCA does not satisfy stability in that $u_1$ extracted by NLPCA using a single bottleneck node is in general not identical to either $u_1$ or $u_2$ extracted from an NLPCA using two bottleneck nodes.[3] Scholz and Vigario (2002) wanted their hierarchical NLPCA to behave in a similar way as PCA, in that the $i$th mode or component accounts for the $i$th highest variance projection.

The objective function in NLPCA is of the form

$$J = E + \text{regularization terms} \qquad (10.45)$$

where $E = \langle \|\mathbf{x} - \mathbf{x}'\|^2 \rangle$ is the MSE. In the hierarchical NLPCA, $E$ is replaced by $E_\mathrm{H}$ in (10.45), where

$$E_\mathrm{H} = E_1 + E_{1,2} \qquad (10.46)$$

for a model with two bottleneck nodes. $E_1$ is the MSE computed from the model output using only the first bottleneck node $u_1$ (i.e. the second bottleneck node $u_2$ is set to 0), while $E_{1,2}$ is the MSE computed using both bottleneck nodes. This way, a two-dimensional manifold solution is found with the first component ($u_1$) explaining more variance than the second component ($u_2$). In the original (non-hierarchical) NLPCA model, $E = E_{1,2}$.

---

[3] Rotated PCA also does not satisfy this stability property (Section 9.2.3)

This hierarchical approach generalizes easily to more than two bottleneck nodes. For instance, with three bottleneck nodes,

$$E_H = E_1 + E_{1,2} + E_{1,2,3} , \tag{10.47}$$

with $E_{1,2,3}$ being the MSE computed using all three bottleneck nodes. The three-dimensional manifold solution has $u_1$ explaining more variance than $u_2$, which explains more variance than $u_3$.

~~~~~~~~~~~~~~~~~~~~~~~~~~~~~~~~~~~~~~~~~~~~~~~~~~~

Let us illustrate the NLPCA method by applying it to the tropical Pacific climate system. The tropical Pacific contains the El Niño–Southern Oscillation (ENSO) (Section 9.1.5), the dominant interannual variability in the global climate system, giving rise to anomalously warm sea surface temperatures (SST) in the eastern equatorial Pacific during El Niño episodes and cool SST in the central equatorial Pacific during La Niña episodes. Hsieh (2001b) used the tropical Pacific SST data (1950–1999) to make a three-way comparison between NLPCA, rotated PCA (RPCA) and PCA. The tropical Pacific SST anomaly (SSTA) data (i.e. the monthly SST data with the climatological seasonal cycle removed) were pre-filtered by PCA (for dimensionality reduction), with only the three leading modes retained. The first three PCs (PC1, PC2 and PC3) were used as the input $\mathbf{x}$ for the NLPCA network (with a single bottleneck node).

The data are shown as dots in a scatter plot in the PC1–PC2 plane in Fig. 10.9, where the cool La Niña states lie in the upper left corner and the warm El Niño states in the upper right corner. The NLPCA solution is a ∪-shaped curve linking the La Niña states at one end (low $u$) to the El Niño states at the other end (high $u$), similar to that found originally by Monahan (2001).

The first PCA eigenvector describes a somewhat unphysical oscillation, as there are no data (dots) close to either end of the horizontal arrow. For the second PCA eigenvector, there are dots close to the bottom of the vertical arrow, but almost no dots near the top of the arrow, that is, one phase of the mode 2 oscillation is realistic, but the opposite phase is not. Thus, if the underlying data have a non-linear structure but we are restricted to finding linear solutions using PCA, the energy of the non-linear oscillation is scattered into multiple PCA modes, many of which represent unphysical linear oscillations.

For comparison, a varimax rotation of the E-frame was applied to the first three PCA eigenvectors, as presented in Section 9.2. The resulting first RPCA eigenvector $\tilde{\mathbf{e}}_1$ spears through the cluster of El Niño states in the upper right corner (Fig. 10.9), thereby yielding a more accurate description of the SSTA during El Niño (Fig. 10.10(c)) than the first PCA mode (Fig. 10.10(a)), which did not fully represent the intense warming of Peruvian waters (just off the west coast of South America). The second RPCA eigenvector $\tilde{\mathbf{e}}_2$ (Fig. 10.9) did not improve much on the second PCA mode, with the PCA spatial pattern shown in Fig. 10.10(b) and the RPCA pattern in Fig. 10.10(d). In terms of variance explained, the first NLPCA mode explained 56.6% of the variance, versus 51.4% by the first PCA mode and 47.2% by the first RPCA mode.

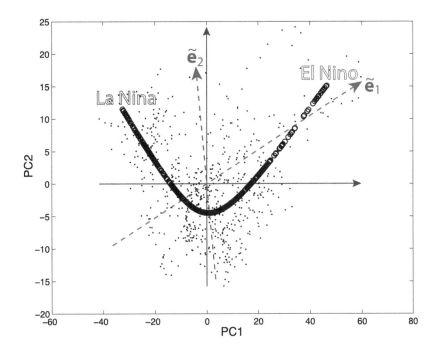

Figure 10.9 Scatter plot of the SST anomaly (SSTA) data (shown as dots) in the PC1–PC2 plane, with the El Niño states lying in the upper right corner and the La Niña states in the upper left corner. The PC2 axis is stretched relative to the PC1 axis for better visualization. The NLPCA first mode approximation to the data is shown by the (overlapping) circles, which trace out a boomerang-shaped curve. The first PCA eigenvector is oriented along the horizontal arrow and the second eigenvector along the vertical arrow. The rotated PCA (RPCA) eigenvectors $\tilde{\mathbf{e}}_1$ and $\tilde{\mathbf{e}}_2$ from a varimax rotation are indicated by the dashed arrows. [Adapted from Hsieh (2001b).]

With the NLPCA, for a given value of the NLPC $u$, one can map from $u$ to the three PCs using the decoder. Each of the three PCs can be multiplied by its associated PCA (spatial) eigenvector and the three added together to yield the spatial pattern for that particular value of $u$. Unlike PCA, which gives the same spatial anomaly pattern except for changes in the amplitude as the PC varies, the NLPCA spatial pattern generally varies continuously as the NLPC changes. Figure 10.10(e) and (f) show, respectively, the spatial anomaly patterns when $u$ has its maximum value (corresponding to the strongest El Niño) and when $u$ has its minimum value (strongest La Niña). Clearly the asymmetry between El Niño and La Niña, that is, the cool anomalies during La Niña episodes (Fig. 10.10(f)) are observed to centre much further west of the warm anomalies during El Niño (Fig. 10.10(e)) (Hoerling et al., 1997), is well captured by the

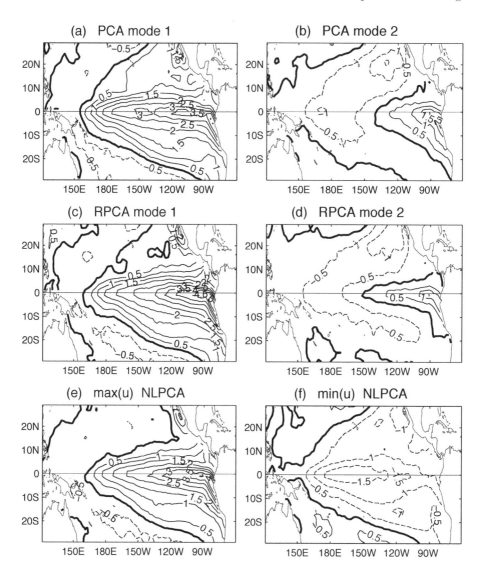

Figure 10.10 The SSTA patterns (in °C) of the PCA, RPCA and the NLPCA. The first and second PCA spatial modes are shown in (a) and (b), respectively (both with their corresponding PCs at maximum value). The first and second varimax RPCA spatial modes are shown in (c) and (d), respectively (both with their corresponding RPCs at maximum value). The anomaly pattern as the NLPC $u$ of the first NLPCA mode varies from (e) its maximum (strong El Niño) to (f) its minimum (strong La Niña). With a contour interval of 0.5°C, the positive contours are shown as solid curves, negative contours as dashed curves and the zero contour as a thick curve. [Reproduced from Hsieh (2004).]

first NLPCA mode. In contrast, the PCA mode 1 gives a La Niña that is simply the mirror image of the El Niño (Fig. 10.10(a)). While El Niño has been known by Peruvian fishermen for many centuries due to its strong SSTA off the coast of Peru and its devastation of the Peruvian fishery, the La Niña, with its weak manifestation in the Peruvian waters, was not appreciated until the last two decades of the twentieth century.

In summary, PCA is used for two main purposes: (i) to reduce the dimensionality of the dataset and (ii) to extract features or patterns from the dataset. Here, both RPCA and NLPCA took the PCs from PCA as input. RPCA eigenvectors (from a varimax E-frame rotation) tend to point more towards local data clusters and are, therefore, more representative of physical states than the PCA eigenvectors but with RPCA mode 1 explaining less variance than PCA mode 1 (Section 9.2).

With a linear approach, it is generally impossible to have a mode 1 solution simultaneously (a) explaining maximum global variance of the dataset and (b) approaching local data clusters, hence the dichotomy between PCA and RPCA (E-frame rotation), with PCA aiming for objective (a) and RPCA for (b). With the more flexible NLPCA method, both objectives (a) and (b) may be attained together (Hsieh, 2001b).

The tropical Pacific SST example illustrates that with a complicated oscillation like the El Niño–La Niña phenomenon, using a linear method such as PCA results in the non-linear mode being scattered into several linear modes (in fact, all three leading PCA modes are related to this phenomenon). In the study of climate variability, the use of PCA methods has tended to produce a slightly misleading view that our climate is dominated by a number of spatially fixed oscillatory patterns, which is in fact an artifact of the linear method. Applying NLPCA to the tropical Pacific SSTA, we found no spatially fixed oscillatory patterns but an oscillation evolving in space as well as in time.

From a broader perspective, extracting an NLPCA mode is a special application of autoencoders, which are in turn a special case of encoder–decoder networks applied to unsupervised learning. In recent years, the advent of convolutional neural network (CNN) models such as the U-net (Section 15.3.1) has made obsolete the pre-filtering step of reducing the input dimension prior to performing NLPCA. The CNN architecture can handle 2-D images/grids using far fewer weights than the traditional MLP NN architecture (Section 15.2), thereby obviating the need for input dimension reduction by the linear PCA method. In general, the interest in extracting non-linear modes from data has declined as it is usually difficult to identify robust modes beyond the first mode. However, the use of a bottleneck NN architecture for dimension reduction and noise reduction has found a myriad of new applications. See Section 15.3 for the modern extension of autoencoder and encoder–decoder methods to deep neural networks (i.e. deep learning).

The NLPCA has been applied to the Lorenz (1963) three component chaotic system (Monahan, 2000; Hsieh, 2001b). For the tropical Pacific climate variability, the NLPCA has been used to study the sea surface temperature (SST) field (Monahan, 2001; Hsieh, 2001b, 2007) and the sea level pressure (SLP) field (Monahan, 2001). In remote sensing, Del Frate and Schiavon (1999) applied NLPCA to the inversion of radiometric data to retrieve atmospheric profiles of temperature and water vapour. PCA has been used to impute missing data (Section 9.1.12); for non-linearly structured data, Scholz, F. Kaplan, et al. (2005) proposed using a variant of NLPCA to impute missing data.

Since PCA can be applied to *complex variables* (Section 9.1.3), NLPCA can also be extended to complex variables (Hsieh, 2009, section 10.5). The *nonlinear complex PCA* (NLCPCA) model introduced by Rattan and Hsieh (2004) and Rattan and Hsieh (2005) uses basically the same architecture as the NLPCA model, except that all the input variables and the weight and offset parameters are now complex-valued. The tropical Pacific wind anomalies $(u, v)$ (expressed as a complex variable $w = u + iv$) have been analysed by NLCPCA in Rattan and Hsieh (2004). Complex PCA has been developed to study fields of scalar time series in the frequency domain by Wallace and Dickinson (1972). A nonlinear generalization of this method has been used to study the time evolution of the tropical Pacific sea surface temperature anomalies (Rattan and Hsieh, 2005) and the offshore propagation of sandbars (Rattan, Ruessink, et al., 2005).

In the following subsections, we examine the problem of overfitting in NLPCA (Section 10.6.1) and the choice of extracting a closed curve NLPCA solution instead of an open curve solution (Section 10.6.2).

## 10.6.1   Overfitting  Ⓒ☺

Unsupervised learning is actually a much harder problem than supervised learning. In supervised learning, such as non-linear regression, the presence of noise in the data can lead to overfitting, that is, the model trying to fit the noise in the data. However, when plentiful data are available (i.e. far more observations than model weights/parameters), overfitting is not a problem when performing non-linear regression on noisy data (Section 8.2). In contrast, in unsupervised learning (e.g. clustering and NLPCA), overfitting remains a problem even with plentiful data. In cluster analysis, the optimal number of clusters to use is an inexact science (Christiansen, 2007a) (Section 10.1.2). In NLPCA, how much flexibility to allow in the curve (or hyperplane) solution is also an inexact science (Christiansen, 2005; Monahan and J. C. Fyfe, 2007; Christiansen, 2007b; Hsieh, 2007).

As illustrated in Fig. 10.11, for a Gaussian-distributed data cloud, a nonlinear model for NLPCA, with enough flexibility, will find the zigzag solution of Fig. 10.11(b) as having a smaller MSE than the linear solution in Fig. 10.11(a). Since the distance between the point $A$ and $a$, its projection on the NLPCA curve, is smaller in Fig. 10.11(b) than the corresponding distance in Fig. 10.11(a), it is easy to see that the more zigzags there are in the curve, the smaller is the MSE. However, the two neighbouring points $A$ and $B$, on opposite sides of an 'ambiguity' line (Malthouse, 1998), are projected far apart on the NLPCA curve in Fig. 10.11(b). Thus, simply searching for the solution which gives the smallest MSE does not guarantee that NLPCA will find a meaningful solution in a noisy dataset.

Figure 10.11 Illustrating how overfitting can occur in NLPCA (even in the limit of infinite sample size). (a) PCA solution for a Gaussian data cloud (shaded ellipse), with two neighbouring points $A$ and $B$ shown projecting to the points $a$ and $b$ on the PCA straight line solution. (b) A zigzag NLPCA solution found by a flexible enough non-linear model, with a smaller MSE than that in (a). Dashed lines illustrate 'ambiguity' lines where neighbouring points (e.g. $A$ and $B$) on opposite sides of these lines are projected to $a$ and $b$, far apart on the NLPCA curve. [Adapted from Hsieh (2007).]

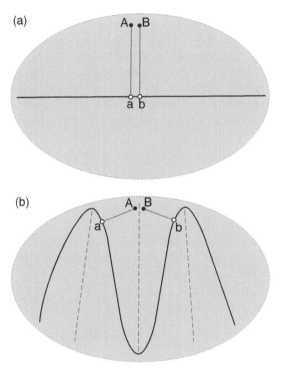

As the earlier use of the NLPCA model did not include regularization terms in the objective function (Kramer, 1991; Monahan, 2000; Monahan, 2001), Hsieh (2001b) added a weight penalty to the NLPCA model to eliminate excessively wiggly solutions. With a large enough weight penalty hyperparameter $P$ in (10.43), the model flexibility of the NLPCA can be reduced to even give the linear PCA solution. If the overfitting arises as in Fig. 10.11(b), using independent data to validate the MSE from various models with different $P$ is not a viable method for choosing the optimal $P$, since independent data from the Gaussian data cloud will still find the validation MSE to be lower in the wiggly model than in the straight line model. Instead, Hsieh (2007) proposed an *inconsistency index* for detecting the projection of neighbouring points to distant parts of the NLPCA curve and used the index to choose the appropriate $P$.

The index is calculated as follows: for each data point $\mathbf{x}$, find its nearest neighbour $\tilde{\mathbf{x}}$. The NLPC for $\mathbf{x}$ and $\tilde{\mathbf{x}}$ are $u$ and $\tilde{u}$, respectively. With $\operatorname{cor}(u, \tilde{u})$ denoting the Pearson correlation between all the pairs $(u, \tilde{u})$, the inconsistency index $I$ is defined as

$$I = 1 - \operatorname{cor}(u, \tilde{u}). \tag{10.48}$$

If for some nearest neighbour pairs, $u$ and $\tilde{u}$ are assigned very different values, $\mathrm{cor}(u, \tilde{u})$ would have a lower value, leading to a larger $I$, indicating greater inconsistency in the NLPC mapping. With $u$ and $\tilde{u}$ standardized to having zero mean and unit standard deviation, (10.48) is equivalent to

$$I = \tfrac{1}{2} \left\langle (u - \tilde{u})^2 \right\rangle, \tag{10.49}$$

(Exercise [10.4]). The $I$ index plays a similar role as the topographic error in self-organizing maps (SOM) (Section 10.4).

In statistics, various criteria, often in the context of linear models, have been developed to select the right amount of model complexity so neither overfitting nor underfitting occurs. These criteria are often called 'information criteria' (IC) (Section 8.11). An IC is typically of the form

$$\mathrm{IC} = \mathrm{MSE} + \text{complexity term}, \tag{10.50}$$

where MSE is evaluated over the training data and the complexity term is larger when a model has more free parameters. The IC is evaluated over a number of models with different numbers of free parameters, and the model with the minimum IC is selected as the best. As the presence of the complexity term in the IC penalizes models that use an excessive number of free parameters to attain low MSE, choosing the model with the minimum IC would rule out complex models with overfitted solutions.

In Hsieh (2007), a holistic IC to deal with the type of overfitting arising from the broad data distribution (Fig. 10.11(b)) was introduced as

$$H = \mathrm{MSE} + \text{inconsistency term}, \tag{10.51}$$

$$= \mathrm{MSE} - \mathrm{cor}(u, \tilde{u}) \times \mathrm{MSE} = \mathrm{MSE} \times I, \tag{10.52}$$

where MSE and cor were evaluated over all (training and validation) data, inconsistency was penalized and the model run with the smallest $H$ value was selected as the best.

A different approach to validating NLPCA was introduced by Scholz (2012), where the absence of target data in unsupervised learning was dealt with by artificially creating target data from randomly removing elements of the input data vector and using a variant of the NLPCA to predict the missing data values for model validation.

## 10.6.2   Closed Curves   [C]☺

While the NLPCA is capable of finding a continuous open curve solution, there are many phenomena involving waves or quasi-periodic fluctuations, which call for a continuous closed curve solution. The limitation of NLPCA to open curves (Malthouse, 1998) was removed by Kirby and Miranda (1996), who introduced an NLPCA with a circular node at the network bottleneck (henceforth referred to as the NLPCA(cir)). The non-linear principal component (NLPC), as represented by the circular node, is an angular variable $\theta$, and the NLPCA(cir) is capable of approximating the data by a closed continuous curve. Figure 10.12 shows the NLPCA(cir) network, which is almost identical to the NLPCA of Fig. 10.8, except at the bottleneck, where there are now two nodes $p$ and $q$ constrained to lie on a unit circle in the $p$–$q$ plane, so there is only one free angular variable $\theta$, the NLPC.

Figure 10.12 NLPCA(cir), the NLPCA model with a circular node at the bottleneck. Instead of having one bottleneck node $u$, there are now two nodes $p$ and $q$ constrained to lie on a unit circle in the $p$–$q$ plane, so there is only one free angular variable $\theta$, the NLPC. This network is suited for extracting a closed curve solution. [Adapted from Hsieh (2001b).]

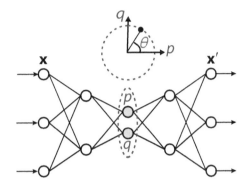

At the bottleneck in Fig. 10.12, analogous to $u$ in (10.41), the pre-states $p_o$ and $q_o$ are computed as

$$p_o = \hat{\mathbf{w}}^{(x)} \cdot \mathbf{h}^{(x)} + \hat{b}^{(x)}, \quad \text{and} \quad q_o = \tilde{\mathbf{w}}^{(x)} \cdot \mathbf{h}^{(x)} + \tilde{b}^{(x)}, \tag{10.53}$$

where $\hat{\mathbf{w}}^{(x)}, \tilde{\mathbf{w}}^{(x)}$ are weight vectors, and $\hat{b}^{(x)}$ and $\tilde{b}^{(x)}$ are offset parameters. Let

$$r = (p_o^2 + q_o^2)^{1/2}, \tag{10.54}$$

then the circular node is defined with

$$p = p_o/r, \quad \text{and} \quad q = q_o/r, \tag{10.55}$$

satisfying the unit circle equation $p^2 + q^2 = 1$. Thus, although there are two variables $p$ and $q$ at the bottleneck, there is only one angular degree of freedom from $\theta$ (Fig. 10.12), due to the circle constraint.

When implementing NLPCA(cir), Hsieh (2001b) found that there are two possible configurations: (i) a restricted configuration where the constraints $\langle p \rangle = 0 = \langle q \rangle$ are applied and (ii) a general configuration without the constraints. With (i), the constraints can be satisfied approximately by adding the extra terms $\langle p \rangle^2$ and $\langle q \rangle^2$ to the objective function. If a closed curve solution is sought, then (i) is better than (ii) as it has effectively two fewer parameters. However, (ii), being more general than (i), can more readily model open curve solutions like a regular NLPCA. The reason is that if the input data mapped onto the $p$–$q$ plane cover only a segment of the unit circle instead of the whole circle, then the inverse mapping from the $p$–$q$ space to the output space will yield a solution resembling an open curve. Hence, given a dataset, (ii) may yield either a closed curve or an open curve solution.

The inconsistency index $I$ in (10.48) is modified to

$$I = 1 - \tfrac{1}{2}\big[\mathrm{cor}(p, \tilde{p}) + \mathrm{cor}(q, \tilde{q})\big], \tag{10.56}$$

where $p$ and $q$ are from the bottleneck, with $\tilde{p}$ and $\tilde{q}$ the corresponding nearest neighbour values, and the information criterion $H = \mathrm{MSE} \times I$ (Hsieh, 2007).

~~~~~~~~~~~~~~~~~~~~~~~~~~~~~~~~~~~~~~~~~~~~~~~~~~

For an application of NLPCA(cir) on real data, consider the quasi-biennial oscillation (QBO), which dominates over the annual cycle or other variations in the equatorial stratosphere, with the period of oscillation varying between 22 and 32 months. Average zonal (i.e. westerly component of) winds at 70, 50, 40, 30, 20, 15 and 10 hPa (i.e. from about 20 to 30 km altitude) during 1956–2006 were studied. After the 51-year means were removed, the zonal wind anomalies $U$ at seven vertical levels in the stratosphere became the seven inputs to the NLPCA(cir) network (Hamilton and Hsieh, 2002; Hsieh, 2007). Since the data were not very noisy (Fig. 10.13), a rather complex model was used, with $l$ (the number of nodes in the encoding layer) ranging from 5 to 9, and weight penalty $P = 10^{-1}, 10^{-2}, 10^{-3}, 10^{-4}, 10^{-5}, 0$. The smallest $H$ occurred when $l = 8$ and $P = 10^{-5}$, with the closed curve solution shown in Fig. 10.13.

By choosing a relatively large $l$ and a small $P$, the $H$ IC justified having considerable model complexity, including the wiggly behaviour seen in the 70 hPa wind (Fig. 10.13(c)). The wiggly behaviour can be understood by viewing the phase–pressure contour plot of the zonal wind anomalies (Fig. 10.14): as the easterly wind anomaly descends with time (i.e. as phase increases), wavy behaviour is seen in the 40, 50 and 70 hPa levels at $\theta_{\mathrm{weighted}}$ around 0.4–0.5. This example demonstrates the benefit of having an IC to decide objectively on how smooth or wiggly the fitted curve should be.

The observed strong asymmetries between the easterly and westerly phases of the QBO (Hamilton, 1998; Baldwin et al., 2001) are captured by this NLPCA(cir) mode – for example, the much more rapid transition from easterlies to westerlies than the reverse transition and the much deeper descent of the easterlies than the westerlies (Fig. 10.14). For comparison, Hamilton and Hsieh (2002) constructed a *linear* model of $\theta$, which was unable to capture the observed strong asymmetry between easterlies and westerlies.

The NLPCA(cir) model has also been used to study the tropical Pacific climate variability (Hsieh, 2001b; An, Hsieh, et al., 2005; An, Z. Q. Ye, et al., 2006; J. Choi et al., 2009), the non-sinusoidal propagation of underwater sandbars off beaches in the Netherlands and Japan (Ruessink et al., 2004), the tidal cycle off the German North Sea coast (Herman, 2007) and the Madden-Julian oscillation (MJO) (Jenkner et al., 2011), where MJO is the dominant component of the intraseasonal (30–90 days) variability in the tropical atmosphere (C. Zhang, 2005).

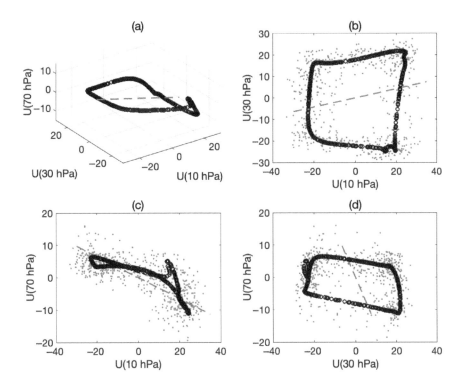

Figure 10.13 The NLPCA(cir) mode 1 solution for the equatorial stratospheric zonal wind anomalies. For comparison, the PCA mode 1 solution is shown by the dashed line. Only three out of seven dimensions are shown, namely the zonal velocity anomaly $U$ at the top, middle and bottom levels (10, 30 and 70 hPa). (a) A 3-D view. (b)–(d) 2-D views. [Reproduced from Hsieh (2007).]

# 10.7   Other Non-linear Dimensionality Reduction Methods  ⌈C⌉☺

Besides autoencoders, there are many other methods for finding a low-dimensional manifold in a high-dimensional data space (J. A. Lee and Verleysen, 2007; Hastie, Tibshirani, et al., 2009, chapter 14; van der Maaten, Postma, et al., 2009). Many of these are local non-parametric methods, for example locally linear embedding (LLE) (Roweis and Saul, 2000) and isomap (Tenenbaum et al., 2000). With a node for each training data point, local graphs are constructed from nearest neighbours, and the general shape of the manifold is obtained by local interpolation. These methods have the disadvantage of requiring a very large number of training data points if the underlying manifold is not smooth (Goodfellow, Bengio, et al., 2016, p. 509). Although these methods are

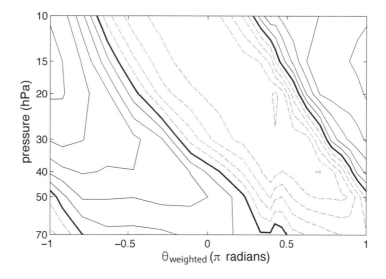

Figure 10.14 Contour plot of the NLPCA(cir) mode 1 zonal wind anomalies as a function of pressure and phase $\theta_{\text{weighted}}$, where $\theta_{\text{weighted}}$ is $\theta$ weighted by the histogram distribution of $\theta$ (see Hamilton and Hsieh, 2002). Thus, $\theta_{\text{weighted}}$ is more representative of actual time during a cycle than $\theta$. Contour interval is 5 m s$^{-1}$, with westerly winds indicated by solid lines easterlies by dashed lines and zero contours by thick lines. [Reproduced from Hsieh (2007).]

successful on artificial datasets, they are often not very successful at visualizing real, high-dimensional data (van der Maaten and Hinton, 2008).

To help visualizing high-dimensional data, van der Maaten and Hinton (2008) developed the t-SNE (t-distributed stochastic neighbour embedding) method, which maps each high-dimensional data point to a point in a 2-D (or 3-D) space. For several test problems, t-SNE is seen to display more clearly separated clusters than methods like LLE, isomap, and so on. This method has a 'perplexity' hyperparameter, with typical values between 5 and 50. According to Wattenberg et al. (2016), t-SNE has become popular as it is very flexible, often finding structure while other dimensionality-reduction algorithms fail; however, that very flexibility makes t-SNE tricky to interpret, as very different cluster patterns can be found in the low-dimensional space by using different perplexity values or different numbers of iterative steps. In addition, the distance separating two clusters in the low-dimensional space may have no relation with their separation in the high-dimensional space. Sonnewald et al. (2020) applied t-SNE to determine global marine ecological provinces from plankton community structure and nutrient flux data.

# 10.8 Non-linear Canonical Correlation Analysis $\boxed{\text{C}}$ ☺

In Section 9.4, canonical correlation analysis (CCA) was presented as the generalization of correlation to the multivariate case. CCA is widely used but is limited by being a linear model. A number of different approaches have been proposed to generalize CCA non-linearly (Lai and C. Fyfe, 1999; Lai and C. Fyfe, 2000; Hsieh, 2000; Suykens et al., 2002; Hardoon et al., 2004; Shawe-Taylor and Cristianini, 2004).

To perform non-linear CCA (NLCCA), a simple way is to use three multilayer perceptron (MLP) NNs, where the linear mappings in the CCA are replaced by the non-linear mapping functions of MLP NNs with one or more hidden layers.

Consider two vector variables $\mathbf{x}$ and $\mathbf{y}$, each with $N$ observations. CCA looks for linear combinations

$$u = \mathbf{f}^{\mathrm{T}}\mathbf{x}, \quad \text{and} \quad v = \mathbf{g}^{\mathrm{T}}\mathbf{y}, \tag{10.57}$$

where the canonical variates $u$ and $v$ have maximum correlation, that is, the weight vectors $\mathbf{f}$ and $\mathbf{g}$ are chosen such that $\text{cor}(u, v)$, the Pearson correlation coefficient between $u$ and $v$, is maximized (Section 9.4.1). For NLCCA, the linear maps $\mathbf{f}$ and $\mathbf{g}$, and their inverse maps, are replaced below by non-linear mapping functions using MLP NNs (Hsieh, 2000, 2001a).

The mappings from $\mathbf{x}$ to $u$ and $\mathbf{y}$ to $v$ are represented by the double-barrelled NN in Fig. 10.15(a). By minimizing the objective function $J = -\text{cor}(u, v)$, one finds the parameters that maximize the correlation $\text{cor}(u, v)$. After the forward mapping with the double-barrelled NN has been solved, inverse mappings from the canonical variates $u$ and $v$ to the original variables, as represented by the two standard MLP NNs in Fig. 10.15(b) and (c), are to be solved.

To have the canonical variates normalized, that is, $\langle u \rangle = 0 = \langle v \rangle$, and $\langle u^2 \rangle = 1 = \langle v^2 \rangle$, the objective function $J$ is modified to

$$J = -\text{cor}(u, v) + \langle u \rangle^2 + \langle v \rangle^2 + \left( \langle u^2 \rangle^{1/2} - 1 \right)^2 + \left( \langle v^2 \rangle^{1/2} - 1 \right)^2$$
$$+ \text{ weight penalty term.} \tag{10.58}$$

The MLP NN mapping from $u$ to the output $\mathbf{x}'$ is solved by minimizing the objective function $J_1$, where

$$J_1 = \left\langle \|\mathbf{x}' - \mathbf{x}\|^2 \right\rangle + \text{ weight penalty term.} \tag{10.59}$$

Similarly, the mapping from $v$ to $\mathbf{y}'$ is solved by minimizing $J_2$, where

$$J_2 = \left\langle \|\mathbf{y}' - \mathbf{y}\|^2 \right\rangle + \text{ weight penalty term.} \tag{10.60}$$

In prediction problems, one may treat $\mathbf{x}$ as the predictors and $\mathbf{y}$ as the response variables, that is, $\hat{\mathbf{y}} = f(\mathbf{x})$, where values of the canonical variate $\hat{v}$ must be predicted from values of the canonical variate $u$. The procedure

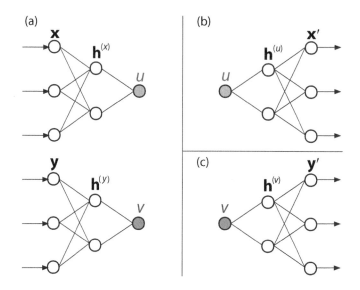

Figure 10.15 The three MLP NNs used to perform NLCCA. (a) The double-barrelled NN maps from the inputs **x** and **y** to the canonical variates $u$ and $v$. The objective function $J$ forces the correlation between $u$ and $v$ to be maximized. (b) The NN maps from $u$ to the output layer **x'**, where the objective function $J_1$ basically minimizes the MSE of **x'** relative to **x**. (c) The NN maps from $v$ to the output layer **y'**, where the objective function $J_2$ basically minimizes the MSE of **y'** relative to **y**. [Reproduced from Hsieh (2001a), ©American Meteorological Society. Used with permission.]

is as follows: first the NLCCA model is trained from some training data and $\text{cor}(u, v)$ is obtained, then from the predictor data **x** (which can either be from the training dataset or new data), corresponding $u$ values are computed by the model. For canonical variates normalized to unit variance and zero mean, the linear least-squares regression solution for estimating $\hat{v}$ from $u$ is given by

$$\hat{v} = u \, \text{cor}(u, v), \tag{10.61}$$

(von Storch and Zwiers, 1999, p. 325). From $\hat{v}$, one obtains a value for $\hat{\mathbf{y}}$ via the inverse mapping NN.

As the Pearson correlation is not a robust measure of association between two variables, a robust NLCCA model was developed by Cannon and Hsieh (2008) using the robust biweight midcorrelation (Section 2.11.5) instead of the Pearson correlation in $J$. For the inverse mappings, the objective functions $J_1$ and $J_2$ can use the robust MAE instead of the non-robust MSE.

~~~~~~~~~~~~~~~~~~~~~~~~~~~~~~~~~~~~~~~~~~~~

NLCCA has been used to analyse the relation between the tropic Pacific sea level pressure anomalies (SLPA) and sea surface temperature anomalies (SSTA)

(Hsieh, 2001a). Relations between the tropical Pacific wind stress anomaly (WSA) and SSTA fields have also been studied using the NLCCA (Wu and Hsieh, 2002; Wu and Hsieh, 2003), where interdecadal changes of ENSO behaviour before and after the mid-1970s climate regime shift were found, with greater non-linearity found during 1981–1999 than during 1961–1975.

Relations between the tropical Pacific SST anomalies and the Northern Hemisphere mid-latitude winter atmospheric variability simulated in an atmospheric general circulation model (GCM) have also been explored using the NLCCA (Wu, Hsieh, and Zwiers, 2003). Monthly 500 hPa geopotential height (Z500) and surface air temperature (SAT) data were produced by the CCCma (Canadian Centre for Climate Modelling and Analysis) GCM2. An ensemble of six 47-year runs of GCM2 were carried out, in which each integration started from minor different initial conditions and was forced by the observed SST. The advantage of using GCM results is that there are multiple runs in the ensemble, which provide far more data than the observed records, thereby allowing successful extraction of non-linear atmospheric teleconnection patterns by NLCCA.

The five leading SSTA PCs and the five leading PCs of atmospheric anomalies from Z500 or SAT (during January 1950 to November 1994) were the inputs to the NLCCA model. Here, minimum $u$ and maximum $u$ are chosen to represent the La Niña states and the El Niño states, respectively. For the Z500 anomalies (Z500A) field, instead of showing the spatial anomaly patterns during minimum $v$ and maximum $v$, patterns are shown for the values of $v$ when minimum $u$ and maximum $u$ occurred. As $u$ takes its minimum value, the SSTA field presents a La Niña with negative anomalies (about $-2.0°$C) over the central-western equatorial Pacific (Fig. 10.16(a)). The corresponding field of Z500A is a negative *Pacific-North American* (PNA) pattern (Horel and Wallace, 1981) with a positive anomaly centred over the North Pacific, a negative anomaly centred over western Canada and a positive anomaly centred over eastern USA.

As $u$ takes its maximum value, the SSTA field presents a fairly strong El Niño with positive anomalies (about 2.5–3.0°C) over the central-eastern Pacific (Fig. 10.16(b)). The SSTA warming centre shifts eastward by 30–40° longitude relative to the cooling centre in Fig. 10.16(a). The warming in Fig. 10.16(b) does not display peak warming off Peru, in contrast to the NLPCA mode 1 of the SSTA (Fig. 10.10(e)), where the El Niño peak warming occurred just off Peru. This difference between the NLPCA and the NLCCA mode implies that warming confined solely to the eastern equatorial Pacific waters does not have a corresponding strong mid-latitude atmospheric response, in agreement with Hamilton (1988).

The corresponding field of Z500A is a PNA pattern (Fig. 10.16(b)), roughly opposite to that shown in Fig. 10.16(a) but with a notable eastward shift. The zero contour of the North Pacific anomaly is close to the western coastline of North America during the El Niño period (Fig. 10.16(b)), while it is about 10–15° further west during the La Niña period (Fig. 10.16(a)). The positive anomaly over eastern Canada and USA in Fig. 10.16(a) becomes a negative anomaly shifted southeastward in Fig. 10.16(b). The amplitude of the Z500

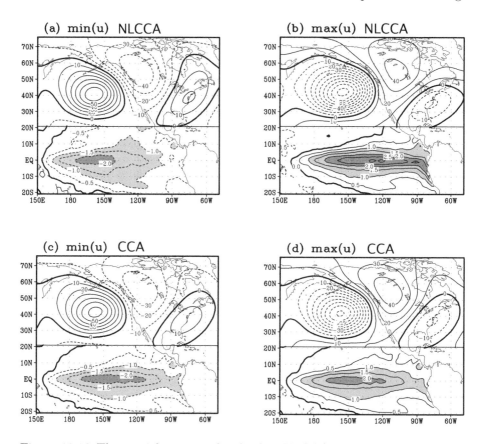

Figure 10.16 The spatial patterns for the first NLCCA mode between the winter Z500A and the tropical Pacific SSTA as the canonical variate $u$ takes its (a) minimum value and (b) maximum value. The Z500A with contour intervals of 10 m are shown north of 20°N. SSTA with contour intervals of 0.5°C are displayed south of 20°N. The SSTA greater than +1°C or less than –1°C are shaded, and more darkly shaded if greater than +2°C or less than –2°C. The linear CCA mode 1 is shown in panels (c) and (d) for comparison. Negative contours are dashed and the zero contour thickened. [Reproduced from Wu, Hsieh, and Zwiers (2003), ©American Meteorological Society. Used with permission.]

anomaly over the North Pacific is stronger during El Niño than La Niña, but the anomaly over western Canada and USA is weaker during El Niño than La Niña (Fig. 10.16(a) and (b)).

For comparison, the spatial patterns extracted by the CCA mode 1 are shown in Fig. 10.16(c) and 10.16(d), where the atmospheric response to La Niña is exactly opposite to that for El Niño, though the El Niño pattens have somewhat stronger amplitudes. The SSTA patterns are also completely antisymmetrical between the two extremes.

The NLCCA method was then applied to the SAT anomalies over North America and the tropical Pacific SSTA, with Fig. 10.17(a) and (b) showing the spatial anomaly patterns for both the SST and SAT associated with La Niña and El Niño, respectively. As $u$ takes its minimum value, positive SAT anomalies (about $1°C$) appear over the southeastern USA, while much of Canada and northwestern USA are dominated by negative SAT anomalies. The maximum cooling centre ($-4°C$) is located around northwestern Canada and Alaska (Fig. 10.17(a)). As $u$ takes its maximum value (Fig. 10.17(b)), the warming centre ($3°C$) is shifted to the southeast of the cooling centre in Fig. 10.17(a), with warming over almost the whole of North America except southeastern USA. The CCA mode 1 between the SAT and SSTA is also shown for reference (Fig. 10.17(c) and (d)).

## Exercises

### 10.1

From the given dataset containing daily $\log_{10}$(streamflow) and the seven-day moving-averaged values of the daily minimum air temperature and precipitation observed at Stave River, British Columbia, perform $K$-means clustering on the three standardized variables and try to explain the reason for the clusters.

### 10.2

From $t = [-1 + (j/30)]\pi$, $(j = 0, \ldots, 60)$, generate the signal $s_1 = \sin t$ and $s_2 = t$. Add Gaussian noise to the signal, that is, let

$$x_1 = s_1 + \alpha \sigma(s_1) \mathcal{N}(0, 1),$$
$$x_2 = s_2 + \alpha \sigma(s_2) \mathcal{N}(0, 1),$$

where $\mathcal{N}(0, 1)$ denotes a random variable from the standard Gaussian distribution, $\sigma(s_1)$ and $\sigma(s_2)$ are the standard deviations of $s_1$ and $s_2$, respectively, and $\alpha$ is a constant specifying the noise level. With low noise, that is, $\alpha = 0.2$, apply a 1-D self-organizing map (SOM) with (a) 6 nodes and (b) 12 nodes to the dataset. Repeat with high noise, that is, $\alpha = 0.5$ for SOM with (c) 6 nodes and (d) 12 nodes. For (e), again use low noise ($\alpha = 0.2$), but with the signal changed to $s_1 = \sin t$ and $s_2 = t/\pi$. (f) Can the problem encountered in (e) be corrected by having more data points, that is, repeat with $t = [-1 + (j/100)]\pi$, $(j = 0, \ldots, 200)$?

### 10.3

Use the dataset that gave the SOM model trouble in part (e) of Exercise 10.2, that is, from $t = [-1 + (j/30)]\pi$, $(j = 0, \ldots, 60)$, generate the signal $s_1 = \sin t$ and $s_2 = t/\pi$, then add Gaussian noise to the signal, giving

$$x_1 = s_1 + 0.2\,\sigma(s_1) \mathcal{N}(0, 1),$$
$$x_2 = s_2 + 0.2\,\sigma(s_2) \mathcal{N}(0, 1),$$

where $\mathcal{N}(0, 1)$ denotes a random variable from the standard Gaussian distribution and $\sigma(s_1)$ and $\sigma(s_2)$ are the standard deviations of $s_1$ and $s_2$, respectively.

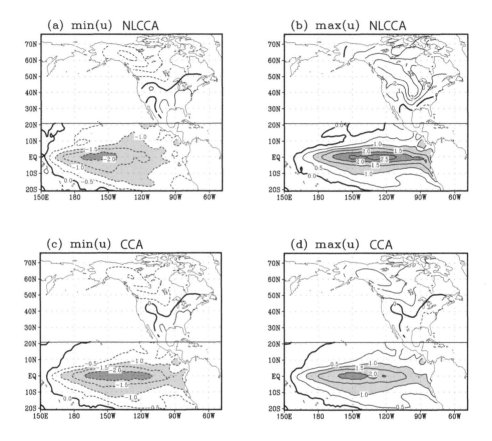

Figure 10.17 Similar to Fig. 10.16, but for the NLCCA mode 1 between the surface air temperature (SAT) anomalies over North America and the tropical SSTA. The contour interval for the SAT anomalies is 1°C. [Reproduced from Wu, Hsieh, and Zwiers (2003), ©American Meteorological Society. Used with permission.]

Perform non-linear principal component analysis (NLPCA) on the dataset of $(x_1, x_2)$ using an autoencoder neural network.

10.4

Prove that with $u$ and $\tilde{u}$ standardized to having zero mean and unit standard deviation, (10.49) is equivalent to (10.48).

10.5

In NLCCA, the objective function $J$ in (10.58) includes terms such as $(\langle u^2 \rangle^{1/2} - 1)^2$ so that the canonical variate $u$ approximately satisfies the normalization condition $\langle u^2 \rangle = 1$. If instead the simpler term $(\langle u^2 \rangle - 1)^2$ is used in $J$, then sometimes the converged solution unexpectedly has $\langle u^2 \rangle \approx 0$. With $\langle u^2 \rangle = (\sum_{i=1}^{N} u_i^2)/N$, where $N$ is the number of observations, use $\partial f / \partial u_i$ to explain the different convergence properties at $\langle u^2 \rangle = 0$ for $f = (\langle u^2 \rangle - 1)^2$ and $f = (\langle u^2 \rangle^{1/2} - 1)^2$.

# 11

# Time Series

A time series is a series of data values observed at successive times, usually with equal time intervals between observations. As data points are often collected sequentially as time progresses, many datasets contain time series. While many of the data techniques we have studied so far, for example linear regression, can be applied to time series data, these techniques do not take advantage of the fact that the data points are ordered in time.

Time series analysis involves using methods that utilize the information contained in the time order of the data points. Methods for time series analysis are divided into two groups: frequency-domain methods and time-domain methods.

This chapter starts from the frequency domain: after introducing Fourier analysis (Section 11.1) and the effects of using different windows for observation (Section 11.2), we come to the major topic of spectral analysis (Section 11.3) – that is, how the energy or variance of a time series is distributed as a function of frequency. With two time series, their relation can be revealed by cross-spectral analysis (Section 11.4). Time series data can be filtered (Section 11.5) to select signals in particular frequency bands while eliminating others. Time-averaging data (Section 11.6) also involves applying a filter, with some unexpected consequences. Principal component analysis can be converted to a multivariate spectral technique called singular spectrum analysis (Section 11.7). Turning to the time domain, we study auto-regressive processes (Section 11.8) and Box-Jenkins models (Section 11.9), a large class of stochastic time series models.

## 11.1 Fourier Analysis [A]☺

With time series data, information is presented in the time domain. Alternatively, one can present the information in the frequency domain, as first introduced by Jean-Baptiste Joseph Fourier in the early nineteenth century. Given a function $y(t)$ defined on the interval $[0, T]$, or $y(t)$ periodic with period $T$, one can write out a corresponding infinite sum of cosine and sine functions of various frequencies, that is, the fluctuations in $y(t)$ can be decomposed into a series of separate frequency components. From Fourier series, the subject is broadened into *Fourier analysis*, which encompasses Fourier series (FS), discrete Fourier transform (DFT), continuous-time Fourier transform (CTFT) and discrete-time Fourier transform (DTFT). Table 11.1 lists the characteristics of these four types of Fourier analysis, where continuous/discrete time in column 2 is paired with infinite/finite frequency range in column 5, and finite/infinite time duration (column 3) is paired with discrete/continuous frequency (column 4). Of the four types, only DFT and its inverse can be performed on a computer as they only need summation over finite number of terms – in contrast, the others involve infinite sums or integrals. The next four subsections look at the four types individually.

Table 11.1 Characteristics of the four types of Fourier analysis: Fourier series (FS), discrete Fourier transform (DFT), continuous-time Fourier transform (CTFT) and discrete-time Fourier transform (DTFT). In FS and DFT, time duration is finite or the function is extended periodically outside the finite duration.

| type | time | time duration | frequency | frequency range |
|------|------|---------------|-----------|-----------------|
| FS | continuous | finite | discrete | infinite |
| DFT | discrete | finite | discrete | finite |
| CTFT | continuous | infinite | continuous | infinite |
| DTFT | discrete | infinite | continuous | finite |

## 11.1.1   Fourier Series  [A]☺

For a function $y(t)$ defined on the interval $[0, T]$, or $y(t)$ being periodic with period $T$, the Fourier series representation for $y(t)$ is

$$\breve{y}(t) = \frac{a_0}{2} + \sum_{m=1}^{\infty} \left[ a_m \cos(\omega_m t) + b_m \sin(\omega_m t) \right], \tag{11.1}$$

with the *angular frequency* $\omega_m$ given by

$$\omega_m = \frac{2\pi m}{T}, \qquad m = 1, 2, \ldots \tag{11.2}$$

and the Fourier coefficients $a_m$ and $b_m$ given by

$$a_m = \frac{2}{T} \int_0^T y(t) \cos(\omega_m t) \, dt, \quad m = 1, 2, \ldots \tag{11.3}$$

$$b_m = \frac{2}{T} \int_0^T y(t) \sin(\omega_m t) \, dt, \quad m = 1, 2, \ldots \tag{11.4}$$

For $m = 0$, (11.2) gives $\omega_0 = 0$, $\cos(\omega_0 t) = \cos 0 = 1$ and (11.3) reduces to

$$a_0 = \frac{2}{T} \int_0^T y(t)\, \mathrm{d}t, \qquad (11.5)$$

hence

$$a_0/2 = \bar{y}, \qquad (11.6)$$

the mean of $y$. If $y(t)$ is differentiable at $t$, then (11.1) has $\breve{y}(t) \to y(t)$, that is, $\breve{y}(t)$ converges to $y(t)$ as the number of terms in the Fourier series approaches infinity. If $y$ has a jump discontinuity at $t$, then $\breve{y}(t) \to [y(t+)+y(t-)]/2$, where $t+$ is infinitesimally larger than $t$, and $t-$ is infinitesimally smaller than $t$.

## 11.1.2   Discrete Fourier Transform  [A]☺

For a *discrete* time series containing $N$ data points, $y(t)$ is replaced by $y(t_n) \equiv y_n$, $(n = 0, \cdots, N - 1)$. With a sampling interval $\Delta t = T/N$, the observations are made at time $t_n = n\Delta t$. The discrete Fourier series representation is given by

$$\breve{y}_n = \frac{a_0}{2} + \sum_{m=1}^{M} \left[ a_m \cos(\omega_m t_n) + b_m \sin(\omega_m t_n) \right], \qquad (11.7)$$

where $M$ is the largest integer $\leq N/2$,

$$\omega_m = 2\pi m/T, \qquad m = 1, 2, \ldots M, \qquad (11.8)$$

and the Fourier coefficients are given by

$$a_m = \frac{2}{N} \sum_{n=0}^{N-1} y_n \cos(\omega_m t_n), \qquad m = 0, 1, 2, \ldots, M, \qquad (11.9)$$

$$b_m = \frac{2}{N} \sum_{n=0}^{N-1} y_n \sin(\omega_m t_n), \qquad m = 1, 2, \ldots, M. \qquad (11.10)$$

If $N$ is even, $b_M = 0$, so the number of non-trivial Fourier coefficients is still $N$.

For fast computation of $a_m$ and $b_m$, the complex-valued *discrete Fourier transform* (DFT) is used (Prabhu, 2014, section 1.5), where

$$\hat{y}_k = \sum_{n=0}^{N-1} y_n\, \mathrm{e}^{-\mathrm{i}\omega_k t_n}, \qquad k = 0, \ldots, N - 1, \qquad (11.11)$$

with $\omega_k = 2\pi k/T$ and the inverse discrete Fourier transform being

$$y_n = \frac{1}{N} \sum_{k=0}^{N-1} \hat{y}_k\, \mathrm{e}^{\mathrm{i}\omega_k t_n}. \qquad (11.12)$$

A slight simplification in notation can be achieved by choosing the time unit so that $\Delta t = 1$, resulting in $t_n = n\Delta t = n$ in the above two equations. In Fourier analysis, a transform can be multiplied by an arbitrary constant $\alpha$ and its inverse transform by $\alpha^{-1}$. For instance, some authors would define DFT by $1/N$ times the right hand side of (11.11) and the inverse DFT by $N$ times the RHS of (11.12).

From Euler's formula $e^{ix} = \cos x + i \sin x$ $(e^{-ix} = \cos x - i \sin x)$, we obtain

$$\hat{y}_k = \sum_{n=0}^{N-1} y_n \cos(\omega_k t_n) - i \sum_{n=0}^{N-1} y_n \sin(\omega_k t_n), \tag{11.13}$$

$$= \frac{N}{2}(a_k - i\,b_k). \tag{11.14}$$

Thus, $a_k$ and $b_k$ can be retrieved from the real and imaginary parts of $\hat{y}_k$, and

$$|\hat{y}_k|^2 = \frac{N^2}{4}\left(a_k^2 + b_k^2\right). \tag{11.15}$$

## 11.1.3 Continuous-Time Fourier Transform  B☺

If $t$ is a continuous variable and the function $y(t)$ is not restricted to a finite domain $[0, T]$, nor to being periodic with period $T$, the Fourier series approach in Section 11.1.1 is replaced by the *Fourier transform*, more precisely the *continuous-time Fourier transform* (CTFT) approach (Prabhu, 2014, section 1.2):

$$\hat{y}(\omega) = \int_{-\infty}^{\infty} y(t)\, e^{-i\omega t}\, dt, \tag{11.16}$$

where $\hat{y}(\omega)$ is the Fourier transform of $y(t)$ and is in general complex. The inverse CTFT is

$$y(t) = \frac{1}{2\pi} \int_{-\infty}^{\infty} \hat{y}(\omega)\, e^{i\omega t}\, d\omega. \tag{11.17}$$

It is also common to use the *frequency* $\nu$ instead of the angular frequency $\omega$, where

$$\nu = \frac{\omega}{2\pi}. \tag{11.18}$$

The Fourier transform pair then becomes

$$\hat{y}(\nu) = \int_{-\infty}^{\infty} y(t)\, e^{-i2\pi\nu t}\, dt, \tag{11.19}$$

$$y(t) = \int_{-\infty}^{\infty} \hat{y}(\nu)\, e^{i2\pi\nu t}\, d\nu, \tag{11.20}$$

which has a more symmetrical appearance than (11.16) and (11.17), as the constant $1/(2\pi)$ disappears from the inverse transform due to $d\nu = d\omega/(2\pi)$.

### 11.1.4   Discrete-Time Fourier Transform  [B]☺

With discrete time, $t = t_n = n\Delta t$, $y(t)$ is replaced by $y(t_n)$ or $y_n$. The *discrete-time Fourier transform* (DTFT) (Prabhu, 2014, section 1.3) is

$$\hat{y}(\omega) = \sum_{n=-\infty}^{+\infty} y_n\, e^{-i\omega n \Delta t}, \tag{11.21}$$

with the inverse DTFT being

$$y_n = \frac{1}{2\pi} \int_{-\pi}^{\pi} \hat{y}(\omega)\, e^{i\omega n \Delta t}\, d\omega. \tag{11.22}$$

One must not confuse DTFT with the discrete Fourier transform (DFT) in Section 11.1.2. In DFT, there is only a finite number of $n$ values ($n = 0, \ldots, N-1$), whereas in DTFT, $-\infty < n < \infty$. Hence, the frequency is discrete in DFT but continuous in DTFT (Table 11.1). The relation between DTFT and DFT is given in Prabhu (2014, section 1.5).

## 11.2   Windows  [A]☺

The data record for an observed variable $y$ is always of finite duration $T$. When applying the Fourier transform to a finite record,[1] periodicity is assumed for $y$, that is, the data record is repeated periodically, so $y$ is extended over the domain $(-\infty, \infty)$. Figure 11.1 illustrates a signal

$$Y = \sin(0.5\,t) + 0.5\sin(0.8\,t), \tag{11.23}$$

composed of two sine functions defined over the domain $(-\infty, \infty)$, the observed data $y$ taken from $Y$ during $t = 0$ to $T$ and the observed data extended periodically to the domain $(-\infty, \infty)$. Unless $y(0)$ and $y(T)$ are of the same value, the periodicity assumption creates jump discontinuities at the ends of the original record. The Fourier representation of a jump discontinuity requires the use of many spectral components, that is, spurious energy is leaked to many frequency bands.

Another way to relate $y$ to $Y$ is to regard $y(t)$ as a product of $Y(t)$ and a rectangular *window* function $w(t)$, that is, $y = Yw$, where

$$w(t) = \begin{cases} 1 & \text{for } 0 \le t \le T, \\ 0 & \text{elsewhere.} \end{cases} \tag{11.24}$$

In other words, $y$ is the signal $Y$ observed over the finite window of $0 \le t \le T$. Outside this window of observation, $y$ contains no information.

---

[1] For simplicity, the discrete nature of the observed data is ignored for now and we assume $y$ is observed continuously in $t$ for a finite duration.

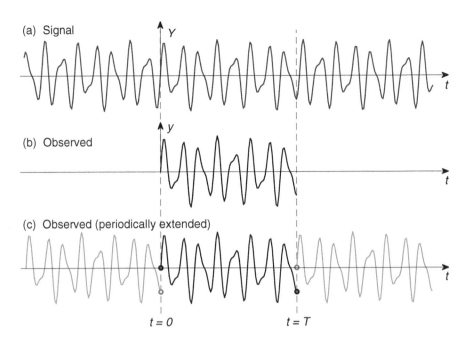

Figure 11.1 A troublesome consequence of the periodicity assumption in Fourier spectral analysis. (a) The true signal $Y$ is over the domain $(-\infty, \infty)$, but (b) the observations $y$ are made during $t = 0$ to $T$. For Fourier spectral analysis, the observed record is assumed to repeat itself periodically, thereby extending the domain to $(-\infty, \infty)$. (c) The periodicity assumption leads to jump discontinuities at $t = 0$ and $t = T$.

If $\hat{Y}(\omega)$ and $\hat{w}(\omega)$ are the (continuous-time) Fourier transforms (see Section 11.1.3) of $Y(t)$ and $w(t)$ over $(-\infty, \infty)$, then the Fourier transform of the product $Yw$ is the *convolution* of $\hat{Y}$ and $\hat{w}$, denoted by $\hat{Y} * \hat{w}$, where

$$\hat{Y} * \hat{w} \equiv \int_{-\infty}^{\infty} \hat{Y}(\tilde{\omega})\hat{w}(\omega - \tilde{\omega})\, d\tilde{\omega}, \tag{11.25}$$

(Prabhu, 2014, sections 1.1.4 and 1.2.1). The convolution operation can be viewed as a weighted average of the function $\hat{Y}(\tilde{\omega})$, where the weighting is given by the function $\hat{w}(-\tilde{\omega})$ simply shifted by amount $\omega$, that is, $\hat{w}(\omega - \tilde{\omega})$.

For the rectangular window, $\hat{w}$ has many strong sidelobes (Fig. 11.2(b)), that is, $|\hat{w}|$ has many strong side peaks besides the main peak at $\omega \approx 0$, thus the convolution of $\hat{w}$ and $\hat{Y}$ leads to spurious energy leakage into other frequency bands (Prabhu, 2014, chapter 3). On the other hand, if a window can be designed to have minimal sidelobes, that is, $\hat{w}(\omega - \tilde{\omega})$ is strongly concentrated at $\tilde{\omega} \approx \omega$, then (11.25) gives $(\hat{Y} * \hat{w})(\omega) \approx \hat{Y}(\omega)$, that is, the Fourier transform of $Yw$ is very close to the Fourier transform of $Y$, hence the window has not introduced much distortion.

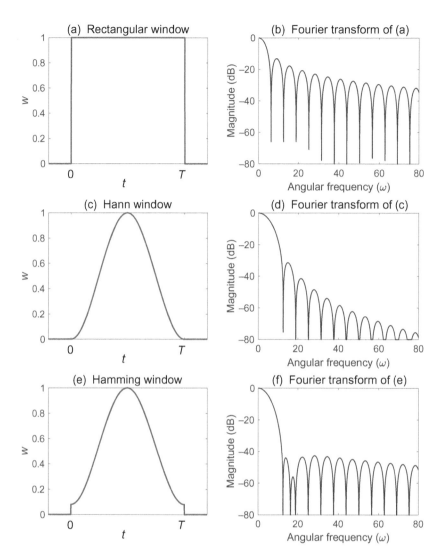

Figure 11.2 Windows and their Fourier transform: (a) and (b) rectangular window, (c) and (d) Hann window, and (e) and (f) Hamming window. The Fourier transforms are plotted only for positive $\omega$, as the function is symmetric about 0, and the magnitude of the Fourier transform is in units of decibel. [The *decibel* (dB) is a common unit for presenting a value (divided by a reference value) on a logarithmic scale. For the magnitude $|y|$ of a variable, the value in dB is given by $20 \log_{10} \left( |y|/|y_0| \right)$, whereas for the power $P$ or variance of $y$, the value in dB is given by $10 \log_{10} \left( P/P_0 \right)$. Here, the maximum value is used as the reference value $y_0$.]

If the ends of the window are tapered (e.g. by a cosine-shaped taper) to alleviate the jump discontinuities, the size of the sidelbes can be much reduced, thereby reducing the spurious energy leakage (Harris, 1978; Prabhu, 2014). A simple *cosine-tapered window* is of the form

$$
w(t) = \begin{cases} c - (1-c)\cos\left(\frac{2\pi t}{T}\right), & 0 \le t \le T \\ 0, & \text{otherwise,} \end{cases} \tag{11.26}
$$

where $c$ is a constant. Two commonly used windows are of this form: the *Hann window* (named after Julius von Hann and often incorrectly called the Hanning window) uses $c = 0.5$, while the *Hamming window* (proposed by Richard W. Hamming) uses $c = 0.54$. Thus, the Hann window is

$$
w(t) = \begin{cases} 0.5\left[1 - \cos\left(\frac{2\pi t}{T}\right)\right], & 0 \le t \le T \\ 0, & \text{otherwise,} \end{cases} \tag{11.27}
$$

while the Hamming window is

$$
w(t) = \begin{cases} 0.54 - 0.46\cos\left(\frac{2\pi t}{T}\right), & 0 \le t \le T \\ 0, & \text{otherwise.} \end{cases} \tag{11.28}
$$

The Hamming window is an optimized version of the Hann window, whereby the first sidelobe is minimized (Fig. 11.2). However, $w(t)$ is 0 at $t = 0$ and $t = T$ for the Hann window but not for the Hamming window, so the fall-off rate of the other sidelobes is slower for the Hamming window. Both windows have much smaller sidelobes than the rectangular window. In Section 11.3, the spectrum for the time series in Fig. 11.1 is computed using the rectangular window, the Hann window and the Hamming window (Fig. 11.3), showing the advantages of the Hann and Hamming windows over the rectangular window.

Both the Hann and Hamming windows are commonly set as the default window for spectral analysis in various software packages. The reader interested in going beyond these two basic windows can try the more advanced windows mentioned in Harris (1978) and (Prabhu, 2014, section 4.3).

## 11.3 Spectrum $\boxed{\text{A}}$ ☺

The variance of the time series $y$ can be written in terms of its Fourier coefficients from (11.9) and (11.10):

$$
\text{var}(y) = \frac{1}{N}\sum_{n=0}^{N-1}(y_n - \bar{y}_n)^2 = \frac{1}{N}\sum_{n=0}^{N-1}\left(y_n - \frac{a_0}{2}\right)^2
$$
$$
= \frac{1}{N}\sum_{n=0}^{N-1}\left[\sum_{m=1}^{M}\left(a_m\cos(\omega_m t_n) + b_m\sin(\omega_m t_n)\right)\right]^2. \tag{11.29}
$$

From the orthogonality properties of the sinusoidal functions in (11.136), var($y$) can be expressed in terms of the Fourier coefficients,

$$\text{var}(y) = \frac{1}{2} \sum_m \left(a_m^2 + b_m^2\right). \tag{11.30}$$

We want to partition var($y$) over a number of frequency bands of width $\Delta\omega$, that is,

$$\text{var}(y) = \sum_m S_m \, \Delta\omega, \tag{11.31}$$

where $S_m \Delta\omega$ is the variance or energy contained in the $m$th frequency band, with bandwidth

$$\Delta\omega = \frac{2\pi}{T} = \frac{2\pi}{N\Delta t}. \tag{11.32}$$

$S_m$ is called the *spectrum, autospectrum, power spectrum* or *periodogram*.[2] When one plots the spectrum, that is, with $\omega$ as the abscissa and $S_m$ as the ordinate, $S_m \, \Delta\omega$ is the area of a rectangular bar with height $S_m$ and width $\Delta\omega$ and (11.31) simply says that summing all these rectangular bars (i.e. the variance contained within individual frequency bands) gives the total variance var($y$).

Matching (11.30) and (11.31) gives

$$S_m = \frac{N\Delta t}{4\pi}\left(a_m^2 + b_m^2\right), \qquad m = 1, \ldots M. \tag{11.33}$$

When $S_m$ is plotted as a function of $\omega_m$, peaks in $S_m$ reveal the frequencies where the energy is relatively high.

The ability to resolve neighbouring spectral peaks is controlled by $\Delta\omega$, which is proportional to $1/T$. Thus, a longer record $T$ would yield sharper spectral peaks, thereby allowing resolution of two signals with close-by frequencies as two distinct peaks in the spectrum. In contrast, a shorter record would merge the two signals into a single spectral peak (see Exercise 11.3).

If frequency $\nu$ is used instead of the angular frequency $\omega$ ($\omega = 2\pi\nu$), the spectrum $S_m^{(\nu)}$ and the variance are related by

$$\text{var}(y) = \sum_m S_m^{(\nu)} \, \Delta\nu, \tag{11.34}$$

with

$$S_m^{(\nu)} = S_m \, \Delta\omega/\Delta\nu = 2\pi \, S_m. \tag{11.35}$$

## 11.3.1   Effects of Window Functions  A☺

It is common to use a window function $w(t)$ with tapered ends (e.g. the Hann and Hamming windows in Section 11.2) to reduce energy leakage via the sidelobes in

---

[2] The term *periodogram* refers to this original and basic form of the spectrum, as more sophisticated methods were developed later to improve estimation of the spectrum.

spectra analysis. However, the tapering of the time series by the window leads to a reduction in the spectral energy or variance, so a *variance compensation factor* $Q$ is commonly used to correct for the reduced variance due to the tapered window (Prabhu, 2014, chapter 4). $Q$ is obtained from the window function $w(t)$, with

$$Q = \frac{1}{T} \int_0^T w^2(t)\mathrm{d}t. \tag{11.36}$$

$Q = 1$ for the rectangular window, but $0 < Q < 1$ for a tapered window. To perform variance compensation, simply replace the raw spectrum $S_m$ by $S_m/Q$.

For the time series composed of two frequencies in (11.23) over the duration $0 \le t \le T$ (Fig. 11.1(b)), its spectra computed using three different windows – the rectangular window (equivalent to no window used in the spectral computation), the Hann window and the Hamming window – are compared in Fig. 11.3. The time series has energy in only two frequencies, yet the spectrum for the rectangular window has much energy leaked to other frequencies. The Hann window shows the least leakage for frequencies away from the two spectral peaks, as expected from the rapid drop-off in its sidelobes (Fig. 11.2(d)). However, near the two peaks, the Hamming window shows less leakage (i.e. having narrower peaks) than the Hann window, as expected from the smaller first sidelobe of the Hamming window (Fig. 11.2(f)).

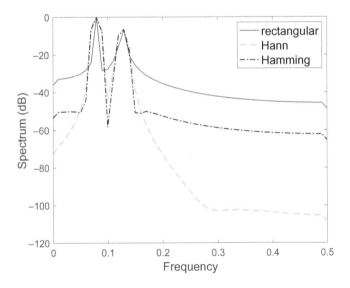

Figure 11.3 The spectrum (in decibels) for the data in Fig. 11.1(b) is computed using the rectangular window (solid curve), the Hann window (dashed) and the Hamming window (dot-dashed). The frequency displayed is $\nu = \omega/(2\pi)$.

## 11.3.2   Trend Removal   $\boxed{\text{A}}$☺

The lowest frequency in the spectrum, known as the *fundamental frequency*, is

$$\omega_1 = \frac{2\pi}{T}, \quad \text{or} \quad \nu_1 = \frac{1}{T}. \tag{11.37}$$

Often a time series displays a *trend*, that is, a positive or negative slope in the data over the time record. As the frequency associated with a trend is lower than the fundamental frequency, the information from the trend cannot be presented properly in spectral analysis. The presence of a trend leaks spurious energy to the fundamental frequency and neighbouring low frequency bands, thereby distorting the low frequency part of the spectrum. Therefore, the usual approach is to remove the trend from the time series before performing spectral analysis.

To *detrend* the data, first find a linear regression relation between $t$ and $y$, that is,

$$y_{\text{trend}} = a_0 + a_1 t. \tag{11.38}$$

Then subtract the regression line from the time series data, that is,

$$y_{\text{detrended}}(t) = y(t) - y_{\text{trend}}(t). \tag{11.39}$$

Spectral analysis is then performed using $y_{\text{detrended}}$.

## 11.3.3   Nyquist Frequency and Aliasing   $\boxed{\text{A}}$☺

The highest resolvable frequency from (11.8) is $\omega = 2\pi M/T$. With $M \approx N/2$ and $T = N\Delta t$, we have $M/T \approx 1/(2\Delta t)$. Thus, the highest resolvable frequency, called the *Nyquist frequency* or the *folding frequency*, is defined to be

$$\omega_{\text{N}} = \frac{\pi}{\Delta t}, \quad \text{or} \quad \nu_{\text{N}} = \frac{1}{2\Delta t}. \tag{11.40}$$

To resolve a wave of period $\tau$, we need at least two data points to cover the period $\tau$, that is, $\tau = 2\Delta t = 1/\nu_{\text{N}}$. *Aliasing* arises when $\Delta t$ is too large to resolve the highest frequency oscillations in the data. Figure 11.4 illustrates a signal measured too infrequently, resulting in an aliased signal of much lower frequency.[3] In a spectrum, aliasing causes signals with frequencies above the Nyquist frequency to be reflected across the Nyquist frequency into the frequency bands below the Nyquist frequency, thereby transferring spurious energy to those frequency bands (Bendat and Piersol, 2010, section 10.2.3).

To illustrate aliasing, consider the following signal with four frequency components,

$$y(t) = \sin(2\pi\nu_1 t) + 0.03\sin(2\pi\nu_2 t) + 0.3\sin(2\pi\nu_3 t) + 0.1\sin(2\pi\nu_4 t), \tag{11.41}$$

---

[3] In movies, one sometimes notices aliasing, where fast spinning wheels and propellors appear to rotate backwards as the frame rate (the number of frames or images displayed per second) is not high enough for the high-frequency phenomenon.

Figure 11.4 Illustrating the phenomenon of aliasing. The sampling
time interval is $\Delta t$, but the signal
(solid curve) is oscillating too quickly
to be resolved by the sampling. From
the observations (dots), an incorrect
signal (dashed curve) of much lower
frequency is inferred. [Source: Hsieh
(2009)]

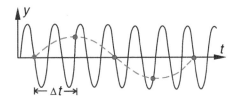

for $t \in [0, 100]$. The frequencies are: $\nu_1 = 0.1$, $\nu_2 = 0.4$, $\nu_3 = 0.65$ and $\nu_4 = 0.8$.
First, we sample twice per time unit, so the Nyquist frequency $\nu_N = 1$. The
four frequencies $\nu_1, \ldots, \nu_4$ are all below the Nyquist frequency, so there is no
aliasing in the computed spectrum (using rectangular window) in Fig. 11.5(a).

Figure 11.5 The (a)
non-aliased spectrum
where all four signals
have frequencies below
the Nyquist frequency
$\nu_N = 1$ and (b) the
aliased spectrum where
by sampling at half
the rate, $\nu_N = 0.5$
and two of the higher
frequency signals in (a)
are reflected or folded
back across the vertical
dashed line at $\nu_N = 0.5$,
creating spurious peaks
at the frequencies of 0.2
and 0.35.

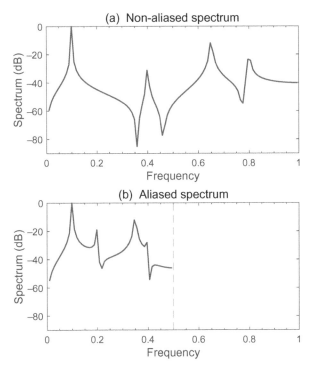

Next, we sample only once per time unit, so $\nu_N = 0.5$. Now $\nu_3$ and $\nu_4$
are above $\nu_N$, so aliasing occurs for these two frequency components and their
spectral peaks are reflected across $\nu_N = 0.5$ into the lower frequency part of the
spectrum. The signal at $\nu_3 = 0.65$ now appears at $\nu = 0.5 - (0.65 - 0.5) = 0.35$,
almost masking the signal from $\nu_2 = 0.4$. The signal at $\nu_4 = 0.8$ now appears
at $\nu = 0.5 - (0.8 - 0.5) = 0.2$.

The aliasing phenomenon reminds us that it is important to filter out contributions from frequencies above $\nu_N$ in the time series data before computing the spectrum. For example, suppose one wants to compute the spectrum using monthly values of air temperature and one is provided with daily observations. One can simply pick the observation from the 15th day of each month to form a monthly time series, or one can average the daily values in each month to get a monthly time series. The latter approach would largely eliminate the contributions from weather systems passing through every few days, whereas the former would not, thereby yielding an aliased spectrum.

## 11.3.4   Smoothing the Spectrum   A☺

Observed data are usually noisy, so the raw spectrum $S_m$ calculated from (11.33) can look terribly noisy. There are two common approaches for smoothing the spectrum: (a) band averaging and (b) ensemble averaging.

In approach (a), a *moving average* filter (a.k.a. running mean filter) (Section 11.6.1) is applied to the raw spectrum, that is,

$$\tilde{S}_m = \frac{1}{(2K+1)} \sum_{k=-K}^{K} S_{m+k}, \tag{11.42}$$

where $\tilde{S}_m$ is the smoothed spectrum resulting from averaging the raw spectrum over $2K + 1$ frequency bands. This is called the *Daniell* method.

In approach (b), the data record is divided into $J$ blocks of equal length $L = T/J$. The periodogram is computed for each block to get $S_m^{(j)}$ ($j = 1, \ldots, J$). The spectrum $S_m$ is then the ensemble average over the $J$ periodograms:

$$S_m = \frac{1}{J} \sum_{j=1}^{J} S_m^{(j)}. \tag{11.43}$$

This is called the *Bartlett* method.

Approach (b) has an advantage over (a) when there are data gaps – in (b), the data gaps do not pose a serious problem since the data record is to be chopped into $J$ blocks anyway, whereas in (a), the data gaps may have to be filled with interpolated values or zeros. The disadvantage of (b) is that the lowest resolvable frequency is $\nu_1 = 1/L = J/T$, hence there is a loss of low frequency information when the record is chopped up.

The Bartlett method has been improved upon by the *Welch* method, which is now the most popular of the three methods. In the Welch method, the time series is divided into overlapping blocks, that is, the data record is split up into $J$ blocks of length $L$, overlapping by $M$ points. For instance, for $J = 8$ and $M = L/2$, one divides the data record into 9 approximately equal segments. The first block contains the data from segment 1 and segment 2, the second block from segments 2 and 3, ... and the eighth block from segments 8 and 9. A window such as a Hann or Hamming window is used in computing the periodogram of each block, and the final spectrum is the ensemble average over the 8 blocks. If $M = 0$, the Welch method reduces to the Bartlett method. The advantage of the Welch method over the Bartlett method is that for similar block length, the Welch method uses more ensemble members for estimating the spectrum. As data near the ends of a block are under-utilized relative to data near the centre of the block by the window function, the Welch method, by allowing blocks to overlap, alleviates this problem associated with the window function.

There is a trade-off between the variance of the spectrum $S$ and the band width. Increasing the band width (by increasing $K$ or $J$) leads to a less noisy $S$, but spectral peaks are broadened, so that neighbouring spectral peaks may merge together, resulting in a loss of resolution.

## 11.3.5 Confidence Interval [B]☺

Next, consider estimating confidence intervals for the spectrum (von Storch and Zwiers, 1999, section 12.3.10; Shumway and Stoffer, 2017, p. 190). In (11.33), $S_m$ is computed from summing two squares, $a_m^2$ and $b_m^2$. Furthermore, the spectral estimate is obtained either by ensemble averaging over $J$ periodograms (in the Bartlett method or Welch method) or by band averaging over $J$ $(= 2K + 1)$ frequency bands (Daniell method). Assuming the total summation is over $2J$ squares of independent Gaussian variables, one gets the asymptotic distribution

$$\frac{2J\,S_m}{\hat{S}_m} \sim \chi_{2J}^2, \tag{11.44}$$

where $\hat{S}$ is the population spectrum and $\chi_{2J}^2$ is the chi-squared distribution (Section 3.16) with $2J$ degrees of freedom.

An estimate of the $100(1-\alpha)\%$ confidence interval for the spectrum is given by

$$\frac{2J\,S_m}{\chi_{2J}^2(1-\alpha/2)} \leq \hat{S}_m \leq \frac{2J\,S_m}{\chi_{2J}^2(\alpha/2)}, \tag{11.45}$$

where $\chi_{2J}^2$ is evaluated at $1 - \alpha/2$ and $\alpha/2$. When plotting the spectrum in logarithmic or decibel scales,

$$\log\left(\frac{2J}{\chi_{2J}^2(1-\alpha/2)}\right) + \log(S_m) \leq \log(\hat{S}_m) \leq \log\left(\frac{2J}{\chi_{2J}^2(\alpha/2)}\right) + \log(S_m), \tag{11.46}$$

so the width of the confidence interval is independent of $S_m$.

## 11.3.6 Examples [B]☺

For pedagogical reasons, the examples so far have used synthetic time series without noise. Time series from the real world are most likely to be quite noisy, so let us look at the spectra of some observed time series. Consider the monthly Niño 3.4 index (sea surface temperature anomalies of the Niño 3.4 region in the equatorial Pacific from 1870 to 2019), seen previously in Fig. 9.3 and Figure 9.4. Figure 11.6(a) shows the spectrum of the Niño 3.4 index. For comparison, Fig. 11.6(b) shows the spectrum of the monthly zonal wind in the equatorial stratosphere at the 30 hPa level during January 1956–May 2020 (Section 10.6.2).

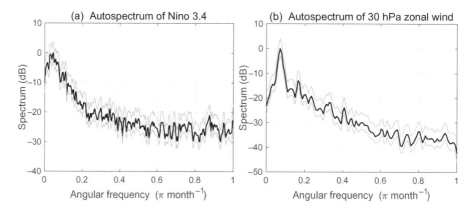

Figure 11.6  Autospectra of (a) the Niño 3.4 index and (b) the equatorial zonal wind at the 30 hPa level, with the 95% confidence interval given by the thin lines. The Welch method (with 8 blocks, 50% overlap and the Hamming window) was used to compute the autospectra after detrending.

With an overall less noisy appearance, the equatorial zonal wind shows a narrower and more prominent main spectral peak than the Niño 3.4 index. The main peak of the 30 hPa wind occurs at $\omega = 0.070\pi$ month$^{-1}$, corresponding to a period of 28.4 months, hence the name quasi-biennial oscillation (QBO). In contrast, the low frequency peak for Niño 3.4 is not sharply defined and is composed of a number of finer peaks. The highest one occurs at $\omega = 0.047\pi$ month$^{-1}$, corresponding to a period of 42.7 months, while the next highest one occurs at $\omega = 0.031\pi$ month$^{-1}$, corresponding to a period of 64.0 months. The decline in spectral energy away from the main peak is more dramatic for the 30 hPa wind, falling by more than 40 dB versus less than 30 dB for Niño 3.4.

## 11.3.7    Fast Fourier Transform  [B]☺

*Fast Fourier transform* (FFT) is the efficient algorithm used to compute the discrete Fourier transform (DFT) in (11.11) to obtain the Fourier coefficients $a_m$ and $b_m$ needed for the spectrum (Bendat and Piersol, 2010, section 11.2). FFT became popular after a paper by James W. Cooley and John W. Tukey in 1965, but it was later discovered that Carl Friedrich Gauss had actually discovered FFT around 1805 (Heideman et al., 1985). FFT is particularly fast when the number of data points is a power of 2. *Zero-padding* is the technique of adding zeros at the end of a data record so the number of data points reaches the next power of 2 (Prabhu, 2014, section 2.5). For instance, if there are 200 data points, one can pad the time series with zeros until there are $2^8 = 256$ data points. As zero padding increases $T$, the bandwidth decreases. Thus, the spectrum is smoother, but the resolution of spectral peaks is not improved.

## 11.3.8 Relation with Auto-covariance B☺

There is an important relation between the spectrum and the *auto-covariance* function. For the complex-valued discrete Fourier transform (DFT)

$$\hat{y}_m = \sum_{n=0}^{N-1} y_n \, e^{-i\omega_m t_n}, \tag{11.47}$$

(11.15) and (11.33) give

$$S_m = \frac{\Delta t}{N\pi} |\hat{y}_m|^2. \tag{11.48}$$

Assume $\{y_n\}$ is stationary and the mean $\bar{y}$ has been subtracted from the data, then

$$S_m = \frac{\Delta t}{N\pi} \left[ \sum_n y_n \, e^{-i\omega_m t_n} \right] \left[ \sum_j y_j \, e^{i\omega_m t_j} \right] \tag{11.49}$$

$$= \frac{\Delta t}{\pi} \sum_{l=-(N-1)}^{N-1} \left[ \frac{1}{N} \left( \sum_{j-n=l} y_n y_j \right) \right] e^{i\omega_m t_l}. \tag{11.50}$$

In general, the *auto-covariance* function with lag $l$ is defined by

$$C_l = \frac{1}{N} \sum_{n=0}^{N-1-l} (y_n - \bar{y})(y_{n+l} - \bar{y}). \tag{11.51}$$

Here (with $\bar{y} = 0$), we have the important relation

$$S_m = \frac{\Delta t}{\pi} \sum_{l=-(N-1)}^{N-1} C_l \, e^{i\omega_m t_l}, \tag{11.52}$$

that is, the spectrum is related to the auto-covariance function by a Fourier transformation.

## 11.3.9 Rotary Spectrum for 2-D Vectors C☺

The winds and the ocean currents are essentially two-dimensional vectors lying in the horizontal plane. One can write the 2-D velocity vector $\mathbf{u}$ in its two Cartesian components $(u, v)$ and, after subtracting off their means $(\bar{u}, \bar{v})$, perform spectral analysis on the two components separately. This approach is not ideal since motion in the atmosphere is often composed of cyclonic circulation[4] (associated with low pressure cells) and anti-cyclonic circulation (high pressure cells), and similarly in the oceans. Furthermore, the orientation of the Cartesian coordinate system may not be well defined. For instance, in the coastal ocean, one may want $u$ to point offshore and $v$ alongshore. However, coastlines and depth contours are not straight, so there are no unique offshore and alongshore directions.

If one works with the complex velocity vector $w = u + iv$, one can decompose each frequency component into a clockwise and an anti-clockwise rotating component, hence the approach is called *rotary spectral analysis* (Gonella, 1972; Mooers, 1973).

With $\bar{u}, \bar{v}$ removed, (11.7) gives the discrete Fourier series representation

$$u(t_n) = \sum_{m=1}^{M} u_{1m} \cos(\omega_m t_n) + u_{2m} \sin(\omega_m t_n), \tag{11.53}$$

$$v(t_n) = \sum_{m=1}^{M} v_{1m} \cos(\omega_m t_n) + v_{2m} \sin(\omega_m t_n). \tag{11.54}$$

With $w = u + iv$, we can write

$$w(t_n) = \sum_{m=1}^{M} w_m(t_n), \tag{11.55}$$

$$w_m(t_n) = (u_{1m} + iv_{1m}) \cos(\omega_m t_n) + (u_{2m} + iv_{2m}) \sin(\omega_m t_n). \tag{11.56}$$

Let

$$w_m(t_n) = w_m^+(t_n) + w_m^-(t_n) = c_m^+ e^{i\omega_m t_n} + c_m^- e^{-i\omega_m t_n}, \tag{11.57}$$

where the first term rotates anti-clockwise with time in the complex plane and the second term clockwise, and $c_m^+$ and $c_m^-$ are the complex Fourier coefficients. Upon matching the terms in (11.56) and (11.57) and using Euler's formula $e^{i\phi} = \cos\phi + i\sin\phi$, we have

$$c_m^+ = \tfrac{1}{2}\left[(u_{1m} + v_{2m}) + i(v_{1m} - u_{2m})\right],$$
$$c_m^- = \tfrac{1}{2}\left[(u_{1m} - v_{2m}) + i(v_{1m} + u_{2m})\right]. \tag{11.58}$$

The complex Fourier coefficients $c_m^+$ and $c_m^-$ can be expressed as in polar form

$$c_m^+ = A_m^+ e^{i\epsilon_m^+}, \qquad c_m^- = A_m^- e^{i\epsilon_m^-}, \tag{11.59}$$

where $A_m^+$ and $A_m^-$ are their magnitudes and $\epsilon_m^+$ and $\epsilon_m^-$ are their phases (or arguments).

---

[4] The *cyclonic* circulation associated with low pressure cells in the atmosphere or ocean is anti-clockwise in the Northern Hemisphere and clockwise in the Southern Hemisphere, due to the change in the Coriolis force. The *anti-cyclonic* circulation associated with high pressure cells is opposite in direction to the cyclonic circulation.

Similar to (11.30) and (11.31), we have

$$\text{var}(u) + \text{var}(v) = \tfrac{1}{2} \sum_m (u_{1m}^2 + u_{2m}^2 + v_{1m}^2 + v_{2m}^2), \tag{11.60}$$

and

$$\text{var}(u) + \text{var}(v) = \sum_m \left[ S_m^+(\omega_m)\Delta\omega + S_m^-(\omega_m)\Delta\omega \right], \tag{11.61}$$

where $S_m^+$ and $S_m^-$ are the power spectrum (autospectrum) associated with anti-clockwise and clockwise rotation at the $\omega_m$ frequency band.

From (11.58), (11.59), we have

$$(A_m^+)^2 + (A_m^-)^2 = \tfrac{1}{2}\left(u_{1m}^2 + u_{2m}^2 + v_{1m}^2 + v_{2m}^2\right). \tag{11.62}$$

Thus,

$$S_m^+ \Delta\omega = (A_m^+)^2, \qquad S_m^- \Delta\omega = (A_m^-)^2, \tag{11.63}$$

with $\Delta\omega = 2\pi/(N\Delta t)$. A rotary autospectrum is usually plotted with the clockwise part $S_m^-$ over negative frequencies $-\omega_m$ ($\omega_m = 2\pi m/T$) and the anti-clockwise part $S_m^+$ over positive frequencies $\omega_m$.

Combining the anti-clockwise and clockwise components for a particular $\omega_m$, the general motion of the velocity vector traces out an ellipse over time (Fig. 11.7), with semi-major axis $A_m^+ + A_m^-$ and semi-minor axis $|A_m^+ - A_m^-|$ (Thomson and Emery, 2014, section 5.4.4). If $A_m^+ = A_m^-$, the ellipse narrows to a straight line, that is, rectilinear motion. The ellipse turns into anti-clockwise circular motion if $A_m^- = 0$ and clockwise circular motion if $A_m^+ = 0$. The orientation of the semi-major axis relative to the $u$ direction is given by the angle $\theta$ (Gonella, 1972; Thomson and Emery, 2014),

$$\theta = \tfrac{1}{2}\left(\epsilon_m^+ + \epsilon_m^-\right). \tag{11.64}$$

Figure 11.7 In general, in the $(u, v)$ plane, as the complex velocity $w_m$ rotates with angular frequency $\omega_m$, the tip of the velocity vector traces out an ellipse. Here it is shown rotating clockwise as the clockwise component $A_m^-$ happens to be larger than the anti-clockwise component $A_m^+$. If $A_m^+ = A_m^-$, the ellipse narrows to a straight line.

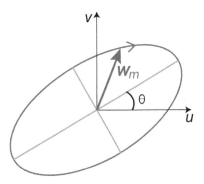

When there are two time series, cross-spectral analysis is used to find their relation at various frequencies (Section 11.4). With two 2-D vector time series, rotary cross-spectral analysis was introduced by Mooers (1973) (with typographical errors corrected by Middleton (1982)).

O'Brien and Pillsbury (1974) computed rotary wind spectra along and off the coast of Oregon in a study of sea breeze. Y. Hayashi (1979) developed space-time rotary spectral analysis to study travelling vortices in the atmosphere. Currents on the continental shelf off Oregon were studied by rotary spectral and cross-spectral analysis (Hsieh, 1982).

## 11.3.10   Wavelets  ⓒ☺

Many oscillations in observed data do not display constant frequency nor constant amplitude over the data record. For instance, for the tropical Pacific sea surface temperature anomalies in the Niño 3 region (Fig. 9.3), Fig. 11.8(a) shows oscillations with higher frequency around 1962–1978 and weaker amplitude around 1923–1962, that is, the El Niño-Southern Oscillation (ENSO) phenomenon changes over time. Modelling such non-stationary behaviour with Fourier analysis, where the time series is represented by a sum of sinusoidal oscillations, is not entirely satisfactory, since the oscillations in the sinusoidal functions do not change with time. Wavelet analysis has emerged in the last few decades to address this limitation in Fourier analysis (Fugal, 2009; Mallat, 2009).

A *wavelet* is a localized wave-like oscillation, that is, the amplitude starts at zero, increases with time, then decreases back to zero, so the oscillations do not last forever as for a sinusoidal function. Wavelets are commonly seen in seismographs. Wavelet analysis represents a time series by a sum of wavelet functions instead of sinusoidal functions. The basic wavelet function $\psi$ is called the '*mother*' wavelet, and the time series is represented by a sum of 'daughter' wavelets, which are simply shifted and scaled versions of the mother wavelet.

There are many types of wavelets. A commonly one used in continuous wavelet transforms is the *Morlet* wavelet (Fig. 11.9(a)),

$$\psi(t) = \pi^{-1/4} \, e^{-t^2/2} \, e^{-i\omega_c t}. \tag{11.65}$$

The Morlet (also called Gabor) wavelet approach differs from the Fourier approach by having a Gaussian window $e^{-t^2/2}$, which damps the oscillations to zero when far from the centre $t = 0$. The central frequency $\omega_c$ was chosen to be 6 in Torrence and Compo (1998), satisfying the condition for being an 'admissible' wavelet. The wavelet $\psi(t)$ is shifted to the right by $\tau$ in $\psi(t-\tau)$, and scaled by a factor $s$ in $\psi(t/s)$. Both shifting and scaling are carried out in $\psi\big((t-\tau)/s\big)$ (Fig. 11.9).

Given a time series $y_n = y(t_n) = y(n\Delta t)$ $(n = 0, \ldots, N-1)$, the continuous wavelet transform of $y_n$ is defined as the convolution of $y_n$ and the shifted and scaled wavelet $\psi$, that is,

$$W_n(s) = \sum_{n'=0}^{N-1} y_{n'} \, \psi^* \left[ \frac{(n' - n)\Delta t}{s} \right], \tag{11.66}$$

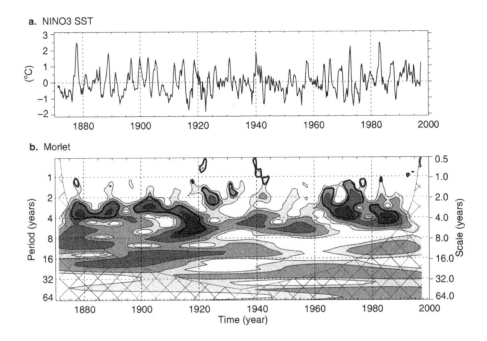

Figure 11.8 (a) The time series of the SST anomalies in the Niño 3 region. (b) The local wavelet power spectrum of the Niño 3 time series using the Morlet wavelet. The left axis is the period in years, corresponding to the wavelet scale on the right axis. The shaded contours are at normalized variances of 1, 2, 5 and 10, with thick contours enclosing regions above 95% confidence for a lag-1 red noise process (see Section 11.8.2). Cross-hatched regions on either end indicate the 'cone of influence', where edge effects become important. [Reproduced from Torrence and Compo (1998, figure 1), ©American Meteorological Society. Used with permission.]

where the asterisk denotes the complex conjugate. It is cheaper to compute $W_n(s)$ using the convolution theorem,[5]

$$W_n(s) = \sum_{k=0}^{N-1} \hat{y}_k \, \hat{\psi}^*(s\,\omega_k) \, \mathrm{e}^{\mathrm{i}\omega_k n \Delta t}, \tag{11.67}$$

where following the notation of Torrence and Compo (1998), $\hat{y}_k$ is the DFT of $y_n$ and the CTFT of $\psi(t/s)$ is $\hat{\psi}(s\omega)$, with

$$\omega_k = \begin{cases} \frac{2\pi k}{N\Delta t}, & 0 \le k \le \frac{N}{2}, \\[2mm] \frac{-2\pi k}{N\Delta t}, & \frac{N}{2} < k \le N-1. \end{cases} \tag{11.68}$$

---

[5] The *convolution theorem* states that under suitable conditions the Fourier transform of a convolution of two signals equals the pointwise product of their Fourier transforms. The theorem holds for CTFT, DTFT and DFT (Mallat, 2009).

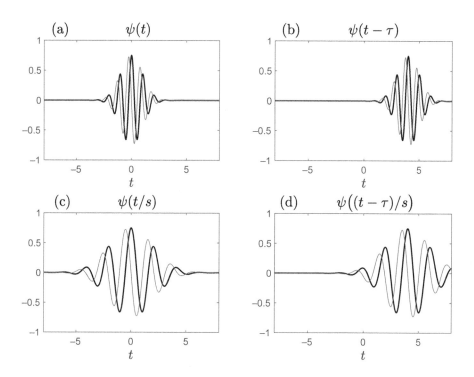

Figure 11.9 (a) The Morlet wavelet with real part (thick line) and imaginary part (thin line), (b) wavelet shifted to the right by $\tau$, (c) wavelet scaled by the factor $s = 2$ and (d) wavelet shifted and scaled.

Before we can plot the wavelet power spectrum $|W_n(s)|^2$, we need to choose a discrete set of values for the scale $s$. Higher $s$ values correspond to lower frequencies or longer periods, as seen in Figs. 11.9(a) and (c). Furthermore, $\hat{\psi}$ needs to be normalized so that $\sum_{k=0}^{N-1} |\hat{\psi}(s\,\omega_k)|^2$ is independent of $s$, so that $|W_n(s)|^2$ can be compared across different values of $s$ (Torrence and Compo, 1998). Then we can produce the wavelet power spectrum plot of $|W_n(s)|^2$ as a function of time and frequency or period. Figure 11.8(b) shows the changes in Niño 3 SST anomalies over time, such as oscillations with higher frequency around 1962–1978 and weaker amplitude around 1923–1962.

With two time series $x_n$ and $y_n$, wavelet cross-spectrum has also been developed to analyse their relation. Additional examples of wavelet power spectrum and cross-spectrum can be found in Thomson and Emery (2014, section 5.7) on coastal wave height and ocean currents, and in Torrence and Webster (1999) on the interdecadal changes in the ENSO-monsoon system. Common software packages carry wavelet programs, with a number of wavelet families (e.g. Morlet, Morse, etc.) to choose from. Wavelet functions have also been used in place of sigmoidal functions in neural network models (Alexandridis and Zapranis, 2013).

## 11.4   Cross-Spectrum  $\boxed{\text{B}}$ ☺

Consider two time series, $x_1, \ldots, x_N$ and $y_1, \ldots, y_N$, with respective Fourier transforms $\hat{x}_m$ and $\hat{y}_m$, which are in general complex numbers. The *cross-spectrum*

$$C_m = \frac{N\Delta t}{4\pi} \hat{x}_m \hat{y}_m^*, \tag{11.69}$$

where the asterisk denotes complex conjugation, so $C_m$ is in general complex. If $\hat{y}_m = \hat{x}_m$, $C_m$ reduces to $S_m$, which is real.

$C_m$ can be split into a real part and an imaginary part,

$$C_m = R_m + i I_m, \tag{11.70}$$

where $R_m$ is the *co-spectrum* and $I_m$ is the *quadrature spectrum*. Note that $C_m$ can also be expressed in polar form,

$$C_m = A_m \, e^{i\theta_m}, \tag{11.71}$$

where $A_m$ is the *amplitude spectrum* and $\theta_m$, the *phase spectrum*.

A useful quantity is the *squared coherence spectrum* (where the word 'squared' is often omitted for brevity):

$$r_m^2 = \frac{A_m^2}{S_m^{(x)} S_m^{(y)}}, \tag{11.72}$$

where $S_m^{(x)}, S_m^{(y)}$ are the autospectrum for the $x$ and $y$ time series, respectively. One can interpret $r_m^2$ as the square of the correlation between $x$ and $y$ in the $m$th frequency band. However, if one does not perform band averaging or ensemble averaging, then $r_m^2 = 1$, that is, perfect correlation at all frequency bands! To see this, let

$$\hat{x}_m = a_m \, e^{i\alpha_m} \quad \text{and} \quad \hat{y}_m = b_m \, e^{i\beta_m}. \tag{11.73}$$

Equation (11.69) becomes

$$C_m = \frac{N\Delta t}{4\pi} a_m \, b_m \, e^{i(\alpha_m - \beta_m)}. \tag{11.74}$$

Thus,

$$A_m = \frac{N\Delta t}{4\pi} a_m \, b_m \quad \text{and} \quad \theta_m = \alpha_m - \beta_m. \tag{11.75}$$

Also,

$$S_m^{(x)} = \frac{N\Delta t}{4\pi} a_m^2, \quad \text{and} \quad S_m^{(y)} = \frac{N\Delta t}{4\pi} b_m^2. \tag{11.76}$$

Substituting these into (11.72) yields $r_m^2 = 1$. The reason is that in a single frequency band, the $x$ and $y$ signals are simply sinusoidals of the same frequency, which are perfectly correlated (other than a possible phase shift between the two).

Suppose there is no real relation between $\hat{x}_m$ and $\hat{y}_m$, then the phase $\alpha_m - \beta_m$ tends to be random. Consider ensemble averaging, with

$$C_m = \frac{1}{J} \sum_{j=1}^{J} C_m^{(j)}. \tag{11.77}$$

With random phase, the $C_m^{(j)}$ vectors are randomly oriented in the complex plane, so summing of the $C_m^{(j)}$ vectors tends not to produce a $C_m$ vector with large magnitude $A_m$. In general, for large $J$, $A_m^2 \ll S_m^{(x)} S_m^{(y)}$, resulting in a small value for $r_m^2$, as desired. Thus, some form of ensemble averaging or band averaging is essential for computing the squared coherence spectrum – without the averaging, even random noise has $r_m^2$ equal to unity.

It can also be shown that the cross-spectrum is the Fourier transform of the *cross-covariance* $\gamma$, where

$$\gamma = \frac{1}{N} \sum_{n=0}^{N-1-l} (x_n - \overline{x})(y_{n+l} - \overline{y}). \tag{11.78}$$

Cross-spectral analysis between the equatorial zonal wind at the 30 hPa and at the 10 hPa levels in the stratosphere yielded the squared coherence spectrum (Fig. 11.10(a)) and the phase spectrum (Fig. 11.10(b)). The phase is positive when the wind at the first time series leads that at the second. The highest squared coherence occurs at $\omega = 0.070\,\pi$ month$^{-1}$ (i.e. at the same $\omega$ found for the main spectral peak in Fig. 11.6(b)), corresponding to a period of 28.4 months, due to the dominant quasi-biennial oscillation (QBO) phenomenon. At this $\omega$, the phase is $-0.481\,\pi$ radian or $-87$ degrees. The negative sign of the phase means the second time series is leading the first series, that is, the zonal wind at the 10 hPa level leads the wind at the 30 hPa level, in agreement with Fig. 10.14, as the zonal wind anomalies propagates downward with time from the 10 hPa level to the 30 hPa level.

As the phase is a cyclic variable, plotting the phase spectrum presents some technical difficulty. For instance, suppose the phase is plotted within $[-\pi, \pi)$; if two neighbouring phase values are $0.99\pi$ and $1.01\pi$, the first value will appear near the top of the plot while the second jumps to the bottom as $-0.99\pi$. One may have to experiment with various intervals, for example $[0, 2\pi)$, $[-\pi/2, 3\pi/2)$, and so on, to see which one displays the fewest jumps between the top and bottom of the plot. For small values of the squared coherence, the phase usually fluctuates wildly, hence it may be best to avoid plotting the phase.

Methods for estimating the confidence intervals for the squared coherence spectrum and the phase spectrum are given in von Storch and Zwiers (1999, sections 12.5.5 and 12.5.6).

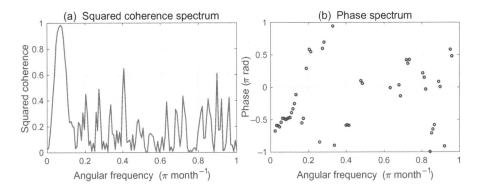

Figure 11.10 (a) The squared coherence and (b) phase from the cross-spectrum between the equatorial zonal wind at the 30 hPa and at the 10 hPa levels. The phase is only plotted when the squared coherence value is $\geq 0.2$. The phase is positive if the wind at the 30 hPa level leads that at the 10 hPa level and negative if vice versa. After detrending, the Welch method (with 8 blocks, 50% overlap and the Hamming window) was used to compute the cross-spectrum.

# 11.5 Filtering $\boxed{\text{A}}$☺

In environmental sciences, one often encounters time series containing strong periodic signals, for example the seasonal cycle, the diurnal (i.e. daily) cycle or tidal cycles. While these periodic signals are important, it is often the non-periodic signals that have the most impact on humans, as they produce the unexpected events. One usually removes the strong periodic signals from the time series first before analysing the non-periodic signals.

In mathematics and statistics, a strict *stationary process* is a stochastic process where its unconditional joint probability distribution remains unchanged when shifted in time. Less strict definitions include first-oder stationarity where the means do not change over time, while second-oder stationarity (a.k.a. weak stationarity) have constant mean, variance and auto-covariance over time.

Since stationarity is an assumption underlying many data methods used in time series analysis, non-stationary data are often transformed to become approximately stationary. Trend removal (Section 11.3.2) is commonly performed to alleviate a linear drift in the mean over time.

Periodic signals can also render the data non-stationary, thereby violating the assumption of stationarity made by many data methods. Removing the periodic signals tend to make the non-stationary data closer to being stationary. Section 11.5.1 looks at the removal of the seasonal cycle, the diurnal cycle and the tidal cycles.

## 11.5.1   Periodic Signals   [A] ☺

**Seasonal and diurnal cycles**

Suppose one has monthly data for a variable $x$, and one would like to extract the seasonal cycle. Average all $x$ values in January to get $\bar{x}_{\mathrm{Jan}}$, and similarly for the other months. The *climatological seasonal cycle* is then given by

$$\bar{x}_{\mathrm{seasonal}} = [\bar{x}_{\mathrm{Jan}}, \, \ldots \, , \bar{x}_{\mathrm{Dec}}]. \qquad (11.79)$$

The filtered time series is obtained by subtracting this climatological seasonal cycle from the raw data – that is, all January values of $x$ will have $\bar{x}_{\mathrm{Jan}}$ subtracted, and similarly for the other months. With daily data, the climatological seasonal cycle would contain the mean of all data from 1 January, the mean of all data from 2 January, and so on.

Furthermore, the variance can also be periodic, for example in mid-latitudes, winter storms often produce larger variance in the winter data than in the summer data, thereby introducing non-stationarity. One can similarly compute the sample standard deviation $s_{\mathrm{seasonal}}$ using data separately from January, February, ... , December. One can then introduce a standardized variable by subtracting the seasonal mean and dividing by the seasonal standard deviation.

The diurnal cycle (daily cycle) can be removed in a similar manner, for example the climatological diurnal cycle can be obtained by separately averaging the values for each hour of the day.

**Tidal cycles**

For tidal cycles, *harmonic analysis* is commonly used to extract the tidal cycles from a record of duration $T$. The method was first developed by Lord Kelvin (William Thomson) around 1867. The tidal frequencies $\omega_m$ are known from astronomy, and one assumes the tidal signals are sinusoidal functions of amplitude $A_m$ and phase $\theta_m$. The best fit of the tidal cycle to the data is obtained by a least squares fit, that is, minimize

$$\int_0^T \left[ x(t) - \sum_m A_m \cos(\omega_m t + \theta_m) \right]^2 \mathrm{d}t \qquad (11.80)$$

by finding the optimal values of $A_m$ and $\theta_m$ for a number of tidal frequencies $\omega_m$. If $T$ is short, then tidal components with closely related frequencies cannot be separately resolved. A time series with the tides filtered is given by

$$y(t) = x(t) - \sum_m A_m \cos(\omega_m t + \theta_m). \qquad (11.81)$$

For more details on tidal harmonic analysis, see Thomson and Emery (2014, section 5.9). Software for tidal analysis are also available in MATLAB (Pawlowicz et al., 2002) and in FORTRAN (Foreman et al., 2009).

## 11.5.2 Ideal Filters $\boxed{\text{A}}$ ☺

Often, one would like to perform digital filtering on time series data. For instance, one may want a smoother time series, or want to concentrate on the low-frequency or high-frequency signals or a frequency band associated with a particular phenomenon (e.g. the El Niño). Consider a time series $x(t)$, $(-\infty < t < \infty)$. Assuming discrete time, $t = t_n = n\Delta t$, $x(t_n) \equiv x_n$, $n$ being an integer $(-\infty < n < \infty)$.

From the DTFT and its inverse (Section 11.1.4),

$$x_n = \frac{1}{2\pi} \int_{-\pi}^{\pi} \hat{x}(\omega)\, \mathrm{e}^{\mathrm{i}\omega n \Delta t} \mathrm{d}\omega, \tag{11.82}$$

where the Fourier coefficient

$$\hat{x}(\omega) = \sum_{n=-\infty}^{\infty} x_n\, \mathrm{e}^{-\mathrm{i}\omega n \Delta t}. \tag{11.83}$$

A filtered time series is given by

$$y_n = \frac{1}{2\pi} \int_{-\pi}^{\pi} f(\omega)\, \hat{x}(\omega)\, \mathrm{e}^{\mathrm{i}\omega n \Delta t} \mathrm{d}\omega, \tag{11.84}$$

where $f(\omega)$ is the *filter response function*. Figure 11.11 illustrates several ideal filters for low-pass, high-pass and band-pass filtering.

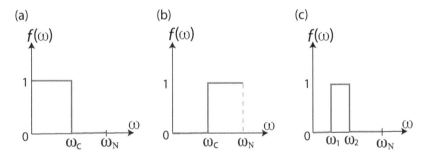

Figure 11.11 Ideal filters: (a) Low-pass, (b) high-pass and (c) band-pass, where $f(\omega)$ is the filter response function and $\omega_N$ is the Nyquist frequency. In (a), frequencies below the cutoff frequency $\omega_c$ are allowed to pass through the filter, while frequencies above $\omega_c$ are eliminated. In (b), the situation is reversed, while in (c), only frequencies within a selected band $(\omega_1 \leq \omega \leq \omega_2)$ are allowed to pass through the filter. In these ideal filters, jump discontinuities in $f(\omega)$ give infinitely sharp transitions, which are not attainable in practice.

In the real world, observed time series are of finite duration, that is, for $x_n$, $n = 0, \ldots, N-1$. At first thought, it seems straightforward to filter a time series in the frequency domain by (i) performing a discrete Fourier transform (DFT) on the time series, (ii) setting its Fourier coefficients to zero at frequencies where the ideal filter response $f(\omega)$ is zero, then (iii) do inverse DFT to get the filtered time series. Unfortunately, the number of observations, $N$, being finite implies a rectangular window has been used to observe the true signal (Section 11.2). This introduces spurious 'ringing' (i.e. oscillations) in the filtered time series, especially at the two ends (Forbes, 1988; Thomson and Emery, 2014, section 6.10). In practice, filtering in the frequency domain is done using filters with more gradual transitions than the infinitely sharp transitions in the ideal filters to alleviate spurious oscillations (Thomson and Emery, 2014, section 6.10).

## 11.5.3   Finite Impulse Response Filters  $\boxed{\text{B}}$☺

Alternatively, one can perform digital filtering in the time domain instead of in the frequency domain. This second approach is actually much more commonly used, largely from historical heritage. In the early days of computing, performing a DFT (Section 11.1.2) was computationally very expensive. The time domain approach was favoured as it avoided the DFT needed in the frequency domain approach.

In time domain filtering, the *finite impulse response* (FIR) filter approach is commonly used. The filtered time series

$$y_n = \sum_{k=0}^{K} h_k \, x_{n-k} \tag{11.85}$$

is the discrete convolution between $h_k$, the impulse response of the filter, and the input time series $x_n$. The $K$th order FIR filter has $K+1$ terms for $h_k$. The filter is called a finite impulse response filter as there is only a finite number of terms in $h_k$. Given the impulse response $h_k$, how would one find the filter response function $f(\omega)$?

Comparing the the inverse DTFT (11.22) and (11.84), we obtain

$$\hat{y}(\omega) = f(\omega)\, \hat{x}(\omega), \tag{11.86}$$

that is, the Fourier coefficient of the filtered time series is simply the Fourier coefficient of the original time series multiplied by the filter response function.

With

$$f(\omega) = \frac{\hat{y}(\omega)}{\hat{x}(\omega)}, \tag{11.87}$$

applying the DTFT (Section 11.1.4) to (11.85) and noting that the DTFT of $x_{n-k}$ is simply the DTFT of $x_n$ multiplied by the phase factor $\mathrm{e}^{-\mathrm{i}\omega k \Delta t}$ (see Exercise $\boxed{11.4}$), one obtains

$$f(\omega) = \frac{\sum_{k=0}^{K} h_k \, \mathrm{e}^{-\mathrm{i}\omega k \Delta t} \, \hat{x}(\omega)}{\hat{x}(\omega)} = \sum_{k=0}^{K} h_k \, \mathrm{e}^{-\mathrm{i}\omega k \Delta t}. \tag{11.88}$$

Hence, from the filter's impulse response $h_k$, one can derive the filter response $f(\omega)$. In (11.88), the filter response $f(\omega)$ can be regarded as the DTFT of the impulse response $h_k$, truncated to $K + 1$ terms from the infinite sum of the DTFT (11.21).

To align the filtered time series $y_n$ and the original time series $x_n$ in time, we can shift $y_n$ in (11.85) back by $K/2$ time steps ($K$ assumed to be even), that is,

$$y_{n-K/2} = \sum_{k=0}^{K} h_k \, x_{n-k}. \tag{11.89}$$

One can also express this centred filter as

$$y_n = \sum_{l=-L}^{L} w_l \, x_{n+l}, \tag{11.90}$$

where $L = K/2$ and the filter weights $w_l = h_{l+L}$. Similar to the derivation of (11.88), using the fact that the DTFT of $x_{n+l}$ is simply the DTFT of $x_n$ multiplied by the phase factor $e^{i\omega l \Delta t}$ gives the filter response $f(\omega)$ in terms of the weights $w_l$ as

$$f(\omega) = \sum_{l=-L}^{L} w_l \, e^{i\omega l \Delta t}. \tag{11.91}$$

For *symmetric filters* (i.e. filters with $w_{-l} = w_l$, $l = 1, \ldots, L$), $f(\omega)$ is real, because $w_{-l} e^{-i\omega l \Delta t}$ and $w_l e^{i\omega l \Delta t}$ are pairs of complex conjugates.

A common way to design a practical FIR filter is by using the *windowing method* or window design method (Prabhu, 2014, sections 5.5.1 and 7.6; Wanhammar and Saramäki, 2020, section 5.4): First, choose an ideal filter (Fig. 11.11), with filter response $f_{\text{ideal}}(\omega)$. From the DTFT and its inverse, we have

$$f_{\text{ideal}}(\omega) = \sum_{n=-\infty}^{\infty} h_{\text{ideal}}(n) \, e^{-i\omega n \Delta t}, \tag{11.92}$$

$$h_{\text{ideal}}(n) = \frac{1}{2\pi} \int_{-\infty}^{\infty} f_{\text{ideal}}(\omega) \, e^{i\omega n \Delta t} \, d\omega. \tag{11.93}$$

As there is an infinite number of terms in $h_{\text{ideal}}(n)$, we need to truncate the filter to only a finite number of terms by multiplying $h_{\text{ideal}}$ with a finite-length window function (Section 11.2). The rectangular window is the simplest but has large sidelobes, so windows with smaller sidelobes are used, such as the Hann or Hamming window. The resulting filter impulse function $h_k$ for the FIR filter has a finite number of terms (with $k = 0, \ldots, K$). Most software packages for digital filtering can automatically generate the FIR filter once the user specifies the desired ideal filter, the filter order $K$ and the window function.

As an example, suppose we want a band-pass filter (Fig. 11.11(c)), allowing transmission for $0.3\pi \leq \omega \leq 0.6\pi$. Figure 11.12 compares the filter response $f(\omega)$ when the filter order $K = 40$ and $K = 200$, and when the Hamming and rectangular windows are used.  Using a higher-order filter helps to reduce the sidelobes and produce a sharper transition at $\omega = 0.3\pi$ and $0.6\pi$.  Compared to the rectangular window, the Hamming window greatly reduces the sidelobes, though it has a more gradual transition at $\omega = 0.3\pi$ and $0.6\pi$.

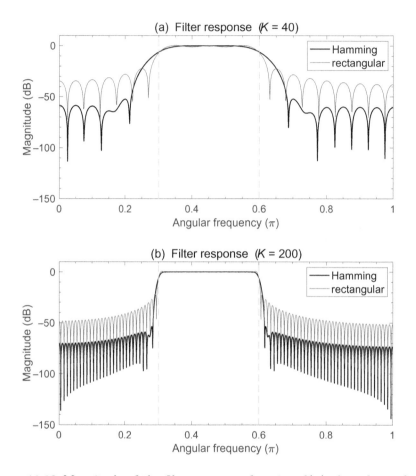

Figure 11.12 Magnitude of the filter response function $f(\omega)$ plotted as a function of the angular frequency $\omega$ (in units of $\pi$), using the Hamming window (thick line) and the rectangular window (thin line), with the filter order (a) $K = 40$ and (b) $K = 200$.  The ideal filter has infinitely sharp transitions at $\omega_1 = 0.3\pi$ and $\omega_2 = 0.6\pi$, as marked by the vertical dashed lines.

In summary, since an ideal filter has an infinite number of terms for the impulse response $h_k$, it is not usable in practice.  A practical filter needs to truncate $h_k$ from the ideal filter to a finite number of terms, $k = 0, \ldots, K$.  Truncation is equivalent to applying a rectangular window, introducing undesirable sidelobes in the frequency response $f(\omega)$ and more gradual transitions than the infinitely sharp transitions in the ideal filters. A tapered window (e.g. a Hann or Hamming window) is commonly used to alleviate the sidelobes. Increasing $K$ allows $f(\omega)$ to more closely approximate the filter response of the ideal filter. On the other hand, as the filtered time series $y_n$ contains only $N - K$ data points compared to the original time series $x_n$ with $N$ data points, increasing $K$ loses data points for $y_n$. The user chooses the $K$ value by balancing the need for a good approximation to the ideal filter and not losing too many data points for $y_n$.

Let us illustrate how an FIR filter works on an observed time series. Suppose we would like to isolate the El Niño-Southern Oscillation (ENSO) phenomenon in the Niño 3.4 index, the monthly sea surface temperature anomalies in the Niño 3.4 region in the equatorial Pacific (Section 9.1.5). Since ENSO is usually considered a 2–7 year period oscillatory phenomenon, we select an ideal filter allowing frequencies corresponding to periods between 24 and 84 months to pass through, and apply a Hamming window. For the filter order $K$, we can try a few values and plot out the filter response $f(\omega)$ to ensure it is acceptable. The time series was detrended and band-pass filtered with $K = 100$. The band-passed filtered time series allows us to see the ENSO phenomenon more clearly by removing oscillations from other phenomena (Fig. 11.13). The filtered time series is shorter than the original time series, having lost $K/2 = 50$ months of data at either end.

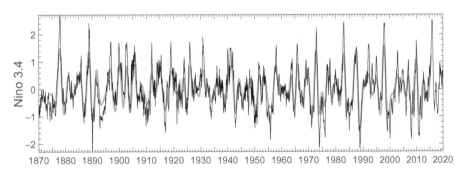

Figure 11.13 The monthly Niño 3.4 time series, unfiltered (thin line) and band-pass filtered (thick line). The grid mark for a year marks the January of that year.

# 11.6  Averaging  $\boxed{\text{A}}$☺

In environmental sciences, the ubiquitous time-averaging operation is performed on hourly data to get daily data, on daily data to get monthly data, and on monthly data to get seasonal and annual data. It is used to remove high-frequency noise or to reduce the amount of data when one is not interested in the higher frequency domain. Averaging is also performed in the spatial domain to provide a smoother spatial field. An unintended consequence of averaging is the filtering of non-linear relations in the data (Section 11.6.3). The averaging operation is equivalent to applying a filter to the time series, so let us examine its filtering effects.

### 11.6.1 Moving Average Filters  [A]☺

The centred *moving average* (MA) *filter* (a.k.a. running mean filter),

$$y_n = \frac{1}{2L+1} \sum_{l=-L}^{L} x_{n+l}, \tag{11.94}$$

is simply the average over $2L+1$ terms of the original time series. It is a rather crude low-pass filter and is a special case of the centred FIR filter in (11.90) where all the weights $w_l$ are set to $1/(2L+1)$.

The shortest MA filter (with $L = 1$) is the three-point moving average filter

$$y_n = \tfrac{1}{3} x_{n-1} + \tfrac{1}{3} x_n + \tfrac{1}{3} x_{n+1}, \tag{11.95}$$

that is, average over the immediate neighbours, commonly used when converting monthly data to seasonal data.

The 12-month moving average converts monthly values to annual values. However, it is awkward to perform 12-month moving average using a centred MA filter, as one needs to average the central monthly value $x_n$ and 5.5 months of data to either side of $x_n$. Hence, in the literature, one can find the 11-month centred MA filter (with $L = 5$) and the 13-month centred MA filter (with $L = 6$). The closest to a 12-month centred MA filter is the 'annual symmetric moving average filter' (Shumway and Stoffer, 2017, p. 212),

$$y_n = \frac{1}{12} \sum_{l=-5}^{5} x_{n+l} + \frac{1}{24} \big( x_{n-6} + x_{n+6} \big), \tag{11.96}$$

which averages the central monthly value $x_n$ and 5.5 months of data to either side of $x_n$ by weighting the values of $x_{n-6}$ and $x_{n+6}$ half as much as the other 11 months. Figure 11.14(a) shows that this filter eliminates signals at the annual frequency $\omega_{\text{annual}}$ better than the 11-month and 13-month MA filters. This filter is also better in filtering out signals in the domain $\omega_{\text{annual}} < \omega$.

### 11.6.2 Grid-Scale Noise  [B]☺

Discretization of space and time in numerical models tend to generate grid-scale noise, that is, noise at the smallest scale allowed in the numerical model. Grid-scale noise can be manifested in time as oscillations at the Nyquist frequency or in space (e.g. a 2-D anomaly field manifesting a checkerboard pattern of alternating $+-$ signs at adjacent grid points). One would like to smooth out the grid-scale noise, and the three-point MA filter seems a handy tool.

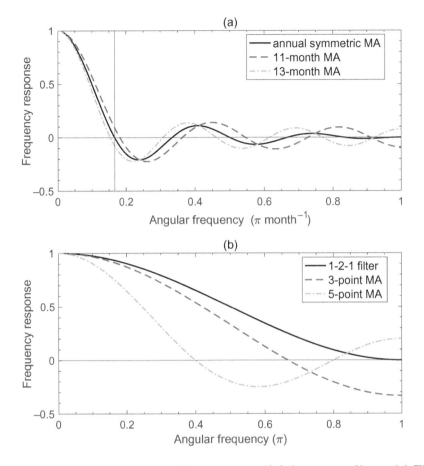

Figure 11.14 Comparison of the filter response $f(\omega)$ for various filters: (a) The annual symmetric moving average (MA) filter and the 11-month and 13-month MA filters and (b) the 1-2-1 filter and the three-point and five-point MA filters. The angular frequency $\omega$ is in units of $\pi$ and the vertical line in (a) marks the frequency $\omega_{\mathrm{annual}}$ for the annual cycle.

However, a better filter to remove grid-scale noise is the *three-point triangular filter*, a.k.a. the *1-2-1 filter* from the central weight being double that of its two neighbours, that is,

$$y_n = \tfrac{1}{4}\, x_{n-1} + \tfrac{1}{2}\, x_n + \tfrac{1}{4}\, x_{n+1}. \tag{11.97}$$

Figure 11.14(b) shows that at the Nyquist frequency $\omega = \pi$, the filter response is zero for the 1-2-1 filter but non-zero for the three-point and five-point MA filters, that is, the 1-2-1 filter eliminates the grid-scale noise but the MA filters can only reduce the noise (see Exercise 11.6 and Exercise 11.7).

## 11.6.3  Linearization from Time-Averaging  B☺

An unintended consequence of time-averaging is the filtering of non-linear relations in the dataset. In a study of the non-linear relation between the precipitation rate (the response variable) and ten other atmospheric variables (the predictors) from the *NCEP/NCAR Reanalysis* dataset,[6] Yuval and Hsieh (2002) analysed the daily, weekly and monthly averaged data by performing non-linear multiple regression using the MLP neural network model and discovered that the non-linear relations found in the daily data by the NN model became dramatically weakened by time-averaging to the almost linear relations found in the monthly data. For instance, off the west coast of British Columbia, Canada (50°N, 130°W), for the daily, weekly and monthly data, the ratio of the RMSE (root mean squared error) of the NN model to that of MLR (multiple linear regression) was 0.88, 0.92 and 0.96, respectively, while the correlation skill of NN to MLR was 1.10, 1.02 and 1.00, respectively. Similar conclusions were found for data in the Middle East (33°N, 35°E) and northeastern China (50°N, 123°E).

To explain this phenomenon, Yuval and Hsieh (2002) invoked the well-known *central limit theorem* from statistics. For simplicity, consider the relation between two variables $x$ and $y$. If $y = f(x)$ is a non-linear function, then even if $x$ is a Gaussian variable, $y$ will, in general, not have a Gaussian distribution. Now consider the effects of time-averaging on the $(x, y)$ data. The bivariate central limit theorem (Bickel and Doksum, 1977, theorem 1.4.3) says that if $(x_1, y_1), \ldots, (x_m, y_m)$ are independent and identically distributed random vectors with finite second moments, then $(\tilde{x}, \tilde{y})$, obtained from averaging $(x_1, y_1), \ldots, (x_m, y_m)$, will, as $m \to \infty$, approach a bivariate Gaussian distribution $N(\mu_1, \mu_2, \sigma_1^2, \sigma_2^2, \rho)$, where $\mu_1$ and $\mu_2$ are the mean of $\tilde{x}$ and $\tilde{y}$, respectively, $\sigma_1^2$ and $\sigma_2^2$ are the corresponding variance, and $\rho$ the correlation between $\tilde{x}$ and $\tilde{y}$.

From the bivariate Gaussian distribution, the conditional probability distribution of $\tilde{y}$ (given $\tilde{x}$) is also a Gaussian distribution (Bickel and Doksum, 1977, theorem 1.4.2), with mean

$$\mathrm{E}[\tilde{y} \,|\, \tilde{x}] = \mu_2 + (\tilde{x} - \mu_1)\rho\,\sigma_2/\sigma_1. \tag{11.98}$$

This linear relation in $\tilde{x}$ explains why time-averaging tends to linearize the relationship between the two variables. With more variables, the bivariate Gaussian distribution readily generalizes to the multivariate Gaussian distribution.

Figure 1.3 uses synthetic data generated from the simple quadratic relation (1.1) to illustrate how time-averaging transforms a non-Gaussian distribution to a near Gaussian distribution. With real data, there is serial correlation in the time series, so the monthly data will be effectively averaging over far fewer than 30 independent observations as done in this synthetic dataset.

---

[6] By incorporating observations and numerical weather prediction (NWP) model output, the NCEP/NCAR Reanalysis dataset is a continually updated (1948–present) globally gridded dataset representing the state of the Earth's atmosphere (Kalnay, Kanamitsu, et al., 1996). It is a joint product from the National Centers for Environmental Prediction (NCEP) and the National Center for Atmospheric Research (NCAR).

If the data have strong serial correlation, so that the time scale over which there is positive autocorrelation is not small compared to the time-averaging window, then the central limit theorem does not apply as one is not averaging over many independent observations. For instance, the eastern equatorial Pacific sea surface temperature anomalies have positive autocorrelation of almost a year (because the dominant ENSO phenomenon is an interannual oscillation), hence non-linear relations can be detected from monthly or seasonal data, as found by non-linear principal component analysis (Fig. 10.9) and non-linear canonical correlation analysis (Section 10.8). In contrast, the mid-latitude weather variables are often positively autocorrelated over 2–5 days, so monthly averaged data would have effectively averaged over about 6–15 independent observations, and seasonal data over 18–45 independent observations, so the influence of the central limit theorem cannot be ignored.

While time-averaging tends to reduce the non-linear signal, it also smooths out the noise. Depending on the type of noise and the type of non-linear signal, it is possible that time-averaging may nevertheless enhance detection of a non-linear signal above the noise for some datasets. For instance, if the noise is autocorrelated over a short time scale and the non-linear signal over a longer time scale, then averaging will be effective in filtering out the noise but not too much of the non-linear signal. In short, researchers should be aware that time-averaging is a double-edged sword – it could have a major impact on their modelling or detection of non-linear empirical relations, and that non-linear machine learning methods often outperform a linear method in weather applications, but may fail to do so in climate applications.

Time-averaging is no longer the only statistic used to convert weather data to climate data. In the last couple of decades, there has been growing interest in the *climate of extreme weather events* (simply called '*climate extremes*'), as global climate change may affect the extremes even more than the averages. There is now a long list of climate extreme statistics derived from daily data, for example the annual number of frost days, the maximum number of consecutive days when precipitation is < 1 mm, and so on (Peterson and coauthors, 2001; X. B. Zhang et al., 2011). By circumventing time-averaging and its linearization effect, non-linear ML methods can model/predict the non-linear relations in the daily data directly, then the daily results from the ML model can be compiled to give the indices for climate extremes (Gaitan, Hsieh, et al., 2014).

# 11.7 Singular Spectrum Analysis B☺

If a multivariate dataset contains observations taken at various spatial locations, principal component analysis (PCA) applied to the dataset gives eigenvectors containing spatial information on the variability of the dataset (Section 9.1). It is possible to use the PCA method to incorporate time information into the eigenvectors. This approach is known as *singular spectrum analysis* (SSA) (Elsner and Tsonis, 1996; Ghil et al., 2002).

For simplicity, start with a single time series $x_j = x(t_j)$ $(j = 1, \ldots, n)$. Lagged copies of the time series are stacked to form the *augmented data matrix* $\mathbf{X}$,

$$
\mathbf{X} = \begin{bmatrix} x_1 & x_2 & \cdots & x_L \\ x_2 & x_3 & \cdots & x_{L+1} \\ \vdots & \vdots & \ddots & \vdots \\ x_{n-L+1} & x_{n-L+2} & \cdots & x_n \end{bmatrix}. \tag{11.99}
$$

This matrix has the same form as the data matrix produced by $L$ variables, each being a time series of length $n - L + 1$ presented as a column in $\mathbf{X}$. $\mathbf{X}$ can also be viewed as composed of its row vectors $\mathbf{x}_{(l)}^{\mathrm{T}}$, that is,

$$
\mathbf{X} \equiv \begin{bmatrix} \mathbf{x}_{(0)}^{\mathrm{T}} \\ \mathbf{x}_{(1)}^{\mathrm{T}} \\ \vdots \\ \mathbf{x}_{(n-L)}^{\mathrm{T}} \end{bmatrix}, \tag{11.100}
$$

where
$$
\mathbf{x}_{(l)}^{\mathrm{T}} = \begin{bmatrix} x_{l+1} & x_{l+2} & \cdots & x_{l+L} \end{bmatrix}, \quad l = 0, \ldots, n - L. \tag{11.101}
$$

The vector space spanned by $\mathbf{x}_{(l)}$ is called the *delay coordinate space*. The number of lags $L$ is usually taken to be at most $1/4$ of the total record length.

The standard PCA can be performed on $\mathbf{X}$, resulting in

$$
\mathbf{x}_{(l)} = \mathbf{x}(t_l) = \sum_j a_j(t_l) \, \mathbf{e}_j, \tag{11.102}
$$

where $a_j$ is the $j$th principal component (PC), a time series of length $n - L + 1$, and $\mathbf{e}_j$ is the $j$th eigenvector (or loading vector) of length $L$. Together, $a_j$ and $\mathbf{e}_j$, represent the $j$th SSA mode. This method is called singular spectrum analysis, as it studies the ordered set (spectrum) of singular values (the square roots of the eigenvalues).

SSA has become popular in the field of dynamical systems (including chaos theory), where time-delay coordinates are commonly used (Bradley and Kantz, 2015). By lagging a time series, one is providing information on the first-order differencing of the discrete time series – with the first-order difference on discrete variables being the counterpart of the derivative. In the delay coordinate approach, repeated lags are supplied, thus information on the time series and its higher-order differences (hence derivatives) is provided.

The first SSA *reconstructed component* (RC) is the approximation of the original time series $x(t)$ by the first SSA mode. As the eigenvector $\mathbf{e}_1$ contains the loading over a range of lags, the first SSA mode, that is, $a_1(t_l)\,\mathbf{e}_1$, provides an estimate for the $x$ values over a range of lags starting from the time $t_l$. For instance, at time $t_L$, estimates of $x(t_L)$ can be obtained from any one of the delay coordinate vectors $\mathbf{x}_{(0)}, \ldots, \mathbf{x}_{(L-1)}$. Hence, each value in the reconstructed RC time series $\tilde{x}$ at time $t_i$ involves averaging over the contributions at $t_i$ from the

$L$ delay coordinate vectors that provide estimates of $x$ at time $t_i$. Near the beginning or end of the record, the averaging will involve fewer than $L$ terms. The RCs for higher modes are similarly defined.

Let us compare SSA with Fourier spectral analysis. Unlike Fourier spectral analysis, SSA does not in general assume the time series to be periodic. Therefore, there is no need to taper the ends of the time series as commonly done in Fourier spectral analysis. As the wave forms extracted from the SSA eigenvectors are not restricted to sinusoidal shapes, the SSA can in principle capture an anharmonic wave more efficiently than the Fourier method. However, the SSA eigenvectors may turn out to have shapes resembling sinusoidal functions, even when the signal is as anharmonic as a sawtooth wave or a square wave (Hsieh, 2009, figures 3.5 and 3.6).

In the real world, many signals do not have a precise frequency. For instance, the El Niño-Southern Oscillation (ENSO) (Section 9.1.5) occurs every 2–10 years. When the frequency is not precise, the energy is scattered over many frequency bands in the Fourier approach, while the SSA approach captures more of the energy or variance in fewer modes (Hsieh, 2009, pp. 72–73).

Another advantage of SSA over the Fourier approach lies in the multivariate situation – the Fourier approach does not generalize naturally to large multivariate datasets, whereas the SSA, based on the PCA method, does.

In the next two subsections, we cover multivariate SSA (a.k.a. multichannel SSA) and non-linear SSA.

## 11.7.1 Multichannel Singular Spectrum Analysis B☺

In a multivariate dataset, there are $m$ variables $x_k$ ($k = 1, \ldots, m$) and $n$ observations in time, that is, $x_k(t_j) \equiv x_{jk}$, ($j = 1, \ldots, n$; $k = 1, \ldots, m$). We define $L$ time-lagged data matrices

$$\mathbf{X}_{(l)} = \begin{bmatrix} x_{l+1,1} & \cdots & x_{l+1,m} \\ \vdots & \ddots & \vdots \\ x_{n-L+l+1,1} & \cdots & x_{n-L+l+1,m} \end{bmatrix}, \tag{11.103}$$

with $l = 0, 1, 2, \ldots, L-1$. Again, treat the time lagged data as extra variables, so the augmented data matrix is

$$\mathbf{X} = \begin{bmatrix} \mathbf{X}_{(0)} & | & \mathbf{X}_{(1)} & | & \mathbf{X}_{(2)} & | \cdots \mathbf{X}_{(L-1)} \end{bmatrix}, \tag{11.104}$$

that is,

$$\mathbf{X} = \begin{bmatrix} x_{11} & \cdots & x_{1m} & \cdots & x_{L1} & \cdots & x_{Lm} \\ \vdots & \ddots & \vdots & \ddots & \vdots & \ddots & \vdots \\ x_{n-L+1,1} & \cdots & x_{n-L+1,m} & \cdots & x_{n1} & \cdots & x_{nm} \end{bmatrix}. \tag{11.105}$$

PCA is applied to the augmented data matrix $\mathbf{X}$ to get the SSA modes.

When the method is applied to multiple spatial variables, it is called the *space–time PCA* (ST–PCA), or *multichannel singular spectrum analysis* (MSSA).

In this book, we will, for brevity, use the term SSA to denote both the univariate and multivariate cases. The term *extended empirical orthogonal function* (EEOF) analysis is also used in the literature, especially when the number of lags $(L - 1)$ is small. So far, we have assumed the time series was lagged one time step at a time. To save computational time, larger lag intervals can be used, that is, lags can be taken over several time steps at a time. ST–PCA can have degenerate modes for some choices of the lag interval (Monahan, Tangang, et al., 1999).

To illustrate the SSA applied to a multivariate dataset, let us consider the tropical Pacific monthly sea surface temperature anomalies (SSTA) data from 1950 to 2000, where the climatological seasonal cycle and the linear trend have been removed from the SST data to give the SSTA. To resolve the ENSO variability, a window of 73 months was chosen (Hsieh and Wu, 2002). With a lag interval of 3 months, the original plus 24 lagged copies of the SSTA data formed the augmented SSTA dataset (note that if a lag interval of 1 month were used instead, then to cover the window of 73 months, the original plus 72 copies of the SSTA data would have produced a much bigger augmented data matrix). The first eight SSA modes respectively explain 12.4%, 11.7%, 7.1%, 6.7%, 5.4%, 4.4%, 3.5% and 2.8% of the total variance of the augmented dataset.

The first six SSA modes are shown in Fig. 11.15, where the contour plots display the space–time eigenvectors (i.e. loading patterns). In a separate panel beneath each contour plot, the principal component (PC) of each SSA mode is plotted as a time series (with the time of the PC synchronized to the lag time of 0 month in the space–time eigenvector).

The first two modes have eigenvectors showing an oscillatory time scale of about 48 months, comparable to the ENSO time scale, with the mode 1 anomaly pattern occurring about 12 months before a very similar mode 2 pattern, that is, the two patterns are in quadrature.[7] The PC time series also show similar time scales for modes 1 and 2. Modes 3 and 5 show longer time scale fluctuations, while modes 4 and 6 show shorter time scale fluctuations – around the 30 month time scale.

Since rotation can be performed on PCA (Section 9.2), and SSA is based on PCA, rotation can also be applied to SSA, for example Groth and Ghil (2011) applied varimax rotation on the SSA eigenvectors.

### Non-linear singular spectrum analysis

In Section 10.6, we saw that an autoencoder neural network model can be used to perform non-linear PCA. Similarly, an autoencoder can be used to perform non-linear singular spectrum analysis (NLSSA). When applied to the tropical Pacific SSTA, the NLSSA improves on the SSA solution in several ways (Hsieh and Wu, 2002; Hsieh, 2009, section 10.6):

---

[7] A pair of periodic signals are '*in quadrature*' when they are 90 degrees out-of-phase.

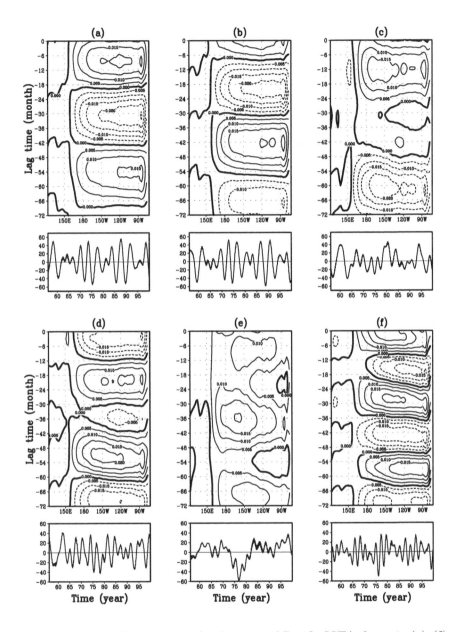

Figure 11.15 The SSA modes 1–6 for the tropical Pacific SSTA shown in (a)–(f), respectively. The contour plots show the SSTA along the equator as a function of the lag. The zero contour is marked by the thick curve, and positive and negative anomalies by solid and dashed curves, respectively. The PC time series is shown beneath each contour plot. [Reproduced from Hsieh and Wu (2002).]

(1) The presence of warm anomalies for 24 months followed by cool anomalies for 24 months in the first two SSA modes (Fig. 11.15) is replaced by warm anomalies for 18 months followed by cool anomalies for about 33 months in the NLSSA mode 1. Although the cool anomalies can be quite mild for long periods, they can develop into full La Niña cool episodes.

(2) The El Niño warm episodes are strongest near the eastern boundary, while the La Niña episodes are strongest near the central equatorial Pacific in the NLSSA mode 1, an asymmetry not found in the individual SSA modes.

(3) The magnitude of the peak positive anomalies is much larger than that of the peak negative anomalies in the NLSSA mode 1, again an asymmetry not found in the individual SSA modes.

## 11.8  Auto-regressive Process  $\boxed{\text{B}}$ ☺

Many physical processes $x(t)$ can be modelled by first- or second-order linear ordinary differential equations with constant coefficients, that is,

$$a_2 \frac{\mathrm{d}^2 x}{\mathrm{d}t^2} + a_1 \frac{\mathrm{d}x}{\mathrm{d}t} + a_0\, x = f(t), \qquad (11.106)$$

where the coefficients $a_0$, $a_1$ and $a_2$ are constants and $f(t)$ is an external forcing function. Differentiation can be approximated by finite-differences, for example

$$\frac{\mathrm{d}x}{\mathrm{d}t} \approx \frac{x(t) - x(t - \Delta t)}{\Delta t}, \qquad (11.107)$$

$$\frac{\mathrm{d}^2 x}{\mathrm{d}t^2} \approx \frac{x(t) - 2\,x(t - \Delta t) + x(t - 2\Delta t)}{(\Delta t)^2}. \qquad (11.108)$$

Taking $\Delta t = 1$, (11.106) can be approximated by the finite-difference equation

$$a_2(x_t - 2x_{t-1} + x_{t-2}) + a_1(x_t - x_{t-1}) + a_0\, x_t = f_t, \qquad (11.109)$$

where $x_t = x(t)$, $x_{t-1} = x(t - 1)$, and so on. This equation can be rewritten as

$$x_t = \phi_1 x_{t-1} + \phi_2\, x_{t-2} + \epsilon_t, \qquad (11.110)$$

where

$$\phi_1 = \frac{a_1 + 2a_2}{a_0 + a_1 + a_2}, \quad \phi_2 = \frac{-a_2}{a_0 + a_1 + a_2} \quad \text{and} \quad \epsilon_t = \frac{f_t}{a_0 + a_1 + a_2}. \qquad (11.111)$$

If $f_t$ or $\epsilon_t$ is a white noise process,[8] (11.110) is called a *second-order auto-regressive process* or AR(2) process. Auto-regressive refers to the fact that $x_t$ depends on $x$ at previous time steps, with the order being the number of time steps back, that is, for a $p$th order process, $x_t$ depends on $x_{t-1}, \ldots, x_{t-p}$. AR models (von Storch and Zwiers, 1999; Wilks, 2011; Shumway and Stoffer, 2017) are relatively simple models in a large class known as Box–Jenkins models, after their popularization by the seminal book by Box and Jenkins (1970).

---

[8] '*White noise*' is a random signal with equal intensity at all frequencies, that is, its power spectral density is constant over all frequencies.

## 11.8.1 AR($p$) Process B☺

In general, an auto-regressive process of order $p$, that is, an AR($p$) process, is defined by

$$x_t = \sum_{k=1}^{p} \phi_k\, x_{t-k} + \phi_0 + \epsilon_t, \tag{11.112}$$

where $\phi_k$ $(k = 0, \ldots, p)$ are real constants (with $\phi_p \neq 0$), and $\epsilon_t$ is a white noise process. Note that $x_t$ depends on $\epsilon_t$, but $x_{t-k}$ $(k = 1, \ldots, p)$ are independent of $\epsilon_t$. The most common AR processes are AR(1) and AR(2). For AR(0), (11.112) reduces to $x_t = \phi_0 + \epsilon_t$, that is, AR(0) is simply a white noise process.

Taking the expectation of (11.112) and writing $\mathrm{E}[x_t] = \mu = \mathrm{E}[x_{t-k}]$ and $\mathrm{E}[\epsilon_t] = 0$, we have

$$\mu = \sum_{k=1}^{p} \phi_k\, \mu + \phi_0, \quad \text{i.e.} \quad \phi_0 = \mu\left(1 - \sum_{k=1}^{p} \phi_k\right). \tag{11.113}$$

Thus, (11.112) can be rewritten as

$$x_t - \mu = \sum_{k=1}^{p} \phi_k\, (x_{t-k} - \mu) + \epsilon_t. \tag{11.114}$$

To obtain the variance of $x_t$, we take the expectation of the product between $(x_t - \mu)$ and (11.114), giving

$$\mathrm{E}\big[(x_t - \mu)^2\big] = \sum_{k=1}^{p} \phi_k\, \mathrm{E}\big[(x_t - \mu)(x_{t-k} - \mu)\big] + \mathrm{E}\big[(x_t - \mu)\epsilon_t\big],$$

$$\mathrm{var}(x_t) = \sum_{k=1}^{p} \phi_k\, \rho_k\, \mathrm{var}(x_t) + \mathrm{var}(\epsilon_t), \tag{11.115}$$

where the lag-$k$ autocorrelation (Section 2.11.2)

$$\rho_k = \frac{\mathrm{E}[(x_t - \mu)(x_{t-k} - \mu)]}{\mathrm{var}(x_t)}, \tag{11.116}$$

and from substituting in (11.114),

$$\mathrm{E}\big[(x_t - \mu)\epsilon_t\big] = \mathrm{E}\!\left[\left(\textstyle\sum_{k=1}^{p} \phi_k\, (x_{t-k} - \mu) + \epsilon_t\right)\epsilon_t\right] = \mathrm{E}\big[\epsilon_t^2\big] = \mathrm{var}(\epsilon_t), \tag{11.117}$$

as $x_{t-k}$ and $\epsilon_t$ are independent. From 11.115, the relation between the variance of $x$ and the variance of the input noise $\epsilon$ is seen as

$$\mathrm{var}(x_t) = \frac{\mathrm{var}(\epsilon_t)}{1 - \sum_{k=1}^{p} \phi_k\, \rho_k}. \tag{11.118}$$

If the mean $\mu$ has been subtracted from $x$, we can set $\mu = 0$ (and $\phi_0 = 0$), thereby simplifying (11.114) to

$$x_t = \sum_{k=1}^{p} \phi_k\, x_{t-k} + \epsilon_t. \tag{11.119}$$

## 11.8.2  AR(1) Process  ☐B☺

For an AR(1) process, (11.119) reduces to

$$x_t = \phi\,x_{t-1} + \epsilon_t, \tag{11.120}$$

with only a single model parameter $\phi$ ($\equiv \phi_1$). For the process to be stationary,[9] $|\phi| < 1$ is required (von Storch and Zwiers, 1999, section 10.3.5). For $0 < \phi < 1$, $\phi\,x_{t-1}$ is a damping term. For $-1 < \phi < 0$, as $\phi\,x_{t-1}$ tries to reverse the sign between $x_{t-1}$ and $x_t$, it is an oscillatory term.

Taking the expectation of the product between $x_{t-1}$ and (11.120), we have

$$\mathrm{E}[x_{t-1}x_t] = \phi\mathrm{E}[x_{t-1}x_{t-1}] + \mathrm{E}[x_{t-1}\epsilon_t]. \tag{11.121}$$

The last term vanishes since $x_{t-1}$ and $\epsilon_t$ are independent, thus

$$\phi = \frac{\mathrm{E}[x_{t-1}x_t]}{\mathrm{E}[x_{t-1}x_{t-1}]} = \frac{\mathrm{E}[x_{t-1}x_t]}{\mathrm{var}(x_t)} = \rho_1, \tag{11.122}$$

where $\rho_1$ is the lag-1 autocorrelation. Thus, the AR(1) model parameter $\phi$ can be obtained from the lag-1 autocorrelation coefficient estimated from the time series data $x_t$. Furthermore, one can show that

$$\rho_k = \phi^k \tag{11.123}$$

(see Exercise ☐11.9☐).

The autospectrum or power spectrum of an AR(1) process (von Storch and Zwiers, 1999, section 11.2.5) is

$$S(\omega) = \frac{\mathrm{var}(\epsilon_t)}{2\pi(1 + \phi^2 - 2\phi\cos\omega)}, \tag{11.124}$$

where $\omega$ is the angular frequency. $S(\omega)$ decreases as $\omega$ increases towards the Nyquist frequency, hence the spectrum is 'red' (i.e. having more power at lower frequencies).

Figure 11.16 shows the spectrum of the monthly Niño 3.4 time series and that of the AR(1) model of the time series. The AR(1) model provides a basic benchmark, allowing us to focus on the spectral peaks that are stronger than the background red noise spectrum. AR(1) models can also provide significance tests against red noise, for example Torrence and Compo (1998) used the chi-squared distribution to estimate the 95% confidence interval of the AR(1) wavelet spectrum as a 5% significance test for the wavelet spectrum in Fig. 11.8.

---

[9] A *stationary process* is a stochastic process where the statistical parameters (e.g. mean and variance) do not change over time. Time series with trends and seasonal cycles are not stationary.

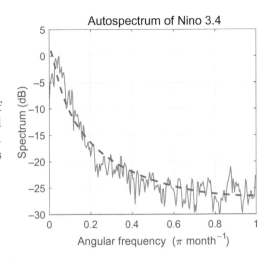

Figure 11.16 Autospectrum of the Niño 3.4 index (solid line) and the AR(1) model (dashed line). The AR(1) model parameter $\phi$ is 0.92.

### 11.8.3  AR(2) Process C☺

As given in (11.110), the AR(2) process, with two model parameters $\phi_1$ and $\phi_2$, can model a broader range of behaviour than the AR(1) model. Taking the expectation of the product between $x_{t-1}$ and (11.110), we get

$$\mathrm{E}[x_{t-1}x_t] = \phi_1\mathrm{E}[x_{t-1}x_{t-1}] + \phi_2\mathrm{E}[x_{t-1}x_{t-2}] + \mathrm{E}[x_{t-1}\epsilon_t]. \tag{11.125}$$

With $\mathrm{E}[x_{t-1}\epsilon_t] = 0$, substituting in (11.116) gives

$$\rho_1 = \phi_1 + \phi_2\rho_1. \tag{11.126}$$

Similarly, taking the expectation of the product between $x_{t-2}$ and (11.110) yields

$$\rho_2 = \phi_1\rho_1 + \phi_2. \tag{11.127}$$

These two equations are known as the *Yule–Walker* equations, named after G. Udny Yule and Sir Gilbert Walker (discoverer of the Southern Oscillation). Hence

$$\phi_1 = \frac{\rho_1(1 - \rho_2)}{1 - \rho_1^2}, \qquad \phi_2 = \frac{\rho_2 - \rho_1^2}{1 - \rho_1^2}, \tag{11.128}$$

that is, the two model parameters can be estimated from the lag-1 and lag-2 autocorrelations of the time series. Conversely, if $\phi_1$ and $\phi_2$ are given, one can solve for the lag-1 and lag-2 autocorrelations of the AR(2) model from the Yule–Walker equations.

An AR(2) process is stationary if

$$\phi_1 + \phi_2 < 1,$$
$$\phi_2 - \phi_1 < 1,$$
$$|\phi_2| < 1 \tag{11.129}$$

(von Storch and Zwiers, 1999, section 10.3.5), with the conditions for stationarity satisfied in a triangular domain in the $\phi_1$–$\phi_2$ plane.

## 11.9   Box–Jenkins Models   C☺

Box and Jenkins (1970) popularized a large class of time series forecasting models, such as the auto-regressive moving average (ARMA) model, the auto-regressive integrated moving average (ARIMA) model, and so on, with the latest edition of the book being Box, Jenkins, et al. (2015). The autoregressive (AR) model studied in the previous section is a small subset of the Box–Jenkins models. Another subset is the moving average (MA) models. Box–Jenkins models were the dominant state-of-the-art stochastic time series forecasting tool for about two decades, until newer ML methods emerged.

Box–Jenkins models have been applied to many areas related to the environment. Examples include fishery dynamics (Mendelssohn, 1980), hydrology (Salas and Obeysekera, 1982), air pollution (Milionis and T. D. Davies, 1994), energy consumption forecasting (Saab et al., 2001) and drought forecasting from remote sensing data (Han et al., 2010).

A relatively brief introduction to Box–Jenkins models is given below. For more details, see Brockwell and R. A. Davis (1991) or Box, Jenkins, et al. (2015).

### 11.9.1   Moving Average (MA) Process   C☺

Moving average (MA) processes or models – not to be confused with moving average filters in Section 11.6.1 – have the form

$$x_t = \mu + \epsilon_t + \sum_{l=1}^{q} \theta_l \, \epsilon_{t-l}, \qquad (11.130)$$

where $\mu$ is the expectation of $x_t$, $\epsilon_t$ is a white noise process and $\theta_l$ $(l = 1, \ldots, q)$ are the model parameters. This simple model is excited by a weighted sum of the white noise over the time steps during $t-q, \ldots, t$. Unlike the AR process, an MA process cannot be solved by linear least squares, because the right hand side of (11.130) contains a weighted sum of the unknown $\epsilon$ terms. Thus, iterative non-linear least squares is needed to solve for the parameters of the MA model.

### 11.9.2   Auto-regressive Moving Average (ARMA) Model   C☺

An auto-regressive moving average (ARMA) model combines the AR and MA processes into

$$x_t = \mu + \epsilon_t + \sum_{k=1}^{p} \phi_k \, x_{t-k} + \sum_{l=1}^{q} \theta_l \, \epsilon_{t-l}, \qquad (11.131)$$

where $\mu$ is the expectation of $x_t$, $\epsilon_t$ is a white noise process and $\phi_k$ $(k = 1, \ldots, p)$ and $\theta_l$ $(l = 1, \ldots, q)$ are the model parameters. This model is referred to as an ARMA$(p, q)$ model. Non-linear least squares is needed to solve for the ARMA model parameters. An ARMA$(p, 0)$ model is simply an AR$(p)$ model, while an ARMA$(0, q)$ is an MA$(q)$ model.

For finding appropriate values for $p$ and $q$, or choosing the best model among several runs with different $p$ and $q$, Box, Jenkins, et al. (2015, chapter 6) recommends examining the estimated autocorrelation and partial autocorrelation[10] functions of the time series, as well as using information criteria such as the Akaike information criterion (AIC) (Section 8.11.2) or the Bayesian information criterion (BIC) (Section 8.11.1).

## 11.9.3 Auto-regressive Integrated Moving Average (ARIMA) Model ☐☺

For time series that are not stationary, for example having a trend, differencing of the time series is used to improve stationarity. The differenced time series $x'$ is defined by

$$x'_t = x_t - x_{t-1}. \tag{11.132}$$

Trends are removed by differencing because the difference operation is a high-pass filter (see Exercise 11.10). The differencing operation can be repeated, for example

$$x''_t = x'_t - x'_{t-1} = x_t - 2x_{t-1} + x_{t-2}, \tag{11.133}$$

and the operation can be repeated $d$ times.

Let $x'$ denote the time series obtained from performing differencing $d$ times on the original time series $x_t$. The auto-regressive integrated moving average ARIMA($p$, $d$, $q$) model is simply the ARMA($p$, $q$) model applied to $x'$, that is,

$$x'_t = \mu' + \epsilon_t + \sum_{k=1}^{p} \phi_k \, x'_{t-k} + \sum_{l=1}^{q} \theta_l \, \epsilon_{t-l}, \tag{11.134}$$

where $\mu'$ is the expectation of $x'_t$, $\epsilon_t$ is a white noise process and $\phi_k$ ($k = 1, \ldots, p$) and $\theta_l$ ($l = 1, \ldots, q$) are the model parameters. The term 'integrated' or 'I' in the name ARIMA comes from the fact that a time series that needs to be differenced to become stationary can be considered the 'integrated' version of a stationary series. With $d = 0$, that is, no differencing, ARIMA($p$, 0, $q$) reduces to ARMA($p$, $q$).

For data with a seasonal cycle, *seasonal differencing* is more relevant, that is,

$$x'_t = x_t - x_{t-s}, \tag{11.135}$$

where $s$ is the duration of the season. For instance, for monthly data with an annual cycle, $s = 12$, that is, subtracting the monthly value one year ago.

---

[10] The partial autocorrelation of lag $j$ is the autocorrelation between $x_t$ and $x_{t+j}$ with the linear dependence of $x_t$ on $x_{t+1}$ through $x_{t+j-1}$ removed, that is, it is the autocorrelation between $x_t$ and $x_{t+j}$ not accounted for by lags 1 through $j - 1$.

# Exercises

### 11.1

Prove that cosine and sine functions have the following orthogonality properties (as used in Section 11.3):

$$\sum_{n=0}^{N-1} \cos(\omega_l t_n) \cos(\omega_m t_n) = \frac{N}{2}\delta_{lm},$$

$$\sum_{n} \sin(\omega_l t_n) \sin(\omega_m t_n) = \frac{N}{2}\delta_{lm},$$

$$\sum_{n} \cos(\omega_l t_n) \sin(\omega_m t_n) = 0, \qquad (11.136)$$

where $\delta_{lm}$ is the Kronecker delta function.

### 11.2

In Section 11.2, the window functions were defined to be zero outside the interval $[0, T]$. What are the corresponding forms of (11.26), (11.27) and (11.28) if the windows are to be zero outside the interval $[-T/2,\ T/2]$ instead?

### 11.3

Fourier spectral analysis is performed on hourly sea level height data. The main tidal period is around 12 hours, but there are actually two components, $M_2$ from the moon at 12.42 hours, and $S_2$ from the sun at 12.00 hours. What is the minimum length of the data record required in order to see two distinct peaks around 12 hours in your spectrum due to the $M_2$ and $S_2$ tidal components?

### 11.4

With $x_n \equiv x(t_n)$, $t_n = n\Delta t$, show that the discrete-time Fourier transform (DTFT) of $x_{n-k}$ is simply the DTFT of $x_n$ multiplied by the phase factor $e^{-i\omega k \Delta t}$.

### 11.5

Analyse the data file provided on the book website using Fourier spectral analysis. (a) Compute the autospectrum for the time series $x_1$ (with the time series $t$ giving the time of the observation in days). (b) Compute the autospectrum for time series $x_2$, and compare with that in (a).

### 11.6

From (11.91), derive the filter response $f(\omega)$ for the three-point moving average filter (11.95) and the 1-2-1 filter (11.97). Check that your plot agrees with Fig. 11.14(b).

### 11.7

If the data are dominated by grid-scale noise, for example the time series data simply flip signs between adjacent time points as $-1, +1, -1, +1, -1, +1, \ldots,$

show that one application of the 1-2-1 filter (11.97) eliminates the grid-scale noise completely, whereas one application of the three-point moving-average filter (11.95) does not.

**11.8**

Using the same time series $x_1$ and $x_2$ from Exercise 11.5, perform singular spectrum analysis (SSA) on (a) $x_1$ and (b) $x_2$, with the time series lagged by an extra 50 days in SSA (so that there is a total of 51 time series). Briefly discuss the difference between the SSA result and the Fourier spectral result in Exercise 11.5 – in particular, which method is more efficient in capturing the variance?

**11.9**

Show that for the AR(1) model (Section 11.8.2),

$$\rho_k = \phi^k, \tag{11.137}$$

starting with the lag $k = 2$.

**11.10**

Show that the frequency response function of the difference operation in (11.132) corresponds to a high-pass filter.

# 12

# Classification

Although classification problems are less commonly encountered than regression problems in environmental sciences, they do arise in many applications. For instance, a classification model is needed to decide whether an evacuation order should be issued for an approaching hurricane, tsunami, wildfire or flood. From satellite remote sensing (Lillesand et al., 2015; Emery and Camps, 2017; Chuvieco, 2020), pixels in a satellite image need to be classified into ocean, cloud, sea ice, various types of land cover, and so on (Maxwell et al., 2018). With recent advances in ML for image recognition, classification of moving atmospheric circulation patterns such as cyclones and atmospheric rivers from numerical weather prediction model output is being pursued (Section 15.2).

The basic concepts of classification have been introduced previously in Sections 1.4.1 and 2.15. This chapter continues with classical parametric statistical models – linear discriminant analysis (LDA) (and the Fisher LDA variant) (Section 12.1), logistic regression (Section 12.2) and naive Bayes (Section 12.3) – followed by the non-parametric $K$-nearest neighbours (KNN) model (Section 12.4). The simplest neural network (NN) classifier using the extreme learning machine (ELM) approach (Section 12.5) is followed by the standard multi-layer perceptron (MLP) NN classifier (Section 12.7). Cross-entropy is often used as an alternative to the mean squared error in the objective function for classification problems (Section 12.6). The difficult problem of having imbalanced class distributions (i.e. one or more classes have relatively few data points for training) is discussed in Section 12.8.

More classification models are presented in later chapters – support vector machines (Section 13.7), classification and regression trees (CART) (Section 14.1), where these trees can be grown into random forests (Section 14.2), boosting (Section 14.3) and deep learning NN models (Chapter 15). Many performance scores are available for classification models (Chapter 16). The different economic and/or human costs associated with different class predictions can be taken into account by minimizing the expected loss (Section 16.7).

# 12.1   Linear Discriminant Analysis  $\boxed{\text{A}}$ ☺

Recall from Section 2.14, given a $D$-dimensional feature (i.e. predictor) vector $\mathbf{x}$ and classes $C_1, \ldots, C_k$, Bayes theorem gives

$$P(C_i|\mathbf{x}) = \frac{p(\mathbf{x}|C_i)P(C_i)}{\sum_i p(\mathbf{x}|C_i)P(C_i)}, \qquad i = 1, \ldots, k, \tag{12.1}$$

where $P(C_i|\mathbf{x})$ is the posterior probability given observation $\mathbf{x}$, $P(C_i)$, the prior probability and $p(\mathbf{x}|C_i)$, the likelihood of $\mathbf{x}$ given the event is of class $C_i$.

Assume $P(C_i|\mathbf{x})$ to have a multivariate Gaussian distribution (Section 3.6),

$$p(\mathbf{x}|C_i) = \frac{1}{(2\pi)^{D/2}\,|\mathbf{C}_i|^{1/2}}\,\exp\left[-\frac{1}{2}(\mathbf{x}-\boldsymbol{\mu}_i)^{\mathrm{T}}\mathbf{C}_i^{-1}(\mathbf{x}-\boldsymbol{\mu}_i)\right], \tag{12.2}$$

where the $D$-dimensional vector $\boldsymbol{\mu}_i$ is the expectation of $\mathbf{x}$ for class $C_i$, $\mathbf{C}_i = \mathrm{E}\left[(\mathbf{x}-\boldsymbol{\mu}_i)(\mathbf{x}-\boldsymbol{\mu}_i)^{\mathrm{T}}\right]$ is the $D \times D$ covariance matrix (Section 2.6) for class $i$ and $|\mathbf{C}_i|$ its determinant. *Linear discriminant analysis* (LDA) makes the simplifying assumption that all the covariance matrices $\mathbf{C}_i$ are the same, that is, $\mathbf{C}_i = \mathbf{C}$.

Given $\mathbf{x}$, we choose the class $C_j$ having the highest posterior probability, that is,

$$P(C_j|\mathbf{x}) > P(C_i|\mathbf{x}), \quad \text{for all } i \neq j. \tag{12.3}$$

When comparing $P(C_j|\mathbf{x})$ with $P(C_i|\mathbf{x})$, one can simply look at their log ratio, that is, *log odds*, where by invoking (12.1),

$$\log\left[\frac{P(C_j|\mathbf{x})}{P(C_i|\mathbf{x})}\right] = \log\left[\frac{p(\mathbf{x}|C_j)}{p(\mathbf{x}|C_i)}\right] + \log\left[\frac{P(C_j)}{P(C_i)}\right]. \tag{12.4}$$

Using (12.2) with $\mathbf{C}_i = \mathbf{C}$ and $\mathbf{x}^{\mathrm{T}}\mathbf{C}^{-1}\boldsymbol{\mu}_i = \boldsymbol{\mu}_i^{\mathrm{T}}\mathbf{C}^{-1}\mathbf{x}$ from $\mathbf{C}$ being Hermitian, one can show that

$$\log\left[\frac{P(C_j|\mathbf{x})}{P(C_i|\mathbf{x})}\right] = \log\left[\frac{P(C_j)}{P(C_i)}\right] - \frac{1}{2}\boldsymbol{\mu}_j^{\mathrm{T}}\mathbf{C}^{-1}\boldsymbol{\mu}_j + \frac{1}{2}\boldsymbol{\mu}_i^{\mathrm{T}}\mathbf{C}^{-1}\boldsymbol{\mu}_i + \mathbf{x}^{\mathrm{T}}\mathbf{C}^{-1}(\boldsymbol{\mu}_j - \boldsymbol{\mu}_i). \tag{12.5}$$

$P(C_j|\mathbf{x}) = P(C_i|\mathbf{x})$ gives a decision boundary, where one side of the boundary has $P(C_j|\mathbf{x}) > P(C_i|\mathbf{x})$ and the other side has $P(C_j|\mathbf{x}) < P(C_i|\mathbf{x})$. At the decision boundary, the left hand side of (12.5) becomes zero, and the right hand side is a linear equation of $\mathbf{x}$. Thus, the decision boundaries in LDA are defined by linear functions of $\mathbf{x}$.[1]

An alternative is to make decisions using the *discriminant functions* $y_1(\mathbf{x}), \ldots, y_k(\mathbf{x})$, where a feature vector $\mathbf{x}$ is assigned to class $C_j$ if

$$y_j(\mathbf{x}) > y_i(\mathbf{x}), \quad \text{for all } i \neq j. \tag{12.6}$$

---

[1] If one does not make the assumption $\mathbf{C}_i = \mathbf{C}$, then the log odds would contain quadratic terms, leading to *quadratic discriminant analysis* (QDA), with quadratic decision boundaries (Hastie, Tibshirani, et al., 2009, section 4.3). Friedman (1989) introduced *regularized discriminant analysis*, a compromise between LDA and QDA.

From (12.5), we can choose the linear discriminant functions to be

$$y_i(\mathbf{x}) = \log P(C_i) - \frac{1}{2}\boldsymbol{\mu}_i^{\mathrm{T}}\mathbf{C}^{-1}\boldsymbol{\mu}_i + \mathbf{x}^{\mathrm{T}}\mathbf{C}^{-1}\boldsymbol{\mu}_i, \tag{12.7}$$

for $i = 1, \ldots, k$.

From the training dataset with $N$ data points, we can estimate the prior probability and the Gaussian distribution parameters:

[1] If there are $N_i$ data points belonging to class $C_i$, the sample prior is estimated by

$$\hat{P}(C_i) = N_i/N. \tag{12.8}$$

[2] $\boldsymbol{\mu}_i$ is estimated by the sample mean computed using only data points $\mathbf{x}^{(l)}$ with $l$ belonging to class $C_i$, that is,

$$\hat{\boldsymbol{\mu}}_i = \frac{1}{N_i} \sum_{l \in C_i} \mathbf{x}^{(l)}. \tag{12.9}$$

[3] The sample covariance matrix is estimated by

$$\hat{\mathbf{C}} = \frac{1}{N-k} \sum_{i=1}^{k} \sum_{l \in C_i} \left(\mathbf{x}^{(l)} - \hat{\boldsymbol{\mu}}_i\right)\left(\mathbf{x}^{(l)} - \hat{\boldsymbol{\mu}}_i\right)^{\mathrm{T}}. \tag{12.10}$$

The linear discriminant functions in (12.7) can be rewritten in the form

$$y_i(\mathbf{x}) = \mathbf{w}_i^{\mathrm{T}}\mathbf{x} + w_{i0}, \tag{12.11}$$

where

$$\mathbf{w}_i = \mathbf{C}^{-1}\boldsymbol{\mu}_i, \tag{12.12}$$

$$w_{i0} = \log P(C_i) - \frac{1}{2}\boldsymbol{\mu}_i^{\mathrm{T}}\mathbf{C}^{-1}\boldsymbol{\mu}_i. \tag{12.13}$$

The decision boundary between class $C_j$ and $C_i$ is obtained from setting $y_j(\mathbf{x}) = y_i(\mathbf{x})$, yielding a hyperplane decision boundary described by

$$(\mathbf{w}_j - \mathbf{w}_i)^{\mathrm{T}}\mathbf{x} + (w_{j0} - w_{i0}) = 0. \tag{12.14}$$

In the environmental sciences, one of the most common applications of classifiers is in the classification of satellite data. As an example, consider the problem of using satellite data to classify land surface in a forested area in Japan (B. Johnson et al., 2012). Class $C_1$ contains Japanese cedar planted forest, $C_2$, Japanese cypress planted forest, $C_3$, mixed deciduous broadleaf natural forest, and $C_4$ contains all other land types (agriculture, roads, buildings, etc.). Satellite data were from the 15 m spatial resolution multispectral Advanced Spaceborne Thermal Emission and Reflection Radiometer (ASTER). Of the 27 predictor variables available, only the first two are used in this example for easy visualization (see Exercise 12.1). The training dataset with 198 data points are shown in a scatter plot (Fig. 12.1(a)), and the classification resulted from using LDA, in Fig. 12.1(b).[2] The four classes have, respectively, 59, 48, 54 and 37 training data points, and the percentage misclassified by the LDA is 6.8, 10.4, 7.4 and 40.5 %, respectively. The much greater misclassification for $C_4$ is not surprising since the class contains numerous very different land types and is widely scattered in Fig. 12.1(a). Of all training data, 14.1% were misclassified, while for a separate test dataset with 325 points, 19.7% were misclassified.

## 12.1.1 Fisher Linear Discriminant $\boxed{\text{B}}$ ☺

Linear discriminant analysis was first introduced by Fisher (1936), though the original approach by Fisher did not use the above posterior probability derivation assuming Gaussian distribution with all $\mathbf{C}_i = \mathbf{C}$. Instead, Fisher proposed a 'sensible' rule for separating classes: find a linear combination of the original variables, that is, $y = \mathbf{w}^{\mathrm{T}}\mathbf{x}$, which maximizes the ratio of the between-class variance to the within-class variance.

The ratio $J$ to be maximized is given by

$$J = \frac{\text{var}_{\text{between}}}{\text{var}_{\text{within}}} = \frac{\mathbf{w}^{\mathrm{T}}\mathbf{S}_{\mathrm{B}}\,\mathbf{w}}{\mathbf{w}^{\mathrm{T}}\mathbf{S}_{\mathrm{W}}\mathbf{w}}, \tag{12.15}$$

where $\mathbf{S}_{\mathrm{B}}$ is the between-class sum of squares and $\mathbf{S}_{\mathrm{W}}$, the within-class sum of squares. With two classes,

$$\mathbf{S}_{\mathrm{B}} = (\boldsymbol{\mu}_2 - \boldsymbol{\mu}_1)(\boldsymbol{\mu}_2 - \boldsymbol{\mu}_1)^{\mathrm{T}}, \tag{12.16}$$

$$\mathbf{S}_{\mathrm{W}} = \sum_{l \in C_1}\left(\mathbf{x}^{(l)} - \boldsymbol{\mu}_1\right)\left(\mathbf{x}^{(l)} - \boldsymbol{\mu}_1\right)^{\mathrm{T}} + \sum_{l \in C_2}\left(\mathbf{x}^{(l)} - \boldsymbol{\mu}_2\right)\left(\mathbf{x}^{(l)} - \boldsymbol{\mu}_2\right)^{\mathrm{T}}, \tag{12.17}$$

(Bishop, 2006, section 4.1.4).

Fisher's LDA approach has been generalized to $k$ classes, where

$$\mathbf{S}_{\mathrm{B}} = \sum_{i=1}^{k}(\boldsymbol{\mu}_i - \boldsymbol{\mu})(\boldsymbol{\mu}_i - \boldsymbol{\mu})^{\mathrm{T}}, \tag{12.18}$$

with $\boldsymbol{\mu}$ being the mean of $\mathbf{x}$ over all classes, and

$$\mathbf{S}_{\mathrm{W}} = \sum_{i=1}^{k}\sum_{l \in C_i}\left(\mathbf{x}^{(l)} - \boldsymbol{\mu}_i\right)\left(\mathbf{x}^{(l)} - \boldsymbol{\mu}_i\right)^{\mathrm{T}}. \tag{12.19}$$

Since $\mathbf{w}$ can be scaled to keep $\mathbf{w}^{\mathrm{T}}\mathbf{S}_{\mathrm{W}}\mathbf{w} = 1$, maximizing $J$ in (12.15) is equivalent to

$$\max_{\mathbf{w}}\left(\mathbf{w}^{\mathrm{T}}\mathbf{S}_{\mathrm{B}}\mathbf{w}\right) \text{ subject to } \mathbf{w}^{\mathrm{T}}\mathbf{S}_{\mathrm{W}}\mathbf{w} = 1. \tag{12.20}$$

---

[2] The decision regions in Fig. 12.1 (b), (c) and (d) were found by applying the classification models to a high resolution 2-D grid of $(x_1, x_2)$ values.

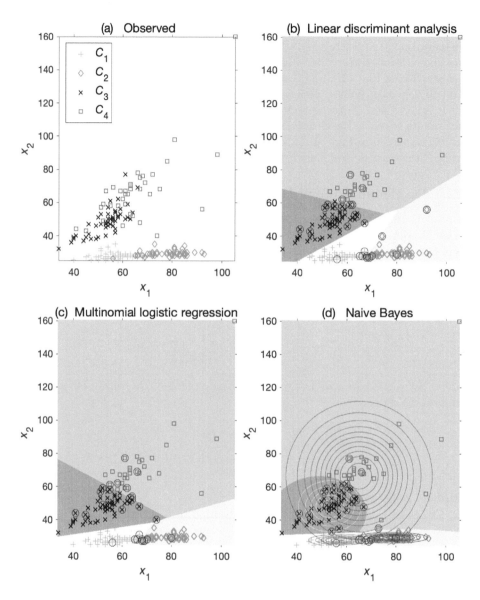

Figure 12.1 (a) Training dataset with two predictor variables $x_1$ and $x_2$ from two spectral bands and four classes $(C_1, \ldots, C_4)$ of land surface and (b) the same data classified by LDA. Decision regions for the four classes are shown by different background shading and misclassified data points are circled. (c) Data classified by multinomial logistic regression and (d) data classified by naive Bayes, where the Gaussian distributions for the four classes are shown by contours extending out to $4\,\sigma_{ij}$. [Data source: B. Johnson et al. (2012), UCI Machine Learning Repository http://archive.ics.uci.edu/ml, Irvine, CA: University of California, School of Information and Computer Science.]

The optimization in (12.20) is a generalized eigenvalue problem. If $\mathbf{S}_\mathrm{W}$ is invertible, the problem can be converted to a regular eigenvalue problem,

$$\mathbf{M}\mathbf{w} = \lambda\mathbf{w}, \tag{12.21}$$

where $\mathbf{M} = \mathbf{S}_\mathrm{W}^{-1}\mathbf{S}_\mathrm{B}$ (Mardia et al., 1979, section 11.5). The eigenvector $\mathbf{w}_1$ corresponds to the largest eigenvalue $\lambda_1$ of $\mathbf{M}$.

Once $\mathbf{w}_1$ has been solved, an observation $\mathbf{x}$ can be classified based on its discriminant score $y = \mathbf{w}_1^\mathrm{T}\mathbf{x}$: Assign $\mathbf{x}$ to class $C_j$ if

$$|\mathbf{w}_1^\mathrm{T}(\mathbf{x} - \boldsymbol{\mu}_j)| \leq |\mathbf{w}_1^\mathrm{T}(\mathbf{x} - \boldsymbol{\mu}_i)| \quad \text{for all } i \neq j, \tag{12.22}$$

that is, assign $\mathbf{x}$ to the class $C_j$ where its mean discriminant score $\mathbf{w}_1^\mathrm{T}\boldsymbol{\mu}_j$ is closest to $\mathbf{w}_1^\mathrm{T}\mathbf{x}$.

A more sophisticated criterion uses more than one eigenvector. With $x \in \mathcal{R}^D$, $\mathbf{M}$ has eigenvalues $\{\lambda_l\}$ and eigenvectors $\{\mathbf{w}_l\}$, $l = 1, \ldots, L$, where $L = \min(k - 1, D)$. One then assigns $\mathbf{x}$ to the class $C_j$ where

$$\sum_{l=1}^{L} \left[\mathbf{w}_l^\mathrm{T}(\mathbf{x} - \boldsymbol{\mu}_j)\right]^2 \leq \sum_{l=1}^{L} \left[\mathbf{w}_l^\mathrm{T}(\mathbf{x} - \boldsymbol{\mu}_i)\right]^2 \quad \text{for all } i \neq j, \tag{12.23}$$

(Wilks, 2011, section 14.3.1). Dropping all terms with $l > 1$ reduces (12.23) to (12.22). When $D > k - 1 = L$, there is a benefit of dimensional reduction when viewing the $L$-dimensional $\mathbf{w}_l^\mathrm{T}\mathbf{x}$ space instead of the $D$-dimensional $\mathbf{x}$ space.

When there are only two classes, there is a simpler solution (K. P. Murphy, 2012, section 8.6.3.1). From (12.16), we have

$$\mathbf{S}_\mathrm{B}\mathbf{w} = (\boldsymbol{\mu}_2 - \boldsymbol{\mu}_1)(\boldsymbol{\mu}_2 - \boldsymbol{\mu}_1)^\mathrm{T}\mathbf{w} = (\boldsymbol{\mu}_2 - \boldsymbol{\mu}_1)(m_2 - m_1), \tag{12.24}$$

that is, $\mathbf{S}_\mathrm{B}\mathbf{w}$ is the vector $(\boldsymbol{\mu}_2 - \boldsymbol{\mu}_1)$ multiplied by a scale factor $(m_2 - m_1)$. From (12.21), we obtain

$$\lambda\mathbf{w} = \mathbf{S}_\mathrm{W}^{-1}(\boldsymbol{\mu}_2 - \boldsymbol{\mu}_1)(m_2 - m_1), \tag{12.25}$$

$$\mathbf{w} \propto \mathbf{S}_\mathrm{W}^{-1}(\boldsymbol{\mu}_2 - \boldsymbol{\mu}_1). \tag{12.26}$$

As we are only interested in the direction of $\mathbf{w}$ and not its magnitude, we can simply take

$$\mathbf{w} = \mathbf{S}_\mathrm{W}^{-1}(\boldsymbol{\mu}_2 - \boldsymbol{\mu}_1) \tag{12.27}$$

as the solution.

(Mardia et al., 1979, section 11.5) showed that, with two classes, Fisher's approach gives the same solution as the previous approach in (12.5) using posterior probabilities (with the same prior for both classes). Since Fisher's approach did not make the Gaussian assumption, LDA may work quite well even if the data are not close to being Gaussian. For $k \geq 3$ classes, the Fisher approach in general will not be the same as the LDA based on posterior probabilities.

## 12.2 Logistic Regression [A]☺

The basic logistic regression method uses a logistic function to model the probability of a binary dependent variable, and the method has been extended to multiclass dependent variables. A history of its development is given in Cramer (2002). The *logistic function*, also known as the logistic sigmoidal function,

$$s(x) = \frac{1}{1 + e^{-x}}, \tag{12.28}$$

often used as an activation function in neural network models (6.5), was originally introduced in three papers by Pierre-François Verhulst between 1838 and 1847 to model biological populations where the unbounded growth of the exponential model was unrealistic (Cramer, 2002). It turns out that the logistic function arises naturally from Bayes decision with two classes $C_1$ and $C_2$.

With two classes, Bayes theorem in (12.1) gives

$$P(C_1|\mathbf{x}) = \frac{p(\mathbf{x}|C_1)P(C_1)}{p(\mathbf{x}|C_1)P(C_1) + p(\mathbf{x}|C_2)P(C_2)} \tag{12.29}$$

$$= \frac{1}{1 + \frac{p(\mathbf{x}|C_2)P(C_2)}{p(\mathbf{x}|C_1)P(C_1)}} \tag{12.30}$$

$$= \frac{1}{1 + e^{-u}}, \tag{12.31}$$

which has the form of a logistic sigmoidal function, with

$$u = \log\left[\frac{p(\mathbf{x}|C_1)P(C_1)}{p(\mathbf{x}|C_2)P(C_2)}\right]. \tag{12.32}$$

We can also write

$$u = \log\left[\frac{P(C_1|\mathbf{x})}{P(C_2|\mathbf{x})}\right], \tag{12.33}$$

therefore $u$ is called the *log odds*. The logistic regression model assumes a linear relation between $u$ and the features $x_j$ $(j = 1, \dots, D)$, that is,

$$u = \sum_j w_j x_j + w_0 = \mathbf{w}^{\mathrm{T}}\mathbf{x} + w_0. \tag{12.34}$$

The name logistic regression came from performing linear regression between $\mathbf{x}$ and $u$, followed by a logistic transformation of $u$. The posterior probability in (12.31) is modelled by

$$P(C_1|\mathbf{x}) = \frac{1}{1 + \exp[-(\mathbf{w}^{\mathrm{T}}\mathbf{x} + w_0)]}. \tag{12.35}$$

Since $P(C_1|\mathbf{x}) + P(C_2|\mathbf{x}) = 1$, we have

$$P(C_2|\mathbf{x}) = \frac{\exp[-(\mathbf{w}^{\mathrm{T}}\mathbf{x} + w_0)]}{1 + \exp[-(\mathbf{w}^{\mathrm{T}}\mathbf{x} + w_0)]}. \tag{12.36}$$

The logistic regression model can be used as a classifier, for example

$$\begin{cases} \text{if } P(C_1|\mathbf{x}) \geq P(C_2|\mathbf{x}), & \text{choose } C_1, \\ \text{otherwise}, & \text{choose } C_2. \end{cases} \tag{12.37}$$

The logistic regression model (12.35) can be viewed as a single-layer percep-tron neural network model, that is, NN with no hidden layers. The target data $y$ takes on the value 1 if it belongs to class $C_1$ and 0 if $C_2$. The inputs $\mathbf{x}$ are fed through a logistic sigmoidal activation function before reaching the single output node $\hat{y}$, that is,

$$\hat{y} = \frac{1}{1 + \exp[-(\mathbf{w}^T\mathbf{x} + w_0)]}, \tag{12.38}$$

which is the same as (12.35) with $\hat{y} = P(C_1|\mathbf{x})$, and $P(C_2|\mathbf{x}) = 1 - P(C_1|\mathbf{x}) = 1 - \hat{y}$. With no hidden layers, the NN model is limited to giving linear decision boundaries.

The model parameters $\mathbf{w}$ and $w_0$ are usually estimated using maximum likelihood. Unlike linear regression with Gaussian errors, there is no closed-form expression for the optimal parameter values, so one needs to use an iterative scheme, such as the iteratively reweighted least squares method (Bishop, 2006, section 4.3.3; Hastie, Tibshirani, et al., 2009, section 4.4.1), to estimate the parameters.

As an example, suppose a biological oceanographer is collecting data for the occurrence of red tide, that is, algal blooms with the water colour turning reddish, which tends to occur during days of warmer water temperatures. $C_1$ are the days when red tide occurred and $C_2$, when no red tide occurred. The target data are assigned the value 1 for $C_1$ and 0 for $C_2$. Figure 12.2 illustrates the logistic regression model applied to this problem. For a given temperature input, the model outputs a posterior probability for the occurrence of red tide.

## 12.2.1 Multiclass Logistic Regression B☺

Logistic regression has been generalized to problems with $k > 2$ classes. *Multi-nomial logistic regression* is used when the dependent variable is *nominal*, that is, the classes are not ordered. For instance, colour (red, green, blue) is nominal, and so is the type of forest cover as used in the example in Fig. 12.1.

The log odds in (12.33) and (12.34) is now generalized to a total of $k-1$ log odds, all modelled by linear relations with $\mathbf{x}$:

$$\log\left[\frac{P(C_1|\mathbf{x})}{P(C_k|\mathbf{x})}\right] = \mathbf{w}_1^T\mathbf{x} + w_{10},$$

$$\log\left[\frac{P(C_2|\mathbf{x})}{P(C_k|\mathbf{x})}\right] = \mathbf{w}_2^T\mathbf{x} + w_{20},$$

$$\vdots \tag{12.39}$$

$$\log\left[\frac{P(C_{k-1}|\mathbf{x})}{P(C_k|\mathbf{x})}\right] = \mathbf{w}_{k-1}^T\mathbf{x} + w_{(k-1)0},$$

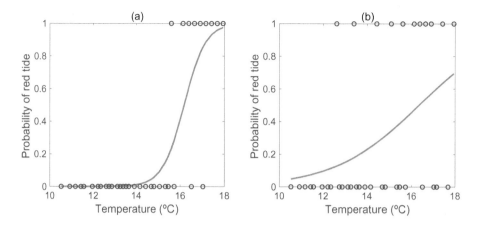

Figure 12.2 Logistic regression fit to synthetic datasets representing the occurrence of red tide for various water temperatures, with data points marked by circles. The transition between classes is relatively sharp in (a) and gradual in (b).

Exponentiating these log odds and using the fact that the probabilities sum to one,

$$P(C_1|\mathbf{x}) + \cdots + P(C_{k-1}|\mathbf{x}) + P(C_k|\mathbf{x}) = 1, \tag{12.40}$$

one can write

$$P(C_i|\mathbf{x}) = \frac{\exp(\mathbf{w}_i^{\mathrm{T}}\mathbf{x} + w_{i0})}{1 + \sum_{j=1}^{k-1} \exp(\mathbf{w}_j^{\mathrm{T}}\mathbf{x} + w_{j0})}, \quad i = 1, \ldots, k-1, \tag{12.41}$$

$$P(C_k|\mathbf{x}) = \frac{1}{1 + \sum_{j=1}^{k-1} \exp(\mathbf{w}_j^{\mathrm{T}}\mathbf{x} + w_{j0})}. \tag{12.42}$$

The model parameters $\mathbf{w}_i$ and $w_{i0}$ $(i = 1, \ldots, k-1)$ are again estimated by an iterative scheme, such as the iteratively reweighted least squares method (Bishop, 2006, section 4.3.3; Hastie, Tibshirani, et al., 2009, section 4.4.1). To use the model as a classifier, choose class $C_l$ with $P(C_l|\mathbf{x}) \geq P(C_i|\mathbf{x})$ for all $i \neq l$. The result for binary classification in (12.35) and (12.36) is of course a special case of the multiclass result in (12.41) and (12.42) (but with all the weights multiplied by a negative sign due to a different sign convention).

Figure 12.1(c) shows the classification of forest cover by the multinomial logistic regression method with linear decision boundaries. The misclassification rate was 13.6% for the training data and 20.6% for the test data, very similar to the rates attained by LDA (14.1% for the training data and 19.7% for the test data).

Since both LDA (12.5) and multinomial logistic regression (12.40) have the log odds modelled by a linear function of $\mathbf{x}$, it seems that the two models are identical. Hastie, Tibshirani, et al. (2009, section 4.4.5) provided some insight:

Although they have exactly the same form, the difference lies in the way the linear coefficients are estimated. The logistic regression model is more general, in that it makes less assumptions.

In practice these assumptions are never correct ... It is generally felt that logistic regression is a safer, more robust bet than the LDA model, relying on fewer assumptions. It is our experience that the models give very similar results, even when LDA is used inappropriately, such as with qualitative predictors.

Logistic regression has also been been extended to multiclass *ordinal* variables, where there is a natural ordering of the classes (McCullagh, 1980). For instance, temperature categories (cold, normal, warm) and weather types (sunny, cloudy, rainy) are ordinal variables.

# 12.3   Naive Bayes Classifier   $\boxed{B}$☺

Naive Bayes is a simple, inexpensive classifier also known by the even more discouraging name 'Idiot's Bayes'. However, it is a good classifier for many problems, often outperforming more sophisticated and computationally more costly methods (Domingos and Pazzani, 1997; Hand and K. M. Yu, 2001; Rish, 2001) and has been widely used in spam filtering (Brunton, 2013, chapter 3). Of course, if one has plentiful data and computing resources, there are more sophisticated models that can outperform naive Bayes (Caruana and Niculescu-Mizil, 2006).

For the posterior probability from Bayes theorem (12.1), naive Bayes makes a simplifying assumption on $p(\mathbf{x}|C_i)$, the likelihood of $\mathbf{x}$ given the event is of class $C_i$. The assumption is that the features $x_j$ $(j = 1, \ldots, D)$ are conditionally independent, that is,

$$p(\mathbf{x}|C_i) = p(x_1, \ldots, x_D|C_i) = \prod_{j=1}^{D} p(x_j|C_i). \tag{12.43}$$

While the assumption of conditionally independent features is 'naive', it makes the model simple with only $O(kD)$ parameters (for $k$ classes and $D$ features), rendering the model quite immune to overfitting, as well as computationally cheap (K. P. Murphy, 2012, section 3.5). Naive Bayes is often used as a benchmark, since sophisticated, expensive models may actually underperform naive Bayes.

For continuous feature variables, the Gaussian distribution is usually assumed, that is,

$$p(x_j|C_i) = \frac{1}{\sqrt{2\pi}\sigma_{ij}} \exp\left[-\frac{(x_j - \mu_{ij})^2}{2\sigma_{ij}^2}\right], \tag{12.44}$$

with model parameters $\mu_{ij}$ and $\sigma_{ij}$. For discrete feature variables, for example in spam filtering, the Bernoulli distribution is usually used for binary variables and the multinomial distribution for categorical features with $k$ classes (K. P. Murphy, 2012, section 3.5).

Once the model parameters have been estimated from the training data, the naive Bayes classifier assigns a feature vector $\mathbf{x}$ to class $C_l$ when the numerator of (12.1) is maximal for class $C_l$ (the denominator, being a normalization factor, is ignored), that is, the model outputs $\hat{y} = C_l$, where

$$P(C_l) \prod_{j=1}^{D} p(x_j|C_l) \geq P(C_i) \prod_{j=1}^{D} p(x_j|C_i), \quad \text{for all } i \neq l. \tag{12.45}$$

As an example, let us apply naive Bayes to the same forest classification dataset in Fig. 12.1(a). Classification of the training data by naive Bayes assuming Gaussian distribution is shown in Fig. 12.1(d). The assumption that the features are conditionally independent restricts the Gaussian ellipses to have their semi-major and semi-minor axes oriented parallel to the $x_1$ and $x_2$ axes. The misclassification rates are 13.1% for the training data and 20.9% for the test data, quite comparable to the LDA values of 14.1% (training) and 19.7% (test) from the previous section.

## 12.4 *K*-nearest Neighbours  B☺

In the previous two sections, we have presented two simple parametric classifiers, namely LDA and naive Bayes, which fit parametric models to the training data. *Parametric models* are defined by a finite number of parameters, whereas the number of parameters in *non-parametric models* grow with the amount of training data. Parametric models tend to be simpler and faster than non-parametric models, but are constrained by the assumed functional form.

In this section, we will study a simple non-parametric model for classification and regression called the *K-nearest neighbours* (KNN),[3] first proposed by Cover and Hart (1967). As a *memory-based* classifier, KNN stores all the training data in memory and does not fit any model to the training data. Being non-parametric, KNN is better than parametric models when the decision boundaries are complicated, but can be very expensive with large datasets.

The basic idea of KNN is extremely simple: given a $D$-dimensional feature vector $\mathbf{x}$, one finds $K$ feature vectors closest to $\mathbf{x}$ in the training dataset based on some distance measure (e.g. the Euclidean distance). For classification, the class $\hat{y}$ to assign to $\mathbf{x}$ is decided by simple voting from the classes in the $K$ nearest neighbours. For example, assume $\mathbf{x}$ is a feature vector containing various meteorological measurements and $y$ is the weather class of the next day, with three classes, 'dry', 'rainy' and 'snowy'. Suppose KNN with $K = 7$ is used to classify the dataset in Fig. 12.3. Among the seven nearest neighbours around the point $\mathbf{x}$, four are labelled 'dry', one 'rainy' and two 'snowy'; thus, 'dry' receives 4 votes, 'rainy' 1 and 'snowy' 2 votes, so the model output is 'dry'. If

---

[3] Be careful not to confuse the classification method $K$-nearest neighbours with $K$-means clustering, a clustering method under unsupervised learning (Section 10.2.1).

the nearest neighbours happen to be 3 'dry', 1 'rainy' and 3 'snowy', we have a tie in the voting. How do we break the tie? One option is to just choose randomly among the tied top classes. Another option is to use the closest neighbour as the tie-breaker. To avoid ties, KNN is often run with $K$ being an odd integer.

For regression, KNN outputs the mean of the $y$ values of the $K$ nearest neighbours.

More formally, for the KNN model, we can define a posterior probability

$$P(C_i|\mathbf{x}) = \frac{1}{K} \sum_{j \in N_K(\mathbf{x})} \delta(y_j, C_i), \tag{12.46}$$

where the $K$ nearest neighbours in the set $N_K(\mathbf{x})$ have class labels $\{y_j\}$ and

$$\delta(u, v) = \begin{cases} 1 & \text{if } u = v \\ 0 & \text{if } u \neq v. \end{cases} \tag{12.47}$$

In other words, the summation in (12.46) simply counts the number of data points among the $K$ nearest neighbours that belong to class $C_i$. KNN outputs $\hat{y} = C_l$ where $P(C_l|\mathbf{x}) \geq P(C_i|\mathbf{x})$ for all $i \neq l$.

Figure 12.3 Illustrating seven nearest neighbours (within the dashed circle) around a particular feature vector $\mathbf{x}$ in a 2-D feature space for the KNN classifier with $K = 7$. The three classes represent 'dry', 'rainy' and 'snowy' conditions. Since 'dry' has the most votes in this neighbourhood, the model outputs 'dry'.

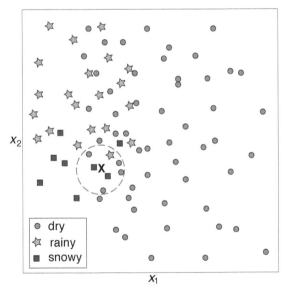

There are many choices for the distance measure. When using the Euclidean distance, it is commonly recommended that one standardizes each feature variable, especially if different features are measured in different units.

For KNN to work well, one needs to choose an optimal value for the hyperparameter $K$. A common 'rule of thumb' is to choose $K = \sqrt{N}$, with $N$ being the sample size of the training dataset. Lall and Sharma (1996) supported this choice for $1 \leq D \leq 6$ and $N \geq 100$. To avoid having to choose one value for $K$, Hassanat et al. (2014) proposed using an ensemble of KNN models, with $K = 1, 3, 5, \ldots, K_{\max}$, where $K_{\max}$ is the largest odd integer $\leq \sqrt{N}$. Each ensemble member votes for a particular class, and the ensemble model gathers all the votes from its members and chooses the class with the most votes. Cross-validation can also be used to find the optimal value for $K$.

When the classes are imbalanced, for example, there are far more 'dry' days than 'snowy' days, KNN tends to ignore the infrequent class. For instance in Fig. 12.3, the two closest neighbours to $\mathbf{x}$ are 'snowy', but the KNN outputs 'dry' since there are four 'dry' points among the seven nearest neighbours. One way to deal with imbalanced classes is to use weighted voting, that is, when summing up the votes, multiply each vote by a weight factor proportional to the inverse of the distance between $\mathbf{x}$ and the nearest neighbour.

Figure 12.4 shows the results from KNN for the forest classification dataset using three different values for $K$. The decision boundaries in KNN are complex when compared with those in LDA, logistic regression and naive Bayes (Fig. 12.1). In general, using a larger $K$ tends to produce smoother decision boundaries. The misclassification rates for the training data are 4.0% ($K = 1$), 13.6% ($K = 11$), 13.6% ($K = 31$) and, for the test data, 25.9% ($K = 1$), 18.2% ($K = 11$), 21.5% ($K = 31$). Using $K = 1$ overfitted the training data, leading to poorer performance on the test data, while $K = 31$ also underperformed $K = 11$ on the test data.

Let us summarize the pros and cons for KNN:

KNN is very easy to understand and simple to implement. As there is no model to be trained, when new data become available there is no need to retrain the model. Being non-parametric, KNN makes no assumption about the underlying distribution of the data.

On the negative side, KNN needs to store all data in memory and can be expensive when making predictions. KNN suffers from the 'curse of dimensionality' (Section 1.5) when $D$, the dimension of the feature vector, grows. One might try to reduce $D$ by some dimension reduction technique such as principal component analysis.

## 12.5   Extreme Learning Machine Classifier  $\boxed{\text{A}}$ ☺

The extreme learning machine (ELM), a non-linear neural network model with random weights in its hidden layer, which requires only the solution of a linear least squares problem, has been introduced in Section 6.4 for regression problems. Here, we will use ELM for classification. The model is run identical to the regression problem except that the target data are coded using *one-hot encoding*. For example, consider the forest cover dataset containing four classes of land cover, $C_1, \ldots, C_4$. If an input $\mathbf{x}$ belongs to class $C_1$, the target is coded by the binary vector [1 0 0 0], whereas [0 1 0 0] is used for $C_2$, [0 0 1 0] for $C_3$ and [0 0 0 1] for $C_4$. In general, with $k$ classes, instead of the target being an

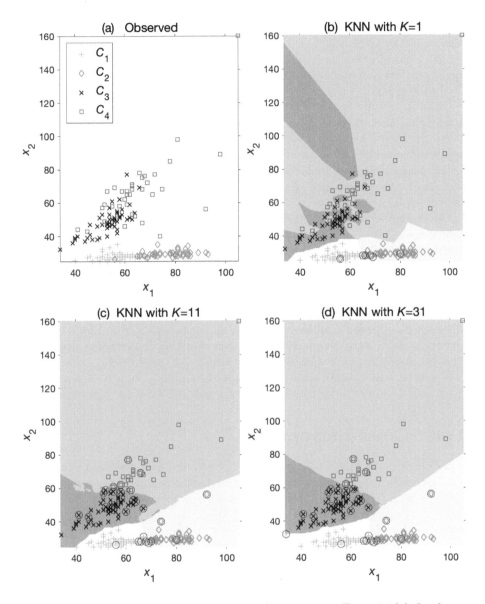

Figure 12.4 (a) Given the same training dataset as in Fig. 12.1(a) for forest classification, the data are classified by the method of $K$-nearest neighbours, with (b) $K = 1$, (c) $K = 11$ and (d) $K = 31$. Decision regions for the four classes are shown by different background shading, and misclassified data points are circled.

integer with $k$ possible values $(1, \ldots, k)$, under one-hot encoding, the target is represented by a vector with $k$ binary elements (with values 0 or 1), where only the $j$th element is non-zero, indicating class $C_j$. For a given input $\mathbf{x}$, the ELM model outputs real numbers $y_i$ $(i = 1, \ldots, k)$. The output class is chosen to be $C_j$ if $y_j \geq y_i$ for all $i \neq j$.

ELM is usually run as an ensemble of $M$ models with different random weight and offset parameters in the hidden layer. Unlike ELM for regression, where the ensemble output is computed as the average of the outputs from the individual members, ELM for classification uses *voting* to determine the output for the ensemble. The predicted class $C_j$ by an ensemble members counts as one vote for that class, and the class $C_i$ receiving the most votes is chosen as the output for the ensemble.

An ensemble of 99 ELM models (with the logistic activation function in the hidden layer) was applied to the forest cover dataset (Fig. 12.5). The hyperparameter $L$ (number of hidden nodes) was estimated by a 6-fold cross-validation (Section 8.5) of the 99-member ensemble, where the misclassification rate over the validation data was smallest when $L = 13$ (Fig. 12.5(b)). Using $L = 13$, the ELM ensemble was again trained using all the data from the training set, with the classification results shown in Fig. 12.5(a). The decision boundaries are more complicated than those found by earlier methods (Fig. 12.1). In areas without training data, the model can be seen assigning classes quite arbitrarily. This is a disadvantage of using non-linear functions to model data, as non-linear functions can extrapolate strangely outside the training domain. The misclassification rate is 15.2% for the training data and 18.2% for the test data.

## 12.6  Cross-Entropy  $\boxed{\text{A}}$ ☺

In Section 8.1, the MSE objective function was found from the maximum likelihood principle assuming the target (i.e. the observed response) to be continuous variables with Gaussian noise. For classification problems with two classes ($C_1$ and $C_2$), the targets are binary variables. While the MSE objective function is widely used in classification problems, there is an alternative objective function, based on cross-entropy (Section 2.17.3), which is designed specifically for classification problems. Its usage in multi-layer perceptron (MLP) NN models has increased, especially since the advent of deep NN models (Goodfellow, Bengio, et al., 2016, section 6.2.1.2).

For a two-class problem, we consider a model with a single output $\hat{y}$ modelling the posterior probability $P(C_1|\mathbf{x})$. The target data $y$ equals 1 if the input $\mathbf{x}$ belongs to class $C_1$, and equals 0 if $C_2$. We can combine the two expressions

$$P(C_1|\mathbf{x}) = \hat{y}, \qquad\qquad (12.48)$$
$$P(C_2|\mathbf{x}) = 1 - \hat{y} \qquad\qquad (12.49)$$

into a single convenient expression

$$P(y|\mathbf{x}) = \hat{y}^y \, (1 - \hat{y})^{1-y}, \qquad\qquad (12.50)$$

Figure 12.5 Given the training dataset in Fig. 12.1(a) for forest classification, the data are classified by an ensemble of 99 ELM models with (a) $L = 13$ hidden nodes. Class $C_1$ is indicated by $+$, $C_2$ by $\Diamond$, $C_3$ by $\times$ and $C_4$ by $\Box$. Decision regions for the four classes are shown by different background shading, and misclassified data points are circled. (b) Misclassification rate for the training and validation data under a 6-fold cross-validation scheme, with the number of hidden nodes ranging from 3 to 50. The minimum validation error occurred when $L = 13$. This estimated value for $L$ can vary considerably if different random weights are used, since the validation error is quite noisy due to the small sample size of 198 data points in the training dataset.

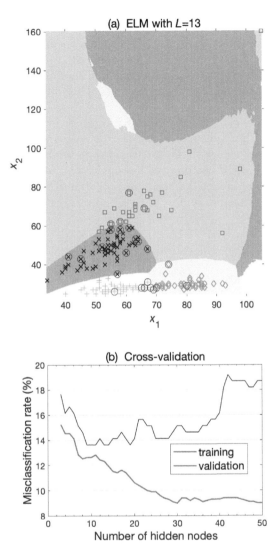

since changing from $y = 1$ to $y = 0$ in (12.50) switches from (12.48) to (12.49).

For a dataset with $n = 1, \dots, N$ independent observations, the likelihood function is then

$$L = \prod_{n=1}^{N} \hat{y}_n^{y_n} \left(1 - \hat{y}_n\right)^{1 - y_n}. \tag{12.51}$$

As in Section 8.1, the objective function $J$ is taken to be the negative natural logarithm of the likelihood function, that is,

$$J = -\ln L = -\sum_{n=1}^{N} \left[ y_n \ln \hat{y}_n + (1 - y_n) \ln(1 - \hat{y}_n) \right], \tag{12.52}$$

which is called the *cross-entropy* error function or objective function. An optimization algorithm is used to find the model weights/parameters that minimize $J$. This error function also applies to problems where the target $y$ is not a binary variable but is a real variable in the interval $[0, 1]$, representing the probability of the input $\mathbf{x}$ belonging to $C_1$ (Bishop, 1995, section 6.8).

Next, we derive the cross-entropy error function for *multiclass classification*, with classes $C_1, \ldots, C_k$, $k > 2$. Assume that the model has one output $\hat{y}_i$ for each class $C_i$, and the target data $y_i$ are coded using one-hot encoding, that is, $y_i = 1$ if the input $\mathbf{x}$ belongs to $C_i$, and 0 otherwise. The model output $\hat{y}_i$ is to represent the posterior probability $P(C_i|\mathbf{x})$. We can combine all $i = 1, \ldots, k$ expressions of $\hat{y}_i = P(C_i|\mathbf{x})$ into a single expression

$$P(\mathbf{y}|\mathbf{x}) = \prod_{i=1}^{k} \hat{y}_i^{y_i}, \tag{12.53}$$

where $\mathbf{y} = [y_1, \ldots, y_k]^{\mathrm{T}}$. With $n = 1, \ldots, N$ independent observations, the likelihood function is then

$$L = \prod_{n=1}^{N} \prod_{i=1}^{k} \hat{y}_{ni}^{y_{ni}}. \tag{12.54}$$

Again, the objective function $J$ is taken to be the negative logarithm of the likelihood function, that is,

$$J = -\ln L = -\sum_{n=1}^{N} \sum_{i=1}^{k} y_{ni} \ln \hat{y}_{ni}, \tag{12.55}$$

giving the cross-entropy error function. This error function also applies to the case where the target $y_i$ is not a binary variable but is a real variable in the interval $[0, 1]$, representing the probability of the input $\mathbf{x}$ belonging to $C_i$ (Bishop, 1995, section 6.10).

## 12.7   Multi-layer Perceptron Classifier  [A] ☺

The multi-layer perceptron (MLP) NN model for non-linear regression, presented in Section 6.3, can easily be modified for classification problems. If there are only two classes $C_1$ and $C_2$, we can take the target data $y$ to be a binary variable, with $y = 1$ denoting $C_1$ and $y = 0$ denoting $C_2$. Since the output is bounded in classification problems, instead of using a linear activation function in the output layer as in the case of MLP regression, we now use the logistic sigmoidal function $s$ in (12.28). The MLP network, with one layer of hidden neurons $h_j$, has the output $\hat{y}$ given by

$$\hat{y} = s\left( \sum_j \tilde{w}_j h_j + \tilde{b} \right), \quad \text{with } h_j = s(\mathbf{w}_j \cdot \mathbf{x} + b_j), \tag{12.56}$$

where $\tilde{w}_j$ and $\mathbf{w}_j$ are weights and $\tilde{b}$ and $b_j$ are offset or bias parameters.

In MLP regression problems, the objective function $J$ minimizes the mean squared error (MSE). The MSE objective function is also used for classification problems, though the alternative, the cross-entropy function (Section 12.6), designed specifically for classification problems, is becoming more widely used. Regularization (i.e. weight penalty or decay terms) (Section 8.4) can also be added to $J$ to control overfitting.

To classify the MLP output $\hat{y}$ as either 0 or 1, we invoke the *indicator function $I$*, where

$$I(x) = \begin{cases} 1 & \text{if} \quad x > 0, \\ 0 & \text{if} \quad x \leq 0. \end{cases} \tag{12.57}$$

The classification is then given by

$$f(\mathbf{x}) = I\big(\hat{y}(\mathbf{x}) - 0.5\big), \tag{12.58}$$

that is, assign class $C_1$ if $\hat{y}(\mathbf{x}) > 0.5$, otherwise assign $C_2$. While posterior probabilities have not been invoked in this MLP classifier, $\hat{y}$, the output from a logistic sigmoidal function, can be interpreted as a posterior probability, as was shown in Section 12.2 for the logistic regression model, which can be regarded as a single-layer perceptron model. With the extra hidden layers in MLP, complicated, non-linear decision boundaries can be modelled, in contrast to the linear decision boundaries in the logistic regression model.

We next turn to classification problems where there are $k$ classes $(k > 2)$. The target data typically use a one-hot encoding scheme, as in Section 12.5. If we are interested in the outputs giving the posterior probability of each class, we will need a generalization of the logistic sigmoidal function. Since posterior probabilities are non-negative, one can model non-negative outputs by exponential functions such as $\exp(a_i)$ for the $i$th model output. Next, we require the posterior probabilities to sum to 1, so the $i$th model output $\hat{y}_i$ is

$$\hat{y}_i = \frac{\exp(a_i)}{\sum_{i'} \exp(a_{i'})}, \tag{12.59}$$

which satisfies $\sum_i \hat{y}_i = 1$. This normalized exponential activation function is commonly referred to as the *softmax* activation function.[4] The name softmax comes about because the function is a smooth form of a 'maximum' function – for example, if $a_j \gg a_i$, for all $i \neq j$, then $\hat{y}_j \approx 1$ and all other $\hat{y}_i \approx 0$.

---

[4] There is a risk of numerical overflow/underflow from the exponentials $\exp(a_i)$ (Goodfellow, Bengio, et al., 2016, p. 78). A numerically stable version of the softmax function is defined by

$$\hat{y}_i = \frac{\exp(a_i - \max_i(a_i))}{\sum_{i'} \exp(a_{i'} - \max_i(a_i))}. \tag{12.60}$$

Since $\exp(-\max_i(a_i))$ in the numerator cancels that in the denominator, there is no change in the analytical value of the softmax function. The possibility of overflow is eliminated as the numerator is now bounded above by 1 and the denominator has at least one term with value 1.

That the softmax activation function is a generalization of the logistic function can be shown readily. Let us try to express $\hat{y}_i$ in the logistic sigmoidal form

$$\hat{y}_i = \frac{1}{1 + \exp(-\alpha_i)}.$$ (12.61)

Equating the right hand side of (12.59) and (12.61), we then cross-multiply to get

$$\sum_{i'} \exp(a_{i'}) = [1 + \exp(-\alpha_i)] \exp(a_i).$$ (12.62)

Solving for $\alpha_i$ yields

$$\alpha_i = a_i - \ln \left[ \sum_{i' \neq i} \exp(a_{i'}) \right].$$ (12.63)

The MLP classifier was applied to the same forest cover dataset in Fig. 12.1(a). An ensemble of 25 MLP models with $L$ hidden neurons was run using tanh as the activation function for the hidden layer, softmax for the output layer, cross-entropy for the objective function and 15% of the training dataset selected randomly as validation data for early stopping (i.e. stopping the optimization early to avoid overfitting). Voting was used to select the output class from the 25 ensemble members. $L$ was varied between 3 and 50, and the lowest misclassification rate over validation[5] data in a 6-fold cross-validation process was used to find the best estimate for $L$ (Fig. 12.6(b)). With the best estimate for $L$, the ensemble was rerun without cross-validation (Fig. 12.6(a)), and the misclassification rate was also computed for the independent test dataset. The misclassification rate was 10.1% (training data) and 21.2% (test data). The use of early stopping alleviates overfitting in Fig. 12.6(b) when the number of hidden nodes exceeds the optimal number, as manifested by the slower divergence between the misclassification rate of the validation data and that of the training data, as compared to that found in Fig. 12.5(b) for ELM, which did not use early stopping nor other regularization techniques.

## 12.8   Class Imbalance   [A] ☺

For the classifiers in this chapter to work well, the classes need to be balanced, that is, the amount of data in each class $C_i$ is similar in magnitude. *Class imbalance* occurs when the number of data points in one or more minority classes are far fewer than the number in the majority classes.

With class imbalance, a performance score such as the misclassification rate can become almost meaningless. For example, consider a dataset contains 99% non-tornado days and 1% tornado days. A trivial model that simply predicts

---

[5] The validation data in the six-fold cross-validation are not the same as the randomly selected validation data from the training portion to stop the optimization algorithm early to avoid overfitting.

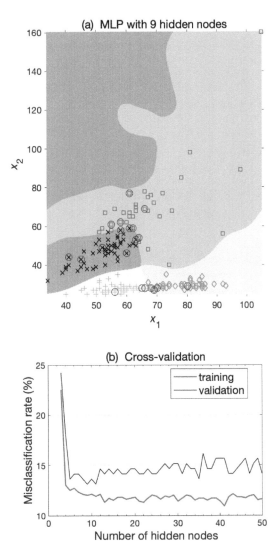

Figure 12.6 Given the training dataset in Fig. 12.1(a) for forest classification, the data are classified by an ensemble of 25 MLP models with (a) $L = 9$ hidden nodes. Class $C_1$ is indicated by $+$, $C_2$ by $\Diamond$, $C_3$ by $\times$ and $C_4$ by $\square$. Decision regions for the four classes are shown by different background shading, and misclassified data points are circled. (b) Misclassification rate for the training and validation data under a 6-fold cross-validation scheme, with the number of hidden nodes ranging from 3 to 50. The minimum validation error occurred when $L = 9$. This estimated value for $L$ can vary considerably if different random weights are used, since the validation error is quite noisy due to the small sample size of 198 data points in the training dataset.

'non-tornado' every day will have an impressive misclassification rate of only 1% while failing to predict any tornado! More meaningful performance scores are provided in Chapter 16.

With class imbalance, the learning algorithm may tend to ignore the minority classes, treating them as noise, hence the model will be unable to predict the rare classes accurately. Minority classes are often the main interest in a classification problem – for example fraud, spam, anomaly, rare disease, extreme weather, natural disaster, and so on.

There are some ways to improve the model's ability to learn the minority classes. One approach is to add extra copies of the data for the minority classes to the training dataset. The disadvantage is that this could increase the likelihood of overfitting, that is, overfitting to the noise in the minority class data. More advanced options involve ensemble approaches such as bagging and boosting (Chapter 14) (Galar et al., 2012).

# Exercises

## 12.1

Use linear discriminant analysis to classify the forest types in the given dataset. Train using (a) the first 2 predictors, (b) the first 4 predictors, (c) the first 6 predictors and (d) all 27 predictors. Examine the misclassification rates for the training data and for the test data.

## 12.2

Use naive Bayes (with Gaussian distribution) to classify the forest types in the given dataset. Train using (a) the first 2 predictors, (b) the first 4 predictors, (c) the first 6 predictors and (d) all 27 predictors. Examine the misclassification rates for the training data and for the test data and compare them with the rates obtained from using linear discriminant analysis in Exercise 12.1.

## 12.3

Evaluate the resistance to outliers for the three classification methods, linear discriminant analysis, multinomial logistic regression and naive Bayes, using the given forest cover dataset and only the first two predictors. Add an outlier data point to the training dataset, where $(x_1, x_2) = (40, 400)$ and $y$ being class (a) $C_1$, (b) $C_2$, (c) $C_3$ and (d) $C_4$. Examine the misclassification rates for the training data and for the test data and compare them with the rates obtained without adding the outlier.

## 12.4

Use a non-linear NN classifier, such as ELM or MLP, to classify the forest types in the given dataset. Train using (a) the first 2 predictors and (b) all 27 predictors. Examine the misclassification rates for the training data and for the test data and compare them with the rates obtained from using linear discriminant analysis in Exercise 12.1. Does increasing the complexity of the problem by using 27 predictors instead of 2 predictors improve the performance of the non-linear classifier relative to the linear classifier?

## 12.5

Develop a prediction model for 'Rain Tomorrow' for Sydney Airport in Australia using the available predictors in the given dataset. Use data from 2009–2015 for training and 2016–2017 for testing. How are the wind direction variables input into the model? One can ignore the variables with many missing data values marked by 'NA', and for the remaining variables, one can omit days with 'NA'. Since the number of rainy days are far fewer than non-rainy days, there is class imbalance, and some of the performance scores in Chapter 16 are more meaningful.

# 13

# Kernel Methods

Neural network methods became popular in the mid to late 1980s, but by the mid to late 1990s, kernel methods also became popular in the field of machine learning. At the beginning of the twenty-first century, in ML conferences, kernel methods were the 'stars', while neural networks appeared relatively old and stagnant. Kernel methods were winning ML problem-solving contests and their cleaner mathematical structures, with more rigorous proofs available than NN methods (Shawe-Taylor and Cristianini, 2004), were also winning the hearts of theoreticians. It seemed the old lion could no longer fend off the younger challenger. But amazingly neural networks bounced back with new breakthroughs, leading to the advent of deep learning (i.e. neural networks with many hidden layers) around 2006. The amazing comeback story of NN continues in Chapter 15.

The first kernel methods were non-linear classifiers called *support vector machines* (SVM). Soon, almost all linear statistical methods had been non-linearly generalized by the kernel approach, examples being ridge regression, linear discriminant analysis (LDA), principal component analysis (PCA), canonical correlation analysis (CCA), and so on. The kernel method has also been extended to probabilisitic models, for example, *Gaussian processes* (GP).

Section 13.1 tries to conceptually bridge from NN to kernel methods. Sections 13.2–13.4 present the mathematical foundation of the kernel method. Since the mathematics behind kernel methods is more sophisticated than that for NN methods, Section 13.5 tries to summarize the main ideas behind kernel methods, as well as their advantages and disadvantages. The pre-image problem, a disadvantage of kernel methods, is discussed in Section 13.6. The most common kernel method, the support vector machine (SVM), available for classification and regression problems, is presented in Section 13.7. Section 13.8 covers Gaussian process regression, while Section 13.9 covers kernel principal component analysis.

# 13.1 From Neural Networks to Kernel Methods $\boxed{\text{B}}$ ☺

Recall multiple linear regression (Section 5.2), where

$$y = \sum_{l=1}^{m} a_l x_l + a_0 + \epsilon, \qquad (13.1)$$

with $x_l$ ($l = 1, \ldots, m$) the predictors or inputs, $y$ the response variable, $\epsilon$ the noise or residual and $a_l$ and $a_0$, the regression coefficients. When a multi-layer perceptron (MLP) NN with a single hidden layer is used for non-linear regression, the mapping does not proceed directly from the inputs $\mathbf{x}$ to the output $\hat{y}$, but passes through an intermediate layer of variables $\mathbf{h}$, that is, the hidden neurons,

$$h_j = \tanh\left(\sum_l w_{jl} x_l + b_j\right), \qquad (13.2)$$

where the mapping from $\mathbf{x}$ to $\mathbf{h}$ is through a non-linear activation function such as the hyperbolic tangent (tanh). The next stage is to map from $\mathbf{h}$ to $\hat{y}$, and is most commonly done via a linear mapping:

$$\hat{y} = \sum_j \tilde{w}_j h_j + \tilde{b}, \qquad y = \hat{y} + \epsilon. \qquad (13.3)$$

Since (13.3) is formally identical to (13.1), one can think of the NN model as first mapping non-linearly from the input space to a 'hidden' space (spanned by the hidden neurons), that is, $\phi : \mathbf{x} \to \mathbf{h}$, then finding a linear regression between the hidden space variables $\mathbf{h}$ and the response variable $y$ (Fig. 13.1).

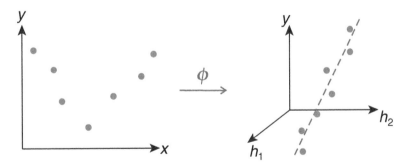

Figure 13.1 Illustrating the effect of the non-linear mapping $\phi$ from the input space to the hidden space, where a non-linear relation between the input $x$ and the response $y$ becomes a linear relation (dashed line) between the hidden variables $\mathbf{h}$ and $y$.

If the relation between $\mathbf{x}$ and $\mathbf{y}$ is highly non-linear, then one must use more hidden neurons, that is, increase the dimension of the hidden space, before one can find a good linear regression fit between the hidden space variables and the response variable. Since the parameters $w_{jl}$, $\tilde{w}_j$, $b_j$ and $\tilde{b}$ are solved together in the MLP model, non-linear optimization is involved, leading to local minima.

The *kernel methods* follow a somewhat similar procedure: in the first stage, a non-linear function $\phi$ maps from the input space to a hidden space, called the *'feature'* space. In the second stage, one performs linear regression from the feature space to the output space. Instead of linear regression, one can also perform linear classification, PCA, CCA, and so on during the second stage. Since only linear methods are used in the feature space, this means rigorous mathematical proofs are generally available for kernel methods (Shawe-Taylor and Cristianini, 2004), in contrast to NN methods, which are often heuristic. Hence, when kernel methods came out, the more mathematically inclined researchers quickly switched their interest from NN to kernel methods.

Like the extreme learning machine (ELM) (Section 6.4) and the radial basis function NN with non-adaptive basis functions (Section 6.5), the optimization in stage two of a kernel method is independent of stage 1, so only linear optimization is involved, and there are no local minima – a main advantage over the MLP NN model (some kernel methods, e.g. Gaussian processes, use non-linear optimization to find the hyperparameters, and thus have the local minima problem). However, the feature space can be of very high (or even infinite) dimension. This disadvantage is eliminated by the use of a *kernel trick*, which manages to avoid direct evaluation of the high-dimensional function $\phi$ altogether. Hence, many methods that were previously buried due to the 'curse of dimensionality' (Section 1.5) have been revived during the kernel renaissance.

## 13.2 Primal and Dual Solutions for Linear Regression Ⓑ ☺

To illustrate the kernel method, we will start with linear regression. We need to find a way to solve the regression problem that will compute efficiently under the kernel framework. From Section 5.2, the multiple linear regression problem is

$$y_i = w_0 + \sum_{l=1}^{m} x_{il} w_l + \epsilon_i, \qquad i = 1, \ldots, n, \tag{13.4}$$

where there are $n$ observations and $m$ predictors $x_l$ for the response $y$, with $\epsilon_i$ being the errors or residuals, and $w_0$ and $w_l$ the regression coefficients. By introducing $x_{i0} \equiv 1$, we can place $w_0$ as the first element of the vector $\mathbf{w}$, which also contains $w_l$, $(l = 1, \ldots, m)$, so the regression problem can be expressed in vector notation as

$$\mathbf{y} = \mathbf{X}\mathbf{w} + \boldsymbol{\epsilon}, \tag{13.5}$$

where $\mathbf{X}$ is an $n \times m$ design matrix (Section 5.2). The solution is

$$\mathbf{w} = (\mathbf{X}^{\mathrm{T}}\mathbf{X})^{-1}\mathbf{X}^{\mathrm{T}}\mathbf{y}, \tag{13.6}$$

where $(\mathbf{X}^{\mathrm{T}}\mathbf{X})$ is an $m \times m$ matrix, with $(\mathbf{X}^{\mathrm{T}}\mathbf{X})^{-1}$ taking $\mathrm{O}(m^3)$ operations to compute. Let us rewrite the equation as

$$\mathbf{w} = (\mathbf{X}^{\mathrm{T}}\mathbf{X})(\mathbf{X}^{\mathrm{T}}\mathbf{X})^{-1}(\mathbf{X}^{\mathrm{T}}\mathbf{X})^{-1}\mathbf{X}^{\mathrm{T}}\mathbf{y}. \tag{13.7}$$

If we introduce *dual variables* $\boldsymbol{\alpha}$, with

$$\boldsymbol{\alpha} = \mathbf{X}(\mathbf{X}^{\mathrm{T}}\mathbf{X})^{-1}(\mathbf{X}^{\mathrm{T}}\mathbf{X})^{-1}\mathbf{X}^{\mathrm{T}}\mathbf{y}, \tag{13.8}$$

then

$$\mathbf{w} = \mathbf{X}^{\mathrm{T}}\boldsymbol{\alpha}. \tag{13.9}$$

Linear regression when performed with a weight penalty term is called *ridge regression* (Section 5.7), where the objective function to be minimized becomes

$$J = \sum_{i=1}^{n} \left( y_i - \sum_{l=0}^{m} x_{il} w_l \right)^2 + \lambda \|\mathbf{w}\|^2, \tag{13.10}$$

with $\lambda$ being the weight penalty parameter. This can be rewritten as

$$J = (\mathbf{y} - \mathbf{X}\mathbf{w})^{\mathrm{T}}(\mathbf{y} - \mathbf{X}\mathbf{w}) + \lambda \mathbf{w}^{\mathrm{T}}\mathbf{w}. \tag{13.11}$$

Setting the gradient of $J$ with respect to $\mathbf{w}$ to zero, we have

$$- \mathbf{X}^{\mathrm{T}}(\mathbf{y} - \mathbf{X}\mathbf{w}) + \lambda \mathbf{w} = 0, \tag{13.12}$$

that is,

$$(\mathbf{X}^{\mathrm{T}}\mathbf{X} + \lambda \mathbf{I})\,\mathbf{w} = \mathbf{X}^{\mathrm{T}}\mathbf{y}, \tag{13.13}$$

where $\mathbf{I}$ is the identity matrix. This yields

$$\mathbf{w} = (\mathbf{X}^{\mathrm{T}}\mathbf{X} + \lambda \mathbf{I})^{-1}\,\mathbf{X}^{\mathrm{T}}\mathbf{y}, \tag{13.14}$$

which requires $\mathrm{O}(m^3)$ operations to compute.

From (13.12),

$$\mathbf{w} = \lambda^{-1}\mathbf{X}^{\mathrm{T}}(\mathbf{y} - \mathbf{X}\mathbf{w}). \tag{13.15}$$

Invoking (13.9), we have

$$\boldsymbol{\alpha} = \lambda^{-1}(\mathbf{y} - \mathbf{X}\mathbf{w}) = \lambda^{-1}(\mathbf{y} - \mathbf{X}\mathbf{X}^{\mathrm{T}}\boldsymbol{\alpha}). \tag{13.16}$$

Multiplying by $\lambda$ and rewriting as

$$(\mathbf{X}\mathbf{X}^{\mathrm{T}} + \lambda \mathbf{I})\,\boldsymbol{\alpha} = \mathbf{y}, \tag{13.17}$$

we obtain

$$\boldsymbol{\alpha} = (\mathbf{G} + \lambda \mathbf{I})^{-1}\mathbf{y}, \tag{13.18}$$

with the $n \times n$ matrix $G \equiv \mathbf{X}\mathbf{X}^{\mathrm{T}}$ called the *Gram matrix*. Using this equation to solve for $\boldsymbol{\alpha}$ takes $\mathrm{O}(n^3)$ operations. If $n \ll m$, solving $\mathbf{w}$ from (13.9) via

(13.18), called the *dual solution*, will be much faster than solving from (13.14), the *primal solution*.

If a new datum $\tilde{\mathbf{x}}$ becomes available, and if one wants to map from $\tilde{\mathbf{x}}$ to the output $\tilde{y}$, then using

$$\tilde{y} = \tilde{\mathbf{x}}^{\mathrm{T}}\mathbf{w} \tag{13.19}$$

requires $O(m)$ operations if $\mathbf{w}$ from the primal solution (13.14) is used.

If the dual solution is used, we can obtain $\tilde{y}$ from

$$\tilde{y} = \mathbf{w}^{\mathrm{T}}\tilde{\mathbf{x}} = \boldsymbol{\alpha}^{\mathrm{T}}\mathbf{X}\tilde{\mathbf{x}} = \sum_{i=1}^{n} \alpha_i\, \mathbf{x}_i^{\mathrm{T}}\tilde{\mathbf{x}}, \tag{13.20}$$

where $\mathbf{x}_i^{\mathrm{T}}$ is the $i$th row of $\mathbf{X}$. With the dimensions of $\boldsymbol{\alpha}^{\mathrm{T}}$, $\mathbf{X}$ and $\tilde{\mathbf{x}}$ being $1 \times n$, $n \times m$ and $m \times 1$, respectively, computing $\tilde{y}$ via the dual solution requires $O(nm)$ operations, more expensive than via the primal solution.

Letting $\mathbf{k} \equiv \mathbf{X}\tilde{\mathbf{x}}$ (with $k_i = \mathbf{x}_i^{\mathrm{T}}\tilde{\mathbf{x}}$), we have

$$\tilde{y} = \boldsymbol{\alpha}^{\mathrm{T}}\mathbf{k}. \tag{13.21}$$

Substituting in (13.18), we have

$$\tilde{y} = \left((\mathbf{G} + \lambda\mathbf{I})^{-1}\mathbf{y}\right)^{\mathrm{T}}\mathbf{k} = \mathbf{y}^{\mathrm{T}}(\mathbf{G} + \lambda\mathbf{I})^{-1}\mathbf{k}, \tag{13.22}$$

where we have used the fact that $\mathbf{G} + \lambda\mathbf{I}$ is symmetric.

## 13.3   Kernels  B☺

In the kernel method, a feature map $\boldsymbol{\phi}(\mathbf{x})$ maps from the input space $X$ to the feature space $F$. For instance, if $\mathbf{x} \in X = \mathbb{R}^m$, then $\boldsymbol{\phi}(\mathbf{x}) \in F \subseteq \mathbb{R}^M$, where $F$ is the feature space, and

$$\boldsymbol{\phi}^{\mathrm{T}}(\mathbf{x}) = [\phi_1(\mathbf{x}), \ldots, \phi_M(\mathbf{x})]. \tag{13.23}$$

In problems where it is convenient to incorporate a constant element $x_{i0} \equiv 1$ into the vector $\mathbf{x}_i$ (e.g. to deal with the constant weight $w_0$ in (13.4)), one also tends to add a constant element $\phi_0(\mathbf{x})$ ($\equiv 1$) to $\boldsymbol{\phi}(\mathbf{x})$. With appropriate choice of the non-linear mapping functions $\boldsymbol{\phi}$, the relation in the high-dimensional space $F$ becomes linear.

In Section 13.2, we had $\mathbf{G} = \mathbf{X}\mathbf{X}^{\mathrm{T}}$ and $k_i = \mathbf{x}_i^{\mathrm{T}}\tilde{\mathbf{x}}$. Now, as we will be performing regression in the feature space, we need to work with

$$G_{ij} = \boldsymbol{\phi}^{\mathrm{T}}(\mathbf{x}_i)\boldsymbol{\phi}(\mathbf{x}_j), \tag{13.24}$$

$$k_i = \boldsymbol{\phi}^{\mathrm{T}}(\mathbf{x}_i)\boldsymbol{\phi}(\tilde{\mathbf{x}}). \tag{13.25}$$

If we assume $\boldsymbol{\phi}(\mathbf{x})$ requires $O(M)$ operations, then $\boldsymbol{\phi}^{\mathrm{T}}(\mathbf{x}_i)\boldsymbol{\phi}(\mathbf{x}_j)$ is still of $O(M)$, but (13.24) has to be done $n^2$ times as $\mathbf{G}$ is an $n \times n$ matrix, thus requiring a total of $O(n^2 M)$ operations. To get $\boldsymbol{\alpha}$ from (13.18), computing the inverse of the $n \times n$ matrix $(\mathbf{G} + \lambda \mathbf{I})$ takes $O(n^3)$ operations, thus a total of $O(n^2 M + n^3)$ operations is needed.

With new datum $\tilde{\mathbf{x}}$, (13.20) now becomes

$$\tilde{y} = \sum_{i=1}^{n} \alpha_i \, \phi^{\mathrm{T}}(\mathbf{x}_i)\phi(\tilde{\mathbf{x}}), \qquad (13.26)$$

which requires $O(nM)$ operations.

To save computation costs, instead of evaluating $\phi$ explicitly, we introduce a *kernel function $K$* to evaluate the inner product (i.e. dot product) in the feature space,

$$K(\mathbf{x}, \mathbf{z}) \equiv \phi^{\mathrm{T}}(\mathbf{x})\phi(\mathbf{z}) = \sum_{l} \phi_l(\mathbf{x})\phi_l(\mathbf{z}), \qquad (13.27)$$

for all $\mathbf{x}, \mathbf{z}$ in the input space. Since $K(\mathbf{x}, \mathbf{z}) = K(\mathbf{z}, \mathbf{x})$, $K$ is a *symmetric function*. The key to the *kernel trick* is that if an algorithm in the input space can be formulated involving only inner products, then the algorithm can be solved in the feature space with the kernel function evaluating the inner products. Although the algorithm may only be solving for a linear problem in the feature space, it is equivalent to solving a non-linear problem in the input space.

As an example, for $\mathbf{x} = (x_1, x_2) \in \mathbb{R}^2$, consider the feature map

$$\phi(\mathbf{x}) = (x_1^2, \, x_2^2, \, \sqrt{2}\,x_1 x_2) \in \mathbb{R}^3. \qquad (13.28)$$

Linear regression performed in $F$ is then of the form

$$y = a_0 + a_1 x_1^2 + a_2 x_2^2 + a_3 \sqrt{2}\,x_1 x_2, \qquad (13.29)$$

so quadratic relations with the inputs become linear relations in the feature space. For this $\phi$,

$$\phi^{\mathrm{T}}(\mathbf{x})\phi(\mathbf{z}) = x_1^2 z_1^2 + x_2^2 z_2^2 + 2x_1 x_2 z_1 z_2 = (x_1 z_1 + x_2 z_2)^2 = (\mathbf{x}^{\mathrm{T}}\mathbf{z})^2. \qquad (13.30)$$

Thus,

$$K(\mathbf{x}, \mathbf{z}) = (\mathbf{x}^{\mathrm{T}}\mathbf{z})^2. \qquad (13.31)$$

Although $K$ is defined via $\phi$ in (13.27), we can obtain $K$ from (13.31) without explicitly involving $\phi$. Note that for

$$\phi(\mathbf{x}) = (x_1^2, \, x_2^2, \, x_1 x_2, \, x_2 x_1) \in \mathbb{R}^4, \qquad (13.32)$$

we would get the same $K$, that is, a kernel function $K$ is not uniquely associated with one feature map $\phi$.

Next, generalize to $\mathbf{x} \in \mathbb{R}^m$. Function $K$ in (13.31) now corresponds to the feature map $\phi(\mathbf{x})$ with elements $x_k x_l$, where $k = 1, \ldots, m$, $l = 1, \ldots, m$. Hence $\phi(\mathbf{x}) \in \mathbb{R}^{m^2}$, that is, the dimension of the feature space is $M = m^2$. Computing $\phi^{\mathrm{T}}(\mathbf{x})\phi(\mathbf{z})$ directly takes $O(m^2)$ operations, but computing this inner product through $K(\mathbf{x}, \mathbf{z})$ using (13.31) requires only $O(m)$ operations. This example clearly illustrates the advantage of using the kernel function to avoid direct computations with the high-dimensional feature map $\phi$.

Given observations $\mathbf{x}_i$, $(i = 1, \ldots, n)$ in the input space, and a feature map $\phi$, the $n \times n$ *kernel matrix* $\mathbf{K}$ is defined with elements

$$K_{ij} = K(\mathbf{x}_i, \mathbf{x}_j) = \phi^{\mathrm{T}}(\mathbf{x}_i)\phi(\mathbf{x}_j) \equiv G_{ij}, \qquad (13.33)$$

where $(G_{ij})$ is simply the Gram matrix.

A matrix $\mathbf{M}$ is *positive semi-definite* if for any vector $\mathbf{v} \in \mathbb{R}^n$, $\mathbf{v}^{\mathrm{T}}\mathbf{M}\mathbf{v} \geq 0$. A symmetric function $f : X \times X \longrightarrow \mathbb{R}$ is said to be a positive semi-definite function if any matrix $\mathbf{M}$ [with elements $M_{ij} = f(\mathbf{x}_i, \mathbf{x}_j)$], formed by restricting $f$ to a finite subset of the input space $X$, is a positive semi-definite matrix.

A special case of *Mercer theorem* from functional analysis guarantees that a function $K : X \times X \longrightarrow \mathbb{R}$ is a kernel associated with a feature map $\phi$ via

$$K(\mathbf{x}, \mathbf{z}) = \phi^{\mathrm{T}}(\mathbf{x})\phi(\mathbf{z}), \qquad (13.34)$$

if and only if $K$ is a positive semi-definite function. To show that a kernel function $K$ is positive semi-definite, consider any kernel matrix $\mathbf{K}$ derived from $K$, and any vector $\mathbf{v}$:

$$\mathbf{v}^{\mathrm{T}}\mathbf{K}\mathbf{v} = \sum_i \sum_j v_i v_j K_{ij} = \sum_i \sum_j v_i v_j \phi(\mathbf{x}_i)^{\mathrm{T}}\phi(\mathbf{x}_j)$$

$$= \left(\sum_i v_i \phi(\mathbf{x}_i)\right)^{\mathrm{T}} \left(\sum_j v_j \phi(\mathbf{x}_j)\right) = \left\|\sum_i v_i \phi(\mathbf{x}_i)\right\|^2 \geq 0, \qquad (13.35)$$

so $K$ is indeed positive semi-definite. The proof for the converse is more complicated, and the reader is referred to Schölkopf and A. J. Smola (2002) or Shawe-Taylor and Cristianini (2004).

Let us now consider what operations can be performed on kernel functions that will yield new kernel functions. For instance, if we multiply a kernel function $K_1$ by a positive constant $c$, then $K = cK_1$ is a kernel function, since for any kernel matrix $\mathbf{K}_1$ derived from $K_1$ and any vector $\mathbf{v}$,

$$\mathbf{v}^{\mathrm{T}}(c\mathbf{K}_1)\mathbf{v} = c\,\mathbf{v}^{\mathrm{T}}\mathbf{K}_1\mathbf{v} \geq 0. \qquad (13.36)$$

Next, consider adding two kernel functions $K_1$ and $K_2$, which also yields a kernel since

$$\mathbf{v}^{\mathrm{T}}(\mathbf{K}_1 + \mathbf{K}_2)\mathbf{v} = \mathbf{v}^{\mathrm{T}}\mathbf{K}_1\mathbf{v} + \mathbf{v}^{\mathrm{T}}\mathbf{K}_2\mathbf{v} \geq 0. \qquad (13.37)$$

Below is a list of operations on kernel functions ($K_1$ and $K_2$) that will yield new kernel functions $K$:

(1)  $K(\mathbf{x}, \mathbf{z}) = cK_1(\mathbf{x}, \mathbf{z}), \quad$ with  $c > 0$. $\qquad (13.38)$

(2)  $K(\mathbf{x}, \mathbf{z}) = K_1(\mathbf{x}, \mathbf{z}) + K_2(\mathbf{x}, \mathbf{z})$. $\qquad (13.39)$

(3) $K(\mathbf{x}, \mathbf{z}) = K_1(\mathbf{x}, \mathbf{z}) K_2(\mathbf{x}, \mathbf{z}).$ (13.40)

(4) $K(\mathbf{x}, \mathbf{z}) = \dfrac{K_1(\mathbf{x}, \mathbf{z})}{\sqrt{K_1(\mathbf{x}, \mathbf{x}) K_1(\mathbf{z}, \mathbf{z})}} = \dfrac{\phi_1(\mathbf{x})^{\mathsf{T}} \phi_1(\mathbf{z})}{\|\phi_1(\mathbf{x})\| \|\phi_1(\mathbf{z})\|}.$ (13.41)

(5) $K(\mathbf{x}, \mathbf{z}) = K_1(\psi(\mathbf{x}), \psi(\mathbf{z}))$, with $\psi$ a real function. (13.42)

If $f$ is a real function, $\mathbf{M}$ a symmetric positive semi-definite $m \times m$ matrix and $p$ a positive integer, we also obtain kernel functions from the following operations:

(6) $K(\mathbf{x}, \mathbf{z}) = f(\mathbf{x}) f(\mathbf{z}).$ (13.43)

(7) $K(\mathbf{x}, \mathbf{z}) = \mathbf{x}^{\mathsf{T}} \mathbf{M} \mathbf{z}.$ (13.44)

(8) $K(\mathbf{x}, \mathbf{z}) = \big(K_1(\mathbf{x}, \mathbf{z})\big)^p,$ (13.45)

where a special case is the popular *polynomial kernel*

$$K(\mathbf{x}, \mathbf{z}) = \big(1 + \mathbf{x}^{\mathsf{T}} \mathbf{z}\big)^p.$$ (13.46)

For example, suppose the true relation is

$$y(x_1, x_2) = b_1 x_1 + b_2 x_2^2 + b_3 x_1 x_2.$$ (13.47)

With a $p = 2$ polynomial kernel, $\phi(\mathbf{x})$ in the feature space contains the elements $x_1, x_2, x_1^2(\equiv x_3), x_1 x_2(\equiv x_4), x_2^2(\equiv x_5)$. Linear regression in $F$ is

$$y = a_0 + a_1 x_1 + a_2 x_2 + a_3 x_3 + a_4 x_4 + a_5 x_5,$$ (13.48)

so the true non-linear relation (13.47) can be extracted by a linear regression in $F$.

Kernels can also be obtained through exponentiation:

(9) $K(\mathbf{x}, \mathbf{z}) = \exp\big(K_1(\mathbf{x}, \mathbf{z})\big).$ (13.49)

(10) $K(\mathbf{x}, \mathbf{z}) = \exp\left(-\dfrac{\|\mathbf{x} - \mathbf{z}\|^2}{2\sigma^2}\right).$ (13.50)

The *Gaussian kernel* or *radial basis function* (RBF) *kernel* (13.50) is the most commonly used kernel. As the exponential function can be approximated arbitrarily accurately by a high-degree polynomial with positive coefficients, the exponential kernel is, therefore, a limit of kernels. To extend further from the exponential kernel (13.49) to the Gaussian kernel (13.50), we first note that

$\exp(\mathbf{x}^{\mathrm{T}}\mathbf{z}/\sigma^2)$ is a kernel according to (13.49). Normalizing this kernel by (13.41) gives the new kernel

$$\frac{\exp(\mathbf{x}^{\mathrm{T}}\mathbf{z}/\sigma^2)}{\sqrt{\exp(\mathbf{x}^{\mathrm{T}}\mathbf{x}/\sigma^2)\exp(\mathbf{z}^{\mathrm{T}}\mathbf{z}/\sigma^2)}} = \exp\left(\frac{\mathbf{x}^{\mathrm{T}}\mathbf{z}}{\sigma^2} - \frac{\mathbf{x}^{\mathrm{T}}\mathbf{x}}{2\sigma^2} - \frac{\mathbf{z}^{\mathrm{T}}\mathbf{z}}{2\sigma^2}\right)$$

$$= \exp\left(\frac{-\mathbf{x}^{\mathrm{T}}\mathbf{x} + \mathbf{z}^{\mathrm{T}}\mathbf{x} + \mathbf{x}^{\mathrm{T}}\mathbf{z} - \mathbf{z}^{\mathrm{T}}\mathbf{z}}{2\sigma^2}\right)$$

$$= \exp\left(\frac{-(\mathbf{x}-\mathbf{z})^{\mathrm{T}}(\mathbf{x}-\mathbf{z})}{2\sigma^2}\right)$$

$$= \exp\left(\frac{-\|\mathbf{x}-\mathbf{z}\|^2}{2\sigma^2}\right). \tag{13.51}$$

Thus, the Gaussian kernel is a normalized exponential kernel. For more details on operations on kernels, see Shawe-Taylor and Cristianini (2004).

## 13.4   Kernel Ridge Regression  $\boxed{\text{B}}$ ☹

For the first application of the kernel method, let us describe *kernel ridge regression* as a technique for performing non-linear regression. We will use the Gaussian kernel $K$ and ridge regression. The Gaussian kernel (13.50) has the parameter $\sigma$ governing the width of the Gaussian function, while ridge regression has a weight penalty parameter $\lambda$ in the objective function (13.10). These two parameters are usually called *hyperparameters*, as we need to perform a search for their optimal values above a search for the basic model parameters like the regression coefficients. Kernel ridge regression is also called *least squares support vector regression*, as it replaces the standard error function in support vector regression (Section 13.7.5) by the mean squared error (MSE) function.

The procedure for kernel regression is set out below:

(1) Set aside some data for validation later; use the remaining observations $\mathbf{x}_i$, $y_i$, $(i = 1, \ldots, n)$ for training.

(2) Choose values for $\sigma$ and $\lambda$.

(3) Calculate the kernel matrix $\mathbf{K}$ with $K_{ij} = K(\mathbf{x}_i, \mathbf{x}_j)$.

(4) Solve the dual problem for ridge regression in (13.18) with $\mathbf{K}$ replacing $\mathbf{G}$, that is, $\boldsymbol{\alpha} = (\mathbf{K} + \lambda\mathbf{I})^{-1}\mathbf{y}$.

(5) For validation data $\tilde{\mathbf{x}}$, compute $\tilde{y}$ using (13.21), that is, $\tilde{y} = \boldsymbol{\alpha}^{\mathrm{T}}\mathbf{k}$, with $k_i = K(\mathbf{x}_i, \tilde{\mathbf{x}})$.

(6) Calculate the MSE of $\tilde{y}$ over the validation data.

(7) Go to (2) and repeat the calculations with different values of $\sigma$ and $\lambda$.

(8) Choose the $\sigma$ and $\lambda$ values with the smallest MSE over the validation data as the optimal solution.

In the environmental sciences, examples of kernel ridge regression include forecasting monthly streamflow (Samsudin et al., 2011), retrieving atmospheric profiles from satellite infrared sounding data (Camps-Valls, Munoz-Mari, et al., 2012) and predicting solar insolation for solar power generation (Ekici, 2014).

## 13.5 Advantages and Disadvantages B☺

Since the mathematical formulation of the last three sections may be difficult for some readers, a summary of the main ideas of kernel methods and their advantages and disadvantages is presented in this section.

First, consider the multiple linear regression problem with a single output variable $\hat{y}$,

$$\hat{y} = \sum_{l=1}^{m} a_l x_l + a_0 , \qquad (13.52)$$

with $x_l$ the predictor variables, and $a_l$ and $a_0$ the regression parameters or coefficients. In the MLP NN approach to non-linear regression (Section 6.3), non-linear *adaptive* basis functions $h_j$ (i.e. hidden neurons) are introduced, so the linear regression is between $\hat{y}$ and $h_j$,

$$\hat{y} = \sum_{j} a_j h_j(\mathbf{x}; \mathbf{w}) + a_0, \qquad (13.53)$$

with a sigmoidal function such as tanh commonly used for $h_j$, that is,

$$h_j(\mathbf{x}; \mathbf{w}) = \tanh \left( \sum_{l} w_{jl} x_l + w_{j0} \right). \qquad (13.54)$$

Since $\hat{y}$ depends on $\mathbf{w}$ non-linearly (due to the non-linear function tanh), the resulting optimization is non-linear, with multiple minima in general.

What happens if instead of adaptive basis functions $h_j(\mathbf{x}; \mathbf{w})$, we use *non-adaptive* basis functions $\phi_j(\mathbf{x})$, that is, the basis functions do not have adjustable parameters $\mathbf{w}$? In this situation,

$$\hat{y} = \sum_{j} a_j \, \phi_j(\mathbf{x}) + a_0. \qquad (13.55)$$

For instance, Taylor series expansion with two predictors $(x_1, x_2)$ would have $\{\phi_j\} = x_1, x_2, x_1^2, x_1 x_2, x_2^2, \ldots$ The advantage of using non-adaptive basis functions is that $y$ does not depend on any parameter non-linearly, so only linear optimization is involved with no multiple minima problem. The disadvantage is that one usually needs a large number of non-adaptive basis functions compared to relatively few adaptive basis functions to model a non-linear relation, hence the dominance of MLP NN models despite the local minima problem.

The 'curse of dimensionality' is brought under control with the kernel trick, that is, although $\phi$ is a very high (or even infinite) dimensional vector function, as long as the solution of the problem can be formulated to involve only inner products like $\phi^{\mathrm{T}}(\mathbf{x}')\phi(\mathbf{x})$, then a kernel function $K$ can be introduced

$$K(\mathbf{x}', \mathbf{x}) = \phi^{\mathrm{T}}(\mathbf{x}')\phi(\mathbf{x}). \tag{13.56}$$

The solution of the problem now involves working only with a very manageable kernel function $K(\mathbf{x}', \mathbf{x})$ instead of the unmanageable $\phi$. From Section 13.4, the kernel ridge regression solution is

$$\hat{y} = \sum_{i=1}^{n} \alpha_i K(\mathbf{x}_i, \mathbf{x}), \tag{13.57}$$

where there are $i = 1, \ldots, n$ data points $\mathbf{x}_i$ in the training dataset. If a Gaussian kernel function is used, $\hat{y}$ is simply a linear combination of Gaussian functions. If a polynomial kernel function is used, then $\hat{y}$ is a linear combination of polynomial functions.

The kernel approach has an elegant architecture with simple modularity, which allows easy interchange of parts. The modularity, as illustrated in Fig. 13.2, is as follows: after the input data have been gathered in stage 1, stage 2

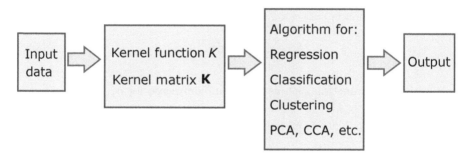

Figure 13.2 The modular architecture of the kernel method. [Follows (Shawe-Taylor and Cristianini, 2004, figure 2.4).]

consists of choosing a kernel function $K$ and calculating the kernel matrix $\mathbf{K}$, stage 3 consists of using a pattern analysis algorithm to perform regression, classification, principal component analysis (PCA), canonical correlation analysis (CCA) or other tasks in the feature space, whereby the extracted information is given as output in stage 4. For example, if at stage 3, one decides to switch from ridge regression to a different regression algorithm such as support vector regression, the other stages require no major adjustments to accommodate this switch. Similarly, we can switch kernels in stage 2 from, say, the Gaussian kernel to a polynomial kernel without having to make significant modifications to the other stages. The pattern analysis algorithms in stage 3 may be limited to working with vectorial data, yet with cleverly designed kernels in stage 2, kernel methods have been used to analyse data with very different structures, for example strings (used in text and DNA analyses) and trees (Shawe-Taylor and Cristianini, 2004).

In summary, with the kernel method, one can analyse the structure in a high-dimensional feature space with only moderate computational costs thanks to the kernel function, which gives the inner product in the feature space without having to evaluate the feature map $\phi$ directly. The kernel method is applicable to all pattern analysis algorithms expressed only in terms of inner products of the inputs. In the feature space, linear pattern analysis algorithms are applied, with no local minima problems since only linear optimization is involved. Although only linear pattern analysis algorithms are used, the fully non-linear patterns in the input space can be extracted due to the non-linear feature map $\phi$.

The main disadvantage of the kernel method is the lack of an easy way to map inversely from the feature space back to the input data space – a difficulty commonly referred to as the *pre-image* problem. This problem arises, for example, in kernel principal component analysis (kernel PCA, see Section 13.9), where one wants to find the pattern in the input space corresponding to a principal component in the feature space. In kernel PCA, where PCA is performed in the feature space, the eigenvectors are expressed as a linear combination of the data points in the feature space, that is,

$$\mathbf{v} = \sum_{i=1}^{n} \beta_i \, \phi(\mathbf{x}_i). \tag{13.58}$$

As the feature space $F$ is generally a much higher dimensional space than the input space $X$ and the mapping function $\phi$ is non-linear, one may not be able to find an $\mathbf{x}$ (i.e. the 'pre-image') in the input space, such that $\phi(\mathbf{x}) = \mathbf{v}$, as illustrated in Fig. 13.3. We will look at methods for finding an approximate pre-image in the next section.

## 13.6   Pre-image Problem   $\boxed{\text{C}}$ ☺

Since in general an exact pre-image may not exist (Fig. 13.3), various methods have been developed to find an approximate pre-image. In Fig. 13.4, we illustrate the situation where we want to find a pre-image for a point $\mathbf{p}(\phi(\mathbf{x}))$. Here $\mathbf{p}$ can represent, for instance, the projection of $\phi(\mathbf{x})$ onto the first PCA eigenvector in the feature space. Mika et al. (1999) proposed finding a pre-image $\mathbf{x}'$ by minimizing the squared distance between $\phi(\mathbf{x}')$ and $\mathbf{p}(\phi(\mathbf{x}))$, that is,[1]

$$\mathbf{x}' = \arg\min_{\mathbf{x}'} \|\phi(\mathbf{x}') - \mathbf{p}(\phi(\mathbf{x}))\|^2. \tag{13.60}$$

---

[1] The operator

$$\arg\min_{\mathbf{x}} f(\mathbf{x}) \tag{13.59}$$

returns the argument (i.e. an $\mathbf{x}$ value) for which the function $f(\mathbf{x})$ attains the minimum value. For example, with $f(x) = x^2 - 1$, the minimum value of $f(x)$ is $-1$, which occurs when $x = 0$, so $\arg\min f(x) = 0$.

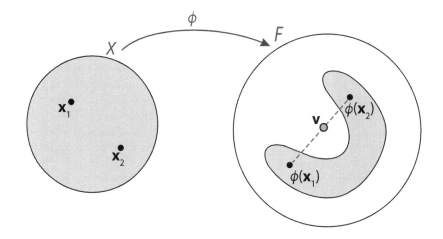

Figure 13.3 Illustrating the pre-image problem in kernel methods. The input space $X$ is mapped by $\phi$ to the shaded area in the much larger feature space $F$. Two data points $\mathbf{x}_1$ and $\mathbf{x}_2$ are mapped to $\phi(\mathbf{x}_1)$ and $\phi(\mathbf{x}_2)$, respectively, in $F$. Although $\mathbf{v}$ is a linear combination of $\phi(\mathbf{x}_1)$ and $\phi(\mathbf{x}_2)$, it lies outside the shaded area in $F$, hence there is no 'pre-image' $\mathbf{x}$ in $X$, such that $\phi(\mathbf{x}) = \mathbf{v}$. [Source: Hsieh (2009)]

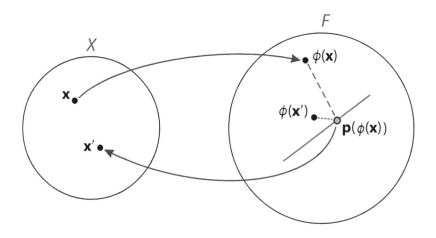

Figure 13.4 Illustrating the approach used by Mika et al. (1999) to extract an approximate pre-image in the input space $X$ for a point $\mathbf{p}(\phi(\mathbf{x}))$ in the feature space $F$. Here, for example, $\mathbf{p}(\phi(\mathbf{x}))$ is shown as the projection of $\phi(\mathbf{x})$ onto the direction of the first PCA eigenvector (solid line). The optimization algorithm looks for $\mathbf{x}'$ in $X$ that minimizes the squared distance between $\phi(\mathbf{x}')$ and $\mathbf{p}(\phi(\mathbf{x}))$. [Source: Hsieh (2009)]

This is a non-linear optimization problem, hence susceptible to finding local minima. It turns out the method is indeed quite unstable, and alternatives have been proposed.

In the approach of Kwok and I. W.-H. Tsang (2004), the distances between $\mathbf{p}(\phi(\mathbf{x}))$ and its $K$ nearest neighbours, $\phi(\mathbf{x}_1) \ldots, \phi(\mathbf{x}_K)$ in $F$ are used. Noting that there is usually a simple relation between feature-space distance and input-space distance for many commonly used kernels, and borrowing an idea from multi-dimensional scaling (MDS), they were able to use the corresponding distances among $\mathbf{x}_1, \ldots, \mathbf{x}_K$ to pinpoint the approximate pre-image $\mathbf{x}'$ for $\mathbf{p}(\phi(\mathbf{x}))$ (Fig. 13.5), analogous to the use of global positioning system (GPS) satellites to

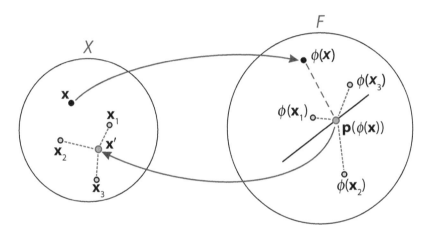

Figure 13.5 Illustrating the approach used by Kwok and I. W.-H. Tsang (2004) to extract an approximate pre-image in the input space $X$ for a point $\mathbf{p}(\phi(\mathbf{x}))$ in the feature space $F$. The distance information in $F$ between $\mathbf{p}(\phi(\mathbf{x}))$ and its several nearest neighbours (e.g. $\phi(\mathbf{x}_1), \phi(\mathbf{x}_2), \ldots$), and the relationship between distance in $X$ and distance in $F$ are exploited to allow $\mathbf{x}_1, \mathbf{x}_2, \ldots$ to pinpoint the desired approximate pre-image $\mathbf{x}'$ in $X$. [Source: Hsieh (2009)]

pinpoint the location of an object. Since the solution required only linear algebra and is non-iterative, there are no numerical instability and local minima problems. Note that the method can be used to find the pre-image for any point $\tilde{\mathbf{p}}$ in $F$. Other applications beside kernel PCA include kernel clustering, for example $\tilde{\mathbf{p}}$ can be the mean position of a cluster in $F$ obtained from a $K$-means clustering algorithm (Section 10.2.1). Alternative approaches to finding an approximate pre-image include Bakir et al. (2004) and B. Tang and Mazzoni (2006).

# 13.7 Support Vector Machines (SVM) $\boxed{\text{B}}$ ☹

Since the mid-1990s, *support vector machines* (SVM) have become popular in non-linear classification and regression problems. A kernel-based method, SVM for classification provides the decision boundaries but not posterior probabilities. While various individual features of SVM have been discovered earlier, B. E. Boser et al. (1992) and Cortes and V. Vapnik (1995) were the first papers to assemble the various ideas into an elegant new method for classification. V. N. Vapnik (1995) then extended the method to support vector regression (SVR) for the non-linear regression problem. For unsupervised learning, ideas from SVM has been used in support vector clustering (Ben-Hur et al., 2001). Books covering SVM include V. N. Vapnik (1998), Cristianini and Shawe-Taylor (2000), Schölkopf and A. J. Smola (2002), Shawe-Taylor and Cristianini (2004), Bishop (2006), and Cherkassky and Mulier (2007).

As SVM is best known for classification, the focus of this section will be on the SVM classifier, with SVR mentioned briefly in Section 13.7.5. The development of the SVM classifier for the two-class problem proceeds naturally in three stages. The basic *maximum margin classifier* is first introduced to problems where the two classes can be separated by a linear decision boundary (a hyperplane). Next, the classifier is modified to tolerate misclassification in problems not separable by a linear decision boundary. Finally, the classifier is non-linearly generalized by using the kernel method.

In the environmental sciences, the SVM classifier has been used on satellite data to identify various types of land cover (C. Huang et al., 2002; Camps-Valls and Bruzzone, 2005) and to predict ground-level ozone (Lu and D. Wang, 2008). Reviews of SVM for classification and regression are available in remote sensing (Mountrakis et al., 2011) and in hydrology (Raghavendra and Deka, 2014).

A caveat on the comparison between SVM and MLP NN models: a main advantage of SVM models is that they do not have the local minima problem found in the MLP models. Because of local minima in MLP, comparing a single SVM model against a single MLP model (with one to two hidden layers) allows many papers to claim SVM outperforms MLP. In reality, MLP is typically run as an *ensemble* of models to remove the local minima disadvantage, so the performance of the SVM and that of the ensemble MLP (with one to two hidden layers) are usually found to be very comparable in regression problems (Aguilar-Martinez and Hsieh, 2009; A. R. Lima et al., 2015). With enough data, modern deep NN models (Chapter 15) do tend to outperform SVM.

## 13.7.1  Linearly Separable Case  B ☹

First we consider the two-class problem and assume the data from the two classes can be separated by a hyperplane decision boundary. The separating hyperplane (Fig. 13.6) is given by the equation

$$\hat{y}(\mathbf{x}) = \mathbf{w}^{\mathrm{T}}\mathbf{x} + w_0 = 0. \tag{13.61}$$

Any two points $\mathbf{x}_1$ and $\mathbf{x}_2$ lying on the hyperplane $\hat{y} = 0$ satisfy

$$\mathbf{w}^{\mathrm{T}}(\mathbf{x}_1 - \mathbf{x}_2) = 0, \tag{13.62}$$

hence the unit vector

$$\hat{\mathbf{w}} = \frac{\mathbf{w}}{\|\mathbf{w}\|} \tag{13.63}$$

is normal to the $\hat{y} = 0$ hyperplane surface.

Figure 13.6 The hyperplane $\hat{y} = 0$, with the vector $\mathbf{w}$ perpendicular to this hyperplane. In this example with $\mathbf{x}$ being two-dimensional, the hyperplane $\hat{y} = 0$ reduces to a straight line in the $x_1$–$x_2$ plane. The component of the vector $\mathbf{x} - \mathbf{x}_0$ projected onto the $\mathbf{w}$ direction is shown by the dot-dashed line.

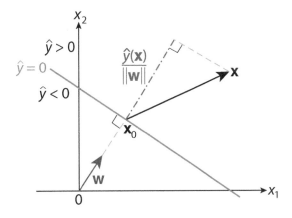

Any point $\mathbf{x}_0$ on the $\hat{y} = 0$ surface satisfies $\mathbf{w}^\mathrm{T}\mathbf{x}_0 = -w_0$. In Fig. 13.6, the component of the vector $\mathbf{x} - \mathbf{x}_0$ projected onto the $\hat{\mathbf{w}}$ direction is given by

$$\hat{\mathbf{w}}^\mathrm{T}(\mathbf{x} - \mathbf{x}_0) = \frac{\mathbf{w}^\mathrm{T}(\mathbf{x} - \mathbf{x}_0)}{\|\mathbf{w}\|} = \frac{\mathbf{w}^\mathrm{T}\mathbf{x} + w_0}{\|\mathbf{w}\|} = \frac{\hat{y}(\mathbf{x})}{\|\mathbf{w}\|}. \tag{13.64}$$

Thus, $\hat{y}(\mathbf{x})$ is proportional to the normal distance between $\mathbf{x}$ and the hyperplane $\hat{y} = 0$, and the sign of $\hat{y}(\mathbf{x})$ indicates on which side of the hyperplane $\mathbf{x}$ lies.

Let the training dataset be composed of predictors $\mathbf{x}_n$ and target data $y_n$ $(n = 1, \ldots, N)$. Since there are two classes, $y_n$ takes on the value of $-1$ or $+1$. When a new predictor vector $\mathbf{x}$ becomes available, it is classified according to the sign of the function

$$\hat{y}(\mathbf{x}) = \mathbf{w}^\mathrm{T}\mathbf{x} + w_0. \tag{13.65}$$

The normal distance from a point $\mathbf{x}_n$ to the decision boundary $\hat{y} = 0$ is, according to (13.64), given by

$$\frac{y_n\,\hat{y}(\mathbf{x}_n)}{\|\mathbf{w}\|} = \frac{y_n\,(\mathbf{w}^\mathrm{T}\mathbf{x}_n + w_0)}{\|\mathbf{w}\|}, \tag{13.66}$$

where $y_n$ contributes the correct sign ($+1$ or $-1$) to ensure that the distance given by (13.66) is non-negative.

The margin (Fig. 13.7(a)) is given by the distance of the closest point(s) $\mathbf{x}_n$ in the dataset. A *maximum margin classifier* determines the decision boundary by maximizing the margin $l$, through searching for the optimal values of $\mathbf{w}$ and $w_0$. The optimization problem is then

$$\max_{\mathbf{w}, w_0} l \quad \text{subject to} \quad \frac{y_n\,(\mathbf{w}^\mathrm{T}\mathbf{x}_n + w_0)}{\|\mathbf{w}\|} \geq l, \quad (n = 1, \ldots, N), \tag{13.67}$$

where the constraint simply ensures that no data points lie inside the margins. Obviously the margin is determined by relatively few points in the dataset, and these points circled in Fig. 13.7(a) are called *support vectors*.

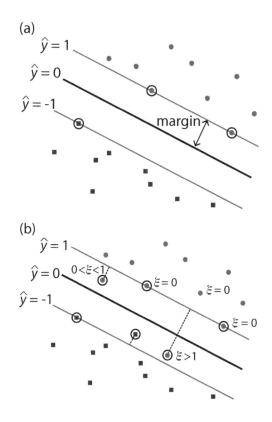

Figure 13.7 (a) A dataset containing two classes (shown by solid circles and squares) separable by a hyperplane decision boundary $\hat{y} = 0$. The margin is maximized. Support vectors, that is, data points used in determining the margins $\hat{y} = \pm 1$, are circled. (b) A dataset not separable by a hyperplane boundary. Slack variables $\xi_n \geq 0$ are introduced, with $\xi_n = 0$ for data points lying on or within the correct margin, $\xi_n > 1$ for points lying to the wrong side of the decision boundary. Support vectors are circled.

Since the distance (13.66) is unchanged if we multiply $\mathbf{w}$ and $w_0$ by an arbitrary scale factor $s$, we are free to choose $\|\mathbf{w}\|$. If we choose $\|\mathbf{w}\| = l^{-1}$, then the constraint in (13.67) becomes

$$y_n \left(\mathbf{w}^{\mathrm{T}} \mathbf{x}_n + w_0\right) \geq 1, \quad (n = 1, \ldots, N), \tag{13.68}$$

and maximizing the margin $l$ becomes equivalent to minimizing $\|\mathbf{w}\|$, which is equivalent to solving

$$\min_{\mathbf{w}, w_0} \tfrac{1}{2} \|\mathbf{w}\|^2, \tag{13.69}$$

subject to the constraint (13.68). This optimization of a quadratic function subject to constraints is referred to as a *quadratic programming* problem (P. E. Gill et al., 1981).

Karush (1939) and Kuhn and Tucker (1951) (KKT) have solved this type of optimization with inequality constraint by a generalization of the Lagrange multiplier method. With a Lagrange multiplier $\lambda_n \geq 0$ introduced for each of the $N$ constraints in (13.68) (see Appendix B), the Lagrange function $L$ takes the form

$$L(\mathbf{w}, w_0, \boldsymbol{\lambda}) = \frac{1}{2} \|\mathbf{w}\|^2 - \sum_{n=1}^{N} \lambda_n \left[y_n (\mathbf{w}^{\mathrm{T}} \mathbf{x}_n + w_0) - 1\right], \tag{13.70}$$

with $\boldsymbol{\lambda} = (\lambda_1, \ldots, \lambda_N)^{\mathrm{T}}$. Setting the derivatives of $L$ with respect to $\mathbf{w}$ and $w_0$ to zero yields, respectively,

$$\mathbf{w} = \sum_{n=1}^{N} \lambda_n y_n \mathbf{x}_n, \tag{13.71}$$

$$0 = \sum_{n=1}^{N} \lambda_n y_n. \tag{13.72}$$

Substituting these into (13.70) allows us to express $L$ solely in terms of the Lagrange multipliers, that is,

$$L_{\mathrm{D}}(\boldsymbol{\lambda}) = \sum_{n=1}^{N} \lambda_n - \frac{1}{2} \sum_{n=1}^{N} \sum_{j=1}^{N} \lambda_n \lambda_j y_n y_j \mathbf{x}_n^{\mathrm{T}} \mathbf{x}_j, \tag{13.73}$$

subject to $\lambda_n \geq 0$ and (13.72), with $L_{\mathrm{D}}$ referred to as the *dual Lagrangian*. If $m$ is the dimension of $\mathbf{x}$ and $\mathbf{w}$, then the original primal optimization problem of (13.69) is a quadratic programming problem with about $m$ variables. In contrast, the dual problem (13.73) is also a quadratic programming problem with $N$ variables from $\boldsymbol{\lambda}$. The main reason we want to work with the dual problem instead of the primal problem is that, in the next stage, to generalize the linear classifier to a non-linear classifier, SVM will invoke kernels, that is, $\mathbf{x}$ will be replaced by $\phi(\mathbf{x})$ in a feature space of dimension $M$; thus, $m$ will be replaced by $M$, which is usually much larger or even infinite.

Our constrained optimization problem satisfies the KKT conditions (Appendix B):

$$\lambda_n \geq 0, \tag{13.74}$$

$$y_n(\mathbf{w}^{\mathrm{T}}\mathbf{x}_n + w_0) - 1 \geq 0, \tag{13.75}$$

$$\lambda_n \left[ y_n(\mathbf{w}^{\mathrm{T}}\mathbf{x}_n + w_0) - 1 \right] = 0. \tag{13.76}$$

Combining (13.65) and (13.71), we have the following formula for classifying a new data point $\mathbf{x}$:

$$\hat{y}(\mathbf{x}) = \sum_{n=1}^{N} \lambda_n y_n \mathbf{x}_n^{\mathrm{T}} \mathbf{x} + w_0, \tag{13.77}$$

where the class ($+1$ or $-1$) is decided by the sign of $\hat{y}(\mathbf{x})$.

From (13.65), the KKT condition (13.76) can be rewritten as

$$\lambda_n \left[ y_n \hat{y}(\mathbf{x}_n) - 1 \right] = 0. \tag{13.78}$$

Thus, for every data point $\mathbf{x}_n$, either $\lambda_n = 0$ or $y_n \hat{y}(\mathbf{x}_n) = 1$. However, the data points with $\lambda_n = 0$ will be omitted in the summation in (13.77), hence will not contribute in the classification of a new data point $\mathbf{x}$. Data points with $\lambda_n > 0$ are the *support vectors* and they satisfy $y_n \hat{y}(\mathbf{x}_n) = 1$, that is, they lie exactly on the maximum margins (Fig. 13.7(a)). Although only support vectors determine the margins and the decision boundary, the entire training dataset was used to locate the support vectors. However, once the model has been trained, classification of new data by (13.77) requires only the support vectors, and the remaining data (usually the majority of data points) can be ignored. The desirable property of being able to ignore the majority of data points is called *sparsity* or *sparseness*, which greatly reduces computational costs.

To obtain the value for $w_0$, we use the fact that for any support vector $\mathbf{x}_j$, $y_j \hat{y}(\mathbf{x}_j) = 1$, that is,

$$y_j \left( \sum_{n=1}^{N} \lambda_n y_n \mathbf{x}_n^{\mathrm{T}} \mathbf{x}_j + w_0 \right) = 1. \tag{13.79}$$

As $y_j = \pm 1$, $y_j^{-1} = y_j$, so

$$w_0 = y_j - \sum_{n=1}^{N} \lambda_n y_n \mathbf{x}_n^{\mathrm{T}} \mathbf{x}_j, \tag{13.80}$$

(where the summation need only be over $n$ for which $\lambda_n$ is non-zero). Since each support vector $\mathbf{x}_j$ gives an estimate of $w_0$, all the estimates are usually averaged together to provide the most accurate estimate of $w_0$.

## 13.7.2   Linearly Non-separable Case  $\boxed{\text{B}}$☹

So far, the maximum margin classifier has only been applied to the case where the two classes are separable by a linear decision boundary (a hyperplane). Next, we move to the case where the two classes are not separable by a hyperplane, that is, no matter how hard we try, one or more data points will end up on the wrong side of the hyperplane and end up being misclassified. Thus, our classifier needs to be generalized to tolerate misclassification.

We introduce *slack variables*, $\xi_n \geq 0$, $n = 1, \ldots, N$, that is, each training data point $\mathbf{x}_n$ is assigned a slack variable $\xi_n$. The slack variables are defined by $\xi_n = 0$ for all $\mathbf{x}_n$ lying on or within the correct margin boundary, and $\xi_n = |y_n - \hat{y}(\mathbf{x}_n)|$ for all other points. Thus, the slack variable measures the distance a data point protrudes beyond the correct margin (Fig. 13.7(b)). A data point that lies right on the decision boundary $\hat{y}(\mathbf{x}_n) = 1$ will have $\xi_n = 1$, while $\xi_n > 1$ corresponds to points lying to the wrong side of the decision boundary, that is, misclassified. Points with $0 < \xi_n \leq 1$ protrude beyond the correct margin but not enough to cross the decision boundary to be misclassified.

The constraint (13.68) is modified to allow for data points extending beyond the correct margin, that is,

$$y_n(\mathbf{w}^{\mathrm{T}} \mathbf{x}_n + w_0) \geq 1 - \xi_n, \quad (n = 1, \ldots, N). \tag{13.81}$$

The minimization problem (13.69) is modified to

$$\min_{\mathbf{w}, w_0} \left( \frac{1}{2} \|\mathbf{w}\|^2 + C \sum_{n=1}^{N} \xi_n \right). \tag{13.82}$$

The objective of the optimization is to maximize the margin while allowing some slack for misclassification. $C$ is a hyperparameter, with a smaller value allowing more slack. If we divide the expression in (13.82) by $C$, the second term can be viewed as an error term, while the first term can be viewed as a weight penalty term, with $C^{-1}$ as the weight penalty parameter – analogous to the regularization of NN models in Section 8.4 where we have a mean squared error (MSE) term plus a weight penalty term in the objective function. The effect of misclassification on the objective function is only linearly related to $\xi_n$ in (13.82), in contrast to the MSE term, which is quadratic. Since any misclassified point has $\xi_n > 1$, $\sum_n \xi_n$ can be viewed as providing an upper bound on the number of misclassified points.

To optimize (13.82) subject to constraint (13.81) and $\xi_n \geq 0$, we again turn to the method of Lagrange multipliers (Appendix B), where the Lagrange function is now .

$$L = \frac{1}{2}\|\mathbf{w}\|^2 + C \sum_{n=1}^{N} \xi_n - \sum_{n=1}^{N} \lambda_n \left[ y_n(\mathbf{w}^\mathrm{T}\mathbf{x}_n + w_0) - 1 + \xi_n \right] - \sum_{n=1}^{N} \mu_n \xi_n, \quad (13.83)$$

with $\lambda_n \geq 0$ and $\mu_n \geq 0$ $(n = 1, \ldots, N)$ the Lagrange multipliers.

Setting the derivatives of $L$ with respect to $\mathbf{w}$, $w_0$ and $\xi_n$ to 0 yields, respectively,

$$\mathbf{w} = \sum_{n=1}^{N} \lambda_n y_n \mathbf{x}_n, \quad (13.84)$$

$$0 = \sum_{n=1}^{N} \lambda_n y_n, \quad (13.85)$$

$$\lambda_n = C - \mu_n. \quad (13.86)$$

Substituting these into (13.83) again allows $L$ to be expressed solely in terms of the Lagrange multipliers $\boldsymbol{\lambda}$, that is,

$$L_\mathrm{D}(\boldsymbol{\lambda}) = \sum_{n=1}^{N} \lambda_n - \frac{1}{2} \sum_{n=1}^{N} \sum_{j=1}^{N} \lambda_n \lambda_j y_n y_j \mathbf{x}_n^\mathrm{T} \mathbf{x}_j, \quad (13.87)$$

with $L_\mathrm{D}$ the dual Lagrangian. Note that $L_\mathrm{D}$ has the same form as that in the separable case, but the constraints are somewhat changed. As $\lambda_n \geq 0$ and $\mu_n \geq 0$, (13.86) implies

$$0 \leq \lambda_n \leq C. \quad (13.88)$$

Furthermore, there are the constraints from the KKT conditions (Appendix B):

$$y_n(\mathbf{w}^\mathrm{T}\mathbf{x}_n + w_0) - 1 + \xi_n \geq 0, \quad (13.89)$$

$$\lambda_n \left[ y_n(\mathbf{w}^\mathrm{T}\mathbf{x}_n + w_0) - 1 + \xi_n \right] = 0, \quad (13.90)$$

$$\mu_n \xi_n = 0, \quad (13.91)$$

for $n = 1, \ldots, N$.

In the computation for $\mathbf{w}$ in (13.84), only data points with $\lambda_n \neq 0$ contributed in the summation, so these points (with $\lambda_n > 0$) are the *support vectors*. Some support vectors lie exactly on the margin (i.e. $\xi_n = 0$), while others may protrude beyond the margin ($\xi_n > 0$) (Fig. 13.7(b)). The value for $w_0$ can be obtained from (13.90) using any of the support vectors lying on the margin ($\xi_n = 0$), that is, (13.80) again follows, and usually the values obtained from the individual vectors are averaged together to give the best estimate.

Optimizing the dual problem (13.87) is a simpler quadratic programming problem than the primal (13.83) and can be readily solved using standard methods (see e.g. P. E. Gill et al., 1981). The hyperparameter $C$ is not solved by the optimization. Instead, one usually reserves some validation data to test the performance of various models trained with different values of $C$ to determine the best value to use for $C$. With new data $\mathbf{x}$, the classification is, as before, based on the sign of $\hat{y}(\mathbf{x})$ from (13.77).

### 13.7.3   Non-linear Classification by SVM  B☺

As our classifier is still restricted to linearly non-separable problems, the final step is to extend the classifier from being linear to non-linear. This is achieved by performing linear classification not with the input $\mathbf{x}$ data but with the $\phi(\mathbf{x})$ data in a feature space, where $\phi$ is the non-linear function mapping from the original input space to the feature space (Section 13.3), that is,

$$\hat{y}(\mathbf{x}) = \mathbf{w}^{\mathrm{T}}\phi(\mathbf{x}) + w_0. \tag{13.92}$$

The dual Lagrangian $L_{\mathrm{D}}$ for this new problem is obtained from (13.87) simply through replacing $\mathbf{x}$ by $\phi(\mathbf{x})$, giving

$$L_{\mathrm{D}}(\boldsymbol{\lambda}) = \sum_{n=1}^{N} \lambda_n - \frac{1}{2} \sum_{n=1}^{N} \sum_{j=1}^{N} \lambda_n \lambda_j y_n y_j \phi^{\mathrm{T}}(\mathbf{x}_n)\phi(\mathbf{x}_j). \tag{13.93}$$

Classification is based on the sign of $\hat{y}(\mathbf{x})$, with $\hat{y}(\mathbf{x})$ modified from (13.77) to

$$\hat{y}(\mathbf{x}) = \sum_{n=1}^{N} \lambda_n \, y_n \, \phi^{\mathrm{T}}(\mathbf{x}_n)\phi(\mathbf{x}) + w_0. \tag{13.94}$$

Since the dimension of the feature space can be very high or even infinite, computations involving the inner product $\phi^{\mathrm{T}}(\mathbf{x})\phi(\mathbf{x}')$ are only practicable because of the *kernel trick* (Section 13.3), where a *kernel function* $K$

$$K(\mathbf{x}, \mathbf{x}') \equiv \phi^{\mathrm{T}}(\mathbf{x})\phi(\mathbf{x}') \tag{13.95}$$

is introduced to obviate direct computation of the inner product, with the most commonly used kernel being the *Gaussian or radial basis function* (RBF) *kernel*,

$$K(\mathbf{x}, \mathbf{x}') = \exp\left(-\frac{\|\mathbf{x} - \mathbf{x}'\|^2}{2\sigma^2}\right). \tag{13.96}$$

Under the kernel approach, (13.94) becomes

$$y(\mathbf{x}) = \sum_{n=1}^{N} \lambda_n y_n K(\mathbf{x}_n, \mathbf{x}) + w_0 \tag{13.97}$$

and (13.93) becomes

$$L_D(\boldsymbol{\lambda}) = \sum_{n=1}^{N} \lambda_n - \frac{1}{2} \sum_{n=1}^{N} \sum_{j=1}^{N} \lambda_n \lambda_j y_n y_j K(\mathbf{x}_n, \mathbf{x}_j). \tag{13.98}$$

Optimizing the dual problem (13.98) is again a quadratic programming problem (P. E. Gill et al., 1981). Since the objective function is only quadratic, and with the constraints being linear, there is no local minima problem, that is, unlike MLP NN models, the minimum found by SVM is the *global minimum*. This formulation of the SVM is sometimes called the $C$-SVM, as there is an alternative formulation, the $\nu$-SVM, by Schölkopf, A. Smola, et al. (2000).

The hyperparameters $C$ and $\sigma$ (assuming the RBF kernel (13.96) is used) are not obtained from the optimization. To determine the values of these two hyperparameters, one usually trains multiple models with different values of $C$ and $\sigma$,[2] and from their classification performance over validation data, determines the best values for $C$ and $\sigma$. An alternative is to use Bayesian optimization to solve for the hyperparameters (Snoek, Larochelle, et al., 2012).

## 13.7.4 Multi-class Classification by SVM C☺

Support vector machine was originally developed for two-class (i.e. binary) problems. The simplest extension to multi-class problems is via the *one-versus-the-rest* approach: that is, given $c$ classes, we train $c$ SVM binary classifiers $\hat{y}^{(k)}(\mathbf{x})$ $(k = 1, \ldots, c)$ where the target data for class $k$ are set to $+1$ and for all other classes to $-1$. New input data $\mathbf{x}$ are then classified as belonging to class $j$ if $\hat{y}^{(j)}(\mathbf{x})$ attains the highest value among $\hat{y}^{(k)}(\mathbf{x})$ $(k = 1, \ldots, c)$.

There are two criticisms of this heuristic approach: (a) as the $c$ binary classifiers were trained on different problems, there is no guarantee that the outputs $\hat{y}^{(k)}$ are on the same scale for fair comparison between the $c$ binary classifiers and (b) if there are many classes, then the $+1$ target data can be greatly outnumbered by the $-1$ target data.

A better approach is *pairwise classification* or *one-versus-one* (C. W. Hsu and C. J. Lin, 2002, with errata in C. J. Lin (2002)). A binary classifier is trained for each possible pair of classes, so there is a total of $c(c-1)/2$ binary classifiers, with each classifier using training data from only class $k$ and class $l$ $(k \neq l)$. With new data $\mathbf{x}$, all the binary classifiers are applied, and if class $j$ gets the most 'yes votes' from the classifiers, then $\mathbf{x}$ is classified as belonging to class $j$.

---

[2] Usually a 2-D grid search is performed over $C$ and $\sigma$. One could first perform a coarse grid search over a broad domain, for example $C = 2^{-5}, 2^{-3}, \ldots, 2^{15}$ and $\sigma = 2^{-15}, 2^{-13}, \ldots, 2^{3}$, then focus on a smaller domain by performing a fine grid search.

As an example, consider the forest cover dataset used earlier in Fig. 12.1(a) and apply the one-versus-one SVM classifier with Gaussian kernel on this dataset (Fig. 13.8). Out of 198 training data points, 49 are support vectors (Fig. 13.8(b)), which are used to determine the boundaries of the decision regions. The misclassification rate is 11.6% for the training data and 20.3% for the test data.

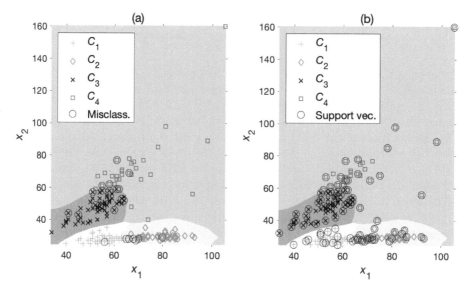

Figure 13.8 One-versus-one classification by SVM on the training dataset in Fig. 12.1(a) for types of forest cover. Class $C_1$ is indicated by $+$, $C_2$ by $\Diamond$, $C_3$ by $\times$ and $C_4$ by $\square$. Decision regions for the four classes are shown by different background shading. Circled data points in (a) are the misclassified data and in (b) the support vectors.

### 13.7.5  Support Vector Regression  Ⓒ☺

We now briefly describe support vector regression (SVR), the extension of the SVM approach to regression problems. In SVR, the objective function to be minimized is

$$J = C \sum_{n=1}^{N} E\big[\hat{y}(\mathbf{x}_n) - y\big] + \frac{1}{2}\|\mathbf{w}\|^2, \tag{13.99}$$

where $C$ is the inverse weight penalty parameter, $E$ is an error function and the second term is the weight penalty term. To retain the sparseness property of the SVM classifier, $E$ is usually taken to be of the form

$$E_\epsilon(z) = \begin{cases} |z| - \epsilon, & \text{if } |z| > \epsilon, \\ 0, & \text{otherwise.} \end{cases} \tag{13.100}$$

This is an *ϵ-insensitive error function*, as it ignores errors of size smaller than $\epsilon$ (Fig. 13.9). For large $z$, as $E_\epsilon(z)$ behaves like the MAE function – very different from the MSE function used in most NN models – SVR is more resistant to outliers in the training data than NN. Kernel ridge regression (Section 13.4) can be regarded as SVR but with its $E_\epsilon(z)$ error function replaced by the MSE.

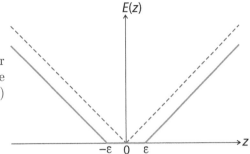

Figure 13.9 The $\epsilon$-insensitive error function $E_\epsilon(z)$ shown by solid line and the mean absolute error (MAE) function by dashed line.

Details of the SVR method can be found in A. J. Smola and Schölkopf (2004) and in various books (Shawe-Taylor and Cristianini, 2004, section 7.3; Bishop, 2006, section 7.1.4; Cherkassky and Mulier, 2007, sections 9.7–9.9; Hsieh, 2009, section 9.1). As $\epsilon$ is an additional hyperparameter, SVR has three hyperparameters versus two in SVM classifiers. If the hyperparameters are to be found by a 3-D grid search in SVR instead of a 2-D grid search in SVM classifiers, the computational burden is much greater for SVR. To avoid the full 3-D grid search, practical guidelines for estimating the hyperparameters are given in Cherkassky and Ma (2004) and Cherkassky and Mulier (2007, section 9.8)

One disadvantage of SVR is that there is only a single response or output variable $\hat{y}$, whereas MLP NN can have multiple response output $\hat{\mathbf{y}}$. If there are multiple response variables, one has to perform SVR on each response separately. If there is relation among the response variables, NN with multiple output nodes can utilize the relation, whereas separate SVR models cannot.

In the environmental sciences, SVR has been introduced to many fields. Examples of earlier applications include forecasting air pollution (Osowski and Garanty, 2007), space-based estimation of moisture transport in the marine atmosphere (X. Xie et al., 2008) and forecasting tropical Pacific sea surface temperatures (Aguilar-Martinez and Hsieh, 2009).

## 13.8  Gaussian Processes  B☺

The kernel trick (Section 13.3) can also be applied to probabilistic models, leading to a new class of kernel methods known as *Gaussian processes* (GP) in the late 1990s. Historically, regression using Gaussian processes has actually been known for a long time in geostatistics as *kriging* (after D. G. Krige, the South African mining engineer), which evolved from the need to perform spatial interpolation of mining exploration data. The new GP method is on a Bayesian framework. Among Bayesian methods (e.g. Bayesian NN), the Bayesian inference generally requires either approximations or extensive numerical computations using Markov chain Monte Carlo methods; however, GP regression has the advantage of assuming Gaussian distributions, so the Bayesian inference at the first level (i.e. obtaining the posterior distribution of the model parameters) is analytically tractable and, therefore, exact. The Gaussian process method is covered in texts such as Rasmussen and Williams (2006), Bishop (2006), and K. P. Murphy (2012). In the environmental sciences, examples of GP regression include estimating chlorophyll concentration in subsurface waters from remote sensing data (Pasolli et al., 2010), forecasting monthly streamflow (A. Y. Sun, D. Wang, et al., 2014) and predicting wind speed for wind power generation (Hoolohan et al., 2018). GP can also be used in classification problems (Bishop, 2006, sections 6.4.5–6.4.6).

In GP regression, the response variable $\hat{y}$ is a linear combination of $M$ fixed basis functions $\phi_l(\mathbf{x})$ (analogous to the radial basis function NN in Section 6.5)

$$\hat{y}(\mathbf{x}) = \sum_{l=1}^{M} w_l \phi_l(\mathbf{x}) = \boldsymbol{\phi}^{\mathrm{T}}(\mathbf{x})\,\mathbf{w}, \tag{13.101}$$

where $\mathbf{x}$ is the input vector and $\mathbf{w}$ the weight vector.

The Gaussian distribution over the $M$-dimensional space of $\mathbf{w}$ vectors is

$$\mathcal{N}(\mathbf{w}|\boldsymbol{\mu}, \mathbf{C}_w) \equiv \frac{1}{(2\pi)^{M/2}|\mathbf{C}_w|^{1/2}}\,\exp\left[-\frac{1}{2}(\mathbf{w}-\boldsymbol{\mu})^{\mathrm{T}}\mathbf{C}_w^{-1}(\mathbf{w}-\boldsymbol{\mu})\right], \quad (13.102)$$

where $\boldsymbol{\mu}$ is the mean, $\mathbf{C}_w$ the $M \times M$ covariance matrix and $|\mathbf{C}_w|$ the determinant of $\mathbf{C}_w$. In GP, we assume the prior distribution of $\mathbf{w}$ to be an isotropic Gaussian with zero mean and covariance $\mathbf{C}_w = \alpha^{-1}\mathbf{I}$ ($\mathbf{I}$ being the identity matrix), that is,

$$p(\mathbf{w}) = \mathcal{N}(\mathbf{w}|\,\mathbf{0}, \alpha^{-1}\mathbf{I}), \tag{13.103}$$

where $\alpha$ is a hyperparameter, with $\alpha^{-1}$ governing the variance of the distribution. This distribution in $\mathbf{w}$ leads to a probability distribution for the functions $\hat{y}(\mathbf{x})$ through (13.101), with

$$\mathrm{E}\big[\hat{y}(\mathbf{x})\big] = \boldsymbol{\phi}^{\mathrm{T}}(\mathbf{x})\,\mathrm{E}[\mathbf{w}] = 0, \tag{13.104}$$

$$\begin{aligned}
\mathrm{cov}\big[\hat{y}(\mathbf{x}_i), \hat{y}(\mathbf{x}_j)\big] &= \mathrm{E}\big[\boldsymbol{\phi}^{\mathrm{T}}(\mathbf{x}_i)\,\mathbf{w}\,\mathbf{w}^{\mathrm{T}}\boldsymbol{\phi}(\mathbf{x}_j)\big] \\
&= \boldsymbol{\phi}^{\mathrm{T}}(\mathbf{x}_i)\,\mathrm{E}\big[\mathbf{w}\,\mathbf{w}^{\mathrm{T}}\big]\,\boldsymbol{\phi}(\mathbf{x}_j) = \boldsymbol{\phi}^{\mathrm{T}}(\mathbf{x}_i)\,\mathbf{C}_w\,\boldsymbol{\phi}(\mathbf{x}_j) \\
&= \alpha^{-1}\boldsymbol{\phi}(\mathbf{x}_i)^{\mathrm{T}}\boldsymbol{\phi}(\mathbf{x}_j) \equiv K_{ij} \equiv K(\mathbf{x}_i, \mathbf{x}_j), \tag{13.105}
\end{aligned}$$

where $K(\mathbf{x}, \mathbf{x}')$ is the kernel function and (given $n$ training input data points $\mathbf{x}_1, \ldots, \mathbf{x}_n$) $K_{ij}$ are the elements of the $n \times n$ covariance or kernel matrix $\mathbf{K}$.

To be a Gaussian process, the function $\hat{y}(\mathbf{x})$ evaluated at $\mathbf{x}_1, \ldots, \mathbf{x}_n$ must have the probability distribution $p(\hat{y}(\mathbf{x}_1), \ldots, \hat{y}(\mathbf{x}_n))$ obeying a joint Gaussian distribution. Thus, the joint distribution over $\hat{y}(\mathbf{x}_1), \ldots, \hat{y}(\mathbf{x}_n)$ is completely specified by second-order statistics (i.e. the mean and covariance). Since the mean is zero, the GP is completely specified by the covariance, that is, the kernel function $K$. There are many choices for kernel functions (see Section 13.3). A popular choice is the (isotropic) Gaussian kernel function

$$K(\mathbf{x}, \mathbf{x}') = a \exp\left(-\frac{\|\mathbf{x} - \mathbf{x}'\|^2}{2\sigma_K^2}\right), \tag{13.106}$$

with two hyperparameters, $a$ and $\sigma_K$.

When GP is used for regression, the target data $y$ are the underlying relation $\hat{y}(\mathbf{x})$ plus Gaussian noise with variance $\sigma^2$, that is, the distribution of $y$ conditional on $\hat{y}$ is a Gaussian distribution $\mathcal{N}$ with mean $\hat{y}$ and variance $\sigma^2$, that is,

$$p(y|\hat{y}) = \mathcal{N}(y|\hat{y}, \sigma^2). \tag{13.107}$$

With $n$ data points $\mathbf{x}_i$ ($i = 1, \ldots, n$), we write $\hat{y}_i = \hat{y}(\mathbf{x}_i)$, $\hat{\mathbf{y}} = (\hat{y}_1, \ldots, \hat{y}_n)^{\mathrm{T}}$ and $\mathbf{y} = (y_1, \ldots, y_n)^{\mathrm{T}}$. Assuming that the noise is independent for each data point, the joint distribution of $\mathbf{y}$ conditional on $\hat{\mathbf{y}}$ is given by an isotropic Gaussian distribution,

$$p(\mathbf{y}|\hat{\mathbf{y}}) = \mathcal{N}(\mathbf{y}|\hat{\mathbf{y}}, \beta^{-1}\mathbf{I}), \tag{13.108}$$

where the hyperparameter $\beta = \sigma^{-2}$.

From (13.101), we note that $\hat{y}$ is a linear combination of the $w_l$ variables, that is, a linear combination of Gaussian distributed variables that again gives a Gaussian distributed variable, so upon invoking (13.104) and (13.105),

$$p(\hat{\mathbf{y}}) = \mathcal{N}(\hat{\mathbf{y}}|\mathbf{0}, \mathbf{K}). \tag{13.109}$$

Since $p(\mathbf{y}|\hat{\mathbf{y}})$ and $p(\hat{\mathbf{y}})$ are both Gaussians, $p(\mathbf{y})$ can be evaluated analytically (Bishop, 2006, Section 6.4.2) to give another Gaussian distribution

$$p(\mathbf{y}) = \int p(\mathbf{y}|\hat{\mathbf{y}}) \, p(\hat{\mathbf{y}}) \, d\hat{\mathbf{y}} = \mathcal{N}(\mathbf{y}|\mathbf{0}, \mathbf{C}), \tag{13.110}$$

where the $n \times n$ covariance matrix $\mathbf{C}$ has elements

$$C_{ij} = K(\mathbf{x}_i, \mathbf{x}_j) + \beta^{-1}\delta_{ij}, \tag{13.111}$$

with $\delta_{ij}$ the Kronecker delta function. These two terms indicate two sources of randomness, with the first term coming from $p(\hat{\mathbf{y}})$ and the second from $p(\mathbf{y}|\hat{\mathbf{y}})$ (i.e. from the noise in the target data).

Suppose we have built a GP regression model from the training set containing input data $\{\mathbf{x}_1, \ldots, \mathbf{x}_n\}$ and target data $\{y_1, \ldots, y_n\}$. Next, we want to make predictions with this model, that is, given a new predictor point $\mathbf{x}_{n+1}$, what can we deduce about the distribution of the target variable $y_{n+1}$?

To answer this, we need to find the conditional distribution $p(y_{n+1}|\mathbf{y})$, (where for notational brevity, the conditional dependence on the predictor data has been omitted). First, we start with the joint distribution $p(\mathbf{y}_{n+1})$, where $\mathbf{y}_{n+1} = (\mathbf{y}^{\mathrm{T}}, y_{n+1})^{\mathrm{T}} = (y_1, \ldots, y_n, y_{n+1})^{\mathrm{T}}$. From (13.110), it follows that

$$p(\mathbf{y}_{n+1}) = \mathcal{N}(\mathbf{y}_{n+1}|\mathbf{0}, \mathbf{C}_{n+1}). \tag{13.112}$$

The $(n+1) \times (n+1)$ covariance matrix $\mathbf{C}_{n+1}$ given by (13.111) can be partitioned into

$$\mathbf{C}_{n+1} = \begin{bmatrix} \mathbf{C}_n & \mathbf{k} \\ \mathbf{k}^{\mathrm{T}} & c \end{bmatrix}, \tag{13.113}$$

where $\mathbf{C}_n$ is the $n \times n$ covariance matrix, the column vector $\mathbf{k}$ has its $i$th element $(i = 1, \ldots, n)$ given by $K(\mathbf{x}_i, \mathbf{x}_{n+1})$ and the scalar $c = K(\mathbf{x}_{n+1}, \mathbf{x}_{n+1}) + \beta^{-1}$.

Since the joint distribution $p(\mathbf{y}_{n+1})$ is a Gaussian, it can be shown that the conditional distribution $p(y_{n+1} | \mathbf{y})$ is also a Gaussian (Bishop, 2006, section 6.4.2), with its mean and variance given by

$$\mu(\mathbf{x}_{n+1}) = \mathbf{k}^{\mathrm{T}} \mathbf{C}_n^{-1} \mathbf{y}, \tag{13.114}$$

$$\sigma^2(\mathbf{x}_{n+1}) = c - \mathbf{k}^{\mathrm{T}} \mathbf{C}_n^{-1} \mathbf{k}. \tag{13.115}$$

The main computational burden in GP regression is the inversion of the $n \times n$ matrix $\mathbf{C}_n$, requiring $O(n^3)$ operations. The method becomes prohibitively costly for large sample size $n$. One can simply use a random subset of the original dataset to train the model, though this is wasteful of the data. A number of other approximations are given in Rasmussen and Williams (2006). An alternative is to solve the regression problem (13.101) with $M$ basis functions directly without using the kernel trick. This would require $O(M^3)$ operations instead of $O(n^3)$ operations. The advantage of the kernel approach in GP regression is that it allows us to use kernel functions representing an infinite number of basis functions.

### 13.8.1   Learning the Hyperparameters   B ☺

Before we can compute the solutions in (13.114) and (13.115), we need to know the values of the hyperparameters, that is, $\beta$, and $a$ and $\sigma_K$ if the Gaussian kernel (13.106) is used. Let $\boldsymbol{\theta}$ denote the vector of hyperparameters. Since parameters like $\beta$, and $a$ and $\sigma_K$ are all positive and we do not want to impose a bound on $\boldsymbol{\theta}$ during optimization, we let

$$\boldsymbol{\theta} = \begin{bmatrix} \ln \beta & \ln a & \ln \sigma_K \end{bmatrix}^{\mathrm{T}}. \tag{13.116}$$

A common way to find the optimal $\boldsymbol{\theta}$ is to maximize the likelihood function $p(\mathbf{y} | \boldsymbol{\theta})$. From (13.110),

$$p(\mathbf{y} | \boldsymbol{\theta}) = \mathcal{N}(\mathbf{y} | \mathbf{0}, \mathbf{C})$$

$$= \frac{1}{(2\pi)^{n/2} |\mathbf{C}|^{1/2}} \exp\left[ -\frac{1}{2} \mathbf{y}^{\mathrm{T}} \mathbf{C}^{-1} \mathbf{y} \right]. \tag{13.117}$$

In practice, we maximize the logarithm of the likelihood function,

$$\ln p(\mathbf{y} | \boldsymbol{\theta}) = -\frac{1}{2} \ln |\mathbf{C}| - \frac{1}{2} \mathbf{y}^{\mathrm{T}} \mathbf{C}^{-1} \mathbf{y} - \frac{n}{2} \ln(2\pi). \tag{13.118}$$

The non-linear optimization problem will in general have multiple minima; thus, it is best to run the optimization procedure multiple times from different initial conditions and choose the run with the lowest minimum. In contrast to MLP NN models, where multiple minima tend to occur from the non-linear optimization with respect to the model weight and offset parameters, multiple minima may occur in GP during non-linear optimization of the hyperparameters, there being usually far fewer hyperparameters than the MLP weight and offset parameters.

The $\boldsymbol{\theta}$ values obtained from the above optimization can be used in (13.114) and (13.115) to give the GP regression solution. A full Bayesian treatment is also possible, where instead of using a single optimal $\boldsymbol{\theta}$ value, integration over all $\boldsymbol{\theta}$ values is performed. The integration cannot be done analytically but can be computed using Markov chain Monte Carlo methods (Neal, 1999).

So far, the GP regression has been limited to a single response or dependent variable. If one has multiple response variables, one can apply the single-response GP regression to model each response separately, but this ignores any correlation between the responses. Generalization to multiple responses while incorporating correlation between the responses is known as *co-kriging* in geostatistics (Cressie, 1993). Boyle and Frean (2005) have generalized GP to multiple responses.

As an example, Fig. 13.10, illustrates GP regression applied to some simple synthetic data. The signal, as shown by the dashed curve, is

$$y_{\text{signal}} = x \sin \pi x, \quad 0 \le x < 2. \tag{13.119}$$

Gaussian noise with $\frac{1}{2}$ the standard deviation of the signal is added to the signal to give the target data $y$. Using the isotropic Gaussian kernel (13.106), GP regression was performed with the number of data points varying from (a) 8 to (b) 16. The mean and variance from (13.114) and (13.115) are used to draw the thick curve and to shade the 95% prediction interval. In regions where data are lacking, the prediction interval widens.

Figure 13.11 illustrates GP regression with some observed data.

## 13.8.2 Other Common Kernels Ⓒ☺

Besides the isotropic Gaussian kernel (13.106), there are a few other common kernels used in GP regression. A Gaussian kernel that allows *automatic relevance determination* (ARD) is

$$K(\mathbf{x}, \mathbf{x}') = a \exp\left(-\frac{1}{2} \sum_{l=1}^{m} \eta_l (x_l - x_l')^2\right), \tag{13.120}$$

where the hyperparameters $\eta_l$ govern the importance or relevance of each particular input $x_l$ (Bishop, 2006, section 6.4.4). Thus, if a particular hyperparameter $\eta_l$ turns out to be close to zero, then the corresponding input variable $x_l$ is not a relevant predictor for $y$.

Another common class of kernels is the Matérn kernels (Rasmussen and Williams, 2006). With $r = \|\mathbf{x} - \mathbf{x}'\|$, and $r_0$ a hyperparameter, the two commonly used Matérn kernels are

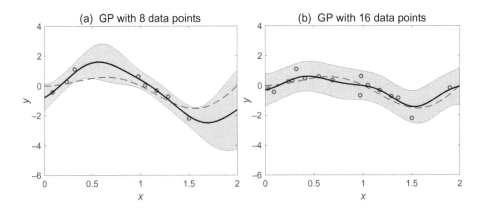

Figure 13.10 GP regression using the isotropic Gaussian kernel, with the number of data points (small circles) being (a) 8 and (b) 16. The thick curve shows the predicted mean, with the two thin curves showing the boundaries of the 95% prediction interval (i.e. ±2 standard deviations). The true underlying signal is indicated by the dashed curve.

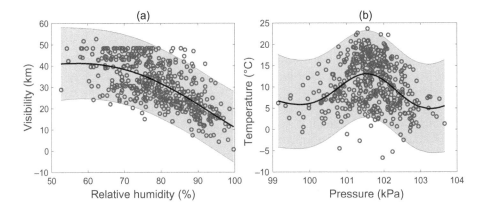

Figure 13.11 GP regression applied to daily weather variables from Vancouver, BC, Canada, with 1/20 the data from 1993 to 2017 used: (a) visibility as a function of relative humidity and (b) temperature as a function of pressure. The mean and variance from (13.114) and (13.115) are used to draw the thick curve and to shade the 95% prediction interval. [Data source: weatherstats.ca based on Environment and Climate Change Canada data.]

$$K_{\nu=3/2}(\mathbf{x}, \mathbf{x}') = a\left(1 + \frac{\sqrt{3}\,r}{r_0}\right)\exp\left(-\frac{\sqrt{3}\,r}{r_0}\right), \tag{13.121}$$

$$K_{\nu=5/2}(\mathbf{x}, \mathbf{x}') = a\left(1 + \frac{\sqrt{5}\,r}{r_0} + \frac{5r^2}{3r_o^2}\right)\exp\left(-\frac{\sqrt{5}\,r}{r_0}\right), \tag{13.122}$$

which are the product of a polynomial kernel with an exponential kernel.

## 13.9 Kernel Principal Component Analysis $\boxed{\text{C}}$ ☺

In Section 10.6 we saw how autoencoder MLP neural network models can be used to perform non-linear principal component analysis. The kernel method allows another approach towards non-linear PCA. Recall that in the kernel method, points in the input space of dimension $m$ are mapped to a 'feature' space by a non-linear mapping function $\phi$. The feature space is of dimension $M$, which is usually much larger than $m$ and can be infinite. Essentially, (linear) PCA is performed in the high-dimensional feature space, which corresponds to finding non-linear modes in the original data space (i.e. the input space).

Note that PCA is meaningful only when applied to centred data, that is, data with zero mean. How to centre data in the feature space is an issue that we will deal with later, but for now assume that data in the feature space have been centred, that is,

$$\frac{1}{n}\sum_{i=1}^{n}\phi(\mathbf{x}_i) = 0, \tag{13.123}$$

where $\mathbf{x}_i$ ($i = 1, \ldots, n$) is the $i$th observation in the input space. The covariance matrix is

$$\mathbf{C} = \frac{1}{n}\sum_{i=1}^{n}\phi(\mathbf{x}_i)\phi^{\mathrm{T}}(\mathbf{x}_i). \tag{13.124}$$

The PCA method involves finding the eigenvalues $\lambda'$ and eigenvectors $\mathbf{v}$ satisfying

$$\mathbf{C}\mathbf{v} = \lambda'\mathbf{v}. \tag{13.125}$$

The eigenvectors can be expressed as a linear combination of the data points in the feature space, that is,

$$\mathbf{v} = \sum_{j=1}^{n}\alpha_j\phi(\mathbf{x}_j). \tag{13.126}$$

Substituting (13.124) and (13.126) into $n$ times (13.125), we get

$$\sum_{i=1}^{n}\sum_{j=1}^{n}\alpha_j\phi(\mathbf{x}_i)K(\mathbf{x}_i,\mathbf{x}_j) = n\lambda'\sum_{j=1}^{n}\alpha_j\phi(\mathbf{x}_j), \tag{13.127}$$

where $K(\mathbf{x}_i,\mathbf{x}_j)$ is an inner-product kernel defined by

$$K(\mathbf{x}_i,\mathbf{x}_j) = \phi^{\mathrm{T}}(\mathbf{x}_i)\phi(\mathbf{x}_j). \tag{13.128}$$

As with other kernel methods, we aim to eliminate $\phi$ (which can be prohibitively costly to compute) by clever use of the kernel trick. Upon left-multiplying both sides of the equation by $\phi^{\mathrm{T}}(\mathbf{x}_k)$, we obtain

$$\sum_{i=1}^{n}\sum_{j=1}^{n}\alpha_j K(\mathbf{x}_k,\mathbf{x}_i)K(\mathbf{x}_i,\mathbf{x}_j) = n\lambda'\sum_{j=1}^{n}\alpha_j K(\mathbf{x}_k,\mathbf{x}_j), \quad k = 1,\ldots,n. \tag{13.129}$$

In matrix notation, this is simply

$$\mathbf{K}^2\boldsymbol{\alpha} = n\lambda'\mathbf{K}\boldsymbol{\alpha}, \tag{13.130}$$

where $\mathbf{K}$ is the $n \times n$ kernel matrix, with $K(\mathbf{x}_i, \mathbf{x}_j)$ as its $(i,j)$th element and $\boldsymbol{\alpha}$ is an $n \times 1$ vector, with $\alpha_j$ as its $j$th element. All the solutions of interest for this eigenvalue problem are also found by solving the simpler eigenvalue problem (Schölkopf and A. J. Smola, 2002)

$$\mathbf{K}\boldsymbol{\alpha} = n\lambda'\boldsymbol{\alpha}. \tag{13.131}$$

This eigenvalue equation can be rewritten in the more familiar form

$$\mathbf{K}\boldsymbol{\alpha} = \lambda\boldsymbol{\alpha}, \tag{13.132}$$

where $\lambda = n\lambda'$. Let $\lambda_1 \geq \lambda_2 \geq \cdots \geq \lambda_n$ denote the solution for $\lambda$ in this eigenvalue problem. Suppose $\lambda_p$ is the smallest non-zero eigenvalue. The eigenvectors $\mathbf{v}^{(1)}, \ldots, \mathbf{v}^{(p)}$ are all normalized to unit length, that is,

$$\mathbf{v}^{(k)\mathrm{T}}\mathbf{v}^{(k)} = 1, \quad k = 1, \ldots, p. \tag{13.133}$$

This normalization of the eigenvectors translates into a normalization condition for $\alpha^{(k)}$ $(k = 1, \ldots, p)$ upon invoking (13.126), (13.128) and (13.132):

$$1 = \sum_{i=1}^{n}\sum_{j=1}^{n}\alpha_i^{(k)}\alpha_j^{(k)}\boldsymbol{\phi}(\mathbf{x}_i)^{\mathrm{T}}\boldsymbol{\phi}(\mathbf{x}_i) = \sum_{i=1}^{n}\sum_{j=1}^{n}\alpha_i^{(k)}\alpha_j^{(k)}K_{ij}$$
$$= \boldsymbol{\alpha}^{(k)\mathrm{T}}\mathbf{K}\boldsymbol{\alpha}^{(k)} = \lambda_k\,\boldsymbol{\alpha}^{(k)\mathrm{T}}\boldsymbol{\alpha}^{(k)}. \tag{13.134}$$

Let us return to the problem of centring data in the feature space. In the kernel evaluations, we actually need to work with

$$\tilde{K}(\mathbf{x}_i, \mathbf{x}_j) = \big(\boldsymbol{\phi}(\mathbf{x}_i) - \overline{\boldsymbol{\phi}}\big)^{\mathrm{T}}\big(\boldsymbol{\phi}(\mathbf{x}_j) - \overline{\boldsymbol{\phi}}\big), \tag{13.135}$$

where the mean

$$\overline{\boldsymbol{\phi}} = \frac{1}{n}\sum_{l=1}^{n}\boldsymbol{\phi}(\mathbf{x}_l). \tag{13.136}$$

It can be shown that (see Exercise 13.3 )

$$\tilde{K}(\mathbf{x}_i, \mathbf{x}_j) = K(\mathbf{x}_i, \mathbf{x}_j) - \frac{1}{n}\sum_{l=1}^{n}K(\mathbf{x}_i, \mathbf{x}_l) - \frac{1}{n}\sum_{l=1}^{n}K(\mathbf{x}_j, \mathbf{x}_l)$$
$$+ \frac{1}{n^2}\sum_{l=1}^{n}\sum_{l'=1}^{n}K(\mathbf{x}_l, \mathbf{x}_{l'}). \tag{13.137}$$

The eigenvalue problem is now solved with $\tilde{\mathbf{K}}$ replacing $\mathbf{K}$ in (13.132).

For any test point $\mathbf{x}$, with a corresponding point $\phi(\mathbf{x})$ in the feature space, we project $\phi(\mathbf{x}) - \overline{\phi}$ onto the eigenvector $\mathbf{v}^{(k)}$ to obtain the $k$th principal component or *feature*:

$$
\begin{aligned}
\mathbf{v}^{(k)\mathrm{T}}\left(\phi(\mathbf{x}) - \overline{\phi}\right) &= \sum_{j=1}^{n} \alpha_j^{(k)} \left(\phi(\mathbf{x}_j) - \overline{\phi}\right)^{\mathrm{T}} \left(\phi(\mathbf{x}) - \overline{\phi}\right) \\
&= \sum_{j=1}^{n} \alpha_j^{(k)} \tilde{K}(\mathbf{x}_j, \mathbf{x}), \quad k = 1, \dots, p.
\end{aligned}
\tag{13.138}
$$

In summary, the basic steps of kernel PCA are as follows: first, having chosen a kernel function, we compute the kernel matrix $\tilde{K}(\mathbf{x}_i, \mathbf{x}_j)$ where $\mathbf{x}_i$ and $\mathbf{x}_j$ are among the data points in the input space. Next, the eigenvalue problem is solved with $\tilde{\mathbf{K}}$ in (13.132). The eigenvector expansion coefficients $\boldsymbol{\alpha}^{(k)}$ are then normalized by (13.134). Finally, the PCs are calculated from (13.138). As the kernel has an adjustable parameter (e.g. the standard deviation $\sigma$ controlling the shape of the Gaussian in the case of the Gaussian kernel), one searches over various values of the kernel parameter for the optimal one, that is, the one explaining the most variance in the feature space, as determined by the magnitude of the leading PC(s) in (13.138).

Suppose $n$, the number of observations, exceeds $m$, the dimension of the input space. With PCA, no more than $m$ PCs can be extracted. In contrast, with kernel PCA, where PCA is performed in the feature space of dimension $M$ (usually much larger than $m$ and $n$), up to $n$ PCs or features can be extracted, that is, the number of features that can be extracted by kernel PCA is determined by the number of observations. The NLPCA method by autoencoder neural networks involves non-linear optimization, hence local minima problems, whereas kernel PCA, which only performs linear PCA in the feature space, does not involve non-linear optimization, hence no local minima problem. On the other hand, with NLPCA, the inverse mapping from the non-linear PCs to the original input space is entirely straightforward, while for kernel PCA, there is no obvious inverse mapping from the feature space to the input space. This is the pre-image problem common to kernel methods, and only an approximate solution can be found, as discussed in Section 13.6.

In the environmental sciences, kernel PCA has been used to help cluster analysis for classification and regionalization of daily sea level pressure over North America and over Europe (Richman and Adrianto, 2010). Kernel PCA has also been applied to satellite data for cloud removal (Greeshma et al., 2016) and for the retrieval of ultraviolet diffuse attenuation coefficients from ocean colour (K. Sun et al., 2020).

# Exercises

13.1

Prove that new kernels result from the operations in (13.40) and (13.41).

**13.2**

From the Taylor series expansion, show that for any kernel $K$, $\exp(K)$ is also a kernel.

**13.3**

Prove (13.137) for the resulting kernel from centring data in (13.135).

# 14

# Decision Trees, Random Forests and Boosting

A *decision tree* is a decision support tool based on a tree-like model of decisions and their consequences. With their flowchart structure, decision trees are commonly used in operations research for decision analysis, for example, finding the best path through the flowchart to reach an objective. They are also widely used in statistics and machine learning on non-linear classification and regression problems (Section 14.1).

While decisions trees are interpretable, they are not stable to noise in the data. An ensemble of decision trees, called a 'forest', has been developed to improve accuracy and robustness. Introducing elements of randomness to the trees turned out to be beneficial, hence the approach is named *random forest* (a.k.a. random decision trees) (Section 14.2).

Since decision trees are simple models, they are considered to be '*weak learners*' relative to 'strong learners', that is, more complex and more accurate models such as neural networks or support vector machines. One-on-one, a weak learner is no match against a strong learner. In the animal world, simple organisms such as ants form large colonies to compete successfully, for example, army ants foraging the forest floor overwhelm much larger adversaries, such as spiders and scorpions. Similarly, by using a large ensemble of weak learners, ML methods such as random forests can compete well against strong learners such as neural networks.

Another common ensemble approach using weak learners is *boosting* (Section 14.3). While random forest constructs all the trees independently, boosting constructs one tree at a time. At each step, boosting tries to a build a weak learner that improves on the previous one. In a random forest, all trees are weighted equally, whereas in boosting, the trees are not weighted equally. Comparing the two methods, boosting tends to outperform the simpler random forest method on most problems, but random forest generally remains competitive (Caruana and Niculescu-Mizil, 2006; Hastie, Tibshirani, et al., 2009, chapters 10 and 15).

The disadvantage of boosting is that it has more hyperparameters to tune and can overfit on very noisy data.

# 14.1 Classification and Regression Trees (CART) [A] ☺

Decision tree methods partition or split the predictor $\mathbf{x}$-space into rectangular regions and fit a simple function $\hat{y} = f(\mathbf{x})$ to the response variable in each region. The most common decision tree method is *classification and regression tree* (CART) (Breiman, Friedman, et al., 1984), which fits $f(\mathbf{x}) = $ constant in each region, so there is a step at the boundary between two regions. The CART method may seem crude but is useful for two main reasons: (i) CART gives an intuitive display of how the response variable broadly depends on the predictors. (ii) When there are many predictors, it provides a computationally inexpensive way to select a smaller number of relevant predictors, which can then be used in more accurate but computationally expensive models such as MLP NN, SVM, and so on (Burrows, 1999), although as noted later in this section, CART is quite unstable on noisy data.

While CART can be used for both non-linear classification and regression problems, we will first focus on the regression problem, with $y$ being the target data. For simplicity, suppose there are only two predictor variables $x_1$ and $x_2$. We look for the partition or split point $x_1^{(1)}$ where the step function $f(\mathbf{x}) = c_1$ for $x_1 < x_1^{(1)}$, and $f(\mathbf{x}) = c_2$ for $x_1 \geq x_1^{(1)}$ gives the best fit to the target data. If the fit is determined by the mean squared error (MSE), then the constants $c_1$ and $c_2$ are simply given by the mean of $y$ over the two separate regions split by $x_1^{(1)}$. A similar search for a split point $x_2^{(1)}$ is performed in the $x_2$ direction. We decide on whether our first split should be at $x_1^{(1)}$ or $x_2^{(1)}$ based on whichever split yields the smaller MSE. In Fig. 14.1(a), the split is made along $x_1^{(1)}$, and there are now two regions. The process is repeated, that is, the next split is made in either the $x_1$ or the $x_2$ dimension. The splitting process is repeated until some stopping criterion is met. In Fig. 14.1(b), the second split is along $x_1 = x_1^{(2)}$, and the third split along $x_2 = x_2^{(3)}$, resulting in the predictor space being split into four regions and the model output described by the four constants over the four regions.

Let us illustrate CART with a dataset containing the daily maximum of the hourly-averaged ozone reading (in ppm) at Los Angeles, with high ozone level indicating poor air quality. The downloaded dataset was prepared by Leo Breiman, and was similar to that used in Breiman and Friedman (1985). The dataset also contained nine predictor variables for the ozone. Among the nine predictors, there are temperature measurements $T_1$ and $T_2$ (in °F) at two stations, visibility (in miles) measured at Los Angeles airport and the pressure gradient (in mm Hg) between the airport and another station. In the nine-dimensional predictor space, CART made the first split at $T_1 = 63.05$°F, the

(a)

(b)

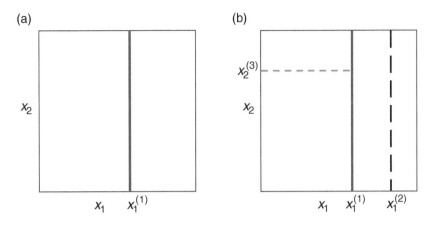

Figure 14.1 Illustrating the partitioning or splitting of the predictor **x**-space by CART. (a) First split at $x_1 = x_1^{(1)}$ yields two regions, each with a constant value for the output $\hat{y}$. (b) Second split at $x_1 = x_1^{(2)}$ (long dash line) is followed by a third split at $x_2 = x_2^{(3)}$ (short dash), yielding four regions of constant $\hat{y}$ values.

second split at $T_1 = 70.97°$F and the third split at $T_2 = 58.50°$F. This sequence of splits can be illustrated by the tree in Fig. 14.2(a). The partitioned regions are also shown schematically in Fig. 14.1(b), with $x_1$ and $x_2$ representing $T_1$ and $T_2$, respectively.

If one continues splitting, the tree grows further. In Fig. 14.2(b), there is now a fourth split at pressure gradient $= -13$ mm Hg, and a fifth split at visibility $= 75$ miles. Thus, CART tells us that the most important predictors, in decreasing order of importance, are $T_1$, $T_2$, pressure gradient and visibility. The tree now has six terminal or leaf nodes, denoting the six regions formed by the splits. Each region is associated with a constant ozone value, the highest being 27.8 ppm (attained by the second leaf from the right). From this leaf node, one can then retrace the path towards the root, which tells us that this highest ozone leaf node was reached after satisfying first $63.05°$F $\leq T_1$, then $70.97°$F $\leq T_1$ and finally visibility $<75$ miles, that is, the highest ozone conditions tend to occur at high temperature and low visibility. The lowest ozone value of 5.61 ppm was attained by the leftmost leaf node, which satisfies $T_1 < 63.05°$F and $T_2 < 58.5°$F, indicating that the lowest ozone values tend to occur when both stations record low temperatures. The CART method also gives the number of data points in each partitioned region, for example 15 points belong to the highest ozone leaf node versus 88 to the lowest ozone node, out of a total of 203 data points. After training is done, when a new value of the predictor $\times$ becomes available, one starts from the top of the tree and descends until reaching a leaf node, with the predicted $\hat{y}$ being the mean value of $y$ for that leaf node.

The intuitive interpretation of the tree structure contributes to CART's popularity in the medical field. Browsing over the American Medical Association's

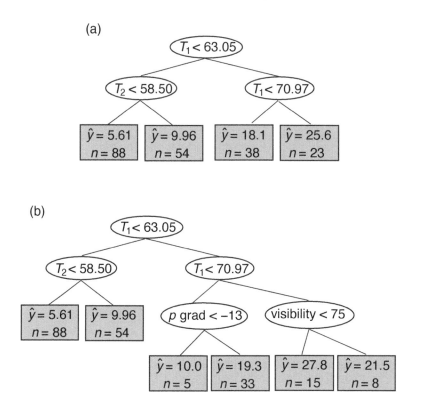

Figure 14.2 Regression tree from CART where the output $\hat{y}$ is the Los Angeles ozone level (in ppm), and there are nine predictor variables. The 'tree' is plotted upside down, with the 'leaves' (i.e. terminal nodes) drawn as rectangular boxes at the bottom and the non-terminal nodes (i.e. internal nodes) as ellipses. (a) The tree after three splits has four leaf nodes. (b) The tree after five splits has six leaf nodes. In each ellipse, a condition is given. Starting from the top ellipse, if the condition is satisfied, proceed along the left branch down to the next node; if not, proceed along the right branch. Continue until a leaf node is reached. In each rectangular box, the constant value of model output $\hat{y}$ (computed from the mean of the target data $y$) in the partitioned region associated with the particular leaf node is given, as well as $n$, the number of data points in that region. Among the nine predictor variables, the most relevant are the temperatures $T_1$ and $T_2$ (in °F) at two stations, $p$ grad (pressure gradient in mm Hg) and visibility (in miles).

*Encyclopedia of Medicine*, one can find many tree-structured flow charts for patient diagnosis. For instance, the questions asked are: Is the body temperature above normal? Is the patient feeling pain? Is the pain in the chest area? The terminal nodes are the likely diseases, for example influenza, heart attack, food poisoning, and so on. Thus, the tree-structured logic in CART is indeed the type of reasoning used by doctors.

How big a tree should one grow? It is not a good idea to stop the growing process after encountering a split that gave little improvement in the MSE, because a further split may lead to a large drop in the MSE. Instead, one grows the tree to a large size, then uses regularization (i.e. weight penalty) to prune the tree down to the optimal size. Suppose $L$ is the number of leaf nodes.[1] A regularized objective function is

$$J(L) = E(L) + \lambda L, \tag{14.1}$$

where $E(L)$ is the MSE for the tree with $L$ leaf nodes, and $\lambda$ is the weight penalty parameter, penalizing trees with excessive leaf nodes. The process to generate a sequence of trees with varying $L$ is as follows: (a) start with the full tree, remove the internal node the demise of which leads to the smallest increase in MSE, and continue until the tree has only one internal node; (b) from this sequence of trees with a wide range of $L$ values, one chooses the tree with the smallest $J(L)$, thus selecting the tree with the optimal size for a given $\lambda$. The best value for $\lambda$ is determined from cross-validation, where multiple runs with different $\lambda$ values are made, and the run with the smallest cross-validated MSE is chosen as the best.

CART can also be used for *classification*. The constant value for $\hat{y}$ over a region is now given by the class $C_k$ ($k = 1, \ldots, K$) to which the largest number of $y$ data belong. During the growth phase of the tree, the term $E$ to be minimized is no longer the MSE. Instead, $E$ denotes *impurity* in classification, which is to be minimized.[2] With $p_{lk}$ denoting the fraction of data in region $l$ belonging to class $k$, $E$ is usually the *entropy* impurity (a.k.a. cross-entropy or *deviance*),

$$E(L) = \sum_{l=1}^{L} \sum_{k=1}^{K} -p_{lk} \log p_{lk}, \tag{14.2}$$

(with log often denoting $\log_2$). Another common choice for $E$ is the *Gini impurity index*,

$$E(L) = \sum_{l=1}^{L} \sum_{k=1}^{K} p_{lk}(1 - p_{lk}). \tag{14.3}$$

During the pruning phase, the *misclassification rate* (i.e. fraction of misclassified data) is usually used for $E$ in (14.1) (Hastie, Tibshirani, et al., 2009, section

---

[1] A tree with $L$ leaf or terminal nodes also has $L - 1$ internal nodes (i.e. non-terminal nodes).

[2] In a classification tree, the ideal is to have purity in the leaf nodes, that is, each leaf node containing a single class. Hence the minimization of impurity.

9.2.3). For binary classification ($K = 2$), and for a single leaf node (with $p$ denoting the fraction of data belonging to class 1), the various choices for $E$ are shown as a function of $p$ in Fig. 14.3 (see Exercise 14.1). When $p$ is close to 0 or 1, one class dominates, so $E$ is small. As the entropy impurity and the Gini impurity are differentiable, they are preferred over the misclassification rate in numerical optimization. Besides CART, other popular classification tree models include C4.5 and the commercial version C5.0 (Quinlan, 1993).

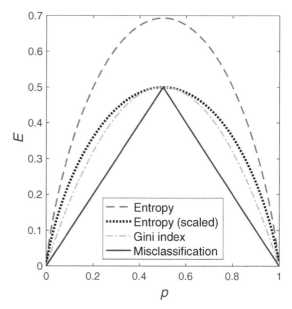

Figure 14.3 The error $E$ for a leaf node in binary classification, where $p$ is the fraction of data belonging to class 1. $E$ is taken to be the entropy impurity (dashed), the Gini impurity (dot-dashed) and the misclassification rate (solid). The dotted line shows the entropy scaled to have the same maximum value as the Gini index to facilitate comparison.

Applications of CART in the environmental sciences include the use of CART classification models to predict lake-effect snow (Burrows, 1991), CART regression to predict ground level ozone concentration (Burrows, Benjamin, et al., 1995), CART regression to predict ultraviolet radiation at the ground in the presence of cloud and other environmental factors (Burrows, 1997), CART to select large-scale atmospheric predictors for surface marine wind prediction by a neuro-fuzzy system (Faucher et al., 1999) and CART regression to predict lightning probability over Canada and northern USA (Burrows, C. Price, et al., 2005). In general, the response variables are not forecast by the numerical weather prediction models. Instead, the numerical weather prediction models provide the predictors for the CART models.

What are the disadvantages of CART? While the tree structure has the advantage of being interpretable, it is unstable to noise perturbations. Small noise in the data is often enough to cause the tree to make a different split. The error in the splitting is then propagated down to all the splits below, resulting in a very different tree structure. To make the method more stable, one grows many trees to form a 'forest', as in Section 14.2 on the method of *random forests*.

As the CART output $\hat{y}$ is constant over a partitioned region in $\mathbf{x}$, this means $\hat{y}$ takes a discrete step when crossing partitioned regions. For classification problems, this is not an issue since the response variable is discrete anyway, but for regression problems, the CART output is not ideal, because CART models the continuous response variable by discrete steps. To extend beyond decision trees such as CART, which is considered a zero-order model since it uses piecewise constant functions to represent the non-linear relation $\hat{y} = f(\mathbf{x})$, *model trees*, for example the *M5 model tree* (Quinlan, 1992), have been developed. M5 is a first-order model, using piecewise linear functions to represent the non-linear relation. The final stage involves applying a smoothing process to smooth out the sharp discontinuities between adjacent linear models at the boundaries of the partitioned regions. The M5 model tree has been used to predict river runoff from rainfall (Solomatine and Dulal, 2003) and for flood forecasting (Solomatine and Xue, 2004).

So far, the splits are of the form $x_j < s$, thus restricting the split boundaries to lie parallel to the axes in $\mathbf{x}$-space. If a decision boundary in the $x_1$–$x_2$ plane is oriented at $45°$ to the $x_1$ axis, then it would take many parallel-axes splits to approximate such a decision boundary. Some versions of CART allow splits of the form $\sum_j a_j x_j < s$, which are not restricted to be parallel to the axes, but the easy interpretability of CART is lost.

## 14.1.1 Relative Importance of Predictors B☺

A CART model can be used to find out the relative importance of predictors (Breiman, Friedman, et al., 1984; Hastie, Tibshirani, et al., 2009, section 10.13.1). For regression, the MSE at an internal node $t$ is given by

$$E_t = \frac{1}{N_t} \sum_{i \in S_t} (y_i - \bar{y}_t)^2, \tag{14.4}$$

where the summation is over all the $N_t$ data points in node $t$, with $S_t$ denoting the set of data points contained in this node and $\bar{y}_t$ the mean of $y_i$ in node $t$. If node $t$ is split into two nodes, we can similarly compute $E_{t\mathrm{L}}$, the MSE for the data points in the left descendent node and $E_{t\mathrm{R}}$, for the right descendent node. The reduction in MSE by the split at node $t$ is then given by

$$\Delta E_t = E_t - E_{t\mathrm{L}} - E_{t\mathrm{R}}. \tag{14.5}$$

The importance of predictor $x_j$ is the sum of $\Delta E_t$ over all the internal nodes $(t = 1, \ldots, L - 1)$ where the best split variable $b_t$ happens to be $x_j$, that is,

$$I_j = \sum_{t=1}^{L-1} \Delta E_t \, \mathbb{1}(b_t = x_j), \tag{14.6}$$

where the function $\mathbb{1}$ is zero but becomes one when $x_j$ matches $b_t$. The relative importance of predictor $x_j$ is given by $I_j / \max(I_j) \times 100\%$, that is, the importance of each predictor is measured relative to the strongest predictor.

For classification, instead of using the MSE for $E_t$, one would use the Gini impurity or the entropy impurity evaluated at the internal node $t$.

One problem with using (14.6) is that an important predictor may be masked by a similar but slightly better predictor. For example, suppose $x_1$ and $x_3$ are similar variables, but $x_1$ is slightly better than $x_3$ as a predictor, so $x_3$ is not used in the splits of the tree structure. Using (14.6) would not detect the importance of $x_3$. If $x_1$ is removed as an input, then $x_3$ will appear prominently in the new tree structure and (14.6) will detect its importance. The next subsection on surrogate splits shows how one can avoid masking of important predictors like $x_3$ when estimating the relative importance of predictors.

## 14.1.2   Surrogate Splits   B☺

Missing data occur commonly, and one has a couple of options: (a) impute a missing value by using the mean or median of that variable or (b) use surrogate splits in CART (Breiman, Friedman, et al., 1984).

The basic idea of *surrogate splits* is as follows: choose a measure of similarity between two splits $s$ and $s'$ at a node $t$ in a tree. Suppose the best split of $t$ is $s$, which uses the variable $x_j$ for splitting. Next, find $s'$, the next best split of $t$, using any variable other than $x_j$, based on the greatest similarity between $s$ and $s'$. Call $s'$ the best surrogate for $s$. Similarly, one can repeat this process to find the second best surrogate, and so on.

For example, take the CART model of Fig. 14.2(b) for predicting the ozone level. Suppose when descending the tree, the split involving the variable $T_2$ is missing the $T_2$ value. One may find that the best surrogate to use is a split involving $T_1$. Suppose for a particular day, values of both $T_2$ and $T_1$ are missing; then, one descends the tree using the second best surrogate.

The measure of similarity between splits $s$ and $s'$ is computed as follows (Breiman, Friedman, et al., 1984, section 5.3): suppose $x_j$ and $x_k$ are the predictor variables used in split $s$ ($x_j < u$) and $s'$ ($x_k < v$). Let $P_L$ be the fraction of data in node $t$ with $x_j < u$ and $P_R$ be the fraction with $x_j \geq u$, where the subscripts L and R label the left and right branches from node $t$. Let $P_{LjLk}$ be the fraction of data at node $t$ with $x_j < u$ and $x_k < v$ (i.e. the descend taking the left branch in both splits), and $P_{RjRk}$ be the fraction of data with $x_j \geq u$ and $x_k \geq v$. If the surrogate split is in excellent agreement with the original split, then $P_{LjLk} + P_{RjRk}$ approaches unity.[3] The similarity measure is defined by

$$\lambda(s, s') = \frac{\min(P_L, P_R) - (1 - P_{LjLk} - P_{RjRk})}{\min(P_L, P_R)}. \tag{14.7}$$

Any split $s'$ with a negative value of $\lambda(s, s')$ is rejected, and the split $s'$ with the largest $\lambda(s, s')$ is taken to be the best surrogate.

Surrogate splits are also helpful in determining the importance of predictor variables (Breiman, Friedman, et al., 1984, section 5.3.4). Using surrogate splits

---

[3] If $x_k$ is negatively oriented with respect to $x_j$, then the split $s'$ has to be reversed to $-x_k < v$ to avoid reversing left and right between $s$ and $s'$.

in (14.6) can reveal the importance of masked variables as mentioned in Section 14.1.1.

For the Los Angeles ozone level problem, in the fully grown tree model, the relative importance histogram (Fig. 14.4(a)) shows the dominant predictor to be $T_1$ (temperature at El Monte), when no surrogate splits are included. When the best surrogate split is included in the calculation of the relative importance, $T_2$ (temperature at Sandburg) is a close second to $T_1$ (Fig. 14.4(b)). When two surrogate splits are included, Z500 (the geopotential height at 500 hPa) is shown as the third most important predictor (Fig. 14.4(c)). When all surrogates are included, the third most important predictor is actually the inversion temperature, with Z500 being the fourth most important (Fig. 14.4(d)). A more reliable way to estimate the relative importance of predictors is to use a random forest model, that is, an ensemble of CART models (Fig. 14.7).

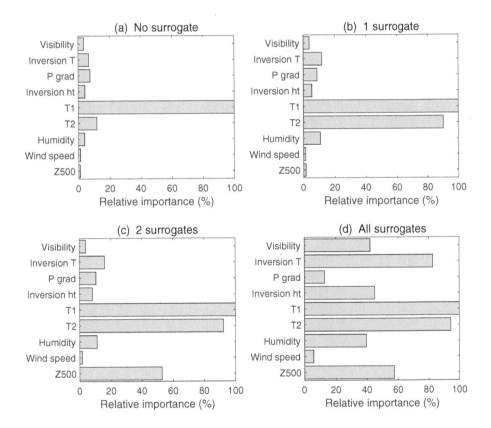

Figure 14.4 Relative importance of the predictors in the CART model for the Los Angeles ozone level, when using (a) no surrogate, (b) one surrogate, (c) two surrogates and (d) all surrogate splits.

## 14.2    Random Forests  $\boxed{\text{A}}$ ☺

The idea of improving the accuracy and stability of a decision tree model by running an ensemble of random tree models was first proposed by Ho (1995), who coined the term 'random forest'. Further improvements of the method were made by several contributors, as mentioned in Hastie, Tibshirani, et al. (2009, p. 602), leading to the widely used version of Breiman (2001a).

*Random forest* (RF) generates an ensemble by applying *bagging* (bootstrap aggregating) (Section 8.7.1) to CART models (Breiman, 2001a). To reduce the variance of a single CART model by an ensemble, simply running the same model on many bootstrap samples may not reduce the variance effectively because the output from the ensemble members can be well correlated. To enhance the diversity or independence of the ensemble members, a subset of the original predictors is randomly selected during the tree construction process. If $N$ is the number of training data points and $m$ the number of predictor variables, the RF algorithm is as follows:

[1] Generate $B$ bootstrap samples (by selecting $N$ data points with replacement from the training dataset and repeating the process $B$ times).

[2] For each bootstrap sample, train a CART model by iterating the following steps for each terminal node (i.e. leaf node) of the tree, till reaching the minimum leaf node size $n_{\min}$ (i.e. the minimum number of data points allowed in a leaf node) or another convergence criterion:

   (a) Randomly select $p$ predictors out of the original $m$ predictors ($1 \leq p < m$).

   (b) Among the $p$ dimensions, choose the best dimension to make the partition, splitting the node into two leaf nodes.

[3] The RF model output $\hat{y}$ is taken to be the mean of the CART output from the $B$ ensemble members in regression problems, or the class $k$ chosen by the largest number of ensemble members in classification problems.

The main tuning hyperparameters in RF are $p$, the number of selected predictors, and $n_{\min}$, the minimum number of data points in a leaf node. For regression problems, the usual default for $p$ is the largest integer $\leq m/3$ and $n_{\min} = 5$. For classification, the default for $p$ is the largest integer $\leq \sqrt{m}$ and $n_{\min} = 1$ (Liaw and Wiener, 2002). In practice, for best performance, these two hyperparameters are tuned using validation. With bootstrapping sampling, about 36.8% of original data are not selected in a bootstrap sample. These unselected data, called out-of-bag (OOB) data, have not been used for model training and can be used as validation data, that is, the error estimated from the OOB data can be used to determine the best values for the two hyperparameters.

For $B$, the number of ensemble members, there is usually no need to spend much effort in tuning this hyperparameter. The reason is that beyond a certain number $B_0$, the OOB error of RF is basically constant for any $B > B_0$, that is, RF does not overfit when using a larger $B$ than is necessary. Typically, one would use $B \sim 100\text{--}200$, or more if computing resources are available. The RF ensemble model can be built sequentially by adding one ensemble member at each step – once the OOB error levels off, one can stop increasing $B$.

An RF model with 200 trees was applied to the same forest cover dataset from Japan in Fig. 12.1(a). As only two predictors were used ($m = 2$), the default value of $p$ being the largest integer $\leq \sqrt{m}$ gave $p = 1$. From repeated runs using $n_{\min}$ in the range $1, 2, \ldots, 12$, the optimal value $n_{\min} = 3$ was chosen based on the smallest misclassification rate (14.1%) on the OOB data. Misclassification rate was 9.1% on the training data and 23.1% on separate test data (not shown). The decision regions found by RF (Fig. 14.5) tend to have boundaries oriented parallel to the $x_1$ and $x_2$ axes, which is not unexpected as each ensemble member in RF is a CART model, with decision boundaries aligned parallel to the $x_1$ and $x_2$ axes.

For the next example, consider the regression problem in Fig. 6.7, where an ensemble of MLP NN models was applied to a synthetic dataset consisting of the signal $y_{\text{signal}} = \cos(x - 1)$ and 121 values of $x$ spaced equally in the interval $[-6, 6]$. Gaussian noise with a quarter of the standard deviation of $y_{\text{signal}}$ was added to give the target data $y$. Here, we use an RF model for regression with 200 trees. The optimal $n_{\min}$ hyperparameter was determined to be seven from the minimum MSE error computed using the OOB data. Compared to the MLP ensemble-averaged solution, the RF solution is less smooth, since each regression tree changes by discrete steps when crossing decision regions. In Fig. 14.6, the greatest difference with MLP occurs near the boundaries of the training domain of $x$. For $x \gtrsim 6$, the RF output is simply the mean of the seven or more $y$ values closest to the boundary of the training domain. The flat extrapolation of RF beyond the training domain, while appropriate for classification problems, is somewhat unrealistic for regression problems. For instance at $x = 6$, the gradient of the response function (dashed curve) is non-zero, but is set to zero by RF. However, this extrapolation by a constant value can be a blessing in disguise, since MLP and many other non-linear regression models can issue extrapolated values with unrealistically large amplitude (Fig. 6.7).

Since RF is an ensemble of CART models, the approach (14.6) used in CART for estimating the *relative importance of predictors* carries over to RF. For RF, the importance $I_j$ for predictor $x_j$ is simply the mean value over all $B$ ensemble members, that is,

$$I_j = \frac{1}{B} \sum_{l=1}^{B} I_j^{(l)}, \tag{14.8}$$

with $I_j^{(l)}$ the importance of predictor $x_j$ in the $l$th ensemble member. There is also the option of including surrogate splits in the CART models.

An alternative method for estimating the relative importance of predictors was given in the original paper by Breiman (2001a, section 10). First, the RF model error is computed using the OOB data. For the $j$th predictor variable, the

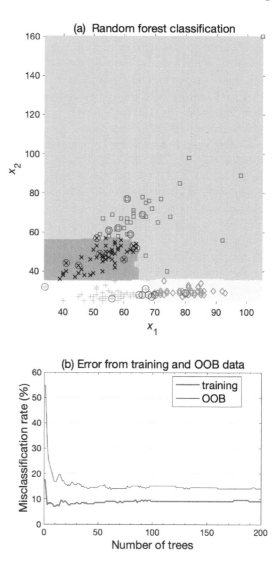

Figure 14.5 Given the training dataset in Fig. 12.1(a) for forest classification, the data are classified by a random forest model with 200 trees in (a), where class $C_1$ is indicated by $+$, $C_2$ by $\Diamond$, $C_3$ by $\times$ and $C_4$ by $\square$. Decision regions for the four classes are shown by different background shading, and misclassified data points are circled. (b) Misclassification rate for the training data and the OOB data as the number of trees increases from 1 to 200.

OOB data are permuted (i.e. the order of the OOB data for the $j$th predictor are randomly shuffled), and the RF model error is again computed. $I_j$, a measure of the importance of the $j$th predictor, is defined to be the difference between the error from the permuted OOB data and the original OOB error. The relative importance of predictor $x_j$ is given by $I_j / \max(I_j) \times 100\%$, that is, the importance of each predictor is measured relative to the strongest predictor. This relative importance measure is not the same as that obtained from (14.8), though the two measures are usually consistent (Hastie, Tibshirani, et al., 2009, section 15.3.2).

For example, consider the Los Angeles ozone level problem from Section 14.1. An RF regression model was run with 200 trees, where each tree was trained

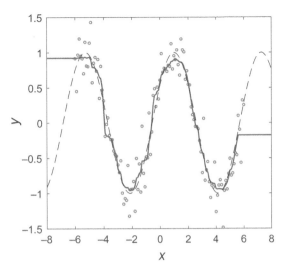

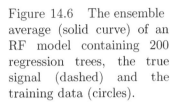

Figure 14.6 The ensemble average (solid curve) of an RF model containing 200 regression trees, the true signal (dashed) and the training data (circles).

on a different bootstrap sample and a random selection of $p = 3$ predictors from the 9 predictors was used to split each terminal node. The optimal $n_{\min}$ hyperparameter was determined to be five from the minimum MSE error computed using the OOB data. The relative importance of predictors, as estimated from the permuted OOB data, is plotted in Fig. 14.7. The difference between not using surrogate splits and using all surrogate splits is much smaller than using a single CART model (Fig. 14.4). As only three out of nine predictors are randomly selected at each terminal node in RF, the odds of having both predictors $T_1$ and $T_2$ selected (with $T_1$ masking $T_2$) is low; thus, even without using surrogate splits, there is less masking of predictors in RF than in a single CART. The relative importance of predictors as determined by the RF model is much more accurate than that determined from a single CART model.

Previously, we have studied *quantile regression* (Section 5.9) and neural network models for (non-linear) quantile regression (Section 6.7). It turns out RF models for regression can be easily adapted to perform quantile regression, as first proposed by Meinshausen (2006).

In RF, a leaf node on a tree contains at least $n_{\min}$ data points. For a leaf node, only the mean value of the data points is used in RF regression, so there is much untapped information remaining. Quantile estimates (Section 2.9) are easily obtained from the conditional (cumulative) distribution function

$$F(y|\mathbf{x}) \equiv P(Y \leq y|\mathbf{x}) \qquad (14.9)$$

for a real-valued random variable $Y$.

In *quantile regression forest* (QRF), the data in a leaf node are utilized to estimate the quantile functions (Meinshausen, 2006). The method is as follows: First train an RF model. For a given value of $\mathbf{x}$, one can go down a trained tree to a leaf node. The data points in the leaf node provide information on $F(y|\mathbf{x})$.

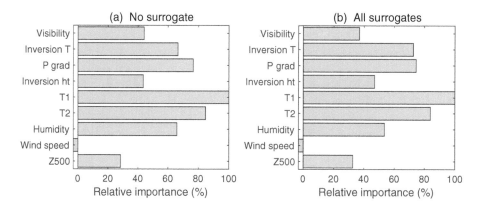

Figure 14.7 Relative importance of the predictors in the random forest regression model for the Los Angeles ozone level, when using (a) no surrogate and (b) all surrogate splits. The most important predictors are the temperatures $T_1$ and $T_2$, the pressure gradient and the inversion temperature.

By combining this information from all the trees in the RF model, one gets an accurate estimate of $F(y \mid \mathbf{x})$, hence the quantile function $q_\alpha(\mathbf{x})$. For instance, a 95% *prediction interval* $I(\mathbf{x})$ can be provided for the RF model output, with

$$I(\mathbf{x}) = [\, q_{0.025}(\mathbf{x}),\, q_{0.975}(\mathbf{x})].  \tag{14.10}$$

QRF works well even when the data are heteroscedastic, for example the variance or skewness of the response variable changes with time.

   In the environmental sciences, RF has been applied to many areas. It has been used to produce probabilistic quantitative precipitation forecast using numerical weather prediction (NWP) model output (Gagne, D. J., II, McGovern, A., and Xue, M., 2014) and to estimate the atmospheric pollutant ($PM_{2.5}$) concentration in in the conterminous United States (X. Hu et al., 2017). Sub-grid scale processes for climate models have been parameterized using RF (J. Yuval and O'Gorman, 2020). In hydrology, RF has been used for flood hazard risk assessment (Z. Wang et al., 2015) and in the ranking of hydrological signatures, that is, indices characterizing hydrologic behaviour (Addor et al., 2018). For solar-based electricity generation, Gagne, McGovern, et al. (2017) used RF and other ensemble decision tree methods to give gridded forecasts of solar irradiance from NWP models, while H. Sun et al. (2016) used RF to assess the potential for estimating daily solar radiation from air pollution index and meteorological data. In marine studies, RF has been used to predict the spatial distribution of sediment pollution in an estuarine system (E. S. Walsh et al., 2017) and to retrieve temperature anomalies in the global subsurface and deeper ocean waters from satellite observations (Hua Su, W. Li, et al., 2018). For agriculture, accurate prediction of sugarcane yield has been developed using RF (Everingham et al., 2016). Invasive plant species have been detected from aerial survey images taken by unmanned aerial systems (UAS) using RF (Michez et al., 2016). RF has been used to forecast the concentration of allergy-inducing pollen at city and regional scales using meteorological and vegetation conditions as input (Lo et al., 2021).

### 14.2.1   Extremely Randomized Trees (Extra Trees)  B☺

Even more randomization can be introduced to the RF approach. *Extremely randomized trees*, commonly referred to by its abbreviated name, *extra trees* (ET) (Geurts et al., 2006), differ from RF in two ways:

[1]  ET uses the entire dataset for training instead of bootstrap samples as in RF.

[2]  Instead of finding an optimal split point for each of the $p$ randomly selected predictors at a terminal node as in RF, ET *randomly* selects a split point for each of the $p$ predictors. ET uses a score to choose the best of the $p$ split points, then performs a split of that node.

Since the split points are not optimized, ET makes more splits than RF, leading to a larger tree structure. Like RF, the main hyperparameters in ET are $p$, the number of selected predictors, and $n_{min}$, the minimum number of data points in a leaf node, which can both be tuned using cross-validation.

While ET and RF usually have similar accuracy, ET is computationally faster because it simply randomly chooses a split point instead of performing an optimization. Another advantage is that for regression problems, the ET output is much smoother than the RF output (Geurts et al., 2006, figure 10). Also, when there are many predictors containing essentially noise, ET may be more accurate than RF.

ET models have been used to forecast atmospheric pollution ($PM_{2.5}$ concentration) over China using satellite and meteorological data as predictors (Wei et al., 2020). Using satellite data to forecast the probability of rapid intensity change of tropical cyclones, Hui Su et al. (2020) combined ET, RF and other ML methods to outperform the National Hurricane Center forecasts.

## 14.3   Boosting  A☺

Along with deep neural networks, boosting is among the most powerful methods to emerge from ML in the last two decades or so. Boosting refers to a family of algorithms that convert weak learners (a.k.a. base learners) to strong learners. Random forests (RF) also convert weak learners to strong learners. However, in contrast to RF where the weak learners (decision trees like CART) are built independently of each other, boosting builds a sequence of weak learners, with each trying to improve on the weakness of its predecessor.

Let the sequence of weak learners be $f_1(\mathbf{x})$, $f_2(\mathbf{x})$, ..., $f_M(\mathbf{x})$, that is, the boosting ensemble has $M$ members, where for input $\mathbf{x}$ the $j$th member gives the output $f_j(\mathbf{x})$ $(j = 1, \ldots, M)$. Assume predictor data $\mathbf{x}_i$ and target data $y_i$ are provided for $i = 1, \ldots, N$. After member $f_{j-1}(\mathbf{x})$ has been built, boosting trains the next member $f_j(\mathbf{x})$ by putting more effort into improving the performance at the data points $\mathbf{x}_i$ where the error between $f_{j-1}(\mathbf{x}_i)$ and the target $y_i$ have been large, that is, boosting tries to improve the fit to data points which have caused trouble for the $(j-1)$th member. How boosting can build a sequence of weak leaners into a strong learner is illustrated in Fig. 14.8 for binary classification, where the final output is the majority vote from $f_1(\mathbf{x})$, $f_2(\mathbf{x})$, ..., $f_M(\mathbf{x})$. For regression problems, the final output is the mean of $f_1(\mathbf{x})$, $f_2(\mathbf{x})$, ..., $f_M(\mathbf{x})$.

Early work on boosting started appearing around 1990 (Hastie, Tibshirani, et al., 2009, pp. 380–384), with the first widely-used boosting model being *AdaBoost*, developed by Freund and Schapire (1997). A comparison of 10 supervised learning algorithms applied to 11 binary classification problems (Caruana and Niculescu-Mizil, 2006) found that boosting was the best method, followed closely by random forest, then support vector machines and MLP NN models. Of course, this ranking order is now dated since this comparison in 2006 was made before the advent of deep NN and more modern gradient boosting methods.

Breiman (1997) pointed out that boosting can be interpreted as the optimization of an objective function. A gradient descent approach to optimization was then used to improve the solution $f_j(\mathbf{x})$ at the $j$th stage from its predecessor $f_{j-1}(\mathbf{x}_i)$ in *gradient boosting machines* (GBM) (a.k.a. *gradient tree boosting*) (Friedman, 2001), now the most common approach to boosting. GBM has been further improved by newer variants, such as XGBoost (T. Chen and Guestrin, 2016), LightGBM (Ke et al., 2017) for large datasets and CatBoost (Prokhorenkova et al., 2018) for datasets containing categorical input variables.

## 14.3.1   Gradient Boosting  B☺

Gradient boosting, used in both regression and classification problems, improved on the earlier boosting models such as AdaBoost by using gradient descent to upgrade the solution at stage $j$ from that at stage $j-1$ (Friedman, 2001). Let $\mathbf{x}_i$ and $y_i$ $(i = 1, \ldots, N)$ be the given data for the predictors and the target response, respectively. The model output $f(\mathbf{x}_i)$ is compared against the target data using an objective or loss function $J$. For instance, for regression problems, $J$ can be the mean squared error,

$$J(\mathbf{f}) \equiv \sum_{i=1}^{N} J\big(y_i, f(\mathbf{x}_i)\big) = \sum_{i=1}^{N} \big(y_i - f(\mathbf{x}_i)\big)^2, \qquad (14.11)$$

where

$$\mathbf{f} = \big(f(\mathbf{x}_1), \ldots, f(\mathbf{x}_N)\big). \qquad (14.12)$$

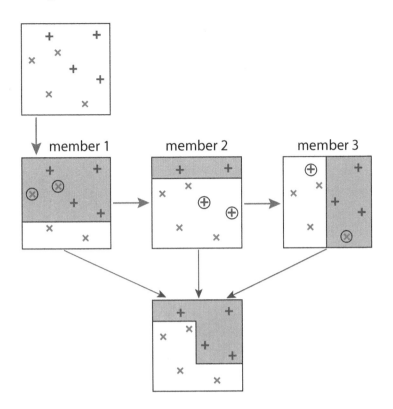

Figure 14.8 How boosting works in classification. Top box contains eight input data points belonging to two classes, '+' and '×'. Next, ensemble member 1 is built using a weak learner, a simple decision tree which splits the domain into two regions, with the shaded region predicting class '+' and the white region predicting class '×'. The shaded region contains two '×' data points, which are circled to indicate their being misclassified. More effort is devoted to improving the two misclassified points when building member 2, so they are classified correctly here, but now two '+' points are misclassified in member 2. With more effort, these two points are correctly classified in member 3, but there are another two misclassified points. Finally, majority voting by the three members gives the more complicated decision regions in the bottom box, where all eight data points are correctly classified.

We can treat **f** as a vector of adjustable parameters for minimizing the objective function, with the optimal **f** being

$$\hat{\mathbf{f}} = \arg\min_{\mathbf{f}} J(\mathbf{f}). \tag{14.13}$$

Instead of the MSE, another objective function, for example entropy or the Gini index, can be used for $J$ in classification problems.

At stage $j$, let $\mathbf{g}_j$ be the gradient of $J(\mathbf{f})$ evaluated at $\mathbf{f} = \mathbf{f}_{j-1}$, that is,

$$g_{ij} = \left[ \frac{\partial J(y_i, f(\mathbf{x}_i))}{\partial f(\mathbf{x}_i)} \right]_{f(\mathbf{x}_i)=f_{j-1}(\mathbf{x}_i)} , \qquad (i = 1, \ldots, N). \qquad (14.14)$$

Functional gradient descent is used to update from stage $j - 1$ to stage $j$, with

$$\mathbf{f}_j = \mathbf{f}_{j-1} - \gamma_j \mathbf{g}_j, \qquad (14.15)$$

where $\gamma_j$ is the step length determined from a line search,

$$\gamma_j = \arg\min_\gamma J(\mathbf{f}_{j-1} - \gamma \mathbf{g}_j). \qquad (14.16)$$

This present solution is actually not useful since it only optimizes $f$ at a set of $N$ training data points. We need a function that can generalize when given new values of $\mathbf{x}$. To achieve generalization ability, we use a weak learner, such as a CART model, to approximate the negative gradient function. The weak learner $\phi(\mathbf{x}_i; \boldsymbol{\theta})$ has adjustable model parameters $\boldsymbol{\theta}$, with optimal values given by

$$\boldsymbol{\theta}_j = \arg\min_{\boldsymbol{\theta}} \sum_{i=1}^{N} \left( -g_{ij} - \phi(\mathbf{x}_i; \boldsymbol{\theta}) \right)^2. \qquad (14.17)$$

To summarize, the gradient boosting algorithm is as follows:

[1] Initialization:
   Fit a constant function $f_0(\mathbf{x}) = c$ to the $y_i$ data, that is,
   $f_0(\mathbf{x}) = \arg\min_c \sum_{i=1}^{N} J(y_i, c)$.

[2] for $j = 1, \ldots, M$:

   (a) Compute the gradient $g_{ij}$ from (14.14);

   (b) Use (14.17) to solve for the optimal parameters $\boldsymbol{\theta}_j$ of the weak learner $\phi$ used to approximate $-g_{ij}$;

   (c) Determine step length $\gamma_j$ from a line search:
   $\gamma_j = \arg\min_\gamma \sum_{i=1}^{N} J\big(y_i, f_{j-1}(\mathbf{x}_i) + \gamma \phi(\mathbf{x}_i; \boldsymbol{\theta}_j)\big)$;

   (d) Update $f_j(\mathbf{x}) = f_{j-1}(\mathbf{x}) + \nu \gamma_j \phi(\mathbf{x}; \boldsymbol{\theta}_j)$,
   with $\nu$ a learning rate hyperparameter.

[3] Finally, model output $f(\mathbf{x}) = f_M(\mathbf{x})$.

Often the weak learner $\phi$ is a CART decision tree model with $L$ terminal nodes or leaf nodes. With the domain for $\mathbf{x}$ partitioned into $L$ regions, Friedman

(2001) proposed that step [2](c) for finding the step length $\gamma_j$ be changed to finding $\gamma_{lj}$ for each partitioned region $l$, $(l = 1, \ldots, L)$, where

$$\gamma_{lj} = \arg\min_{\gamma} \sum_{\mathbf{x}_i \in R_{lj}} J\big(y_i,\, f_{j-1}(\mathbf{x}_i) + \gamma\big), \tag{14.18}$$

with $R_{lj}$ the $l$th partitioned region during the $j$th iteration stage.

The number of leaf nodes is a hyperparameter – $L = 2$ is usually insufficient, while $L > 10$ is likely not needed. Hastie, Tibshirani, et al. (2009, p. 363) recommended small trees with $4 \leq L \leq 8$. Since the final model output $f_M(\mathbf{x})$ is computed from a linear combination of decision trees, the relative importance of predictors can be estimated similarly to the RF model in (14.8).

The learning rate hyperparameter $\nu$ controls the step size taken during the gradient descent optimization process. Although in theory $\nu = 1$, in practice, taking a smaller step size (called *shrinkage*), with $0 < \nu < 1$, gives better results than without shrinkage ($\nu = 1$). Hastie, Tibshirani, et al. (2009, p. 365) recommended small learning rates, for example $\nu < 0.1$. With small step size, one would need to increase $M$, the number of steps (i.e. stages). The strategy recommended by Hastie, Tibshirani, et al. (2009) is to set the learning rate to a small value ($\nu < 0.1$), then choose $M$ by early stopping.

Recall that RF uses bagging, that is, different bootstrap samples to train the various ensemble members. Similarly, gradient boosting can use *subsampling*, that is, randomly choose (without replacement) a fraction $\eta$ of the training data to train the tree model at each stage of the boosting process; a typical value is $\eta = 0.5$. However, for subsampling to work well, it needs to be applied together with shrinkage (i.e. small $\nu$) (Hastie, Tibshirani, et al., 2009, p. 365). Also, similar to bagging, subsampling has data points not used for building the next weak learner, so these unused data can provide an out-of-bag validation error to help choose the best values of the hyperparameters. Compared to the simpler RF method, gradient boosting can have higher accuracy, but it also has more hyperparameters to tune, namely, $L$, $M$, $\nu$ and $\eta$. Hastie, Tibshirani, et al. (2009, p. 367) recommended doing some early explorations to determine suitable values of $L$, $\nu$ and $\eta$, leaving only $M$ as the main hyperparameter to tune accurately.

Since around 2015, *XGBoost* (eXtreme Gradient Boosting), a new variant of the gradient boosting machine developed by T. Chen and Guestrin (2016), has gained popularity over the original version by winning data science competitions, for example those sponsored by Kaggle www.kaggle.com, and from being a fast algorithm. XGBoost has added the second order Hessian matrix information in addition to the gradient information in the optimization process, and weight penalty in the objective function. To further prevent overfitting, it also uses shrinkage and randomly selects a subset of predictors.

Gradient boosting methods have been applied to the environmental sciences: Lawrence et al. (2004) used gradient boosting to classify landcover from remote sensed imagery at three locations in the USA. In hydrology, Rice et al. (2015)

used gradient boosting to examine the influence of watershed characteristics on streamflow trend magnitudes over the continental US Comparing four tree-based models (random forest, M5 model tree, gradient boosting and XGBoost), extreme learning machine (ELM) and support vector machine (SVM) on estimating daily reference evapotranspiration from meteorological data at eight sites over China, J. Fan et al. (2018) found ELM and SVM to have the highest accuracy followed closely by XGBoost and gradient boosting, with XGBoost being computationally the fastest. For air pollution, the $PM_{2.5}$ concentration over China has been estimated from satellite data using XGBoost ( Z.-Y. Chen et al., 2019). From surface remote sensing observations, Hua Su, Xin Yang, et al. (2019) used XGBoost to estimate the subsurface thermohaline structure of the global ocean, that is, the subsurface temperature anomaly and the subsurface salinity anomaly in the upper 2,000 m of the global ocean.

# Exercises

**14.1**

For binary classification, and for a single leaf node (with $p$ denoting the portion of data belonging to class 1), derive the theoretical expressions for the error $E(p)$ as plotted in Fig. 14.3 where $E$ is the (a) entropy impurity, (b) the Gini impurity and (c) the misclassification rate.

**14.2**

In the given wilt dataset from satellite data (with 4,339 data points for training and 500 for testing), there are two classes for the response variable, 'w' for wilted (i.e. diseased) trees and 'n' for non-wilted land cover, and five predictors (B. A. Johnson et al., 2013). There are few training data points for the wilted class (74) relative to the 'n' class (4265). Build (a) a decision tree model and (b) a random forest model for predicting wilt. Determine the relative importance of the predictors. [Optional: (c) Build a gradient boosting model.]

**14.3**

Develop a classification model using random forest or boosting to predict 'Rain Tomorrow' for Sydney Airport in Australia using the available predictors in the given dataset. Use data from 2009–2015 for training and 2016–2017 for testing (see Exercise 12.5).

**14.4**

(a) Build a regression model using random forest to predict the daily precipitation amount (only for days with positive precipitation) for the given dataset (used previously in Exercise 6.6), with sea-level pressure, 700 hPa specific humidity and 500 hPa geopotential height as the three predictors (Cannon, 2011b). Determine the relative importance of the predictors. Compute the RMSE and Pearson correlation between the observed and predicted response variable over the test data. [Note: the given response variable is actually the fourth root of the prediction amounts.] [Optional: (b) Build a gradient boosting model.]

14.5

Bike-sharing systems help to reduce traffic congestion and improve the environment and health (Fanaee-T and Gama, 2014). Weather conditions affect the demand of bikes on a given day. For the given dataset, (a) build a random forest regression model to predict the count of rental bikes, and determine the relative importance of the predictors. Reserve the data from January, April, July and October of 2012 for testing, and train with the rest of the data. [Optional: (b) Build a gradient boosting model.]

# 15

# Deep Learning

The traditional multi-layer perceptron (MLP) neural network (NN) model seen previously in Section 6.3 is commonly run with only one layer of hidden neurons, because using two hidden layers often fails to significantly improve the model accuracy on test data. Adding even more hidden layers usually leads to a drop in model accuracy, so in practice the traditional MLP NN is mostly limited to one or two hidden layers. This was very frustrating to the NN researchers, since biological neural networks are certainly not limited to so few layers. The problem lies in using gradient-based back-propagation for optimization, where the gradients (error signals) can become vanishingly small after propagating through many layers of neurons. This is known as the *vanishing gradient problem*, first explained by Sepp Hochreiter in his diplom thesis of 1991 (written in German). Without a solution in sight, NN research stalled while newer methods – kernel methods (e.g. support vector machines) emerging from the mid-1990s and random forests from 2001 – seriously challenged NN's dominant position in ML.

NN models with more hidden layers than the traditional NN are referred to as *deep neural network* (DNN) or *deep learning* (DL) models. Reviews of DL are given by LeCun, Bengio, et al. (2015), Schmidhuber (2015), and Ponti et al. (2017). The Venn diagram in Fig. 15.1 illustrates the relation between DL, NN, ML, AI, and so on, with  DL $\subset$ NN $\subset$ ML $\subset$ AI.

Interest in deep feedforward networks was revived around 2006 by a group of researchers brought together by the Canadian Institute for Advanced Research (CIFAR) (LeCun, Bengio, et al., 2015). As DNN models were difficult to train with back-propagation, the early DL models first pre-trained the network using unsupervised learning so the network weights were already in a reasonable range before using back-propagation (Hinton, Osindero, et al., 2006). These early deep networks, called *deep belief networks*, have since fallen out of favour but were the first NN models demonstrating that deep networks can be trained successfully (Goodfellow, Bengio, et al., 2016, section 20.3). The deep belief networks have been replaced by newer DNN models that do not require pre-training by unsupervised learning.

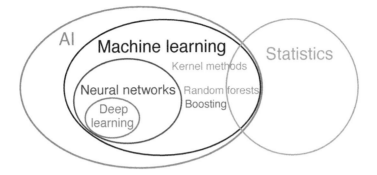

Figure 15.1 A schematic Venn diagram illustrating the relation between AI, statistics, machine learning, neural networks and deep learning, as well as kernel methods (Chapter 13), random forests and boosting (Chapter 14). [Reproduced from Hsieh (2022).]

The ImageNet project is a large database of images for use in computer vision research. During 2010–2017, the ImageNet project ran a series of annual ML contests, the ImageNet Large Scale Visual Recognition Challenge (ILSVRC), where various ML models competed in classifying and detecting objects and scenes from images (Russakovsky et al., 2015).

The first time DNN made a monumental impact was in the ILSVRC 2012 competition, where the AlexNet DNN model with eight layers (Krizhevsky et al., 2012), submitted by the Supervision group from the University of Toronto, won the competition, surpassing previous year's records by wide margins. This breakthrough started the exponential growth in DNN research worldwide. By ILSVRC 2014, the winning DNN model, GooLeNet from Google, had 22 layers. By ILSVRC 2015, the winning DNN model, from the Microsoft Research Lab in Beijing, had 152 layers (He et al., 2016). In theory, an MLP NN with a single hidden layer is enough to represent any continuous function (Section 6.3.2); in practice, a model with more hidden layers (and modern training approach) tends to outperform one with fewer layers, as explained by Montufar et al. (2014) and Goodfellow, Bengio, et al. (2016, section 6.4.1).

A common question on terminology: How many layers of neurons does an NN model need in order to be considered a *deep* NN? Most leading researchers consider DNN models as having $\gtrsim 5$ layers of mapping functions with adjustable weights (i.e. $\gtrsim 4$ or more hidden layers) (LeCun, Bengio, et al., 2015, p. 438; Montufar et al., 2014). Unfortunately, the term 'deep NN' or 'deep learning' has been watered down by some researchers to describe any NN model with more than one hidden layer. The author has been frustrated from attending conference presentations where the 'deep learning' advertised in the abstract turned out to be a traditional two-hidden layer MLP NN!

A number of technical advances made DNN possible:

[1] The vanishing gradient problem found in NN models using sigmoidal-shaped activation functions (e.g. the hyperbolic tangent function) was overcome by new types of activation functions, for example the *rectified linear unit* (ReLU). The ReLU function $f$, defined by

$$f(x) = \max(0, x) \qquad (15.1)$$

(Fig. 6.3(d)), has been found to be much better than the sigmoidal-shaped activation functions in the training of DNN models (Nair and Hinton, 2010; Glorot et al., 2011). The hyperbolic tangent function has gradients in the range $(1, 1)$, and as $x \to \pm\infty$, the gradient of $\tanh x \to 0$. As back-propagation computes gradients using the chain rule, in a network with many layers, this could lead to multiplying many numbers of small magnitude to obtain the gradients of the early layers. In contrast, the ReLU function has gradient $= 1$ for $x > 0$, allowing it to circumvent the vanishing gradient problem in DNN.[1]

There are now several activation functions which have evolved from the original ReLU function, for example the leaky ReLU, the parametric ReLU, the softplus (a smooth version of ReLU), the swish function, and so on, with a comparison of their performance given by Ramachandran et al. (2017). The best performer seems to be the *swish function* (Fig. 6.3(d)), defined by

$$f(x) = x \cdot \text{sigmoid}(\beta x) = \frac{x}{1 + e^{-\beta x}}, \qquad (15.2)$$

where $\beta$ can be simply set to the constant 1 or be treated as a tuneable hyperparameter.

[2] The traditional MLP NN has each neuron in one layer connected to all the neurons in the preceding layer. When working with image data, the MLP uses a huge number of weights – for example mapping an input $100 \times 100$ image to just one neuron in the first hidden layer requires 10,000 weights! Since in nature, neurons are connected only to neighbouring neurons, to have every neuron in one layer of an NN model connected to all the neurons in the preceding layer appears unnatural and very wasteful of computing resources. Inspired by biological studies in visual cortexes, convolutional layers were first introduced in the 'neocognitron' model (Fukushima, 1975; Fukushima, 1980). In *convolutional layers*, a neuron is only connected to a small patch of neurons in the preceding layer, thereby resulting in a drastic reduction of weights compared to the traditional fully-connected

---

[1] A caveat is that ReLU is designed for deep NN – in a shallow NN, the non-smooth ReLU activation function is likely to underperform the smooth sigmoidal-shaped activation functions. In NN research, ReLU dates back to at least Fukushima (1975) but has not been able to compete against sigmoidal-shaped activation functions in shallow NN models (Goodfellow, Bengio, et al., 2016, p. 219).

layers, making deep NN models practical.[2] The modern *convolutional neural network* (CNN), first developed by LeCun, B. Boser, et al. (1989), is widely used in DL.

[3] The regularization technique *dropout* was introduced by Hinton, Srivastava, et al. (2012) to alleviate overfitting in DNN (Section 8.8).

[4] *Batch normalization* (Ioffe and Szegedy, 2015), which normalizes (i.e. standardizes) the inputs at each layer during training, was found to stabilize and accelerate DNN training.

With all the impressive breakthroughs in DL, the reader may start to believe DNN models to be invincible against all other ML models. That belief is not correct, since the best ML model is very problem dependent. For instance, a farmer would find a rifle perfect for defending against wolves or bears, but useless against locusts or caterpillars. If a dataset has only a small number of data points, it is possible for even linear regression to be the best model.

There are two main types of datasets, structured and unstructured. *Structured* datasets have a tabular format, for example like an Excel spreadsheet, with the variables listed in columns. In contrast, *unstructured* datasets include images, videos, audio, text, and so on. For unstructured datasets, DL has indeed been dominant. For structured datasets, however, 'shallow' ML methods, mainly gradient boosting methods such as XGBoost, have often beaten DL methods in competitions, for example those organized by Kaggle (www.kaggle .com).

How can 'shallow' boosting beat deep NN in structured data? Typically, in a structured dataset, the predictors are quite inhomogeneous (e.g. pressure, temperature, humidity, etc.), whereas in an unstructured dataset, the predictors are more homogeneous (e.g. temperature at various pixels in an image or at various grid points in a numerical model). Boosting is based on decision trees (Chapter 14), where the effects of the predictors are treated independently of each other, as the path through a decision tree is controlled by questions like: is '$x_1 > a$?', '$x_2 > b$?', and so on. In contrast, NN models combine the predictors by a linear combination ($\sum_i w_i x_i$) before passing it though an activation function. With inhomogeneous predictors, for example temperature and pressure, treating the two separately as in decision trees makes more sense than trying to add the two together by a linear combination.

In environmental science, both structured and unstructured datasets are common. Examples of unstructured datasets include satellite images and 2-D or 3-D gridded output from weather/climate models. In the last few years, there has been rapid growth in the applications of DNN models to satellite images and weather/climate model output. Review/perspective papers covering DL in

---

[2] Prior to the advent of the convolutional layer, one usually had to rely on the linear technique of principal component analysis to compress the large number of input variables from images or 2-D grids, then supply a modest number of leading principal components as input to the NN to greatly reduce the number of model weights.

environmental science are found in C. Shen (2018), Reichstein et al. (2019), A. Y. Sun and Scanlon (2019) and in the edited book by Camps-Valls, Tuia, et al. (2021).

# 15.1   Transfer Learning   [A]☺

ML models, especially DNN models, generally require a large number of data points to train, which may not be available. Fortunately, in the *transfer learning* approach, ML models trained on a dataset with large sample size can transfer their learning to a different problem hampered by a relatively small sample size. For example, consider the task of developing facial recognition software for a rare species of monkeys with fewer than 100 images available. If one trains a DNN model with the small dataset, it would perform poorly. However, training the same model using a million or more human facial images would give a good model for human facial recognition. In the transfer learning approach, one trains this human model further with the small number of monkey images, and gets a much better model for monkey recognition. Basically, after the model has learned to recognize human faces well, its weights only need a moderate amount of tuning using the monkey images for it to adapt to the monkey problem. Transfer learning started as early as 1976 (Bozinovski, 2020), with a 1997 special issue of Machine Learning devoted to the subject (Pratt and Thrun, 1997).

Since DNN models usually require a large amount of data and computer time to train, many pre-trained DNN models developed by leading DL groups have been made available to the public. Such models have already been trained on large datasets and have learned to recognize patterns/objects. Researchers would then further train the pre-trained DNN with data particular to their areas of interest. Thus, transfer learning is able to greatly reduce the amount of training time and data for DNN models, thereby popularizing the use of DNN models. For instance, to detect *beluga whales* from passive acoustics monitoring, Zhong et al. (2020) used transfer learning to train CNN models to successfully classify input acoustic signals as belonging to the belugas or not. As the images used for transfer learning were from objects (e.g. dogs and cars) unrelated to acoustic signals, the model was able to learn how to extract features from these images, before it was further trained with relevant acoustic data.

In environmental science, the application of ML methods to climate problems has been impeded by relatively short observational records, as the long timescales in climate problems preclude a large number of independent data points in time. However, there are many climate models (a.k.a. general circulation models or dynamical models) that can simulate climate on a computer for centuries, millennia or longer. These large climate model datasets offer an exciting opportunity for transfer learning, that is, ML models can be first trained using the climate model data and then further trained using the short observational data records.

In interannual variability, prediction of the El Niño-Southern Oscillation (ENSO) by a CNN model was made by Ham et al. (2019). Transfer learning was used to overcome the limitation of short observational records, which had hampered earlier NN models (Tangang et al., 1997; B. Tang, Hsieh, et al., 2000; Wu, Hsieh, and B. Tang, 2006). For each target month of the year, the CMIP5 climate model data for 2,961 months (from the Coupled Model Intercomparison Project (CMIP), World Climate Research Programme) were first used to train the CNN, followed by additional training using 103 months of reanalysis data (between 1871 and 1973). The prediction model was tested using the reanalysis data from 1984 to 2017, where the 10-year gap between the final year in the training period and the earliest year in the test period was to prevent oceanic memory from the training period influencing the test period. During the test period, the all-season correlation skill of the Niño 3.4 index of the CNN model was much higher than those of current state-of-the-art dynamical forecast systems. In other words, by using a CNN to learn the ENSO behaviour from the dynamical models, then transfer the learning to observed data (i.e. reanalysis data), this transfer learning approach yielded a CNN with better accuracy in ENSO prediction than the dynamical models.

## 15.2 Convolutional Neural Network ☒☺

Biological studies in visual cortexes pioneered by David H. Hubel and Torsten N. Wiesel in 1959 (which led to the 1981 Nobel Prize in Physiology or Medicine) inspired the development of the convolutional layer, first introduced as the neocognitron (Fukushima, 1975, 1980), later evolving into the convolutional neural network (CNN) in DL (LeCun, B. Boser, et al., 1989). In convolutional layers, a neuron is only connected to a small patch of neurons in the preceding layer, greatly reducing the number of weights compared to the traditional fully-connected layers in MLP NN models. For recognizing objects in images, Urban et al. (2017) empirically found that shallow NN and shallow CNN were unable to attain the accuracy of deep CNN models.

### 15.2.1 Convolution Operation ☒☺

We have encountered the convolution operation previously – (11.25) in Section 11.2 on windows under time series analysis. In CNN, *discrete* convolution is performed by applying a *filter* (a.k.a. *kernel* or *mask*) $\mathbf{F}$ to an array $\mathbf{A}$, yielding an output array $\mathbf{B}$, with elements

$$B(i,j) = (A * F)(i,j) = \sum_k \sum_l A(k,l)\, F(i-k, j-l), \qquad (15.3)$$

with the summation indices starting from $k = 0$ and $l = 0$. Equivalently, as the convolution operation is commutative $(\mathbf{A} * \mathbf{F} = \mathbf{F} * \mathbf{A})$,

$$B(i,j) = (F * A)(i,j) = \sum_k \sum_l F(k,l)\, A(i-k, j-l). \qquad (15.4)$$

It is sometimes more convenient to express the convolution operation using a *correlation* operator,

$$B(i, j) = (F \star A)(i, j) = \sum_{k} \sum_{l} F'(k, l) A(i + k, j + l),$$   (15.5)

where $\mathbf{F}'$ is the *flipped filter* or flipped kernel, obtained by reversing the order of the rows and the columns of $\mathbf{F}$. For example, in Fig. 15.2, $\mathbf{F}$ and $\mathbf{F}'$ are given by

$$\mathbf{F} = \begin{pmatrix} 0 & 0 & 1 \\ -2 & 0 & 2 \\ 0 & -1 & 0 \end{pmatrix}, \qquad \mathbf{F}' = \begin{pmatrix} 0 & -1 & 0 \\ 2 & 0 & -2 \\ 1 & 0 & 0 \end{pmatrix}$$   (15.6)

(note $\mathbf{F}'$ is not the same as the transpose of $\mathbf{F}$).

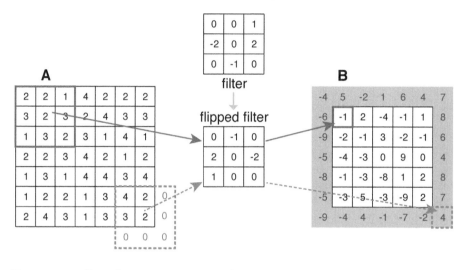

Figure 15.2 Convolution operation as illustrated by applying a $3 \times 3$ filter to a $7 \times 7$ matrix. The filter $\mathbf{F}$ is first flipped (both rows and columns) to give the flipped filter $\mathbf{F}'$. Nine elements of the input matrix $\mathbf{A}$ on the left are multiplied by the nine elements of the flipped filter, then summed and placed in the output matrix $\mathbf{B}$ on the right. If $\mathbf{B}$ is to remain the same size as $\mathbf{A}$, zeros must be padded outside the boundary of $\mathbf{A}$ to produce the extra elements shaded in grey (e.g. the dashed pixel).

In Fig. 15.2, $\mathbf{A}$ is a $7 \times 7$ matrix and $\mathbf{F}$ and $\mathbf{F}'$ are $3 \times 3$ matrices. Using (15.5), the convolution is simply the dot product between the nine points of $\mathbf{F}'$ and a patch of nine data points in $\mathbf{A}$. For the top left $3 \times 3$ patch in $\mathbf{A}$, the convolution gives

$$0 \times 2 - 1 \times 2 + 0 \times 1 +$$
$$2 \times 3 + 0 \times 2 - 2 \times 3 +$$
$$1 \times 1 + 0 \times 3 + 0 \times 2 = -2 + 6 - 6 + 1 = -1.$$   (15.7)

To generate the rest of **B**, the nine-point patch in **A** is shifted both horizontally and vertically, one grid or pixel at a time (i.e. using a *stride* of 1), and a dot product is taken with **F′**. **B** has a smaller size than **A** – here the $7 \times 7$ matrix **A** is reduced to a $5 \times 5$ matrix **B**. In general, for an **A** matrix of width $m$ using a filter of width $f$, the output **B** has width $m - f + 1$. For example, If we use a $5 \times 5$ filter in Fig. 15.2, **B** would be reduced to a $3 \times 3$ matrix.

The stride $s$ is a hyperparameter. If $s = 2$ is used, the patch over **A** to be operated on by **F′** is shifted by two pixels or grids, as in Fig. 15.3. The width of the matrix **B** becomes $(m-f)/s+1$, so increasing $s$ can greatly reduce the output matrix size. In many DNN applications, the input consists of high-dimensional images, but the output is of much lower dimension, for example 'storm' or 'no storm', so it is useful to reduce the dimension as the signal propagates through the layers of a DNN.

At the boundary of **A**, one has the option of padding with zeros beyond the boundary, so matrix **B** can remain the same size as **A** (Fig. 15.2). If $z$ number of zeros are padded beyond a boundary, the width of the matrix **B** is

$$\frac{m - f + 2z}{s} + 1. \tag{15.8}$$

After **B** is generated by the convolution operation, there are two more steps: the same bias or offset is added to each element of **B**, and the adjusted value is passed through an activation function, usually the rectified linear unit (ReLU), onto the next layer of neurons. The next layer can be another convolutional layer or a different type of layer. Details of the computations in CNN models are provided by Dumoulin and Visin (2018).

A 2-D image of length $m_L$ and width $m_W$ often uses three colour *channels* (RGB), so the input is a 3-D array ($m_L \times m_W \times 3$). This third dimension is called the *depth* of the layer – not to be confused with the *depth of the model*, which is the number of layers (with adjustable weights) in the CNN. The filter is then also 3-D, $f \times f \times 3$. In general, the input **A** can have $m_D$ channels, that is, **A** is of dimension $m_L \times m_W \times m_D$. The output **B** can have $m_D'$ channels. The convolution operation in (15.5) is generalized to

$$B(i, j, d') = \sum_d \sum_k \sum_l F'(k, l, d, d') \, A(i + k, j + l, d), \tag{15.9}$$

where $d$ and $d'$ are the indices for the channels in **A** and **B**, respectively. The number of weights used to get **B** is

$$n_{\mathbf{B}} = f \times f \times m_D \times m_D'. \tag{15.10}$$

The weights in a filter are found by optimizing an objective function. Usually, the same filter is used for all the elements of **B**, which allows a very economic sharing of weights compared to the traditional MLP NN. It is possible to not

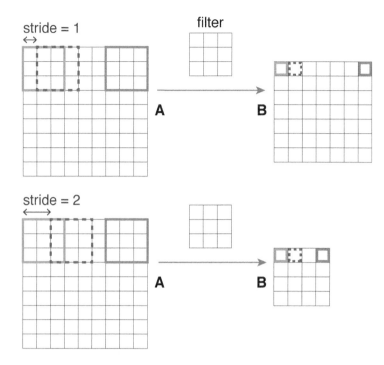

Figure 15.3 Effect of moving the (flipped) filter over the input matrix $\mathbf{A}$ at different stride $s$. $\mathbf{A}$ is of size $9 \times 9$ and the filter $3 \times 3$, while the output $\mathbf{B}$ is of size $7 \times 7$ (for $s = 1$) and $4 \times 4$ ($s = 2$).

share the same filter for all elements of $\mathbf{B}$ (Goodfellow, Bengio, et al., 2016, section 9.5), which leads to more weights but also more flexibility. Even with the same filter weights used for all elements of $\mathbf{B}$, the model is flexible as one can use more sets of filters, that is, increasing the channel depth $m'_{\mathrm{D}}$.

## 15.2.2 Pooling  B☺

Pooling is another operation used to *downsample*, that is, reduce the dimension of an array $\mathbf{A}$, by extracting a single number from a patch of size $f \times f$ in $\mathbf{A}$. The most common pooling used is maximum pooling or *max pooling*, where the filter simply extracts the maximum value from the patch. *Average pooling*, which returns the mean value of the patch, was used historically. While it has largely been replaced by max pooling, Weyn, Durran, and Caruana (2020) switched to average pooling from their earlier use of max pooling (Weyn, Durran, and Caruana, 2019).

Max pooling is most commonly performed over a $2 \times 2$ patch (Fig. 15.4), which reduces the width of $\mathbf{A}$ by half. When counting the number of layers in a CNN model, max pooling is not counted as a layer since it has no adjustable weights.

Max pooling can also be done with a stride $s \neq f$, with the width of the output **B** being $(m - f)/s + 1$. For example, if the input width is $m = 55$ pixels, $f = 3$ and $s = 2$, the output width is 27 (Krizhevsky et al., 2012, figure 2). With $s < f$, a pooling patch overlaps with its neighbouring patch, so this is called overlapping pooling, in contrast to the non-overlapping pooling in Fig. 15.4 with $s = f = 2$.

Figure 15.4 Example of a $4 \times 4$ input array undergoing a max pooling operation, where the output is the maximum value from each $2 \times 2$ patch. Here, the filter width $f = 2$ and the stride $s = 2$.

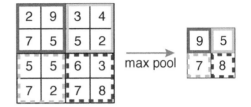

### 15.2.3 AlexNet B☺

The first DNN to have a major impact was AlexNet, which won the ImageNet Large Scale Visual Recognition Challenge (ILSVRC) 2012 competition (Krizhevsky et al., 2012). The eight-layer AlexNet CNN model is composed of five convolutional layers followed by three *dense layers* (i.e. the traditional fully connected layers) (Fig. 15.5). Input images have $227 \times 227$ pixels (not $224 \times 224$ as given in Krizhevsky et al. (2012)), with three colour channels (RGB). Applying convolution using an $11 \times 11$ filter with a stride of $s = 4$ led to an array of $55 \times 55$ nodes in layer 1. Using 96 such filters led to 96 channels, with the number of weights in layer 1 being $11 \times 11 \times 3 \times 96 = 34,848$. There are an additional 96 weights from the bias or offset (one for each channel), so the total number of weights in layer $1 = 34,944$. Increasing the number of channels in a CNN layer increases the number of weights at that layer, thereby increasing the model complexity. There are about 62 million weights/parameters in AlexNet, and the model used roughly 1.2 million training images, 50,000 validation images, and 150,000 testing images. Since the number of weights greatly exceeded the number of available images, data augmentation (Section 15.2.5) and dropout were used to avoid overfitting.

How does CNN actually detect an object from an image?

> The learned features in the first layer of representation typically represent the presence or absence of edges at particular orientations and locations in the image. The second layer typically detects motifs by spotting particular arrangements of edges, regardless of small variations in the edge positions. The third layer may assemble motifs into larger combinations that correspond to parts of familiar objects, and subsequent layers would detect objects as combinations of these parts. The key aspect of deep learning is that these layers of features are not designed by human engineers: they are learned from data

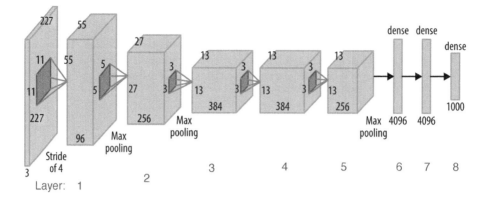

Figure 15.5 The eight-layer AlexNet CNN model has input images of 227 × 227 pixels and three colour channels (RGB), with the number of channels indicated by the depth (i.e. thickness) of the input block. Convoluting the input by a 11 × 11 filter with a stride of $s = 4$ led to an array of 55 × 55, and the use of 96 such filters led to 96 channels in layer 1. After max pooling, convoluting by a 5 × 5 filter (with $z = 2$ for zero padding) led to the layer 2 array of size 27 × 27, with 256 channels. All max pooling operations are done using a 3 × 3 filter with $s = 2$. Again max pooling, then convoluting by a 3 × 3 filter (with $z = 1$) led to the layer 3 array of 13 × 13, with 384 channels. Convoluting with a 3 × 3 filter (with $z = 1$) led to the layer 4 array of 13 × 13, with 384 channels, and convoluting again led to layer 5 of 13 × 13, with 256 channels. After max pooling, the resulting 6 × 6 array with 256 channels is reshaped into a 1-D array of 4,096 nodes and is fully or densely connected to layer 6 with 4,096 nodes, which is fully connected to layer 7 with 4,096 nodes. Layer 7 is fully connected to layer 8 with 1,000 nodes, with the softmax activation function indicating which one of the 1,000 classes (e.g. cats, dogs, cars, etc.) the output belongs to. [Adapted from *TensorFlow for Deep Learning*, by Bharath Ramsundar and Reza Bosagh Zadeh. Copyright © 2018 Reza Zadeh, Bharath Ramsundar. Published by O'Reilly Media, Inc. Used with permission.]

using a general-purpose learning procedure. (LeCun, Bengio, et al., 2015)

The reason why CNN can use a filter with only a modest number of weights on an entire image is because the local statistics and signals in images are usually invariant to location. If a motif or pattern can appear in one part of the image, it could appear elsewhere, hence one can use the same filter weights for every node in the array, which allows CNN to greatly reduce the number of weights compared to the traditional fully connected NN.

Progress in CNN skills in image recognition continued: ZFNet, a modified AlexNet, won the ILSVRC 2013 competition (Zeiler and Fergus, 2014). For the first layer, instead of a $11 \times 11$ filter and a stride of 4 in AlexNet, ZFNet used a $7 \times 7$ filter and a stride of 2. Using a CNN with 22 layers, GoogLeNet (from Google, with 'LeNet' imbedded in the name to pay tribute to the original LeNet CNN from LeCun, B. Boser, et al. (1989)) won the ILSVRC 2014 competition (Szegedy et al., 2015). The first runner-up at this competition was VGGNet, from the Visual Geometry Group (VGG) at the University of Oxford (Simonyan and Zisserman, 2015). VGGNet has variants using 16 or 19 layers.

### 15.2.4 Residual Neural Network (ResNet) [B]☺

For the winners in the ILSVRC competitions, as the number of layers in the CNN increased from 8 in 2012 to 22 in 2014, eventually the problem of vanishing gradients again put an upper limit on the number of layers. The *residual neural network* (ResNet) from Microsoft introduced *skip connections* (i.e. shortcuts to jump over some layers) to overcome this problem. Using models with up to 152 layers, ResNet came first in the ILSVRC 2015 competition (He et al., 2016). The ResNet architecture even allowed a model with 1,202 layers to perform fairly well.

Figure 15.6 illustrates a skip connection, where the output from layer $l$ is directly connected to the input of layer $l + 3$. Layer $l + 3$ thus receives the output from layer $l + 2$ and from layer $l$. Bigger steps are also allowed in skip connections, mapping layer $l$ directly to layer $l + k$ ($k \geq 3$). The use of skip connections is further extended in *densely connected convolutional networks* (DenseNet) (G. Huang, Z. Liu, et al., 2017).

Figure 15.6 Unlike the standard CNN architecture on the left, the ResNet architecture allows skip connections (dot-dashed line) to connect the output from layer $l$ directly to layer $l + 3$. The basic building block for residual learning (dashed) is repeated to give a deep network structure.

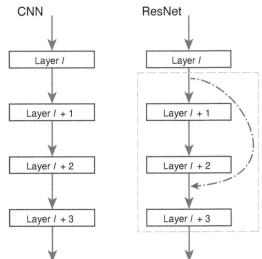

The term 'residual' comes from using the NN layers to model the residual $\mathbf{F}(\mathbf{x}) = \mathbf{H}(\mathbf{x}) - \mathbf{x}$, where $\mathbf{H}(\mathbf{x})$ is the desired underlying mapping. In other words, one tries to remove the linear signal from $\mathbf{H}(\mathbf{x})$, so the NN layers can focus on modelling the non-linear residual signal, as it is inefficient for non-linear NN layers to model a linear signal. The skip connection allows the linear signal to bypass[3] the non-linear NN layers in the residual learning block (Fig. 15.6).

## 15.2.5   Data Augmentation   [C]☺

In many applications, the number of available training images is quite limited, too few to properly train the large number of weights in deep CNN models. We have seen the use of transfer learning in Section 15.1 to alleviate this problem. Furthermore, one can generate more training images by various *data augmentation* techniques. For instance, translation, rotation, flipping, geometric scaling, colour modification, cropping, noise injection and random erasing can be used to generate more training images (Shorten and Khoshgoftaar, 2019). One should use only transformations that give physically realistic images. For instance, shifting a tropical cyclone to polar latitudes or flipping the cyclone (i.e. reversing the anticlockwise circulation to clockwise circulation in the Northern Hemisphere) would supply irrelevant training data for the CNN model.

A more advanced approach to data augmentation is to use generative adversarial networks (GAN) (Section 15.5) to produce highly realistic fake images. The GAN model has two adversarial components: the generative network creates fake images, while the adversarial network tries to determine if the image is fake or genuine. The competition between the adversarial components leads to increasingly more realistic fake images.

## 15.2.6   Applications in Environment Science   [B]☺

In the last few years, there have been many papers applying CNN models to environmental science. To detect extreme weather patterns from 2-D images of atmospheric variables, CNN models composed of two convolutional layers followed by two fully connected layers were used by Y. Liu, Racah, et al. (2016), with three separate models built to estimate the posterior probability of three types of extreme events – tropical cyclones, atmospheric rivers and weather fronts. A deep CNN model with six convolutional layers followed by two fully connected layers was used by Lagerquist, Ryan and McGovern, Amy and Gagne, II, D. J. (2019) to detect synoptic-scale weather fronts, where the three output nodes with softmax activation functions gave the probabilities for cold front, warm front and no front. For next-hour tornado prediction, with radar images as input, Lagerquist et al. (2020) used a deep CNN model to output the probability of a tornado.

For automated detection of weather fronts, Baird and Kunkel (2019) used a four-layer CNN model, where three convolutional layers followed by one fully connected layer using the softmax action function (with all layers having the same spatial dimension) gave the posterior probability for five frontal categories (cold, warm, stationary, occluded and none) at each pixel (unlike the previous

---

[3] Linear bypass NN architecture has also been used in shallow NN models (Section 6.4.2).

two studies where the output had no spatial resolution). Assigning a class label to each pixel of an image is called *semantic segmentation*. Most CNN models for semantic segmentation use an encoder–decoder structure (Section 15.3).

To predict the El Niño-Southern Oscillation (ENSO) phenomenon, Ham et al. (2019) used CNN to predict the Niño 3.4 sea surface temperature (SST) anomaly index. The input were the 2-D anomaly maps of SST and upper ocean heat content at time $t$, $t-1$ and $t-2$ months (i.e. six input channels of $24 \times 72$ pixel images). The four layers were three convolutional layers followed by a fully connected layer, with the output 1-D vector containing Niño 3.4 at time $t$, $t+1$, ..., $t+22$ months. The model was pre-trained using long records from CMIP5 climate model output as described in Section 15.1 on transfer learning.

Training deep CNN with past weather data to predict future weather has been explored (Weyn, Durran, and Caruana, 2019; Weyn, Durran, and Caruana, 2020). For short- to medium-range forecasting, CNN has been able to significantly outperform persistence, climatology and a coarse-resolution dynamical numerical weather prediction (NWP) model, though underperforming a high-resolution state-of-the-art operational NWP system. The model has also been extended to sub-seasonal forecasting (Weyn, Durran, Caruana, and Cresswell-Clay, 2021).

For researching the occurrence and movement of the *humpback whale*, A. N. Allen et al. (2021) used a ResNet CNN to automatically detect humpback whale song in a passive acoustic dataset with 13 monitoring sites in the North Pacific over a 14-year period.

# 15.3 Encoder–Decoder Network $\boxed{\text{B}}$ ☺

The encoder–decoder NN architecture has served as the foundation in autoencoders (and its applications to non-linear principal component analysis) (Sections 10.5 and 10.6) under unsupervised learning. The arrival of the CNN architecture has greatly advanced encoder–decoder networks and autoencoders. The encoder–decoder structure is also used in *supervised* learning, for example in semantic segmentation, where a class label is assigned to each pixel of an output image, or in regression problems with real values assigned to each output pixel. The encoder structure funnels the high-resolution input image(s) to a bottleneck, thereby extracting the important information but ignoring the unimportant details in the input. The decoder structure then expands from the low-resolution bottleneck back to higher-resolution image(s). Proceeding from lower resolution to higher resolution is called *upsampling*, opposite to downsampling operations such as max pooling.

Nearest neighbour upsampling is the simplest, where an input pixel value is reproduced over multiple pixels in the higher-resolution output (Fig. 15.7). To avoid the blocky appearance, bilinear interpolation upsampling is an alternative.

Figure 15.7 Example of a $2 \times 2$ input array using nearest neighbour upsampling to generate values for a $4 \times 4$ grid.

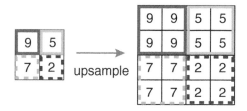

### 15.3.1   U-net   B ☺

However, too much detail can be lost by going through the bottleneck in the encoder–decoder, so the high-resolution output image may lack details. Following the earlier work of Long et al. (2015), Ronneberger et al. (2015) introduced the U-Net, which has *skip connections* linking the layers in the encoder to the corresponding layers in the decoder, transmitting details from the earlier layers to avoid the loss of details in the output (Fig. 15.8). Using only around 30 training images plus data augmentation, the U-net was able to perform accurate semantic segmentation, where a class label was assigned to each pixel of an output image, allowing individual cells to be identified and outlined in microscopic images. The networks in Long et al. (2015) and Ronneberger et al. (2015) are also referred to as *fully convolutional networks* (FCN), because they do not contain dense layers like the earlier CNN (e.g. AlexNet in Fig. 15.5).

The U-net has recently been applied to environmental science problems: using a U-net type CNN model, Chapman et al. (2019) improved the forecasts for atmospheric rivers in the NCEP Global Forecast System's integrated vapour transport forecast field. From microwave satellite data, Y. Choi and Kim (2020) used the U-net for rain-type classification. The input $40 \times 40$ images have 14 channels, consisting of nine microwave channels at several frequency bands, four channels of polarized corrected temperature and the surface type (ocean, land, coastal and in-land water), while the output $40 \times 40$ images have four channels for the classes (no-rain, stratiform, convective and others).

A system for Precipitation Estimation from Remotely Sensed Information using Artificial Neural Networks (PERSIANN) has undergone many upgrades since 1997 (K. L. Hsu, Gao, et al., 1997; Ashouri et al., 2015; Nguyen et al., 2018; Q. Sun et al., 2018). The latest version used a U-net model with satellite infrared images and latitude–longitude as input and precipitation as output (Sadeghi et al., 2020).

*Cloud cover nowcasting* (i.e. very short-term forecasting) is useful for photovoltaic energy production forecast. Using images from visible and infrared channels of the Meteosat Second Generation (MSG) geostationary satellite as input, Berthomier et al. (2020) compared U-net and several methods in classifying 16 cloud types and found that U-net outperformed other NN methods – CNN, long-short term memory (LSTM) and recurrent neural network (RNN) – as well as AROME, a non-hydrostatic very-high-resolution model from Météo-France, and EXIM, an image extrapolation tool from EUMETSAT (European Organisation for the Exploitation of Meteorological Satellites).

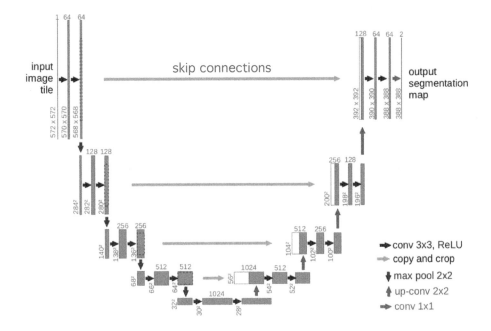

Figure 15.8 In this U-net model, the input image of 572 × 572 pixels passes through two convolutional layers (each with 64 channels) using 3 × 3 filters and the ReLU activation function, then undergoes 2 × 2 max pooling to a 284 × 284 layer with 64 channels. Descending the left arm of the 'U' structure is the encoding part, where the spatial resolution decreases but the number of channel increases, reaching a 30 × 30 layer with 1,024 channels at the bottom of the 'U'. From this bottleneck, ascending the right arm of the 'U' structure is the decoding part, where up-convolution (i.e. upsampling followed by 2 × 2 convolution) increases the spatial resolution but decreases the number of channels. Skip connections linking layers in the encoder to the corresponding layers in the decoder are used to avoid the loss of details in the output. At the final layer, a 1 × 1 convolution is used to map from the 64 channels to the desired number of classes (two classes in this example). In total, the network has 23 convolutional layers. [Adapted from Ronneberger et al. (2015, figure 1).]

As *missing data* are common in climate datasets, *imputation* of missing values is vital for producing a useful dataset. In Section 9.1.12, principal component analysis (PCA) has been shown to be effective in imputing missing values. Neural networks have led to autoencoders (Section 10.5), which can be viewed as non-linear PCA (Section 10.6). With its encoder–decoder architecture, U-net is a further extension of autoencoders and it has been used to impute missing data. With *partial convolutional* layers introduced to work with missing data, G. Liu et al. (2018) used a modified U-net to impute the missing values from images. Using this approach, Kadow et al. (2020) took sea surface temperature

(SST) data from two sources, the Twentieth-Century Reanalysis (20CR) from NOAA and the historical experiment of the Coupled Model Intercomparison Project Phase 5 (CMIP5), to train two separate models, both of which outperformed PCA and kriging methods in imputing missing SST values. The models were used to fill missing global SST data in the HadCRUT4 dataset during 1850–2018.

A somewhat related problem is *super-resolution*, that is, from a low-resolution input image, generate a higher-resolution output image. Often there are far more low-resolution images available than high-resolution images, so ML is used to convert the low-resolution ones to higher resolution. Geiss and Hardin (2020) applied a modified U-net to generate super-resolution radar images. High-resolution images were first converted to low-resolution images by taking either $8 \times 8$ or $4 \times 4$ pixel averages. The modified U-net was then trained using low-resolution images as input and high-resolution images as target. The trained model can then be used to convert other low-resolution images to higher resolution. A different approach to super-resolution is through conditional generative adversarial networks (CGAN) (Section 15.5).

More recent models with encoder–decoder architecture include UNet++, UNet 3+, attention U-net and DeepLabv3. In UNet++ (Zhou et al., 2018), the encoder and decoder sub-networks are connected through a series of nested, dense skip connections. UNet++ has been used to downscale precipitation from numerical weather prediction (Sha et al., 2020b) (see Section 16.11 on downscaling). UNet 3+ (H. Huang et al., 2020) redesigned the skip connections in UNet++, resulting in fewer weights. Attention U-net (Oktay et al., 2018) added attention gates to the skip connections for focusing the model attention on more important regions of the image. DeepLabv3 (L.-C. Chen et al., 2018) has been used for mapping retrogressive thaw slumps (a type of thawing permafrost) in the Tibetan Plateau from CubeSat satellite images (L. Huang et al., 2020).

## 15.4   Time Series   Ⓒ☺

CNN was developed mainly for image recognition and semantic segmentation. For time series, a number of different deep NN methods have also been introduced, which we examine in this section. In Section 15.4.1, the recurrent NN is much enhanced by the long short-term memory (LSTM) network. Convolutional layers can be used in LSTM, leading to the convolutional LSTM network. Although LSTM models have been widely used, CNN adapted for temporal modelling, that is, the temporal convolutional network (TCN), has recently been found to be more accurate and computationally faster than LSTM (Section 15.4.2).

### 15.4.1   Long Short-Term Memory (LSTM) Network   Ⓒ☺

For time series modelling and prediction, retaining earlier events/conditions in memory is important. The *recurrent neural network* (RNN) adds memory to the

MLP NN model. In Fig. 15.9(a), the one-hidden layer MLP NN with input $\mathbf{x}(t)$ and output $\mathbf{y}(t)$ is modified to an RNN by adding a feedback loop, allowing the hidden neurons $\mathbf{h}(t)$ to be submitted as additional input at the next time step, that is, at time $t+1$, the input are $\mathbf{x}(t+1)$ and $\mathbf{h}(t)$. This arrangement provides $\mathbf{h}(t+1)$ with information from its previous state $\mathbf{h}(t)$. The loop structure can be unfolded in time to give the equivalent network structure in Fig. 15.9(b). A major limitation of RNN is that memory fades away quite quickly with this method of memory storage.

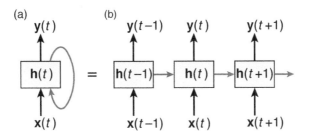

Figure 15.9  (a) Recurrent neural network (RNN) with one hidden layer $\mathbf{h}(t)$. The network can be unfolded to give the equivalent structure in (b), where $\mathbf{h}(t+1)$ receives $\mathbf{x}(t+1)$ and $\mathbf{h}(t)$ as input and the output is $\mathbf{y}(t+1)$.


The LSTM was proposed by Hochreiter and Schmidhuber (1997) to supplement short-term memory storage in RNN with additional long-term memory storage. A crucial 'forget gate' was added to LSTM by Gers et al. (2000), because it is important to be able to 'forget' or remove the long-term memory when it becomes obsolete, with the forget gate deciding when it is time to forget. An LSTM block is compared with a corresponding RNN hidden layer block in Fig. 15.10.

RNN and LSTM models are inherently deep in time, since the hidden state (in $\mathbf{h}$ and $\mathbf{C}$) is a function of the hidden state from all previous time. However, one can still increase the depth of the RNN and LSTM models for deep learning. Analogous to adding more hidden layers in an MLP NN to increase the depth of the network, one can also add more hidden layers to the RNN (Fig. 15.9(a)) each with a feedback loop. Similarly, one can add more LSTM blocks to increase the depth of the LTSM model for deep learning, leading to deep LSTM (a.k.a. stacked LSTM) models.

LSTM has led to major improvements in speech recognition and other time series applications. There are many variants of the LSTM model – fortunately, a comparison of eight variants by Greff et al. (2017) found that they all had similar performance.

LSTM has been applied to environmental sciences (Shi et al., 2015; Kratzert, Klotz, Brenner, et al., 2018). In hydrology, the relation between streamflow and precipitation is very complex, because water from precipitation is affected by the type of soil and vegetation in the watershed, or locked into snow or ice, before

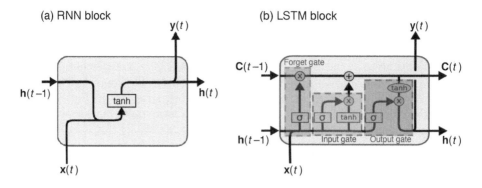

Figure 15.10 (a) A hidden layer block from the the RNN in Fig. 15.9(b) has the combined input from $\mathbf{x}(t)$ and $\mathbf{h}(t-1)$ passing through an activation function (e.g. tanh) to give $\mathbf{h}(t)$, which is further passed on to the output node $\mathbf{y}(t)$ and as input to the next block at time $t+1$. (b) A corresponding block from LSTM where the main difference is the addition of the memory cell vector $\mathbf{C}$, which stores the long-term memory to supplement the short-term memory stored in $\mathbf{h}$. There are three components inside the block: the forget gate, which decides whether to clear the long-term memory from $\mathbf{C}$; the input gate, which updates $\mathbf{C}$; and the output gate, which outputs $\mathbf{h}(t)$. Logistic sigmoidal functions $\sigma$ provide smooth switching between on and off, while $\otimes$ and $\oplus$ denote element-wise multiplication and addition. The equations for the LSTM block are given in Ki et al. (2020) and Kratzert, Klotz, Brenner, et al. (2018).

it eventually feeds into the streamflow. Since 3-D information on the subsurface physical properties of a watershed is difficult to obtain, models based on physical processes are rarely used in in operational streamflow forecasting; instead, NN models have been commonly used (Maier and Dandy, 2000; Dawson and Wilby, 2001; Abrahart, Anctil, et al., 2012).

Kratzert, Klotz, Brenner, et al. (2018) applied LSTM for streamflow prediction in the USA using meteorological data as input (precipitation, shortwave downward radiation, maximum and minimum temperature and humidity). The LSTM has two LSTM layers (each with 20 neurons) followed by a fully connected layer to the single output for the streamflow. Kratzert, Klotz, Shalev, et al. (2019) showed that a single LSTM trained on data from 531 basins (using meteorological time series and static catchment attributes as input) outperformed LSTMs individually trained for each basin.

### Convolutional LSTM network

With image data, the fully connected LSTM model has a very large number of weights/parameters. Shi et al. (2015) introduced the *convolutional LSTM* (ConvLSTM) model, which replaced fully connected layers with convolutional layers in the LSTM to greatly reduce the number of weights. The model was applied to precipitation *nowcasting* (i.e. very short-term forecasting at 0–6 hour lead time) using radar data, with input radar images of $100 \times 100$ pixels. The model has two LSTM layers, each with 64 hidden neurons and a $3 \times 3$ kernel.

To predict the El Niño-Southern Oscillation (ENSO) phenomenon, using transferred learning from global climate model as in Ham et al. (2019), Mahesh et al. (2019) encoded the spatial information of each global surface temperature grid using a six-layer CNN, then fed the encoded information into an LSTM to learn the temporal behaviour of ENSO.

## 15.4.2 Temporal Convolutional Network Ⓒ☺

More recently, CNN has been adapted for time series problems. Lea et al. (2017) proposed two very different *temporal convolutional network* (TCN) models. The first, called the *encoder–decoder TCN* (ED-TCN), uses temporal convolutions, pooling and upsampling in an encoder–decoder structure. The second, called the *dilated TCN*, uses a deep stack of dilated temporal convolutions. Bai et al. (2018) modified the dilated TCN and compared it against several benchmark models, including the LSTM. Among 11 tests using different datasets, the dilated TCN outperformed the LSTM in 10 of the 11 tests (Bai et al., 2018, table 1).

From the 2-D convolution in (15.4), the 1-D *dilated causal convolution* used for temporal convolution is written as

$$B(i) = (F *_d A)(i) = \sum_{k=0}^{f-1} F(k)A(i - kd), \qquad (15.11)$$

where $\mathbf{A}$ is the input, $\mathbf{B}$ the output, $f$ is the number of elements in the 1-D filter $\mathbf{F}$, $d$ (a positive integer) is the dilation factor and the summation index $k$ is restricted to be non-negative, so no elements of $\mathbf{A}$ from future time are used in the summation to satisfy the causality requirement.

The basic dilated TCN from Bai et al. (2018) (Fig. 15.11) shows input time series $x_0, x_1, \ldots, x_t$ passing through a dilated causal convolutional layer with $d = 1$, followed by additional hidden layers with $d = 2$ and $d = 4$. There are two ways to increase the receptive field of the TCN: (i) increasing the dilation factor $d$ and (ii) increasing the filter size $f$. Adding more hidden layers with $d = 8, 16, 32, \ldots$ increases the long-term memory of the model. More details of the model structure (including optional skip connections) are given in Bai et al. (2018). The time window size, that is, number of time points in $x_0, x_1, \ldots, x_t$, is often chosen based on the autocorrelation or partial autocorrelation (Duan et al., 2020).

TCN has computational advantage over RNN models such as LSTM: in an RNN, the predictions for later time steps must wait for their predecessors to complete, whereas convolutions can be performed in parallel as the same filter is used in each layer. Hence, during both training and evaluation, a long input sequence can be processed as a whole in TCN, instead of sequentially in RNN, leading to much faster computation in TCN.

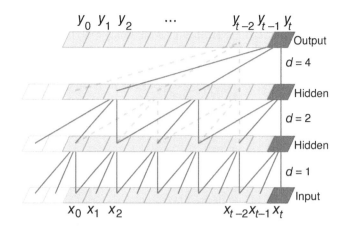

Figure 15.11 Dilated causal convolution layers with dilation factors $d = 1, 2, 4$ and filter size $f = 3$. [Adapted from Bai et al. (2018, figure 1)].

For multiple input time series, R. Wan et al. (2019) introduced a multivariate TCN model. Each input time series goes through a separate dilated TCN sub-model. The output from the sub-models are concatenated into a vector, which then passes through a fully connected layer to give the final output. The method was applied to predict the Beijing hourly $PM_{2.5}$ concentration using $PM_{2.5}$ and meteorological time series (dew point, temperature, atmospheric pressure, combined wind direction, cumulated wind speed, hours of snow and hours of rain) as input. In a second application, the New England hourly electric demand was predicted, using electric demand and dry-bulb temperature time series as input. The multivariate TCN model outperformed a number of benchmarks, including the LSTM and ConvLSTM models.

In hydrology, Duan et al. (2020) applied a dilated TCN to daily streamflow prediction in California. Having three dilated causal convolutional layers (with $d = 1, 6, 12$ and $f = 7$) and a time window size of five months, the model used time series of precipitation, temperature and solar radiation as input and streamflow as output. K. Lin et al. (2020) developed a TCN encoder–decoder model with time series of precipitation, evaporation and runoff (i.e. streamflow) as input and runoff as output.

## 15.5 Generative Adversarial Network  ⒞☺

In many games, from chess to tennis, having two individuals playing against each other enhances the skill level of both. In some situations, the game is asymmetrical – one individual attacks while the other defends, for example a goal scorer practices against a goalkeeper. The generative adversarial network (GAN), introduced by Goodfellow, Pouget-Abadie, et al. (2014), has two sub-models, the generator and the discriminator, playing as adversaries. The goal is to produce realistic fake data. Given a random input vector, the generator outputs a set of fake data. The discriminator receives either real data or fake data as input and classifies them as either 'real' or 'fake' (Fig. 15.12). If a fake dataset is correctly classified, the generator's model weights are updated, whereas if the fake is mistaken to be real, the discriminator's model weights are updated. The skill level of both players improves until at the end the discriminator can only identify fake data from the generator about 50% of the time. After training is done, the discriminator is discarded while the generator is retained to produce new fake datasets.

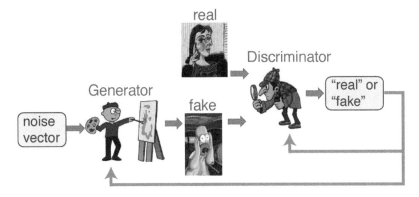

Figure 15.12 Generative adversarial network (GAN) with the generator creating a fake image (e.g. a fake Picasso painting) from random noise input, and the discriminator classifying images as either real or fake. Whether the discriminator classifies a fake image rightly or wrongly leads, respectively, to further training for the generator or for the discriminator. [Reproduced from Hsieh (2022).]

The generator $G$ inputs a noise vector $\mathbf{z}$ and outputs a set of fake data $\mathbf{y}'$, that is, $\mathbf{y}' = G(\mathbf{z})$. The discriminator $D$ can input either (i) a set of real data $\mathbf{y}$ or (ii) a set of fake data $\mathbf{y}'$. $D$ outputs the probability that the data is real. The GAN objective can be written as

$$J(G, D) = E_{\mathbf{y}}[\log D(\mathbf{y})] + E_{\mathbf{z}}[\log(1 - D(G(\mathbf{z})))], \qquad (15.12)$$

and the optimization problem is to solve for

$$\min_{G} \max_{D} J(G, D). \qquad (15.13)$$

On the right hand side of (15.12), the first expectation term is large if $D$ is good in identifying real data, while the second expectation is also large if $D$ is good in detecting fake data from $G$. Thus, the optimization searches for model weights of $D$ to maximize $J$, but for the weights of $G$ to minimize $J$, as $G$ is an adversary of $D$.

GAN commonly works with image data, so both the generator and the discriminator are usually deep CNN models. GAN can suffer from 'mode collapse', that is, the final generator fails to produce some variety/class of data, for example the generator produces only 'no storm' images. The stability of GAN has been improved in more recent models, for example Radford et al. (2016) and the Wasserstein GAN model (Arjovsky et al., 2017; Gulrajani et al., 2017). With GAN, one has little control over the fake images generated, since the input for the generator is a random vector. To have more control over the type of fake image wanted, additional non-random input is needed.

*Conditional* GAN (CGAN), introduced by Mirza and Osindero (2014), supplies additional input to both the generator and the discriminator. For instance, the additional input can be a class label, such as 'day' or 'night', so the generator produces fake images for daytime or nighttime, as specified by the user. There can be more than two classes, for example the four seasons, so the generator can produce fake images for a given season.

Besides class labels, CGAN can also be conditioned on other types of input, for example an image **x**, whereby CGAN can be used for image-to-image translation tasks (Isola et al., 2017). Examples include translating a line drawing to a photo image (Fig. 15.13), a map to a satellite map and vice versa.

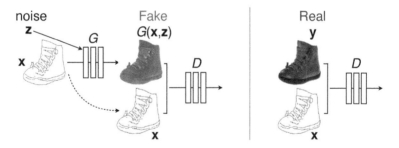

Figure 15.13 Conditional generative adversarial network (CGAN) where the generator $G$ receives an image **x** and a random noise vector **z** as input. The discriminator $D$ receives **x** plus either a fake image from $G$ or a real image **y** as input. Here, a line drawing is converted to a photo image; similarly, a photo image can be converted to a line drawing. [Adapted from Isola et al. (2017, figure 2).]

The CGAN objective function is modified from that of GAN by having the extra input **x**, that is,

$$J(G, D) = \mathrm{E}_{\mathbf{x},\mathbf{y}}[\log D(\mathbf{x}, \mathbf{y})] + \mathrm{E}_{\mathbf{x},\mathbf{z}}[\log(1 - D(\mathbf{x}, G(\mathbf{x}, \mathbf{z})))]. \qquad (15.14)$$

CGAN has been applied to environmental science problems: in atmospheric remote sensing, Leinonen et al. (2019) used CGAN to generate cloud structures in a 2-D vertical plane in the satellite's along-track direction. The real

2-D images **y** are radar scenes from the CloudSat satellite, and the input **x** are 1-D along-track satellite observations from the Moderate-Resolution Imaging Spectroradiometer (MODIS), providing data for cloud top pressure, cloud optical depth, effective radius, cloud water path and a binary cloud mask indicating whether cloud was detected by MODIS in a column. Their model was usually able to generate reasonable guesses of the cloud structure and infer complex structures (e.g. multilayer clouds) from MODIS data. As an alternative to U-net (Section 15.3.1) in super-resolution applications, CGAN has been used by Pashaei et al. (2020) to convert low-resolution unmanned aircraft system (UAS) images to high-resolution images. The potential of using CGAN to estimate *sub-grid processes* in dynamical models has been investigated by Gagne, Christensen, et al. (2020).

For global satellite precipitation estimation at high spatiotemporal resolution, infrared (IR) data suffer from low detection accuracy while passive microwave data, which estimate precipitation accurately, suffer from low spatiotemporal coverage, as the microwave data are discontinuous in space and time. In their PrecipGAN model, C. Wang et al. (2021) used a type of conditional GAN to merge incomplete microwave precipitation estimates with the complete IR precipitation estimates.

# Exercises

### 15.1

Determine the dimensions and number of weights of each layer in AlexNet (Fig. 15.5) to arrive at the total number of weights for the model.

### 15.2

Try a variety of deep NN models using the MeteoNet data provided by METEO FRANCE (Larvor et al., 2020), available on the Kaggle website (www.kaggle.com/katerpillar/meteonet) [with the full dataset on the MeteoNet website (https://meteonet.umr-cnrm.fr/)]. Exercises include rainfall and cloud cover nowcasting, and time series prediction.

# 16

# Forecast Verification and Post-processing

In environmental science, forecasting has great socio-economic benefits for society. Forecasts for hurricanes, tornados, Arctic blizzards, floods and tsunami save lives and properties, while forecasts for seasonal climate variability, such as drought, have economic value. This chapter covers a number of topics related to forecasting, for both weather and climate problems.

Once a forecast model has been built, it is important that we evaluate the quality of its forecasts, a process known as *forecast verification*, forecast evaluation or, in ML terminology, forecast testing. Sections 16.1–16.6 cover forecast verification for a variety of problems: binary classes, multiple classes, continuous variables and probabilistic forecasts (for binary classes, multiple classes and continuous variables). Economic and human costs can be taken into account when issuing forecasts (Section 16.7). Spurious skills (Section 16.8) and extrapolation (Section 16.9) are important considerations, as non-linear NN models can underperform linear models due to extrapolation with new data.

Weather forecasting and climate prediction are heavily dependent on numerical models built on physics/dynamics. The raw numerical model output are not practical due to model bias, coarse spatial resolution, and so on. Statistical and ML methods are used to post-process and downscale numerical model output (Sections 16.10 and 16.11).

Forecast *lead time* is defined as 'the length of time between the issuance of a forecast and the occurrence of the phenomena that were predicted' (Glossary of Meteorology, American Meteorology Society, http://glossary.ametsoc.org/wiki/Forecast_lead_time). If forecasts are issued on 1 May, then the one-day lead time forecast will be for 2 May, the two-day lead time forecast for 3 May, and so on. However, for seasonal forecasts, the lead time has been not been uniquely defined in the literature. Suppose at the end of spring (1 March–31 May); a forecast was made for the next season (1 June–31 August); some researchers considered this forecast to have a lead time of one season. However, the time gap between 31

May when the forecast was made and 1 June (the start of the period forecasted) was zero season, so many researchers considered this forecast to have a lead time of zero season (Barnston, van den Dool, et al., 1994). This latter naming scheme has gained wide acceptance in recent decades, though the former naming scheme was commonly found in the earlier literature (Kirtman et al., 2001).

# 16.1 Binary Classes $\boxed{\text{B}}$ ☺

A considerable number of statistical measures have been developed to evaluate how accurate forecasts are compared with observations (Jolliffe and Stephenson, 2012). Let us start with forecasts for two classes or categories, where class 1 is for an event (e.g. tornado) and class 2 for a non-event (e.g. non-tornado). Model forecasts and observed data can be compared and arranged in a $2\times2$ *contingency table* (Table 16.1). The number of events forecast and indeed observed are called 'hits' and are placed in entry $a$ of the table. In our tornado example, this would correspond to forecasts for tornados which turned out to be correct. Entry $b$ is the number of false alarms, that is, tornados forecast but never materialized. Entry $c$ is the number of misses, that is, tornados observed under non-tornado forecasts, while entry $d$ is the number of correct negatives, that is, non-tornado forecasts which turned out to be correct.

Table 16.1 A $2\times2$ contingency table used in the forecast verification of a binary-class problem. The number of forecast 'yes' and 'no', and the number of observed 'yes' and 'no' are the entries in the table. The marginal totals are obtained by summing over the rows or the columns.

|  |  | Observed | | |
|---|---|---|---|---|
|  |  | Yes | No | Total |
| Forecast | Yes | $a$ (hits) | $b$ (false alarms) | $a + b$ (forecast yes) |
|  | No | $c$ (misses) | $d$ (correct negatives) | $c + d$ (forecast no) |
|  | Total | $a + c$ (observed yes) | $b + d$ (observed no) | $a + b + c + d = N$ |

An alternative nomenclature scheme has:

hits = true positives (TP),     false alarms = false positives (FP),

misses = false negatives (FN),   correct negatives = true negatives (TN).

Marginal totals are also listed in the table, for example, the top row sums to $a + b$, the total number of tornados forecast, whereas the first column sums to $a + c$, the total number of tornados observed. Finally, the total number of cases $N$ is given by $N = a + b + c + d$. The *base rate*, $s$, defined by the ratio between the total number of tornados observed and the total number of cases,

$$s = \frac{a+c}{N}, \tag{16.1}$$

is a sample estimate of the marginal probability of the event occurring. The *forecast rate*, $r$, defined by the ratio between the total number of tornados forecast and the total number of cases,

$$r = \frac{a+b}{N}, \tag{16.2}$$

is a sample estimate of the marginal probability of the event being forecast.

The simplest measure of accuracy of binary forecasts is the *proportion correct* (PC) or *fraction correct*, that is, the number of correct forecasts divided by the total number of forecasts,

$$\text{PC} = \frac{a+d}{N} = \frac{a+d}{a+b+c+d}, \tag{16.3}$$

where PC ranges between 0 and 1, with 1 being the perfect score. Unfortunately, this measure becomes very misleading if the number of non-events vastly outnumbers the number of events. For instance, if $d \gg a, b, c$, then (16.3) yields $\text{PC} \approx 1$. In Marzban and Stumpf (1996), one NN tornado forecast model has $a = 41$, $b = 31$, $c = 39$ and $d = 1{,}002$, since the vast majority of days has no tornado forecast and none observed. While $a$, $b$ and $c$ are comparable in magnitude, the overwhelming size of $d$ lifts PC to a lofty value of 0.937.

In such situations, where the non-events vastly outnumber the events, including $d$ in the score is rather misleading. Dropping $d$ in both the numerator and the denominator in (16.3) gives the *threat score* (TS) or *critical success index* (CSI),

$$\text{TS} = \text{CSI} = \frac{a}{a+b+c}, \tag{16.4}$$

which is a much better measure of forecast accuracy than PC in such situations. The worst TS is 0 and the best TS is 1.

To see what fraction of the observed events ('yes') were correctly forecast, we compute the *probability of detection* (POD), also known as the *hit rate* ($H$),

$$H = \text{POD} = \frac{\text{hits}}{\text{hits} + \text{misses}} = \frac{a}{a+c}, \tag{16.5}$$

with the worst POD score being 0 and the best score being 1.

Besides measures of forecast accuracy, we also want to know if there is forecast *bias* or *frequency bias*, that is,

$$B = \frac{\text{total 'yes' forecast}}{\text{total 'yes' observed}} = \frac{a+b}{a+c}. \tag{16.6}$$

It is easy to increase the POD if we simply issue many more forecasts of events ('yes'), despite most of them being false alarms. Increasing $B$ would raise concern that the model is forecasting far too many events compared to the number of observed events. While theoretically the ideal score is $B = 1$, there are often economic/safety reasons to tolerate more false alarms than missing on devastating events, that is, having $B > 1$ is common in practice.

To see what fraction of the forecast events ('yes') never materialized, we compute the *false alarm ratio* (FAR)

$$\text{FAR} = \frac{\text{false alarms}}{\text{hits} + \text{false alarms}} = \frac{b}{a+b}, \qquad (16.7)$$

with the worst FAR score being 1 and the best score being 0.

Be careful not to confuse the false alarm ratio (FAR) with the *false alarm rate* ($F$), also known as the *probability of false detection* (POFD). The value of $F$ measures the fraction of the observed 'no' events that were incorrectly forecast as 'yes', that is,

$$F = \text{POFD} = \frac{\text{false alarms}}{\text{false alarms} + \text{correct negatives}} = \frac{b}{b+d}, \qquad (16.8)$$

with the worst $F$ score being 1 and the best score being 0. While $F$ is not as commonly given as FAR and POD, it is one of the axes in the *relative operating characteristic* (ROC) diagram, used widely in probabilistic forecasts (Kharin and Zwiers, 2003; Marzban, 2004).

In an ROC diagram (Fig. 16.1), $F$ (i.e. POFD) is the abscissa and POD, the ordinate. Although our model may be issuing probabilistic forecasts in a two-class problem, we are actually free to choose the decision threshold used in the classification, that is, instead of using a posterior probability of 0.5 as the threshold for deciding whether to issue a 'yes' forecast, we may want to use 0.7 as the threshold if we want fewer false alarms (i.e. lower $F$) (at the expense of a lower POD) or 0.3 if we want to increase our POD (at the expense of increasing $F$ as well). The result of varying the threshold generates a curve in the ROC diagram. The choice of the threshold hinges on the cost associated with missing an event and that with issuing a false alarm. For instance, if we miss forecasting a powerful hurricane hitting a vulnerable coastal city, the cost may be far higher than that from issuing a false alarm, so we would want a low threshold value to increase the POD. If we have a second model, where for a given $F$ it has a higher POD than the first model, then its ROC curve (dashed curve in Fig. 16.1) lies above that from the first model. A model with zero skill (POD $= F$) is shown by the diagonal line in the ROC diagram.

For actual ROC diagrams, consider Marzban and Witt (2001), where an MLP NN classifier was used to predict the size of severe hail, given that severe hail had occurred or was expected to occur. The predictors were four variables from Doppler weather radar and five other environmental variables. For the target, one-hot encoding was used to represent the three classes of severe hail – hailstones of coin size (class 1), golf-ball size (class 2) and baseball size (class 3). The ROC diagrams for the three classes (Fig. 16.2) indicate class 3 forecasts to be the best and class 2 forecasts to be the worst, that is, the model was most accurate in predicting large hailstones and least accurate in mid-sized ones.

Figure 16.1 A schematic relative operating characteristic (ROC) diagram illustrating the trade-off between the false alarm rate ($F$) and the probability of detection (POD) as the classification decision threshold is varied for a given model (solid curve). The dashed curve shows the ROC of a better model while the diagonal line (POD = $F$) indicates a model with zero skill. ROC can be characterized by a single number, the *area under the curve* (AUC), where AUC is larger for the better model.

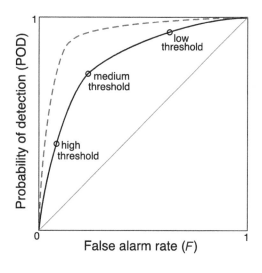

In situations where the non-events vastly outnumber the events, the number of true negatives ($d$) can be very large, rendering the false alarm rate $F$ in (16.8) to be tiny, hence the ROC diagram (plotting POD versus $F$) becomes uninformative. Instead, the following set of three measures are often used, namely precision, recall and their harmonic mean (K. P. Murphy, 2012, pp. 182–183). *Precision* is defined by

$$\text{precision} = \frac{a}{a+c} = \frac{\text{TP}}{\text{TP}+\text{FN}}, \tag{16.9}$$

which is the same as the hit rate $H$ or POD defined in (16.5). The denominator is the number of observed 'yes'.

*Recall* is defined by

$$\text{recall} = \frac{a}{a+b} = \frac{\text{TP}}{\text{TP}+\text{FP}}, \tag{16.10}$$

where the denominator is the number of forecast 'yes'. Both precision and recall lie in the interval $[0, 1]$, with 1 being perfect. In analogy to the ROC diagram, one can draw a curve in a graph with precision along the ordinate and recall along the abscissa by varying the classification decision threshold.

Precision and recall can be conveniently combined into a single measure by taking their harmonic mean, yielding the $F_1$-measure (or simply $F$-measure),

$$F_1 = 2\,\frac{\text{precision} \cdot \text{recall}}{\text{precision} + \text{recall}} = \frac{2a}{2a+b+c}. \tag{16.11}$$

The $F_1$-measure also lies in the interval $[0, 1]$, with 1 being perfect.

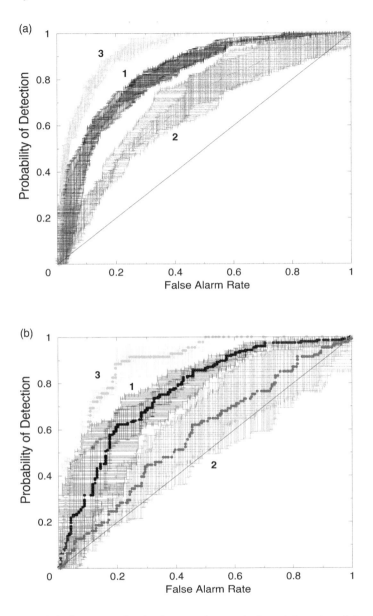

Figure 16.2 ROC diagrams for hailstone classes 1, 2 and 3, using (a) training and (b) validation data. The error bars in the horizontal and vertical directions are the one standard deviation intervals based on bootstrapping. The diagonal line indicates a model with zero skill. [Reproduced from Marzban and Witt (2001, figure 7), ©American Meteorological Society. Used with permission.]

The *odds ratio* is another measure of accuracy, widely used in medicine and introduced into meteorology by Stephenson (2000). The odds is the ratio of a probability to its complementary probability, for example the odds of a 'yes' forecast being right is

$$O_{\text{right}} = \frac{P(f_1|o_1)}{1 - P(f_1|o_1)} = \frac{a/(a+c)}{1 - a/(a+c)} = \frac{a}{c}, \qquad (16.12)$$

where $P(f_1|o_1)$ denotes the probability of forecasting 'yes' given 'yes' is observed. The odds of a 'yes' forecast being wrong is

$$O_{\text{wrong}} = \frac{P(f_1|o_2)}{1 - P(f_1|o_2)} = \frac{b/(b+d)}{1 - b/(b+d)} = \frac{b}{d}, \qquad (16.13)$$

where $P(f_1|o_2)$ denotes the probability of forecasting 'yes' given 'no' is observed. The odds ratio (OR) estimates the ratio of the odds of a 'yes' forecast being right to the odds of a 'yes' forecast being wrong, that is,

$$\text{OR} = \frac{O_{\text{right}}}{O_{\text{wrong}}} = \frac{ad}{bc}. \qquad (16.14)$$

OR can also be expressed in terms of the probability of detection (H) and the probability of false detection (F),

$$\text{OR} = \frac{H/(1-H)}{F/(1-F)}. \qquad (16.15)$$

OR $\in [0, \infty)$, with 1 indicating no skill.

### 16.1.1   Skill Scores for Binary Classes  B☺

Various *skill scores* have been designed to measure the relative accuracy of a set of forecasts, with respect to a set of *reference* or control forecasts. Choices for the reference forecasts include persistence, climatology, random forecasts and forecasts from a standard model. *Persistence forecasts* simply carry forward the anomalies to the future (e.g. tomorrow's weather is forecast to be the same as today's). *Climatology forecasts* simply issue the climatological mean value in the forecast. For instance, to forecast the temperature for 11 November, the mean of all historical temperature values for 11 November is issued as the forecast. In *random forecasts*, events are forecast randomly but in accordance to the model's forecast frequency of such events. For instance, if the model forecasts tornados 2% of the time, then random forecast also only forecasts tornados 2% of the time. Finally, the reference model can be a standard model, as the researcher is trying to determine if the new model is better.

For a particular model performance score $A$, the skill score (SS) is defined generically by

$$\text{SS} = \frac{A - A_{\text{ref}}}{A_{\text{perfect}} - A_{\text{ref}}}, \tag{16.16}$$

where $A_{\text{perfect}}$ is the value of $A$ for a set of perfect forecasts, and $A_{\text{ref}}$ is the value of $A$ computed over the set of reference forecasts. Note that if we define $A' = -A$, then SS is unchanged if computed using $A'$ instead of $A$. This shows that SS is unaffected by whether $A$ is positively or negatively oriented (i.e. whether better performance is indicated by a higher or lower value of $A$). Thus SS is always positively oriented, that is, higher SS is better, and the definition of SS in (16.16) is not limited to binary-class problems.

In meteorology, random forecasts may not be the best reference to use in a skill score. For short lead time (i.e. short range) forecasts, serial correlation allows persistence forecasts to outperform random forecasts, so it is more meaningful to test if a new model has higher skills than persistence forecasts, that is, persistence forecasts are used for $A_{\text{ref}}$. For longer lead time, climatology forecasts are generally better than persistence forecasts, so climatology forecasts are commonly used for $A_{\text{ref}}$ (A. H. Murphy, 1992).

The *Heidke skill score* (HSS) (Heidke, 1926) is the skill score (16.16) using the proportion correct (PC) for $A$ and random forecasts as the reference, that is,

$$\text{HSS} = \frac{\text{PC} - \text{PC}_{\text{random}}}{\text{PC}_{\text{perfect}} - \text{PC}_{\text{random}}}. \tag{16.17}$$

Thus, if the forecasts are perfect, HSS $= 1$; if they are only as good as random forecasts, HSS $= 0$ and if they are worse than random forecasts, HSS is negative. From (16.3), PC can be interpreted as the fraction of hits $(a/N)$ plus the fraction of correct negatives $(d/N)$. For $\text{PC}_{\text{random}}$ obtained from random forecasts, the fraction of hits is the product of two probabilities, $P(\text{'yes' forecast})$ and $P(\text{'yes' observed})$, that is, $(a+b)/N$ and $(a+c)/N$, respectively. Similarly, the fraction of correct negatives from random forecasts is the product of $P(\text{'no' forecast})$ and $P(\text{'no' observed})$, that is, $(c + d)/N$ and $(b + d)/N$. With $\text{PC}_{\text{random}}$ being the fraction of hits plus the fraction of correct negatives, we have

$$\text{PC}_{\text{random}} = \left(\frac{a+b}{N}\right)\left(\frac{a+c}{N}\right) + \left(\frac{c+d}{N}\right)\left(\frac{b+d}{N}\right). \tag{16.18}$$

Substituting this into (16.17) and invoking (16.3) and $\text{PC}_{\text{perfect}} = 1$, HSS is then given by

$$\text{HSS} = \frac{(a + d)/N - [(a + b)(a + c) + (b + d)(c + d)]/N^2}{1 - [(a + b)(a + c) + (b + d)(c + d)]/N^2}, \tag{16.19}$$

which can be simplified to (see Exercise 16.1)

$$\text{HSS} = \frac{2(ad - bc)}{(a + c)(c + d) + (a + b)(b + d)}. \tag{16.20}$$

As mentioned above, random forecasts are not too useful as the reference in skill scores in fields like meteorology, where persistence forecasts and climatology forecasts are better reference at short and long lead time, respectively.

The *Peirce skill score* (PSS) (Peirce, 1884), also called the *Hanssen and Kuipers' score* or the *true skill statistics* (TSS), is similar to the HSS, except that the reference used in the denominator of (16.17) is unbiased, that is, $P$('yes' forecast) is set to equal $P$('yes' observed), and $P$('no' forecast) to $P$('no' observed) for $PC_{\text{random}}$ in the denominator of (16.17), from which

$$PC_{\text{random}} = \left(\frac{a+c}{N}\right)^2 + \left(\frac{b+d}{N}\right)^2. \tag{16.21}$$

The PSS is computed from

$$\text{PSS} = \frac{(a+d)/N - [(a+b)(a+c) + (b+d)(c+d)]/N^2}{1 - [(a+c)^2 + (b+d)^2]/N^2}, \tag{16.22}$$

which simplifies to

$$\text{PSS} = \frac{ad - bc}{(a+c)(b+d)}. \tag{16.23}$$

PSS can also be expressed as

$$\text{PSS} = \frac{a}{a+c} - \frac{b}{b+d} = H - F, \tag{16.24}$$

upon invoking (16.5) and (16.8) (see Exercise 16.2). PSS $= H - F$ means PSS is simply the difference between POD and POFD. Again, if the forecasts are perfect, PSS $= 1$; if they are only as good as random forecasts, PSS $= 0$ and if they are worse than random forecasts, PSS is negative.

In situations where the non-events vastly outnumber the events, including $d$ in the score is rather misleading, so the threat score (TS) in (16.4) is usually used instead of PC. The *equitable threat score* (ETS), also known as the *Gilbert skill score* (GSS), uses $a_{\text{random}}/(a+b+c)$ as the reference, where $a_{\text{random}}$ is expected from random forecasts (G. K. Gilbert, 1884; Hogan, Ferro, et al., 2010).

$$\text{ETS} = \frac{a/(a+b+c) - a_{\text{random}}/(a+b+c)}{1 - a_{\text{random}}/(a+b+c)} = \frac{a - a_{\text{random}}}{a - a_{\text{random}} + b + c}, \tag{16.25}$$

where from the first term of (16.18), $a_{\text{random}} = (a+b)(a+c)/N$.

For very rare events, $a$ can be so small that even the TS and ETS are ineffective. Skill scores have been introduced specifically for rare extreme events, for example the extreme dependency score (EDS), the symmetric extreme dependency score (SEDS) and the symmetric extremal dependence index (SEDI) (Hogan and I. B. Mason, 2012).

## 16.2   Multiple Classes  $\boxed{\text{C}}$☺

Next, consider the forecast verification problem with $k$ classes, where $k$ is an integer $> 2$. Examples of $k = 3$ are most abundant, for example temperature being cold, normal or warm, and seasonal rainfall condition being dry, normal or wet. Instead of the $2 \times 2$ matrix in Table 16.1, the contingency table is now a $k \times k$ matrix, where each element $n_{ij}$ is the cell count given by the number of cases where class $C_i$ was forecast and class $C_j$ was observed. The $i$th diagonal element gives the number of correct forecasts for class $C_i$. The relative sample frequency is given by

$$p_{ij} = n_{ij}/N, \qquad (16.26)$$

with $N \ (= \sum_{ij} n_{ij})$ being the sample size. The sample probability of forecasts and observations, $\hat{p}_i$ and $p_j$, respectively, are obtained from the sample marginal distributions,

$$\hat{p}_i = \sum_{j=1}^{k} p_{ij}, \quad i = 1, \ldots, k, \qquad (16.27)$$

$$p_j = \sum_{i=1}^{k} p_{ij}, \quad j = 1, \ldots, k. \qquad (16.28)$$

Among the many measures developed for the two-class problem in Section 16.1, only three of them generalize to the $k$-class problem (Livezey, 2012), namely, proportion correct (PC), bias ($B$) and probability of detection (POD), also known as the hit rate ($H$).

$$\text{PC} = \sum_{i=1}^{k} p_{ii}, \qquad (16.29)$$

$$B_i = \hat{p}_i/p_i, \quad i = 1, \ldots, k, \qquad (16.30)$$

$$\text{POD}_i = H_i = p_{ii}/p_i, \quad i = 1, \ldots, k. \qquad (16.31)$$

PC is simply the sum of the diagonal elements of the contingency matrix divided by $N$, while $B_i$ indiates whether the $i$th class is being over-forecast or under-forecast and $\text{POD}_i$ or $H_i$ gives the hit rate for the $i$th class.

The other measures in Section 16.1 that cannot be generalized from 2 to $k > 2$ classes can still be used (Wilks, 2011, section 8.2.6). The approach is to collapse the $k \times k$ matrix to a $2 \times 2$ matrix. For instance, if the forecast classes are 'cold', 'normal' and 'warm', we can put 'normal' and 'warm' together to form the class of 'non-cold' events. Then we are back to two classes, namely 'cold' and 'non-cold', and the measures for binary classes can be easily applied. Similarly, we can collapse to only 'warm' and 'non-warm' events, or to 'normal' and 'non-normal' events.

For skill scores, the Heidke skill score and Peirce skill score can be generalized to $k > 2$ classes, but are not recommended (Livezey, 2012). Instead, equitable skill scores introduced by Gandin and A. H. Murphy (1992) and Gerrity (1992) are used. Gandin and A. H. Murphy (1992) proposed the use of *equitable skill scores*, where equitability requires that all constant forecasts and random forecasts receive the same expected score.

In many environmental science problems, the classes are *ordinal* variables, that is, there is a natural ordering of the classes. For instance, temperature categories (cold, normal, warm) and weather types (sunny, cloudy, rainy) are ordinal variables. The natural ordering means that a 'cloudy' forecast is not as bad as a 'sunny' forecast when 'rainy' is observed. For ordinal classes, Gerrity (1992) found a family of equitable skill scores as a subset of the Gandin and Murphy scores, which allows convenient computation and gives good performance (Livezey, 2012).

## 16.3   Probabilistic Forecasts for Binary Classes  B☺

In probabilistic forecasts of binary events, the model outputs probabilities. One can perform binary classification based on which event has the larger posterior probability, then compute verification scores for the binary classification problem as in Section 16.1. Alternatively, in this section, we look at scores developed for probabilistic forecasts of binary events. Traditionally, the most widely used score has been the *Brier score* (BS) (Brier, 1950). Formally, this score resembles the MSE, that is,

$$\text{BS} = \frac{1}{N} \sum_{n=1}^{N} (f_n - y_n)^2, \tag{16.32}$$

where there is a total of $N$ pairs of probabilistic forecasts $f_n$ and observations $y_n$. While $f_n$ is a continuous variable within $[0, 1]$, $y_n$ is a binary variable, being 1 if the event occurred and 0 if it did not occur. BS is negatively oriented, that is, the lower the better. Since $|f_n - y_n|$ is bounded between 0 and 1 for each $n$, BS is also bounded between 0 and 1, with 0 being the perfect score.

From (16.16), the Brier skill score (BSS) is then

$$\text{BSS} = \frac{\text{BS} - \text{BS}_{\text{ref}}}{0 - \text{BS}_{\text{ref}}} = 1 - \frac{\text{BS}}{\text{BS}_{\text{ref}}}, \tag{16.33}$$

where the reference forecasts are often taken to be random forecasts based on climatological probabilities. Unlike BS, BSS is positively oriented, with 1 being the perfect score and 0 meaning no skill relative to the reference forecasts.

Based on information theory, the *logarithmic score* or *ignorance score* has also become commonly used, as it has a more robust philosophical justification than the Brier score (Good, 1952; Roulston and L. A. Smith, 2002; Benedetti, 2010). Ignorance is the negative of the logarithm of the forecast probability $f_n$ for the outcome that is also observed, that is,

$$I_n = \begin{cases} -\log f_n & \text{if } y_n = 1, \\ -\log(1 - f_n) & \text{if } y_n = 0, \end{cases} \qquad (16.34)$$

where log can be either $\log_2$ or $\log_e$ in the literature. The ignorance score $I$ is the average of $I_n$ over all forecasts, that is,

$$I = \frac{1}{N} \sum_{n=1}^{N} I_n. \qquad (16.35)$$

The ignorance score is zero for perfect forecasts and is unbounded above since $-\log 0$ is $\infty$, which can occur if the model allows output of $f_n = 1$ or $f_n = 0$. Benedetti (2010) showed that the Brier score is equivalent to the second-order approximation of the ignorance score, which explains why the two scores are similar; however, for extreme probability forecasts ($f_n \approx 0$ or $f_n \approx 1$), where Brier score can be very different from the ignorance score, BS is inadequate as it does not sufficiently discriminate between small changes in forecasts that are significant for rare events.

## 16.3.1 Reliability Diagram B☺

While verification scores such as the Brier score or the ignorance score provide a compact assessment on how good our probabilistic forecasts are, one also turns to diagrams to see more details. The reliability diagram checks if the forecast probability is reliable, that is, if the event actually occurs with an observed relative frequency consistent with the forecast probability. For instance, if the model forecasts a probability of 0.70 for rain on some days, one would like to check over the observed data what fraction of the days with a 0.70 forecast probability actually had rain. Suppose this observed relative frequency is 0.65. For forecast probability of 0.30, the observed relative frequency turns out to be 0.39 (Fig. 16.3(a)). By plotting the observed relative frequency with respect to the forecast probability, we get the *reliability* curve. Details on how to collect the continuous forecast probability values into a finite number of bins are given in Bröcker and L. A. Smith (2007). Examples on how reliability diagrams are used are given in Hartmann et al. (2002) and Weisheimer and T. N. Palmer (2014).

In Fig. 16.3, four types of imperfect models are illustrated, while the dashed diagonal line shows the perfect model (with forecast probability exactly matched by the observed relative frequency). (a) In the overconfident or poor resolution case, the observed relative frequencies are all near the climatological probability with only weak dependence on the forecast probability. The forecast probabilities at the extremes are too extreme, hence 'overconfident'. (b) In the underconfident or good resolution case, the observed relative frequencies are strongly dependent on the forecast probabilities, though at the extremes, the forecast probabilities are not extreme enough, hence 'underconfident'.

(c) Overforecasting occurs when forecast probability tends to be higher than the observed relative frequency. If rain days are being forecast, this means there is a wet bias. (d) Underforecasting in the rain example means a dry bias.

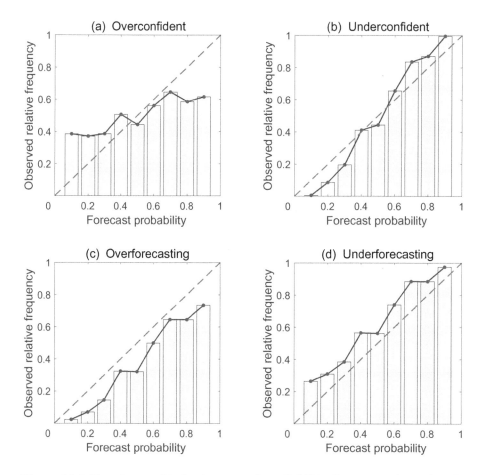

Figure 16.3 Four types of behaviour seen in *reliability diagrams*, where the observed relative frequency is plotted as a function of the forecast probability: (a) overconfident forecasts, (b) underconfident forecasts, (c) overforecasting and (d) underforecasting, with the dashed diagonal line indicating a perfect model.

# 16.4   Probabilistic Forecasts for Multiple Classes   $\boxed{\text{B}}$ ☺

The *Brier score* (BS) generalizes to multiple classes, where for $k$ classes,

$$\text{BS} = \frac{1}{N} \sum_{n=1}^{N} \sum_{i=1}^{k} (f_{ni} - y_{ni})^2, \tag{16.36}$$

where for the $n$th forecast, $f_{ni}$ is the forecast probability for the $i$th class and $y_{ni} = 0$ except $y_{ni} = 1$ for $i = i^*$, the index for the actual observed class at the $n$th forecast.

For example, consider $k = 3$ (with the three classes being 'cold', 'normal' and 'warm') and the $n$th forecast is $f_{n1} = 0.2$, $f_{n2} = 0.3$ and $f_{n3} = 0.5$, that is, 'warm' has the highest forecast probability. However, the observed outcome is 'normal', that is, $i^* = 2$, with $y_{n1} = 0$, $y_{n2} = 1$, and $y_{n3} = 0$.

The *ignorance score* also generalizes to multiple classes, with

$$I = -\frac{1}{N} \sum_{n=1}^{N} \log(f_{ni^*}). \tag{16.37}$$

In our example, $f_{ni^*} = f_{n2} = 0.3$.

For multiple classes, there are two types – *nominal* classes and *ordinal* classes. The former has no natural order for the classes (e.g. land, ocean, ice and clouds) while the latter has a natural order or distance between classes (e.g. cold, normal and warm). Neither BS nor $I$ takes into account the natural order in ordinal classes.

The *ranked probability score* (RPS) adapts the Brier score to ordinal classes. The cumulative forecasts $F$ and cumulative observations $Y$ are defined by

$$F_{nm} = \sum_{i=1}^{m} f_{ni}, \quad m = 1, \ldots, k, \tag{16.38}$$

$$Y_{nm} = \sum_{i=1}^{m} y_{ni}, \quad m = 1, \ldots, k. \tag{16.39}$$

In our example, $F_{n1} = f_{n1} = 0.2$, $F_{n2} = f_{n1} + f_{n2} = 0.2 + 0.3 = 0.5$ and $F_{n3} = f_{n1} + f_{n2} + f_{n3} = 0.2 + 0.3 + 0.5 = 1$, while $Y_{n1} = y_{n1} = 0$, $Y_{n2} = y_{n1} + y_{n2} = 0 + 1 = 1$ and $Y_{n3} = y_{n1} + y_{n2} + y_{n3} = 0 + 1 + 0 = 1$. $Y_{nm}$ is actually a step function, being 0 for $m < i^*$ and 1 for $m \geq i^*$.

The ranked probability score (RPS) is given by

$$\text{RPS} = \frac{1}{N} \sum_{n=1}^{N} \sum_{m=1}^{k} (F_{nm} - Y_{nm})^2, \tag{16.40}$$

which formally resembles BS in (16.36), but with the cumulative values $F$ and $Y$ replacing $f$ and $y$, respectively. All three scores (BS, $I$ and RPS) are negatively oriented, that is, lower is better.

The reliability diagram (Fig. 16.3) generalizes easily to multiple classes. For each class, the reliability curve of the observed relative frequency can be plotted as a function of the forecast probability. For example, in the 'cold', 'normal', warm' problem, three reliability curves can be constructed, one for each class.

## 16.5   Continuous Variables   $\boxed{\text{B}}$ ☺

Next, we turn to the situation where the response variable is continuous instead of discrete. In this section, we consider models issuing forecast values $f_n$, with observed values being $y_n$ ($n = 1, \ldots, N$), where both $f_n$ and $y_n$ are continuous variables. In Section 16.6, we turn to probabilistic forecasts, where the models issue posterior probabilities for the response variable.

### 16.5.1   Forecast Scores   $\boxed{\text{B}}$ ☺

When both $f_n$ and $y_n$ are continuous variables, the most common forecast scores are the mean squared error (MSE) (or its square root, the root mean squared error, RMSE), and the mean absolute error (MAE), where

$$\text{MSE} = \frac{1}{N} \sum_{n=1}^{N} \left( f_n - y_n \right)^2, \tag{16.41}$$

$$\text{MAE} = \frac{1}{N} \sum_{n=1}^{N} |f_n - y_n|. \tag{16.42}$$

The mean error ME

$$\text{ME} = \frac{1}{N} \sum_{n=1}^{N} (f_n - y_n) = \overline{f} - \overline{y}, \tag{16.43}$$

is simply the difference between the mean of the forecast values and that of the observed values, thus providing a measure of the bias in the forecasts.

The MSE can be decomposed into components (A. H. Murphy, 1988). By adding and subtracting $\overline{f}$ and $\overline{y}$ in (16.41), we have

$$\text{MSE} = \frac{1}{N} \sum_{n=1}^{N} \left[ (f_n - \overline{f}) - (y_n - \overline{y}) + (\overline{f} - \overline{y}) \right]^2,$$

$$= (\overline{f} - \overline{y})^2 + s_f^2 + s_y^2 - 2s_{fy}, \tag{16.44}$$

where $s_f^2$ and $s_y^2$ are the sample variance of the forecasts and observations, respectively, and $s_{fy}$ is the sample covariance, that is,

$$s_f^2 = \frac{1}{N} \sum_{n=1}^{N} (f_n - \overline{f})^2, \quad s_y^2 = \frac{1}{N} \sum_{n=1}^{N} (y_n - \overline{y})^2, \tag{16.45}$$

$$s_{fy} = \frac{1}{N} \sum_{n=1}^{N} (f_n - \overline{f})(y_n - \overline{y}). \tag{16.46}$$

With the Pearson correlation coefficient $r_{fy} = s_{fy}/(s_f s_y)$, we can write

$$\text{MSE} = (\overline{f} - \overline{y})^2 + s_f^2 + s_y^2 - 2 s_f s_y r_{fy}, \qquad (16.47)$$

where the first term is simply ME squared. If we set ME to zero and $r_{fy}$ to one, then MSE reduces to

$$s_f^2 + s_y^2 - 2 s_f s_y = (s_f - s_y)^2, \qquad (16.48)$$

which drops to zero only when $s_f = s_y$. Thus, there are three components affecting MSE: (i) the ME, (ii) difference between the sample variance of the forecasts and of the observations and (iii) the correlation between forecasts and observations. Although increasing the correlation lowers the MSE, correlation is obviously not as good a forecast score as MSE, since it ignores (i) and (ii).

If one is working with standardized variables $f$ and $y$, then the MSE and the correlation are simply related by

$$\text{MSE} = 2(1 - r_{fy}) \qquad (16.49)$$

(see Exercise 16.4).

To assess if the forecast score of model A is significantly different from that of model B, some form of the *t*-test for paired samples (Section 4.2.3) is often used. The *cross-validated* *t*-test was recommended by Dietterich (1998) and was used to compare different models for seasonal prediction (B. Tang, Hsieh, et al., 2000). In general, serial correlation needs to be taken into account in the *t*-tests to avoid excessive Type-I errors (Geer, 2016).

## 16.5.2 Skill Scores B☺

Skill scores (SS) of the form (16.16) can be constructed for MSE, MAE, and so on. There are various choices for the reference forecasts, for example climatology forecasts and persistence forecasts. With *climatology forecasts*, the value forecast $f_n$ is simply the observed climatology value $\overline{y}$, so (16.41) becomes

$$\text{MSE}_{\text{clim}} = \frac{1}{N} \sum_{n=1}^{N} (\overline{y} - y_n)^2. \qquad (16.50)$$

While $\overline{y}$ can be just the constant sample mean, it is common to have $\overline{y}$ being the climatological seasonal cycle (i.e. replace the overall mean by the mean over individual months or seasons). The MSE SS from (16.16) using climatology forecasts as reference is

$$\text{SS}_{\text{clim}} = \frac{\text{MSE} - \text{MSE}_{\text{clim}}}{0 - \text{MSE}_{\text{clim}}} = 1 - \frac{\text{MSE}}{\text{MSE}_{\text{clim}}}, \qquad (16.51)$$

since the MSE for a perfect model is 0.

With *persistence forecasts* at lead time $l$, the latest observed value $y_{n-l}$ is used as the forecast $f_n$, yielding

$$\text{MSE}_{\text{persist}} = \frac{1}{N} \sum_{n=1}^{N} (y_{n-l} - y_n)^2. \tag{16.52}$$

The MSE SS using persistence forecasts as reference is given by

$$\text{SS}_{\text{persist}} = 1 - \frac{\text{MSE}}{\text{MSE}_{\text{persist}}}. \tag{16.53}$$

For short lead time forecasts, persistence forecasts are commonly used, while for longer lead time forecasts, climatology forecasts are preferred. A. H. Murphy (1992) showed that when the lag-$l$ autocorrelation of the observed $y$ data is $> 0.5$, persistence forecasts are more accurate than climatology forecasts; otherwise, climatology forecasts are better. A. H. Murphy (1992) also showed that an optimal linear combination of persistence and climatology produces better forecasts than either alone and can be used as the reference in computing SS.

The MSE SS using climatology forecasts as reference, that is, (16.51), called the *Nash-Sutcliffe efficiency* $E$ in the hydrological literature, is widely used as a measure of model performance (Nash and Sutcliffe, 1970; McCuen et al., 2006; Gupta et al., 2009), where from (16.51) and (16.50),

$$E = 1 - \left[ \sum_n (f_n - y_n)^2 \Big/ \sum_n (\bar{y} - y_n)^2 \right]. \tag{16.54}$$

There have been improvements to the Nash–Sutcliffe efficiency, for example by replacing MSE with the more robust MAE in the *Legates–McCabe efficiency* (Legates and McCabe, 1999)

$$E_1 = 1 - \left[ \sum_n |f_n - y_n| \Big/ \sum_n |\bar{y} - y_n| \right]. \tag{16.55}$$

In other words, $E$ is the MSE SS while $E_1$ is the MAE SS, both using climatology forecasts as reference. As skill scores, both $E$ and $E_1$ are positively oriented (i.e. higher value means better performance). A common disadvantage of $E$ and $E_1$ is that they have a range of $(-\infty, 1]$. To have an index with a range of $[-1, 1]$, Willmott, Robeson, and Matsuura (2012) introduced a positively oriented 'refined' *index of agreement* (refined from an earlier 1982 version),

$$d_r = \begin{cases} 1 - \left[ \sum_n |f_n - y_n| / 2 \sum_n |\bar{y} - y_n| \right], & \text{if } \sum_n |f_n - y_n| \le 2 \sum_n |\bar{y} - y_n| \\ \left[ 2 \sum_n |\bar{y} - y_n| / \sum_n |f_n - y_n| \right] - 1, & \text{if } \sum_n |f_n - y_n| > 2 \sum_n |\bar{y} - y_n|. \end{cases} \tag{16.56}$$

A comparison of the three indices ($E$, $E_1$ and $d_r$) is given by Willmott, Robeson, Matsuura, and Ficklin (2015).

## 16.6 Probabilistic Forecasts for Continuous Variables $\boxed{\text{B}}$ ☺

In Section 16.3 and Section 16.4, we studied verification of probabilistic forecasts for binary classes and multiple classes, respectively. In this section, the response variable $y$ is a continuous variable. Two common scores for evaluating probabilistic forecasts for a continuous $y$ are the continuous ranked probability score and the ignorance score (Gneiting et al., 2005).

The ranked probability score (RPS) in (16.40) generalizes to the *continuous ranked probability score* (CRPS), which is the integral of the Brier scores at all possible threshold forecast values $f$, and is defined by

$$
\begin{aligned}
\text{CRPS} &= \frac{1}{N} \sum_{n=1}^{N} \text{crps}\,(F_n, y_n) \\
&= \frac{1}{N} \sum_{n=1}^{N} \left( \int_{-\infty}^{\infty} \left[ F_n\,(f) - H\,(f - y_n) \right]^2 \mathrm{d}f \right),
\end{aligned}
\tag{16.57}
$$

where for the $n$th forecast, the cumulative probability $F_n\,(f) \equiv p\left( \tilde{f} \leq f \right)$, and $H$ is the Heaviside step function, with $H\,(f - y_n)$ giving the value 0 when $f < y_n$, and 1 when $f \geq y_n$.

The *ignorance score* is the average of the negative logarithm of the predictive density $p$ evaluated at the observed values $y_n$, that is,

$$
I = -\frac{1}{N} \sum_{n=1}^{N} \log p(y_n).
\tag{16.58}
$$

Both scores are negatively oriented. If the predictive distribution is Gaussian (with mean $\mu$ and standard deviation $\sigma$), the analytical forms of the scores can be derived (Gneiting et al., 2005). Rasp and Lerch (2018) used the CRPS for a Gaussian distribution as the objective function in their NN model for postprocessing ensemble weather forecasts.

For a Gaussian distribution, the key difference between these two scores is that CRPS grows linearly with the normalized prediction error $(y - \mu)\,/\sigma$, but $I$ grows quadratically. Note that CRPS can be interpreted as a generalized version of the MAE (Gneiting et al., 2005). As the ignorance score assigns harsh penalties to particularly poor probabilistic forecasts, it can be very sensitive to outliers and extreme events. Hence Gneiting et al. (2005) preferred the more robust CRPS over $I$.

## 16.7 Minimizing Loss $\boxed{\text{C}}$ ☺

In probabilistic forecasts for $k$ classes, given an input $\mathbf{x}$, the model gives the posterior probability $P(C_i|\mathbf{x})$ for class $C_i$, $i = 1, \ldots, k$. The forecast class is usually chosen as the class $C_j$ with the largest posterior probability. In the real world, there are economic or safety reasons to choose the class differently. For instance, for natural disasters such as tornados, flash floods and tsunamis, the economic/human cost from a missed event can be so great that the forecasters would rather have a few false alarms than a missed event. One can define a loss matrix $\mathbf{L}$ where its element $L_{li}$ is the loss or cost if $f_l$ is forecast and $y_i$ is observed. For example, a city has the choice of issuing a disaster forecast and the disaster may or may not materialize. If a disaster is forecast and the inhabitants make preparations, the loss $L_{11}$ is \$10 million if the disaster hits, whereas the cost $L_{12}$ is \$1 million if there is no disaster. If no forecast is issued and disaster strikes, $L_{21}$ is \$50 million, whereas $L_{22} = \$0$ if there is no disaster. Thus, the loss matrix is

$$\mathbf{L} = \begin{bmatrix} 10 & 1 \\ 50 & 0 \end{bmatrix}. \tag{16.59}$$

We want to find the forecast that minimizes the expected loss, that is,

$$\min_l \sum_i L_{li} P(C_i|\mathbf{x}), \tag{16.60}$$

which is trivial to solve since we know $L_{li}$ and $P(C_i|\mathbf{x})$ (Bishop, 2006, section 1.5.6).

Suppose on a particular day, the posterior probability for a disaster is $P(C_1|\mathbf{x})$ = 0.1 and for no disaster is $P(C_2|\mathbf{x})$ = 0.9. For $l = 1$, $\sum_i L_{1i}P(C_i|\mathbf{x})$ = $10 \times 0.1 + 1 \times 0.9 = 1.9$, whereas $\sum_i L_{2i}P(C_i|\mathbf{x}) = 50 \times 0.1 + 0 \times 0.9 = 5$. Hence the expect loss is \$1.9 million if a disaster warning to the public is issued and \$5.0 million if no warning is issued. Of course, getting an accurate estimate of $\mathbf{L}$ can be difficult.

## 16.8 Spurious Skills $\boxed{\text{A}}$ ☺

In the early applications of methods such as MLP NN to environmental problems, researchers often did not fully appreciate the problem of overfitting using these very flexible models, and the need to test/verify using as many independent data as possible, for example by using 'cross-validation' for validation and for testing (i.e. verification) when the data record is not long (Section 8.5). Hence, the control of overfitting and the testing/verification process in the early papers may seem primitive by latter day standards.

Even with proper testing/verification, there is still room for spurious skills to slip into publications. In many papers, the authors propose a new ML/statistical method and compare its performance with other standard methods. In our present peer-reviewed publication system, a new method will have difficulty getting published in a high-standard journal if it cannot be shown to be better in some way than the established methods. Authors tend to present their new methods in the best possible light to enhance their chance of passing the peer review. For instance, an author might test his new method A against the traditional or benchmark method B on two different datasets $D_1$ and $D_2$, and finds A outperforming B on $D_1$ but B outperforming A on $D_2$. The author convinces himself why $D_2$ is not entirely suitable for testing his new method, then writes up a journal paper describing his method and the comparison of methods A and B on dataset $D_1$.

Publications biased in favour of new methods can arise even without the authors doing 'cherry-picking'. Suppose author X applied method A to a dataset $D_1$ and found A performing better than the traditional method B, while authors Y and Z applied the same method A to different datasets $D_2$ and $D_3$ but found A worse than B. The paper by author X was published but the paper by the unlucky author Y was rejected and the discouraged author Z never bothered to submit his work for publication. Ironically, it is actually the high-standard journals that have unwittingly done the 'cherry-picking'!

Sometimes, the comparison between the new method A and the benchmark B in a publication was simply unfair. Many of the non-linear methods can give better performance if the user is willing to devote more resources, for example running larger ensembles or spending more computing time searching for better hyperparameters. Often minimal computer resources were spent on method B than on method A. The literature is full of papers claiming a new method has beaten the neural network model, where a single MLP NN (instead of an ensemble of NN models) is used. In the real world, no experienced data scientist would run a single NN model with its well-known local minima problem.

Nowadays, it is common to have 'data challenge' competitions where multiple groups submit their solutions for a given data problem using different methods, and the winning entry is determined by the organizer of the competition. While one has more confidence in the winning entry method W determined by an impartial organizer than method A published by an author, it is still not certain that the method W is superior to the others used in the competition. First, there is always luck that W came out first in the particular dataset used in the competition. Next, real datasets often have problems of missing data, outliers, and so on, and the competitors would likely deal with these problems differently before applying their ML/statistical methods. Furthermore, there may be other sources of information, for example physics, which can be utilized in a competition. For instance, for wind power forecasting, an astute competitor would take the observed wind speed $w$ and based on physics (wind power being proportional to $w^3$) supply $w^3$ as a predictor. Hence, method W winning the competition still does not guarantee it is the best ML/statistical method.

In practice, a simple recommendation like 'method A is better than method B' is not meaningful for many reasons. For example: (i) method A outperforms B when the dataset has high signal-to-noise ratio, but B beats A when the ratio is low; (ii) A beats B when there are many independent observations but vice versa when there are few observations; (iii) A beats B when there are few predictor variables, but vice versa when there are many predictors; (iv) A beats B when the noise is Gaussian but vice versa when the noise is non-Gaussian, and so on. In short, *caveat emptor*!

In Section 8.5, subtle sources of spurious model skills in *cross-validation* have been discussed. Using 1,000 years of simulated coupled GCM data to test

the prediction of seasonal air temperature over Canada by canonical correlation
analysis (CCA), Shabbar and Kharin (2007) compared the model skill found
by cross-validation over 50 years and the model skill found in the subsequent
50 years, and found that the skill in the subsequent 50 years were often lower
than the cross-validated skill. While this seems to imply cross-validation gives
inflated skills, a more likely explanation is that the presence of low-frequency
multi-decadal variability in the climate system leads to the decline of skill in
the subsequent decades. For instance, for the tropical Pacific sea surface temperature in Fig. 9.4, the Niño 3.4 index from 1900 to 1970 shows fluctuations
with smaller amplitude and higher frequency than those from 1970 to 2000.
Thus, applying a model developed under cross-validation to subsequent decades
will give lower skills due to the low-frequency changes in the climate system.
Changes in the amplitude of the predictors over the subsequent period could
lead to extrapolation, which is particularly bad for many non-linear NN models
(Section 16.9). Thus, the decline of skill after a model has been built can be
more rapid for a non-linear NN model than a linear statistical model, as seen in
Barnston, Tippett, et al. (2012). Therefore, frequent updates of non-linear NN
models may be needed to avoid the decline in skill from low-frequency variability
in the climate system.

An alternative to cross-validation is *retroactive validation*, which is much
closer to how operational forecast models are actually run (S. J. Mason and
Baddour, 2008). For example, under retroactive validation, a model (or models)
is trained using data from 1950 to 1980 (cross-validation can be used to develop
the model for this period). After the model forecast for 1981 is saved, the
model is retrained with data from 1950 to 1981 and the forecast for 1982 saved,
then the model is retrained for 1950–1982, and so on. The forecasts from 1981
onward are then the test/verification data to determine the model forecast skill.
Instead of repeating every year, the process can be repeated every $k$ years –
for example for $k = 5$, the model trained from 1950 to 1980 is used to forecast
1981–1985, then the model is retrained using data from 1950 to 1985, and so
on. With efficient online learning models, the model can be retrained every
day or every hour (Sections 5.4 and 6.4.1). The retroactive validation has the
advantage that it follows closely how operational forecast models are actually
run, but has the disadvantage that data from the early period (1950–1980) is
not used for forecast verification and the earlier models have been trained with
shorter data records than the models used to forecast the later period.

## 16.9   Extrapolation   $\boxed{\text{A}}$ ☺

It is commonly assumed that an ML or statistical model that has been properly regularized to avoid overfitting (e.g. by cross validation) will have similar
prediction accuracy with new data as with the original training data. However,
merely focusing on not overfitting is not enough for environmental scientists
using unbounded continuous input data, where new input can easily lie well

outside the domain of the input data used for training. Extrapolation, especially with non-linear ML models such as NN models, has the potential to inflict terrible damage to model performance on new data.

Let us illustrate the effects of extrapolation using a toy problem, where the 'true' signal is $y = x + 0.2\,x^2$, with $x$ a standard Gaussian random variable (Hsieh, 2009, pp. 303–306; 2020). Gaussian noise with twice the standard deviation of the $y$ signal was added to generate the training data. Linear regression (LR) and non-linear regression (NLR) models were trained using a training dataset of 100 points. The extreme learning machine (ELM) NN model is run 100 times to produce a 100-member ensemble for NLR, with three types of activation functions tested, the logistic sigmoidal function $1/[1 + \exp(-x)]$, the Gaussian or radial basis function $\exp(-x^2)$ and the *softplus* function $\log(1+\mathrm{e}^x)$ (which is a smooth version of the ReLU function).

Within the domain of the training data, there is very little difference between the true signal, the individual ensemble members and the ensemble average of the NLR members in Figs. 16.4(a), (b) and (c). However, outside the training domain, there is great divergence among the true signal, the individual members and the ensemble average. The NLR (ensemble averaged) models using the three different activation functions are very similar within the training domain but are quite different outside. Within the training domain, the NLR model is closer to the true signal than LR; however, that may not be the case when outside the training domain (e.g. when $x > 3$ in Fig. 16.4(b)).

By following the gradient of the NLR function at the boundary of the training domain, linear extrapolation of the NLR model beyond the training domain was performed. The linear extrapolation of the NLR model is closer to the true solution than the NLR model in Fig. 16.4(a), (b) and (c). The three activation functions have very different asymptotic behaviour (Fig. 16.4(d)): as the sigmoidal activation function is bounded as $x \to \pm\infty$, the NN model output, being a linear combination of the activation functions, is also bounded asymptotically. The radial basis function asymptotes to zero, so the model output also asymptotes to zero. The softplus function asymptotes linearly as $x \to +\infty$, hence the model output asymptotes linearly. Thus, the particular choice of an activation function affects the extrapolation of the NLR model as noted in Hsieh (2009, pp. 303–305).

Six environmental datasets were used to check the performance of NLR versus LR on test data partitioned into 'non-outliers' and 'outliers' based on the Mahalanobis distance of the predictors (Hsieh, 2020). The general outperformance of NLR over LR on non-outliers tended to turn into underperformance on outliers. Wild extrapolation was found in predicting the Beijing air quality ($PM_{2.5}$ concentration) when data from 2013 to 2015 were used for training and 2010–2012 were used for testing. Cumulated precipitation, one of the important predictors, had a maximum value of 51 mm in the training data, but a maximum of 223 mm in the test data! Wild extrapolation was also found in streamflow prediction at Stave River, British Columbia, where a predictor, the accumulated snow depth, had a maximum value of 64 mm in the training data but a maximum of 231 mm in the test data. How much damage an outlier in a

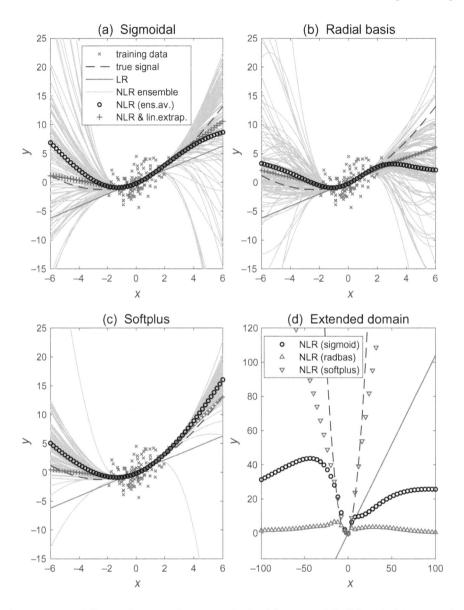

Figure 16.4 Effects of extrapolation with the (a) sigmoidal, (b) radial basis and (c) softplus activation function used in the ELM NLR model. *Linear* extrapolation of the NLR model beyond the training domain is marked by '+'. (d) shows extrapolation of the NLR model over an extended domain for the three activation functions, as well as the true signal (dashed) and LR (solid line). [Adapted from Hsieh (2020, figure 1)].

particular predictor can cause depends on how strongly the NLR model weights this predictor – that is, an outlier in an important predictor would do far more damage than that in an unimportant predictor.

Underperformance on input outliers in the test data is worrying, as such outliers often occur during extreme environmental events, which exert major impact on human health, safety and economy. The following procedure is recommended to alleviate the effects of extrapolation (Hsieh, 2020): (a) New input data are screened for outliers relative to the training input data. (b) For strong or extreme input outliers, avoid the non-linear extrapolations from the NLR model. Instead, either (i) use LR, (ii) linearly extrapolate using the NLR model within the training domain or (iii) issue a warning that no reliable prediction can be made.

The climate system has low-frequency variability, which changes the amplitude for some of the predictor variables over time (Section 16.8). This leads to a decline in the model performance over time, more so for non-linear NLR models vulnerable to wild extrapolations than linear models. It is therefore important to update the NLR models frequently with new data, for example by using online learning methods (Section 6.4.1).

In general, online learning helps to alleviate extrapolation effects in the following way: Suppose the training data of a predictor $x$ covers the range $[x_{\min}, x_{\max}]$, and a new extreme event arrives with values $x^{(1)}$, $x^{(2)}$ and $x^{(3)}$ over three consecutive days, where $x_{\max} < x^{(1)} < x^{(2)} < x^{(3)}$. With a batch model, there would be strong extrapolation effects when making forecasts at day 3 as $x_{\max} \ll x^{(3)}$, but with an online learning model updating daily, $x_{\max}$ would have been updated to $x_{\max} = x^{(2)}$, so the extrapolation would be much less drastic.

This discussion on extrapolation has been focused on NN models, which extrapolate using non-linear functions. Tree-based models such as random forest (RF) and gradient boosting do not extrapolate wildly, as seen in Fig. 14.6. Hence, when coupling ML models to physical/dynamical models, NN models can lead to instability while RF models remain stable (Brenowitz, Beucler, et al., 2020).

# 16.10   Post-processing [C]☺

In many areas of environmental science, physical/dynamical models have better prediction accuracy than statistical models. For instance, in *numerical weather prediction* (NWP), the governing equations of the atmosphere are solved numerically by finite-difference or spectral methods on supercomputers (Kalnay, 2003; Warner, 2011). Such dynamical models can be integrated forward in time to give weather forecasts. Nevertheless, *post-processing* of the dynamical model output with statistical techniques is performed to improve the raw forecasts (Wilks, 2011, section 7.5.2). The reason is that the variables in the dynamical model usually have poor spatial resolution and are often too idealized. Presently, typical spatial resolution is 10–30 km for global NWP models (and about an order of magnitude smaller for regional NWP models), too coarse to resolve cumulus clouds with a scale of $< 1$ km. The lowest temperature level in the model may be some considerable distance above the ground and the fine-structure in local topography may be completely missing in the model. Thus, it would be difficult to directly use the output from such a dynamical model to predict the ground temperature at a village located in a valley. Furthermore, some local variables such as atmospheric pollutant concentrations or precipitation may not even be variables carried in the dynamical model.

Let $\mathbf{x}(t)$ and $y(t)$ be the observed variables and $\mathbf{x}_{\mathrm{dm}}(t)$ be the dynamical model output for $\mathbf{x}(t)$. A common post-processing method is the *perfect prog* (abbreviation for 'perfect prognosis') scheme, which computes a multiple linear regression (MLR) relation using the historical observed data, that is,

$$y(t) = \mathbf{x}(t)^{\mathrm{T}}\mathbf{a} + \epsilon(t), \qquad (16.61)$$

where $y$ is the response variable, $\mathbf{x}^{\mathrm{T}} = \begin{bmatrix} 1 & x_1 & \dots & x_m \end{bmatrix}$ are the predictors, $\mathbf{a} = \begin{bmatrix} a_0 & a_1 & \dots & a_m \end{bmatrix}^{\mathrm{T}}$ are the regression coefficients and $\epsilon$ is the error or residual. For example, $\mathbf{x}^{\mathrm{T}}$ can be meteorological observations, while $y$ is the concentration of an atmospheric pollutant or the streamflow of a river. When making a forecast at time $t$, one wants to estimate $y$ at the future time $t + \Delta t$. Since $\mathbf{x}(t + \Delta t)$ is not yet observed at time $t$, the dynamical model is run to time $t + \Delta t$, yielding $\mathbf{x}_{\mathrm{dm}}(t + \Delta t)$. Thus, in perfect prog, $y(t + \Delta t)$ is estimated by

$$\hat{y}(t + \Delta t) = \mathbf{x}_{\mathrm{dm}}(t + \Delta t)^{\mathrm{T}}\mathbf{a}, \qquad (16.62)$$

with $\mathbf{a}$ from (16.61). The problem with this scheme is that while the regression model was developed or trained using historical observed data for $\mathbf{x}$, the actual forecasts used the dynamical model forecasts $\mathbf{x}_{\mathrm{dm}}$. Bias in the dynamical model forecasts $\mathbf{x}_{\mathrm{dm}}$ relative to the real data $\mathbf{x}$ has not been taken into account – that is, zero bias (perfect prognosis) is assumed, hence the name of this scheme.

In contrast, a better approach is the *model output statistics* (MOS) scheme, where the dynamical model forecasts have been archived, so the regression relation is obtained using $y(t)$ from the data archive and $\mathbf{x}_{\mathrm{dm}}(t)$ from the dynamical model forecast archive, that is,

$$y(t) = \mathbf{x}_{\mathrm{dm}}(t)^{\mathrm{T}}\mathbf{a}_{\mathrm{MOS}} + \epsilon(t), \qquad (16.63)$$

with $\mathbf{a}_{\mathrm{MOS}}$ containing the regression coefficients. When making a forecast at time $t$ for future time $t + \Delta t$, we let

$$\hat{y}(t + \Delta t) = \mathbf{x}_{\mathrm{dm}}(t + \Delta t)^{\mathrm{T}}\mathbf{a}_{\mathrm{MOS}}. \qquad (16.64)$$

Since $\mathbf{x}_{\mathrm{dm}}$ is from the dynamical model forecasts during both model training in (16.63) and actual forecasting in (16.64), the model bias in the perfect prog scheme has been eliminated. While MOS is more accurate than perfect prog, it is considerably more cumbersome to implement. Operational forecast models are frequently upgraded, and a slight modification of the dynamical model would require regeneration of the dynamical model forecast archive and recalculation of the regression relations. As the recalculation of the regression relations (when the model has been updated or new data have become available) can be costly, *online learning* methods have been devised to frequently update MOS at relatively low cost, for example *updatable* MOS (UMOS) using MLR (Wilson and Vallée, 2002; Wilson and Vallée, 2003).

In more recent decades, besides MLR, non-linear ML/statistical methods (e.g. NN) have been used in post-processing (Casaioli et al., 2003; Marzban, 2003), with reviews given by McGovern, Elmore, et al. (2017) and Vannitsem et al. (2021). Methods for non-linear updatable MOS have also been developed (Yuval and Hsieh, 2003; A. R. Lima et al., 2016; A. R. Lima, Hsieh, et al., 2017). The convolutional NN model, with its proficiency in working with geospatial data, has been used for MOS with non-local spatial data as input (Steininger et al., 2020).

# 16.11   Downscaling  ⃝C ☺

The goal of downscaling is to infer higher-resolution information from low-resolution data. The topic of downscaling overlaps somewhat with the previous topic of post-processing in numerical weather prediction (NWP) (Section 16.10), since post-processing is often used to estimate local variables from dynamical model output, which are not capable of resolving the local variables. However, downscaling is used in many applications beyond NWP.

*General circulation models*, (a.k.a. *global climate models*) (GCM) are the main tools for estimating future climate conditions under increasing concentrations of atmospheric greenhouse gases (Gettelman and Rood, 2016). As GCMs need to be integrated over a much longer time duration than NWP models, the GCM spatial resolution is currently about 100–200 km, about a factor of ten coarser than the resolution used in global NWP models. The GCM resolution is far too coarse to resolve climate change at the local scale. Thus, downscaling the GCM model output is crucial for estimating climate change for local scale and local variables (Maraun et al., 2010; Ekström et al., 2015). Downscaling methods are also used in remote sensing to improve the spatial resolution of satellite images (Atkinson, 2013).

There are two main approaches to climate downscaling – *dynamical downscaling* (a.k.a. process-based downscaling) and *statistical downscaling* (a.k.a. empirical downscaling) (Hewitson and Crane, 1996). Dynamical downscaling typically involves *nesting* (i.e. imbedding) a *regional climate model* (RCM) within a global climate model, where the RCM is a higher resolution numerical model of a limited region, run with boundary conditions provided by the GCM (Giorgi and Gutowski, 2015).

For statistical climate downscaling, Wilby and Wigley (1997) grouped the methods into three categories: regression methods, weather pattern-based approaches and stochastic weather generators (Wilks and Wilby, 1999), with regression methods being the most popular due to ease of implementation and low computational cost. Besides linear regression, non-linear regression methods have also been widely used in statistical downscaling.

Hewitson and Crane (1996) used MLP NN for precipitation forecasts with predictors from the GCM atmospheric data over southern Africa and the surrounding ocean. The six leading principal components (PCs) of the sea level pressure field and the seven leading PCs of the 500 hPa geopotential height field from the GCM were used as inputs to the NN, and the response variables were the $1° \times 1°$ gridded daily precipitation data over southern Africa. The GCM was also run at double the present amount of $CO_2$ in the atmosphere (called the $2 \times CO_2$ experiment), and the model output fed into the previously derived NN to predict precipitation in the $2 \times CO_2$ atmosphere. The assumption is that the empirical relation derived between the local precipitation and the synoptic-scale atmospheric circulation for the $1 \times CO_2$ atmosphere remains valid for the $2 \times CO_2$ atmosphere. This assumption is risky since extreme weather events are intensified under climate change (P. Stott, 2016) and extrapolation with non-linear ML/statistical models can cause havoc (Section 16.9).

For precipitation downscaling, Olsson et al. (2004) found it advantageous to use two MLP NN models in sequence – the first performs classification into 'rain' or 'no-rain', and if there is rain, the second NN predicts the intensity. Using MLP, McGinnis (1997) downscaled five-day averaged snowfall, while Schoof and Pryor (2001) downscaled daily minimum temperature ($T_{min}$) and maximum temperature ($T_{max}$), daily precipitation and monthly precipitation. Also, the radial basis function (RBF) NN was compared against a linear statistical model by Weichert and Bürger (1998) in downscaling daily temperature, precipitation and vapour pressure. For probabilistic multi-site precipitation downscaling, Cannon (2008) developed a conditional density network (CDN) model, where the MLP model outputs are the parameters of the Bernoulli-gamma distribution, and the objective function has constraint terms forcing the predicted between-site covariances to match the observed covariances. The quantile regression neural network (QRNN) model has also been applied to precipitation downscaling (Cannon, 2011b).

In the last couple of decades, interests in climate change have broadened from assessments of average behaviour to extreme weather events, as there is great concern that climate change can increase the occurrence of extreme events (Tebaldi et al., 2006; Bürger et al., 2012). Many indices have been introduced to measure the climate of extreme weather events. Changes in the value of these indices would indicate how the *climate of extremes* are affected under climate change. For example, some of the indices for *extreme temperature* listed in X. B. Zhang et al. (2011, table 1) include: (a) monthly maximum value of daily maximum temperature, (b) monthly maximum value of daily minimum temperature, (c) monthly minimum value of daily max temperature, (d) monthly minimum value of daily min temperature, (e) monthly mean difference between daily max and min temperature (i.e. diurnal temperature range), and so on. Some of the indices for *extreme precipitation* include: (a) monthly maximum one-day precipitation, (b) monthly maximum consecutive five-day precipitation, (c) annual count of days with precipitation $> 20$ mm, (d) maximum number of consecutive dry days ($< 1$ mm precipitation), (e) maximum number of consecutive wet days ($\geq 1$ mm), and so on.

Haylock et al. (2006) compared six statistical and two dynamical downscaling models with regard to their ability to downscale several seasonal indices of heavy precipitation for two station networks in northwest and southeast England. The skill among the eight downscaling models was high for those indices and seasons that had greater spatial coherence, that is, winter generally showed the highest downscaling skill and summer the lowest. The six statistical models used included a canonical correlation analysis (CCA) model, three MLP NN models in different setups, a radial basis function (RBF) NN model and SDSM (a statistical downscaling method using a two-step conditional resampling) (Wilby, Dawson, et al., 2002). The ranking of the models based on their correlation skills (with Spearman rank correlation used) revealed the NN models performing well. However, as the NN models were designed to reproduce the conditional mean precipitation for each day, there was a tendency to underestimate extremes. The rainfall indices indicative of rainfall occurrence were better modelled than those indicative of intensity.

Six of the models were then applied to the Hadley Centre global circulation model HadAM3P forced by emissions according to two different scenarios for the projected period of 2071–2100. As the inter-model differences between the future changes in the downscaled precipitation indices were large, Haylock et al. (2006) cautioned against interpreting the output from a single model or a single type of model (e.g. regional climate models) and emphasized the advantage of including as many different types of downscaling model, global model and emission scenario as possible when developing climate-change projections at the local scale.

The lack of observations to verify future climate projections can be partially addressed by using RCM model output as pseudo-observations, that is, use GCM output as predictors and RCM output as response (Vrac et al., 2007) in both model training and future climate projections. Gaitan, Hsieh, et al. (2014) compared several non-linear and linear ML/statistical models in future climate projections as measured by a number of indices for extreme precipitation over southern Ontario and Quebec in Canada. Overall, NN models and tree ensembles (i.e. random forests) outscored the linear models and simple non-linear models in terms of precipitation occurrences, without performance deteriorating in future climate (2041–2070) based on pseudo-observations provided by an RCM. In contrast, for the precipitation amounts and related climate indices, the performance of downscaling models deteriorated in future climate.

Climate downscaling is not limited to temperature and precipitation. Other environmental variables for climate downscaling include, for example, wind speed, streamflow, and so on. Downscaling *wind speed* is important for assessing how wind power generation may be affected by climate change (Sailor, T. Hu, et al., 2000; Pryor et al., 2005; Sailor, M. Smith, et al., 2008). Using RCM as pseudo-observations to verify future wind speed projections, Gaitan and Cannon (2013) downscaled GCM output for climate of extremes, as measured by indices relevant for wind power generation, for example number of days in the year with wind speed < cut-in speed (i.e. minimum wind speed required to generate power), number of days with wind speed > cut-out speed (maximum speed allowed for operation), and so on.

Downscaling is also needed to link climate change modelling to hydrological systems (Fowler et al., 2007). There are several approaches for downscaling *streamflow*: (a) statistically downscale from GCM output to local streamflow (Cannon and Whitfield, 2002); (b) statistically downscale GCM output for temperature and precipitation, which are then fed into a physical-based hydrological model to obtain streamflow (Dibike and Coulibaly, 2005) or (c) dynamically downscale GCM by RCM, then statistically downscale RCM output for temperature and precipitation, which are fed into a physical-based hydrological model (Wood et al., 2004).

Recently, deep NN models have been used in downscaling. For weather forecasting, Sha et al. (2020a) used a modified U-net (Section 15.3.1), called a 'UNet-Autoencoder', to downscale daily maximum and minimum temperatures (Tmax and Tmin) over the western continental United States. Input are the low-resolution (LR) image of Tmax/Tmin, LR and HR (high-resolution) images of the topographic elevation, while the output are the HR Tmax/Tmin and HR elevation. Since HR elevation is used as both input and target output, this part of the network has an autoencoder architecture. For precipitation downscaling, Sha et al. (2020b) used a nested U-net model, called UNet++ (Zhou et al., 2018).

## 16.11.1   Reduced Variance   Ⓒ☺

A common problem with a statistically downscaled variable is that its variance is generally smaller than the observed variance (Zorita and von Storch, 1999). This is because the influence of the large-scale variables can only account for a portion of the variance of the local variable. Various methods have been proposed to boost the variance of the downscaled variable to match the observed variance:

[1] The simplest method is *inflation*, which linearly rescales the downscaled variable to match the observed variance (Karl et al., 1990). von Storch (1999) criticized the underlying assumption of this method – namely all local variability can be traced back to large-scale variability – as being invalid. The inflation method increases the RMSE (see Exercise 16.5). Neighbouring local variables would also be more strongly correlated as the influence of the large-scale variables has been inflated.

[2] *Quantile mapping*, where quantiles from the model output are mapped to the corresponding observed quantiles, still suffers from the problems of inflation (Maraun, 2013).

[3] An alternative is to treat the local variability not explainable by the large-scale variables as random noise. *Randomization* adds some ran-

dom noise to the downscaled variable (von Storch, 1999). This method also increases the RMSE (see Exercise 16.5) and has other undesirable properties (Bürger et al., 2012, appendix A).

[4] *Expanded downscaling* adds a constraint term to the objective function, forcing the predicted variance towards the observed variance during optimization (Bürger, 1996; Bürger et al., 2012). If expanded downscaling is for variables at multiple sites, then the constraint term is for the covariance matrix of the predicted variables to match that of the observations.

# Exercises

## 16.1

Derive expression (16.20) for the Heidke skill score (HSS).

## 16.2

Derive expression (16.24) for the Peirce skill score (PSS).

## 16.3

Denote the reference Brier score using climatological forecasts by $B_{\text{ref}}$. If we define $s$ as the climatological probability for the occurrence of the event (i.e. using observed data, count the number of occurrences of the event, then divide by the total number of observations), show that $B_{\text{ref}} = s(1 - s)$.

## 16.4

From (16.47), show that when one is working with standardized variables (having zero mean and unit variance), the MSE and correlation are related by (16.49). What are the values of the MSE when the correlation is 1, 0.5, 0 and -1?

## 16.5

For statistical downscaling, compare the increase in the MSE when using the inflation method versus the randomization method. Consider downscaling by a simple linear regression model relating the observed response variable $y$ and the predictor $x$,

$$y = \alpha x + \epsilon, \tag{16.65}$$

where $\epsilon$ is Gaussian noise with zero mean and standard deviation $\sigma$, the variables have been scaled with $\text{var}(x) = \text{var}(y) = 1$ and the regression coefficient $0 < \alpha < 1$.
(a) Show that $\sigma^2 = 1 - \alpha^2$.

(b) The downscaling model outputs

$$\hat{y} = \alpha x, \tag{16.66}$$

with $\mathrm{var}(\hat{y}) = \alpha^2$. With the inflation method, the inflated output

$$\tilde{y} = \sqrt{\frac{\mathrm{var}(y)}{\mathrm{var}(\hat{y})}} \, \hat{y}. \tag{16.67}$$

Show that the MSE for the inflation method is

$$\mathrm{var}(\tilde{y} - y) = (1 - \alpha)^2 + 1 - \alpha^2 = 2(1 - \alpha), \tag{16.68}$$

larger than the MSE of the original output $\hat{y}$.

(c) For the randomization method,

$$y^* = \hat{y} + \delta, \tag{16.69}$$

where $\delta$ is Gaussian noise with standard deviation chosen so that $\mathrm{var}(y^*) = \mathrm{var}(y) = 1$. Show that the MSE is

$$\mathrm{var}(y^* - y) = 2(1 - \alpha^2) = 2(1 - \alpha)(1 + \alpha). \tag{16.70}$$

(d) Over the range $0 < \alpha < 1$, does inflation or randomization result in a larger MSE? Plot, as a function of $\alpha$, the MSE of the original output $\hat{y}$, the MSE for the inflation method and for the randomization method.

# 17

# Merging of Machine Learning and Physics

In environmental science, ML methods were originally used as non-linear statistical tools, without direct relation to models based on physics.[1] The clear separation between physical/dynamical models and ML models is becoming blurry, as the two types of models have been merging. Interestingly, this parallels the evolution of science fiction, where the term 'robot' first appeared in 1920, followed by 'cyborg' in 1960, with the robot being a machine but the more intriguing cyborg being a combination of a living organism and a machine. Strictly speaking, a man with a wooden leg could be considered a cyborg, but what fascinated the science fiction audience were cyborgs attaining super-human ability by having some damaged body parts replaced by machines (e.g. the Bionic Man, RoboCop, Darth Vader, Iron Man, etc.). By combining the best qualities from humans and machines, cyborgs have tended to upstage robots in science fiction. The merging of ML and physical/dynamical models could be the next stage in the evolution of ML in environmental science.

In this chapter, we look at three areas where ML and physics have been merging: (a) Physical models with computationally expensive components can have these components replaced by inexpensive ML models, giving rise to hybrid models (Section 17.1). (b) ML models can be solved satisfying the laws of physics, e.g. conservation of energy, mass, and so on. (Section 17.2). (c) In forecasting, ML models can be combined with physical/dynamical models under data assimilation (Section 17.3).

---

[1] The term 'physics' in this chapter is used in the broadest sense, that is, for some problems, it can mean chemistry, biology, and so on.

# 17.1    Physics Emulation and Hybrid Models  ☐C☺

Traditionally, statistical and ML methods have been used to find relations between observed data $\{\mathbf{x}, \mathbf{y}\}$, for example $\mathbf{y} = \mathbf{f}(\mathbf{x})$. In physics *emulation*, $\{\mathbf{x}, \mathbf{y}\}$ are simulated data from models based on physics. Why would one want to emulate physics with an ML model when one has the actual equations of physics? The reason is that direct computation using the equations of physics can be very expensive, but ML can often emulate the physics accurately and cheaply. Hence some of the physics in numerical models can be replaced by ML emulation, leading to hybrid models.

## 17.1.1    Radiation in Atmospheric Models  ☐C☺

The earth radiates primarily in the infrared frequency band. This outgoing radiation is called the *longwave radiation* (LWR), as the wavelengths are long relative to those of the incoming solar radiation. Greenhouse gases, (e.g. carbon dioxide, water vapour, methane and nitrous oxide) absorb certain wavelengths of the LWR, adding heat to the atmosphere and, in turn, causing the atmosphere to emit more radiation. Some of this radiation is directed back towards the Earth, hence warming the Earth's surface. The Earth's radiation balance is very closely achieved since the outgoing LWR very nearly equals the absorbed incoming *shortwave radiation* (SWR) from the sun (primarily as visible, near-ultraviolet and near-infrared radiation). Atmospheric general circulation models (GCM) typically spend a major part of their computational resources on calculating the LWR and SWR fluxes through the atmosphere.

A GCM computes the net LWR heat flux $F(p)$, where the pressure $p$ serves as a vertical coordinate. The cooling rate $C_{\mathrm{r}}(p)$ is simply proportional to $\partial F/\partial p$. Besides being a function of $p$, $F$ is also a function of $\mathbf{S}$, variables at the Earth's surface, $\mathbf{T}$, the vertical temperature profile, $\mathbf{V}$, vertical profiles of chemical concentrations (e.g. $CO_2$ concentration), and $\mathbf{C}$, cloud variables.

Chevallier, Cheruy, et al. (1998) and Chevallier, Morcrette, et al. (2000) developed MLP NN models to replace LWR fluxes in GCMs. For the flux $F$ at the discretized pressure level $p_j$, the original GCM computed

$$F = \sum_{i} a_i(\mathbf{C}) F_i(\mathbf{S}, \mathbf{T}, \mathbf{V}), \qquad (17.1)$$

where the summation over $i$ is from the Earth's surface to the level $p_j$, and $F_i$ is the flux at level $p_i$ without the cloud correction factor $a_i$. Neural network models $N_i(\mathbf{S}, \mathbf{T}, \mathbf{V})$ were developed to replace $F_i(\mathbf{S}, \mathbf{T}, \mathbf{V})$. This 'NeuroFlux' model was highly accurate, and has been implemented in the European Centre for Medium-Range Weather Forecasts (ECMWF) global atmospheric model, as it ran eight times faster than the original LWR code (Chevallier, Morcrette, et al., 2000).

In an alternative approach by Krasnopolsky, Fox-Rabinovitz, and Chalikov (2005a), MLP NN models of the more general form $N(\mathbf{S}, \mathbf{T}, \mathbf{V}, \mathbf{C})$ were developed to replace the cooling rates $C_{\mathrm{r}}$ at levels $p_j$ and several radiation fluxes in the GCM. Discussions on the relative merits of the two approaches were given in Chevallier (2005) and Krasnopolsky, Fox-Rabinovitz, and Chalikov (2005a). This alternative NN approach is also highly accurate and has been implemented in the moderate-resolution NCAR Community Atmospheric Model (CAM) and in the NASA Natural Seasonal-to-Interannual Predictability Program (NSIPP) GCM (Krasnopolsky, Fox-Rabinovitz, and Chalikov, 2005b; Krasnopolsky and Fox-Rabinovitz, 2006; Krasnopolsky, 2007). For the NCAR CAM model, the NN model has 220 inputs and 33 outputs, and a single hidden layer with 50 neurons was found to be enough, giving an NN LWR code capable of running 150 times faster than the original LWR code (Krasnopolsky, 2007).

Similarly, this NN approach has been applied to replace the SWR codes in the NCAR CAM GCM (Krasnopolsky and Fox-Rabinovitz, 2006; Krasnopolsky, 2007). For the CAM3 model, the NN SWR model has 451 inputs and 33 outputs, and a single hidden layer with 55 neurons was found to be enough, yielding an NN SWR code capable of running about 20 times faster than the original SWR code (Krasnopolsky, 2007).

In Krasnopolsky, Fox-Rabinovitz, Hou, et al. (2010), NN models for LWR and SWR were implemented into the higher complexity National Centers for Environmental Prediction (NCEP) coupled Climate Forecast System (CFS), with the NN codes for LWR and SWR running, respectively, 16 and 60 times faster than the original radiation codes. Comparison of parallel decadal climate simulations and seasonal predictions performed with the original NCEP model and with the NN emulations showed differences within the observation errors and uncertainties of reanalysis.

## 17.1.2   Clouds  Ⓒ☺

With a broad range of types and sizes, clouds exert important influence on the hydrological and energy cycles in the global climate system. Unfortunately, global climate models (GCM) cannot resolve the scales of air movements driving individual clouds (typically tens of meters). Cloud parameterization schemes have been used in GCMs to model cloud effects without resolving the full complexity of cloud motions and processes. How clouds respond to anthropogenic global warming is the greatest source of uncertainty in climate projections (Schneider et al., 2017). Thus, modelling the effects of clouds in GCMs is a vital and challenging area of research.

*Cloud resolving models* (CRM) are physics-based numerical models with much finer spatial scales (several kilometres) and temporal scales (seconds to minutes) than a GCM. Global CRMs can be run for months but are too expensive for the multi-decadal runs needed for climate simulation. There is great interest in having an ML model learn from a CRM, then using the ML model to supply moist convection parameterization to a GCM (Fig. 17.1).

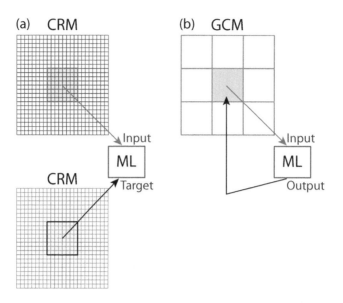

Figure 17.1 Using ML to learn parameterization from a high-resolution numerical model such as a cloud resolving model (CRM). In stage (a), high-resolution data from the CRM are coarse-grained (i.e. averaged over a number of grids to match the coarser GCM grid size), then supplied as input data and target data for training the ML model. In stage (b), the trained ML model is coupled to the GCM, with the ML output supplying moist convection and/or other parameterization to the GCM.

Krasnopolsky, Fox-Rabinovitz, and Belochitski (2013) ran a CRM with a 1-km horizontal solution and 96 vertical layers for 120 days to model a domain of $256 \times 256$ km in the tropical Pacific. An ensemble of 10 MLP NN models, with 36 inputs (temperature and atmospheric moisture), 55 outputs (apparent heat source, apparent moist sink, precipitation and cloudiness) and 5 neurons in a single hidden layer were used to learn the CRM model simulation. The NN approach yields a stochastic parameterization scheme, as different ensemble members tend to converge to different local minima. This NN parameterization scheme was found to give good results when used diagnostically with given GCM data (but was not actually used in a coupled run with the GCM).

When an NN parameterization model developed from CRM is coupled to a GCM, numerical instability can emerge (Brenowitz and Bretherton, 2019). This is not surprising in view of the wild extrapolation behaviour of NN models seen in Section 16.9. Interactions between gravity waves and the parameterized heating (Brenowitz, Beucler, et al., 2020) is believed to cause the coupled simulation to drift beyond the boundary of the training domain for the NN model, whereby extrapolation by the NN model ultimately leads to numerical instability and the model crashing within a few days. Brenowitz and Bretherton (2019) were able to avoid numerical instability by not using the humidity and temperature variables higher in the atmosphere as input to the NN, as well as minimizing the NN objective function over several predicted time steps (Brenowitz and Bretherton, 2018).

Deep NN parameterization models have also been developed by learning from CRMs (Gentine et al., 2018; Rasp, M. S. Pritchard, et al., 2018). An NN model having 9 fully connected layers with 256 neurons per layer was stable when coupled to a GCM, and remained stable even when simulating a climate with the tropical SST perturbed by zonal anomalies, though the accuracy dropped when the anomaly amplitude was $\gtrsim$ 3K.

Besides NN, random forest (RF) has also been used for moist convection parameterization (O'Gorman and Dwyer, 2018; J. Yuval and O'Gorman, 2020). Since the RF output are the means over subsets of the training data, the output satisfy conservation of energy and non-negativity of surface precipitation and do not give rise to wild extrapolation (Section 16.9) nor numerical instability when coupled to a GCM.

On the other hand, NN models require substantially less memory and can run much faster than RF models (J. Yuval, O'Gorman, and Hill, 2021). Beucler, M. Pritchard, Gentine, et al. (2020) and Beucler, M. Pritchard, Rasp, et al. (2021) showed that conservation laws or physical constraints can be implemented in an NN model either approximately by adding a regularization term in the objective function or exactly by adding architectural constraints in the NN model (see physics-informed neural network (PINN) models in Section 17.2).

## 17.1.3 Turbulent Fluxes Ⓒ☺

In numerical modelling of fluid motion, as the small-scale turbulent motion is generally not resolved, parameterization is needed in the numerical models. In the momentum equations, the turbulent fluxes are called *Reynolds stresses*, which are the momentum fluxes exerted by the turbulence on the larger scale flow (a.k.a. mean or time-averaged flow) (Davidson, 2015, chapter 4). Traditionally, parameterization terms known as eddy viscosity are added to the momentum equations as a crude way to incorporate the turbulent or eddy effects on the mean flow.

In *direct numerical simulation* (DNS), the equations of fluid motion (e.g. the Navier–Stokes equations) are numerically solved at very high resolution without any turbulence parameterization. This is computationally extremely expensive as one needs to resolve the whole range of spatial and temporal scales of the turbulence, down to the smallest dissipative scales.

J.-X. Wang et al. (2017) trained a random forest (RF) model to correct crude Reynolds stresses used in a low-resolution model by learning from DNS data. The inputs to the RF model are data for 10 flow features from the low-resolution model, for example pressure gradient, streamline curvature and so on. The output targets are the discrepancy between the Reynolds stresses from the low-resolution model and the stresses from DNS. After training, the RF model provides improved parameterization for the Reynolds stresses when running the low-resolution model.

Besides momentum fluxes, turbulence also contributes *heat fluxes* to the mean flow. C. Xie et al. (2019) used MLP NN (with two hidden layers) to learn heat fluxes from DNS data. The trained NN model provides improved turbulent heat flux parameterization when running the low-resolution model.

Turbulence has an important role in building the surface mixed layer of the ocean. The horizontally-averaged temperature equation is

$$\frac{\partial}{\partial t}\overline{T}(z,t) = -\frac{\partial}{\partial z}\left[\overline{w'T'} - \kappa\frac{\partial\overline{T}}{\partial z}\right], \tag{17.2}$$

where $z$ is the vertical coordinate, the overline denotes horizontal averaging, $w'$ and $T'$ denote the turbulent vertical velocity and temperature, respectively, and $\kappa\frac{\partial\overline{T}}{\partial z}$ denotes the vertical diffusive heat flux with $\kappa$ the diffusivity. Learning from high-resolution 3-D simulation data, an MLP NN (with two hidden layers) takes $\overline{T}$ at 32 vertical grid points and the prescribed surface heat flux $Q_T$ as input and outputs the vertical turbulent heat flux $\overline{w'T'}$ at 33 levels in the water column (Ramadhan et al., 2020). The horizontally-averaged temperature equation becomes

$$\frac{\partial}{\partial t}\overline{T}(z,t) = -\frac{\partial}{\partial z}\left[\text{NN}(\overline{T},Q_T) - \kappa\frac{\partial\overline{T}}{\partial z}\right]. \tag{17.3}$$

In other words, the NN model provides a 1-D parameterization of the vertical turbulent heat flux in the water column based on high-resolution 3-D simulations. This approach can be used in realistic 3-D ocean simulation of a gyre by resolving vertical mixing at every column of the 3-D model.

## 17.1.4   Hybrid Coupled Atmosphere–Ocean Modelling   C☺

In numerical modelling of the coupled atmosphere–ocean system, the large density difference between air and water leads to very different dynamical behaviour and computational requirements for the two fluids. As there are fast waves in the atmosphere, numerical integration with small time steps are needed to keep the atmospheric model stable, hence large computational cost.

In the tropical Pacific, a hybrid approach has been used to replace the costly dynamical atmosphere in the coupled ocean–atmosphere system with a statistical or ML model, that is, a dynamical ocean model is coupled to an inexpensive statistical/ML atmospheric model. Originally, only linear statistical models were used to predict wind stress from upper ocean variables, with the predicted wind stress in turn driving the ocean (Barnett, Latif, et al., 1993; Balmaseda et al., 1994; Syu and Neelin, 2000).

Y. Tang, Hsieh, et al. (2001) used MLP NN to build a non-linear regression model for the tropical wind stress anomalies. A six-layer dynamical ocean model of the tropical Pacific was driven by observed wind stress. Leading principal components (PC) of the model upper ocean heat content anomalies and PCs of the observed wind stress anomalies were computed, and NN models were built using the heat content PCs as inputs and the wind stress PCs as outputs, so that given a heat content anomaly field in the ocean model, a simultaneous wind stress anomaly field can be predicted.

Next, the empirical atmospheric model using NN was coupled to the dynamical ocean model, with the upper ocean heat content giving an estimate of the wind stress via the NN model, and the wind stress in turn driving the dynamical ocean model (Y. Tang, 2002). When used for forecasting the El Niño-Southern Oscillation (ENSO), the ocean model was first driven by the observed wind stress to time $t_1$, then to start forecasting at $t_1$, the NN atmospheric model was coupled to the ocean model (Y. Tang and Hsieh, 2002). Adding data assimilation further improved the forecasting capability of this hybrid coupled model (Y. Tang and Hsieh, 2003).

In the hybrid coupled model of S. Li et al. (2005), the dynamical ocean model from Zebiak and Cane (1987) was coupled to an NN model of the wind stress. The original ocean model had a rather simple parameterization of $T_{sub}$ (the ocean temperature anomaly below the mixed layer) in terms of the thermocline depth anomaly $h$. In S. Li et al. (2005), the parameterization was improved by using an NN model to estimate $T_{sub}$ from $h$. A similar hybrid coupled model was used to study how ENSO properties changed when the background climate state changed (Z. Ye and Hsieh, 2006).

## 17.1.5 Wind Wave Modelling Ⓒ☺

Incoming water waves with periods of several seconds observed by sunbathers on a beach are called *wind waves*, as they are surface gravity waves (LeBlond and Mysak, 1978; A. E. Gill, 1982) generated by the wind. As the ocean surface allows efficient propagation of wind waves, they are usually generated by distant storms, often thousands of kilometres away from the calm sunny beach.

The wave energy spectrum $F$ on the ocean surface is a function of the two-dimensional horizontal wave vector $\mathbf{k}$. The wavelength is $2\pi\|\mathbf{k}\|^{-1}$, while the wave's phase propagation is along the direction of $\mathbf{k}$. The evolution of $F(\mathbf{k})$ is described by

$$\frac{dF}{dt} = S_{in} + S_{nl} + S_{ds} + S_{sw}, \qquad (17.4)$$

where $S_{in}$ is the input source term, $S_{nl}$ the non-linear wave–wave interaction term, $S_{ds}$ the dissipation term, and $S_{sw}$ is the term incorporating shallow-water effects, with $S_{nl}$ being the most complicated one. $S_{nl}$ can be computed numerically by a six-dimensional integration (S. Hasselmann and K. Hasselmann, 1985), which requires $10^3$–$10^4$ more computation than all other terms in the model. An approximation, for example the discrete interaction approximation (DIA) (S. Hasselmann, K. Hasselmann, et al., 1985), has to be made to reduce the amount of computation for operational wind wave forecast models.

Krasnopolsky, Chalikov, et al. (2002) and Tolman et al. (2005) proposed the use of MLP NN to map from $F(\mathbf{k})$ to $S_{nl}(\mathbf{k})$ using training data generated from the six-dimensional integration. Since $\mathbf{k}$ is two-dimensional, both $F$ and $S_{nl}$ are 2-D fields, containing of the order of $10^3$ grid points in the $\mathbf{k}$-space. The computational burden is reduced by using PCA on $F(\mathbf{k})$ and $S_{nl}(\mathbf{k})$ and retaining only the leading PCs (about 20–50 for $F$ and 100–150 for $S_{\mathrm{nl}}$) before training the NN model. Once the NN model has been trained, computation of $S_{\mathrm{nl}}$ from $F$ can be obtained from the NN model in place of the costly six-dimensional integration. The NN model is about $10^5$ times faster than the original approach and only seven times slower than DIA (but nearly ten times more accurate than DIA) (Krasnopolsky, 2007).

## 17.2 Physics-Informed Machine Learning $\boxed{\text{C}}$☺

A common problem with applying ML models to environmental science is that the available data are not abundant. However, there are often equations of physics governing the environmental system that can provide constraints for the ML models, thereby allowing ML models to be built with much smaller datasets. The other problem with ML models is that the model solutions do not in general obey the laws of physics, for example conservation of energy, mass, and so on.

*Physics-informed machine learning* (PIML) (Karniadakis et al., 2021), also known as theory-guided data science (Karpatne et al., 2017), uses equations of physics as constraints in ML models. When neural network models are used, PIML is also referred to as *physics-informed neural network* (PINN) (Raissi et al., 2019). There are two main approaches to implementing physics constraints – '*soft*' constraints implemented as an extra regularization term in the objective function of the NN model and '*hard*' constraints implemented by fixed conservation layers that enforce the constraints to machine precision. With PIML or PINN, one can be ensured that the model solution either satisfies the conservation laws approximately, in the case of soft constraints, or exactly (i.e. to machine precision), in the case of hard constraints.

### 17.2.1 Soft Constraint $\boxed{\text{C}}$☺

Let us first consider the soft constraint approach. For a simple example, consider a field $u(x, t)$ obeying the *viscous Burgers' equation*,

$$\frac{\partial u}{\partial t} + u\frac{\partial u}{\partial x} - \nu\frac{\partial^2 u}{\partial x^2} = 0, \tag{17.5}$$

with $\nu$ a constant diffusion coefficient (a.k.a. kinematic viscosity in fluid mechanics), and with suitable initial and boundary conditions (Karniadakis et al., 2021). In the PINN approach, an NN model is used to represent $u$, that is, the NN has $x$ and $t$ as input and $u$ as output. The objective or loss function

$$J = w_{\mathrm{data}}J_{\mathrm{data}} + w_{\mathrm{phys}}J_{\mathrm{phys}}, \tag{17.6}$$

where

$$J_{\text{data}} = \frac{1}{N_{\text{data}}} \sum_{i=1}^{N_{\text{data}}} \left( u(x_i, t_i) - u_i \right)^2, \tag{17.7}$$

$$J_{\text{phys}} = \frac{1}{N_{\text{phys}}} \sum_{j=1}^{N_{\text{phys}}} \left[ \frac{\partial u}{\partial t} + u \frac{\partial u}{\partial x} - \nu \frac{\partial^2 u}{\partial x^2} \right]^2 \Bigg|_{(x_j, t_j)}. \tag{17.8}$$

There are two sets of points: $\{x_i, t_i\}$ sampled at the initial and boundary locations, and $\{x_j, t_j\}$ sampled in the entire domain. Observations $u_i$ at the points $(x_i, t_i)$, $i = 1, \ldots, N_{\text{data}}$ are compared with the corresponding NN model output $u(x_i, t_i)$ in (17.7). The physics constraint term in (17.8) is evaluated at the points $(x_j, t_j)$, $j = 1, \ldots, N_{\text{phys}}$. *Automatic differentiation* (Baydin et al., 2018) is used to compute the partial derivatives of $u$ in (17.8). When minimizing $J$, one is trying to (i) minimize the MSE between the NN output $u(x_i, t_i)$ and the observed values $u_i$ via $J_{\text{data}}$ and (ii) fit the physics equation via $J_{\text{phys}}$. The hyperparameters $w_{\text{data}}$ and $w_{\text{phys}}$ can be used to weight the two terms $J_{\text{data}}$ and $J_{\text{phys}}$ differently. In other words, the objective can emphasize fitting the model output to the target data more closely or satisfying the physics equations more accurately. What makes PINN different from a traditional NN approach is the addition of the physics constraint term $J_{\text{phys}}$, which serves as a regularization term in the objective function.

Next, we look at a fluid mechanics problem with more variables and more than one spatial dimension. The *Navier–Stokes equation* is a fundamental equation governing many types of fluid flows. In two spatial dimensions, its momentum equations give

$$\begin{aligned} u_t + \lambda_1 (u u_x + v u_y) &= -p_x + \lambda_2 (u_{xx} + u_{yy}), \\ v_t + \lambda_1 (u v_x + v v_y) &= -p_y + \lambda_2 (v_{xx} + v_{yy}), \end{aligned} \tag{17.9}$$

where $u(x, y, t)$ denotes the $x$-component of the velocity field, $v(x, y, t)$ the $y$-component, $p(x, y, t)$ the pressure field, and $\lambda_1$ and $\lambda_2$ denote two unknown parameters to be determined by optimizing an objective function (Raissi et al., 2019). Burgers' equation in (17.5) can be considered a special case of the one-dimensional Navier–Stokes equation with the pressure gradient term neglected.

For an incompressible fluid, conservation of mass gives the continuity equation

$$u_x + v_y = 0. \tag{17.10}$$

A *stream function* $\psi$ is introduced where

$$u \equiv \psi_y, \qquad v \equiv -\psi_x, \tag{17.11}$$

as the continuity equation is automatically satisfied by this assumed form for $u$ and $v$.

An NN model has input $x$, $y$ and $t$ and two outputs $\psi$ and $p$. The two terms in the objective function $J$ in (17.6) now become

$$J_{\text{data}} = \frac{1}{N_{\text{data}}} \sum_{i=1}^{N_{\text{data}}} \left[ \Big(u(x_i, y_i, t_i) - u_i\Big)^2 + \Big(v(x_i, y_i, t_i) - v_i\Big)^2 \right], \quad (17.12)$$

$$J_{\text{phys}} = \frac{1}{N_{\text{phys}}} \sum_{j=1}^{N_{\text{phys}}} \left[ \Big(f(x_j, y_j, t_j)\Big)^2 + \Big(g(x_j, y_j, t_j)\Big)^2 \right], \quad (17.13)$$

where

$$f \equiv u_t + \lambda_1(uu_x + vu_y) + p_x - \lambda_2(u_{xx} + u_{yy}),$$
$$g \equiv v_t + \lambda_1(uv_x + vv_y) + p_y - \lambda_2(v_{xx} + v_{yy}). \quad (17.14)$$

Minimizing $J$ solves for all the weights in the NN model as well as the parameters $\lambda_1$ and $\lambda_2$. Raissi et al. (2019) used a (fully connected) NN model with 9 layers and 20 neurons per layer. Even without supplying any observed values for $p$, the pressure field was accurately predicted (up to an arbitrary constant as (17.9) is unaffected by adding a constant to $p$).

## 17.2.2   Hard Constraint  C☺

As an alternative to the above soft constraint approach (which only imposes the physics constraints approximately), the hard constraint approach of Beucler, M. Pritchard, Rasp, et al. (2021) imposes the physics constraints exactly (i.e. down to machine precision) via architectural constraint on the NN model output.

Given appropriately non-dimensionalized input variables $\mathbf{x} = [x_1, \ldots, x_m]^{\text{T}}$ and output variables $\mathbf{y} = [y_1, \ldots, y_p]^{\text{T}}$, there are $n$ constraint equations from physics ($n < m + p$). For now, consider linear constraints

$$\mathbf{C}\,[\mathbf{x}\ \ \mathbf{y}]^{\text{T}} = 0, \quad (17.15)$$

where $\mathbf{C}$ is a matrix with $n$ rows and $m + p$ columns.

The *architecture-constrained* NN (ACnet) model is built as illustrated in Fig. 17.2. The output variables are divided into two groups, direct outputs, $y_1, \ldots, y_{p-n}$, from the NN model, and the remaining 'residual' outputs, $y_{p-n+1}$, $\ldots$, $y_p$, which have been constrained by the physics by passing through $n$ 'constraint layers'. The NN model is optimized by minimizing the MSE between the outputs (all direct and residual outputs) and corresponding target data.

If the constraint relations are *non-linear*, the above scheme will still work by redefining the input and output variables of the NN model. Let the original input and output variables be $\mathbf{x}^{(0)}$ and $\mathbf{y}^{(0)}$, respectively. The non-linear constraint relations are $\mathbf{c}(\mathbf{x}^{(0)}, \mathbf{y}^{(0)}, \mathbf{z}) = 0$, with $\mathbf{z}$ containing additional parameters involved in the physics constraints. Beucler, M. Pritchard, Rasp, et al. (2021) showed that by introducing new input variables $\mathbf{x}$ and output variables $\mathbf{y}$ for

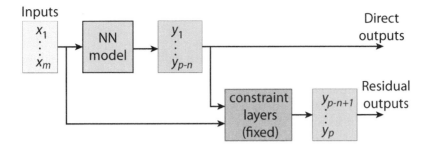

Figure 17.2 The ACnet: With predictors $x_1, \ldots, x_m$, the neural network model generates the direct outputs $y_1, \ldots, y_{p-n}$. The constraint layers take in $x_1, \ldots, x_m$ and $y_1, \ldots, y_{p-n}$, then use the physics constraints to give the residual outputs $y_{p-n+1}, \ldots, y_p$. The NN model weights are optimized by minimizing the MSE between the outputs $y_1, \ldots, y_p$ and corresponding target data. [Follows Beucler, M. Pritchard, Rasp, et al. (2021), figure 2).]

the NN model, where $\mathbf{x}$ is dependent only on $\mathbf{x}^{(0)}$, and $\mathbf{y}$ is dependent on $\mathbf{x}^{(0)}$, $\mathbf{y}^{(0)}$ and $\mathbf{z}$, the non-linear constraints can be rewritten as

$$\mathbf{c}(\mathbf{x}^{(0)}, \mathbf{y}^{(0)}, \mathbf{z}) = \mathbf{C}\,[\mathbf{x} \;\; \mathbf{y}]^{\mathrm{T}} = 0, \qquad (17.16)$$

that is, the constraints are again of the same linear form as in (17.15). The choice of $\mathbf{x}$, $\mathbf{y}$ and $\mathbf{C}$ is not unique.

Upon training on data from high-resolution climate simulations, the ACnet has been applied to emulate the effect of sub-grid scale convective processes in coarse-resolution climate models (Beucler, M. Pritchard, Gentine, et al., 2020); (Beucler, M. Pritchard, Rasp, et al., 2021). The goal for the ACnet is to predict the rate at which subgrid convection vertically redistributes heat and water based on the current large-scale thermodynamic state. The input $\mathbf{x}$ contains 304 variables, among which are five vertical profiles (with 30 levels each) for the specific humidity $\mathbf{q}_v$, the liquid water concentration $\mathbf{q}_l$, the ice concentration $\mathbf{q}_i$, the absolute temperature $\mathbf{T}$ and the north–south velocity $\mathbf{v}$. The output $\mathbf{y}$ contains 216 variables, among which are seven vertical profiles with 30 levels for the time tendencies $\dot{\mathbf{q}}_v$, $\dot{\mathbf{q}}_l$, $\dot{\mathbf{q}}_i$, $\dot{\mathbf{T}}$, and so on. The NN model has 5 hidden layers, each with 512 neurons (using the leaky ReLU activation function), resulting in about 1.3 million model weights (Beucler, M. Pritchard, Rasp, et al., 2021). The four physics constraints ensure the conservation of four quantities: column-integrated energy, mass, long-wave radiation and short-wave radiation. As simulating a warmer climate remains problematic even with physical constraints implemented in the NN model, Beucler, M. Pritchard, Gentine, et al. (2020) alleviated the problem by rescaling variables – for example replacing specific humidity by relative humidity, as relative humidity is expected to change much less than specific humidity in a warmer climate.

# 17.3   Data Assimilation and ML   $\boxed{\text{C}}$ ☺

Even with accurate numerical models, there are many applications where observed data are needed to run the numerical models: (a) In numerical weather prediction (Bauer et al., 2015), where dynamical equations are integrated forward in time, observed data are used to estimate the initial conditions. (b) Some numerical models need to have boundary conditions supplied from observed data, for example an ocean model of a bay with an open boundary. (c) As numerical models usually need parameterization to cover processes that cannot be properly resolved, observed data are needed to estimate the values for these model parameters. How to utilize observed data in numerical models is called *data assimilation*, also referred to as *inverse modelling* (Daley, 1991; Bennett, 2005; Lahoz et al., 2010; Carrassi et al., 2018).

Another important use of data assimilation is to use a numerical model to interpolate observed data. Observations are often not taken over a regular grid in the spatial domain and they can be sparse at the early part of the historical record or in some regions of the world. Assimilating the observed data into a numerical model essentially uses the numerical model to interpolate the data onto regular grids. The output data satisfy conservation laws based on the physics in the numerical model, in contrast to statistical interpolation schemes where conservation laws are not satisfied. Besides the observed variables, the numerical model also provides 'interpolation' for the unobserved variables, for example inferring winds by assimilating satellite-observed radiances, which are sensitive to relative humidity in the upper troposphere (Peubey and McNally, 2009). *Reanalysis* combines past short-range weather forecasts with observations via data assimilation to reconstruct the observed history of the atmosphere (Kalnay, Kanamitsu, et al., 1996; Uppala et al., 2005).

There are many methods used in data assimilation, for example smoothing splines, Kriging, optimal interpolation, successive corrections, constrained initialization, Kalman filter and variational methods, which are all related (Lorenc, 1986). In this section, we focus on the variational methods, especially the widely used 4D-Var method (Kalnay, 2003, chapter 5; Talagrand, 2010; Bonavita et al., 2016).

To illustrate the basic idea of *variational data assimilation*, let us start with the simplest case. Suppose $y_1$ and $y_2$ are two different ways to observe or estimate the true state $x_{\text{true}}$, that is,

$$y_1 = x_{\text{true}} + \epsilon_1, \qquad y_2 = x_{\text{true}} + \epsilon_2, \tag{17.17}$$

where the errors $\epsilon_1$ and $\epsilon_2$ have zero mean and variance $\sigma_1^2$ and $\sigma_2^2$, respectively, that is,

$$\langle \epsilon_1 \rangle = 0, \qquad \langle \epsilon_2 \rangle = 0, \tag{17.18}$$

$$\langle \epsilon_1^2 \rangle = \sigma_1^2, \qquad \langle \epsilon_2^2 \rangle = \sigma_2^2, \tag{17.19}$$

and zero covariance,

$$\langle \epsilon_1 \epsilon_2 \rangle = 0. \tag{17.20}$$

We want to linearly combine $y_1$ and $y_2$ for a better estimate of the true state $x_{\text{true}}$, that is, we let

$$x_{\text{a}} = w_1 y_1 + w_2 y_2. \tag{17.21}$$

If we require $x_{\text{a}}$ to be unbiased, that is, $\langle x_{\text{a}} \rangle = x_{\text{true}}$, then the weights satisfy

$$w_1 + w_2 = 1, \tag{17.22}$$

and

$$x_{\text{a}} = w_1 y_1 + (1 - w_1) y_2. \tag{17.23}$$

Next, find the value of $w_1$ which minimizes the error of $x_{\text{a}}$. In other words, minimize the mean squared error

$$E = \langle (x_{\text{a}} - x_{\text{true}})^2 \rangle \tag{17.24}$$

by setting $dE/dw_1 = 0$, which gives the optimal value of the weights as

$$w_1 = \frac{\sigma_2^2}{\sigma_1^2 + \sigma_2^2}, \qquad w_2 = \frac{\sigma_1^2}{\sigma_1^2 + \sigma_2^2} \tag{17.25}$$

(see Exercise 17.3). Thus, the *best linear unbiased estimator* (BLUE) for combining $y_1$ and $y_2$ is

$$x_{\text{a}} = \frac{\sigma_2^2 \, y_1 + \sigma_1^2 \, y_2}{\sigma_1^2 + \sigma_2^2}. \tag{17.26}$$

Alternatively, the same expression for $x_{\text{a}}$ can be found by minimizing $J$, the objective function (a.k.a. *cost function* in the data assimilation literature), where

$$J(x) = \frac{(x - y_1)^2}{\sigma_1^2} + \frac{(x - y_2)^2}{\sigma_2^2}, \tag{17.27}$$

with $x_{\text{a}} = \arg\min_x (J(x))$ (see Exercise 17.4). Minimizing $J$ amounts to doing a least squares fit between $x$ and the two observations/estimates $y_1$ and $y_2$, weighted, respectively, by $\sigma_1^{-2}$ and $\sigma_2^{-2}$, that is, the reciprocal of their error variance. For instance, if $y_1$ is less accurate than $y_2$, that is, $\sigma_1^2 > \sigma_2^2$, then $y_1$ will be weighted less than $y_2$ in the least squares fit.

Next, generalize from two observations/estimates, $y_1$ and $y_2$, to $m$ observations, $\mathbf{y} = (y_1, y_2, \ldots, y_m)^{\text{T}}$. The error variances generalizes to the error covariance matrix $\mathbf{C}$, and the objective function to

$$J(x) = (x\mathbf{1} - \mathbf{y})^{\text{T}} \mathbf{C}^{-1} (x\mathbf{1} - \mathbf{y}), \tag{17.28}$$

with $\mathbf{1}$ being the $m$-dimensional vector with all elements being 1. If $\mathbf{C}$ is diagonal and $\mathbf{y} = (y_1, y_2)^{\text{T}}$, this equation reduces to (17.27).

## 17.3.1   3D-Var  Ⓒ☹

In numerical weather prediction (NWP), a dynamical model is integrated forward in time to predict future weather.[2] The success of NWP depends critically on being able to supply accurate initial conditions for integrating the dynamical model. The process of estimating the initial conditions in NWP is called *analysis*. The analysis $\mathbf{x}_a$ is obtained from combining two sources of information: (i) a short-term forecast $\mathbf{x}_b$ made earlier and (ii) recent observed data $\mathbf{y}$. In other words, the short-term forecast $\mathbf{x}_b$ (also referred to as the *background state*) needs to be corrected by the recent observed data $\mathbf{y}$ to come up with an improved estimate $\mathbf{x}_a$, which serves as the initial condition for integrating the dynamical model forward in time to produce a new forecast.

The objective function for the *3D-Var* method is

$$J(\mathbf{x}) = (\mathbf{x} - \mathbf{x}_b)^T \mathbf{B}^{-1}(\mathbf{x} - \mathbf{x}_b) + \left[\mathcal{H}(\mathbf{x}) - \mathbf{y}\right]^T \mathbf{R}^{-1}\left[\mathcal{H}(\mathbf{x}) - \mathbf{y}\right], \quad (17.29)$$

where $\mathbf{B}$ is the background error covariance, that is, the error covariance matrix of the forecast variables $\mathbf{x}_b$, and $\mathbf{R}$ is the error covariance matrix of the observations $\mathbf{y}$. Since the location of the observations may not coincide with the grid points of the numerical model, the *observation operator* $\mathcal{H}(\mathbf{x})$ represents an interpolation of the analysis $\mathbf{x}$ to the exact location of $\mathbf{y}$.[3] The two terms in $J$ correspond to fitting $\mathbf{x}$ to the background short-term forecast $\mathbf{x}_b$ and to the observed data $\mathbf{y}$. This approach uses variational data assimilation to solve for the three-dimensional structure of the atmosphere or ocean, hence the name 3D-Var.

It is worth noting the similarity between $J$ in (17.29) and $J$ in (8.29) for MLP NN models. The second term on the right hand side in (17.29) is a measure of the (squared) error between the dynamical model $\mathcal{H}(\mathbf{x})$ and the observation $\mathbf{y}$, whereas the first term in (8.29) measures the error between the NN model output $\hat{y}$ and the observation $y$. The first term in (17.29) is a regularization term to ensure $\mathbf{x}$ is not too far from the background $\mathbf{x}_b$, analogous to the second term in (8.29) which regularizes the weight $w$ to be not too far from 0.

To find $\mathbf{x}_a = \arg\min_{\mathbf{x}}(J(\mathbf{x}))$, the minimization of $J$ requires the gradient of $J$, that is,

$$\nabla J(\mathbf{x}) = 2\mathbf{B}^{-1}(\mathbf{x} - \mathbf{x}_b) + 2\mathbf{H}^T\mathbf{R}^{-1}\left[\mathcal{H}(\mathbf{x}) - \mathbf{y}\right], \quad (17.30)$$

where

$$\mathbf{H} = [H_{ij}] = \left[\frac{\partial \mathcal{H}_i}{\partial x_j}\right] \quad (17.31)$$

is the *Jacobian* matrix of $\mathcal{H}$, and $\mathbf{H}^T$ is the transpose or *adjoint* of $\mathbf{H}$. The gradient descent optimization techniques in Chapter 7, for example the conjugate gradient method (Section 7.7) and the quasi-Newton method (Section 7.8), have been used for data assimilation in operational weather forecasting.

---

[2] Since time is discretized in a numerical model, integration amounts to time-stepping.

[3] In modern data assimilation, the observation operator $\mathcal{H}$ performs more than just interpolation. It has been designed to assimilate complex observation types such as satellite radiances. The direct assimilation of satellite radiances has much advanced weather forecasting since the late 1990s (Eyre et al., 2020).

## 17.3.2   4D-Var  C ☹

An extension of 3D-Var is the *4D-Var* method, which assimilates observed data over a short time window, typically of 6-hour or 12-hour duration (Fig. 17.3). With observations available at times $t_n$ $(n = 0, 1, \ldots, N)$ during the time window, the objective function is

$$J(\mathbf{x}(t_0)) = \big[\mathbf{x}(t_0) - \mathbf{x}_\mathrm{b}(t_0)\big]^\mathrm{T} \mathbf{B}^{-1}\big[\mathbf{x}(t_0) - \mathbf{x}_\mathrm{b}(t_0)\big]$$
$$+ \sum_{n=0}^{N} \big[\mathcal{H}\big(\mathbf{x}(t_n)\big) - \mathbf{y}(t_n)\big]^\mathrm{T} \mathbf{R}_n^{-1}\big[\mathcal{H}\big(\mathbf{x}(t_n)\big) - \mathbf{y}(t_n)\big], \qquad (17.32)$$

where

$\mathbf{x}(t_0)$ is the model state at time $t_0$, which gives the analysis when $J(\mathbf{x}(t_0))$ is at a minimum;

$\mathbf{x}_\mathrm{b}(t_0)$ is the background state at $t_0$ (usually a short-term forecast from the previous analysis);

$\mathbf{B}$ is the background error covariance matrix;

$\mathbf{y}(t_n)$ is the observation vector at time $t_n$ $(n = 0, 1, \ldots, N)$;

$\mathbf{R}_n$ is the observation error covariance matrix at time $t_n$;

$\mathcal{H}\big(\mathbf{x}(t_n)\big)$ is the observation operator at $t_n$ and

$\mathbf{x}(t_n)$ is the model state at $t_n$, obtained by integrating the dynamical model $\mathcal{M}$ from time $t_0$ to $t_n$.

A more compact notation for $J$ is

$$J(\mathbf{x}(t_0)) = \big\|\mathbf{x}(t_0) - \mathbf{x}_\mathrm{b}(t_0)\big\|^2_{\mathbf{B}^{-1}} + \sum_{n=0}^{N} \big\|\mathcal{H}\big(\mathbf{x}(t_n)\big) - \mathbf{y}(t_n)\big\|^2_{\mathbf{R}_n^{-1}}, \qquad (17.33)$$

where the matrix norm notation $\|\mathbf{v}\|^2_\mathbf{M} \equiv \mathbf{v}^\mathrm{T}\mathbf{M}\mathbf{v}$.

If the summation has only one term $(n = 0)$ in (17.32), that is, only observations at $t_0$ are assimilated, the 4D-Var objective function reduces to that of the 3D-Var in (17.29). Details on how to find the optimal initial condition $\mathbf{x}_\mathrm{a}(t_0)$ by minimizing $J$ using the gradient $\nabla J$ are given in Kalnay (2003, appendix B) and (Talagrand, 2010, section 3). Although it is most common to use variational assimilation to estimate initial conditions of dynamical models, the method can also be used to estimate other model parameters, or both initial conditions and model parameters.

4D-Var is heavily used in many forecasting centres, for example ECMWF, Météo-France and the UK Met Office (Hatfield et al., 2021).

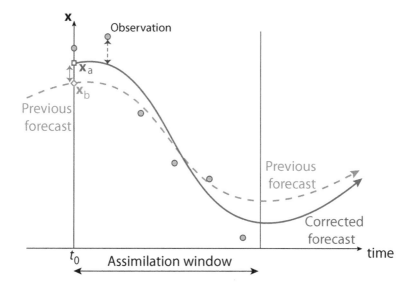

Figure 17.3 In 4D-Var, observations are assimilated over a time window starting at $t_0$. The solid curve, generated by integrating the dynamical model, is fitted to the observations, as well as to the background forecast $\mathbf{x}_b$ at $t_0$, by minimizing the objective or cost function $J$. The optimally estimated $\mathbf{x}_a$ at $t_0$ serves as the initial condition for integrating the dynamical model forward in time to generate future forecasts.

### 17.3.3   Neural Networks in 4D-Var   [C]☺

Neural network modelling and variational data assimilation, with their common approach of minimizing an objective function, are structurally very similar despite using completely different jargons (Hsieh and B. Tang, 1998). For instance, the NN jargon for 'back-propagation' in MLP models is analogous to 'backward integration of the adjoint model' in variational data assimilation. Their similar structure allows an NN model to be coupled to a dynamical model by minimizing a single objective function $J$.

Suppose we have a dynamical model $\mathcal{M}$ with governing equations in discrete time,

$$\mathbf{u}(t + \delta t) = \mathcal{M}(\mathbf{u}, \mathbf{v}, \mathbf{p}, t),  \tag{17.34}$$

where $\mathbf{u}$ denotes the vector of state variables in the dynamical model, $\mathbf{v}$ denotes the vector of variables not modelled by the dynamical model and $\mathbf{p}$ denotes a vector of model parameters and/or initial conditions. Suppose the $\mathbf{v}$ variables, which could not be modelled or forecast well by a dynamical model with reasonable computing resources, could be forecast by an MLP NN model, that is,

$$\mathbf{v}(t + \Delta t) = \mathcal{N}(\mathbf{u}, \mathbf{v}, \mathbf{q}, t),  \tag{17.35}$$

where the NN model $\mathcal{N}$ has inputs $\mathbf{u}$ and $\mathbf{v}$, and parameters or weights $\mathbf{q}$. A single objective function can be used for variational data assimilation for this coupled dynamical and NN system (Hsieh and B. Tang, 1998, appendix). This approach contrasts with the traditional approach of using post-processing (Section 16.10) to estimate the variables $\mathbf{v}$ not computed by the dynamical model.

A proof-of-concept study was carried out by Y. Tang and Hsieh (2001) using the simple Lorenz (1963) three-component non-linear system, renowned in the development of chaos theory (Gleick, 1987). Describing a two-dimensional fluid layer uniformly warmed from below and cooled from above, the Lorenz system contains three differential equations:

$$\frac{\mathrm{d}x_1}{\mathrm{d}t} = \sigma(x_2 - x_1), \tag{17.36}$$

$$\frac{\mathrm{d}x_2}{\mathrm{d}t} = rx_1 - x_2 - x_1x_3, \tag{17.37}$$

$$\frac{\mathrm{d}x_3}{\mathrm{d}t} = x_1x_2 - bx_3, \tag{17.38}$$

where $x_1$ is proportional to the rate of convection, $x_2$ to the horizontal temperature variation and $x_3$ to the vertical temperature variation, and the three parameters, $\sigma$, $r$, $b$, are proportional to the Prandtl number, Rayleigh number and certain physical dimensions of the fluid layer, respectively.

Assuming the third equation is unavailable, Y. Tang and Hsieh (2001) replaced (17.38) by an MLP NN model $\mathcal{N}$

$$\frac{\mathrm{d}x_3}{\mathrm{d}t} = \mathcal{N}(x_1, x_2, x_3). \tag{17.39}$$

Thus, the three-component system contained two dynamical equations (17.36) and (17.37) and one neural network equation (17.39), with a single objective function used in 4D-Var data assimilation. Three separate assimilation experiments were run to estimate: (i) the NN model weights (26 weights), (ii) two dynamical parameters ($\sigma$ and $r$) and three initial conditions for $x_1$, $x_2$ and $x_3$ and (iii) the dynamical parameters, initial conditions and NN weights (28 weights/parameters plus three initial conditions). The results show that dynamical equations and NN equations can be run together under variational assimilation with a single objective function.

Geer (2021) presented a *Bayesian* perspective on data assimilation with dynamical and ML models. Both data assimilation and ML solve an inverse problem. The corresponding forward problem is of the form

$$\mathbf{y} = \mathcal{F}(\mathbf{x}, \mathbf{w}), \tag{17.40}$$

where a function (or model) $\mathcal{F}$, controlled by some parameters/weights $\mathbf{w}$, maps from a state $\mathbf{x}$ to an observation $\mathbf{y}$. The inverse problem is to find the state $\mathbf{x}$ and/or parameters $\mathbf{w}$ from the observed data $\mathbf{y}$.

A joint probability distribution can be written in terms of conditional probabilities using the chain rule of probability. For instance, $p(\mathbf{y}, \mathbf{x}, \mathbf{w})$ can be written as

$$p(\mathbf{y}, \mathbf{x}, \mathbf{w}) = p(\mathbf{y}|\mathbf{x}, \mathbf{w}) \, p(\mathbf{x}|\mathbf{w}) \, p(\mathbf{w}), \tag{17.41}$$

and alternatively as

$$p(\mathbf{y}, \mathbf{x}, \mathbf{w}) = p(\mathbf{x}|\mathbf{w}, \mathbf{y}) \, p(\mathbf{w}|\mathbf{y}) \, p(\mathbf{y}). \tag{17.42}$$

Equating the right hand sides in (17.41) and (17.42) and dividing by $p(\mathbf{y})$, we get

$$p(\mathbf{x}|\mathbf{w}, \mathbf{y}) \, p(\mathbf{w}|\mathbf{y}) = \frac{p(\mathbf{y}|\mathbf{x}, \mathbf{w}) \, p(\mathbf{x}|\mathbf{w}) \, p(\mathbf{w})}{p(\mathbf{y})} . \tag{17.43}$$

Replacing $p(\mathbf{x}|\mathbf{w}, \mathbf{y}) \, p(\mathbf{w}|\mathbf{y})$ by $p(\mathbf{x}, \mathbf{w}|\mathbf{y})$ on the left hand side and $p(\mathbf{x}|\mathbf{w}) \, p(\mathbf{w})$ by $p(\mathbf{x}, \mathbf{w})$ on the right hand side, we have

$$p(\mathbf{x}, \mathbf{w}|\mathbf{y}) = \frac{p(\mathbf{y}|\mathbf{x}, \mathbf{w}) \, p(\mathbf{x}, \mathbf{w})}{p(\mathbf{y})} . \tag{17.44}$$

In this equation, $p(\mathbf{y}|\mathbf{x}, \mathbf{w})$ is from the forward model, while $p(\mathbf{x}, \mathbf{w}|\mathbf{y})$ is what we want out of the inverse modelling, that is, estimate $\mathbf{x}$ and $\mathbf{w}$ from observation $\mathbf{y}$. If $\mathbf{x}$ and $\mathbf{w}$ can be considered statistically independent, then $p(\mathbf{x}, \mathbf{w})$ can be replaced by $p(\mathbf{x}) \, p(\mathbf{w})$, giving

$$p(\mathbf{x}, \mathbf{w}|\mathbf{y}) = \frac{p(\mathbf{y}|\mathbf{x}, \mathbf{w}) \, p(\mathbf{x}) \, p(\mathbf{w})}{p(\mathbf{y})} . \tag{17.45}$$

To keep the discussion simple, we follow Geer (2021) in replacing the vectors $\mathbf{y}$, $\mathbf{x}$ and $\mathbf{w}$ by scalars $y$, $x$ and $w$. With the forward model $\mathcal{F}(x, w)$ and assuming a Gaussian distribution of observational errors, the conditional probability of observing a particular value of $y$ is then

$$p(y|x, w) = \frac{1}{c_1} \exp\left(-\frac{1}{2} \frac{[y - \mathcal{F}(x, w)]^2}{\sigma_y^2}\right), \tag{17.46}$$

where $c_1$ is a normalization constant and $\sigma_y^2$ is the variance. Assume the model $\mathcal{F}(x, w)$ to be perfect, so model errors only arise from the error in the model parameter $w$. We also assume Gaussian distributions for $p(x)$ and $p(w)$, that is,

$$p(x) = \frac{1}{c_2} \exp\left(-\frac{1}{2} \frac{(x - x_b)^2}{\sigma_x^2}\right), \quad p(w) = \frac{1}{c_3} \exp\left(-\frac{1}{2} \frac{(w - w_b)^2}{\sigma_w^2}\right). \tag{17.47}$$

The objective function $J$ can be written as

$$J(x, w) = -2 \ln\left[p(x, w|y)\right] + c, \tag{17.48}$$

where $p(x, w|y)$ can be decomposed according to (17.45) and $c$ takes care of all the normalization constants (i.e. $c_1$, $c_2$, $c_3$ and $p(y)$) from $p(x, w|y)$ to give a cleaner form for $J$, that is,

$$J(x, w) = \frac{[y - \mathcal{F}(x, w)]^2}{\sigma_y^2} + \frac{(x - x_b)^2}{\sigma_x^2} + \frac{(w - w_b)^2}{\sigma_w^2} . \tag{17.49}$$

This is of the same form as (17.29) (but for the simplification from vector to scalar) and has added the third term on the right hand side, as we now try to estimate $w$ as well as $x$. Initial estimates $x_b$ and $w_b$ are supplied, for example in NWP, $x_b$ is the background from a previous forecast. This Bayesian approach allows $J$ to be derived from a probabilistic framework and treats $x$ and $w$ the same way, thereby placing the estimation of initial conditions and model parameters/weights on equal footing in variational assimilation. $J$ in (8.29) for MLP NN models can be viewed as a special case of (17.49), which has an extra regularization term $(x - x_b)^2/\sigma_x^2$.

While it is possible to use an NN model to replace the dynamical model entirely in variational data assimilation, as demonstrated with low-order chaotic models by Bocquet et al. (2020), for real-world forecasting problems, dynamical models are likely to continue to play a major role.

In Farchi et al. (2020), an NN model is used to correct the error of the dynamical model $\mathcal{M}$. The 4D-Var objective function $J$ has two terms added to the right hand side of (17.33):

$$+ \sum_{n=0}^{N-1} \left\| \left[ \mathcal{M}\big(\mathbf{x}(t_n)\big) - \mathbf{x}(t_{n+1}) \right] + \mathcal{N}\big(\mathbf{w}, \mathbf{x}(t_n)\big) \right\|_{\mathbf{Q}_n^{-1}}^2 + \mathcal{L}(\mathbf{w}), \qquad (17.50)$$

where the term in the square brackets is the error of the dynamical model to be corrected by the NN model $\mathcal{N}$ with weights $\mathbf{w}$, and $\mathcal{L}(\mathbf{w})$ is a regularization term. Forecasts are obtained from the combined output of the dynamical and NN models, that is,

$$\mathbf{x}_a(t_{n+1}) = \mathcal{M}\big(\mathbf{x}_a(t_n)\big) + \mathcal{N}\big(\mathbf{w}, \mathbf{x}_a(t_n)\big), \qquad (17.51)$$

with $\mathbf{x}_a = \arg\min_{\mathbf{x}}(J(\mathbf{x}))$. The approach was tested numerically in 4D-Var using a two-layer, two-dimensional quasi-geostrophic channel model, where model error was introduced by perturbing parameters in the dynamical model (Farchi et al., 2020). NN models were able to learn a substantial part of the model error, and the resulting hybrid models produced better short- to mid-range forecasts than the original dynamical model.

As NN models can be differentiated trivially, they can greatly lighten the burden of generating the tangent-linear and adjoint models needed for 4D-Var. Hatfield et al. (2021) demonstrated this approach by emulating the non-orographic gravity wave drag parametrization scheme in an atmospheric model using an NN model, and deriving its tangent-linear and adjoint models. They concluded that this approach holds the promise of significantly easing the maintenance of tangent-linear and adjoint codes in weather forecasting centres.

Besides 4D-Var, the *ensemble Kalman filter* is another common method used in data assimilation (Kalnay, H. Li, et al., 2007). NN models are also being incorporated into data assimilation using the ensemble Kalman filter (Brajard et al., 2020, 2021).

## Exercises

### 17.1

For an NN model with original inputs $\mathbf{x}^{(0)} = \left(x_1^{(0)}, x_2^{(0)}\right)$ and output $y^{(0)}$, satisfying the hard constraint $\left(x_1^{(0)}\right)^2 + \left(x_2^{(0)}\right)^2 + \left(y^{(0)}\right)^2 = a^2$, with constant $a$, convert the non-linear constraint relation to a linear constraint by a change of variables as in (17.16).

### 17.2

Consider a biological system governed by the *Lotka–Volterra equations*,

$$\dot{x} = \alpha x - \beta xy,$$
$$\dot{y} = \delta xy - \gamma y, \tag{17.52}$$

where $x$ is the predator population, $y$ the prey population, $\dot{x}$ and $\dot{y}$ are their time derivatives, and $\alpha$, $\beta$, $\delta$ and $\gamma$ are constants.
An NN model with original inputs $x^{(0)}$ and $y^{(0)}$ and outputs $\dot{x}^{(0)}$ and $\dot{y}^{(0)}$ satisfies the hard constraints

$$x_{t+1}^{(0)} - x_t^{(0)} = \alpha\, x_t^{(0)} - \beta\, x_t^{(0)}\, y_t^{(0)},$$
$$y_{t+1}^{(0)} - y_t^{(0)} = \delta\, x_t^{(0)}\, y_t^{(0)} - \gamma\, y_t^{(0)}, \tag{17.53}$$

where the time derivative $\dot{x}$ has been approximated by a forward Euler finite difference approximation, with the subscripts $t+1$ and $t$ indicating time $t + \Delta t$ and $t$, respectively (with $\Delta t$ taken to be 1 for simplicity). Convert the non-linear constraint relations to linear constraints by a change of variables as in (17.16).

### 17.3

Derive the optimal value of the weights in (17.25).

### 17.4

Show that the minimum of the objective function in (17.27) occurs when $x_\mathrm{a}$ is given by (17.26).

# Appendices

## A   Trends in Terminology

Google Trends is a website by Google that compares the relative search volume of various queries over time. Searches were carried out for terms such as 'machine learning', 'artificial intelligence', and so on on the Google Trends website to see which terms are becoming more/less popular in usage over time. Google Trends scales the monthly worldwide search volume data to 100 for the maximum value among the multiple terms searched. The median over the 12 months of each year is shown in Fig. A1. In Fig. A1(a), 'machine learning' has surpassed 'artificial intelligence' in search volume since 2015, while 'data science' has also surpassed 'artificial intelligence' recently. In contrast, 'data mining' and 'informatics' have declining trends. In Fig. A1(b), 'Big data' had a spectacular rise starting in 2011, but has already peaked. Search volumes for 'data driven', 'soft computing' and 'computational intelligence' have been low, especially for for the latter two.

## B   Lagrange Multipliers

Let us consider the problem of finding the maximum of a function $f(\mathbf{x})$ subject to constraints on $\mathbf{x}$ (if we need to find the minimum of $f(\mathbf{x})$, the problem is equivalent to finding the maximum of $-f(\mathbf{x})$, so we need only to consider the case of finding the maximum). We will consider two types of constraint: (a) equality constraints such as $g(\mathbf{x}) = 0$ and (b) inequality constraints such as $g(\mathbf{x}) \geq 0$. Other constraints forms such as $\tilde{g}(\mathbf{x}) \leq 0$ or $\hat{g}(\mathbf{x}) \leq b$ can be converted to (b) by letting $g = -\tilde{g}$ and $g = b - \hat{g}$, respectively.

First, consider type (a), with the constraint $g(\mathbf{x}) = 0$ describing a surface in the $\mathbf{x}$-space (Fig. B1). The gradient vector $\nabla g(\mathbf{x})$ is normal to this surface. Suppose at the point $\mathbf{x}_0$ lying on this surface, the maximum value of $f$ occurs. The gradient vector $\nabla f(\mathbf{x})$ must also be normal to the surface at $\mathbf{x}_0$; otherwise, $f$ can be increased if we move on the surface to a point $\mathbf{x}_1$ slightly to the side of $\mathbf{x}_0$, which contradicts the assumption that $f$ at $\mathbf{x}_0$ is the maximum on the surface. This means $\nabla f$ and $\nabla g$ are parallel to each other at $\mathbf{x}_0$ (but may point in opposite directions), thus

$$\nabla f + \lambda \nabla g = 0, \tag{B.1}$$

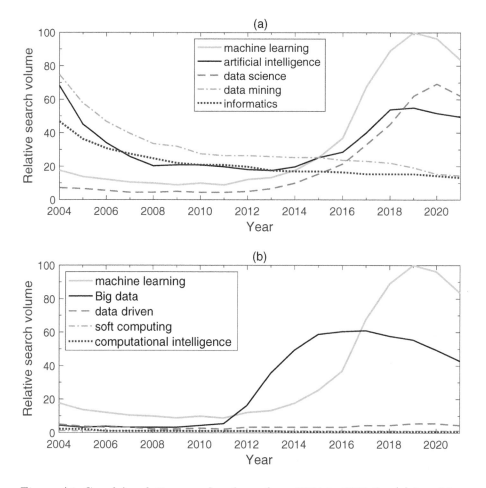

Figure A1 Google's relative search volume from 2004 to 2021 for (a) 'machine learning', 'artificial intelligence', 'data science', 'data mining' and 'informatics' and (b) 'machine learning', 'Big data', 'data driven', 'soft computing' and 'computational intelligence'. [Data source: Google Trends.]

for some $\lambda \neq 0$. This $\lambda$ parameter is called a *Lagrange multiplier* and can have either positive or negative sign.

The *Lagrangian function* is defined by

$$L(\mathbf{x}, \lambda) = f(\mathbf{x}) + \lambda\, g(\mathbf{x}). \tag{B.2}$$

From $\nabla_{\mathbf{x}} L = 0$, we obtain (B.1), while $\partial L / \partial \lambda = 0$ gives the original constraint $g(\mathbf{x}) = 0$. If $\mathbf{x}$ is of dimension $m$, the original constrained maximization problem is solved by finding the stationary point of $L$ with respect to $\mathbf{x}$ and to $\lambda$, that is, use $\nabla_{\mathbf{x}} L = 0$ and $\partial L / \partial \lambda = 0$ to provide $m + 1$ equations for solving the values of the stationary point $\mathbf{x}_0$ and $\lambda$.

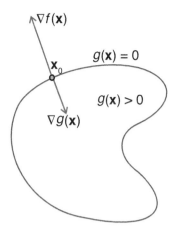

Figure B1 Illustrating the situation where the maximum of $f(\mathbf{x})$ occurs at a point $\mathbf{x}_0$ on the surface described by $g(\mathbf{x}) = 0$. Both gradient vectors $\nabla f(\mathbf{x})$ and $\nabla g(\mathbf{x})$ are normal to the surface at $\mathbf{x}_0$. Here, the interior is assumed to have $g(\mathbf{x}) > 0$, so the gradient vector $\nabla g$ points to the interior.

Next, consider a type (b) constraint, that is, $g(\mathbf{x}) \geq 0$. There are actually two situations to consider. The first situation is when the constrained stationary point lies on the boundary $g(\mathbf{x}) = 0$, while in the second situation, the point lies in the region $g(\mathbf{x}) > 0$. In the first situation, we are back to the previous case of a type (a) constraint. However, this time $\lambda$ is not free to take on either sign. If $f(\mathbf{x})$ is a maximum at a point $\mathbf{x}_0$ on the surface $g(\mathbf{x}) = 0$, its gradient $\nabla f$ must point opposite to $\nabla g$ (Fig. B1) (otherwise $f$ increases in the region $g(\mathbf{x}) > 0$, contradicting that $f$ is maximum on the boundary surface). Thus, $\nabla f(\mathbf{x}) = -\lambda \nabla g(\mathbf{x})$ for some $\lambda > 0$. In the second situation, $g(\mathbf{x})$ does not affect the maximization of $f(\mathbf{x})$, so $\lambda = 0$ and $\nabla_\mathbf{x} L = 0$ gives back $\nabla f(\mathbf{x}) = 0$.

In either situation, $\lambda\, g(\mathbf{x}) = 0$. Thus, the problem of maximizing $f(\mathbf{x}) = 0$ subject to $g(\mathbf{x}) \geq 0$ is solved by finding the stationary point of the Lagrangian (B.2) with respect to $\mathbf{x}$ and $\lambda$, subject to the *Karush–Kuhn–Tucker* (KKT) *conditions* (Karush, 1939; Kuhn and Tucker, 1951)

$$\lambda \geq 0, \tag{B.3}$$

$$g(\mathbf{x}) \geq 0, \tag{B.4}$$

$$\lambda\, g(\mathbf{x}) = 0. \tag{B.5}$$

Next, instead of maximization, *minimization* of $f(\mathbf{x})$ subject to $g(\mathbf{x}) \geq 0$ is sought. In the situation where the stationary point is on the boundary surface $g(\mathbf{x}) = 0$, $\nabla f$ must point in the same direction as $\nabla g$, that is, $\nabla f = \lambda \nabla g$, with $\lambda$ positive. Thus, the Lagrangian function to be used for the minimization problem with inequality constraint is

$$L(\mathbf{x}, \lambda) = f(\mathbf{x}) - \lambda\, g(\mathbf{x}), \tag{B.6}$$

with $\lambda \geq 0$.

Finally, if $f(\mathbf{x})$ is to be maximized subject to multiple constraints, $g_i(\mathbf{x}) = 0$ $(i = 1, \ldots, I)$ and $h_j(\mathbf{x}) \geq 0$ $(j = 1, \ldots, J)$, then the Lagrangian function becomes

$$L(\mathbf{x}, \boldsymbol{\lambda}, \boldsymbol{\mu}) = f(\mathbf{x}) + \sum_i \lambda_i g_i(\mathbf{x}) + \sum_j \mu_j h_j(\mathbf{x}). \tag{B.7}$$

The KKT conditions also give $\mu_j \geq 0$ and $\mu_j h_j(\mathbf{x}) = 0$ for $j = 1, \ldots, J$.

# References

Abraham, B. & Ledolter, J. (1983). *Statistical Methods for Forecasting.* Wiley.

Abrahart, R. J., Anctil, F., Coulibaly, P., Dawson, C. W., Mount, N. J., See, L. M., ... Wilby, R. L. (2012). Two decades of anarchy? Emerging themes and outstanding challenges for neural network river forecasting. *Progress in Physical Geography–Earth and Environment, 36,* 480–513.

Abrahart, R. J., Kneale, P. E., & See, L. M. (Eds.). (2004). *Neural Networks for Hydrological Modelling.* CRC Press.

Addor, N., Nearing, G., Prieto, C., Newman, A. J., Le Vine, N., & Clark, M. P. (2018). A ranking of hydrological signatures based on their predictability in space. *Water Resources Research, 54,* 8792–8812.

Aggarwal, C. C. (2018). *Neural Networks and Deep Learning: A Textbook.* Springer.

Aguilar-Martinez, S. & Hsieh, W. W. (2009). Forecasts of tropical Pacific sea surface temperatures by neural networks and support vector regression. *International Journal of Oceanography, 2009,* Article ID 167239.

Akaike, H. (1973). Information theory and an extension of the maximum likelihood principle. In Petrov, B. N. & Csáki, F. (Eds.), *2nd International Symposium on Information Theory, Tsahkadsor, Armenia, USSR, September 2–8, 1971* (pp. 267–281). Republished in Kotz, S. & Johnson, N. L. (Eds.) (1992), *Breakthroughs in Statistics, I,* Springer-Verlag, pp. 610–624.

Akaike, H. (1974). A new look at the statistical model identification. *IEEE Trans. Automatic Control, 19,* 716–723.

Alexandridis, A. K. & Zapranis, A. D. (2013). Wavelet neural networks: A practical guide. *Neural Networks, 42,* 1–27.

Ali, M. M., Swain, D., & Weller, R. A. (2004). Estimation of ocean subsurface thermal structure from surface parameters: A neural network approach. *Geophysical Research Letters, 31,* L20308.

Allen, A. N., Harvey, M., Harrell, L., Jansen, A., Merkens, K. P., Wall, C. C., ... Oleson, E. M. (2021). A convolutional neural network for automated detection of humpback whale song in a diverse, long-term passive acoustic dataset. *Frontiers in Marine Science, 8,* 165.

Allen, M. R. & Stott, P. A. (2003). Estimating signal amplitudes in optimal fingerprinting, part I: Theory. *Climate Dynamics, 21,* 477–491.

Allen, M. R. & Tett, S. F. B. (1999). Checking for model consistency in optimal fingerprinting. *Climate Dynamics, 15,* 419–434.

Amari, S., Murata, N., Müller, K.-R., Finke, M., & Yang, H. (1996). Statistical theory of overtraining: Is cross validation asymptotically effective? *Advances in Neural Information Processing Systems, 8,* 176–182.

An, S. I., Hsieh, W. W., & Jin, F. F. (2005). A nonlinear analysis of the ENSO cycle and its interdecadal changes. *Journal of Climate, 18,* 3229–3239.

An, S. I., Ye, Z. Q., & Hsieh, W. W. (2006). Changes in the leading ENSO modes associated with the late 1970s climate shift: Role of surface zonal current. *Geophysical Research Letters, 33,* L14609, doi:10.1029/2006GL026604.

Anderson, T. W. & Darling, D. A. (1954). A test of goodness of fit. *Journal of the American Statistical Association, 49,* 765–769.

Argueso, D., Evans, J. P., Fita, L., & Bormann, K. J. (2014). Temperature response to future urbanization and climate change. *Climate Dynamics, 42,* 2183–2199.

Arjovsky M., Chintala, S., & Bottou, L. (2017). Wasserstein generative adversarial networks. In *Proceedings of the 34th International Conference on Machine Learning* (Vol. 70, pp. 214–223). Proceedings of Machine Learning Research.

Arnold, T. B. & Emerson, J. W. (2011). Nonparametric goodness-of-fit tests for discrete null distributions. *The R Journal, 3,* 34–39.

Arpit, D., Zhou, Y., Ngo, H., & Govindaraju, V. (2016). Why regularized auto-encoders learn sparse representation? In *Proceedings of the 33rd International Conference on Machine Learning* (Vol. 48, pp. 136–144). Proceedings of Machine Learning Research.

Arthur, D. & Vassilvitskii, S. (2007). K-means++ : The advantages of careful seeding. In *Proceedings of the Eighteenth An-*

nual *ACM-SIAM Symposium on Discrete Algorithms* (pp. 1027–1035).

Ashouri, H., Hsu, K.-L., Sorooshian, S., Braithwaite, D. K., Knapp, K. R., Ceceil, L. D., ... Prat, O. P. (2015). PERSIAN-NCDR: Daily precipitation climate data record from multisatellite observations for hydrological and climate studies. *Bulletin of the American Meteorological Society*, *96*, 69–83.

Atkinson, P. M. (2013). Downscaling in remote sensing. *International Journal of Applied Earth Observation and Geoinformation*, *22*, 106–114.

Atkinson, P. M. & Tatnall, A. R. L. (1997). Introduction: Neural networks in remote sensing. *International Journal of Remote Sensing*, *18*, 699–709.

Bacao, F., Lobo, V., & Painho, M. (2005). Self-organizing maps as substitutes for k-means clustering. In Sunderam, V. S., VanAlbada, G. D., Sloot, P. M. A., & Dongarra, J. J. (Eds.), *Computational Science – ICCS 2005, Pt 3* (Vol. 3516, 476–483). Lecture Notes in Computer Science. 5th International Conference on Computational Science (ICCS 2005), Atlanta, GA, 22–25 May, 2005.

Badran, F., Thiria, S., & Crépon, M. (1991). Wind ambiguity removal by the use of neural network techniques. *Journal of Geophysical Research*, *96*, 20521–20529.

Bai, S., Kolter, J. Z., & Koltun, V. (2018). An empirical evaluation of generic convolutional and recurrent networks for sequence modeling. *CoRR*, *abs/1803.01271*. arXiv: 1803.01271

Baik, J.-J. & Hwang, H.-S. (1998). Tropical cyclone intensity prediction using regression method and neural network. *Journal of Meteorological Society Japan*, *76*, 711–717.

Baird, J. C. & Kunkel, K. E. (2019). Automated detection of weather fronts using a deep learning neural network. *Advances in Statistical Climatology, Meteorology and Oceanography*, *5*, 147–160.

Bakir, G. H., Weston, J., & Schölkopf, B. (2004). Learning to find pre-images. *Advances in Neural Information Processing Systems*, *16*, 449–456.

Balakrishnan, P. V., Cooper, M. C., Jacob, V. S., & Lewis, P. A. (1994). A study of the classification capabilities of neural networks using unsupervised learning: A comparison with k-means clustering. *Psychometrika*, *59*, 509–525.

Baldwin, M. P., Gray, L. J., Dunkerton, J., Hamilton, K., Haynes, P. H., Randel, W. J., ... Takahashi, M. (2001). The Quasi-Biennial Oscillation. *Reviews of Geophysics*, *39*, 179–229.

Balmaseda, M. A., Anderson, D. L. T., & Davey, M. K. (1994). ENSO prediction using a dynamical ocean model coupled to statistical atmospheres. *Tellus A*, *46*, 497–511.

Bankert, R. L. (1994). Cloud classification of AVHRR imagery in maritime regions using a probabilistic neural network. *Journal of Applied Meteorology*, *33*, 909–918.

Bao, L., Gneiting, T., Grimit, E. P., Guttorp, P., & Raftery, A. E. (2010). Bias correction and Bayesian model averaging for ensemble forecasts of surface wind direction. *Monthly Weather Review*, *138*, 1811–1821.

Bao, M. & Wallace, J. M. (2015). Cluster analysis of Northern Hemisphere winter-time 500-hPa flow regimes during 1920–2014. *Journal of the Atmospheric Sciences*, *72*, 3597–3608.

Barnes, E. A., Toms, B., Hurrell, J. W., Ebert-Uphoff, I., Anderson, C., & Anderson, D. (2020). Indicator patterns of forced change learned by an artificial neural network. *Journal of Advances in Modeling Earth Systems*, *12*, e2020MS002195.

Barnett, T. P., Latif, M., Graham, N., Flugel, M., Pazan, S., & White, W. (1993). ENSO and ENSO-related predictability. Part I: Prediction of equatorial Pacific sea surface temperature with a hybrid coupled ocean-atmosphere model. *Journal of Climate*, *6*, 1545–1566.

Barnett, T. P. & Preisendorfer, R. (1987). Origins and levels of monthly and seasonal forecast skill for United States surface air temperatures determined by canonical correlation analysis. *Monthly Weather Review*, *115*, 1825–1850.

Barnston, A. G. & Livezey, R. E. (1987). Classification, seasonality and persistence of low-frequency atmospheric circulation patterns. *Monthly Weather Review*, *115*(6), 1083–1126.

Barnston, A. G. & Ropelewski, C. F. (1992). Prediction of ENSO episodes using canonical correlation analysis. *Journal of Climate*, *5*, 1316–1345.

Barnston, A. G., Tippett, M. K., L'Heureux, M. L., Li, S. H., & DeWitt, D. G. (2012). Skill of real-time seasonal ENSO model predictions during 2002–11: Is our capability increasing? *Bulletin of the American Meteorological Society, 93*, 631–651.

Barnston, A. G., van den Dool, H. M., Zebiak, S. E., Barnett, T. P., Ji, M., Rodenhuis, D. R., ... Livezey, R. E. (1994). Long-lead seasonal forecasts: Where do we stand? *Bulletin of the American Meteorological Society, 75*, 2097–2114.

Barron, A. R. (1993). Universal approximation bounds for superposition of a sigmoidal function. *IEEE Transactions on Information Theory, 39*, 930–945.

Barros, A. M. G., Pereira, J. M. C., & Lund, U. J. (2012). Identifying geographical patterns of wildfire orientation: A watershed-based analysis. *Forest Ecology and Management, 264*, 98–107.

Bastola, S., Murphy, C., & Sweeney, J. (2011). The role of hydrological modelling uncertainties in climate change impact assessments of Irish river catchments. *Advances in Water Resources, 34*, 562–576.

Bauer, P., Bauer, P., Thorpe, A., & Brunet, G. (2015). The quiet revolution of numerical weather prediction. *Nature, 525* (7567), 47–55.

Baydin, A. G., Pearlmutter, B. A., Radul, A. A., & Siskind, J. M. (2018). Automatic differentiation in machine learning: A survey. *Journal of Machine Learning Research, 18*, 1–43.

Beaton, A. E. & Tukey, J. W. (1974). Fitting of power-series, meaning polynomials, illustrated on band-spectroscopic data. *Technometrics, 16*, 147–185.

Bellman, R. (1961). *Adaptive Control Processes: A Guided Tour*. Princeton University Press.

Belsley, D. A., Luh, E., & Welsch, R. E. (1980). *Regression Diagnostics: Identifying Influential Data and Sources of Collinearity*. Wiley.

Bendat, J. S. & Piersol, A. G. (2010). *Random Data: Analysis and Measurement Procedures* (4th ed.). Wiley.

Benedetti, R. (2010). Scoring rules for forecast verification. *Monthly Weather Review 138*, 203–211.

Benediktsson, J. A., Swain, P. H., & Ersoy, O. K. (1990). Neural network approaches versus statistical methods in classification of multisource remote sensing data. *IEEE Transactions on Geoscience and Remote Sensing, 28*, 540–552.

Bengio, Y. (2012). Practical recommendations for gradient-based training of deep architectures. *CoRR, abs/1206.5533*. arXiv: 1206.5533.

Ben-Hur, A., Horn, D., Siegelmann, H. T., & Vapnik, V. (2001). Support vector clustering. *Journal of Machine Learning Research, 2*, 125–137.

Benjamini, Y. & Hochberg, Y. (1995). Controlling the false discovery rate: A practical and powerful approach to multiple testing. *Journal of the Royal Statistical Society Series B-Statistical Methodology, 57*, 289–300.

Bennett, A. F. (2005). *Inverse Modeling of the Ocean and Atmosphere*. Cambridge University Press.

Bergstra, J. & Bengio, Y. (2012). Random search for hyper-parameter optimization. *Journal of Machine Learning Research, 13*, 281–305.

Berthomier, L., Pradel, B., & Perez, L. (2020). Cloud cover nowcasting with deep learning. *2020 Tenth International Conference on Image Processing Theory, Tools and Applications (IPTA)*, arXiv:2009.11577.

Beucler, T., Pritchard, M., Gentine, P., & Rasp, S. (2020). Towards physically-consistent, data-driven models of convection. In *IGARSS 2020 – 2020 IEEE International Geoscience and Remote Sensing Symposium* (pp. 3987–3990).

Beucler, T., Pritchard, M., Rasp, S., Ott, J., Baldi, P., & Gentine, P. (2021). Enforcing analytic constraints in neural networks emulating physical systems. *Physical Review Letters, 126*, 098302.

Bickel, P. J. & Doksum, K. A. (1977). *Mathematical Statistics: Basic Ideas and Selected Topics*. Oakland: Holden-Day.

Bishop, C. M. (1994). *Mixture density networks*. Technical Report NCRG/94/004. Neural Computing Research Group, Aston University, Birmingham, UK.

Bishop, C. M. (1995). *Neural Networks for Pattern Recognition*. Oxford: Clarendon Press.

Bishop, C. M. (2006). *Pattern Recognition and Machine Learning*. New York: Springer.

Bishop, C. M., Svensen, M., & Williams, C. K. I. (1998). GTM: The generative topographic mapping. *Neural Computation, 10*(1), 215–234.

Blackwell, W. J. & Chen, F. W. (2009). *Neural Networks in Atmospheric Remote Sensing*. Artech House.

Bocquet, M., Brajard, J., Carrassi, A., & Bertino, L. (2020). Bayesian inference of chaotic dynamics by merging data assimilation, machine learning and expectation-maximization. *Foundations of Data Science, 2* (1), 55–80.

Bonavita, M., Holm, E., Isaksen, L., & Fisher, M. (2016). The evolution of the ECMWF hybrid data assimilation system. *Quarterly Journal of the Royal Meteorological Society, 142*, 287–303.

Bondell, H. D., Reich, B. J., & Wang, H. X. (2010). Non-crossing quantile regression curve estimation. *Biometrika, 97*, 825–838.

Boser, B. E., Guyon, I. M., & Vapnik, V. N. (1992). A training algorithm for optimal margin classifiers. In Haussler, D. (Ed.), *Proceedings of the 5th Annual ACM Workshop on Computational Learning Theory* (pp. 144–152). ACM Press.

Botter, G., Basso, S., Rodriguez-Iturbe, I., & Rinaldo, A. (2013). Resilience of river flow regimes. *Proceedings of the National Academy of Sciences of the United States of America, 110*, 12925–12930.

Bove, M. C., Elsner, J. B., Landsea, C. W., Niu, X. F., & O'Brien, J. J. (1998). Effect of El Niño on US landfalling hurricanes, revisited. *Bulletin of the American Meteorological Society, 79*, 2477–2482.

Bowley A. L. (1901). *Elements of Statistics.* London: P.S. King & Sons.

Box, G. E. P. & Cox, D. R. (1964). An analysis of transformations. *Journal of the Royal Statistical Society Series B-Statistical Methodology, 26*, 211–252.

Box, G. E. P. & Jenkins, G. M. (1970). *Time Series Analysis: Forecasting and Control.* San Francisco, CA: Holden-Day.

Box, G. E. P., Jenkins, G. M., Reinsel, G. C., & Ljung, G. M. (2015). *Time Series Analysis: Forecasting and Control* (5th ed.). Wiley.

Boyle, P. & Frean, M. (2005). Dependent Gaussian processes. In Saul, L., Weiss, Y., & Bottou, L. (Eds.), *Advances in Neural Information Processing Systems* (Vol. 17, pp. 217–224). MIT Press.

Bozinovski, S. (2020). Reminder of the first paper on transfer learning in neural networks, 1976. *Informatica, 44*, 291–302.

Boznar, M., Lesjak, M., & Mlakar, P. (1993). A neural-network-based method for short-term predictions of ambient SO2 concentrations in highly polluted industrial-areas of complex terrain. *Atmospheric Environment Part B-Urban Atmosphere, 27*, 221–230.

Bradley, E. & Kantz, H. (2015). Nonlinear time-series analysis revisited. *Chaos, 25*, 097610.

Brajard, J., Carrassi, A., Bocquet, M., & Bertino, L. (2020). Combining data assimilation and machine learning to emulate a dynamical model from sparse and noisy observations: A case study with the Lorenz 96 model. *Journal of Computational Science, 44*, 101171.

Brajard, J., Carrassi, A., Bocquet, M., & Bertino, L. (2021). Combining data assimilation and machine learning to infer unresolved scale parametrization. *Philosophical Transactions of the Royal Society A, 379*, 20200086.

Brajard, J., Jamet, C., Moulin, C., & Thiria, S. (2006). Use of a neuro-variational inversion for retrieving oceanic and atmospheric constituents from satellite ocean colour sensor: Application to absorbing aerosols. *Neural Networks, 19*, 178–185.

Breiman, L. (1996a). Bagging predictions. *Machine Learning, 24*, 123–140.

Breiman, L. (1996b). Stacked regressions. *Machine Learning, 24*, 49–64.

Breiman, L. (1997). *Arcing the Edge.* Technical Report 486, Statistics Department, University of California, Berkeley.

Breiman, L. (2001a). Random forests. *Machine Learning, 45*, 5–32.

Breiman, L. (2001b). Statistical modeling: The two cultures. *Statistical Science 16*, 199–215.

Breiman, L. & Friedman, J. H. (1985). Estimating optimal transformations for multiple regression and correlation. *Journal of the American Statistical Association 80*, 580–598.

Breiman, L. & Friedman, J. H. (1997). Predicting multivariate responses in multiple linear regression. *Journal of the Royal Statistical Society Series B-Methodological, 59* 3–37.

Breiman, L., Friedman, J., Olshen, R. A., & Stone, C. (1984). *Classification and Regression Trees.* New York: Chapman and Hall.

Brenowitz, N. D., Beucler, T., Pritchard, M., & Bretherton, C. S. (2020). Interpreting and stabilizing machine-learning parametrizations of convection. *Journal of the Atmospheric Sciences, 77*, 4357–4375.

Brenowitz, N. D. & Bretherton, C. S. (2018). Prognostic validation of a neural network unified physics parameterization. *Geophysical Research Letters, 45*, 6289–6298.

Brenowitz, N. D. & Bretherton, C. S. (2019). Spatially extended tests of a neural network parametrization trained by coarse-graining. *Journal of Advances in Modeling Earth Systems*, *11*, 2728–2744.

Brent, R. P. (1973). *Algortihms for Minimization without Derivatives*. Englewood Cliffs, New Jersey: Prentice-Hall.

Bretherton, C. S., Smith, C., & Wallace, J. M. (1992). An intercomparison of methods for finding coupled patterns in climate data. *Journal of Climate*, *5*, 541–560.

Bretherton, C. S., Widmann, M., Dymnikov, V. P., Wallace, J. M., & Blade, I. (1999). The effective number of spatial degrees of freedom of a time-varying field. *Journal of Climate*, *12*, 1990–2009.

Brier, W. G. (1950). Verification of forecasts expressed in terms of probabilities. *Monthly Weather Review*, *78*, 1–3.

Bröcker, J. & Smith, L. A. (2007). Increasing the reliability of reliability diagrams. *Weather and Forecasting*, *22*, 651–661.

Brockwell, P. J. & Davis, R. A. (1991). *Time Series: Theory and Methods* (2nd ed.). Springer.

Broomhead, D. S. & Lowe, D. (1988). Multivariable functional interpolation and adaptive networks. *Complex Systems*, *2*, 321–355.

Broyden, C. G. (1970). The convergence of a class of double-rank minimization algorithms. *Journal of the Institute of Mathematics and Its Applications*, *6*, 76–90.

Brunton, F. (2013). *Spam: A Shadow History of the Internet*. MIT Press.

Bühlmann, P. & Künsch, H. R. (1999). Block length selection in the bootstrap for time series. *Computational Statistics & Data Analysis*, *31*, 295–310.

Bürger, G. (1996). Expanded downscaling for generating local weather scenarios. *Climate Research*, *7*, 111–128.

Bürger, G., Murdock, T. Q., Werner, A. T., Sobie, S. R., & Cannon, A. J. (2012). Downscaling extremes: An intercomparison of multiple statistical methods for present climate. *Journal of Climate*, *25*, 4366–4388.

Burnham, K. P. & Anderson, D. R. (2004). Multimodel inference: Understanding AIC and BIC in model selection. *Sociological Methods & Research*, *33* (2), 261–304.

Burnham, K. P. & Anderson, D. R. (2002). *Model Selection and Multimodel Inference: A Practical Information-Theoretic Approach* (2nd ed.). Springer.

Burrows, W. R. (1991). Objective guidance for 0–24-hour and 24–48-hour mesoscale forecasts of lake-effect snow using CART. *Weather and Forecasting*, *6*, 357–378.

Burrows, W. R. (1997). CART regression models for predicting UV radiation at the ground in the presence of cloud and other environmental factors. *Journal of Applied Meteorology*, *36*, 531–544.

Burrows, W. R. (1999). Combining classification and regression trees and the neuro-fuzzy inference system for environmental data modeling. In *18th International Conference of the North American Fuzzy Information Processing Society–NAFIPS* (pp. 695–699). New York, NY.

Burrows, W. R., Benjamin, M., & Beauchamp, S. (1995). CART decision-tree statistical analysis and prediction of summer season maximum surface ozone for the Vancouver, Montreal, and Atlantic regions of Canada. *Journal of Applied Meteorology*, *34*, 1848–1862.

Burrows, W. R., Price, C., & Wilson, L. J. (2005). Warm season lightning probability prediction for Canada and the northern United States. *Weather and Forecasting*, *20*, 971–988.

Calinski, T. & Harabasz, J. (1974). A dendrite method for cluster analysis. *Communications in Statistics: Theory and Methods*, *1* (3), 1–27.

Camps-Valls, G. & Bruzzone, L. (2005). Kernel-based methods for hyperspectral image classification. *IEEE Transactions on Geoscience and Remote Sensing*, *43*, 1351–1362.

Camps-Valls, G., Munoz-Mari, J., Gomez-Chova, L., Guanter, L., & Calbet, X. (2012). Nonlinear statistical retrieval of atmospheric profiles from MetOp-IASI and MTG-IRS infrared sounding data. *IEEE Transactions on Geoscience and Remote Sensing*, *50*, 1759–1769.

Camps-Valls, G., Tuia, D., Zhu, X. X., & Reichstein, M. (Eds.). (2021). *Deep Learning for the Earth Sciences: A Comprehensive Approach to Remote Sensing, Climate Science and Geosciences*. Wiley.

Cannon, A. J. (2008). Probabilistic multi-site precipitation downscaling by an expanded Bernoulli-gamma density network. *Journal of Hydrometeorology*, *9*, 1284–1300.

Cannon, A. J. (2010). A flexible nonlinear modelling framework for nonstationary generalized extreme value analysis in hydroclimatology. *Hydrological Processes*, *24*, 673–685.

Cannon, A. J. (2011a). GEVcdn: An R package for nonstationary extreme value analysis by generalized extreme value conditional density estimation network. *Computers & Geosciences*, *37*, 1532–1533.

Cannon, A. J. (2011b). Quantile regression neural networks: Implementation in R and application to precipitation downscaling. *Computers & Geosciences*, *37*, 1277–1284.

Cannon, A. J. (2012). Neural networks for probabilistic environmental prediction: Conditional Density Estimation Network Creation and Evaluation (CaDENCE) in R. *Computers & Geosciences*, *41*, 126–135.

Cannon, A. J. (2018). Non-crossing nonlinear regression quantiles by monotone composite quantile regression neural network, with application to rainfall extremes. *Stochastic Environmental Research and Risk Assessment*, *32*, 3207–3225.

Cannon, A. J. & Hsieh, W. W. (2008). Robust nonlinear canonical correlation analysis: Application to seasonal climate forecasting. *Nonlinear Processes in Geophysics*, *15*, 221–232.

Cannon, A. J. & Lord, E. R. (2000). Forecasting summertime surface-level ozone concentrations in the Lower Fraser Valley of British Columbia: An ensemble neural network approach. *Journal of the Air and Waste Management Association*, *50*, 322–339.

Cannon, A. J. & Whitfield, P. H. (2002). Downscaling recent streamflow conditions in British Columbia, Canada using ensemble neural network models. *Journal of Hydrology*, *259*, 136–151.

Cao, J. & Lin, Z. (2015). Extreme learning machines on high dimensional and large data applications: A survey. *Mathematical Problems in Engineering*, *2015*, 103796.

Carney, J. G., Cunningham, P., & Bhagwan, U. (1999). Confidence and prediction intervals for neural network ensembles. In *International Joint Conference on Neural Networks, IJCNN'99* (Vol. 2, pp. 1215–1218).

Carrassi, A., Carrassi, A., Bocquet, M., & Evensen, G. (2018). Data assimilation in the geosciences: An overview of methods, issues, and perspectives. *Wiley Interdisciplinary Reviews: Climate Change*, *9*, e535.

Carta, J. A., Bueno, C., & Ramirez, P. (2008). Statistical modelling of directional wind speeds using mixtures of von Mises distributions: Case study. *Energy Conversion and Management*, *49*, 897–907.

Carta, J. A., Ramirez, P., & Velazquez, S. (2009). A review of wind speed probability distributions used in wind energy analysis: Case studies in the Canary Islands. *Renewable & Sustainable Energy Reviews*, *13*, 933–955.

Caruana, R. & Niculescu-Mizil, A. (2006). An empirical comparison of supervised learning algorithms. In *Proceedings of the 23rd International Conference on Machine Learning* (pp. 161–168). Pittsburgh, PA.

Casaioli, M., Mantovani, R., Scorzoni, F. P., Puca, S., Speranza, A., & Tirozzi, B. (2003). Linear and nonlinear postprocessing of numerically forecasted surface temperature. *Nonlinear Processes in Geophysics*, *10*, 373–383.

Casanova, S. & Ahrens, B. (2009). On the weighting of multimodel ensembles in seasonal and short-range weather forecasting. *Monthly Weather Review*, *137*, 3811–3822.

Cavanaugh, J. E. & Neath, A. A. (2019). The Akaike information criterion: Background, derivation, properties, application, interpretation, and refinements. *Wiley Interdisciplinary Reviews: Computational Statistics*, *11*(3). doi:10.1002/wics.1460

Cavazos, T. (1999). Large-scale circulation anomalies conducive to extreme precipitation events and derivation of daily rainfall in northeastern Mexico and southeastern Texas. *Journal of Climate*, *12*, 1506–1523.

Cavazos, T. (2000). Using self-organizing maps to investigate extreme climate events: An application to wintertime precipitation in the Balkans. *Journal of Climate*, *13*(10), 1718–1732.

Chang, A. T. C. & Tsang, L. (1992). A neural network approach to inversion of snow water equivalent from passive microwave measurements. *Nordic Hydrology*, *23*, 173–182.

Chapman, W. E., Subramanian, A. C., Delle Monache, L., Xie, S. P., & Ralph, F. M. (2019). Improving atmospheric river forecasts with machine learning. *Geophysical Research Letters*, *46*, 10627–10635.

Charney, J. G., Fjørtoft, R., & von Newmann, J. (1950). Numerical integration of the barotropic vorticity equation. *Tellus*, *2*, 237–254.

Charrad, M., Ghazzali, N., Boiteau, V., & Niknafs, A. (2014). NbClust: An R package for determining the relevant number of clusters in a data set. *Journal of Statistical Software*, *61*(6), 1–36.

Chen, K. S., Tzeng, Y. C., & Chen, P. C. (1999). Retrieval of ocean winds from satellite scatterometer by a neural network. *IEEE Transactions on Geoscience and Remote Sensing*, *37*, 247–256.

Chen, L., Guo, S. L., Yan, B. W., Liu, P., & Fang, B. (2010). A new seasonal design flood method based on bivariate joint distribution of flood magnitude and date of occurrence. *Hydrological Sciences Journal (Journal Des Sciences Hydrologiques)*, *55*, 1264–1280.

Chen, L.-C., Zhu, Y., Papandreou, G., Schroff, F., & Adam, H. (2018). Encoder-decoder with atrous separable convolution for semantic image segmentation. In *Computer Vision: ECCV 2018* (pp. 833–851). Springer.

Chen, T. & Guestrin, C. (2016). XGBoost: A scalable tree boosting system. In *KDD'16: Proceedings of the 22nd ACM SIGKDD International Conference on Knowledge Discovery and Data Mining* (pp. 785–794).

Chen, Z.-Y., Zhang, T.-H., Zhang, R., Zhu, Z.-M., Yang, J., Chen, P.-Y., ... Guo, Y. (2019). Extreme gradient boosting model to estimate $PM_{2.5}$ concentrations with missing-filled satellite data in China. *Atmospheric Environment*, *202*, 180–189.

Cherkassky, V. & Ma, Y. Q. (2004). Practical selection of SVM parameters and noise estimation for SVM regression. *Neural Networks*, *17*, 113–126.

Cherkassky, V. & Mulier, F. (2007). *Learning from Data* (2nd ed.). Wiley.

Chevallier, F. (2005). Comments on 'New approach to calculation of atmospheric model physics: Accurate and fast neural network emulation of longwave radiation in a climate model'. *Monthly Weather Review*, *133*, 3721–3723.

Chevallier, F., Cheruy, F., Scott, N. A., & Chedin, A. (1998). A neural network approach for a fast and accurate computation of a longwave radiative budget. *Journal of Applied Meteorology*, *37*, 1385–1397.

Chevallier, F., Morcrette, J. J., Cheruy F., & Scott, N. A. (2000). Use of a neural-network-based long-wave radiative-transfer scheme in the ECMWF atmospheric model. *Quarterly Journal of the Royal Meteorological Society*, *126*, 761–776.

Choi, J., An, S.-I., Dewitte, B., & Hsieh, W. W. (2009). Interactive feedback between the tropical Pacific decadal oscillation and ENSO in a coupled general circulation model. *Journal of Climate*, *22*, 6597–6611.

Choi, Y. & Kim, S. (2020). Rain-type classification from microwave satellite observations using deep neural network segmentation. *IEEE Geoscience and Remote Sensing Letters*, *18*, 2137–2141. doi:10.1109/LGRS.2020.3016001

Chong, E. K. P. & Zak, S. H. (2013). *An Introduction to Optimization* (4th ed.). Wiley.

Christiansen, B. (2005). The shortcomings of nonlinear principal component analysis in identifying circulation regimes. *Journal of Climate*, *18*(22), 4814–4823.

Christiansen, B. (2007a). Atmospheric circulation regimes: Can cluster analysis provide the number? *Journal of Climate*, *20* (10), 2229–2250.

Christiansen, B. (2007b). Reply *Journal of Climate*, *20*, 378–379.

Chuvieco E. (2020). *Fundamentals of Satellite Remote Sensing: An Environmental Approach* (3rd ed.). CRC Press.

Clarke, A. J. (2008). *An Introduction to the Dynamics of El Niño and the Southern Oscillation*. Academic Press.

Clarke, B. (2003). Comparing Bayes model averaging and stacking when model approximation error cannot be ignored. *Journal of Machine Learning Research*, *4*, 683–712.

Cleveland, W. S. (2001). Data science: An action plan for expanding the technical areas of the field of statistics. *International Statistical Review*, *69*, 21–26.

Cohen, A. J., Brauer, M., Burnett, R., Anderson, H. R., Frostad, J., Estep, K., ... Forouzanfar, M. H. (2017). Estimates and 25-year trends of the global burden of disease attributable to ambient air pollution: An analysis of data from the Global Burden of Diseases Study 2015. *Lancet*, *389*, 1907–1918.

Coles, S. (2001). *An Introduction to Statistical Modeling of Extreme Values*. Springer.

Compagnucci, R. H. & Richman, M. B. (2008). Can principal component analysis provide atmospheric circulation or teleconnection patterns? *International Journal of Climatology*, *28*, (6), 703–726.

Comrie, A. C. (1997). Comparing neural networks and regression models for ozone forecasting. *Journal of the Air and Waste Management Association*, *47*, 653–663.

Cordisco, E., Prigent, C., & Aires, F. (2006). Snow characterization at a global scale with passive microwave satellite observations. *Journal of Geophysical Research*, *111*, D19102, doi:10.1029/2005JD006773.

Cornford, D., Nabney, I. T., & Bishop, C. M. (1999). Neural network-based wind vector retrieval from satellite scatterometer data. *Neural Computing and Applications, 8*, 206–217.

Cortes, C. & Vapnik, V. (1995). Support vector networks. *Machine Learning, 20*, 273–297.

Cover, T. M. & Hart, P. E. (1967). Nearest neighbor pattern classification. *IEEE Transactions on Information Theory, 13*, 21–27.

Cover, T. M. & Thomas, J. A. (2006). *Elements of Information Theory* (2nd ed.).

Cowan, G. (2007). Data analysis: Frequently Bayesian. *Physics Today, 60*, 82–83.

Cox, D. R., Efron, B., Hoadley, B., Parzen, E., & Breiman, L. (2001). Statistical modeling: The two cultures. Comments and rejoinders. *Statistical Science, 16*, 216–231.

Cox, D. R. & Hinkley D. V. (1974). *Theoretical Statistics*. Chapman and Hall.

Cox, D. T., Tissot, P., & Michaud, P. (2002). Water level observations and short-term predictions including meteorological events for entrance of Galveston Bay, Texas. *Journal of Waterway, Port, Coastal and Ocean Engineering, 128*, 21–29.

Cramer, J. S. (2002). *The Origins of Logistic Regression*. Technical Report TI 2002-119/4, Tinbergen Institute.

Crespo, J. L. & Mora, E. (1993). Drought estimation with neural networks. *Advances in Engineering Software, 18*, 167–170.

Cressie, N. (1993). *Statistics for Spatial Data*. Wiley.

Crevier, D. (1993). *AI: The Tumultuous History of the Search for Artificial Intelligence*. Basic Books.

Cristianini, N. & Shawe-Taylor, J. (2000). *An Introduction to Support Vector Machines and Other Kernel-based Methods*. Cambridge University Press.

Cybenko, G. (1989). Approximation by superpositions of a sigmoidal function. *Mathematics of Control, Signals, and Systems, 2*, 303–314.

Daley R. (1991). *Atmospheric Data Analysis*. Cambridge University Press.

Das, S., Mullick, S. S., & Suganthan, P. N. (2016). Recent advances in differential evolution: An updated survey. *Swarm and Evolutionary Computation, 27*, 1–30.

Das, S. & Suganthan, P. N. (2011). Differential evolution: A survey of the state-of-the-art. *IEEE Transactions on Evolutionary Computation, 15*(1), 4–31.

Dauphin, Y. N., Pascanu, R., Gulcehre, C., Cho, K., Ganguli, S., & Bengio, Y. (2014). Identifying and attacking the saddle point problem in high-dimensional non-convex optimization. In *Advances in Neural Information Processing Systems 27* (pp. 2933–2941).

David, F. N. (1938). *Tables of the Ordinates and Probability Integral of the Distribution of the Correlation Coefficient in Small Samples*. Biometrika Office, London.

Davidon, W. C. (1959). *Variable metric methods for minimization* (A.E.C.Res. and Develop. Report No. ANL-5990). Argonne National Lab.

Davidson, P. A. (2015). *Turbulence: An Introduction for Scientists and Engineers* (2nd ed.). Oxford University Press.

Davies, D. L. & Bouldin, D. W. (1979). Cluster separation measure. *IEEE Transactions on Pattern Analysis and Machine Intelligence, PAMI-1*(2), 224–227.

Davis, J. M., Eder, B. K., Nychka, D., & Yang, Q. (1998). Modeling the effects of meteorology on ozone in Houston using cluster analysis and generalized additive models. *Atmospheric Environment, 32*, 2505–2520.

Davison, A. C. & Hinkley D. V. (1997). *Bootstrap Methods and Their Applications*. New York: Cambridge University Press.

Dawson, C. W. & Wilby, R. L. (2001). Hydrological modelling using artificial neural networks. *Progress in Physical Geography, 25*, 80–108.

de Haan, L. & Ferreira, A. (2006). *Extreme Value Theory*. Springer.

Del Frate, F., Pacifici, F., Schiavon, G., & Solimini, C. (2007). Use of neural networks for automatic classification from high-resolution images. *IEEE Transactions on Geoscience and Remote Sensing 45*, 800–809.

Del Frate, F., Petrocchi, A., Lichtenegger, J., & Calabresi, G. (2000). Neural networks for oil spill detection using ERS-SAR data. *IEEE Transactions on Geoscience and Remote Sensing, 38*, 2282–2287.

Del Frate, F. & Schiavon, G. (1999). Nonlinear principal component analysis for the radiometric inversion of atmospheric profiles by using neural networks. *IEEE Transactions on Geoscience and Remote Sensing, 37*, 2335–2342.

DelSole, T. (2004). Predictability and information theory Part I: Measures of predictability. *Journal of the Atmospheric Sciences, 61*, 2425–2440.

DelSole, T. (2005). Predictability and information theory. Part II: Imperfect forecasts. *Journal of the Atmospheric Sciences, 62,* 3368–3381.

DelSole, T. & Shukla, J. (2009). Artificial skill due to predictor screening. *Journal of Climate, 22,* 331–345.

DelSole, T. & Tippett, M. K. (2007). Predictability: Recent insights from information theory. *Reviews of Geophysics, 45.* doi:10.1029/2006rg000202

DelSole, T., Trenary, L., Yan, X., & Tippett, M. K. (2019). Confidence intervals in optimal fingerprinting. *Climate Dynamics, 52* (7–8), 4111–4126.

DelSole, T., Yang, X. [Xiaosong], & Tippett, M. K. (2013). Is unequal weighting significantly better than equal weighting for multi-model forecasting? *Quarterly Journal of the Royal Meteorological Society, 139,* 176–183.

Dempster, A. P., Laird, N. M., & Rubin, D. B. (1977). Maximum likelihood from incomplete data via EM algorithm. *Journal of the Royal Statistical Society Series B-Methodological, 39,* 1–38.

Dennis, B. & Patil, G. P. (1984). The gamma-distribution and weighted multimodal gamma-distributions as models of population abundance. *Mathematical Biosciences, 68,* 187–212.

Derksen, S. & Keselman, H. J. (1992). Backward, forward and stepwise automated subset-selection algorithms: Frequency of obtaining authentic and noise variables. *British Journal of Mathematical & Statistical Psychology, 45,* 265–282.

Derome, J., Brunet, G., Plante, A., Gagnon, N., Boer, G. J., Zwiers, F. W., ... Ritchie, H. (2001). Seasonal predictions based on two dynamical models. *Atmosphere-Ocean, 39,* 485–501.

Dibike, Y. B. & Coulibaly, P. (2005). Hydrologic impact of climate change in the Saguenay watershed: Comparison of downscaling methods and hydrologic models. *Journal of Hydrology, 307,* 145–163.

Dibike, Y. B., Gachon, P., St-Hilaire, A., Ouarda, T. B. M. J., & Nguyen, V. T. V. (2008). Uncertainty analysis of statistically downscaled temperature and precipitation regimes in Northern Canada. *Theoretical and Applied Climatology, 91,* 149–170.

DiCiccio, T. J. & Efron, B. (1996). Bootstrap confidence intervals. *Statistical Science, 11,* 189–212.

Dietterich, T. G. (1998). Approximate statistical tests for comparing supervised classification learning algorithms. *Neural Computation, 10,* 1895–1923.

Dobson, M. C., Ulaby, F. T., & Pierce, L. E. (1995). Land-cover classification and estimation of terrain attributes using synthetic aperture radar. *Remote Sensing of Environment, 51,* 199–214.

Domingos, P. & Pazzani, M. (1997). On the optimality of the simple Bayesian classifier under zero-one loss. *Machine Learning, 29,* 103–130.

Douglas, E. M., Vogel, R. M., & Kroll, C. N. (2000). Trends in floods and low flows in the United States: Impact of spatial correlation. *Journal of Hydrology, 240,* 90–105.

Draper N. R. & Smith, H. (1981). *Applied Regression Analysis* (2nd ed.). New York: Wiley.

Duan, S., Ullrich, P., & Shu, L. (2020). Using convolutional neural networks for streamflow projection in California. *Frontiers in Water, 2,* 28.

Duda, R. O., Hart, P. E., & Stork, D. G. (2001). *Pattern Classification* (2nd ed.). New York: Wiley.

Dudoit, S. & Fridlyand, J. (2002). A prediction-based resampling method for estimating the number of clusters in a dataset. *Genome Biology, 3*(7), 0036.1.

Dumoulin, V. & Visin, F. (2018). A guide to convolution arithmetic for deep learning. arXiv: 1603.07285

Durbin, J. & Watson, G. S. (1950). Testing for serial correlation in least squares regression. I. *Biometrika, 37,* 409–428.

Durbin, J. & Watson, G. S. (1951). Testing for serial correlation in least squares regression. II. *Biometrika, 38,* 159–178.

Durbin, J. & Watson, G. S. (1971). Testing for serial correlation in least squares regression. III. *Biometrika, 58,* 1–19.

Ebert-Uphoff, I. & Hilburn, K. (2020). Evaluation, tuning, and interpretation of neural networks for working with images in meteorological applications. *Bulletin of the American Meteorological Society, 101,* E2149–E2170.

Ebisuzaki, W. (1997). A method to estimate the statistical significance of a correlation when the data are serially correlated. *Journal of Climate, 10,* 2147–2153.

Efron, B. (1979). Bootstrap methods: Another look at the jackknife. *Annals of Statistics, 7,* 1–26.

Efron, B. (1987). Better bootstrap confidence-intervals. *Journal of the American Statistical Association, 82,* 171–185.

Efron, B., Rogosa, D., & Tibshirani, R. (2015). Resampling methods of estimation. In Wright, J. (Ed.), *International Encyclopedia of the Social & Behavioral Sciences* (2nd ed., Vol. 20, pp. 492–495). New York, NY: Elsevier.

Efron, B. & Tibshirani, R. J. (1993). *An Introduction to the Bootstrap.* Boca Raton, Florida: CRC Press.

Eiben, A. E. & Smith J. E. (2015). *Introduction to Evolutionary Computing* (2nd ed.). Springer.

Ekici, B. B. (2014). A least squares support vector machine model for prediction of the next day solar insolation for effective use of PV systems. *Measurement, 50,* 255–262.

Ekström, M., Ekstrom, M., Grose, M. R., & Whetton, P. H. (2015). An appraisal of downscaling methods used in climate change research. *Wiley Interdisciplinary Reviews: Climate Change, 6,* 301–319.

El Hourany, R., Abboud-Abi Saab, M., Faour, G., Aumont, O., Crépon, M., & Thiria, S. (2019). Estimation of secondary phytoplankton pigments from satellite observations using self-organizing maps (SOMs). *Journal of Geophysical Research: Oceans, 124,* 1357–1378.

Elachi, C. & van Zyl, J. (2021). *Introduction to the Physics and Techniques of Remote Sensing* (3rd ed.). Wiley.

Elsner, J. B. & Schmertmann, C. P. (1994). Assessing forecast skill through cross-validation. *Weather and Forecasting, 9,* 619–624.

Elsner, J. B. & Tsonis, A. A. (1996). *Singular Spectrum Analysis.* New York: Plenum.

Emery W. & Camps, A. (2017). *Introduction to Satellite Remote Sensing: Atmosphere, Ocean, Cryosphere and Land Applications.* Elsevier.

Engmann, S. & Cousineau, D. (2011). Comparing distributions: The two-sample Anderson–Darling test as an alternative to the Kolmogorov–Smirnoff test. *Journal of Applied Quantitative Methods, 6,* 1–17.

Epstein, E. S. (1969). Stochastic dynamic prediction. *Tellus, 21* (6), 739–759.

Escalante, H. J., Escalera, S., Guyon, I., Baro, X., Gucluturk, Y., Guclu, U., & van Gerven, M. (Eds.). (2018). *Explainable and Interpretable Models in Computer Vision and Machine Learning.* Springer.

Everingham, Y., Sexton, J., Skocaj, D., & Inman-Bamber, G. (2016). Accurate prediction of sugarcane yield using a random forest algorithm. *Agronomy for Sustainable Development, 36,* 27.

Eyre, J. R., English, S. J., & Forsythe, M. (2020). Assimilation of satellite data in numerical weather prediction. Part I: The early years. *Quarterly Journal of the Royal Meteorological Society, 146,* 49–68.

Eyring, V., Bony, S., Meehl, G. A., Senior, C. A., Stevens, B., Stouffer, R. J., & Taylor, K. E. (2016). Overview of the Coupled Model Intercomparison Project Phase 6 (CMIP6) experimental design and organization. *Geoscientific Model Development, 9*(5), 1937–1958.

Falls, L. W. (1974). The beta distribution: A statistical model for world cloud cover. *Journal of Geophysical Research, 79,* 1261–1264.

Fan, J., Yue, W., Wu, L., Zhang, F., Cai, H., Wang, X., . . . Xiang, Y. (2018). Evaluation of SVM, ELM and four tree-based ensemble models for predicting daily reference evapotranspiration using limited meteorological data in different climates of China. *Agricultural and Forest Meteorology, 263,* 225–241.

Fanaee-T, H. & Gama, J. (2014). Event labeling combining ensemble detectors and background knowledge. *Progress in Artificial Intelligence, 2,* 113–127.

Farchi, A., Laloyaux, P., Bonavita, M., & Bocquet, M. (2020). Using machine learning to correct model error in data assimilation and forecast applications. arXiv: 2010.12605

Faucher, M., Burrows, W. R., & Pandolfo, L. (1999). Empirical-statistical reconstruction of surface marine winds along the western coast of Canada. *Climate Research, 11,* 173–190.

Feller, W. (1948). On the Kolmogorov–Smirnov limit theorems for empirical distributions. *Annals of Mathematical Statistics, 19,* 177–189.

Feng, X., Porporato, A., & Rodriguez-Iturbe, I. (2013). Changes in rainfall seasonality in the tropics. *Nature Climate Change, 3,* 811–815.

Ferreira, J. A. & Soares, C. G. (1999). Modelling the long-term distribution of significant wave height with the Beta and Gamma models. *Ocean Engineering, 26,* 713–725.

Finnis, J., Cassano, J., Holland, M., Serreze, M., & Uotila, P. (2009a). Synoptically forced hydroclimatology of major Arctic watersheds in general circulation models. Part 1: the Mackenzie River Basin. *International Journal of Climatology, 29* (9), 1226–1243.

Finnis, J., Cassano, J., Holland, M., Serreze, M., & Uotila, P. (2009b). Synoptically forced hydroclimatology of major Arctic watersheds in general circulation models. Part 2: Eurasian watersheds. *International Journal of Climatology, 29*, (9), 1244–1261.

Fischer, H. (2010). *A History of the Central Limit Theorem.* Springer.

Fisher, N. I. (1995). *Statistical Analysis of Circular Data.* Cambridge.

Fisher, R. A. (1915). Frequency distribution of the values of the correlation coefficient in samples from an indefinitely large population. *Biometrika, 10,* 507–521.

Fisher, R. A. (1918). The correlation between relatives on the supposition of Mendelian inheritance. *Philosophical Transactions of the Royal Society of Edinburgh, 52,* 399–433.

Fisher, R. A. (1921). On the 'probable error' of a coefficient of correlation deduced from a small sample. *Metron, 1,* 3–32.

Fisher, R. A. (1936). The use of multiple measurements in taxonomic problems. *Annals of Eugenics, 7,* 179–188.

Fleming, S. W. & Goodbody, A. G. (2019). A machine learning metasystem for robust probabilistic nonlinear regression-based forecasting of seasonal water availability in the US West. *IEEE Access, 7,* 119943–119964.

Fletcher, R. (1970). A new approach to variable metric algorithms. *Computer Journal, 13,* 317–322.

Fletcher, R. & Powell, M. J. D. (1963). A rapidly convergent descent method for minimization. *Computer Journal, 6,* 163–168.

Fletcher, R. & Reeves, C. M. (1964). Function minimization by conjugate gradients. *Computer Journal, 7,* 149–154.

Flom, P. L. & Cassell, D. L. (2007). Stopping stepwise: Why stepwise and similar selection methods are bad, and what you should use. *NESUG.*

Foody G. M. & Mathur, A. (2004). A relative evaluation of multiclass image classification by support vector machines. *IEEE Transactions on Geoscience and Remote Sensing, 42,* 1335–1343.

Forbes, A. M. G. (1988). Fourier-transform filtering: A cautionary note. *Journal of Geophysical Research-Oceans, 93,* (C6), 6958–6962.

Foreman, M. G. G., Cherniawsky J. Y., & Ballantyne, V. A. (2009). Versatile harmonic tidal analysis: Improvements and applications. *Journal of Atmospheric and Oceanic Technology, 26*(4), 806–817.

Foresee, F. D. & Hagan, M. T. (1997). Gauss-Newton approximation to Bayesian regularization. In *Proceedings of the 1997 International Joint Conference on Neural Networks.*

Forsythe, G. E., Malcolm, M. A., & Moler, C. B. (1977). *Computer Methods for Mathematical Computations.* Prentice-Hall.

Fovell, R. G. & Fovell, M. Y. C. (1993). Climate zones of the conterminous United States defined using cluster-analysis. *Journal of Climate, 6,* 2103–2135.

Fowler, H. J., Blenkinsop, S., & Tebaldi, C. (2007). Linking climate change modelling to impacts studies: Recent advances in downscaling techniques for hydrological modelling. *International Journal of Climatology, 27,* 1547–1578.

Fowlkes, E. B. & Mallows, C. L. (1983). A method for comparing 2 hierarchical clusterings. *Journal of the American Statistical Association, 78,* 553–596.

Fraser, R. H. & Li, Z. (2002). Estimating fire-related parameters in boreal forest using SPOT VEGETATION. *Remote Sensing of Environment, 82,* 95–110.

Freund, Y. & Schapire, R. E. (1997). A decision-theoretical generalization of online learning and an application to boosting. *Journal of Computer System Sciences, 55* 119–139.

Friedman, J. H. (1989). Regularized discriminant-analysis. *Journal of the American Statistical Association, 84,* 165–175.

Friedman, J. H. (2001). Greedy function approximation: A gradient boosting machine. *Annals of Statistics, 29,* 1189–1232.

Fugal, D. L. (2009). *Conceptual Wavelets in Digital Signal Processing: An In-Depth Practical Approach for the Non-Mathematician* Space and Signals Technical Publishing.

Fukushima, K. (1975). Cognitron: A self-organizing multilayered neural network. *Biological Cybernetics, 20,* 121–136. Retrieved from https://doi.org/10.1007/BF00342633

Fukushima, K. (1980). Neocognitron: A self-organizing neural network model for a mechanism of pattern-recognition unaffected by shift in position. *Biological Cybernetics, 36,* 193–202.

Gagne, D. J., II, Christensen, H. M., Subramanian, A. C., & Monahan, A. H. (2020). Machine learning for stochastic parameterization: Generative adversarial networks in the Lorenz '96 model. *Journal of Advances in Modeling Earth Systems, 12,* e2019MS001896.

Gagne, D. J., II, Haupt, S. E., Nychka, D. W., & Thompson, G. (2019). Interpretable deep learning for spatial analysis of severe hailstorms. *Monthly Weather Review, 147,* 2827–2845.

Gagne, D. J., II, McGovern, A., Haupt, S. E., & Williams, J. K. (2017). Evaluation of statistical learning configurations for gridded solar irradiance forecasting. *Solar Energy, 150,* 383–393.

Gagne, D. J., II, McGovern, A., & Xue, M. (2014). Machine learning enhancement of storm-scale ensemble probabilistic quantitative precipitation forecasts. *Weather and Forecasting, 29,* 1024–1043.

Gaitan, C. F. & Cannon, A. J. (2013). Validation of historical and future statistically downscaled pseudo-observed surface wind speeds in terms of annual climate indices and daily variability. *Renewable Energy, 51,* 489–496.

Gaitan, C. F., Hsieh, W. W., & Cannon, A. J. (2014). Comparison of statistically downscaled precipitation in terms of future climate indices and daily variability for southern Ontario and Quebec, Canada. *Climate Dynamics, 43,* 3201–3217.

Galar, M., Fernandez, A., Barrenechea, E., Bustince, H., & Herrera, F. (2012). A review on ensembles for the class imbalance problem: Bagging-, boosting-, and hybrid-based approaches. *IEEE Transactions on Systems Man and Cybernetics Part C–Applications and Reviews,* 463–484.

Galton, F. J. (1885). Regression towards mediocrity in hereditary stature. *Journal of the Anthropological Institute, 15,* 246–263.

Gandin, L. S. & Murphy, A. H. (1992). Equitable skill scores for categorical forecasts. *Monthly Weather Review, 120,* 361–370.

Gardner, M. W. & Dorling, S. R. (1999). Neural network modelling and prediction of hourly $NO_x$ and $NO_2$ concentrations in urban air in London. *Atmospheric Environment, 33,* 709–719.

Gardner, M. W. & Dorling, S. R. (2000). Statistical surface ozone models: An improved methodology to account for non-linear behaviour. *Atmospheric Environment, 34,* 21–34.

Garrett, C. & Müller, P. (2008). Supplement to Extreme Events. *Bulletin of the American Meteorological Society, 89,* Retrieved from https://doi.org/10.1175/2008BAMS2566.2

Geer, A. J. (2016). Significance of changes in medium-range forecast scores. *Tellus A, 68,* 30229.

Geer, A. J. (2021). Learning earth system models from observations: Machine learning or data assimilation? *Philosophical Transactions of the Royal Society A, 379,* 20200089.

Geiss, A. & Hardin, J. C. (2020). Radar super resolution using a deep convolutional neural network. *Journal of Atmospheric and Oceanic Technology, 37,* 2197–2207.

Gemmrich, J. & Garrett, C. (2011). Dynamical and statistical explanations of observed occurrence rates of rogue waves. *Natural Hazards and Earth System Sciences 11,* 1437–1446.

Gentine, P., Pritchard, M., Rasp, S., Reinaudi, G., & Yacalis, G. (2018). Could machine learning break the convection parameterization deadlock? *Geophysical Research Letters,* 45(11), 5742–5751.

Gerrity J. P. (1992). A note on Gandin and Murphy's equitable skill score. *Monthly Weather Review, 120,* 2709–2712.

Gers, F. A., Schmidhuber, J., & Cummins, F. (2000). Learning to forget: Continual prediction with LSTM. *Neural Computation, 12,* 2451–2471.

Gettelman, A. & Rood, R. B. (2016). *Demystifying Climate Models: A Users Guide to Earth System Models.* Springer.

Geurts, P., Ernst, D., & Wehenkel, L. (2006). Extremely randomized trees. *Machine Learning, 63,* 3–42.

Ghil, M., Allen, M. R., Dettinger, M. D., Ide, K., Kondrashov, D., Mann, M. E., ... Yiou, P. (2002). Advanced spectral methods for climatic time series. *Reviews of Geophysics, 40,* 1003, doi:10.1029/2000RG000092.

Gilbert, G. K. (1884). Finley's tornado predictions. *American Meteorological Journal, 1,* 166–172.

Gilbert, R. O. (1987). *Statistical Methods for Environmental Pollution Monitoring.* Wiley.

Gill, A. E. (1980). Some simple solutions for heat-induced tropical circulation. *Quarterly Journal of the Royal Meteorological Society, 106*, 447–462.

Gill, A. E. (1982). *Atmosphere-Ocean Dynamics*. Academic Press.

Gill, P. E., Murray W., & Wright, M. H. (1981). *Practical Optimization*. Academic Press.

Gill, P. E., Murray W., & Wright, M. H. (1991). *Numerical Linear Algebra and Optimization*. Addison-Wesley.

Giorgi, F., Bi, X. Q., & Pal, J. (2004). Mean, interannual variability and trends in a regional climate change experiment over Europe. II: Climate change scenarios (2071–2100). *Climate Dynamics, 23*, 839–858.

Giorgi, F. & Gutowski, W. J. (2015). Regional dynamical downscaling and the CORDEX initiative. *Annual Review of Environment and Resources, 40*, 467–490.

Giraud, C. (2015). *Introduction to High-Dimensional Statistics*. CRC Press.

Glazman, R. E. & Greysukh, A. (1993a). Correction to 'Satellite altimeter measurements of surface wind' by Roman E. Glazman and Alexander Greysukh. *98*(C8), 14751.

Glazman, R. E. & Greysukh, A. (1993b). Satellite altimeter measurements of surface wind. *Journal of Geophysical Research, 98*(C2), 2475–2483.

Gleick, J. (1987). *Chaos: Making a New Science*. Viking.

Glorot, X. & Bengio, Y. (2010). Understanding the difficulty of training deep feedforward neural networks. In *13th International Conference on Artificial Intelligence and Statistics (AISTATS)*. Chia Laguna Resort, Sardinia, Italy.

Glorot, X., Bordes, A., & Bengio, Y. (2011). Deep sparse rectifier neural networks. In *Proceedings of the 14th International Conference on Artificial Intelligence and Statistics (AISTATS)* (pp. 315–323). Fort Lauderdale, Florida.

Gneiting, T., Raftery, A. E., Westveld, A. H. I., & Goldman, T. (2005). Calibrated probabilistic forecasting using ensemble model output statistics and minimum CRPS estimation. *Monthly Weather Review, 133*, 1098–1118.

Goddard, L., Mason, S. J., Zebiak, S. E., Ropelewski, C. F., Basher, R., & Cane, M. A. (2001). Current approaches to seasonal-to-interannual climate predictions. *International Journal of Climatology, 21*, 1111–1152.

Goel, N. S. & Strebel, D. E. (1984). Simple beta distribution representation of leaf orientation in vegetation canopies. *Agronomy Journal, 76*, 800–802.

Goldfarb, F. (1970). A family of variable metric methods derived by variational means. *Mathematics of Computation, 24*, 23–26.

Golub, G. H., Heath, M., & Wahba, G. (1979). Generalized cross-validation as a method for choosing a good ridge parameter. *Technometrics, 21*, 215–223.

Gonella, J. (1972). Rotary-component method for analyzing meteorological and oceanographic vector time series. *Deep-Sea Research, 19*, 833–846.

Gong, X. F. & Richman, M. B. (1995). On the application of cluster-analysis to growing-season precipitation data in North America east of the Rockies. *Journal of Climate, 8*(4), 897–931.

Good, I. (1952). Rational decisions. *Journal of the Royal Statistical Society. Series B, 14*, 107–114.

Goodfellow, I., Bengio, Y., & Courville, A. (2016). *Deep Learning*. MIT Press.

Goodfellow, I., Pouget-Abadie, J., Mirza, M., Xu, B., Warde-Farley D., Ozair, S., ... Bengio, Y. (2014). Generative adversarial nets. In *Advances in Neural Information Processing Systems 27* (pp. 2672–2680).

Gopal, S. & Woodcock, C. (1996). Remote sensing of forest change using artificial neural networks. *IEEE Transactions on Geoscience and Remote Sensing, 34*, 398–404.

Gorder, P. F. (2006). Neural networks show new promise for machine vision. *Computing in Science & Engineering, 8*, 4–8.

Gorsuch, R. L. (1983). *Factor Analysis*. Lawrence Erlbaum Associates.

Graham, V. A. & Hollands, K. G. T. (1990). A method to generate synthetic hourly solar radiation globally. *Solar Energy, 44*, 333–341.

Greeshma, N. K., Baburaj, M., & George, S. N. (2016). Reconstruction of cloud-contaminated satellite remote sensing images using kernel PCA-based image modelling. *Arabian Journal of Geosciences, 9*, 239.

Greff, K., Srivastava, R. K., Koutnik, J., Steunebrink, B. R., & Schmidhuber J. (2017). LSTM: A search space odyssey. *IEEE Transactions on Neural Networks and Learning Systems, 28*, 2222–2232.

Grieger, B. & Latif, M. (1994). Reconstruction of the El-Niño attractor with neural networks. *Climate Dynamics, 10*, 267–276.

Groisman, P. Y., Karl, T. R., Easterling, D. R., Knight, R. W., Jamason, P. F., Hennessy, K. J., ... Zhai, P. M. (1999). Changes in the probability of heavy precipitation: Important indicators of climatic change. *Climatic Change*, *42*, 243–283.

Gross, L., Thiria, S., & Frouin, R. (1999). Applying artificial neural network methodology to ocean color remote sensing. *Ecological Modelling*, *120*, 237–246.

Gross, L., Thiria, S., Frouin, R., & Greg, M. B. (2000). Artificial neural networks for modeling the transfer function between marine reflectance and phytoplankton pigment concentration. *Journal of Geophysical Research*, *105*(C2), 3483–3496.

Groth, A. & Ghil, M. (2011). Multivariate singular spectrum analysis and the road to phase synchronization. *Physical Review E*, *84*, 036206.

Gulrajani, I., Ahmed, F., Arjovsky, M., Dumoulin, V., & Courville, A. C. (2017). Improved training of Wasserstein GANs. *CoRR*, *abs/1704.00028*. arXiv: 1704.00028

Gupta, H. V., Kling, H., Yilmaz, K. K., & Martinez, G. F. (2009). Decomposition of the mean squared error and NSE performance criteria: Implications for improving hydrological modelling. *Journal of Hydrology*, *377*, 80–91.

Hahn, G. J. & Shapiro, S. S. (1994). *Statistical Models in Engineering*. Wiley.

Hall, P., Horowitz, J. L., & Jing, B. Y. (1995). On blocking rules for the bootstrap with dependent data. *Biometrika*, *82*, 561–574.

Ham, Y.-G., Kim, J.-H., & Luo, J.-J. (2019). Deep learning for multi-year ENSO forecasts. *Nature*, *573*, 568–572.

Hamed, K. H. (2009). Exact distribution of the Mann–Kendall trend test statistic for persistent data. *Journal of Hydrology*, *365*, 86–94.

Hamed, K. H. & Rao, A. R. (1998). A modified Mann–Kendall trend test for autocorrelated data. *Journal of Hydrology*, *204*, 182–196.

Hamilton, K. (1988). A detailed examination of the extratropical response to tropical El Niño/Southern Oscillation events. *Journal of Climatology*, *8*, 67–86.

Hamilton, K. (1998). Dynamics of the tropical middle atmosphere: A tutorial review. *Atmosphere-Ocean*, *36*, 319–354.

Hamilton, K. & Hsieh, W. W. (2002). Representation of the QBO in the tropical stratospheric wind by nonlinear principal component analysis. *Journal of*

*Geophysical Research*, *107*(D15), 4232, doi:10.1029/2001JD001250.

Han, P., Wang, P. X., Zhang, S. Y., & Zhu, D. H. (2010). Drought forecasting based on the remote sensing data using ARIMA models. *Mathematical and Computer Modelling*, *51*, 1398–1403.

Hand, D. J. & Yu, K. M. (2001). Idiot's Bayes: Not so stupid after all? *International Statistical Review*, *69*, 385–398.

Hannachi, A., Jolliffe, I. T., & Stephenson, D. B. (2007). Empirical orthogonal functions and related techniques in atmospheric science: A review. *International Journal of Climatology*, *27*(9), 1119–1152.

Hansen, P. C., Pereyra, V., & Scherer, G. (2013). *Least Squares Data Fitting with Applications*. Johns Hopkins University Press.

Härdle, W. (1991). *Smoothing Techniques with Implementation in S*. Springer.

Hardman-Mountford, N. J., Richardson, A. J., Boyer, D. C., Kreiner, A., & Boyer, H. J. (2003). Relating sardine recruitment in the Northern Benguela to satellite-derived sea surface height using a neural network pattern recognition approach. *Progress in Oceanography*, *59*, 241–255.

Hardoon, D. R., Szedmak, S., & Shawe-Taylor, J. (2004). Canonical correlation analysis: An overview with application to learning methods. *Neural Computation*, *16*, 2639–2664.

Hardy, D. M. (1977). Empirical eigenvector analysis of vector wind measurements. *Geophysical Research Letters*, *4*, 319–320.

Hardy D. M. & Walton, J. J. (1978). Principal component analysis of vector wind measurements. *Journal of Applied Meteorology*, *17*, 1153–1162.

Harrell, F. E. (2015). *Regression Modeling Strategies: With Applications to Linear Models, Logistic and Ordinal Regression, and Survival Analysis*. Springer.

Harris, F. J. (1978). Use of windows for harmonic-analysis with discrete Fourier-transform. *Proceedings of the IEEE*, *66*, 51–83.

Hartmann, H. C., Pagano, T. C., Sorooshian, S., & Bales, R. (2002). Confidence builders: Evaluating seasonal climate forecasts from user perspectives. *Bulletin of the American Meteorological Society*, *83*, 683–698.

Haskett, J. D., Pachepsky, Y. A., & Acock, B. (1995). Use of the beta distribution for parameterizing variability of soil properties at the regional level for crop yield estimation. *Agricultural Systems*, *48*, 73–86.

Hassanat, A. B., Abbadi, M. A., & Altarawneh, G. A. (2014). Solving the problem of the K parameter in the KNN classifier using an ensemble learning approach. *International Journal of Computer Science and Information Security, 12*, 33–39.

Hasselmann, K. (1993). Optimal fingerprints for the detection of time-dependent climate-change. *Journal of Climate, 6*(10), 1957–1971.

Hasselmann, S. & Hasselmann, K. (1985). Computations and parameterizations of the nonlinear energy-transfer in a gravity-wave spectrum. Part I: A new method for efficient computations of the exact nonlinear transfer integral. *Journal of Physical Oceanography, 15*, 1369–1377.

Hasselmann, S., Hasselmann, K., Allender, J. H., & Barnett, T. P. (1985). Computations and parameterizations of the nonlinear energy-transfer in a gravity-wave spectrum. Part II: parameterizations of the nonlinear energy-transfer for application in wave models. *Journal of Physical Oceanography, 15*, 1378–1391.

Hastie, T. & Stuetzle, W. (1989). Principal curves. *Journal of the American Statistical Association, 84*, 502–516.

Hastie, T., Tibshirani, R., & Friedman, J. (2009). *The Elements of Statistical Learning: Data Mining, Inference, and Prediction* (2nd ed.). Springer.

Hatfield, S., Chantry M., Dueben, P., Lopez, P., Geer, A., & Palmer, T. (2021). Building tangent-linear and adjoint models for data assimilation with neural networks. *Journal of Advances in Modeling Earth Systems, 13*, e2021MS002521.

Haupt, R. L. & Haupt, S. E. (2004). *Practical Genetic Algorithms* (2nd ed.). Wiley.

Haupt, S. E., Gagne, D. J., Hsieh, W. W., Krasnopolsky, V., McGovern, A., Marzban, C., ... Williams, J. K. (2022). The history and practice of AI in the environmental sciences. *Bulletin of the American Meteorological Society, 103*, E1351–E1370.

Haupt, S. E., Pasini, A., & Marzban, C. (Eds.). (2009). *Artificial Intelligence Methods in the Environmental Sciences*. Springer.

Hayashi, C. (1998). What is data science? Fundamental concepts and a heuristic example. In Hayashi, C., Yajima, K., Bock, H.-H., Ohsumi, N., Tanaka, Y., & Baba, Y. (Eds.), *Data Science, Classification, and Related Methods. Proceedings of the Fifth Conference of the International Federation of Classification Societies (IFCS-96)* (pp. 40–51). Kobe, Japan, 27–30 March, 1996: Springer Japan.

Hayashi, Y. (1979). Space-time spectral anaysis of rotary vector series. *Journal of the Atmospheric Sciences, 36*, 757–766.

Haykin S. (1999). *Neural Networks: A Comprehensive Foundation*. New York: Prentice Hall.

Haylock, M. R., Cawley, G. C., Harpham, C., Wilby, R. L., & Goodess, C. M. (2006). Downscaling heavy precipitation over the United Kingdom: A comparison of dynamical and statistical methods and their future scenarios. *International Journal of Climatology, 26*, 1397–1415.

He, K. M., Zhang, X. Y., Ren, S. Q., & Sun, J. (2016). Deep residual learning for image recognition. In *2016 IEEE Conference on Computer Vision and Pattern Recognition* (pp. 770–778).

Hegerl, G. & Zwiers, F. (2011). Use of models in detection and attribution of climate change. *Wiley Interdisciplinary Reviews-Climate Change, 2*, (4), 570–591.

Heideman, M. T., Johnson, D. H., & Burrus, C. S. (1985). Gauss and the history of the fast Fourier transform. *Archive For History Of Exact Sciences, 34*(3), 265–277.

Heidke, P. (1926). Berechnung des Erfolges und der Güte der Windstärkevorhersagen in Sturmwarnungsdienst. *Geografiska Annaler, 8*, 310–349.

Herman, A. (2007). Nonlinear principal component analysis of the tidal dynamics in a shallow sea. *Geophysical Research Letters, 34*.

Hermanns, W. (1983). *Einstein and the Poet: In Search of the Cosmic Man*. Branden Press.

Hertz, J., Krogh, A., & Palmer, R. G. (1991). *Introduction to the Theory of Neural Computation*. Addison-Wesley.

Heskes, T. (1997). Practical confidence and prediction intervals. In Mozer, M. C., Jordan, M. I., & Petsche, T. (Eds.), *Advances in Neural Information Processing Systems* (Vol. 9, pp. 176–182). MIT Press.

Hestenes, M. R. & Stiefel, E. (1952). Methods of conjugate gradients for solving linear systems. *Journal of Research of the National Bureau of Standards, 49*, 409–436.

Hewitson, B. C. & Crane, R. G. (1996). Climate downscaling: Techniques and application. *Climate Research, 7*, 85–95.

Hewitson, B. C. & Crane, R. G. (2002). Self-organizing maps: Applications to synoptic climatology. *Climate Research*, *22*, 13–26.

Hinton, G. E. & Salakhutdinov, R. R. (2006). Reducing the dimensionality of data with neural networks. *Science*, *313*(5786), 504–507.

Hinton, G. E., Srivastava, N., Krizhevsky, A., Sutskever, I., & Salakhutdinov, R. R. (2012). Improving neural networks by pre-venting co-adaptation of feature detectors. *arXiv e-prints*, arXiv:1207.0580.

Hinton, G. E., Osindero, S., & Teh, Y.-W. (2006). A fast learning algorithm for deep belief nets. *Neural Computation*, *18*, 1527–1554.

Hirsch, R. M., Slack, J. R., & Smith, R. A. (1982). Techniques of trend analysis for monthly water-quality data. *Water Resources Research*, *18*, 107–121.

Ho, T. K. (1995). Randon decision forests. In *3rd International Conference on Document Analysis and Recognition* (Vol. 1, pp. 278–282). Montreal, QC.

Hochreiter, S. & Schmidhuber, J. (1997). Long short-term memory. *Neural Computation*, *9*, 1735–1780.

Hodson, R. (2016). The dark universe. *Nature*, *537*, S193.

Hoerl, A. E. & Kennard, R. W. (1970). Ridge regression: Biased estimation for nonorthogonal problems. *Technometrics*, *12*, 55–67.

Hoerling, M. P., Kumar, A., & Zhong, M. (1997). El Niño, La Niña and the nonlinearity of their teleconnections. *Journal of Climate*, *10*, 1769–1786.

Hoeting, J. A., Madigan, D., Raftery, A. E., & Volinsky, C. T. (1999). Bayesian model averaging: A tutorial. *Statistical Science*, *14* (4), 382–401.

Hogan, R. J., Ferro, C. A. T., Jolliffe, I. T., & Stephenson, D. B. (2010). Equitability revisited: Why the 'equitable threat score' is not equitable. *Weather and Forecasting*, *25*, 710–726.

Hogan, R. J. & Mason, I. B. (2012). Deterministic forecasts of binary events. In Jolliffe, I. T. & Stephenson, D. B. (Eds.), *Forecast Verification: A Practitioner's Guide in Atmospheric Science* (2nd ed., Chap. 3, pp. 31–59). Wiley-Blackwell.

Hollander, M., Wolfe, D. A., & Chicken, E. (2014). *Nonparametric Statistical Methods* (3rd ed.). Wiley.

Hoolohan, V., Tomlin, A. S., & Cockerill, T. (2018). Improved near surface wind speed predictions using Gaussian process regression combined with numerical weather predictions and observed meteorological data. *Renewable Energy*, *126*, 1043–1054.

Horel, J. D. (1981). A rotated principal component analysis of the interannual variability of the Northern Hemisphere 500 mb height field. *Monthly Weather Review*, *109*, 2080–2092.

Horel, J. D. & Wallace, J. M. (1981). Planetary-scale atmospheric phenomena associated with the Southern Oscillation. *Monthly Weather Review*, *109*, 813–829.

Hornik, K. (1991). Approximation capabilities of multilayer feedforward networks. *Neural Networks*, *4*, 252–257.

Hornik, K., Stinchcombe, M., & White, H. (1989). Multilayer feedforward networks are universal approximators. *Neural Networks*, *2*, 359–366.

Horstmann, J., Schiller, H., Schulz-Stellenfleth, J., & Lehner, S. (2003). Global wind speed retrieval from SAR. *IEEE Transactions on Geoscience and Remote Sensing*, *41*, 2277–2286.

Hosking, J. R. M. (1990). L-moments: Analysis and estimation of distributions using linear combinations of order statistics. *Journal of the Royal Statistical Society, Series B*, *52*, 105–124.

Hotelling, H. (1931). The generalization of Student's ratio. *Annals of Mathematical Statistics*, *2*, 360–378.

Hotelling, H. (1933). Analysis of a complex of statistical variables into principal components. *Journal of Educational Psychology*, *24* 417–441.

Hotelling, H. (1936). Relations between two sets of variates. *Biometrika*, *28*, 321–377.

Hrachowitz, M., Soulsby, C., Tetzlaff, D., Malcolm, I. A., & Schoups, G. (2010). Gamma distribution models for transit time estimation in catchments: Physical interpretation of parameters and implications for time-variant transit time assessment. *Water Resources Research*, *46*. doi:10.1029/2010wr009148

Hsieh, W. W. (1982). On the detection of continental shelf waves. *Journal of Physical Oceanography*, *12*, 414–427.

Hsieh, W. W. (2000). Nonlinear canonical correlation analysis by neural networks. *Neural Networks*, *13*, 1095–1105.

Hsieh, W. W. (2001a). Nonlinear canonical correlation analysis of the tropical Pacific climate variability using a neural network approach. *Journal of Climate*, *14*, 2528–2539.

Hsieh, W. W. (2001b). Nonlinear principal component analysis by neural networks. *Tellus*, *53A*, 599–615.

Hsieh, W. W. (2004). Nonlinear multivariate and time series analysis by neural network methods. *Reviews of Geophysics*, *42*, RG1003, doi:10.1029/2002RG000112.

Hsieh, W. W. (2007). Nonlinear principal component analysis of noisy data. *Neural Networks*, *20*, 434–443.

Hsieh, W. W. (2009). *Machine Learning Methods in the Environmental Sciences*. Cambridge: Cambridge University Press.

Hsieh, W. W. (2020). Improving predictions by nonlinear regression models from outlying input data. arXiv: 2003.07926.

Hsieh, W. W. (2022). Evolution of machine learning in environmental science: A perspective. *Environmental Data Science*, 1, e3, doi:10.1017/eds.2022.2

Hsieh, W. W. & Cannon, A. J. (2008). Towards robust nonlinear multivariate analysis by neural network methods. In Donner, R. & Barbosa, S. (Eds.), *Nonlinear Time Series Analysis in the Geosciences: Applications in Climatology, Geodynamics, and Solar-terrestrial Physics* (pp. 97–124). Springer.

Hsieh, W. W. & Tang, B. (1998). Applying neural network models to prediction and data analysis in meteorology and oceanography. *Bulletin of the American Meteorological Society*, *79*, 1855–1870.

Hsieh, W. W. & Wu, A. (2002). Nonlinear multichannel singular spectrum analysis of the tropical Pacific climate variability using a neural network approach. *Journal of Geophysical Research*, *107*(C7), 3076, doi:10.1029/2001JC000957.

Hsu, C. W. & Lin, C. J. (2002). A comparison of methods for multiclass support vector machines. *IEEE Transactions on Neural Networks*, *13*, (2), 415–425.

Hsu, K. L., Gao, X. G., Sorooshian, S., & Gupta, H. V. (1997). Precipitation estimation from remotely sensed information using artificial neural networks. *Journal of Applied Meteorology*, *36*, 1176–1190.

Hsu, K. L., Gupta, H. V., & Sorooshian, S. (1995). Artificial neural-network modeling of the rainfall-runoff process. *Water Resources Research*, *31*, 2517–2530.

Hu, X., Belle, J. H., Meng, X., Wildani, A., Waller, L. A., Strickland, M. J., & Liu, Y. (2017). Estimating $PM_{2.5}$ concentrations in the conterminous United States using the random forest approach. *Environmental Science & Technology*, *51*, 6936–6944.

Huang, C., Davis, L. S., & Townshend, J. R. G. (2002). An assessment of support vector machines for land cover classification. *International Journal of Remote Sensing*, *23*, 725–749.

Huang, G., Huang, G.-B., Song, S., & You, K. (2015). Trends in extreme learning machines: A review. *Neural Networks*, *61*, 32–48.

Huang, G., Liu, Z., van der Maaten, L., & Weinberger, K. Q. (2017). Densely connected convolutional networks. In *Proceedings of the IEEE Conference on Computer Vision and Pattern Recognition (CVPR)* (pp. 4700–4708).

Huang, G.-B. (2008). Reply to 'Comments on "The extreme learning machine"' *IEEE Transactions on Neural Networks 19*, 1495–1496.

Huang, G.-B. (2014). An insight into extreme learning machines: Random neurons, random features and kernels. *Cognitive Computation*, *6*, 376–390.

Huang, G.-B., Wang, D. H., & Lan, Y. (2011). Extreme learning machines: A survey. *International Journal of Machine Learning and Cybernetics*, *2*, 107–122.

Huang, G.-B., Zhu, Q.-Y., & Siew, C.-K. (2006). Extreme learning machine: Theory and applications. *Neurocomputing*, *70*, 489–501.

Huang, H., Lin, L., Tong, R., Hu, H., Zhang, Q., Iwamoto, Y., ... Wu, J. (2020). UNet3+: A full-scale connected UNet for medical image segmentation. arXiv: 2004.08790

Huang, L., Luo, J., Lin, Z., Niu, F., & Liu, L. (2020). Using deep learning to map retrogressive thaw slumps in the Beiluhe region (Tibetan Plateau) from CubeSat images. *Remote Sensing of Environment*, *237*, 111534.

Huang, W., Murray, C., Kraus, N., & Rosati, J. (2003). Development of a regional neural network for coastal water level predictions. *Ocean Engineering*, *30*, 2275–2295.

Huber, P. J. (1964). Robust estimation of a location parameter. *The Annals of Mathematical Statistics*, *35*, 73–101.

Hubert, L. & Arabie, P. (1985). Comparing partitions. *Journal of Classification*, *2*, 193–218.

Huth, R. & Beranová, R. (2021). How to recognize a true mode of atmospheric circulation variability. *Earth and Space Science, 8*, e2020EA001275.

Hyndman, R. J. & Fan, Y. N. (1996). Sample quantiles in statistical packages. *American Statistician, 50*, 361–365.

Ioffe, S. & Szegedy, C. (2015). Batch normalization: Accelerating deep network training by reducing internal covariate shift. arXiv:1502.03167

Isola, P., Zhu, J.-Y., Zhou, T., & Efros, A. A. (2017). Image-to-image translation with conditional adversarial networks. In *Proceedings of the IEEE Conference on Computer Vision and Pattern Recognition (CVPR)* (pp. 1125–1134).

Jain, A. K. & Dubes, R. C. (1988). *Algorithms for Clustering Data*. Prentice-Hall.

Jamet, C., Thiria, S., Moulin, C., & Crépon, M. (2005). Use of a neurovariational inversion for retrieving oceanic and atmospheric constituents from ocean color imagery: A feasibility study. *Journal of Atmospheric and Oceanic Technology, 22*, 460–475.

Jaynes, E. T. (2003). *Probability Theory: The Logic of Science*. Bretthorst, G. L. (Ed.). Cambridge: Cambridge University Press.

Jenkner, J., Hsieh, W. W., & Cannon, A. J. (2011). Seasonal modulations of the active MJO cycle characterized by nonlinear principal component analysis. *Monthly Weather Review, 139*(7), 2259–2275.

Johnson, B., Tateishi, R., & Xie, Z. (2012). Using geographically weighted variables for image classification. *Remote Sensing Letters, 3*, 491–499.

Johnson, B. A., Tateishi, R., & Hoan, N. T. (2013). A hybrid pansharpening approach and multiscale object-based image analysis for mapping diseased pine and oak trees. *International Journal of Remote Sensing, 34* (20), 6969–6982.

Johnson, N. L. (1949). Systems of frequency curves generated by methods of translation. *Biometrika, 36*, 149–176.

Johnson, N. C. (2013). How many ENSO flavors can we distinguish? *Journal of Climate, 26*(13), 4816–4827.

Johnson, R. & Wehrly T. (1977). Measures and models for angular-correlation and angular-linear correlation. *Journal of the Royal Statistical Society Series B-Methodological, 39*, (2), 222–229.

Jolliffe, I. T. (1972). Discarding variables in a principal component analysis. 1. Artificial data. *Journal of The Royal Statistical Society Series C-Applied Statistics 21*, 160–173.

Jolliffe, I. T. (1987). Rotation of principal components: Some comments. *Journal of Climatology, 7*, 507–510.

Jolliffe, I. T. (1989). Rotation of ill-defined principal components. *Journal of the Royal Statistical Society Series C-Applied Statistics), 38*, 139–147.

Jolliffe, I. T. (2002). *Principal Component Analysis* (2nd ed.). New York: Springer.

Jolliffe, I. T. & Cadima, J. (2016). Principal component analysis: A review and recent developments. *Philosophical Transactions of the Royal Society, A374*, 20150202.

Jolliffe, I. T. & Stephenson, D. B. (Eds.). (2012). *Forecast Verification: A Practitioner's Guide in Atmospheric Science* (2nd ed.). Wiley-Blackwell.

Justel, A., Pena, D., & Zamar, R. (1997). A multivariate Kolmogorov–Smirnov test of goodness of fit. *Statistics & Probability Letters, 35*, 251–259.

Kadow, C., Hall, D. M., & Ulbrich, U. (2020). Artificial intelligence reconstructs missing climate information. *Nature Geoscience, 13*(6), 408–413.

Kaiser, H. F. (1958). The varimax criterion for analytic rotation in factor analysis. *Psychometrika, 23*, 187–200.

Kalnay E. (2003). *Atmospheric Modeling, Data Assimilation and Predictability*. Cambridge.

Kalnay, E., Kanamitsu, M., Kistler, R., Collins, W., Deaven, D., Gandin, L., ... Joseph, D. (1996). The NCEP/NCAR 40-year reanalysis project. *Bulletin of the American Meteorological Society, 77*, 437–471.

Kalnay E., Li, H., Miyoshi, T., Yang, S.-C., & Ballabrera-Poy J. (2007). 4-D-Var or ensemble Kalman filter? *Tellus A: Dynamic Meteorology and Oceanography, 59*, 758–773.

Kamangir, H., Collins, W., Tissot, P., & King, S. A. (2020). Deep-learning model used to predict thunderstorms within 400 km² of South Texas domains. *Meteorological Applications, 27*, e1905.

Kaplan, A., Kushnir, Y., & Cane, M. A. (2000). Reduced space optimal interpolation of historical marine sea level pressure: 1854–1992. *Journal of Climate, 13*, 2987–3002.

Karl, T. R., Wang, W. C., Schlesinger, M. E., Knight, R. W., & Portman, D. (1990). A method of relating general-circulation model simulated climate to the observed local climate. 1. Seasonal statistics. *Journal of Climate, 3,* 1053–1079.

Karniadakis, G. E., Kevrekidis, I. G., Lu, L., Perdikaris, P., Wang, S., & Yang, L. (2021). Physics-informed machine learning. *Nature Reviews Physics, 3,* 422–440.

Karpatne, A., Atluri, G., Faghmous, J. H., Steinbach, M., Banerjee, A., Ganguly, A., ... Kumar, V. (2017). Theory-guided data science: A new paradigm for scientific discovery from data. *IEEE Transactions on Knowledge and Data Engineering, 29,* 2318–2331.

Karunanithi, N., Grenney, W. J., Whitley, D., & Bovee, K. (1994). Neural networks for river flow prediction. *Journal of Computing in Civil Engineering, 8,* 201–220.

Karush, W. (1939). *Minima of functions of several variables with inequalities as side constraints* (M.Sc. thesis, University of Chicago).

Ke, G., Meng, Q., Finley, T., Wang, T., Chen, W., Ma, W., ... Liu, T.-Y. (2017). Light-GBM: A highly efficient gradient boosting decision tree. In *Advances in Neural Information Processing Systems (NIPS 2017)* (Vol. 30).

Keiner, L. E. & Yan, X.-H. (1998). A neural network model for estimating sea surface chlorophyll and sediments from Thematic Mapper imagery. *Remote Sensing of Environment, 66,* 153–165.

Kelley, H. (1960). Gradient theory of optimal flight paths. *ARS Journal, 30*(10), 947–954.

Kelly, K. (1988). Comment on 'Empirical orthogonal function analysis of advanced very high resolution radiometer surface temperature patterns in Santa Barbara Channel' by G.S.E. Lagerloef and R.L. Bernstein. *Journal of Geophysical Research, 93*(C12), 15, 743–15, 754.

Kendall, M. G. (1938). A new measure of rank correlation. *Biometrika, 30,* 81–93.

Kendall, M. G. (1945). The treatment of ties in ranking problems. *Biometrika, 33,* 239–251.

Kharin, V. V. & Zwiers, F. W. (2003). On the ROC score of probability forecasts. *Journal of Climate, 16,* 4145–4150.

Kharin, V. V., Zwiers, F. W., Zhang, X., & Hegerl, G. C. (2007). Changes in temperature and precipitation extremes in the IPCC ensemble of global coupled model simulations. *Journal of Climate, 20,* 1419–1444.

Kharin, V. V., Zwiers, F. W., Zhang, X., & Wehner, M. (2013). Changes in temperature and precipitation extremes in the CMIP5 ensemble. *Climatic Change, 119,* 345–357.

Ki, S., Jang, I., Cha, B., Seo, J., & Kwon, O. (2020). Restoration of missing pressures in a gas well using recurrent neural networks with long short-term memory cells. *Energies, 13,* 4696.

Kingma, D. P. & Welling, M. (2014). Auto-encoding variational Bayes. arXiv: 1312.6114

Kirby M. J. & Miranda, R. (1996). Circular nodes in neural networks. *Neural Computation, 8,* 390–402.

Kirtman, B. P., Shukla, J., Balmaseda, M., Graham, N., Penland, C., Xue, Y., & Zebiak, S. (2001). *Current status of ENSO forecast skill: A report to the CLIVAR Working Group on seasonal to interannual prediction.* WCRP Informal Report No 23/01, World Climate Research Programme.

Kiviluoto, K. (1996). Topology preservation in self-organizing maps. In *Proceedings of International Conference on Neural Networks (ICNN)* (pp. 294–299).

Kleeman, R. (2002). Measuring dynamical prediction utility using relative entropy. *Journal of the Atmospheric Sciences, 59,* 2057–2072.

Knuth, D. (1998). *The Art of Computer Programming* (3rd ed.). Boston: Addison-Wesley.

Knutti, R., Masson, D., & Gettelman, A. (2013). Climate model genealogy: Generation CMIP5 and how we got there. *Geophysical Research Letters, 40,* (6), 1194–1199.

Knutti, R., Sedlacek, J., Sanderson, B. M., Lorenz, R., Fischer, E. M., & Eyring, V. (2017). A climate model projection weighting scheme accounting for performance and interdependence. *Geophysical Research Letters, 44*(4), 1909–1918.

Koenker, R. & Bassett, G. (1978). Regression quantiles. *Econometrica, 46,* 33–50.

Koenker, R. & Hallock, K. F. (2001). Quantile regression. *Journal of Economic Perspectives, 15,* 143–156.

Kohonen, T. (1982). Self-organizing formation of topologically correct feature maps. *Biological Cybernetics, 43,* 59–69.

Kohonen, T. (2001). *Self-Organizing Maps* (3rd ed.). Springer.

Kolehmainen, M., Martikainen, H., & Ruuskanen, J. (2001). Neural networks and periodic components used in air quality forecasting. *Atmospheric Environment*, *35*, 815–825.

Konishi, S. & Kitagawa, G. (2008). *Information Criteria and Statistical Modeling*. Springer.

Kramer, M. A. (1991). Nonlinear principal component analysis using autoassociative neural networks. *AIChE Journal*, *37*, 233–243.

Kraskov, A., Stogbauer, H., & Grassberger, P. (2004). Estimating mutual information. *Physical Review E*, *69*. doi:10.1103/PhysRevE.69.066138

Krasnopolsky, V. M. (2007). Neural network emulations for complex multidimensional geophysical mappings: Applications of neural network techniques to atmospheric and oceanic satellite retrievals and numerical modeling. *Reviews of Geophysics*, *45*, RG3009, doi:10.1029/2006RG000200.

Krasnopolsky V. M. (2013). *The Application of Neural Networks in the Earth System Sciences: Neural Network Emulations for Complex Multidimensional Mappings*. Springer.

Krasnopolsky, V. M., Breaker, L. C., & Gemmill, W. H. (1995). A neural network as a nonlinear transfer function model for retrieving surface wind speeds from the special sensor microwave imager. *Journal of Geophysical Research*, *100*(C6), 11033–11045.

Krasnopolsky, V. M., Chalikov, D. V., & Tolman, H. L. (2002). A neural network technique to improve computational efficiency of numerical oceanic models. *Ocean Modelling*, *4*, 363–383.

Krasnopolsky, V. M. & Fox-Rabinovitz, M. S. (2006). Complex hybrid models combining deterministic and machine learning components for numerical climate modeling and weather prediction. *Neural Networks*, *19*, 122–134.

Krasnopolsky, V. M., Fox-Rabinovitz, M. S., & Belochitski, A. A. (2013). Using ensemble of neural networks to learn stochastic convection parameterizations for climate and numerical weather prediction models from data simulated by a cloud resolving model. *Advances in Artificial Neural Systems*, *2013*, 485913.

Krasnopolsky, V. M., Fox-Rabinovitz, M. S., & Chalikov, D. V. (2005a). Comments on 'New approach to calculation of atmospheric model physics: Accurate and fast neural network emulation of longwave

radiation in a climate model' – Reply. *Monthly Weather Review*, *133*, 3724–3728.

Krasnopolsky, V. M., Fox-Rabinovitz, M. S., & Chalikov, D. V. (2005b). New approach to calculation of atmospheric model physics: Accurate and fast neural network emulation of longwave radiation in a climate model. *Monthly Weather Review*, *133*, 1370–1383.

Krasnopolsky, V. M., Fox-Rabinovitz, M. S., Hou, Y. T., Lord, S. J., & Belochitski, A. A. (2010). Accurate and fast neural network emulations of model radiation for the NCEP coupled Climate Forecast System: Climate simulations and seasonal predictions. *Monthly Weather Review*, *138*, 1822–1842.

Krasnopolsky V. M., Gemmill, W. H., & Breaker, L. C. (1999). A multiparameter empirical ocean algorithm for SSM/I retrievals. *Canadian Journal of Remote Sensing*, *25*, 486–503.

Kratzert, F., Klotz, D., Brenner, C., Schulz, K., & Herrnegger, M. (2018). Rainfall–runoff modelling using long short-term memory (LSTM) networks. *Hydrology and Earth System Sciences*, *22*, 6005–6022.

Kratzert, F., Klotz, D., Shalev, G., Klambauer, G., Hochreiter, S., & Nearing, G. (2019). Towards learning universal, regional, and local hydrological behaviors via machine learning applied to large-sample datasets. *Hydrology and Earth System Sciences*, *23*, 5089–5110.

Krizhevsky A., Sutskever, I., & Hinton, G. E. (2012). ImageNet classification with deep convolutional neural networks. In *Advances in neural information processing systems* (Vol. 25, pp. 1090–1098).

Kruskal, W. H. (1957). Historical notes on the Wilcoxon unpaired two-sample test. *Journal of the American Statistical Association*, *52*, 356–360.

Kuhn, H. W. & Tucker, A. W. (1951). Nonlinear programming. In *Proceedings of the 2nd Berkeley Symposium on Mathematical Statistics and Probabilities* (pp. 481–492). University of California Press.

Künsch, H. R. (1989). The jackknife and the bootstrap for general stationary observations. *Annals of Statistics*, *17*, 1217–1241.

Kwok, J. T.-Y. & Tsang, I. W.-H. (2004). The pre-image problem in kernel methods. *IEEE Transactions on Neural Networks*, *15*, 1517–1525.

Lagerquist, R. and McGovern, A., & Gagne, D. J., II, (2019). Deep learning for spatially explicit prediction of synoptic-scale fronts. *Weather and Forecasting, 34*, 1137–1160.

Lagerquist, R., McGovern, A., Homeyer, C. R., Gagne, D. J., II, & Smith, T. (2020). Deep learning on three-dimensional multiscale data for next-hour tornado prediction. *Monthly Weather Review*, 2837–2861.

Lahiri, S. N. (2003). *Resampling Methods for Dependent Data*. Springer.

Lahoz, W., Khattatov, B., & Menard, R. (Eds.). (2010). *Data Assimilation*. Springer.

Lai, P. L. & Fyfe, C. (1999). A neural implementation of canonical correlation analysis. *Neural Networks, 12*, 1391–1397.

Lai, P. L. & Fyfe, C. (2000). Kernel and non-linear canonical correlation analysis. *International Journal of Neural Systems, 10*, 365–377.

Lakshmanan, V. (2012). *Automating the Analysis of Spatial Grids: A Practical Guide to Data Mining Geospatial Images for Human & Environmental Applications*. Springer.

Lall, U. & Sharma, A. (1996). A nearest neighbor bootstrap for resampling hydrologic time series. *Water Resources Research, 32*, 679–693.

Lambert, S. J. & Fyfe, J. C. (2006). Changes in winter cyclone frequencies and strengths simulated in enhanced greenhouse warming experiments: Results from the models participating in the IPCC diagnostic exercise. *Climate Dynamics, 26*, 713–728.

Lan, Y., Soh, Y. C., & Huang, G.-B. (2009). Ensemble of online sequential extreme learning machine. *Neurocomputing, 72*, 3391–3395.

Lang, B. (2005). Monotonic multi-layer perceptron networks as universal approximators. In Duch, W., Kacprzyk, J., Oja, E. & Zadrozny, S. (Eds.), *Artificial Neural Networks: Formal Models and Their Applications – ICANN 2005, Pt 2, Proceedings* (Vol. 3697, pp. 31–37). Lecture Notes in Computer Science.

Larvor, G., Berthomier, L., Chabot, V., Pape, B. L., Pradel, B., & Perez, L. (2020). MeteoNet, an open reference weather dataset by METEO FRANCE. https://meteonet.umr-cnrm.fr/.

Lawrence, R., Bunn, A., Powell, S., & Zambon, M. (2004). Classification of remotely sensed imagery using stochastic gradient boosting as a refinement of classification tree analysis. *Remote Sensing of Environment, 90*, 331–336.

Le, N. D. & Zidek, J. V. (2006). *Statistical Analysis of Environmental Space-Time Processes*. Springer.

Le, Q. V., Ngiam, J., Coates, A., Lahiri, A., Prochnow, B., & Ng, A. Y. (2011). On optimization methods for deep learning. In *Proceedings of the 28th International Conference on Machine Learning* (pp. 265–272). ICML'11. Bellevue, Washington, USA: Omnipress.

Lea, C., Flynn, M. D., Vidal, R., Reiter, A., & Hager, G. D. (2017). Temporal convolutional networks for action segmentation and detection. In *Proceedings of the IEEE Conference on Computer Vision and Pattern Recognition (CVPR)* (pp. 156–165).

LeBlond, P. H. & Mysak, L. A. (1978). *Waves in the Ocean*. Elsevier.

LeCun, Y., Bengio, Y., & Hinton, G. (2015). Deep learning. *Nature, 521*, 436–444.

LeCun, Y., Boser, B., Denker, J. S., Henderson, D., Howard, R. E., Hubbard, W., & Jackel, L. D. (1989). Backpropagation applied to handwritten zip code recognition. *Neural Computation, 1*, 541–551.

LeCun, Y., Kanter, I., & Solla, S. A. (1991). Second order properties of error surfaces: Learning time and generalization. In *Advances in Neural Information Processing Systems* (Vol. 3, pp. 918–924). MIT Press.

Lee, A. (2010). Circular data. *Wiley Interdisciplinary Reviews: Computational Statistics, 2*, 477–486.

Lee, J. A. & Verleysen, M. (2007). *Nonlinear Dimensionality Reduction*. Springer.

Lee, J., Weger, R. C., Sengupta, S. K., & Welch, R. M. (1990). A neural network approach to cloud classification. *IEEE Transactions on Geoscience and Remote Sensing, 28*, 846–855.

Legates, D. R. & McCabe, G. J. (1999). Evaluating the use of 'goodness-of-fit' measures in hydrologic and hydroclimatic model validation. *Water Resources Research, 35*, 233–241.

Legler, D. M. (1983). Empirical orthogonal function analysis of wind vectors over the tropical Pacific region. *Bulletin of the American Meteorological Society, 64*, 234–241.

Leinonen, J., Guillaume, A., & Yuan, T. (2019). Reconstruction of cloud vertical structure with a generative adversarial network. *Geophysical Research Letters, 46*, 7035–7044.

Leith, C. E. (1974). Theoretical skill of Monte-Carlo forecasts. *Monthly Weather Review, 102*(6), 409–418.

Lettenmaier, D. P., Wood, E. F., & Wallis, J. R. (1994). Hydro-climatological trends in the continental United States, 1948–88. *Journal of Climate, 7*, 586–607.

Leung, L. Y. & North, G. R. (1990). Information-theory and climate prediction. *Journal of Climate, 3*, 5–14.

Levenberg, K. (1944). A method for the solution of certain non-linear problems in least squares. *Quarterly of Applied Mathematics, 2*, 164–168.

Levine, S. N. & Schindler, D. W. (1999). Influence of nitrogen to phosphorus supply ratios and physicochemical conditions on cyanobacteria and phytoplankton species composition in the Experimental Lakes Area, Canada. *Canadian Journal of Fisheries and Aquatic Sciences, 56*, 451–466.

Li, S., Hsieh, W. W., & Wu, A. (2005). Hybrid coupled modeling of the tropical Pacific using neural networks. *Journal of Geophysical Research, 110*(C9), doi:10.1029/2004JC002595.

Liang, N.-Y., Huang, G.-B., Saratchandran, P., & Sundararajan, N. (2006). A fast and accurate online sequential learning algorithm for feedforward networks. *IEEE Transactions on Neural Networks, 17*, 1411–1423.

Liang, X., Li, S., Zhang, S. Y., Huang, H., & Chen, S. X. (2016). PM$_{2.5}$ data reliability, consistency and air quality assessment in five Chinese cities. *Journal of Geophysical Research Atmosphere, 121*, 10220–10236.

Liaw, A. & Wiener, M. (2002). Classification and regression by randomForest. *R News, 2/3*, 18–22.

Lighthill, J. (1973). *Artificial Intelligence: A General Survey*. Science Research Council.

Lillesand, T. M., Kiefer, R. W., & Chipman, J. W. (2015). *Remote Sensing and Image Interpretation* (7th ed.). Wiley.

Lilliefors, H. W. (1967). On the Kolmogorov–Smirnov test for normality with mean and variance unknown. *Journal of the American Statistical Association, 62*, 399–402.

Lima, A. R., Cannon, A. J., & Hsieh, W. W. (2015). Nonlinear regression in environmental sciences using extreme learning machines: A comparative evaluation. *Environmental Modelling & Software, 73*, 175–188.

Lima, A. R., Cannon, A. J., & Hsieh, W. W. (2016). Forecasting daily streamflow using online sequential extreme learning machines. *Journal of Hydrology, 537*, 431–443.

Lima, A. R., Hsieh, W. W., & Cannon, A. J. (2017). Variable complexity online sequential extreme learning machine, with applications to streamflow prediction. *Journal of Hydrology, 555*, 983–994.

Lin, C. J. (2002). Errata to 'A comparison of methods for multiclass support vector machines'. *IEEE Transactions on Neural Networks, 13*(4), 1026–1027.

Lin, G. F. & Chen, L. H. (2006). Identification of homogeneous regions for regional frequency analysis using the self-organizing map. *Journal of Hydrology, 324*(1–4), 1–9.

Lin, K., Sheng, S., Zhou, Y., Liu, F., Li, Z., Chen, H., ... Guo, S. (2020). The exploration of a temporal convolutional network combined with encoder–decoder framework for runoff forecasting. *Hydrology Research, 51*, 1136–1149.

Littell, J. S., McKenzie, D., Peterson, D. L., & Westerling, A. L. (2009). Climate and wildfire area burned in western US eco-provinces, 1916–2003. *Ecological Applications, 19*, 1003–1021.

Liu, G., Reda, F. A., Shih, K. J., Wang, T.-C., Tao, A., & Catanzaro, B. (2018). Image inpainting for irregular holes using partial convolutions. In *Computer Vision – ECCV 2018* (pp. 89–105).

Liu, Y. G. & Weisberg, R. H. (2005). Patterns of ocean current variability on the West Florida Shelf using the self-organizing map. *Journal of Geophysical Research-Oceans, 110*(C6), C06003, doi:10.1029/2004JC002786.

Liu, Y. G., Weisberg, R. H., & Shay L. K. (2007). Current patterns on the West Florida Shelf from joint self-organizing map analyses of HF radar and ADCP data. *Journal of Atmospheric and Oceanic Technology, 24*, 702–712.

Liu, Y., Racah, E., Prabhat, Correa, J., Khosrowshahi, A., Lavers, D., ... Collins, W. (2016). Application of deep convolutional neural networks for detecting extreme weather in climate datasets. arXiv: 1605.01156

Liu, Y. & Weisberg, R. H. (2011). A review of self-organizing map applications in meteorology and oceanography In Mwasiagi, J. I. (Ed.), *Self Organizing Maps: Applications and Novel Algorithm Design* (pp. 253–272). IntechOpen.

Liu, Y., Weisberg, R. H., & Mooers, C. N. K. (2006). Performance evaluation of the self-organizing map for feature extraction. *Journal of Geophysical Research*, *111*, C05018, doi:10.1029/2005JC003117.

Livezey, R. E. (2012). Deterministic forecasts of multi-category events. In Jolliffe, I. T. & Stephenson, D. B. (Eds.), *Forecast Verification: A Practitioner's Guide in Atmospheric Science* (2nd ed., Chap. 4, pp. 61–75). Wiley-Blackwell.

Livezey, R. E. & Chen, W. Y. (1983). Statistical field significance and its determination by Monte Carlo techniques. *Monthly Weather Review*, *111*, 46–59.

Lo, F., Bitz, C. M., & Hess, J. J. (2021). Development of a random forest model for fore-casting allergenic pollen in North America. *Science of the Total Environment*, *773*, 145590.

Long, J., Shelhamer, E., & Darrell, T. (2015). Fully convolutional networks for semantic segmentation. In *IEEE Conference on Computer Vision and Pattern Recognition (CVPR)* (pp. 3431–3440).

Lorenc, A. C. (1986). Analysis methods for numerical weather prediction. *Quarterly Journal of the Royal Meteorological Society*, *112*, 1177–1194.

Lorenz, E. N. (1956). *Empirical Orthogonal Functions and Statistical Weather Prediction*. Statistical Forecasting Project, Dept. of Meteorology Mass. Inst. Tech.

Lorenz, E. N. (1963). Deterministic nonperiodic flow. *Journal of the Atmospheric Sciences*, *20*, 130–141.

Lu, W.-Z. & Wang, D. (2008). Ground-level ozone prediction by support vector machine approach with a cost-sensitive classification scheme. *Science of the Total Environment*, *395*, 109–116.

Luenberger, D. G. (1984). *Linear and Nonlinear Programming* (2nd ed.). Reading, Massachusetts: Addison-Wesley.

Luenberger, D. G. & Ye, Y. (2016). *Linear and Nonlinear Programming* (4th ed.). Springer.

MacKay, D. J. C. (1992a). Bayesian interpolation. *Neural Computation*, *4*, 415–447.

MacKay, D. J. C. (1992b). A practical Bayesian framework for backpropagation networks. *Neural Computation*, *4*, 448–472.

MacKay D. J. C. (2003). *Information Theory, Inference and Learning Algorithms*. Cambridge: Cambridge University Press.

Maclin, R. & Opitz, D. (1997). An empirical evaluation of bagging and boosting. In *Fourteenth National Conference on Artificial Intelligence* (pp. 546–551). Providence, RI: AAAI Press.

Mahesh, A., Evans, M., Jain, G., Castillo, M., Lima, A., Lunghino, B., ... Brown, P. T. (2019). Forecasting El Niño with convolutional and recurrent neural networks. In *33rd Conference on Neural Information Processing Systems (NeurIPS 2019)*.

Maier, H. R. & Dandy, G. C. (2000). Neural networks for the prediction and forecasting of water resources variables: A review of modelling issues and applications. *Environmental Modelling and Software*, *15*, 101–124.

Mallat, S. (2009). *A Wavelet Tour of Signal Processing: The Sparse Way* (3rd ed.). Elsevier.

Malthouse, E. C. (1998). Limitations of nonlinear PCA as performed with generic neural networks. *IEEE Transactions on Neural Networks*, *9*, 165–173.

Manabe, S. & Bryan, K. (1969). Climate calculations with a combined oceanatmosphere model. *Journal of the Atmospheric Sciences*, *26*, 786–789.

Mann, H. B. (1945). Non-parametric test against trend. *Econometrica*, *13*, 245–259.

Mann, H. B. & Whitney, D. R. (1947). On a test of whether one of two random variables is stochastically larger than the other. *Annals of Mathematical Statistics*, *18*, 50–56.

Maraun, D. (2013). Bias correction, quantile mapping, and downscaling: Revisiting the inflation issue. *Journal of Climate*, *26*, 2137–2143.

Maraun, D., Wetterhall, F., Ireson, A. M., Chandler, R. E., Kendon, E. J., Widmann, M., ... Thiele-Eich, I. (2010). Precipitation downscaling under climate change: Recent developments to bridge the gap between dynamical models and the end user. *Reviews of Geophysics*, *48*, RG3003, doi:10.1029/2009RG000314.

Mardia, K. V., Kent, J. T., & Bibby J. M. (1979). *Multivariate Analysis*. London: Academic Press.

Maronna, R. A. & Zamar, R. H. (2002). Robust estimates of location and dispersion for high-dimensional datasets. *Technometrics*, *44*, 307–17.

Marquardt, D. (1963). An algorithm for least-squares estimation of nonlinear parameters. *SIAM Journal on Applied Mathematics*, *11*, 431–441.

Marsaglia, G. (2004). Evaluating the Anderson–Darling distribution. *Journal of Statistical Software*, *9*, 730–737.

Marzban, C. (2003). Neural networks for post-processing model output: ARPS. *Monthly Weather Review, 131*, 1103–1111.

Marzban, C. (2004). The ROC curve and the area under it as performance measures. *Weather and Forecasting, 19*, 1106–1114.

Marzban, C. & Stumpf, G. J. (1996). A neural network for tornado prediction based on doppler radar-derived attributes. *Journal of Applied Meteorology, 35*, 617–626.

Marzban, C. & Stumpf, G. J. (1998). A neural network for damaging wind prediction. *Weather and Forecasting, 13*, 151–163.

Marzban, C. & Witt, A. (2001). A Bayesian neural network for severe-hail size prediction. *Weather and Forecasting, 16*, 600–610.

Masis, S. (2021). *Interpretable Machine Learning with Python*. Packt Publishing.

Mason, S. J. & Baddour, O. (2008). Statistical modelling. In Troccoli, A., Harrison, M., Anderson, D. L. T. & Mason, S. J. (Eds.), *Seasonal Climate: Forecasting and Managing Risk* (pp. 163–201). Springer.

Masters, T. (1995). *Advanced Algorithms for Neural Networks: A C++ Sourcebook*. New York: Wiley.

Maurer, E. P., Hidalgo, H. G., Das, T., Dettinger, M. D., & Cayan, D. R. (2010). The utility of daily large-scale climate data in the assessment of climate change impacts on daily streamflow in California. *Hydrology and Earth System Sciences, 14*, 1125–1138.

Maxwell, A. E., Warner, T. A., & Fang, F. (2018). Implementation of machine-learning classification in remote sensing: An applied review. *International Journal of Remote Sensing, 39*, 2784–2817.

May, R. J., Dandy, G. C., Maier, H. R., & Nixon, J. B. (2008). Application of partial mutual information variable selection to ANN forecasting of water quality in water distribution systems. *Environmental Modelling & Software, 23*, 1289–1299.

May, R. J., Maier, H. R., Dandy, G. C., & Fernando, T. (2008). Non-linear variable selection for artificial neural networks using partial mutual information. *Environmental Modelling & Software, 23*, 1312–1326.

McCuen, R. H., Knight, Z., & Cutter, A. G. (2006). Evaluation of the Nash–Sutcliffe efficiency index. *Journal of Hydrologic Engineering, 11*, 597–602.

McCullagh, P. (1980). Regression-models for ordinal data. *Journal of the Royal Statistical Society Series B–Methodological 42*, 109–142.

McCulloch, W. S. & Pitts, W. (1943). A logical calculus of the ideas immanent in neural nets. *Bulletin of Mathematical Biophysics, 5*, 115–137.

McGill, R., Tukey J. W., & Larsen, W. A. (1978). Variations of box plots. *American Statistician, 32*, 12–16.

McGinnis, D. L. (1997). Estimating climate-change impacts on Colorado Plateau snow-pack using downscaling methods. *Professional Geographer, 49*, 117–125.

McGovern, A., Elmore, K. L., Gagne, D. J., II, Haupt, S. E., Karstens, C. D., Lagerquist, R., . . . Williams, J. K. (2017). Using artificial intelligence to improve real-time decision-making for high-impact weather. *Bulletin of the American Meteorological Society, 98*, 2073–2090.

McGovern, A., Lagerquist, R., Gagne, D. J., II, Jergensen, G. E., Elmore, K. L., Homeyer, C. R., & Smith, T. (2019). Making the black box more transparent: Understanding the physical implications of machine learning. *Bulletin of the American Meteorological Society, 100*, 2175–2199.

McKendry, I. G. (2002). Evaluation of artificial neural networks for fine particulate pollution ($PM_{10}$ and $PM_{2.5}$) forecasting. *Journal of the Air & Waste Management Association, 52*, 1096–1101.

McLachlan, G. & Krishnan, T. (2008). *The EM Algorithm and Extensions* (2nd ed.). Wiley.

McPhaden, M. J., Zebiak, S. E., & Glantz, M. H. (2006). ENSO as an integrating concept in Earth science. *Science, 314* (5806), 1740–1745.

Meinshausen, N. (2006). Quantile regression forests. *Journal of Machine Learning Research, 7*, 983–999.

Mejia, C., Thiria, S., Tran, N., Crépon, M., & Badran, F. (1998). Determination of the geophysical model function of the ERS-1 scatterometer by the use of neural networks. *Journal of Geophysical Research, 103* (C6), 12853–12868.

Mendelssohn, R. (1980). Using Box–Jenkins-models to forecast fishery dynamics: Identification, estimation, and checking. *Fishery Bulletin, 78*, 887–896.

Meng, X.-L. & van Dyk, D. (1997). The EM algorithm: An old folk-song sung to a fast new tune. *Journal of the Royal Statistical Society Series B-Methodological, 59*, 511–540.

Metropolis, N. (1987). The beginning of the Monte Carlo method. *Los Alamos Science Special Issue, 15*, 125–130.

Miao, C., Duan, Q., Sun, Q., Huang, Y., Kong, D., Yang, T., ... Gong, W. (2014). Assessment of CMIP5 climate models and projected temperature changes over Northern Eurasia. *Environmental Research Letters, 9*(5), 055007.

Michez, A., Piegay H., Jonathan, L., Claessens, H., & Lejeune, P. (2016). Mapping of riparian invasive species with supervised classification of Unmanned Aerial System (UAS) imagery *International Journal of Applied Earth Observation and Geoinformation, 44*, 88–94.

Middleton, J. H. (1982). Outer rotary cross spectra, coherences, and phases. *Deep Sea Research, 29*(10A), 1267–1269.

Mika, S., Schölkopf, B., Smola, A., Müller, K.-R., Scholz, M., & Rätsch, G. (1999). Kernel PCA and de-noising in feature spaces. In Kearns, M., Solla, S., & Cohn, D. (Eds.), *Advances in Neural Information Processing Systems* (Vol. 11, pp. 536–542). MIT Press.

Milionis, A. E. & Davies, T. D. (1994). Regression and stochastic-models for air-pollution: I. review, comments and suggestions. *Atmospheric Environment, 28*, 2801–2810.

Miller, S. W. & Emery, W. J. (1997). An automated neural network cloud classifier for use over land and ocean surfaces. *Journal of Applied Meteorology, 36*, 1346–1362.

Milligan, G. W. & Cooper, M. C. (1985). An examination of procedures for determining the number of clusters in a data set. *Psychometrika, 50*(2), 159–179.

Min, S. K. & Hense, A. (2006). A Bayesian approach to climate model evaluation and multi-model averaging with an application to global mean surface temperatures from IPCC AR4 coupled climate models. *Geophysical Research Letters, 33*(8), L08708.

Min, S. K., Zhang, X. B., Zwiers, F. W., & Hegerl, G. C. (2011). Human contribution to more-intense precipitation extremes. *Nature, 470*, 378–381.

Min, S.-K., Simonis, D., & Hense, A. (2007). Probabilistic climate change predictions applying Bayesian model averaging. *Philosophical Transactions of the Royal Society, A365*, 2103–2116.

Mingoti, S. A. & Lima, J. O. (2006). Comparing SOM neural network with Fuzzy c-means, K-means and traditional hierarchical clustering algorithms. *European Journal of Operational Research, 174*, 1742–1759.

Minka, T. P. (2002). Estimating a gamma distribution. (unpublished manuscript) https://tminka.github.io/papers/minka-gamma.pdf.

Minns, A. W. & Hall, M. J. (1996). Artificial neural networks as rainfall-runoff models. *Hydrological Sciences Journal (Journal Des Sciences Hydrologiques), 41*, 399–417.

Minsky M. & Papert, S. (1969). *Perceptrons.* MIT Press.

Mirza, M. & Osindero, S. (2014). Conditional generative adversarial nets. *CoRR, abs/1411.1784.* arXiv: 1411.1784

Monahan, A. H. (2000). Nonlinear principal component analysis by neural networks: Theory and application to the Lorenz system. *Journal of Climate, 13*, 821–835.

Monahan, A. H. (2001). Nonlinear principal component analysis: Tropical Indo-Pacific sea surface temperature and sea level pressure. *Journal of Climate, 14*, 219–233.

Monahan, A. H. & Fyfe, J. C. (2007). Comment on 'The shortcomings of nonlinear principal component analysis in identifying circulation regimes'. *Journal of Climate, 20*, 375–377.

Monahan, A. H., Fyfe, J. C., Ambaum, M. H. P., Stephenson, D. B., & North, G. R. (2009). Empirical orthogonal functions: The medium is the message. *Journal of Climate, 22*, 6501–6514.

Monahan, A. H., Tangang, F. T., & Hsieh, W. (1999). A potential problem with extended EOF analysis of standing wave fields. *Atmosphere-Ocean, 37*, 241–254.

Montufar, G. F., Pascanu, R., Cho, K., & Bengio, Y. (2014). On the number of linear regions of deep neural networks. In Ghahramani, Z., Welling, M., Cortes, C., Lawrence, N. D., & Weinberger, K. Q. (Eds.), *Advances in Neural Information Processing Systems 27* (pp. 2924–2932). Curran Associates, Inc. Retrieved from http://papers.nips.cc/paper/5422-on-the-number-of-linear-regions-of-deep-neural-networks.pdf

Moody, J. & Darken, C. J. (1989). Fast learning in networks of locally-tuned processing units. *Neural Computation, 1*, 281–294.

Mooers, C. N. K. (1973). A technique for the cross spectrum analysis of pairs of complex-valued time series, with emphasis on properties of polarized components and rotational invariants. *Deep Sea Research, 20*, 1129–1141.

Morellato, L. P. C., Alberti, L. F., & Hudson, I. L. (2010). Applications of circular statistics in plant phenology: A case studies approach. In Hudson, I. & Keatley, M. (Eds.), *Phenological Research* (pp. 339–359). Springer.

Mosteller, F. & Tukey, J. W. (1977). *Data Analysis and Regression: A Second Course in Statistics*. Addison-Wesley.

Mountrakis, G., Im, J., & Ogole, C. (2011). Support vector machines in remote sensing: A review. *ISPRS Journal of Photogrammetry and Remote Sensing, 66*, 247–259.

Murdoch, W. J., Singh, C., Kumbier, K., Abbasi-Asl, R., & Yu, B. (2019). Definitions, methods, and applications in interpretable machine learning. *Proceedings of the National Academy of Sciences, 116*, 22071–22080.

Murphy, A. H. (1988). Skill scores based on the mean square error and their relationships to the correlation coefficient. *Monthly Weather Review, 116*, 2417–2424.

Murphy, A. H. (1992). Climatology, persistence, and their linear combination as standards of reference in skill scores. *Weather and Forecasting, 7*, 692–698.

Murphy K. P. (2012). *Machine Learning: A Probabilistic Perspective*. MIT Press.

Nabney I. T. (2002). *Netlab: Algorithms for Pattern Recognition*. London: Springer.

Naghettini, M. (2017). *Fundamentals of Statistical Hydrology*. Springer.

Nair, V. & Hinton, G. E. (2010). Rectified linear units improve restricted Boltzmann machines. In *Proceedings of the 27th International Conference on Machine Learning (ICML 2010)*.

Najafi, M. R., Moradkhani, H., & Jung, W. (2011). Assessing the uncertainties of hydrologic model selection in climate change impact studies. *Hydrological Processes, 25*(18), 2814–2826.

Nash, J. & Sutcliffe, J. (1970). River flow fore-casting through conceptual models part I: A discussion of principles. *Journal of Hydrology, 10*, 282–290.

Neal, R. M. (1996). *Bayesian Learning for Neural Networks*. Lecture Notes in Statistics. New York: Springer.

Neal, R. M. (1999). Regression and classification using Gaussian process priors. In Bernardo, J. M., Berger, J. O., Dawid, A. P., & Smith, A. F. M. (Eds.), *Bayesian Statistics 6* (475–501). 6th Valencia International Meeting on Bayesian Statistics, Alcoceber, Spain, 6–10 Jun, 1998.

Neelin, J. D., Battisti, D. S., Hirst, A. C., Jin, F. F., Wakata, Y., Yamagata, T., & Zebiak, S. E. (1998). ENSO theory. *Journal of Geophysical Research, 103*(C7), 14261–14290.

New, M., Lister, D., Hulme, M., & Makin, I. (2002). A high-resolution data set of surface climate over global land areas. *Climate Research, 21*, 1–25.

Nguyen, P., Ombadi, M., Sorooshian, S., Hsu, K., AghaKouchak, A., Braithwaite, D., ... Thorstensen, A. R. (2018). The PERSIANN family of global satellite precipitation data: A review and evaluation of products. *Hydrology and Earth System Sciences, 22*, 5801–5816.

Niang, A., Badran, A., Moulin, C., Crépon, M., & Thiria, S. (2006). Retrieval of aerosol type and optical thickness over the Mediterranean from SeaWiFS images using an automatic neural classification method. *Remote Sensing of Environment, 100*, 82–94.

Niermann, S. (2006). Evolutionary estimation of parameters of Johnson distributions. *Journal of Statistical Computation and Simulation, 76*, 185–193.

Nigam, S. & Baxter, S. (2015). Teleconnections. In North, G., Pyle, J. & Zhang, F. (Eds.), *Encyclopedia of Atmospheric Sciences* (2nd ed., Vol. 3, pp. 90–109).

Nilsson, N. J. (2009). *The Quest for Artificial Intelligence*. Cambridge University Press.

Nocedal, J. & Wright, S. J. (2006). *Numerical Optimization*. Springer.

Nolan, P., Lynch, P., McGrath, R., Semmler, T., & Wang, S. Y. (2012). Simulating climate change and its effects on the wind energy resource of Ireland. *Wind Energy, 15*, 593–608.

North, G. R., Bell, T. L., Cahalan, R. F., & Moeng, F. J. (1982). Sampling errors in the estimation of empirical orthogonal functions. *Monthly Weather Review, 110*, 699–706.

O'Brien, J. J. & Pillsbury R. D. (1974). Rotary wind spectra in a sea breeze regime. *Journal of Applied Meteorology, 13*, 820–825.

O'Gorman, P. A. & Dwyer, J. G. (2018). Using machine learning to parameterize moist convection: Potential for modeling of climate, climate change, and extreme events. *Journal of Advances in Modeling Earth Systems, 10*, 2548–2563.

Oja, E. (1982). A simplified neuron model as a principal component analyzer. *Journal of Mathematical Biology, 15*, 267–273.

Oktay, O., Schlemper, J., Folgoc, L. L., Lee, M., Heinrich, M., Misawa, K., ... Rueckert, D. (2018). Attention U-Net: Learning where to look for the pancreas. arXiv: 1804.03999

Olive, D. J. (2004). A resistant estimator of multivariate location and dispersion. *Computational Statistics & Data Analysis, 46*, 93–102.

Olsson, J., Uvo, C. B., Jinno, K., Kawamura, A., Nishiyama, K., Koreeda, N., Nakashima, T., & Morita, O. (2004). Neural networks for rainfall forecasting by atmospheric downscaling. *Journal of Hydrologic Engineering, 9*, 1–12.

*NIST Digital Library of Mathematical Functions* http://dlmf.nist.gov/, Release 1.0.20 of 2018-09-15. Olver, F. W. J., Olde Daalhuis, A. B., Lozier, D. W., Schneider, B. I., Boisvert, R. F., Clark, C. W., Miller, B. R., & Saunders, B. V. (Eds.) Retrieved from http://dlmf.nist.gov/

Osowski, S. & Garanty K. (2007). Forecasting of the daily meteorological pollution using wavelets and support vector machine. *Engineering Applications of Artificial Intelligence, 20*, 745–755.

Ouarda, T. B. M. J., Girard, C., Cavadias, G. S., & Bobee, B. (2001). Regional flood frequency estimation with canonical correlation analysis. *Journal of Hydrology, 254*, 157–173.

Pacifici, F., Del Frate, F., Solimini, C., & Emery W. J. (2007). An innovative neuralnet method to detect temporal changes in high-resolution optical satellite imagery. *IEEE Transactions on Geoscience and Remote Sensing, 45*, 2940–2952.

Palmer, T. (2019). The ECMWF ensemble prediction system: Looking back (more than) 25 years and projecting forward 25 years. *Quarterly Journal of the Royal Meteorological Society, 145* (Suppl. 1), 12–24.

Pao, Y. H., Park, G. H., & Sobajic, D. J. (1994). Learning and generalization characteristics of the random vector functionallink net. *Neurocomputing, 6*, 163–180.

Parviainen, E., Riihimaki, J., Miche, Y., & Lendasse, A. (2010). Interpreting extreme learning machine as an approximation to an infinite neural network. In Fred, A. & Filipe, J. (Eds.), *KDIR 2010: Proceedings of the International Conference on Knowledge Discovery and Information Retrieval* (pp. 65–73).

Pashaei, M., Starek, M. J., Kamangir, H., & Berryhill, J. (2020). Deep learning-based single image super-resolution: An investigation for dense scene reconstruction with UAS photogrammetry. *Remote Sensing, 12*(11), 1757.

Pasolli, L., Melgani, F., & Blanzieri, E. (2010). Gaussian process regression for estimating chlorophyll concentration in subsurface waters from remote sensing data. *IEEE Geoscience and Remote Sensing Letters, 7*, 464–468.

Patton, A., Politis, D. N., & White, H. (2009). Correction to 'Automatic block-length selection for the dependent bootstrap' by D. Politis and H. White. *Econometric Reviews, 28*, 372–375.

Pawlowicz, R., Beardsley, B., & Lentz, S. (2002). Classical tidal harmonic analysis including error estimates in MATLAB using T_TIDE. *Computers & Geosciences, 28*(8), 929–937.

Pearson, K. (1895). Contributions to the mathematical theory of evolution. II. Skew variation in homogeneous material. *Philosophical Transactions of the Royal Society, A186*, 343–414.

Pearson, K. (1901). On lines and planes of closest fit to systems of points in space. *Philosophical Magazine, Ser. 6, 2*, 559–572.

Peirce, C. S. (1884). The numerical measure of the success of predictions. *Science, 4*, 453–454.

Peng H. C., Long, F. H., & Ding, C. (2005). Feature selection based on mutual information: Criteria of max-dependency, max-relevance, and min-redundancy. *IEEE Transactions on Pattern Analysis and Machine Intelligence, 27*, 1226–1238.

Peng, H., Lima, A. R., Teakles, A., Jin, J., Cannon, A. J., & Hsieh, W. W. (2017). Evaluating hourly air quality forecasting in Canada with nonlinear updatable machine learning methods. *Air Quality, Atmosphere & Health, 10*, 195–211.

Pérez, P., Trier, A., & Reyes, J. (2000). Prediction of $PM_{2.5}$ concentrations several hours in advance using neural networks in Santiago, Chile. *Atmospheric Environment, 34*, 1189–1196.

Peterson, T. C. & coauthors. (2001). *Report on the Activities of the Working Group on Climate Change Detection and Related Rapporteurs 1998–2001.* WMO, Rep. WCDMP-47, WMO-TD 1071.

Peubey C. & McNally A. P. (2009). Characterization of the impact of geostationary clear-sky radiances on wind analyses in a 4D-Var context. *Quarterly Journal of the Royal Meteorological Society*, *135*, 1863–1876.

Pfeffer, K., Pebesma, E. J., & Burrough, P. A. (2003). Mapping alpine vegetation using vegetation observations and topographic attributes. *Landscape Ecology*, *18*, 759–776.

Polak, E. (1971). *Computational Methods in Optimization: A Unified Approach*. New York: Academic Press.

Polak, E. & Ribiere, G. (1969). Note sur la convergence de methods de directions conjures. *Revue Francaise d'Informat. et de Recherche Operationnelle*, *16*, 35–43.

Politis, D. N. & White, H. (2004). Automatic block-length selection for the dependent bootstrap. *Econometric Reviews*, *23*, 53–70.

Ponti, M. A., Ribeiro, L. S. F., Nazare, T. S., Bui, T., & Collomosse, J. (2017). Everything you wanted to know about deep learning for computer vision but were afraid to ask. In *2017 30th SIBGRAPI Conference on Graphics, Patterns and Images Tutorials* (pp. 17–41).

Powell, M. J. D. (1987). Radial basis functions for multivariate interpolation: A review. In Mason, J. & Cox, M. (Eds.), *Algorithms for Approximation* (pp. 143–167). Oxford: Clarendon Press.

Prabhu, K. M. M. (2014). *Window Functions and Their Applications in Signal Processing*. CRC Press.

Pratt, L. & Thrun, S. (Eds.). (1997). Machine Learning: Special Issue on Inductive Transfer, *28*(1).

Preisendorfer, R. W. (1988). *Principal Component Analysis in Meteorology and Oceanography*. New York: Elsevier.

Press, W. H., Flannery, B. P., Teukolsky, S. A., & Vetterling, W. T. (1986). *Numerical Recipes*. Cambridge: Cambridge University Press.

Price, K. V., Storn, R. M., & Lampinen, J. A. (2005). *Differential Evolution: A Practical Approach to Global Optimization*. Berlin: Springer.

Prokhorenkova, L., Gusev, G., Vorobev A., Dorogush, A. V., & Gulin, A. (2018). Catboost: Unbiased boosting with categorical features. In *NIPS'18: Proceedings of the 32nd International Conference on Neural Information Processing Systems* (pp. 6639–6649).

Pryor, S. C., Schoof, J. T., & Barthelmie, R. J. (2005). Climate change impacts on wind speeds and wind energy density in northern Europe: Empirical downscaling of multiple AOGCMs. *Climate Research*, *29*, 183–198.

Puth, M. T., Neuhauser, M., & Ruxton, G. D. (2015). On the variety of methods for calculating confidence intervals by bootstrapping. *Journal of Animal Ecology*, *84*, 892–897.

Quinlan, J. R. (1992). Learning with continuous classes. In *Proceedings of AI'92 (5th Australian Joint Conference on Artificial Intelligence)* (pp. 343–348). Singapore.

Quinlan, J. R. (1993). *C4.5: Programs for Machine Learning*. San Mateo: Morgan Kaufmann.

Radford, A., Metz, L., & Chintala, S. (2016). Unsupervised representation learning with deep convolutional generative adversarial networks. arXiv: 1511.06434

Radić, V., Cannon, A. J., Menounos, B., & Gi, N. (2015). Future changes in autumn atmospheric river events in British Columbia, Canada, as projected by CMIP5 global climate models. *Journal of Geophysical Research: Atmospheres*, *120*, 9279–9302.

Radić, V. & Clarke, G. K. C. (2011). Evaluation of IPCC models' performance in simulating late-twentieth-century climatologies and weather patterns over North America. *Journal of Climate*, *24*, 5257–5274.

Raftery, A. E., Gneiting, T., Balabdaoui, F., & Polakowski, M. (2005). Using Bayesian model averaging to calibrate forecast ensembles. *Monthly Weather Review*, *133*(5), 1155–1174.

Raftery A. E., Madigan, D., & Hoeting, J. A. (1997). Bayesian model averaging for linear regression models. *Journal of the American Statistical Association*, *92*(437), 179–191.

Raghavendra, S. N. & Deka, P. C. (2014). Support vector machine applications in the field of hydrology: A review. *Applied Soft Computing*, *19*, 372–386.

Rahman, N. A. (1968). *A Course in Theoretical Statistics*. London: Griffin.

Raissi, M., Perdikaris, P., & Karniadakis, G. E. (2019). Physics-informed neural networks: A deep learning framework for solving forward and inverse problems involving nonlinear partial differential equations. *Journal of Computational Physics*, *378*, 686–707.

Ramachandran, P., Zoph, B., & Le, Q. V. (2017). Searching for activation functions. *CoRR, abs/1710.05941*. arXiv: 1710.05941

Ramadhan, A., Marshall, J., Souza, A., Wagner, G. L., Ponnapati, M., & Rackauckas, C. (2020). Capturing missing physics in climate model parameterizations using neural differential equations. arXiv: 2010.12559

Rand, W. M. (1971). Objective criteria for evaluation of clustering methods. *Journal of the American Statistical Association, 66* 846–850.

Rasmussen, C. E. & Williams, C. K. I. (2006). *Gaussian Processes for Machine Learning.* MIT Press.

Rasp, S. & Lerch, S. (2018). Neural networks for postprocessing ensemble weather fore-casts. *Monthly Weather Review, 146,* 3885–3900.

Rasp, S., Pritchard, M. S., & Gentine, P. (2018). Deep learning to represent subgrid processes in climate models. *Proceedings of the National Academy of Sciences of the United States of America*, 115(39), 9684–9689.

Rattan, S. S. P. & Hsieh, W. W. (2004). Nonlinear complex principal component analysis of the tropical Pacific interannual wind variability. *Geophysical Research Letters, 31*, L21201, doi:10.1029/2004GL020446.

Rattan, S. S. P. & Hsieh, W. W. (2005). Complex-valued neural networks for nonlinear complex principal component analysis. *Neural Networks, 18*, 61–69.

Rattan, S. S. P., Ruessink, B. G., & Hsieh, W. W. (2005). Non-linear complex principal component analysis of nearshore bathymetry. *Nonlinear Processes in Geophysics, 12*, 661–670.

Razali, N. M. & Wah, Y. B. (2011). Power comparisons of Shapiro–Wilk, Kolmogorov–Smirnov, Lilliefors and Anderson–Darling tests. *Journal of Statistical Modeling and Analytics, 2*, 21–33.

Reichstein, M., Camps-Valls, G., Stevens, B., Jung, M., Denzler, J., Carvalhais, N., & Prabhat. (2019). Deep learning and process understanding for data-driven Earth system. *Nature, 566*, 195–204.

Renard, B., Lang, M., Bois, P., Dupeyrat, A., Mestre, O., Niel, H., ... Gailhard, J. (2008). Regional methods for trend detection: Assessing field significance and regional consistency. *Water Resources Research, 44*, W08419, doi:10.1029/2007WR006268.

Reshef, D. N., Reshef, Y. A., Finucane, H. K., Grossman, S. R., McVean, G., Turnbaugh, P. J., ... Sabeti, P. C. (2011). Detecting novel associations in large data sets. *Science, 334*, 1518–1524.

Rey D. & Neuhäuser, M. (2011). Wilcoxon-signed-rank test. In Lovric, M. (Ed.), *International Encyclopedia of Statistical Science* (pp. 1658–1659). Berlin, Heidelberg: Springer.

Reynolds, R. W. & Smith, T. M. (1994). Improved global sea surface temperature analyses using optimum interpolation. *Journal of Climate, 7*, 929–948.

Rice, J. S., Emanuel, R. E., Vose, J. M., & Nelson, S. A. C. (2015). Continental US streamflow trends from 1940 to 2009 and their relationships with watershed spatial characteristics. *Water Resources Research, 51*, 6262–6275.

Richardson, A. J., Risien, C., & Shillington, F. A. (2003). Using self-organizing maps to identify patterns in satellite imagery. *Progress in Oceanography, 59*, 223–239.

Richaume, P., Badran, F., Crépon, M., Mejia, C., Roquet, H., & Thiria, S. (2000). Neural network wind retrieval from ERS-1 scatterometer data. *Journal of Geophysical Research*, 105(C4), 8737–8751.

Richman, M. B. (1986). Rotation of principal components. *Journal of Climatology, 6*, 293–335.

Richman, M. B. (1987). Rotation of principal components: A reply *Journal of Climatology, 7*, 511–520.

Richman, M. B. & Adrianto, I. (2010). Classification and regionalization through kernel principal component analysis. *Physics and Chemistry of the Earth, 35*, 316–328.

Richman, M. B., Trafalis, T. B., & Adrianto, I. (2009). Missing data imputation through machine learning algorithms. In Haupt, S. E., Pasini, A., & Marzban, C. (Eds.), *Artificial Intelligence Methods in the Environmental Sciences* (pp. 153–169). Springer.

Riedmiller, M., Gabel, T., Hafner, R., & Lange, S. (2009). Reinforcement learning for robot soccer. *Auton Robot, 27*, 55–73.

Rish, I. (2001). An empirical study of the naive Bayes classifier. In *IJCAI 2001 Workshop on Empirical Methods in Artificial Intelligence* (Vol. 3, 22 pp. 41–46).

Risovic, D. (1993). 2-component model of sea particle-size distribution. *Deep-Sea Research Part I–Oceanographic Research Papers, 40*, 1459–1473.

Rojas, R. (1996). *Neural Networks: A Systematic Introduction* New York: Springer.

Romao, X., Delgado, R., & Costa, A. (2010). An empirical power comparison of univariate goodness-of-fit tests for normality *Journal of Statistical Computation and Simulation, 80*, 545–591.

Ronneberger O., Fischer, P., & Brox, T. (2015). U-Net: Convolutional networks for biomedical image segmentation. In *Medical Image Computing and Computer-Assisted Intervention – MICCAI 2015* (pp. 234–241). Lecture Notes in Computer Science, vol 9351. Springer.

Rosenblatt, F. (1958). The perceptron: A probabilistic model for information storage and organization in the brain. *Psychological Review, 65*, 386–408.

Rosenblatt, F. (1962). *Principles of Neurodynamics*. New York: Spartan.

Roulston, M. S. & Smith, L. A. (2002). Evaluating probabilistic forecasts using information theory. *Monthly Weather Review, 130*, 1653–1660.

Rousseeuw, P. J. (1987). Silhouettes: A graphical aid to the interpretation and validation of cluster-analysis. *Journal of Computational and Applied Mathematics, 20*, 53–65.

Rousseeuw, P. J. & van Driessen, K. (1999). A fast algorithm for the minimum covariance determinant estimator. *Technometrics, 41*, 212–23.

Roweis, S. T. & Saul, L. K. (2000). Nonlinear dimensionality reduction by locally linear embedding. *Science, 290*, 2323–2326.

Ruessink, B. G., van Enckevort, I. M. J., & Kuriyama, Y. (2004). Non-linear principal component analysis of nearshore bathymetry. *Marine Geology, 203*, 185–197.

Rumelhart, D. E., Hinton, G. E., & Williams, R. J. (1986a). Learning internal representations by error propagation. In Rumelhart, D., McClelland, J., & PDP Research Group (Eds.), *Parallel distributed processing* (Vol. 1, pp. 318–362). MIT Press.

Rumelhart, D. E., Hinton, G. E., & Williams, J. (1986b). Learning representations by back-propagating errors. *Nature, 323*, 533–536.

Russakovsky, O., Deng, J., Su, H., Krause, J., Satheesh, S., Ma, S., ... Fei-Fei, L. (2015). ImageNet large scale visual recognition challenge. *International Journal of Computer Vision, 115*(3), 211–252.

Ruxton, G. D. (2006). The unequal variance *t*-test is an underused alternative to Student's *t*-test and the Mann–Whitney *U* test. *Behavioral Ecology, 17*, 688–690.

Saab, S., Badr, E., & Nasr, G. (2001). Univariate modeling and forecasting of energy consumption: The case of electricity in Lebanon. *Energy, 26*, 1–14.

Sadeghi, M., Phu, N., Hsu, K., & Sorooshian, R. (2020). Improving near real-time precipitation estimation using a U-Net convolutional neural network and geographical information. *Environmental Modelling and Software, 134*, 104856.

Sailor, D. J., Hu, T., Li, X., & Rosen, J. N. (2000). A neural network approach to local downscaling of GCM output for assessing wind power implications of climate change. *Renewable Energy, 19*, 359–378.

Sailor, D. J., Smith, M., & Hart, M. (2008). Climate change implications for wind power resources in the Northwest United States. *Renewable Energy, 33*, 2393–2406.

Salas, J. D. & Obeysekera, J. T. B. (1982). ARMA model identification of hydrologic time-series. *Water Resources Research, 18*, 1011–1021.

Sallenger, A. H., Doran, K. S., & Howd, P. A. (2012). Hotspot of accelerated sea-level rise on the Atlantic coast of North America. *Nature Climate Change, 2*, 884–888.

Samsudin, R., Saad, P., & Shabri, A. (2011). River flow time series using least squares support vector machines. *Hydrology and Earth System Sciences, 15*, 1835–1852.

Samuel, A. L. (1959). Some studies in machine learning using the game of checkers. *IBM Journal of Research and Development, 3*, 210–229.

Sanger, T. D. (1989). Optimal unsupervised learning in a single-layer linear feedforward neural network. *Neural Networks, 2*, 459–473.

Santosa, F. & Symes, W. W. (1986). Linear inversion of band-limited reflection seismograms. *SIAM Journal on Scientific and Statistical Computing, 7*, 1307–1330. doi:10.1137/0907087

Sarachik, E. S. & Cane, M. A. (2010). *The El Niño-Southern Oscillation Phenomenon.* Cambridge University Press.

Schiller, H. & Doerffer, R. (1999). Neural network for emulation of an inverse model: Operational derivation of Case II water properties from MERIS data. *International Journal of Remote Sensing, 20*, 1735–1746.

Schmidhuber, J. (2015). Deep learning in neural networks: An overview. *Neural Networks, 61*, 85–117.

Schmidt, W. F., Kraaijveld, M. A., & Duin, R. P. W. (1992). Feed forward neural networks with random weights. In *11th IAPR International Conference on Pattern Recognition, Proceedings, Vol II: Conference B: Pattern Recognition Methodology and Systems* (pp. 1–4). Int. Assoc. Pattern Recognition.

Schneider, T., Teixeira, J., Bretherton, C. S., Brient, F., Pressel, K. G., Schar, C., & Siebesma, A. P. (2017). Climate goals and computing the future of clouds. *Nature Climate Change, 7*, 3–5.

Schölkopf, B. & Smola, A. J. (2002). *Learning with Kernels: Support Vector Machines, Regularization, Optimization, and Beyond (Adaptive Computation and Machine Learning)*. MIT Press.

Schölkopf, B., Smola, A., Williamson, R., & Bartlett, P. L. (2000). New support vector algorithms. *Neural Computation, 12*, 1207–1245.

Scholz, M. (2012). Validation of nonlinear PCA. *Neural Processing Letters, 36*(1), 21–30.

Scholz, M., Kaplan, F., Guy, C. L., Kopka, J., & Selbig, J. (2005). Non-linear PCA: A missing data approach. *Bioinformatics, 21*(20), 3887–3895.

Scholz, M. & Vigario, R. (2002). Nonlinear PCA: A new hierarchical approach. In *ESANN'2002 Proceedings* (pp. 439–444). Bruges, Belgium.

Schoof, J. T. & Pryor, S. C. (2001). Downscaling temperature and precipitation: A comparison of regression-based methods and artificial neural networks. *International Journal of Climatology, 21*, 773–790.

Schwarz, G. (1978). Estimating the dimension of a model. *Annals of Statistics, 6*, 461–464.

Schwenker, F., Kestler, H. A., & Palm, G. (2001). Three learning phases for radial-basis-function networks. *Neural Networks, 14*, 439–458.

Scott, D. W. (2015). *Multivariate Density Estimation* (2nd ed.). Wiley.

Sha, Y., Gagne, D. J., II, West, G., & Stull, R. (2020a). Deep-learning-based gridded downscaling of surface meteorological variables in complex terrain. Part I: Daily maximum and minimum 2-m temperature. *Journal of Applied Meteorology and Climatology, 59*, 2057–2073.

Sha, Y., Gagne, D. J., II, West, G., & Stull, R. (2020b). Deep-learning-based gridded downscaling of surface meteorological variables in complex terrain. Part II: Daily precipitation. *Journal of Applied Meteorology and Climatology, 59*, 2075–2092.

Shabbar, A. & Barnston, A. G. (1996). Skill of seasonal climate forecasts in Canada using canonical correlation analysis. *Monthly Weather Review, 124*, 2370–2385.

Shabbar, A., Bonsal, B., & Khandekar, M. (1997). Canadian precipitation patterns associated with the Southern Oscillation. *Journal of Climate, 10*, 3016–3027.

Shabbar, A. & Kharin, V. (2007). An assessment of cross validation for estimating skill of empirical seasonal forecasts using a global coupled model simulation. *CLIVAR Exchanges, 12*, 10–12.

Shanno, D. F. (1970). Conditioning of quasi-Newton methods for function minimization. *Mathematics of Computation 24*, 647–657.

Shanno, D. F. (1978). Conjugate-gradient methods with inexact searches. *Mathematics of Operations Research, 3*, 244–256.

Shannon, C. E. (1948a). A mathematical theory of communication. *Bell System Technical Journal, 27*, 379–423.

Shannon, C. E. (1948b). A mathematical theory of communication. *Bell System Technical Journal, 27*, 623–656.

Sharma, A. (2000). Seasonal to interannual rainfall probabilistic forecasts for improved water supply management. Part 1: A strategy for system predictor identification. *Journal of Hydrology, 239*, 232–239.

Shawe-Taylor, J. & Cristianini, N. (2004). *Kernel Methods for Pattern Analysis*. Cambridge University Press.

Sheather, S. J. (2004). Density estimation. *Statistical Science, 19*, 588–597.

Shen, C. (2018). A transdisciplinary review of deep learning research and its relevance for water resources scientists. *Water Resources Research, 54*, 8558–8593.

Shi, X., Chen, Z., Wang, H., Yeung, D.-Y., Wong, W.-k., & Woo, W.-c. (2015). Convolutional LSTM network: A machine learning approach for precipitation nowcasting. In *Proceedings of the 28th International Conference on Neural Information Processing System* (pp. 802–810). arXiv: 1506.04214

Shorten, C. & Khoshgoftaar, T. M. (2019). A survey on image data augmentation for deep learning. *Journal of Big Data, 6*, 60.

Shumway R. H. & Stoffer, D. S. (2017). *Time Series Analysis and Its Applications: With R Examples* (4th ed.). Springer.

Silverman, B. W. (1986). *Density Estimation for Statistics and Data Analysis*. London: Chapman & Hall.

Simon, D. J. (2013). *Evolutionary Optimization Algorithms*. Wiley.

Simonyan, K. & Zisserman, A. (2015). Very deep convolutional networks for large-scale image recognition. In *International Conference on Learning Representations*.

Simpson, J. J. & McIntire, T. J. (2001). A recurrent neural network classifier for improved retrievals of areal extent of snow cover. *IEEE Transactions on Geoscience and Remote Sensing*, *39*, 2135–2147.

Singh, V. P. (1998). *Entropy-Based Parameter Estimation in Hydrology*. Springer.

Sit, M., Demiray, B. Z., Xiang, Z., Ewing, G. J., Sermet, Y., & Demir, I. (2020). A comprehensive review of deep learning applications in hydrology and water resources. *Water Science and Technology*, *82*, 2635–2670.

Smith, T. M., Reynolds, R. W., Livezey, R. E., & Stokes, D. C. (1996). Reconstruction of historical sea surface temperatures using empirical orthogonal functions. *Journal of Climate*, *9*, 1403–1420.

Smola, A. J. & Schölkopf, B. (2004). A tutorial on support vector regression. *Statistics and Computing*, *14*, 199–222.

Snoek, J., Larochelle, H., & Adams, R. P. (2012). Practical Bayesian optimization of machine learning algorithms. In Pereira, F., Burges, C. J. C., Bottou, L., & Weinberger, K. Q. (Eds.), *Advances in Neural Information Processing Systems 25* (pp. 2951–2959). Curran Associates, Inc.

Snoek, J., Rippel, O., Swersky, K., Kiros, R., Satish, N., Sundaram, N., ... Adams, R. P. (2015). Scalable Bayesian optimization using deep neural networks. arXiv: 1502.05700

Solomatine, D. P. & Dulal, K. N. (2003). Model trees as an alternative to neural networks in rainfall-runoff modelling. *Hydrological Sciences Journal*, *48*, 399–411.

Solomatine, D. P. & Xue, Y. P. (2004). M5 model trees and neural networks: Application to flood forecasting in the upper reach of the Huai River in China. *Journal of Hydrologic Engineering*, *9*, 491–501.

Sonnewald, M., Dutkiewicz, S., Hill, C., & Forget, G. (2020). Elucidating ecological complexity: Unsupervised learning determines global marine eco-provinces. *Science Advances*, *6*.

Sorooshian, S., Hsu, K. L., Gao, X., Gupta, H. V., Imam, B., & Braithwaite, D. (2000). Evaluation of PERSIANN system satellite-based estimates of tropical rain-fall. *Bulletin of the American Meteorological Society*, *81*, 2035–2046.

Spall, J. C. (2003). *Introduction to Stochastic Search and Optimization: Estimation, Simulation, and Control*. Wiley.

Spearman, C. (1904). The proof and measurement of association between two things. *American Journal of Psychology*, *15*, 72–101.

Srivastava, N., Hinton, G., Krizhevsky, A., Sutskever, I., & Salakhutdinov, R. (2014). Dropout: A simple way to prevent neural networks from overfitting. *Journal of Machine Learning Research*, *15*, 1929–1958.

Stacey, M. W., Pond, S., & LeBlond, P. H. (1986). A wind-forced Ekman spiral as a good statistical fit to low-frequency currents in a coastal strait. *Science*, *233*, 470–472.

Statheropoulos, M., Vassiliadis, N., & Pappa, A. (1998). Principal component and canonical correlation analysis for examining air pollution and meteorological data. *Atmospheric Environment*, *32*, 1087–1095.

Steiger, J. H. (1980). Tests for comparing elements of a correlation matrix. *Psychological Bulletin*, *87*, 245–251.

Steininger, M., Abel, D., Ziegler, K., Krause, A., Paeth, H., & Hotho, A. (2020). Deep learning for climate model output statistics. arXiv: 2012.10394

Stephens, M. A. (1974). EDF statistics for goodness of fit and some comparisons. *Journal of the American Statistical Association*, *69*, 730–737.

Stephens, M. A. (1986). Tests based on EDF statistics. In D'Agostino, R. & Stephens, M. A. (Eds.), *Goodness-of-Fit Techniques*. New York: Marcel Dekker.

Stephenson, D. B. (2000). Use of the 'odds ratio' for diagnosing forecast skill. *Weather and Forecasting*, *15*, 221–232.

Stogryn, A. P., Butler, C. T., & Bartolac, T. J. (1994). Ocean surface wind retrievals from Special Sensor Microwave Imager data with neural networks. *Journal of Geophysical Research*, *99*(C1), 981–984.

Storn, R. & Price, K. (1997). Differential evolution: A simple and efficient heuristic for global optimization over continuous spaces. *Journal of Global Optimization*, *11*, 341–359.

Stott, P. (2016). How climate change affects extreme weather events. *Science, 352,* 1517–1518.

Strang, G. (2005). *Linear Algebra and Its Applications.* Cengage Learning.

Su, H. [Hua], Li, W., & Yan, X.-H. (2018). Retrieving temperature anomaly in the global subsurface and deeper ocean from satellite observations. *Journal of Geophysical Research-Oceans, 123,* 399–410.

Su, H. [Hua], Yang, X. [Xin], Lu, W., & Yan, X.-H. (2019). Estimating subsurface thermohaline structure of the global ocean using surface remote sensing observations. *Remote Sensing, 11,* 1598.

Su, H. [Hui], Wu, L., Jiang, J. H., Pai, R., Liu, A., Zhai, A. J., ... DeMaria, M. (2020). Applying satellite observations of tropical cyclone internal structures to rapid intensification forecast with machine learning. *Geophysical Research Letters, 47* (17), e2020GL089102.

Sun, A. Y. & Scanlon, B. R. (2019). How can Big Data and machine learning benefit environment and water management: A survey of methods, applications, and future directions. *Environmental Research Letters, 14* (7), 073001.

Sun, A. Y., Wang, D., & Xu, X. (2014). Monthly streamflow forecasting using Gaussian process regression. *Journal of Hydrology, 511,* 72–81.

Sun, H., Gui, D., Yan, B., Liu, Y., Liao, W., Zhu, Y., ... Zhao, N. (2016). Assessing the potential of random forest method for estimating solar radiation using air pollution index. *Energy Conversion and Management, 119,* 121–129.

Sun, K., Zhang, T., Chen, S., Xue, C., Zou, B., & Shi, L. (2020). Retrieval of ultraviolet diffuse attenuation coefficients from ocean color using the kernel principal components analysis over ocean. *IEEE Transactions on Geoscience and Remote Sensing, 59,* 4579–4589.

Sun, Q., Miao, C., Duan, Q., Ashouri, H., Sorooshian, S., & Hsu, K.-L. (2018). A review of global precipitation data sets: Data sources, estimation, and intercomparisons. *Reviews of Geophysics, 56,* 79–107.

Sun, Y. X. (2013). A heteroskedasticity and autocorrelation robust *F* test using an orthonormal series variance estimator. *Econometrics Journal, 16,* 1–26.

Suykens, J. A. K., Van Gestel, T., De Braanter, J., De Moor, B., & Vandewalle, J. (2002). *Least Squares Support Vector Machines.* New Jersey: World Scientific.

Syu, H. H. & Neelin, J. D. (2000). ENSO in a hybrid coupled model. Part II: Prediction with piggyback data assimilation. *Climate Dynamics,* 16(1), 35–48.

Szegedy, C., Liu, W., Jia, Y. Q., Sermanet, P., Reed, S., Anguelov, D., ... Rabinovich, A. (2015). Going deeper with convolutions. In *2015 IEEE Conference on Computer Vision and Pattern Recognition* (pp. 1–9). Boston, MA.

Sztobryn, M. (2003). Forecast of storm surge by means of artificial neural network. *Journal of Sea Research, 49,* 317–322.

Tag, P. M., Bankert, R. L., & Brody L. R. (2000). An AVHRR multiple cloud-type classification package. *Journal of Applied Meteorology, 39,* 125–134.

Takada, Y. (2018). More PRML Errata. Retrieved from https://yousuketakada.github .io/prml_errata/prml_errata.pdf

Talagrand, O. (2010). Variational assimilation. In Lahoz, W., Khattatov, B. & Menard, R. (Eds.), *Data Assimilation* (pp. 41–67). Springer.

Tang, B. Y., Hsieh, W. W., Monahan, A. H., & Tangang, F. T. (2000). Skill comparisons between neural networks and canonical correlation analysis in predicting the equatorial Pacific sea surface temperatures. *Journal of Climate, 13,* 287–293.

Tang, B. & Mazzoni, D. (2006). Multiclass reduced-set support vector machines. In *Proceedings of the 23rd International Conference on Machine Learning (ICML 2006).* Pittsburgh, PA.: New York: ACM.

Tang, Y. (2002). Hybrid coupled models of the tropical Pacific: I. Interannual variability. *Climate Dynamics, 19,* 331–342.

Tang, Y. M., Kleeman, R., & Moore, A. M. (2005). Reliability of ENSO dynamical pre-dictions. *Journal of the Atmospheric Sciences, 62,* 1770–1791.

Tang, Y. & Hsieh, W. W. (2001). Coupling neural networks to incomplete dynamical systems via variational data assimilation. *Monthly Weather Review, 129,* 818–834.

Tang, Y. & Hsieh, W. W. (2002). Hybrid coupled models of the tropical Pacific: II. ENSO prediction. *Climate Dynamics, 19,* 343–353.

Tang, Y. & Hsieh, W. W. (2003). ENSO simulation and prediction in a hybrid coupled model with data assimilation. *Journal of the Meteorological Society of Japan, 81,* 1–19.

Tang, Y., Hsieh, W. W., Tang, B., & Haines, K. (2001). A neural network atmospheric model for hybrid coupled modelling. *Climate Dynamics, 17*, 445–455.

Tangang, F. T., Hsieh, W. W., & Tang, B. (1997). Forecasting the equatorial Pacific sea surface temperatures by neural network models. *Climate Dynamics, 13*, 135–147.

Tangang, F. T., Hsieh, W. W., & Tang, B. (1998). Forecasting the regional sea surface temperatures of the tropical Pacific by neural network models, with wind stress and sea level pressure as predictors. *Journal of Geophysical Research, 103*(C4), 7511–7522.

Tangang, F. T., Tang, B., Monahan, A. H., & Hsieh, W. W. (1998). Forecasting ENSO events: A neural network-extended EOF approach. *Journal of Climate, 11*, 29–41.

Taylor, J. A., Jakeman, A. J., & Simpson, R. W. (1986). Modeling distributions of air pollutant concentrations. 1. Identification of statistical-models. *Atmospheric Environment, 20*, 1781–1789.

Taylor, J. W. (2000). A quantile regression neural network approach to estimating the conditional density of multiperiod returns. *Journal of Forecasting, 19*, 299–311.

Tebaldi, C., Hayhoe, K., Arblaster, J. M., & Meehl, G. A. (2006). Going to the extremes. *Climatic Change, 79*, 185–211.

Tedesco, M., Pulliainen, J., Takala, M., Hallikainen, M., & Pampaloni, P. (2004). Artificial neural network-based techniques for the retrieval of SWE and snow depth from SSM/I data. *Remote Sensing of Environment, 90*, 76–85.

Tenenbaum, J. B., de Silva, V., & Langford, J. C. (2000). A global geometric framework for nonlinear dimensionality reduction. *Science, 290*, 2319–2323.

Tesauro, G. (1994). TD-Gammon, a self-teaching backgammon program, achieves master-level play. *Neural Computation, 6*, 215–219.

Teschl, R., Randeu, W. L., & Teschl, F. (2007). Improving weather radar estimates of rain-fall using feed-forward neural networks. *Neural Networks, 20*, 519–527.

Thiébaux, H. J. & Zwiers, F. W. (1984). The interpretation and estimation of effective sample size. *Journal of Climate and Applied Meteorology, 23*, 800–811.

Thiria, S., Mejia, C., Badran, F., & Crépon, M. (1993). A neural network approach for modeling nonlinear transfer functions: Application for wind retrieval from spaceborne scatterometer data. *Journal of Geophysical Research, 98*(C12), 22827–22841.

Thom, H. C. S. (1958). A note on the gamma distribution. *Monthly Weather Review, 86*, 117–122.

Thomson, R. E. & Emery W. J. (2014). *Data Analysis Methods in Physical Oceanography* (3rd ed.). Elsevier.

Tibshirani, R. (1996). Regression shrinkage and selection via the lasso. *Journal of the Royal Statistical Society. Series B, 58*, 267–288.

Tibshirani, R., Walther, G., & Hastie, T. (2001). Estimating the number of clusters in a data set via the gap statistic. *Journal of the Royal Statistical Society: Series B, 63, Part 2*, 411–423.

Timmermann, A., An, S.-I., Kug, J.-S., Jin, F.-F., Cai, W., Capotondi, A., ... Zhang, X. (2018). El Niño-Southern Oscillation complexity. *Nature, 559*, 535–545.

Tolman, H. L., Krasnopolsky, V. M., & Chalikov, D. V. (2005). Neural network approximations for nonlinear interactions in wind wave spectra: Direct mapping for wind seas in deep water. *Ocean Modelling, 8*, 252–278.

Toms, B. A., Barnes, E. A., & Ebert-Uphoff, I. (2020). Physically interpretable neural networks for the geosciences: Applications to earth system variability. *Journal of Advances in Modeling Earth Systems, 12*, e2019MS002002.

Torrecilla, E., Stramski, D., Reynolds, R. A., Millan-Nunez, E., & Piera, J. (2011). Cluster analysis of hyperspectral optical data for discriminating phytoplankton pigment assemblages in the open ocean. *Remote Sensing of Environment, 115*, 2578–2593.

Torrence, C. & Compo, G. P. (1998). A practical guide to wavelet analysis. *Bulletin of the American Meteorological Society 79*, 61–78.

Torrence, C. & Webster, P. J. (1999). Interdecadal changes in the ENSO-monsoon system. *Journal of Climate, 12*, 2679–2690.

Trenberth, K. E., Branstator, G. W., Karoly, D., Kumar, A., Lau, N.-C., & Ropelewski, C. (1998). Progress during TOGA in understanding and modeling global teleconnections associated with tropical sea surface temperatures. *Journal of Geophysical Research-Oceans, 103*(C7), 14291–14324.

Tukey J. W. (1977). *Exploratory Data Analysis*. Addison-Wesley.

Turing, A. M. (1950). Computing machinery and intelligence. *Mind, 49*, 433–460.

Uppala, S. M., Kallberg, P. W., Simmons, J., Andrae, U., Bechtold, V. D. C., Fiorino,

M., ... Woollen, J. (2005). The ERA-40 re-analysis. *Quarterly Journal of the Royal Meteorological Society, 131*, 2961–3012.

Urban, G., Geras, K. J., Kahou, S. E., Aslan, O., Wang, S., Mohamed, A., ... Caruana, (2017). Do deep convolutional nets really need to be deep and convolutional? arXiv: 1603.05691

Vahatalo, A. V., Aarnos, H., & Mantyniemi, (2010). Biodegradability continuum and biodegradation kinetics of natural organic matter described by the beta distribution. *Biogeochemistry, 100*, 227–240.

van der Maaten, L. & Hinton, G. (2008). Visualizing data using t-SNE. *Journal of Machine Learning Research, 9*(86), 2579–2605.

van der Maaten, L., Postma, E., & van den Herik, J. (2009). *Dimensionality reduction: A comparative review* Technical Report TiCC TR 2009-005, Tilburg Centre for Creative Computing, Tilburg University.

Vannitsem, S., Bremnes, J., Demaeyer, J., Evans, G., Flowerdew, J., Hemri, S., ... Ylhaisi, J. (2021). Statistical postprocessing for weather forecasts: Review, challenges, and avenues in a Big Data world. *Bulletin of the American Meteorological Society, 102*, E681–E699.

Vapnik, V. N. (1995). *The Nature of Statistical Learning Theory*. Berlin: Springer Verlag.

Vapnik, V. N. (1998). *Statistical Learning Theory*. New York: Wiley

Vapnik, V., Golowich, S. E., & Smola, A. (1997). Support vector method for function approximation, regression estimation, and signal processing. In Mozer, M. C., Jordan, M. I. & Petsche, T. (Eds.), *Advances in Neural Information Processing Systems 9: Proceedings of the 1996 Conference* (Vol. 9, pp. 281–287).

Ventura, V., Paciorek, C. J., & Risbey, J. S. (2004). Controlling the proportion of falsely rejected hypotheses when conducting multiple tests with climatological data. *Journal of Climate, 17*, 4343–4356.

Vesterstrøm, J. & Thomsen, R. (2004). A comparative study of differential evolution, particle swarm optimization, and evolutionary algorithms on numerical benchmark problems. In *CEC2004: Proceedings of the 2004 Congress on Evolutionary Computation, Vols 1 and 2* (pp. 1980–1987). Portland, OR.

Villmann, T., Merenyi, E., & Hammer, B. (2003). Neural maps in remote sensing

image analysis. *Neural Networks 16*(3–4), 389–403.

Vincent, P., Larochelle, H., Lajoie, I., Bengio, Y., & Manzagol, P.-A. (2010). Stacked denoising autoencoders: Learning useful representations in a deep network with a local denoising criterion. *Journal of Machine Learning Research, 11*, 3371–3408.

von Storch, H. (1999). On the use of 'inflation' in statistical downscaling. *Journal of Climate, 12*, 3505–3506.

von Storch, H. & Zwiers, F. W. (1999). *Statistical Analysis in Climate Research*. Cambridge: Cambridge University Press.

Vrac, M., Stein, M. L., Hayhoe, K., & Liang, X. Z. (2007). A general method for validating statistical downscaling methods under future climate change. *Geophysical Research Letters, 34*, L18701.

Vrieze, S. I. (2012). Model selection and psychological theory: A discussion of the differences between the Akaike information criterion (AIC) and the Bayesian information criterion (BIC). *Psychological Methods, 17*(2), 228–243.

Wallace, J. M. & Dickinson, R. E. (1972). Empirical orthogonal representation of time series in the frequency domain. Part I: Theoretical considerations. *Journal of Applied Meteorology, 11*, 887–892.

Wallace, J. M. & Gutzler, D. S. (1981). Teleconnections in the geopotential height fields during the northern hemisphere winter. *Monthly Weather Review, 109*, 784–812.

Wallace, J. M., Smith, C., & Bretherton, C. S. (1992). Singular value decomposition of wintertime sea surface temperature and 500-mb height anomalies. *Journal of Climate, 5*, 561–576.

Walsh, E. S., Kreakie, B. J., Cantwell, M. G., & Nacci, D. (2017). A random forest approach to predict the spatial distribution of sediment pollution in an estuarine system. *PLOS ONE, 12*(7), e0179473.

Walsh, J. E. & Richman, M. B. (1981). Seasonality in the associations between surface temperatures over the United-States and the North Pacific Ocean. *Monthly Weather Review, 109*, 767–783.

Wan, R., Mei, S., Wang, J., Liu, M., & Yang, F. (2019). Multivariate temporal convolutional network: A deep neural networks approach for multivariate time series forecasting. *Electronics, 8*, 876.

Wand, M. P. & Jones, M. C. (1995). *Kernel Smoothing*. Chapman & Hall.

Wang, C., Tang, G., & Gentine, P. (2021). PrecipGAN: Merging microwave and infrared data for satellite precipitation estimation using generative adversarial network. *Geophysical Research Letters*, *48*(5), e2020GL092032.

Wang, J.-X., Wu, J.-L., & Xiao, H. (2017). Physics-informed machine learning approach for reconstructing Reynolds stress modeling discrepancies based on DNS data. *Physical Review Fluids*, *2*, 034603.

Wang, L. P. & Wan, C. R. (2008). Comments on 'The extreme learning machine'. *IEEE Transactions on Neural Networks*, *19*, 1494–1495.

Wang, W. M., Li, Z. L., & Su, H. B. (2007). Comparison of leaf angle distribution functions: Effects on extinction coefficient and fraction of sunlit foliage. *Agricultural and Forest Meteorology*, *143*, 106–122.

Wang, X. L. L. (2008). Accounting for autocorrelation in detecting mean shifts in climate data series using the penalized maximal $t$ or $F$ test. *Journal of Applied Meteorology and Climatology*, *47*, 2423–2444.

Wang, Z., Lai, C., Chen, X., Yang, B., Zhao, S., & Bai, X. (2015). Flood hazard risk assessment model based on random forest. *Journal of Hydrology*, *527*, 1130–1141.

Wanhammar, L. & Saramäki, T. (2020). *Digital Filters using MATLAB*. Springer.

Ward, J. H. (1963). Hierarchical grouping to optimize an objective function. *Journal of the American Statistical Association*, *58*, 236–244.

Warner, T. T. (2011). *Numerical Weather and Climate Prediction*. Cambridge.

Wasserman, L. (2000). Bayesian model selection and model averaging. *Journal of Mathematical Psychology*, *44*(1), 92–107.

Wattenberg, M., Viégas, F., & Johnson, I. (2016). How to use t-SNE effectively. *Distill*. Retrieved from http://distill.pub/2016/misread-tsne

Wei, J., Li, Z., Cribb, M., Huang, W., Xue, W., Sun, L., ... Song, Y. (2020). Improved 1 km resolution $PM_{2.5}$ estimates across China using enhanced space-time extremely randomized trees. *Atmospheric Chemistry and Physics*, *20*, 3273–3289.

Weichert, A. & Bürger, G. (1998). Linear versus nonlinear techniques in downscaling. *Climate Research*, *10*, 83–93.

Weigend, A. S. & Gershenfeld, N. A. (Eds.). (1994). *Time Series Prediction: Forecasting the Future and Understanding the Past*. Santa Fe Institute Studies in the Sciences of Complexity, Proceedings vol. XV. Addison-Wesley.

Weisheimer, A. & Palmer, T. N. (2014). On the reliability of seasonal climate forecasts. *Journal of the Royal Society Interface*, *11*, 20131162.

Welch, B. L. (1947). The generalization of 'Student's' problem when several different population variances are involved. *Biometrika*, *34*, 28–35.

Welford, B. P. (1962). Note on a method for calculating corrected sums of squares and products. *Technometrics*, *4*, 419–420.

Weyn, J. A., Durran, D. R., & Caruana, R. (2019). Can machines learn to predict weather? Using deep learning to predict gridded 500-hPa geopotential height from historical weather data. *Journal of Advances in Modeling Earth Systems*, *11*, 2680–2693.

Weyn, J. A., Durran, D. R., & Caruana, R. (2020). Improving data-driven global weather prediction using deep convolutional neural networks on a cubed sphere. *Journal of Advances in Modeling Earth Systems*, *12*, e2020MS002109.

Weyn, J. A., Durran, D. R., Caruana, R., & Cresswell-Clay, N. (2021). Sub-seasonal forecasting with a large ensemble of deep-learning weather prediction models. *Journal of Advances in Modeling Earth Systems*, *13*, e2021MS002502.

Whitfield, P. H. & Cannon, A. J. (2000). Recent variations in climate and hydrology in Canada. *Canadian Water Resources Journal*, *25*, 19–65.

Widrow, B. & Hoff, M. E. (1960). Adaptive switching circuits. In *IRE WESCON Convention Record* (Vol. 4, pp. 96–104). New York.

Wikipedia contributors. (2018). List of category 5 Atlantic hurricanes. Wikipedia, the free encyclopedia. [Online; accessed 6-September-2018]. Retrieved from https://en.wikipedia.org/w/index.php?title=List_of_Category_5_Atlantic_hurricanes&oldid=858173401

Wilamowski, B. M. & Yu, H. (2010). Improved computation for Levenberg–Marquardt training. *IEEE Transactions on Neural Networks*, *21*, 930–937.

Wilby R. L., Dawson, C. W., & Barrow, E. M. (2002). SDSM: A decision support tool for the assessment of regional climate change impacts. *Environmental Modelling & Software*, *17*, 147–159.

Wilby, R. L. & Wigley, T. M. L. (1997). Downscaling general circulation model output: A review of methods and limitations. *Progress in Physical Geography: Earth and Environment*, *21*, 530–548.

Wilcox, R. R. (2004). *Introduction to Robust Estimation and Hypothesis Testing* (2nd ed.). Amsterdam: Elsevier.

Wilcoxon, F. (1945). Individual comparisons by ranking methods. *Biometrics Bulletin*, *1*, 80–83.

Wilks, D. S. (1997). Resampling hypothesis tests for autocorrelated fields. *Journal of Climate*, *10*, 65–82.

Wilks, D. S. (2006). On 'field significance' and the false discovery rate. *Journal of Applied Meteorology and Climatology*, *45*, 1181–1189.

Wilks, D. S. (2011). *Statistical Methods in the Atmospheric Sciences* (3rd ed.). Academic Press.

Wilks, D. S. & Shen K. W. (1991). Threshold relative humidity duration forecasts for plant disease prediction. *Journal of Applied Meteorology*, *30*, 463–477.

Wilks, D. S. & Wilby, R. L. (1999). The weather generation game: A review of stochastic weather models. *23*, 329–357.

Willmott, C. J., Robeson, S. M., & Matsuura, K. (2012). A refined index of model performance. *International Journal of Climatology*, *38*, 2088–2094.

Willmott, C. J., Robeson, S. M., Matsuura, K., & Ficklin, D. L. (2015). Assessment of three dimensionless measures of model performance. *Environmental Modelling & Software*, *73*, 167–174.

Wilson, L. J. & Vallée, M. (2002). The Canadian updateable model output statistics (UMOS) system: Design and development tests. *Weather and Forecasting*, *17*, 206–222.

Wilson, L. J. & Vallée, M. (2003). The Canadian updateable model output statistics (UMOS) system: Validation against perfect prog. *Weather and Forecasting*, *18*, 288–302.

Witten, I. H., Frank, E., & Hall, M. A. (2011). *Data Mining: Practical Machine Learning Tools and Techniques* (3rd ed.). Elsevier.

Wolker, K. (1987). The Southern Oscillation in surface circulation and climate over the tropical Atlantic, eastern Pacific, and Indian Oceans as captured by cluster-analysis. *Journal of Climate and Applied Meteorology*, *26*, 540–558.

Wolpert, D. H. (1992). Stacked generalization. *Neural Networks*, *5*(2), 241–259.

Wood, A. W., Leung, L. R., Sridhar, V., & Lettenmaier, D. P. (2004). Hydrologic implications of dynamical and statistical approaches to downscaling climate model outputs. *Climatic Change*, *62*, 189–216.

Woodruff, S. D., Slutz, R. J., Jenne, R. L., & Steurer, P. M. (1987). A comprehensive ocean-atmosphere data set. *Bulletin of the American Meteorological Society*, *68*, 1239–1250.

Wu, A. & Hsieh, W. W. (2002). Nonlinear canonical correlation analysis of the tropical Pacific wind stress and sea surface temperature. *Climate Dynamics*, *19*, 713–722.

Wu, A. & Hsieh, W. W. (2003). Nonlinear interdecadal changes of the El Niño-Southern Oscillation. *Climate Dynamics*, *21*, 719–730.

Wu, A., Hsieh, W. W., & Tang, B. (2006). Neural network forecasts of the tropical Pacific sea surface temperatures. *Neural Networks*, *19*, 145–154.

Wu, A., Hsieh, W. W., & Zwiers, F. W. (2003). Nonlinear modes of North American winter climate variability detected from a general circulation model. *Journal of Climate*, *16*, 2325–2339.

Xie, C., Li, K., Ma, C., & Wang, J. (2019). Modeling subgrid-scale force and divergence of heat flux of compressible isotropic turbulence by artificial neural network. *Physical Review Fluids*, *4*, 104605.

Xie, X., Liu, W. T., & Tang, B. (2008). Spacebased estimation of moisture transport in marine atmosphere using support vector regression. *Remote Sensing of Environment*, *112*, 1846–1855.

Xu, Q. F., Deng, K., Jiang, C. X., Sun, F., & Huang, X. (2017). Composite quantile regression neural network with applications. *Expert Systems with Applications*, *76*, 129–139.

Xu, R. & Wunsch, D. C., II. (2008). *Clustering*. Wiley-IEEE Press.

Xu, R. & Wunsch, D., II. (2005). Survey of clustering algorithms. *IEEE Transactions on Neural Networks*, *16*(3), 645–678.

Xu, W. C., Hou, Y. H., Hung, Y. S., & Zou, Y. X. (2013). A comparative analysis of Spearman's rho and Kendall's tau in normal and contaminated normal models. *Signal Processing*, *93*, 261–276.

Yacoub, M., Badran, F., & Thiria, S. (2001). A topological hierarchical clustering: Application to ocean color classification. In *Artificial Neural Networks-ICANN 2001, Proceedings. Lecture Notes in Computer*

*Science.* (Vol. 2130, pp. 492–499). Berlin: Springer.

Yadav, B., Ch, S., Mathur, S., & Adamowski, J. (2016). Discharge forecasting using an Online Sequential Extreme Learning Machine (OS-ELM) model: A case study in Neckar River, Germany. *Measurement, 92,* 433–445.

Yao A. Y. M. (1974). A statistical model for the surface relative humidity. *Journal of Applied Meteorology, 13,* 17–21.

Ye, Z. & Hsieh, W. W. (2006). The influence of climate regime shift on ENSO. *Climate Dynamics, 26,* 823–833.

Yeo, I.-K. & Johnson, R. A. (2000). A new family of power transformations to improve normality or symmetry. *Biometrika, 87,* 954–959.

Yhann, S. R. & Simpson, J. J. (1995). Application of neural networks to AVHRR cloud segmentation. *IEEE Transactions on Geoscience and Remote Sensing, 33,* 590–604.

Yi, J. S. & Prybutok, V. R. (1996). A neural network model forecasting for prediction of daily maximum ozone concentration in an industrialized urban area. *Environmental Pollution, 92,* 349–357.

Yue, S. & Wang, C. Y. (2002). The influence of serial correlation on the Mann–Whitney test for detecting a shift in median. *Advances in Water Resources, 25,* 325–333.

Yue, S. & Wang, C. Y. (2004). The Mann–Kendall test modified by effective sample size to detect trend in serially correlated hydrological series. *Water Resources Management, 18,* 201–218.

Yule, G. U. (1912). *An Introduction to the Theory of Statistics.* London: C. Griffin, Ltd.

Yuval. (2000). Neural network training for prediction of climatological time series; regularized by minimization of the generalized cross validation function. *Monthly Weather Review, 128,* 1456–1473.

Yuval. (2001). Enhancement and error estimation of neural network prediction of Niño 3.4 SST anomalies. *Journal of Climate, 14,* 2150–2163.

Yuval, J. & O'Gorman, P. A. (2020). Stable machine-learning parameterization of sub-grid processes for climate modeling at a range of resolutions. *Nature Communications, 11,* 3295.

Yuval, J., O'Gorman, P. A., & Hill, C. N. (2021). Use of neural networks for stable, accurate and physically consistent parameterization of subgrid atmospheric processes with good performance at reduced precision. *Geophysical Research Letters, 48,* e2020GL091363.

Yuval & Hsieh, W. W. (2002). The impact of time-averaging on the detectability of non-linear empirical relations. *Quarterly Journal of the Royal Meteorological Society, 128* 1609–1622.

Yuval & Hsieh, W. W. (2003). An adaptive nonlinear MOS scheme for precipitation forecasts using neural networks. *Weather and Forecasting, 18,* 303–310.

Zebiak, S. E. & Cane, M. A. (1987). A model El Niño-Southern Oscillation. *Monthly Weather Review, 115,* 2262–2278.

Zehna, P. W. (1970). *Probability Distributions and Statistics.* Allyn and Bacon.

Zeiler, M. D. & Fergus, R. (2014). Visualizing and understanding convolutional networks. In *Computer Vision – ECCV 2014* (pp. 818–833). Springer International Publishing.

Zeng, Z., Hsieh, W. W., Shabbar, A., & Burrows, W. R. (2011). Seasonal prediction of winter extreme precipitation over Canada by support vector regression. *Hydrology and Earth System Sciences, 15,* 65–74.

Zhang, C. (2005). Madden-Julian Oscillation. *Reviews of Geophysics, 43,* RG2003.

Zhang, H., Chu, P.-S., He, L., & Unger, D. (2019). Improving the CPC's ENSO fore-casts using Bayesian model averaging. *Climate Dynamics, 53*(5–6), 3373–3385.

Zhang, H. & Zhang, Z. (1999). Feedforward networks with monotone constraints. In *International Joint Conference on Neural Networks* (Vol. 3, pp. 1820–1823).

Zhang, T., Ramakrishnan, R., & Livny M. (1996). BIRCH: An efficient data clustering method for very large databases. In *Proceedings of the 1996 ACM SIGMOD International Conference on Management of Data* (pp. 103–114).

Zhang, X. B., Alexander, L., Hegerl, G. C., Jones, P., Tank, A. K., Peterson, T. C., ... Zwiers, F. W. (2011). Indices for monitoring changes in extremes based on daily temperature and precipitation data. *Wiley Interdisciplinary Reviews-Climate Change, 2,* 851–870.

Zhang, X., Hogg, W. D., & Mekis, E. (2001). Spatial and temporal characteristics of heavy precipitation events over Canada. *Journal of Climate, 14,* 1923–1936.

Zhang, Y. S., Wu, J., Cai, Z. H., Du, B., & Yu, P. S. (2019). An unsupervised parameter learning model for RVFL neural network. *Neural Networks*, *112*, 85–97.

Zhong, M., Castellote, M., Dodhia, R., Ferres, J. L., Keogh, M., & Brewer, A. (2020). Beluga whale acoustic signal classification using deep learning neural network models. *Journal of the Acoustical Society of America*, *147*, 1834–1841.

Zhou, Z., Siddiquee, M. R., Tajbakhsh, N., & Liang, J. (2018). UNet++: A nested U-net architecture for medical image segmentation. *CoRR*, *abs/1807.10165*. arXiv: 1807.10165

Zorita, E. & von Storch, H. (1999). The analog method as a simple statistical down-scaling technique: Comparison with more complicated methods. *Journal of Climate*, *12*, 2474–2489.

Zou, G. Y. (2007). Toward using confidence intervals to compare correlations. *Psychological Methods*, *12*(4), 399–413.

Zounemat-Kermani, M., Matta, E., Cominola, A., Xia, X., Zhang, Q., Liang, Q., & Hinkelmann, R. (2020). Neurocomputing in surface water hydrology and hydraulics: A review of two decades retrospective, current status and future prospects. *Journal of Hydrology*, *588*, 125085.

Zwiers, F. W. & Kharin, V. V. (1998). Changes in the extremes of the climate simulated by CCC GCM2 under $CO_2$ doubling. *Journal of Climate*, *11*, 2200–2222.

Zwiers, F. W. & Von Storch, H. (2004). On the role of statistics in climate research. *International Journal of Climatology*, *24*, 665–680.

Zwiers, F. W. & von Storch, H. (1995). Taking serial correlation into account in tests of the mean. *Journal of Climate*, *8*, 336–351.

# Index